THE MULTILEVEL FAST MULTIPOLE ALGORITHM (MLFMA) FOR SOLVING LARGE-SCALE COMPUTATIONAL ELECTROMAGNETICS PROBLEMS

IEEE Press
445 Hoes Lane
Piscataway, NJ 08854

THE MULTILEVEL FAST MULTIPOLE ALGORITHM (MLFMA) FOR SOLVING LARGE-SCALE COMPUTATIONAL ELECTROMAGNETICS PROBLEMS

Özgür Ergül

Middle East Technical University, Turkey

Levent Gürel

Bilkent University, Turkey

IEEE Antennas and Propagation Society, *Sponsor*

The IEEE Press Series on Electromagnetic Wave Theory
Andreas C. Cangellaris, *Series Editor*

IEEE PRESS

WILEY

This edition first published 2014

Registered office
John Wiley & Sons Ltd, The Atrium, Southern Gate, Chichester, West Sussex, PO19 8SQ, United Kingdom

For details of our global editorial offices, for customer services and for information about how to apply for permission to reuse the copyright material in this book please see our website at www.wiley.com.

Library of Congress Cataloging-in-Publication Data applied for.

ISBN: 9781119977414

10/12pt TimesLTStd by Laserwords Private Limited, Chennai, India

1 2014

Contents

Preface

This book is about a sophisticated algorithm, namely, the multilevel fast multipole algorithm (MLFMA), which has been developed and improved continuously in the last two decades for the fast and accurate solution of real-life electromagnetics problems. When it was first proposed, MLFMA enabled accurate solutions of problems relating to metallic objects that cannot be solved directly on computers. Since then, the algorithm has been extended to more complicated problems involving metallic, dielectric, and composite objects, has been improved for realistic solutions with unprecedented levels of accuracy, and has been parallelized to handle discretizations involving millions of unknowns. Recently, MLFMA has been the core algorithm to solve the largest dense matrix equations involving more than 1 billion unknowns.

Due to high academic and industrial impacts, computational simulations of electromagnetic phenomena have attracted the interest of many researchers. The literature includes excellent studies involving huge efforts to develop novel solvers for more efficient and accurate simulations. Along this direction, MLFMA has been one of the major algorithms that have passed the time test. Since it was proposed, this algorithm has proved its robustness, efficiency, and accuracy in many simulations involving diverse components and parameters. Today, most of the leading research centers now have their own implementations of MLFMA, for sequential and/or parallel platforms, among their code libraries. Unsurprisingly, MLFMA has also attracted the interest of industrial bodies via commercial programs using the high potential of the algorithm. And, more importantly, due to its superior efficiency and robustness, MLFMA will definitely be one of the key algorithms of computational electromagnetics in the future.

This book provides a detailed and instructional overview of MLFMA for post-graduate students, researchers, and academics, studying in the areas of computational electromagnetics, numerical analysis, and computer science.

December 2011

List of Abbreviations

AMLFMA: Approximate multilevel fast multipole algorithm
BCE: Boundary-condition error
BDP: Block-diagonal preconditioner
BiCG: Biconjugate gradient (method)
BiCGStab: Biconjugate gradient stabilized (method)
CFIE: Combined-field integral equation
CG: Conjugate gradient (method)
CGS: Conjugate gradient squared (method)
CMM: Composite metamaterial
CTF: Combined tangential formulation
CNF: Combined normal formulation
DS-CTF: Double-stabilized combined tangential formulation
EFIE: Electric field integral equation
EFVIE: Electric-field volume integral equation
FBS-CTF: Field-based-stabilized combined tangential formulation
FFT: Fast Fourier transform
FGMRES: Flexible generalized minimal residual (method)
FMM: Fast multipole method
GAs: Genetic algorithms
GMRES: Generalized minimal residual (method)
HIE: Hybrid integral equation
ILU: Incomplete LU
IMLFMA: Incomplete multilevel fast multipole algorithm
JMCFIE: Electric and magnetic current combined-field integral equation
LF-MLFMA: Low-frequency multilevel fast multipole algorithm
LHS: Left-hand side
LL: Linear-linear (function)
LOD: Level of distribution
LP: Log-periodic (antennas)
LSQR: Least-squares QR (method)
M-CTF: Modified combined tangential formulation
MFIE: Magnetic-field integral equation
MLFMA: Multilevel fast multipole algorithm
MNMF: Modified normal Müller formulation

MOM:	Method of moments
MPI:	Message passing interface
MVM:	Matrix-vector multiplication
N-:	Normal
NFP:	Near-field preconditioner
NMF:	Normal Müller formulation
NP:	No-preconditioner (case)
OBSF:	Operator-based-stabilized formulation
QMR:	Quasi-minimal residual (method)
PEC:	Perfect electric conductor
PMCHWT:	Poggio-Miller-Chang-Harrington-Wu-Tsai (formulation)
RBC:	Red blood cell
RCS:	Radar cross section
RHS:	Right-hand side
RMS:	Root mean square
RWG:	Rao-Wilton-Glisson (function)
SAI:	Sparse approximate inverse
S-CNF:	Stabilized combined normal formulation
SCS:	Scattering cross section
S-CTF:	Stabilized combined tangential formulation
SRR:	Split-ring resonator
SWG:	Schaubert-Wilton-Glisson (function)
T-:	Tangential
TFQMR:	Transpose-free quasi-minimal residual (method)
2PBDP:	Two-partition block-diagonal preconditioner
4PBDP:	Four-partition block-diagonal preconditioner

1

Basics

This chapter presents some of the preliminaries to the multilevel fast multipole algorithm (MLFMA). Using Maxwell's equations and boundary conditions, surface integral equations are derived to formulate electromagnetics problems involving metallic and dielectric objects. Discretizations of the surface integral equations with basis and testing functions on triangular domains lead to dense matrix equations, which can be solved iteratively via MLFMA. Numerical integrations on triangular domains, different types of excitations, iterative algorithms, and preconditioning are also discussed.

1.1 Introduction

Solving electromagnetics problems is extremely important to analyze electromagnetic interactions of electronic devices with each other and with their environments including living and nonliving objects [1]. A plethora of applications in the areas of antennas [2]–[13], radars [14], optics [15], medical imaging [16], wireless communications [17], nanotechnology [18], metamaterials [19]–[25], photonic crystals [26]–[33], remote sensing, and electronic packaging involve scattering and/or radiation of electromagnetic waves. The following are some examples of popular electromagnetics problems:

- Radiation from antennas, e.g., designing novel antennas and estimating their adverse effects on the human health.
- Indoor and outdoor propagation, e.g., optimizations of wireless electromagnetic sources to cover required areas with minimum power.
- Visualization via optical imaging devices, such as lens systems in microscopy, astronomy, surgery, and biology.
- Transmission through frequency-selective photonic crystals and metamaterials, which are artificial structures having a vast variety of applications, such as invisibility cloaking.
- Detection of ordinary and stealth airborne and naval targets with radar systems using the radar cross section (RCS).

The Multilevel Fast Multipole Algorithm (MLFMA) for Solving Large-Scale Computational Electromagnetics Problems, First Edition. Özgür Ergül and Levent Gürel.

- Scattering from biological structures, such as red blood cells (RBCs), to diagnose various diseases.
- Electromagnetic compatibility of novel electronic devices, such as THz-range circuits.

Solutions of these problems are essential in the design and manufacturing cycles of electronic devices; they enable engineers to explore novel designs and prevent the waste of the resources involved in building prototypes and carrying out laboratory tests.

A quick look at a list of emerging technologies shows that electromagnetic fields will play a major role in the future. Wireless communications, e.g., cellular phones, have already become key components of modern life, with continuously increasing markets. Wireless energy transferring systems allowing unplugged electrical devices are now at the prototype stage, and are a major priority of technology companies. Novel designs in nanotechnology will require better imaging systems with subwavelength focusing abilities that can be achieved with metamaterials. Electromagnetic scattering from human tissues will become more and more popular to design diagnostic tools for noninvasive treatments. Hence, electromagnetics simulations will maintain their importance in the future in testing novel devices based on electromagnetic fields and in reducing their design and manufacturing costs.

Electromagnetics problems can be formulated rigorously with Maxwell's equations. Unfortunately, these equations can only be solved analytically for a few canonical objects, such as a sphere [34]. Recently, computational electromagnetics has become a scientifically popular area, where electromagnetic radiation, scattering, and transmission problems in the aforementioned areas are solved with numerical techniques. Mathematical formulations of physical events lead to a set of equations that can be solved numerically by using computers. Thanks to advances in computer technology and solution algorithms, it has become possible to solve many real-life problems involving complicated structures.

Developing a fast and accurate electromagnetics solver requires a well-designed combination of diverse components in many areas, such as wave theory, integral equations, numerical techniques, fast algorithms, iterative solvers, preconditioners, parallel implementations, high-performance computing, and computer technology. Accurate simulations of real-life problems usually require the solution of numerical problems involving large numbers of unknowns, which cannot be achieved easily, even when using the most powerful computers. Therefore, one needs to develop special acceleration algorithms, such as MLFMA [35], in order to solve those large-scale problems efficiently on relatively inexpensive computing platforms. This book is about MLFMA and its implementation for fast and accurate solutions of electromagnetics problems.

1.2 Simulation Environments Based on MLFMA

Figure 1.1 presents a typical simulation environment based on MLFMA. Surface integral equations are derived from Maxwell's equations without any fundamental approximations and are applied to the modeled geometry represented by surfaces. Simultaneous discretizations of the integral equations and surfaces using a set of basis and testing functions defined on triangles lead to dense matrix equations. These matrix equations can be solved iteratively, where the required matrix-vector multiplications (MVMs) are performed efficiently by MLFMA. Optionally, convergence of iterations can be accelerated with preconditioners. MLFMA is employed on parallel computers, particularly those with distributed-memory architectures, to

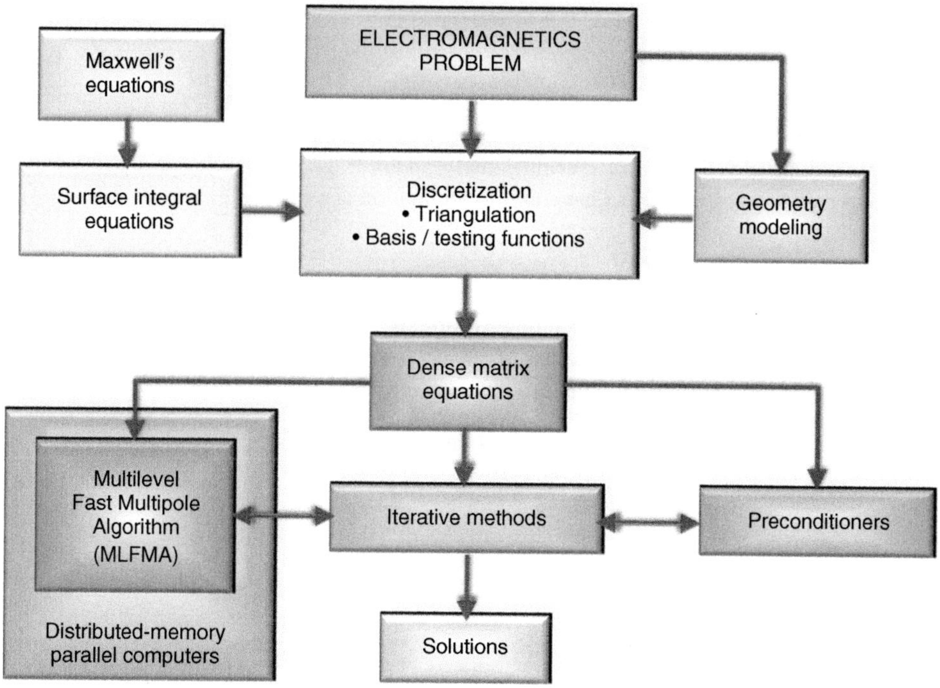

Figure 1.1 A typical simulation environment based on MLFMA.

solve very large problems. Finally, solutions, e.g., scattered or radiated electromagnetic fields, are obtained after a certain number of iterations.

Although Figure 1.1 illustrates the main focus of this book, many different simulation environments can be constructed using alternative components. For example, volume integral equations [36]–[40] discretized with tetrahedra can be used to formulate problems. Instead of MLFMA, one can use the fast Fourier transform (FFT) [41]–[46] or the adaptive integral method [47],[48] to accelerate MVMs. MLFMA can be further accelerated via asymptotic techniques [49],[50]. Besides, in the context of parallelization, it can be employed on shared-memory architectures. Figure 1.1 merely depicts one of the simulation environments that have passed the time test for the accurate and efficient solutions of electromagnetics problems.

1.3 From Maxwell's Equations to Integro-Differential Operators

Maxwell's equations for linear, isotropic, and homogeneous (simple) media can be written as

$$\nabla \times \boldsymbol{E}(\boldsymbol{r}, t) = -\mu \frac{\partial \boldsymbol{H}(\boldsymbol{r}, t)}{\partial t} \tag{1.1}$$

$$\nabla \times \boldsymbol{H}(\boldsymbol{r}, t) = \epsilon \frac{\partial \boldsymbol{E}(\boldsymbol{r}, t)}{\partial t} + \boldsymbol{J}(\boldsymbol{r}, t) \tag{1.2}$$

$$\nabla \cdot \boldsymbol{E}(\boldsymbol{r},t) = \frac{1}{\epsilon}\rho_e(\boldsymbol{r},t) \tag{1.3}$$

$$\nabla \cdot \boldsymbol{H}(\boldsymbol{r},t) = 0, \tag{1.4}$$

where ϵ and μ are the permittivity and permeability of the medium, $\boldsymbol{E}$ and $\boldsymbol{H}$ are the electric and magnetic fields, $\boldsymbol{J}$ is the electric current density, and ρ_e is the electric charge density, which is related to the electric current via the continuity equation as

$$\nabla \cdot \boldsymbol{J}(\boldsymbol{r},t) = -\frac{\partial \rho_e(\boldsymbol{r},t)}{\partial t}. \tag{1.5}$$

In general, electromagnetic fields and sources (currents and charges) depend on position $\boldsymbol{r} = \hat{\boldsymbol{x}}x + \hat{\boldsymbol{y}}y + \hat{\boldsymbol{z}}z$ and time t. Assuming time-harmonic electromagnetic fields and sources with $e^{-i\omega t}$ time dependence, Maxwell's equations can be rewritten as

$$\nabla \times \boldsymbol{E}(\boldsymbol{r}) = i\omega\mu\boldsymbol{H}(\boldsymbol{r}) \tag{1.6}$$

$$\nabla \times \boldsymbol{H}(\boldsymbol{r}) = -i\omega\epsilon\boldsymbol{E}(\boldsymbol{r}) + \boldsymbol{J}(\boldsymbol{r}) \tag{1.7}$$

$$\nabla \cdot \boldsymbol{E}(\boldsymbol{r}) = \frac{1}{\epsilon}\rho_e(\boldsymbol{r}) \tag{1.8}$$

$$\nabla \cdot \boldsymbol{H}(\boldsymbol{r}) = 0, \tag{1.9}$$

where

$$\nabla \cdot \boldsymbol{J}(\boldsymbol{r}) = i\omega\rho_e(\boldsymbol{r}). \tag{1.10}$$

Note that, using the phasor notation,

$$f_s(\boldsymbol{r},t) = \mathrm{Re}\{f_s(\boldsymbol{r})\exp(-i\omega t)\} \tag{1.11}$$

$$\boldsymbol{f}_v(\boldsymbol{r},t) = \mathrm{Re}\{\boldsymbol{f}_v(\boldsymbol{r})\exp(-i\omega t)\} \tag{1.12}$$

$$\frac{\partial f_s(\boldsymbol{r},t)}{\partial t} = -i\omega\mathrm{Re}\{f_s(\boldsymbol{r})\exp(-i\omega t)\} \tag{1.13}$$

for a scalar f_s and a vector $\boldsymbol{f}_v$.

Applying the equivalence principle on dielectric objects introduces magnetic sources, in addition to electric sources. Maxwell's equations can be extended by including the magnetic current density $\boldsymbol{M}$ and the magnetic charge density ρ_m as

$$\nabla \times \boldsymbol{E}(\boldsymbol{r}) = i\omega\mu\boldsymbol{H}(\boldsymbol{r}) - \boldsymbol{M}(\boldsymbol{r}) \tag{1.14}$$

$$\nabla \times \boldsymbol{H}(\boldsymbol{r}) = -i\omega\epsilon\boldsymbol{E}(\boldsymbol{r}) + \boldsymbol{J}(\boldsymbol{r}) \tag{1.15}$$

$$\nabla \cdot \boldsymbol{E}(\boldsymbol{r}) = \frac{1}{\epsilon}\rho_e(\boldsymbol{r}) \tag{1.16}$$

$$\nabla \cdot \boldsymbol{H}(\boldsymbol{r}) = \frac{1}{\mu}\rho_m(\boldsymbol{r}), \tag{1.17}$$

where

$$\nabla \cdot \boldsymbol{M}(\boldsymbol{r}) = i\omega\rho_m(\boldsymbol{r}) \tag{1.18}$$

is the continuity equation for the magnetic sources.

A set of Helmholtz equations for the electric and magnetic fields can be directly derived from Maxwell's equations. Taking the curl of (1.14) and using (1.15), (1.16), and the identity

$$\nabla \times \nabla \times \boldsymbol{f}_v = \nabla\nabla \cdot \boldsymbol{f}_v - \nabla^2 \boldsymbol{f}_v, \tag{1.19}$$

one can obtain

$$\nabla \times \nabla \times \boldsymbol{E}(\boldsymbol{r}) = i\omega\mu\nabla \times \boldsymbol{H}(\boldsymbol{r}) - \nabla \times \boldsymbol{M}(\boldsymbol{r}) \tag{1.20}$$

$$\nabla\nabla \cdot \boldsymbol{E}(\boldsymbol{r}) - \nabla^2 \boldsymbol{E}(\boldsymbol{r}) = i\omega\mu[-i\omega\epsilon\boldsymbol{E}(\boldsymbol{r}) + \boldsymbol{J}(\boldsymbol{r})] - \nabla \times \boldsymbol{M}(\boldsymbol{r}) \tag{1.21}$$

$$\frac{1}{\epsilon}\nabla\rho_e(\boldsymbol{r}) - \nabla^2 \boldsymbol{E}(\boldsymbol{r}) = \omega^2\mu\epsilon\boldsymbol{E}(\boldsymbol{r}) + i\omega\mu\boldsymbol{J}(\boldsymbol{r}) - \nabla \times \boldsymbol{M}(\boldsymbol{r}) \tag{1.22}$$

$$\nabla^2 \boldsymbol{E}(\boldsymbol{r}) + k^2\boldsymbol{E}(\boldsymbol{r}) = -i\omega\mu\boldsymbol{J}(\boldsymbol{r}) + \frac{1}{\epsilon}\nabla\rho_e(\boldsymbol{r}) + \nabla \times \boldsymbol{M}(\boldsymbol{r}), \tag{1.23}$$

where $k = \omega\sqrt{\epsilon\mu} = 2\pi/\lambda$ is the wavenumber and λ is the wavelength. Similarly, taking the curl of (1.15) and using (1.14), (1.17), and (1.19),

$$\nabla \times \nabla \times \boldsymbol{H}(\boldsymbol{r}) = -i\omega\epsilon\nabla \times \boldsymbol{E}(\boldsymbol{r}) + \nabla \times \boldsymbol{J}(\boldsymbol{r}) \tag{1.24}$$

$$\nabla\nabla \cdot \boldsymbol{H}(\boldsymbol{r}) - \nabla^2 \boldsymbol{H}(\boldsymbol{r}) = -i\omega\epsilon[i\omega\mu\boldsymbol{H}(\boldsymbol{r}) - \boldsymbol{M}(\boldsymbol{r})] + \nabla \times \boldsymbol{J}(\boldsymbol{r}) \tag{1.25}$$

$$\frac{1}{\mu}\nabla\rho_m(\boldsymbol{r}) - \nabla^2 \boldsymbol{H}(\boldsymbol{r}) = \omega^2\mu\epsilon\boldsymbol{H}(\boldsymbol{r}) + i\omega\epsilon\boldsymbol{M}(\boldsymbol{r}) + \nabla \times \boldsymbol{J}(\boldsymbol{r}) \tag{1.26}$$

$$\nabla^2 \boldsymbol{H}(\boldsymbol{r}) + k^2\boldsymbol{H}(\boldsymbol{r}) = -i\omega\epsilon\boldsymbol{M}(\boldsymbol{r}) + \frac{1}{\mu}\nabla\rho_m(\boldsymbol{r}) - \nabla \times \boldsymbol{J}(\boldsymbol{r}). \tag{1.27}$$

By writing the fields in terms of the vector and scalar potentials as

$$\boldsymbol{E}(\boldsymbol{r}) = i\omega\boldsymbol{A}_m(\boldsymbol{r}) - \nabla\phi_e(\boldsymbol{r}) - \frac{1}{\epsilon}\nabla \times \boldsymbol{A}_e(\boldsymbol{r}) \tag{1.28}$$

$$\boldsymbol{H}(\boldsymbol{r}) = i\omega\boldsymbol{A}_e(\boldsymbol{r}) - \nabla\phi_m(\boldsymbol{r}) + \frac{1}{\mu}\nabla \times \boldsymbol{A}_m(\boldsymbol{r}) \tag{1.29}$$

and using the Lorentz gauge, i.e.,

$$\nabla \cdot \boldsymbol{A}_m(\boldsymbol{r}) = i\omega\epsilon\mu\phi_e(\boldsymbol{r}) \tag{1.30}$$

$$\nabla \cdot \boldsymbol{A}_e(\boldsymbol{r}) = i\omega\mu\epsilon\phi_m(\boldsymbol{r}), \tag{1.31}$$

a set of Helmholtz equations for the potentials can be derived as

$$\nabla^2\phi_e(\boldsymbol{r}) + k^2\phi_e(\boldsymbol{r}) = -\frac{1}{\epsilon}\rho_e(\boldsymbol{r}) \tag{1.32}$$

$$\nabla^2\phi_m(\boldsymbol{r}) + k^2\phi_m(\boldsymbol{r}) = -\frac{1}{\mu}\rho_m(\boldsymbol{r}) \tag{1.33}$$

$$\nabla^2\boldsymbol{A}_e(\boldsymbol{r}) + k^2\boldsymbol{A}_e(\boldsymbol{r}) = -\epsilon\boldsymbol{M}(\boldsymbol{r}) \tag{1.34}$$

$$\nabla^2\boldsymbol{A}_m(\boldsymbol{r}) + k^2\boldsymbol{A}_m(\boldsymbol{r}) = -\mu\boldsymbol{J}(\boldsymbol{r}). \tag{1.35}$$

The Helmholtz equations in (1.32)–(1.35) can be solved for arbitrary current and charge distributions, leading to

$$\phi_e(\boldsymbol{r}) = \frac{1}{\epsilon}\int d\boldsymbol{r}' g(\boldsymbol{r},\boldsymbol{r}')\rho_e(\boldsymbol{r}') \tag{1.36}$$

$$\phi_m(\boldsymbol{r}) = \frac{1}{\mu}\int d\boldsymbol{r}' g(\boldsymbol{r},\boldsymbol{r}')\rho_m(\boldsymbol{r}') \tag{1.37}$$

$$\boldsymbol{A}_e(\boldsymbol{r}) = \epsilon\int d\boldsymbol{r}' g(\boldsymbol{r},\boldsymbol{r}')\boldsymbol{M}(\boldsymbol{r}') \tag{1.38}$$

$$\boldsymbol{A}_m(\boldsymbol{r}) = \mu\int d\boldsymbol{r}' g(\boldsymbol{r},\boldsymbol{r}')\boldsymbol{J}(\boldsymbol{r}'), \tag{1.39}$$

where

$$g(\boldsymbol{r},\boldsymbol{r}') = \frac{\exp(ikR)}{4\pi R} \qquad (R = |\boldsymbol{r}-\boldsymbol{r}'|) \tag{1.40}$$

is the homogeneous-space Green's function. Using (1.36)–(1.39) in (1.28) and (1.29), electromagnetic fields due to arbitrary current and charge distributions can be written as

$$\begin{aligned}\boldsymbol{E}(\boldsymbol{r}) &= ik\eta\int d\boldsymbol{r}'\left[\boldsymbol{J}(\boldsymbol{r}') + \frac{1}{k^2}\nabla'\cdot\boldsymbol{J}(\boldsymbol{r}')\nabla\right]g(\boldsymbol{r},\boldsymbol{r}')\\ &\quad - \int d\boldsymbol{r}'\nabla g(\boldsymbol{r},\boldsymbol{r}')\times\boldsymbol{M}(\boldsymbol{r}')\end{aligned} \tag{1.41}$$

$$\begin{aligned}\boldsymbol{H}(\boldsymbol{r}) &= ik\eta^{-1}\int d\boldsymbol{r}'\left[\boldsymbol{M}(\boldsymbol{r}') + \frac{1}{k^2}\nabla'\cdot\boldsymbol{M}(\boldsymbol{r}')\nabla\right]g(\boldsymbol{r},\boldsymbol{r}')\\ &\quad + \int d\boldsymbol{r}'\nabla g(\boldsymbol{r},\boldsymbol{r}')\times\boldsymbol{J}(\boldsymbol{r}'),\end{aligned} \tag{1.42}$$

where $\eta = \sqrt{\frac{\mu}{\epsilon}}$ is the intrinsic impedance. Finally, by defining the integro-differential operators $\mathcal{T}$ and $\mathcal{K}$ as

$$\mathcal{T}\{\boldsymbol{X}\}(\boldsymbol{r}) = ik\int d\boldsymbol{r}'\left[\boldsymbol{X}(\boldsymbol{r}') + \frac{1}{k^2}\nabla'\cdot\boldsymbol{X}(\boldsymbol{r}')\nabla\right]g(\boldsymbol{r},\boldsymbol{r}') \tag{1.43}$$

$$\mathcal{K}\{\boldsymbol{X}\}(\boldsymbol{r}) = \int d\boldsymbol{r}'\boldsymbol{X}(\boldsymbol{r}')\times\nabla' g(\boldsymbol{r},\boldsymbol{r}'), \tag{1.44}$$

the expressions for the electromagnetic fields can be rewritten as

$$\boldsymbol{E}(\boldsymbol{r}) = \eta\mathcal{T}\{\boldsymbol{J}\}(\boldsymbol{r}) - \mathcal{K}\{\boldsymbol{M}\}(\boldsymbol{r}) \tag{1.45}$$

$$\boldsymbol{H}(\boldsymbol{r}) = \eta^{-1}\mathcal{T}\{\boldsymbol{M}\}(\boldsymbol{r}) + \mathcal{K}\{\boldsymbol{J}\}(\boldsymbol{r}). \tag{1.46}$$

Equations (1.45) and (1.46) clearly show the dependence of electromagnetic fields on electromagnetic sources in a simple medium.

1.4 Surface Integral Equations

In general, electromagnetics problems involve three-dimensional objects with arbitrary geometries that can be modeled with perfectly conducting and dielectric (lossy or lossless) regions. As depicted in Figure 1.2, the equivalence principle can be used to decompose the original problem into equivalent problems by employing equivalent currents on surfaces. Equivalent problems are defined for all nonmetallic regions, including the host medium extending to infinity. Then, integral equations can be derived by applying the operators on the currents and enforcing the boundary conditions for the tangential components of the electromagnetic fields.

Consider a homogeneous dielectric region D_u bounded by a surface S_u. As depicted in Figure 1.3, this region may extend to infinity. Considering the tangential components of the

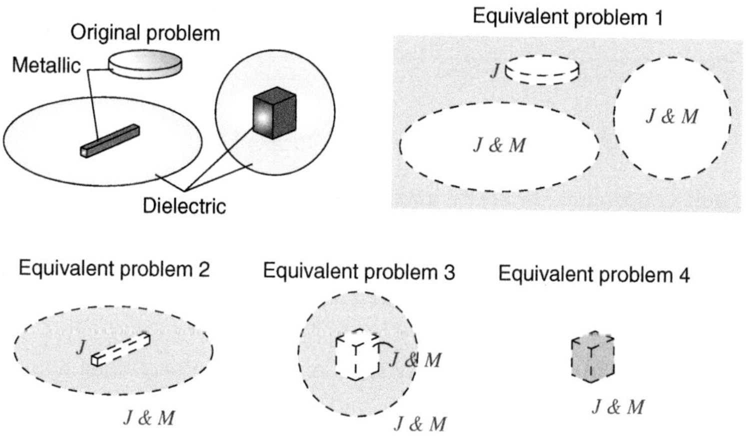

Figure 1.2 Applying the equivalence principle to an electromagnetics problem involving multiple dielectric and metallic regions.

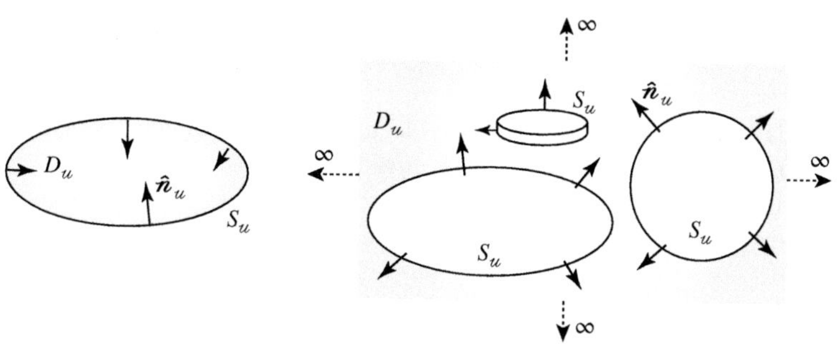

Figure 1.3 A dielectric region enclosed in a surface and a dielectric region extending to infinity. In each case, the region is bounded by a surface S_u, which consists of multiple unconnected parts in the second case.

electric and magnetic fields on S_u, one can derive

$$-\eta_u \hat{\boldsymbol{n}}_u \times \hat{\boldsymbol{n}}_u \times \mathcal{T}_u\{\boldsymbol{J}_u\}(\boldsymbol{r}) + \hat{\boldsymbol{n}}_u \times \hat{\boldsymbol{n}}_u \times \mathcal{K}_u\{\boldsymbol{M}_u\}(\boldsymbol{r}) \\ - \hat{\boldsymbol{n}}_u \times \hat{\boldsymbol{n}}_u \times \boldsymbol{E}_u^{inc}(\boldsymbol{r}) = -\hat{\boldsymbol{n}}_u \times \hat{\boldsymbol{n}}_u \times \boldsymbol{E}_u(\boldsymbol{r}) \tag{1.47}$$

$$-\eta_u^{-1} \hat{\boldsymbol{n}}_u \times \hat{\boldsymbol{n}}_u \times \mathcal{T}_u\{\boldsymbol{M}_u\}(\boldsymbol{r}) - \hat{\boldsymbol{n}}_u \times \hat{\boldsymbol{n}}_u \times \mathcal{K}_u\{\boldsymbol{J}_u\}(\boldsymbol{r}) \\ - \hat{\boldsymbol{n}}_u \times \hat{\boldsymbol{n}}_u \times \boldsymbol{H}_u^{inc}(\boldsymbol{r}) = -\hat{\boldsymbol{n}}_u \times \hat{\boldsymbol{n}}_u \times \boldsymbol{H}_u(\boldsymbol{r}), \tag{1.48}$$

where $\boldsymbol{r} \in S_u$, $\hat{\boldsymbol{n}}_u$ is the inward normal unit vector, and $\boldsymbol{E}_u^{inc}$ and $\boldsymbol{H}_u^{inc}$ are the incident electric and magnetic fields, respectively, created by external sources located in D_u. The equivalent currents equivalent electric and magnetic currents are defined as

$$\boldsymbol{J}_u(\boldsymbol{r}) = \mathcal{I}^{\times n_u}\{\boldsymbol{H}_u\}(\boldsymbol{r}) = \hat{\boldsymbol{n}}_u \times \boldsymbol{H}_u(\boldsymbol{r}) \tag{1.49}$$

$$\boldsymbol{M}_u(\boldsymbol{r}) = -\mathcal{I}^{\times n_u}\{\boldsymbol{E}_u\}(\boldsymbol{r}) = -\hat{\boldsymbol{n}}_u \times \boldsymbol{E}_u(\boldsymbol{r}), \tag{1.50}$$

where

$$\mathcal{I}^{\times n_u}\{\boldsymbol{X}\}(\boldsymbol{r}) = \hat{\boldsymbol{n}}_u \times \mathcal{I}\{\boldsymbol{X}\}(\boldsymbol{r}) = \hat{\boldsymbol{n}}_u \times \boldsymbol{X}(\boldsymbol{r}) \tag{1.51}$$

and $\mathcal{I}$ is the identity operator. The operator $\mathcal{K}_u$ is commonly separated into principal-value and limit parts (see Section 6.1) as

$$\mathcal{K}_u\{\boldsymbol{X}\}(\boldsymbol{r}) = \mathcal{K}_{PV,u}\{\boldsymbol{X}\}(\boldsymbol{r}) - \frac{4\pi - \Omega_u}{4\pi} \mathcal{I}^{\times n_u}\{\boldsymbol{X}\}(\boldsymbol{r}), \tag{1.52}$$

where $0 \le \Omega_u \le 4\pi$ is the solid angle that is nonzero when the observation point $\boldsymbol{r}$ is on the surface. Using (1.52) in (1.47) and (1.48), and rearranging the terms, one can obtain

$$-\eta_u \hat{\boldsymbol{n}}_u \times \hat{\boldsymbol{n}}_u \times \mathcal{T}_u\{\boldsymbol{J}_u\}(\boldsymbol{r}) + \hat{\boldsymbol{n}}_u \times \hat{\boldsymbol{n}}_u \times \mathcal{K}_{PV,u}\{\boldsymbol{M}_u\}(\boldsymbol{r}) \\ - \frac{\Omega_u}{4\pi} \mathcal{I}^{\times n_u}\{\boldsymbol{M}_u\}(\boldsymbol{r}) = \hat{\boldsymbol{n}}_u \times \hat{\boldsymbol{n}}_u \times \boldsymbol{E}_u^{inc}(\boldsymbol{r}) \tag{1.53}$$

$$-\eta_u^{-1} \hat{\boldsymbol{n}}_u \times \hat{\boldsymbol{n}}_u \times \mathcal{T}_u\{\boldsymbol{M}_u\}(\boldsymbol{r}) - \hat{\boldsymbol{n}}_u \times \hat{\boldsymbol{n}}_u \times \mathcal{K}_{PV,u}\{\boldsymbol{J}_u\}(\boldsymbol{r}) \\ + \frac{\Omega_u}{4\pi} \mathcal{I}^{\times n_u}\{\boldsymbol{J}_u\}(\boldsymbol{r}) = \hat{\boldsymbol{n}}_u \times \hat{\boldsymbol{n}}_u \times \boldsymbol{H}_u^{inc}(\boldsymbol{r}). \tag{1.54}$$

Equations (1.53) and (1.54) are known as the electric-field integral equation (EFIE) and the magnetic-field integral equation (MFIE), respectively.

As discussed in Chapter 2, EFIE and MFIE each have two forms. In the forms given in (1.53) and (1.54), the tangential components of the electric and magnetic fields are sampled directly since

$$-\hat{\boldsymbol{n}} \times \hat{\boldsymbol{n}} \times \boldsymbol{f}_\upsilon = \boldsymbol{f}_\upsilon - \hat{\boldsymbol{n}}\hat{\boldsymbol{n}} \cdot \boldsymbol{f}_\upsilon \tag{1.55}$$

for an arbitrary vector $\boldsymbol{f}_\upsilon$. Hence, (1.53) and (1.54) are commonly known as tangential integral equations and denoted with T-EFIE and T-MFIE, respectively. On the other hand, it is also possible to sample a rotated version of the tangential electromagnetic fields. Specifically, (1.53) and (1.54) can be modified as

$$\eta_u \hat{\boldsymbol{n}}_u \times \mathcal{T}_u\{\boldsymbol{J}_u\}(\boldsymbol{r}) - \hat{\boldsymbol{n}}_u \times \mathcal{K}_{PV,u}\{\boldsymbol{M}_u\}(\boldsymbol{r}) + \frac{\Omega_u}{4\pi}\mathcal{I}\{\boldsymbol{M}_u\}(\boldsymbol{r}) = -\hat{\boldsymbol{n}}_u \times \boldsymbol{E}_u^{inc}(\boldsymbol{r}) \tag{1.56}$$

$$\eta_u^{-1} \hat{\boldsymbol{n}}_u \times \mathcal{T}_u\{\boldsymbol{M}_u\}(\boldsymbol{r}) + \hat{\boldsymbol{n}}_u \times \mathcal{K}_{PV,u}\{\boldsymbol{J}_u\}(\boldsymbol{r}) - \frac{\Omega_u}{4\pi}\mathcal{I}\{\boldsymbol{J}_u\}(\boldsymbol{r}) = -\hat{\boldsymbol{n}}_u \times \boldsymbol{H}_u^{inc}(\boldsymbol{r}), \tag{1.57}$$

which are commonly known as normal integral equations and denoted with N-EFIE and N-MFIE, respectively. It should be emphasized that normal integral equations do not involve electromagnetic fields in normal directions; this convention is used throughout the book.

1.5 Boundary Conditions

Solutions of the surface integral equations (EFIE and MFIE) to compute the electric and magnetic currents require the use of appropriate boundary conditions at boundaries. As depicted in Figure 1.4, two scenarios are of particular interest. When two dielectric regions D_u and D_v are separated by a surface S_{uv}, the tangential electric and magnetic fields are continuous across the surface. This can be written as

$$\boldsymbol{J}_u(\boldsymbol{r}) = \hat{\boldsymbol{n}}_u \times \boldsymbol{H}_u(\boldsymbol{r}) = -\boldsymbol{J}_v(\boldsymbol{r}) = -\hat{\boldsymbol{n}}_v \times \boldsymbol{H}_v(\boldsymbol{r}) \tag{1.58}$$

$$\boldsymbol{M}_u(\boldsymbol{r}) = -\hat{\boldsymbol{n}}_u \times \boldsymbol{E}_u(\boldsymbol{r}) = -\boldsymbol{M}_v(\boldsymbol{r}) = \hat{\boldsymbol{n}}_v \times \boldsymbol{E}_v(\boldsymbol{r}), \tag{1.59}$$

or simply

$$\boldsymbol{J}_u(\boldsymbol{r}) = -\boldsymbol{J}_v(\boldsymbol{r}) = \boldsymbol{J}(\boldsymbol{r}) = \hat{\boldsymbol{n}} \times \boldsymbol{H}(\boldsymbol{r}) \tag{1.60}$$

$$\boldsymbol{M}_u(\boldsymbol{r}) = -\boldsymbol{M}_v(\boldsymbol{r}) = \boldsymbol{M}(\boldsymbol{r}) = -\hat{\boldsymbol{n}} \times \boldsymbol{E}(\boldsymbol{r}) \tag{1.61}$$

with $\hat{\boldsymbol{n}} = \hat{\boldsymbol{n}}_u$. Hence, the integral equations derived for D_u and D_v can be combined in different ways to solve for $\boldsymbol{J}$ and $\boldsymbol{M}$.

In many electromagnetics problems, some of the objects can be modeled as perfect electric conductors (PECs). Since electromagnetics fields cannot penetrate into PECs, surface integral equations are not required inside such objects. On a surface separating a dielectric medium

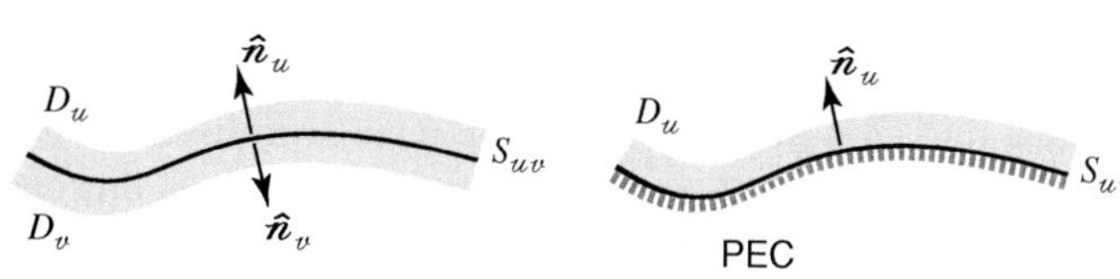

Figure 1.4 Boundaries between two dielectric regions and on a PEC surface.

from a PEC, the magnetic current is identically zero and the electric current corresponds to the induced electric current, i.e.,

$$\boldsymbol{J}_u(\boldsymbol{r}) = \hat{\boldsymbol{n}} \times \boldsymbol{H}(\boldsymbol{r}) = \boldsymbol{J}^{ind}(\boldsymbol{r}) \tag{1.62}$$

$$\boldsymbol{M}_u(\boldsymbol{r}) = 0. \tag{1.63}$$

As discussed in Chapter 2, (1.62) and (1.63) are also valid for an open PEC with zero volume, since an open surface can be considered as a limiting case of a closed surface.

Finally, finding the equivalent currents $\boldsymbol{J}_u$ and $\boldsymbol{M}_u$ on the surface S_u bounding D_u, the secondary (scattered or radiated) electromagnetic fields can be computed everywhere inside D_u (and on S_u) as

$$\boldsymbol{E}_u^{sec}(\boldsymbol{r}) = \eta_u \mathcal{T}_u\{\boldsymbol{J}_u\}(\boldsymbol{r}) - \mathcal{K}_{PV,u}\{\boldsymbol{M}_u\}(\boldsymbol{r}) + \frac{4\pi - \Omega_u}{4\pi} \mathcal{I}^{\times n_u}\{\boldsymbol{M}_u\}(\boldsymbol{r}) \tag{1.64}$$

$$\boldsymbol{H}_u^{sec}(\boldsymbol{r}) = \eta_u^{-1} \mathcal{T}_u\{\boldsymbol{M}_u\}(\boldsymbol{r}) + \mathcal{K}_{PV,u}\{\boldsymbol{J}_u\}(\boldsymbol{r}) - \frac{4\pi - \Omega_u}{4\pi} \mathcal{I}^{\times n_u}\{\boldsymbol{J}_u\}(\boldsymbol{r}). \tag{1.65}$$

1.6 Surface Formulations

For the solution of electromagnetics problems, various surface formulations can be derived by using diverse combinations of surface integral equations. Some of these formulations are known to be stable and provide accurate results [51]–[83], although their performances may vary significantly in terms of efficiency and accuracy. In general, surface formulations can be categorized into three groups, i.e., tangential, normal, and mixed, depending on their contents. Tangential formulations involve T-EFIE and/or T-MFIE, while normal formulations involve N-EFIE and/or N-MFIE. Mixed formulations are obtained by combining the tangential and normal formulations, and they contain at least one tangential equation and at least one normal equation.

For dielectric objects with homogeneous material properties, surface formulations are derived by properly combining the equations obtained for the inner and outer regions. To avoid numerical internal resonances, it is preferable to derive a combined-field integral equation (CFIE) by linearly combining EFIE and MFIE [62]. For example, a combination of T-EFIE and T-MFIE leads to a T-T-CFIE formulation. Similarly, one can obtain T-N-CFIE (T-EFIE+N-MFIE), N-T-CFIE (N-EFIE+T-MFIE), and N-N-CFIE (N-EFIE+N-MFIE) by combining EFIE and MFIE appropriately. In these formulations, the equations obtained for the inner and outer regions are solved simultaneously to obtain the two sets of unknowns, i.e., electric and magnetic currents. One can also derive various other formulations involving triple combinations, such as TN-N-CFIE (T-EFIE+N-EFIE+N-MFIE), for more stable solutions [68].

The CFIE formulations described above are based on a linear combination of EFIE and MFIE in the same way for each medium, but it is also possible to use different combinations for the inner and outer regions [69]. Alternatively, it is quite popular to linearly combine the inner and outer equations while solving EFIE and MFIE simultaneously. In this class of formulations, the Poggio-Miller-Chang-Harrington-Wu-Tsai (PMCHWT) [52]–[54] and Müller [51] formulations are well known and commonly used in the literature. Both tangential and normal versions of these formulations are possible; however, using a Galerkin discretization, only the tangential PMCHWT (T-PMCHWT) formulation and the normal Müller formulation (NMF) are stable, whereas the normal PMCHWT and the tangential Müller formulations are usually unstable. Recently, remarkable efforts have been made to further improve dielectric formulations by devising novel combinations of integral equations. For example, the combined tangential formulation (CTF) [74] is similar to the T-PMCHWT formulation, but it involves a careful (and improved) scaling of T-EFIE and T-MFIE. A similar combination of N-EFIE and N-MFIE leads to the combined normal formulation (CNF) [74]. The modified NMF (MNMF) is obtained by normalizing equations in NMF to produce better-conditioned matrix equations [73]. Finally, a mixed formulation called the electric and magnetic current combined-field integral equation (JMCFIE) is derived by combining all four types of integral equations [71],[78],[80],[83].

Surface formulations of dielectric objects tend to be less accurate as the contrast of the object decreases and electromagnetic properties change slightly across dielectric interfaces. There are various applications that involve scattering from low-contrast objects, such as dielectric photonic crystals [29], RBCs in blood plasma [84]–[92], plastic land mines buried in soil [93], and polymeric materials [94]. When the contrast is low, however, conventional surface formulations encounter stability problems, and scattered fields cannot be calculated accurately with them. Those scattering problems can be solved accurately with volume formulations [36]–[39], which are usually stable independent of the contrast. On the other hand, it is also desirable to extend the applicability of surface integral equations to low-contrast problems to be able to take the advantage of surface formulations, which are usually discretized with fewer unknowns compared to volume formulations. The inaccuracy of surface formulations for the solution of low-contrast problems is due to insufficient modeling of radiating parts of equivalent currents defined on objects [95]. By extracting nonradiating currents and solving modified equations only for radiating currents (similar to volume formulations), scattered fields from low-contrast objects can be calculated accurately [96],[97]. These types of formulations are extensively discussed in Chapter 2, Section 2.2.

When the object is a PEC, T-EFIE, T-MFIE, N-EFIE, and N-MFIE can be solved independently to obtain the induced electric current on the surface. Especially, T-EFIE and N-MFIE are commonly used in the literature [52],[57],[61]. However, for closed conductors, those formulations suffer from internal resonances. Therefore, similar to dielectric formulations, it is essential to combine EFIE and MFIE appropriately to derive various CFIE formulations [55]. Conventional CFIE formulations are usually obtained by convex combinations of EFIE and MFIE. However, there are many studies presenting hybrid formulations, where EFIE and MFIE are coupled in diverse ways. For example, EFIE and MFIE can be applied on different parts of the object to efficiently and accurately solve electromagnetics problems involving complicated structures [59],[64],[65],[76],[82]. This type of hybrid formulations are detailed in Chapter 3, Section 3.6.1.

For PEC objects, T-EFIE suffers from the low-frequency breakdown problem (see Chapter 2, Section 2.3.4) when it is discretized with ordinary basis functions [98],[99]. Specifically, matrix equations obtained from T-EFIE become increasingly ill-conditioned as the discretization is refined. The T-N-CFIE formulation is also affected by the low-frequency breakdown since it contains T-EFIE [100],[101], whereas N-MFIE is usually stable unless the frequency is extremely low [102]. To eliminate the low-frequency breakdown problem, one can apply loop-star or loop-tree decomposition methods [103]–[107]. In addition, current and charge decompositions can be used [108],[109] to derive very stable formulations at low frequencies. Recently, regularization of T-EFIE using Calderon identities has become very popular to tackle the low-frequency breakdown issue [101],[110]–[116].

1.7 Method of Moments and Discretization

Electromagnetics problems involving complicated objects can be discretized and solved numerically. By means of a simultaneous discretization of the geometry and surface integral equations, the equivalent currents are expanded in a series of basis functions. Then, the coefficients of the basis functions are calculated by solving dense matrix equations, which are derived by using the method of moments (MOM) [117],[118].

Let $\mathcal{L}$ be a linear operator applied on an unknown vector function $\boldsymbol{f}(\boldsymbol{r})$ to produce a known vector function $\boldsymbol{g}(\boldsymbol{r})$, i.e.,

$$\mathcal{L}\{\boldsymbol{f}\}(\boldsymbol{r}) = \boldsymbol{g}(\boldsymbol{r}). \tag{1.66}$$

Considering surface formulations, $\mathcal{L}$ can be a combination of the integro-differential $\mathcal{T}$ and $\mathcal{K}$ operators as well as the identity operator $\mathcal{I}$, while the right-hand-side (RHS) function can be a combination of the incident electromagnetic fields created by external sources. Expanding $\boldsymbol{f}(\boldsymbol{r})$ in a series of known basis functions $\boldsymbol{b}_n$ $(n = 1, 2, \ldots, N)$, i.e.,

$$\boldsymbol{f}(\boldsymbol{r}) \approx \sum_{n=1}^{N} \boldsymbol{a}[n]\boldsymbol{b}_n(\boldsymbol{r}), \tag{1.67}$$

and testing (1.66) using a set of testing functions $\boldsymbol{t}_m$ $(m = 1, 2, \ldots, N)$, one can obtain

$$\int d\boldsymbol{r}\boldsymbol{t}_m(\boldsymbol{r}) \cdot \sum_{n=1}^{N} \boldsymbol{a}[n]\mathcal{L}\{\boldsymbol{b}_n\}(\boldsymbol{r}) = \int d\boldsymbol{r}\boldsymbol{t}_m(\boldsymbol{r}) \cdot \boldsymbol{g}(\boldsymbol{r}) \qquad (m = 1, 2, \ldots, N). \tag{1.68}$$

In (1.67) and (1.68), $\boldsymbol{a}[n]$ represents the nth unknown coefficient, i.e., nth element of the unknown vector $\boldsymbol{a}$. Changing the order of the summation and integration, the equation becomes

$$\sum_{n=1}^{N} \boldsymbol{a}[n] \int d\boldsymbol{r}\boldsymbol{t}_m(\boldsymbol{r}) \cdot \mathcal{L}\{\boldsymbol{b}_n\}(\boldsymbol{r}) = \int d\boldsymbol{r}\boldsymbol{t}_m(\boldsymbol{r}) \cdot \boldsymbol{g}(\boldsymbol{r}) \qquad (m = 1, 2, \ldots, N), \tag{1.69}$$

which can be written as a matrix equation, i.e.,

$$\sum_{n=1}^{N} \boldsymbol{a}[n]\overline{\boldsymbol{Z}}[m,n] = \boldsymbol{w}[m] \qquad (m = 1, 2, \ldots, N), \tag{1.70}$$

where

$$\overline{\boldsymbol{Z}}[m,n] = \int d\boldsymbol{r}\boldsymbol{t}_m(\boldsymbol{r}) \cdot \mathcal{L}\{\boldsymbol{b}_n\}(\boldsymbol{r}) \tag{1.71}$$

$$\boldsymbol{w}[m] = \int d\boldsymbol{r}\boldsymbol{t}_m(\boldsymbol{r}) \cdot \boldsymbol{g}(\boldsymbol{r}). \tag{1.72}$$

In the context of electromagnetics problems, $\overline{\boldsymbol{Z}}$ and $\boldsymbol{w}$ are usually called the impedance matrix and the excitation vector, respectively. Furthermore, elements of the impedance matrices (obtained by the discretization of the integral-equation formulations) can be interpreted as electromagnetic interactions between discretization elements, i.e., basis and testing functions.

Using small planar triangles on which basis and testing functions are defined is a popular approach to discretize three-dimensional objects. Some examples of the triangulation of closed and open objects are illustrated in Figures 1.5, 1.6, and 1.7. In general, the size of triangles should be small with respect to wavelength, but the discretization depends on the nature of the problem, e.g., geometry and excitation, as well as the desired level of accuracy. There are actually two error sources that can appear in the triangulation stage. The first one is the approximation of the sources (and fields) with discrete functions. The second one is the modeling of

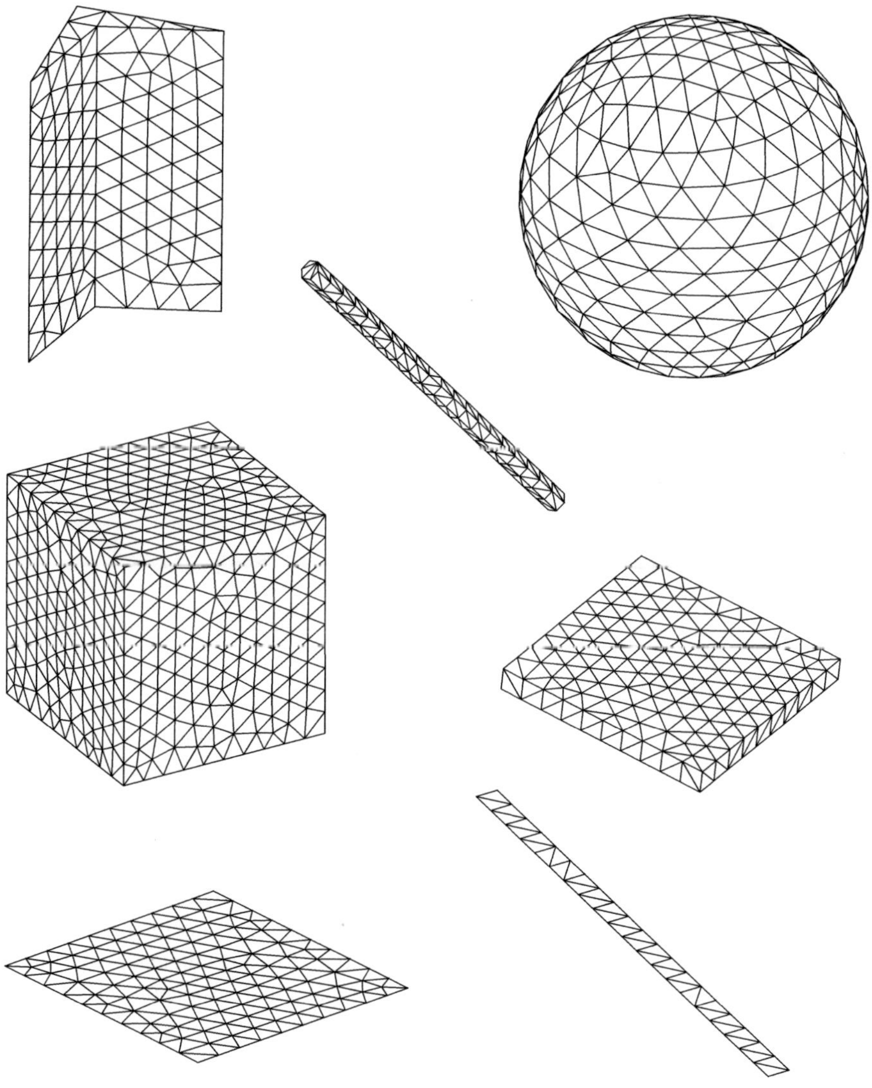

Figure 1.5 Surface models of various canonical geometries with triangular meshing.

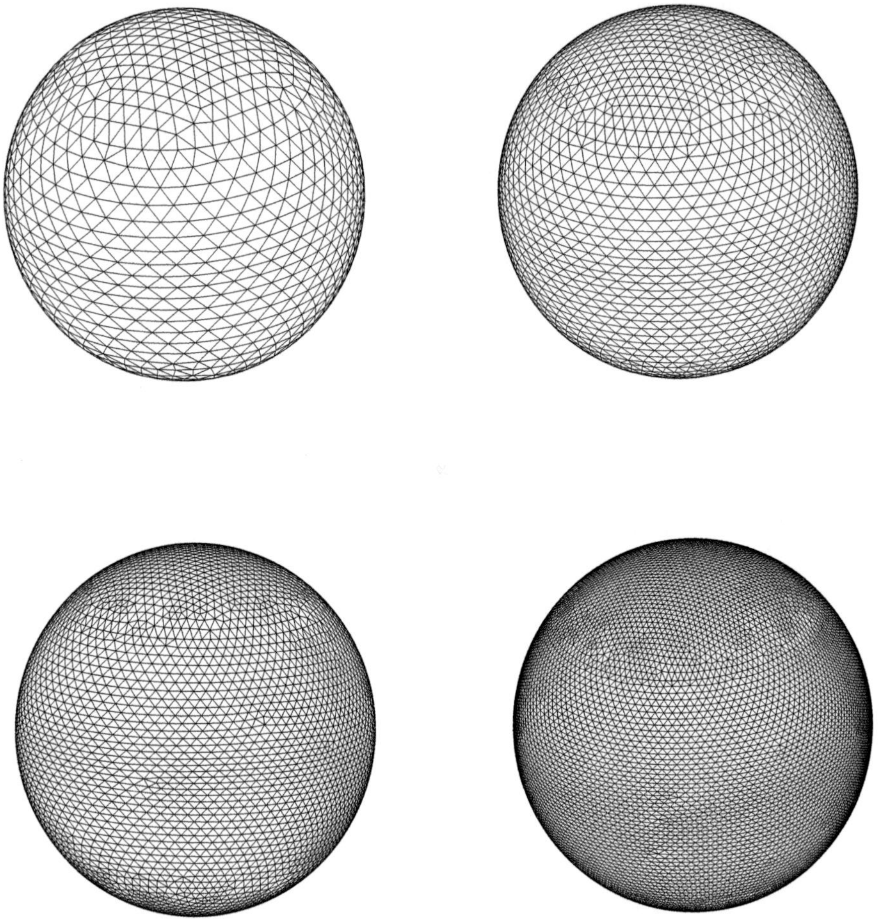

Figure 1.6 Surface models of a sphere with triangular meshing using various mesh sizes.

curved surfaces with planar elements (e.g., see Figure 1.6). Both types of errors may be reduced by decreasing the element size unless limited by low-frequency breakdowns.

Applying a Galerkin scheme for the discretization of surface formulations, i.e., using the same set of functions to expand the current densities (basis functions) and to test the boundary conditions (testing functions), normal and mixed formulations will contain well-tested identity operators. It is well-known that matrix equations involving well-tested identity operators are diagonally dominant and well-conditioned [74],[119],[120]. However, tangential formulations do not contain well-tested identity operators, and their discretizations may lead to ill-conditioned matrix equations. Therefore, for the efficiency of solutions, normal and mixed formulations are usually preferable, especially for large objects discretized with large numbers of unknowns. On the other hand, errors in the discretization of well-tested identity operators may decrease the accuracy of solutions obtained with normal and mixed formulations [120]–[122]. The error level in these formulations compared to tangential formulations can be significant, especially in conventional implementations employing the

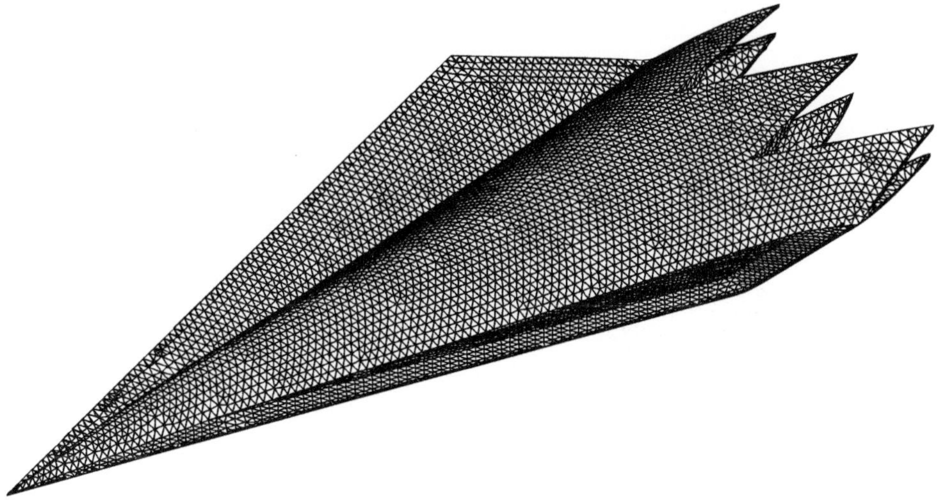

Figure 1.7 Surface model of a stealth airborne target Flamme with triangular meshing.

Rao-Wilton-Glisson (RWG) [57] functions. Investigations on PEC objects show that scattered fields obtained with N-MFIE and T-N-CFIE are significantly inaccurate compared to those obtained with T-EFIE [123]–[129]. Similar observations have been made for the solution of dielectric objects [74],[80]. In such cases, it is helpful to improve the discretizations by employing more appropriate, such as higher-order [130]–[133], basis functions in order to obtain accurate results with normal and mixed formulations [74],[127]–[129],[134]–[139]. Recent studies also show that a regularization of the identity operator improves the accuracy of N-MFIE [140].

1.7.1 Linear Functions

Linear functions are widely used to discretize surface integral equations. Among such functions, the RWG functions [57] are very popular due to their simplicity, flexibility, and usefulness. Over the years, a plethora of electromagnetics problems involving complicated three-dimensional geometries have been solved with the help of the RWG functions. Spatial distributions of the RWG functions are depicted in Figure 1.8, and can be written as

$$\boldsymbol{b}_n^{\mathrm{RWG}}(\boldsymbol{r}) = \begin{cases} \frac{l_n}{2A_{n1}}(\boldsymbol{r}-\boldsymbol{r}_{n1}), & \boldsymbol{r} \in S_{n1} \\ \frac{l_n}{2A_{n2}}(\boldsymbol{r}_{n2}-\boldsymbol{r}), & \boldsymbol{r} \in S_{n2} \\ 0, & \boldsymbol{r} \notin S_n. \end{cases} \tag{1.73}$$

In (1.73), l_n represents the length of the main edge e_n, A_{n1} and A_{n2} are respectively areas of the first (S_{n1}) and the second (S_{n2}) triangles associated with the edge, and $S_n = S_{n1} \bigcup S_{n2}$. Since they have constant normal (CN) and linear tangential (LT) components at the main edge, the RWG functions can be classified as a CN-LT type. Despite their simplicity, they provide the necessary flexibility to model current and field distributions on arbitrarily complicated surfaces with sufficient accuracy. They are also shown to possess the ability to properly model

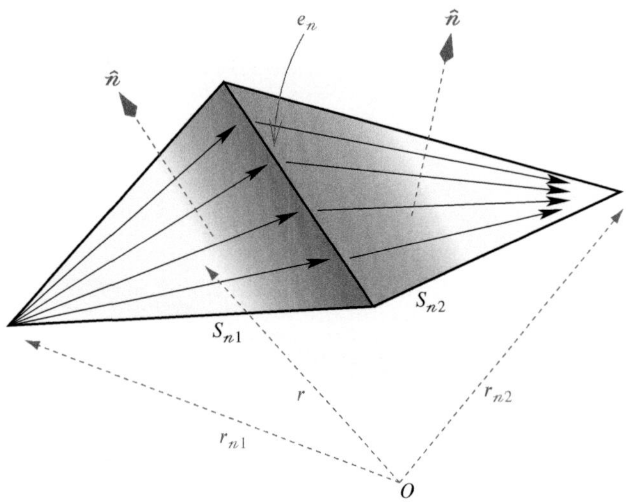

Figure 1.8 Spatial distributions of the RWG functions. Arrows show the direction while the shading indicates the magnitude of the vector distribution. Light and dark colors represent the low and high values, respectively.

the charge distribution, in addition to the current distribution [141]. Therefore, the RWG functions have been extensively used for the discretization of surface integral equations. The RWG functions are divergence conforming and their divergence is finite everywhere, e.g.,

$$\nabla \cdot \boldsymbol{b}_n^{\mathrm{RWG}}(\boldsymbol{r}) = \begin{cases} \dfrac{l_n}{A_{n1}}, & \boldsymbol{r} \in S_{n1} \\ -\dfrac{l_n}{A_{n2}}, & \boldsymbol{r} \in S_{n2} \\ 0, & \boldsymbol{r} \notin S_n. \end{cases} \tag{1.74}$$

This property can be crucial in some formulations, such as T-EFIE. In addition, the total charge distribution associated with an RWG function is identically zero since

$$\int_{S_n} d\boldsymbol{r} \nabla \cdot \boldsymbol{b}_n^{\mathrm{RWG}}(\boldsymbol{r}) = A_{n1}\frac{l_n}{A_{n1}} - A_{n2}\frac{l_n}{A_{n2}} = 0. \tag{1.75}$$

As extensively discussed in Chapter 2, the RWG functions can be replaced with other simple functions to improve the accuracy of normal and mixed formulations. For example, the curl-conforming $\hat{\boldsymbol{n}} \times$RWG functions have a spatial distribution shown in Figure 1.9 that can be expressed as

$$\boldsymbol{b}_n^{\mathrm{nRWG}}(\boldsymbol{r}) = \hat{\boldsymbol{n}} \times \boldsymbol{b}_n^{\mathrm{RWG}}(\boldsymbol{r}), \tag{1.76}$$

where $\boldsymbol{b}_n^{\mathrm{RWG}}(\boldsymbol{r})$ is the associated RWG function. Obviously, the $\hat{\boldsymbol{n}} \times$RWG functions are LN-CT (linear normal, constant tangential) type. Implementations of the $\hat{\boldsymbol{n}} \times$RWG functions can easily be obtained from the implementations of the RWG functions considering the simple rotation relation in (1.76), whereas more efficient implementations can be derived using the curl-conforming property of these functions, as discussed in Chapter 2, Section 2.3.9.

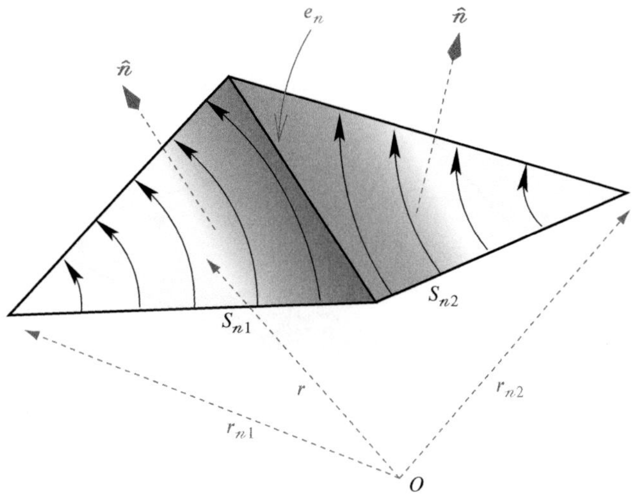

Figure 1.9 Spatial distribution of the $\hat{\boldsymbol{n}}\times$RWG functions. Arrows show the direction while the shading indicates the magnitude of the vector distribution. Light and dark colors represent the low and high values, respectively.

In addition to the $\hat{\boldsymbol{n}}\times$RWG functions obtained via rotation, the RWG functions can be decomposed into the linear-linear (LL) functions [133] to further improve the accuracy of normal and mixed formulations, as detailed in Chapter 2, Section 2.3.10. Spatial distributions of the LL functions are depicted in Figure 1.10, where two kinds of the LL functions are defined on the same edge e_n with the expressions

$$\boldsymbol{b}_n^{\mathrm{LL}(1)}(\boldsymbol{r}) = \begin{cases} \frac{l_n}{4(A_{n1})^2}(\boldsymbol{r}-\boldsymbol{r}_{n1})\cdot\left[(\boldsymbol{r}_{n4}-\boldsymbol{r}_{n1})\times\hat{\boldsymbol{n}}\right](\boldsymbol{r}_{n3}-\boldsymbol{r}_{n1}), & r\in S_{n1} \\ \frac{l_n}{4(A_{n2})^2}(\boldsymbol{r}-\boldsymbol{r}_{n2})\cdot[(\boldsymbol{r}_{n4}-\boldsymbol{r}_{n2})\times\hat{\boldsymbol{n}}](\boldsymbol{r}_{n3}-\boldsymbol{r}_{n2}), & r\in S_{n2} \\ 0, & r\notin S_n \end{cases} \tag{1.77}$$

for the LL function of the first kind, and

$$\boldsymbol{b}_n^{\mathrm{LL}(2)}(\boldsymbol{r}) = \begin{cases} \frac{l_n}{4(A_{n1})^2}(\boldsymbol{r}-\boldsymbol{r}_{n1})\cdot\left[(\boldsymbol{r}_{n3}-\boldsymbol{r}_{n1})\times\hat{\boldsymbol{n}}\right](\boldsymbol{r}_{n4}-\boldsymbol{r}_{n1}), & r\in S_{n1} \\ \frac{l_n}{4(A_{n2})^2}(\boldsymbol{r}-\boldsymbol{r}_{n2})\cdot[(\boldsymbol{r}_{n3}-\boldsymbol{r}_{n2})\times\hat{\boldsymbol{n}}](\boldsymbol{r}_{n4}-\boldsymbol{r}_{n2}), & r\in S_{n2} \\ 0, & r\notin S_n \end{cases} \tag{1.78}$$

for the LL function of the second kind. In Figure 1.10, vertex locations $\boldsymbol{r}_{n1}$, $\boldsymbol{r}_{n2}$, $\boldsymbol{r}_{n3}$, and $\boldsymbol{r}_{n4}$ are labeled as well as the three edges of the triangles e_n, e_p, and e_z depending on the kind of the LL function. Similar to the $\hat{\boldsymbol{n}}\times$RWG functions that are derived from the RWG functions, the curl-conforming $\hat{\boldsymbol{n}}\times$LL functions can be derived from the LL functions as

$$\boldsymbol{b}^{\mathrm{nLL}(1)}(\boldsymbol{r}) = \hat{\boldsymbol{n}}\times\boldsymbol{b}^{\mathrm{LL}(1)}(\boldsymbol{r}) \tag{1.79}$$

$$\boldsymbol{b}^{\mathrm{nLL}(2)}(\boldsymbol{r}) = \hat{\boldsymbol{n}}\times\boldsymbol{b}^{\mathrm{LL}(2)}(\boldsymbol{r}). \tag{1.80}$$

Spatial distributions of the $\hat{\boldsymbol{n}}\times$LL functions are depicted in Figure 1.11.

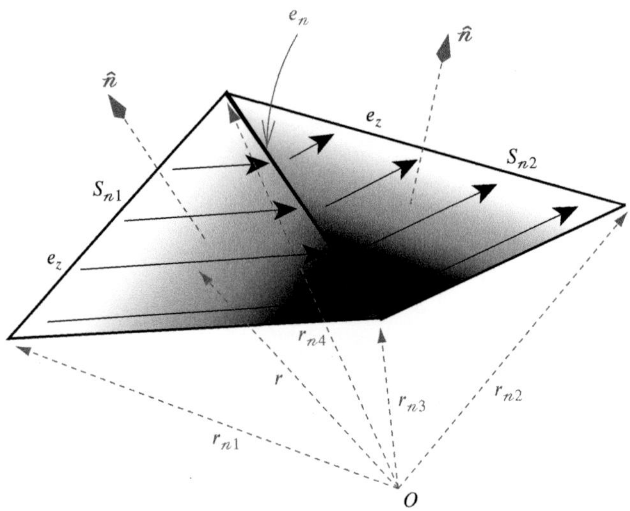

(a)

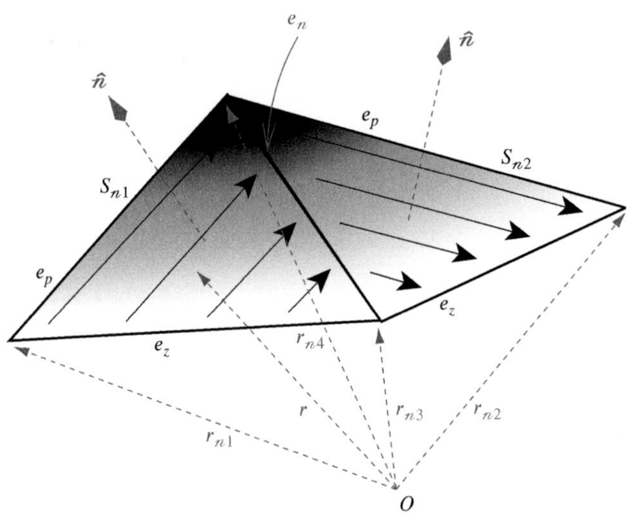

(b)

Figure 1.10 (a) First-kind and (b) second-kind LL functions defined at the nth edge (e_n). Arrows show the direction while the shading indicates the magnitude of the vector distribution. Light and dark colors represent the low and high values, respectively, and white corresponds to zero.

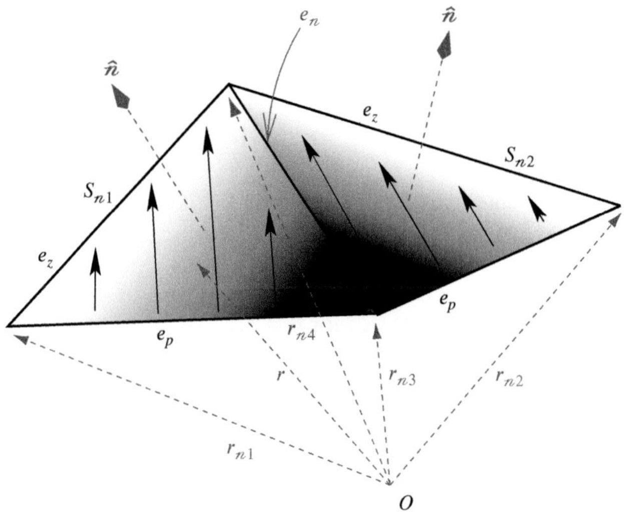

(a)

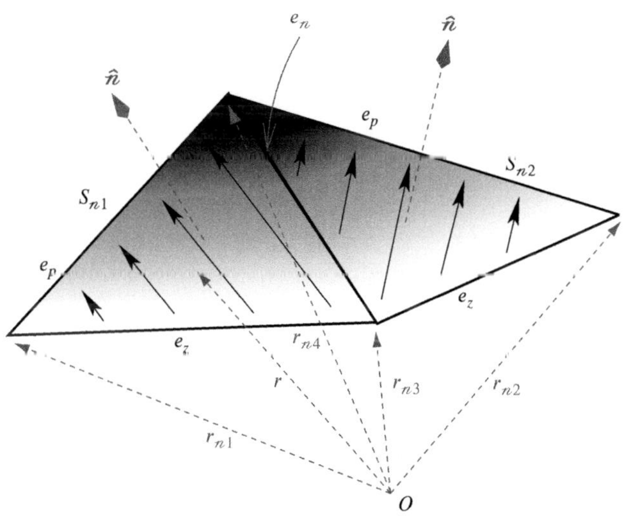

(b)

Figure 1.11 (a) First-kind and (b) second-kind $\hat{\boldsymbol{n}} \times$LL functions defined at the nth edge (e_n). Arrows show the direction while the shading indicates the magnitude of the vector distribution. Light and dark colors represent the low and high values, respectively, and white corresponds to zero.

In order to understand the relation between the LL and RWG functions (or the $\hat{n}\times$LL and $\hat{n}\times$RWG functions), one can consider the general representation of vector functions [130]. The zeroth-order divergence-conforming basis functions for triangular elements, such as the RWG functions, can be written as

$$\boldsymbol{f}_\beta(\boldsymbol{r}) = \frac{1}{J_\beta}[\zeta_{\beta+1}(\boldsymbol{r})\boldsymbol{\ell}_{\beta-1} - \zeta_{\beta-1}(\boldsymbol{r})\boldsymbol{\ell}_{\beta+1}] \qquad (\beta = 1,2,3), \tag{1.81}$$

where J_β indicates the Jacobian that can be derived for the RWG functions as

$$J_\beta = \frac{2A}{|\boldsymbol{\ell}_\beta|} \tag{1.82}$$

and A is the area of the triangle. In (1.81), ζ_β for $\beta = 1,2,3$ represent parent (simplex) coordinates with the dependency relationship

$$\zeta_1(\boldsymbol{r}) + \zeta_2(\boldsymbol{r}) + \zeta_3(\boldsymbol{r}) = 1 \tag{1.83}$$

when $\boldsymbol{r}$ is on the triangle. In addition, $\boldsymbol{\ell}_\beta$ in (1.81) and (1.82) for $\beta = 1,2,3$ represent edge vectors with the dependence

$$\boldsymbol{\ell}_1 + \boldsymbol{\ell}_2 + \boldsymbol{\ell}_3 = 0. \tag{1.84}$$

The RWG functions, which provide three degrees of freedom per triangle, are zeroth-order complete since they can represent any constant vector function on a triangle and their surface divergence is also constant. Given the three RWG functions corresponding to the three edges of a triangle as in (1.81), the set of LL functions associated with the same triangle can be derived as

$$\boldsymbol{f}^{(1)}_\beta(\boldsymbol{r}) = \frac{1}{J_\beta}\zeta_{\beta+1}(\boldsymbol{r})\boldsymbol{\ell}_{\beta-1} \tag{1.85}$$

$$\boldsymbol{f}^{(2)}_\beta(\boldsymbol{r}) = -\frac{1}{J_\beta}\zeta_{\beta-1}(\boldsymbol{r})\boldsymbol{\ell}_{\beta+1} \tag{1.86}$$

with the decomposition

$$\boldsymbol{f}_\beta(\boldsymbol{r}) = \boldsymbol{f}^{(1)}_\beta(\boldsymbol{r}) + \boldsymbol{f}^{(2)}_\beta(\boldsymbol{r}) \qquad (\beta = 1,2,3). \tag{1.87}$$

Note that both of the distributions in (1.81) and (1.85)–(1.86) provide a linear variation for parallel components along the edges β. On the other hand,

$$\nabla\zeta_\beta(\boldsymbol{r})|_{\zeta_\beta(\boldsymbol{r})=0} \cdot \boldsymbol{f}^{(1)}_\beta(\boldsymbol{r}) = \zeta_{\beta+1}(\boldsymbol{r}) \tag{1.88}$$

$$\nabla\zeta_\beta(\boldsymbol{r})|_{\zeta_\beta(\boldsymbol{r})=0} \cdot \boldsymbol{f}^{(2)}_\beta(\boldsymbol{r}) = \zeta_{\beta-1}(\boldsymbol{r}) \tag{1.89}$$

so that normal components of the LL functions in (1.85) and (1.86) also have a linear variation along edges β, while they are constant for the RWG functions in (1.81), i.e.,

$$\nabla\zeta_\beta(\boldsymbol{r})|_{\zeta_\beta(\boldsymbol{r})=0} \cdot \boldsymbol{f}_\beta(\boldsymbol{r}) = \zeta_{\beta+1}(\boldsymbol{r}) + \zeta_{\beta-1}(\boldsymbol{r}) = 1 - \zeta_\beta(\boldsymbol{r}) = 1. \tag{1.90}$$

Any linearly varying vector function on a triangle can be represented by a combination of the LL functions since

$$\frac{\boldsymbol{f}^{(1)}_2(\boldsymbol{r})}{|\boldsymbol{\ell}_2|} + \frac{\boldsymbol{f}^{(2)}_2(\boldsymbol{r})}{|\boldsymbol{\ell}_2|} - \frac{\boldsymbol{f}^{(1)}_3(\boldsymbol{r})}{|\boldsymbol{\ell}_3|} - \frac{\boldsymbol{f}^{(2)}_3(\boldsymbol{r})}{|\boldsymbol{\ell}_3|} = \frac{\boldsymbol{\ell}_1}{2A} \tag{1.91}$$

$$\frac{f_3^{(1)}(\mathbf{r})}{|\ell_3|} + \frac{f_3^{(2)}(\mathbf{r})}{|\ell_3|} - \frac{f_1^{(1)}(\mathbf{r})}{|\ell_1|} - \frac{f_1^{(2)}(\mathbf{r})}{|\ell_1|} = \frac{\ell_2}{2A} \tag{1.92}$$

$$\frac{f_2^{(2)}(\mathbf{r})}{|\ell_2|} - \frac{f_3^{(1)}(\mathbf{r})}{|\ell_3|} = \frac{\zeta_1(\mathbf{r})\ell_1}{2A} \tag{1.93}$$

$$\frac{f_3^{(2)}(\mathbf{r})}{|\ell_3|} - \frac{f_1^{(1)}(\mathbf{r})}{|\ell_1|} = \frac{\zeta_2(\mathbf{r})\ell_2}{2A} \tag{1.94}$$

$$-\frac{f_3^{(2)}(\mathbf{r})}{|\ell_3|} = \frac{\zeta_2(\mathbf{r})\ell_1}{2A} \tag{1.95}$$

$$\frac{f_3^{(1)}(\mathbf{r})}{|\ell_3|} = \frac{\zeta_1(\mathbf{r})\ell_2}{2A}. \tag{1.96}$$

On the other hand, the LL functions are not strictly first-order complete since their divergences

$$\nabla \cdot \mathbf{f}_\beta^{(1)}(\mathbf{r}) = \frac{1}{J_\beta}\ell_{\beta-1} \cdot \nabla\zeta_{\beta+1}(\mathbf{r}) = \frac{1}{J_\beta} = \frac{|\ell_\beta|}{2A} \tag{1.97}$$

$$\nabla \cdot \mathbf{f}_\beta^{(2)}(\mathbf{r}) = -\frac{1}{J_\beta}(\mathbf{r})\ell_{\beta+1} \cdot \nabla\zeta_{\beta-1}(\mathbf{r}) = -\frac{1}{J_\beta} = -\frac{|\ell_\beta|}{2A} \tag{1.98}$$

for $\beta = 1, 2$, or 3 are constants. As indicated in [130], a strictly first-order complete basis requires eight degrees of freedom on the triangle, whereas the LL functions have six degrees of freedom. Nevertheless, the LL functions (and their curl-conforming versions, i.e., the $\hat{\mathbf{n}} \times$LL functions) have the advantage that their implementations do not require higher-order techniques yet they provide significantly higher accuracy than the RWG functions for normal and mixed formulations. This is discussed in Chapter 2, Section 2.3.10.

1.8 Integrals on Triangular Domains

Elements of the matrices derived from the discretization of surface integral equations involve integrals over the basis and testing functions. Outer integrals over the testing functions can usually be performed via numerical integration techniques, such as Gaussian quadratures [142] and adaptive integration methods, while inner integrals over the basis functions may require singularity extractions [143], [144] before such numerical techniques are applied. In general, considering a basis triangle with a surface of S_{nb} lying on the x-y plane, as depicted in Figure 1.12, the integrals to be evaluated can be written as

$$I_{in}^{f} = \int_{S_{nb}} d\mathbf{r}' f(\mathbf{r},\mathbf{r}')g(\mathbf{r},\mathbf{r}') \tag{1.99}$$

$$\mathbf{I}_{in}^{f\nabla'} = \int_{S_{nb}} d\mathbf{r}' f(\mathbf{r},\mathbf{r}')\nabla' g(\mathbf{r},\mathbf{r}'), \tag{1.100}$$

where $f(\mathbf{r},\mathbf{r}') = 1$ or $f(\mathbf{r},\mathbf{r}') = (x' - x)$ or $f(\mathbf{r},\mathbf{r}') = (y' - y)$. Since the Green's function is singular and its value becomes infinite as $R = |\mathbf{r} - \mathbf{r}'|$ goes to zero, singularity extraction is required

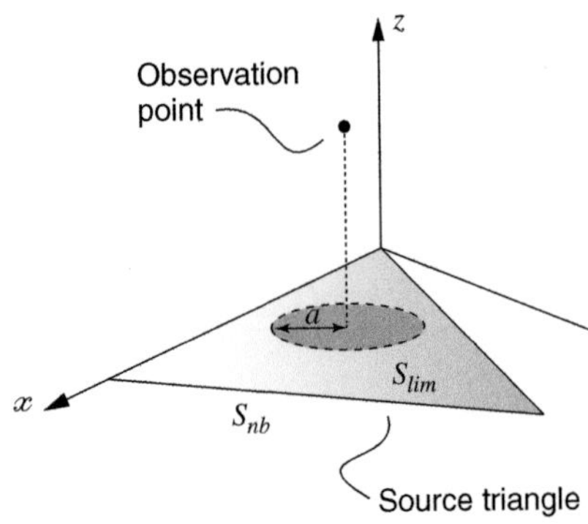

Figure 1.12 A source triangle on the x-y plane and the projection of an observation point.

in (1.99) and (1.100) for accurate computations of the integrals, especially when the observation point is close to the source (basis) triangle. Specifically, the integrals are divided into numerical and analytical parts as

$$I_{in}^{f} = \frac{1}{4\pi}\int_{S_{nb}} d\boldsymbol{r}' f(\boldsymbol{r},\boldsymbol{r}')\left(\frac{\exp(ikR)-1}{R}\right) + \frac{1}{4\pi}\int_{S_{nb}} d\boldsymbol{r}' f(\boldsymbol{r},\boldsymbol{r}')\frac{1}{R} \tag{1.101}$$

$$\begin{aligned}\boldsymbol{I}_{in}^{f\nabla'} &= \frac{1}{4\pi}\int_{S_{nb}} d\boldsymbol{r}' f(\boldsymbol{r},\boldsymbol{r}')\nabla'\left(\frac{\exp(ikR)-1+0.5k^2R^2}{R}\right)\\ &+ \frac{1}{4\pi}\int_{S_{nb}} d\boldsymbol{r}' f(\boldsymbol{r},\boldsymbol{r}')\nabla'\left(\frac{1}{R}\right) - \frac{k^2}{8\pi}\int_{S_{nb}} d\boldsymbol{r}' f(\boldsymbol{r},\boldsymbol{r}')\nabla' R.\end{aligned} \tag{1.102}$$

The numerical integrals in (1.101)–(1.102) can be evaluated accurately by using Gaussian quadratures or adaptive integration methods.

1.8.1 Analytical Integrals

Consider the scenario depicted in Figure 1.12 involving a source triangle lying on the x-y plane and an observation point $\boldsymbol{r}$. The analytical integrals can be further divided as

$$\int_{S_{nb}} d\boldsymbol{r}' f(\boldsymbol{r},\boldsymbol{r}')\frac{1}{R} = \int_{S_{PV,nb}} d\boldsymbol{r}' f(\boldsymbol{r},\boldsymbol{r}')\frac{1}{R} + \int_{S_{lim}} d\boldsymbol{r}' f(\boldsymbol{r},\boldsymbol{r}')\frac{1}{R} \tag{1.103}$$

$$\begin{aligned}\int_{S_{nb}} d\boldsymbol{r}' f(\boldsymbol{r},\boldsymbol{r}')\nabla'\left(\frac{1}{R}\right) &= \int_{S_{PV,nb}} d\boldsymbol{r}' f(\boldsymbol{r},\boldsymbol{r}')\nabla'\left(\frac{1}{R}\right)\\ &+ \int_{S_{lim}} d\boldsymbol{r}' f(\boldsymbol{r},\boldsymbol{r}')\nabla'\left(\frac{1}{R}\right)\end{aligned} \tag{1.104}$$

$$\int_{S_{nb}} d\boldsymbol{r}' f(\boldsymbol{r},\boldsymbol{r}')\nabla' R = \int_{S_{PV,nb}} d\boldsymbol{r}' f(\boldsymbol{r},\boldsymbol{r}')\nabla' R + \int_{S_{lim}} d\boldsymbol{r}' f(\boldsymbol{r},\boldsymbol{r}')\nabla' R, \tag{1.105}$$

where S_{lim} is defined as the infinitesimal circular surface (with radius $a \to 0$) in the vicinity of the projection of the observation point and $S_{PV,nb} = S_{nb} - S_{lim}$. Depending on the location of the observation point, S_{lim} may not exist. In general, the decompositions in (1.103)–(1.105) allow for rigorous analytical evaluations of the integrals in terms of the geometric quantities associated with the basis triangle and the location of the observation point [144]. These quantities are depicted in Figure 1.13 and can be listed for each edge ($i = 1, 2, 3$) as follows:

- $\hat{\boldsymbol{u}}_i$ represents the outward unit vector perpendicular to the ith edge on the x-y plane.
- R_i^+ and R_i^- are the distances between the observation point and the end points of the ith edge. The superscripts "+" and "−", referring to the end points of the edge, are determined by the right-hand rule applied on the triangle in the z direction.
- R_i^0 is the distance from the observation point to the ith edge and can be written as

$$R_i^0 = \sqrt{z^2 + |t_i^0|^2}, \tag{1.106}$$

  where $|t_i^0|$ is the distance between the projection of the observation point on the x-y plane and the ith edge. The sign of t_i^0 is determined by the position of the projection of the observation point with respect to the edge.
- s_i^+ and s_i^- have magnitudes equal to the "+" and "−" segments of the ith edge. These segments are formed by the projection of the observation point on the edge. The signs of s_i^+ and s_i^- are determined by the relative position of this projection compared to the "+" and "−" ends of the edge. In addition, $\hat{\boldsymbol{s}}_i$ is the unit vector in the direction of the edge using the right-hand rule.

In order to evaluate the integrals in the form of (1.103), one can use the identities

$$\nabla'_S \cdot \left(\frac{R}{P}\hat{\boldsymbol{P}}\right) = \frac{1}{R} \tag{1.107}$$

$$\nabla'_S R = \frac{\boldsymbol{\rho}' - \boldsymbol{\rho}}{R} = \hat{\boldsymbol{x}}\frac{(x' - x)}{R} + \hat{\boldsymbol{y}}\frac{(y' - y)}{R}, \tag{1.108}$$

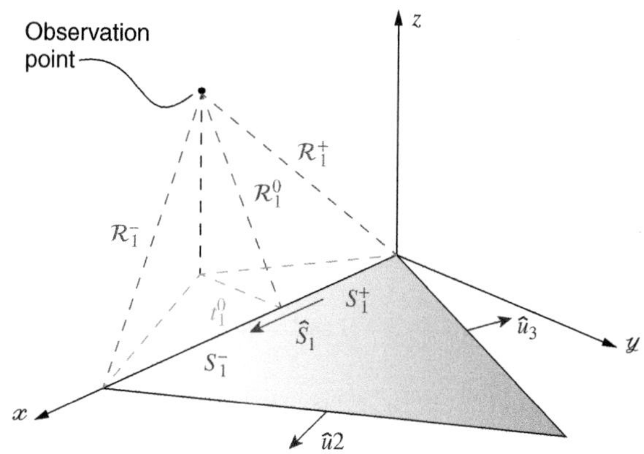

Figure 1.13 Geometric variables introduced to express the results of the analytical integrals. Most of the variables are shown only for the first edge on the x axis.

where ∇'_S represents the surface divergence or gradient in the primed coordinates and

$$P = \hat{\boldsymbol{P}}P = \boldsymbol{\rho} - \boldsymbol{\rho}' = \hat{\boldsymbol{x}}(x - x') + \hat{\boldsymbol{y}}(y - y'). \tag{1.109}$$

Then, using the surface divergence theorem and evaluating the limits (as the infinitesimal surface in Figure 1.12 shrinks to zero), one can obtain

$$\int_{S_{nb}} d\boldsymbol{r}' \frac{1}{R} = \sum_{i=1}^{3} t_i^0 f_{2,i} - |z| \sum_{i=1}^{3} \beta_i \tag{1.110}$$

$$\int_{S_{nb}} d\boldsymbol{r}' \frac{(x' - x)}{R} = \frac{1}{2}\hat{\boldsymbol{x}} \cdot \sum_{i=1}^{3} \hat{\boldsymbol{u}}_i f_{3,i} \tag{1.111}$$

$$\int_{S_{nb}} d\boldsymbol{r}' \frac{(y' - y)}{R} = \frac{1}{2}\hat{\boldsymbol{y}} \cdot \sum_{i=1}^{3} \hat{\boldsymbol{u}}_i f_{3,i}, \tag{1.112}$$

where

$$\beta_i = \tan^{-1} \frac{t_i^0 s_i^+}{(R_i^0)^2 + |z| R_i^+} - \tan^{-1} \frac{t_i^0 s_i^-}{(R_i^0)^2 + |z| R_i^-} \tag{1.113}$$

$$f_{2,i} = \ln \left(\frac{R_i^+ + s_i^+}{R_i^- + s_i^-} \right) \tag{1.114}$$

$$f_{3,i} = s_i^+ R_i^+ - s_i^- R_i^- + (R_i^0)^2 f_{2,i}. \tag{1.115}$$

Similar procedures can be used to evaluate the integrals in the form of (1.104), leading to

$$\int_{S_{nb}} d\boldsymbol{r}' \nabla' \left(\frac{1}{R} \right) = \sum_{i=1}^{3} \hat{\boldsymbol{u}}_i f_{2,i} + \hat{z} \frac{z}{|z|} \sum_{i=1}^{3} \beta_i \tag{1.116}$$

$$\int_{S_{nb}} d\boldsymbol{r}' (x' - x) \nabla' \left(\frac{1}{R} \right) = \hat{\boldsymbol{x}} |z| \sum_{i=1}^{3} \beta_i - \sum_{i=1}^{3} \hat{\boldsymbol{x}} \cdot \hat{\boldsymbol{s}}_i \boldsymbol{f}_{1,i} - \hat{z} z \hat{\boldsymbol{x}} \cdot \sum_{i=1}^{3} \hat{\boldsymbol{u}}_i f_{2,i} \tag{1.117}$$

$$\int_{S_{nb}} d\boldsymbol{r}' (y' - y) \nabla' \left(\frac{1}{R} \right) = \hat{\boldsymbol{y}} |z| \sum_{i=1}^{3} \beta_i - \sum_{i=1}^{3} \hat{\boldsymbol{y}} \cdot \hat{\boldsymbol{s}}_i \boldsymbol{f}_{1,i} - \hat{z} z \hat{\boldsymbol{y}} \cdot \sum_{i=1}^{3} \hat{\boldsymbol{u}}_i f_{2,i}, \tag{1.118}$$

where

$$\boldsymbol{f}_{1,i} = \hat{\boldsymbol{s}}_i t_i^0 f_{2i} - \hat{\boldsymbol{u}}_i (R_i^+ - R_i^-). \tag{1.119}$$

For the integrals in the form of (1.105), $f(\boldsymbol{r}, \boldsymbol{r}')$ is usually unity and one can derive

$$\int_{S_{nb}} d\boldsymbol{r}' \nabla' R = \frac{1}{2} \sum_{i=1}^{3} \hat{\boldsymbol{u}}_i f_{3,i} + \hat{z} z |z| \sum_{i=1}^{3} \beta_i - \hat{z} z \sum_{i=1}^{3} t_i^0 f_{2,i}. \tag{1.120}$$

It is essential to investigate the values of the integrals when the observation point approaches the source triangle. For example,

$$lim_{z\to 0}\sum_{i=1}^{3}\beta_i = \alpha(x,y), \tag{1.121}$$

where

$$\begin{cases}\alpha(x,y)=0, & \boldsymbol{r}\notin S_{nb}\\ \alpha(x,y)=2\pi, & \boldsymbol{r}\in S_{nb}\quad\text{and}\quad \boldsymbol{r}\notin \partial S_{nb}\\ 0<\alpha(x,y)<2\pi, & \boldsymbol{r}\in\partial S_{nb}\end{cases} \tag{1.122}$$

and ∂S_{nb} represents the bounding contour around the triangle. Interestingly, the value of $\alpha(x, y)$ at any edge depends on how the observation point approaches the edge. Specifically, $\alpha \to 2\gamma$ as $\boldsymbol{r} \to \partial S_{nb}$, where γ is the angle between the approach path and the x-y plane, as depicted in Figure 1.14. This angle-dependent factor vanishes in (1.110), (1.117), and (1.118), since $z \to 0$ as the observation point approaches the edge. However, it does not vanish in (1.116). In fact, considering this integral, it can be shown that this factor is directly related to the solid-angle factor in the limit part of the $\mathcal{K}$ operator [125] (see Appendix, Section A.1).

Another important parameter in the analytical integrals is $f_{2,i}$ given in (1.114). The value of this parameter is infinite when the observation point is at one of the edges of the source triangle. In such a case, the value of the integral in (1.116) is also infinite due to the non-vanishing contribution of $f_{2,i}$. Since the singularity is logarithmic and it is mild, it does not cause a serious problem in the numerical integration over the testing triangle, as long as the integration (observation) points are taken inside the testing triangle. It is also possible to apply a singularity-extraction technique for the outer integrals in near-neighboring interactions [124].

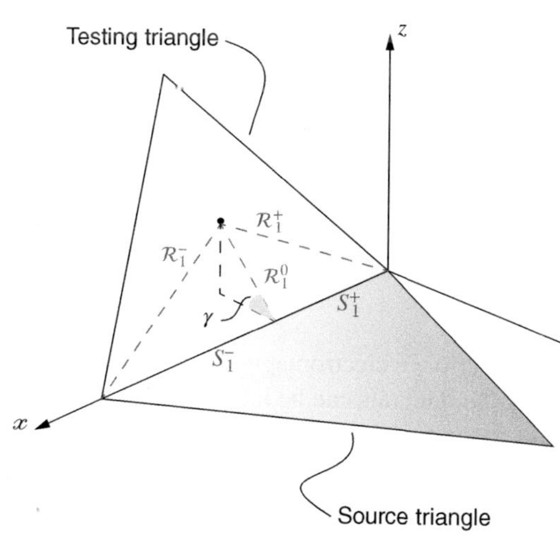

Figure 1.14 A limit case as the observation point approaches the edge of the basis triangle in a near-neighboring interaction.

As an example, consider the scenario in Figure 1.14 and an integral

$$\int_{S_{ma}} d\boldsymbol{r}\, f(x,y,z) f_{2,1} = \int_{S_{ma}} d\boldsymbol{r} f(x,y,z) \ln\left(\frac{R_1^+ + s_1^+}{R_1^- + s_1^-}\right) \tag{1.123}$$

over the testing triangle, where $f(x,y,z)$ is a linear function. The logarithmic singularity due to the first edge of the basis triangle can be extracted as

$$\int_{S_{ma}} d\boldsymbol{r} f(x,y,z) \ln\left(\frac{R_1^+ + s_1^+}{R_1^- + s_1^-}\right) = \int_{S_{ma}} d\boldsymbol{r} \left[f(x,y,z) \ln\left(\frac{R_1^+ + s_1^+}{R_1^- + s_1^-}\right) + 2\ln\,(R_1^0)\right]$$
$$- 2\int_{S_{ma}} d\boldsymbol{r} f(x,y,z) \ln\,(R_1^0), \tag{1.124}$$

where the second integral on the RHS can be evaluated analytically [124].

1.8.2 Gaussian Quadratures

Numerical integrations of nonsingular functions on triangles can be performed accurately and efficiently by using Gaussian quadratures [142]. A symmetrical quadrature rule on a triangle with a spatial support of S_m can be written as

$$\int_{S_m} d\boldsymbol{r} f(\boldsymbol{r}) \approx A_m \sum_{p=1}^{P} w_p f(\boldsymbol{r}_{pm}), \tag{1.125}$$

where A_m is the area of the triangle, P is the number of quadrature points, and w_p represents the integration weight for the pth sample point. Locations of the sample points are determined as

$$\boldsymbol{r}_{pm} = \zeta_{p1}\boldsymbol{r}_{m1} + \zeta_{p2}\boldsymbol{r}_{m2} + \zeta_{p3}\boldsymbol{r}_{m3}, \tag{1.126}$$

where $\boldsymbol{r}_{m1}$, $\boldsymbol{r}_{m2}$, and $\boldsymbol{r}_{m3}$ are node coordinates of the triangle, whereas ζ_{p1}, ζ_{p2}, and ζ_{p3} are the simplex coordinates satisfying $\zeta_{p1} + \zeta_{p2} + \zeta_{p3} = 1$.

1.8.3 Adaptive Integration

In order to reduce the integration error in electromagnetic interactions, the order of the Gaussian quadrature rule applied for the integrals can be increased. Alternatively, low-order Gaussian quadratures can be employed in an adaptive scheme, which can efficiently control the error by adjusting the sampling rate according to the integrand. Specifically, employing an adaptive integration method, the integration points can be selected economically by using more samples at critical locations, where the integrand changes rapidly. Figure 1.15 depicts a simple adaptive integration method based on a three-point Gaussian quadrature rule. The following steps are carried out for the integration of a function f on S_m:

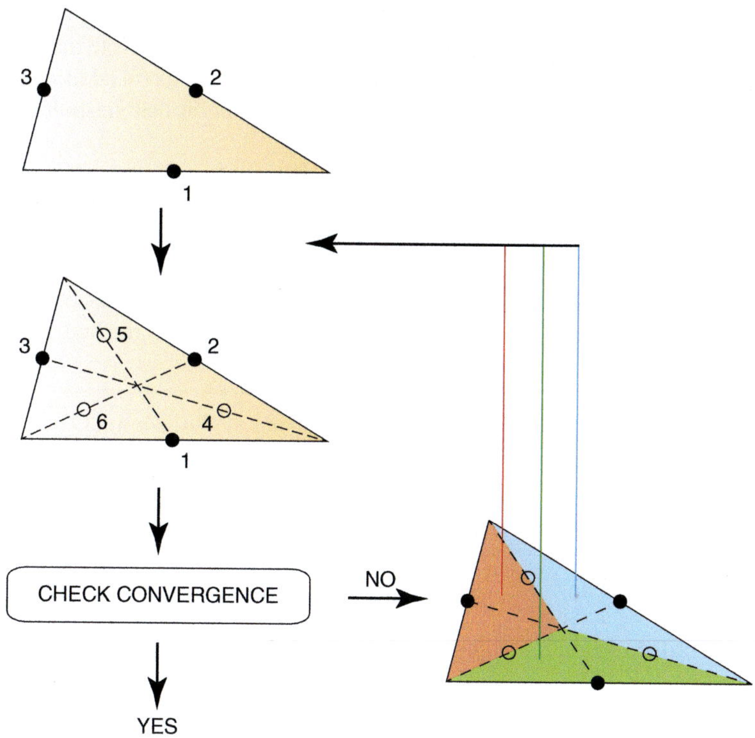

Figure 1.15 An adaptive integration method using a three-point Gaussian quadrature and three subtriangles for each triangle.

- Given a triangle, three integration points are chosen at the middle of the edges. Using a simple Gaussian quadrature rule, the value of the integral is given as

$$\int_{S_m} d\boldsymbol{r} f(\boldsymbol{r}) \approx I_3 = \frac{A_m}{3}[f(\boldsymbol{r}_{p1}) + f(\boldsymbol{r}_{p2}) + f(\boldsymbol{r}_{p3})], \tag{1.127}$$

  where $\boldsymbol{r}_{p1}$, $\boldsymbol{r}_{p2}$, and $\boldsymbol{r}_{p3}$ are the integration points.
- Three additional integration points are chosen on the medians. These points are located at 1/3 of the medians nearer to the vertices of the triangle. Using the three-point Gaussian quadrature rule in (1.127), the value of the integral using six points can be written as

$$\int_{S_m} d\boldsymbol{r} f(\boldsymbol{r}) \approx I_6 = \frac{I_3}{3} + \frac{2A_m}{9}[f(\boldsymbol{r}_{p4}) + f(\boldsymbol{r}_{p5}) + f(\boldsymbol{r}_{p6})], \tag{1.128}$$

  where $\boldsymbol{r}_{p4}$, $\boldsymbol{r}_{p5}$, and $\boldsymbol{r}_{p6}$ are the new integration points on the medians.
- Three-point and six-point integration values are compared. If the error

$$\Delta_{3-6} = \frac{|I_6 - I_3|}{|I_6|} \tag{1.129}$$

is lower than a given threshold, then I_6 is returned as the value of the integral over the triangle. Otherwise, the three subtriangles are considered separately. Note that in each subtriangle, the integration is carried out by sampling three more points (in addition to the three points that are already sampled) and comparing the three-point and six-point integration values.

The steps listed above are performed recursively until the convergence is satisfied for all subtriangles. The final distribution of the samples on the main triangle can be nonuniform since the depth of the recursion varies depending on the integrand.

The adaptive integration method described above works well in many cases, but its efficiency may deteriorate for triangles with large aspect ratios. An alternative method is depicted in Figure 1.16, and can be described as follows:

- Given a triangle, five integration points are chosen. Using the three-point Gaussian quadrature rule in (1.127), the value of the integral over the triangle can be written as

$$\int_{S_m} d\boldsymbol{r} f(\boldsymbol{r}) \approx I_5 = \frac{A_m}{6}[f(\boldsymbol{r}_{p1}) + f(\boldsymbol{r}_{p2}) + f(\boldsymbol{r}_{p3}) + f(\boldsymbol{r}_{p4}) + 2f(\boldsymbol{r}_{p5})]. \tag{1.130}$$

Note that two of the integration points are located at the longest edge of the triangle.

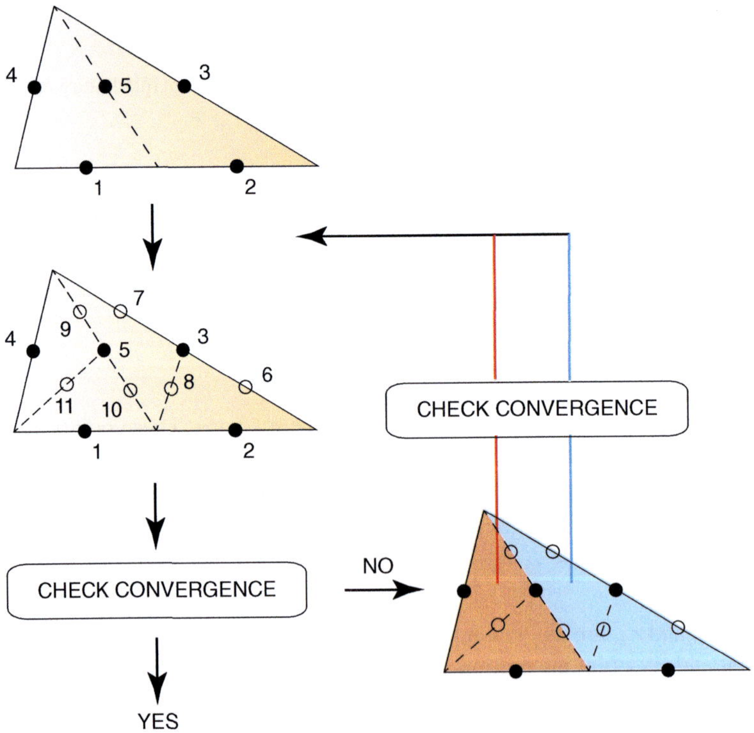

Figure 1.16 An adaptive integration method using a three-point Gaussian quadrature and two subtriangles for each triangle.

- Six additional integration points are chosen. In order to determine the location of these points, the medians are drawn to the longest edges of both subtriangles. The value of the integral using ten points can be written as

$$\int_{S_m} d\mathbf{r} f(\mathbf{r}) \approx I_{10} = \frac{A_m}{12}[f(\mathbf{r}_{p1}) + f(\mathbf{r}_{p2}) + f(\mathbf{r}_{p4}) + f(\mathbf{r}_{p5}) + f(\mathbf{r}_{p6}) + f(\mathbf{r}_{p7})$$
$$+ 2f(\mathbf{r}_{p8}) + f(\mathbf{r}_{p9}) + f(\mathbf{r}_{p10}) + 2f(\mathbf{r}_{p11})]. \tag{1.131}$$

- The five-point and ten-point integration values are compared. If the error

$$\Delta_{10-5} = \frac{|I_{10} - I_5|}{|I_{10}|} \tag{1.132}$$

is lower than a given threshold, then I_{10} is returned as the value of the integral over the triangle. Otherwise, the two subtriangles are considered separately.
- Before starting the next recursion cycle, the three-point and five-point integration values are compared for each subtriangle. Note that the required integration points were already sampled in the previous recursion step. The recursion continues for a subtriangle if the convergence is not satisfied.

The adaptive integration method shown in Figure 1.16 is more robust than the method in Figure 1.15. Although there is an additional cost of finding the longest edge of the triangle, this cost can be considered negligible compared to the gain from a more efficient distribution of the integration points. As an example, Figure 1.17 presents the integration of a function on a triangle with node coordinates $(x_1, y_1) = (0.0, 0.0)$, $(x_2, y_2) = (0.083, 0.0)$, and $(x_3, y_3) = (0.042, 0.048, 0.0)$. The integrand is depicted in Figure 1.17(a) and the integral over the triangle is evaluated adaptively with 1% error. Figures 1.17(b) and 1.17(c) illustrate the integration points when the methods in Figures 1.15 and 1.16 are used, respectively. It can be observed that, using the first method (Figure 1.15), the integration points are not distributed efficiently and there is an accumulation near one of the edges of the triangle. In contrast, using the second method (Figure 1.16), the integration points are distributed efficiently and fewer points are used to compute the integral, but with the same level of accuracy.

1.9 Electromagnetic Excitation

Let $\overline{\mathbf{Z}} \cdot \mathbf{a} = \mathbf{w}$ be an $N \times N$ matrix equation constructed using MOM. The excitation vector $\mathbf{w}$ is obtained by testing the incident electric and magnetic fields, i.e., it contains integrals in the form of

$$v[m] = \int_{S_m} d\mathbf{r} \left\{ \begin{matrix} \mathbf{t}_m(\mathbf{r}) \cdot \\ \mathbf{t}_m(\mathbf{r}) \cdot \hat{\mathbf{n}} \times \end{matrix} \right\} \left\{ \begin{matrix} \eta^{-1} \mathbf{E}^{inc}(\mathbf{r}) \\ \mathbf{H}^{inc}(\mathbf{r}) \end{matrix} \right\}, \tag{1.133}$$

where S_m is the spatial support of the testing function $\mathbf{t}_m$ for $m = 1, 2, \ldots, N$. Different types of excitations can be used, as described below.

1.9.1 *Plane-Wave Excitation*

A common excitation used in numerical simulations, e.g., RCS computations, is the plane-wave illumination. A plane wave propagating in the $\hat{\mathbf{k}}$ direction with the electric field

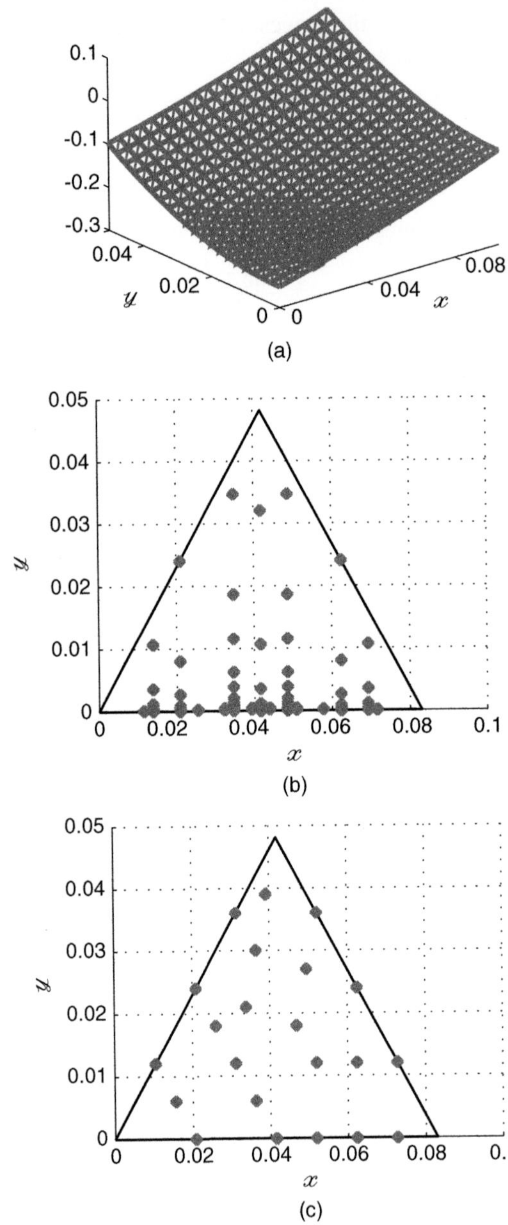

Figure 1.17 Numerical integration of a function on a triangle using the adaptive integration methods depicted in Figures 1.15 and 1.16. Figures show (a) the function and the integration points to evaluate the integral with 1% error using (b) the method in Figure 1.15 and (c) the method in Figure 1.16.

polarized in the $\hat{e}$ direction ($\hat{e} \perp \hat{k}$) can be written as

$$E^{inc}(r) = \hat{e}E_a \exp(ik\hat{k} \cdot r) \tag{1.134}$$

$$H^{inc}(r) = \frac{1}{\eta}\hat{k} \times E^{inc}(r) = \hat{k} \times \hat{e}\frac{E_a}{\eta} \exp(ik\hat{k} \cdot r), \tag{1.135}$$

where E_a is the amplitude of the plane wave.

1.9.2 Hertzian Dipole

A Hertzian (ideal) dipole can be used as a source in many simulations. The electric and magnetic fields of a Hertzian dipole with dipole moment I_{DM} located at r_d can be written as

$$\begin{aligned} E^{inc}(r) = i\omega\mu\frac{\exp(ik|r - r_d|)}{4\pi|r - r_d|}\Bigg[I_{DM}\left(1 + \frac{i}{k|r - r_d|} - \frac{1}{k^2|r - r_d|^2}\right) \\ -(r - r_d)\frac{I_{DM} \cdot (r - r_d)}{|r - r_d|^2}\left(1 + \frac{3i}{k|r - r_d|} - \frac{3}{k^2|r - r_d|^2}\right)\Bigg] \end{aligned} \tag{1.136}$$

and

$$H^{inc}(r) = I_{DM} \times (r - r_d)\frac{\exp(ik|r - r_d|)}{4\pi|r - r_d|^2}\left(\frac{1}{|r - r_d|} - ik\right), \tag{1.137}$$

respectively. Singularity extraction methods are required to evaluate the integrals in (1.133) when the dipole is close to the observation point and $|r - r_d|$ is small.

1.9.3 Complex-Source-Point Excitation

In order to obtain an electromagnetic beam, which satisfies Maxwell's equations exactly, an ideal dipole can be located at $r_d = r_{d,R} + ir_{d,I}$ in complex coordinates [145]. Then, the expressions in (1.136) and (1.137) can be used to compute the electric and magnetic fields at any point r, where

$$|r - r_d| = \sqrt{(r - r_d) \cdot (r - r_d)^*}. \tag{1.138}$$

The real part of the source location ($r_{d,R}$) determines the position of the beam in real coordinates, whereas the imaginary part ($r_{d,I}$) determines the direction of the beam and the width of the beam waist.

Figures 1.18 and 1.19 present some experiments involving a Hertzian dipole with $I_{DM} = \hat{z}$ in complex coordinates. Figure 1.18 depicts the normalized electric field, i.e., $|E(r)| \exp(-k|r_{d,I}|)$, on the x-y plane, when $r_d = i\lambda\hat{x}$ and $r_d = i\lambda\hat{y}$. Since $r_{d,R} = 0$, both beams are located at the origin. However, the beam propagates in the positive x and y directions when $r_{d,I} = \lambda\hat{x}$ and $r_{d,I} = \lambda\hat{y}$, respectively. Figure 1.19 presents two different beams in the x direction with $r_d = 2i\lambda\hat{x}$ and $r_d = 10i\lambda\hat{x}$. Note that increasing the value of $|r_{d,I}|$ increases the width of the beam waist.

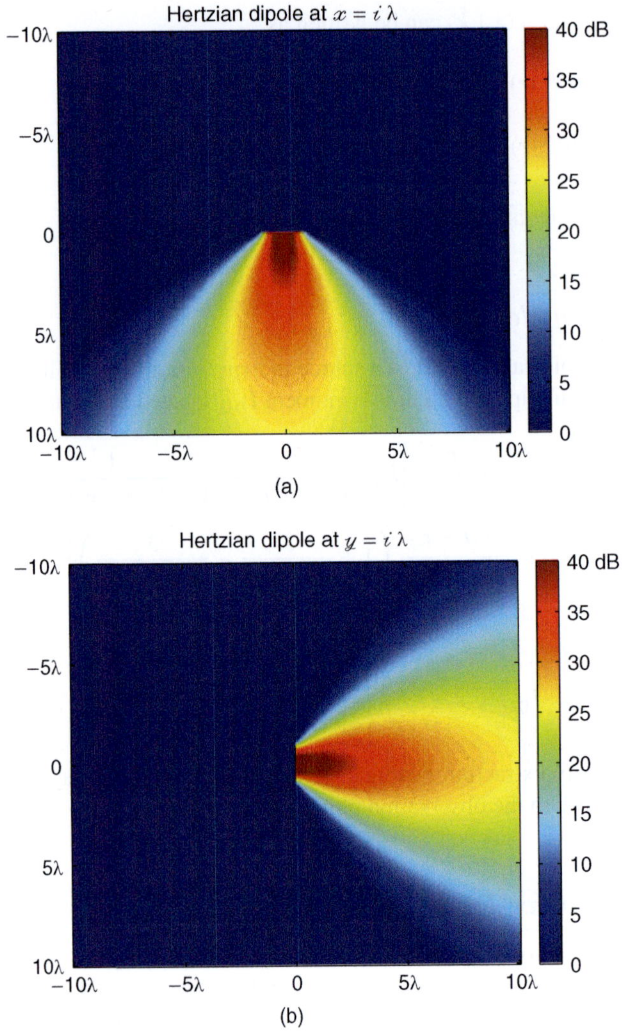

Figure 1.18 The normalized electric field on the x-y plane when a Hertzian dipole with $\boldsymbol{I}_{DM} = \hat{\boldsymbol{z}}$ is located at (a) $\boldsymbol{r}_d = i\lambda\hat{\boldsymbol{x}}$ and (b) $\boldsymbol{r}_d = i\lambda\hat{\boldsymbol{y}}$.

1.9.4 Delta-Gap Excitation

In some applications involving metallic objects, voltage sources are required as external excitations. A voltage source can be modeled as a delta-gap excitation, as depicted in Figure 1.20. A local electric field is defined inside an infinitely narrow opening between two coplanar triangles of the discretization, i.e.,

$$\boldsymbol{E}^{inc}(\boldsymbol{r}) = I_n lim_{\mathrm{d}\to 0}\hat{\boldsymbol{u}}\delta(\boldsymbol{r}, \boldsymbol{r}_n)/d, \tag{1.139}$$

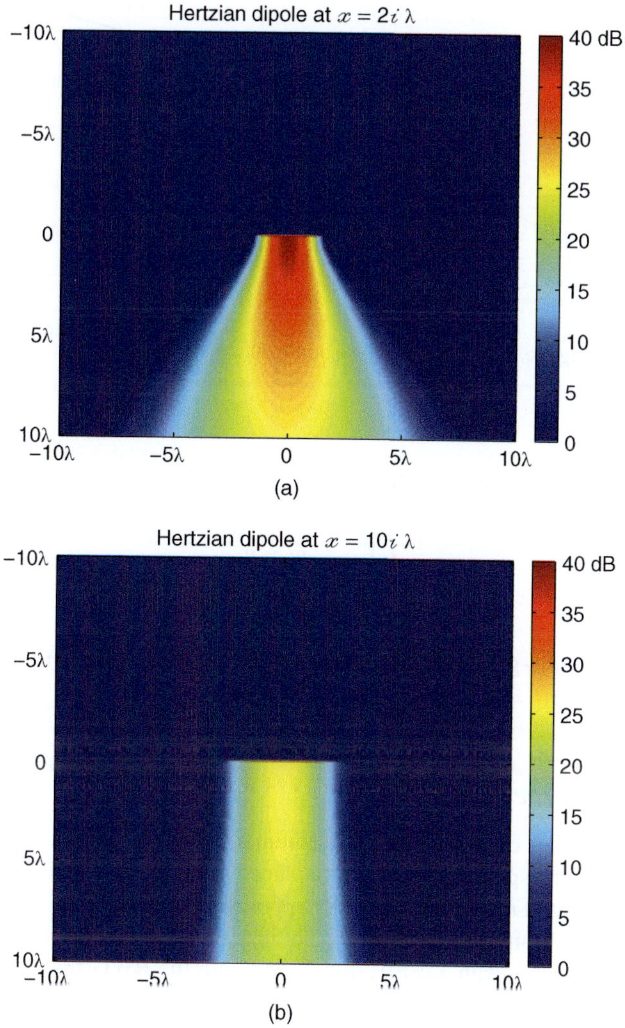

Figure 1.19 The normalized electric field on the x-y plane when a Hertzian dipole with $\boldsymbol{I}_{DM} = \hat{z}$ is located at (a) $\boldsymbol{r}_d = 2i\lambda\hat{\boldsymbol{x}}$ and (b) $\boldsymbol{r}_d = 10i\lambda\hat{\boldsymbol{x}}$.

where $\boldsymbol{r}_n$ represents any point at the nth edge, $\hat{\boldsymbol{u}}$ is the unit vector perpendicular to the edge in the plane of the triangles, I_n is a complex coefficient to represent the strength of the feed, and d is the width of the theoretical gap. In the limit $d \to 0$, the gap shrinks to the edge e_n. The Dirac delta function $\delta(\boldsymbol{r}, \boldsymbol{r}_n)$ in (1.139) indicates that $\boldsymbol{E}^{inc}(\boldsymbol{r})$ is zero outside the small gap.

As an example, consider T-EFIE and let a delta-gap excitation at the nth edge be tested by the RWG functions. Using (1.139), one can obtain

$$\boldsymbol{w}^{\text{T-EFIE}}[m] = -\eta^{-1}\boldsymbol{v}_E^T[m] = -\eta^{-1}\int_{S_m} d\boldsymbol{r}\boldsymbol{t}_m(\boldsymbol{r}) \cdot \boldsymbol{E}^{inc}(\boldsymbol{r}) = \pm\frac{I_n l_n}{\eta}\delta[m,n], \tag{1.140}$$

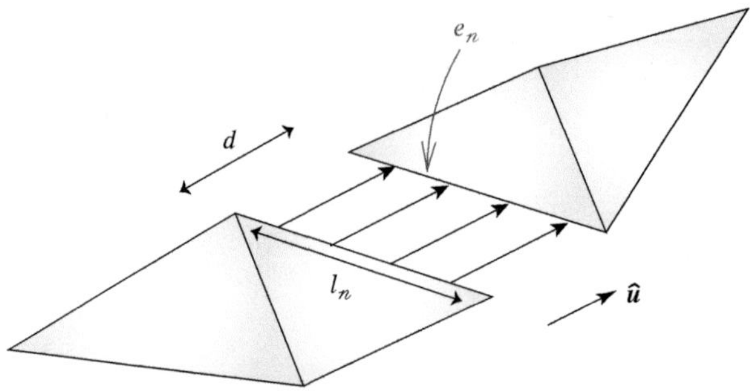

Figure 1.20 A delta-gap source defined at the nth edge.

where l_n represents the length of the edge and

$$\delta[m,n] = \begin{cases} 1, & m = n \\ 0, & m \neq n. \end{cases} \tag{1.141}$$

When using a single delta-gap excitation, only one element ($m = n$) of the excitation vector (1.133) is nonzero.

1.9.5 Current-Source Excitation

Similar to voltage sources, current sources are commonly used to excite metallic structures. A current source and a corresponding sink can be placed at two different edges, with an electrical connection established mathematically. As opposed to the delta-gap excitation, a current source can be placed at a physical edge that is not necessarily shared by two triangles.

As an example, consider again T-EFIE discretized with the RWG functions, and let a current source and a current sink be placed at two physical edges n_1 and n_2, respectively. Then, one can define two half RWG functions at n_1 and n_2. The resulting matrix equation can be written as

$$\sum_{n=1}^{N+2} \bar{\boldsymbol{Z}}^{\text{T-EFIE}}[m,n]\boldsymbol{a}[n] = \boldsymbol{w}^{\text{T-EFIE}}[m] = 0 \qquad (m = 1, \ldots, N), \tag{1.142}$$

where the dimension of the equation is increased to $N \times (N+2)$ due to the extra half functions. Also note that $\boldsymbol{w}^{\text{T-EFIE}}[m] = 0$ since there is no incident-field excitation in this case and the integral for the RHS evaluates to zero. On the other hand, the coefficients corresponding to n_1 and n_2 are fixed, i.e.,

$$\boldsymbol{a}[n_1] = I_n \tag{1.143}$$

$$\boldsymbol{a}[n_2] = -I_n. \tag{1.144}$$

In other words, the expansion coefficients $\boldsymbol{a}[n_1]$ and $\boldsymbol{a}[n_2]$ are forced to be $\pm I_n$ to simulate the source and sink. By setting the two coefficients as above, one can solve the system

$$\sum_{\substack{n=1 \\ n\neq n_1,n_2}}^{N+2} \overline{\boldsymbol{Z}}^{\text{T-EFIE}}[m,n]\boldsymbol{a}[n] = -I_n\overline{\boldsymbol{Z}}^{\text{T-EFIE}}[m,n_1] + I_n\overline{\boldsymbol{Z}}^{\text{T-EFIE}}[m,n_2] \qquad (m=1,\ldots,N), \tag{1.145}$$

to determine the coefficients $\boldsymbol{a}[n]$ for $n \neq n_1, n_2$.

1.10 Multilevel Fast Multipole Algorithm

The fast multipole method (FMM) was developed for efficient solutions of radiation and scattering problems in electromagnetics [146],[147]. Discretizations of integral-equation formulations lead to dense matrix equations involving $\mathcal{O}(N)$ unknowns that can be solved iteratively via a Krylov-subspace algorithm. Iterative solutions require MVMs, i.e., $\boldsymbol{y} = \overline{\boldsymbol{Z}} \cdot \boldsymbol{x}$, which can be performed directly with $\mathcal{O}(N^2)$ complexity. For large-scale problems, direct multiplications render the solution impossible with limited computational resources. Despite that, FMM provides an efficient way of performing the MVMs required by the iterative algorithms in $\mathcal{O}(N^{3/2})$ time using $\mathcal{O}(N^{3/2})$ memory. By reducing the computational complexity from $\mathcal{O}(N^2)$ to $\mathcal{O}(N^{3/2})$, FMM enables the solution of large-scale problems on relatively inexpensive computing platforms [148],[149]. The clustering idea of FMM can be extended and applied in a recursive manner, leading to MLFMA [35], which enables the solution of even larger problems by reducing the complexity of MVMs to $\mathcal{O}(N\log^2 N)$ [150] or $\mathcal{O}(N\log N)$ [1], [68], [151]–[157].

In conventional implementations of MLFMA, a tree structure of $L = \mathcal{O}(\log N)$ levels is constructed by placing the object in a cubic box and recursively dividing the computational domain into subdomains. This recursive clustering is illustrated in Figure 1.21, where a box is divided into eight identical boxes and one of those smaller boxes is further divided into eight boxes, etc. Practically, only nonempty boxes are considered and divided into subboxes, whereas empty boxes created at any stage are simply omitted. The recursion depth depends on the problem as well as the desired accuracy; but in general, the size of the smallest boxes at the lowest level is in the order of the wavelength. As depicted in Figure 1.22, nonempty boxes and the recursive relationship between them are used to construct a multilevel tree structure. In this structure, each box at the lowest level contains some of the discretization elements and each box at the higher levels contains maximum eight subboxes. In order to calculate the far-field interactions, radiated and incoming fields are defined and sampled on the unit sphere.

MLFMA calculates the interactions between the radiating (basis) and receiving (testing) elements that are far from each other in a group-by-group manner consisting of three stages, namely, aggregation, translation, and disaggregation. In each MVM, these stages are performed on the tree structure in a multilevel scheme. The group-by-group calculation of the interactions is based on the factorization and diagonalization of the Green's function, as detailed in Chapter 3, Section 3.1. At this point, it is essential to define the near and far zones for the boxes. A common approach is to use the one-box-buffer scheme illustrated in Figure 1.23. Considering the red box at an arbitrary level in this two-dimensional picture, the blue boxes (which are touching the red box) are assumed to be in the near-zone. In a three-dimensional clustering, there are 27 near-zone boxes, some of which may be empty; all

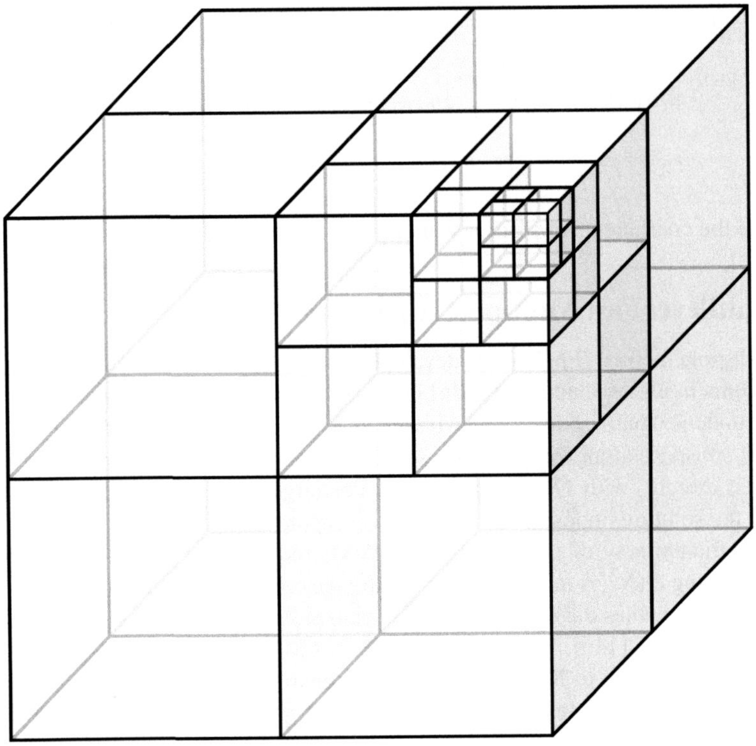

Figure 1.21 Recursive clustering based on dividing each box into eight subboxes.

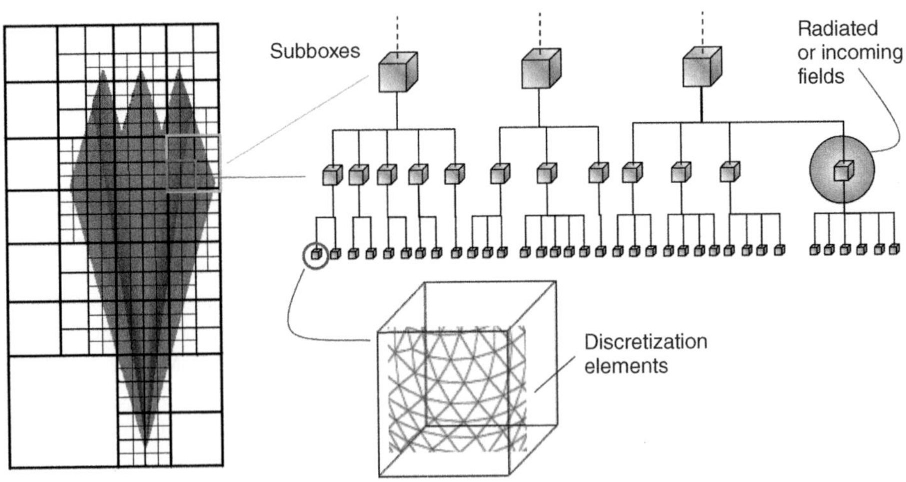

Figure 1.22 Recursive clustering of an arbitrary object and the construction of the multilevel tree structure.

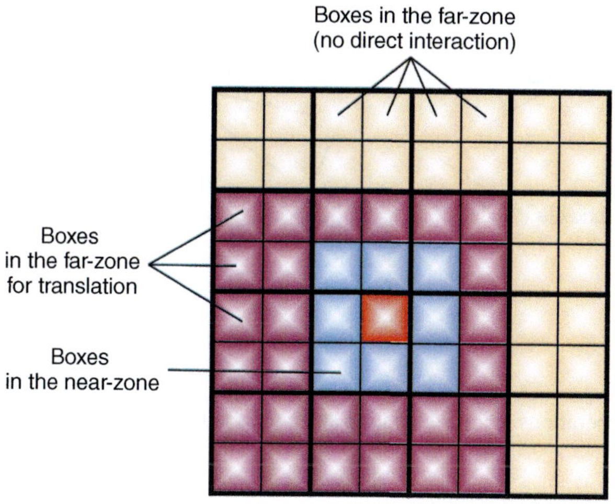

Figure 1.23 Boxes in the near and far zones for a given box (red) according to the one-box-buffer scheme.

other boxes are considered to be in the far zone. Among these far-zone boxes, only some of them (i.e., the magenta boxes, whose parents are in the near zone of the parent of the red box) are directly interacting with the red box through translations, while the remaining interactions are implicitly included at the higher levels.

The major operations that are performed in the three stages of MLFMA are illustrated in Figures 1.24–1.26, and can be summarized as follows.

- Aggregation (Figure 1.24): Radiated fields of boxes are calculated from the lowest level of the tree structure to the highest level. At the lowest level (bottom), the radiated field of a box is obtained by combining radiation patterns of the basis functions inside the box. At the higher levels, the radiated field of a box is the combination of the radiated fields of its subboxes.
- Translation (Figure 1.25): Radiated fields computed during the aggregation stage are translated into incoming fields. For each box at any level, there are $\mathcal{O}(1)$ boxes to translate the radiated field to. Using identical cubic boxes and the one-box-buffer scheme, there is a maximum of $7^3 - 3^3 = 316$ different translations at each level, independent of the number of boxes [158]. Examples of translations performed at two consecutive levels are depicted.
- Disaggregation (Figure 1.26): Total incoming fields at the box centers are calculated from the top of the tree structure to the bottom. As depicted, the total incoming field for a box is obtained by combining the incoming fields due to translations and the incoming field from the center of its parent box, if it exists. At the lowest level, incoming fields are received by testing functions.

In MLFMA, radiated and incoming fields are sampled on the unit sphere with respect to spherical coordinates θ and ϕ. The number of samples required for each box is proportional to the size of the box, as measured by the wavelength [159]. Therefore, to match different sampling rates of consecutive levels, interpolation and transpose interpolation (anterpolation)

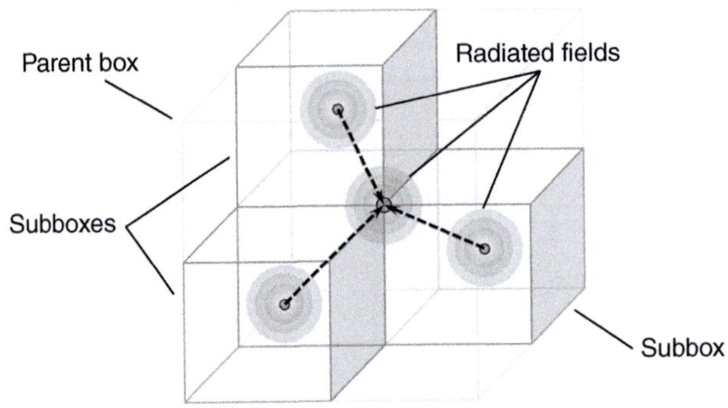

Figure 1.24 Operations in an aggregation stage for a parent box involving three subboxes.

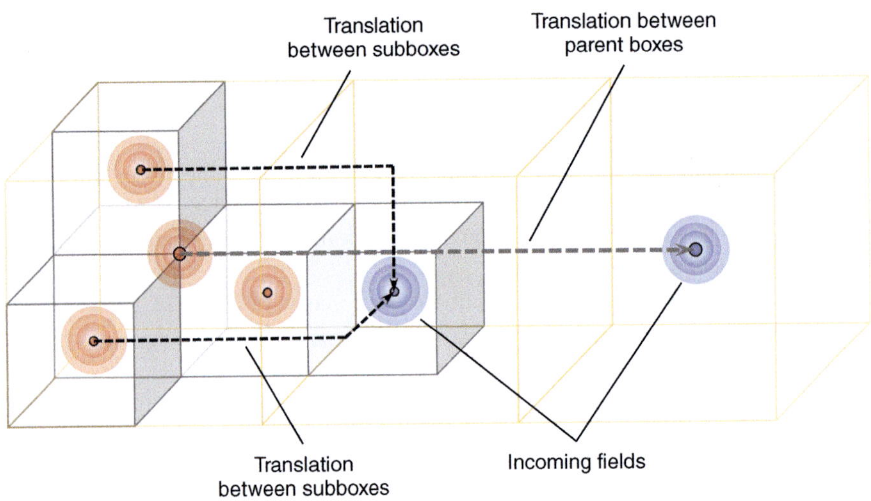

Figure 1.25 Operations in a translation stage for some boxes in two consecutive levels.

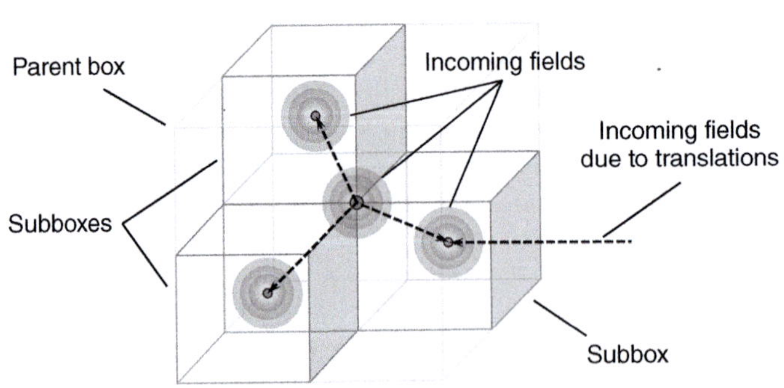

Figure 1.26 Operations in a disaggregation stage for a parent box involving three subboxes.

[160] are required during the aggregation and disaggregation stages, respectively. There are two major ways of implementing interpolations (and anterpolations), namely, through global and local interpolation methods. Global interpolations are usually based on FFT along the ϕ direction and the Legendre transform along the θ direction, performed via one-dimensional FMM [150],[161]. Using a uniform sampling, FFT can also be used along the θ direction [162]. Such an MLFMA implementation has $\mathcal{O}(N \log^2 N)$ time complexity, while interpolations/anterpolations are performed without error, provided that the Nyquist criterion is applied for the sampling rate. On the other hand, local interpolation methods introduce controllable errors [163]–[165], and such methods lead to more efficient MLFMA implementations with $\mathcal{O}(N \log N)$ complexity.

To be employed repeatedly, translation operators must usually be calculated and stored in memory before iterations. A direct calculation of these operators requires $\mathcal{O}(N^{3/2})$ operations, and the processing time for their setup may become substantial as the problem size grows. As a remedy, a two-step computation can be used based on the interpolation of the translation operator [166],[167]. First, the translation operator is expressed as a band-limited function and it is sampled at maximum $\mathcal{O}(N)$ points. Second, the operator is evaluated at the required points by interpolation from the previous samples. With an efficient interpolation algorithm, the processing time for the calculation of the translation operators can be reduced to $\mathcal{O}(N)$, as detailed in Chapter 3, Section 3.3.4.

In MLFMA, there are also $\mathcal{O}(N)$ near-field interactions, which are calculated directly in the setup stage of the program and stored in memory to be used multiple times during iterations. These interactions are between the basis and testing functions that are near to each other, i.e., in the same box or in two touching boxes at the lowest level using the one box-buffer scheme. Near-field interactions can be calculated accurately by using singularity extraction techniques [57],[143],[168],[169], adaptive integration methods, and Gaussian quadratures [142] (see this chapter, above, Section 1.8).

In order to achieve increased efficiency and reduced complexity, MLFMA employs several ways of gradual elimination of redundancies in the computations. Due to their gradual nature, such techniques may be interpreted as controllable error sources [1],[159],[166],[170]–[173]. For example, truncation of an infinite series and the angular integration over the unit sphere are two error sources of controllable error, which are also common to FMM. Interpolation and anterpolation operations constitute the third error source introduced by MLFMA. In addition, other errors arise due to MOM, such as the modeling of the geometry, discretization of integral equations, and numerical integrations on spatial basis and testing functions.

1.11 Low-Frequency Breakdown of MLFMA

MLFMA is an efficient algorithm to solve large-scale electromagnetics problems discretized with large numbers of unknowns. However, it suffers from the low-frequency breakdown, i.e., MLFMA boxes that are small compared to the wavelength cause errors. This is because MLFMA is based on the plane wave expansion, which becomes invalid for short distances. Therefore, MLFMA becomes inefficient for problems involving small objects discretized with large numbers of unknowns. Specifically, when ordinary MLFMA is applied to those low-frequency problems, the lowest-level boxes may involve many discretization elements. This significantly increases the processing time and memory required for the near-field interactions, which must be calculated directly. Even the complexity of the MLFMA implementation can be more than $\mathcal{O}(N \log N)$ due to excessively large numbers of near-field interactions.

In the literature, there are two common approaches to solve low-frequency problems efficiently via (modified versions of) MLFMA. The first approach is based on a spectral representation of the Green's function, where radiated fields are divided into propagating and evanescent parts [174]–[177]. Evanescent waves are employed to compute the interactions between small boxes separated by short distances. The second approach is based on the multipole representation of radiated fields [178]–[183]. As also detailed in Chapter 3, Section 3.7, the Green's function is factorized in a series of multipoles, but the multipoles are not converted into plane waves. In both approaches, box size is not restricted, and one can divide the object into boxes that can be much smaller than the wavelength. There are also new studies using alternative expansion methods for stable implementations at low frequencies [184],[185].

Obviously, diagonalization is the major source for the low-frequency breakdown of MLFMA. As a heuristic explanation, plane waves introduced by diagonalization cannot accurately represent electromagnetic fields of localized sources in small volumes (e.g., boxes). For a rigorous analysis, one should consider the factorized and diagonalized forms of the Green's function. As detailed in Chapter 3, Section 3.1, the factorization of the homogeneous-space Green's function is represented by an infinite summation involving the spherical Bessel and Hankel functions of the first kind, i.e., $j_t(r)$ and $h_t^{(1)}(r)$. In the factorized form, $j_t(r)$ and $h_t^{(1)}(r)$ balance each other, even for localized sources. Further stabilization is also possible by scaling $j_t(r)$ and $h_t^{(1)}(r)$ with the size of the source [183]. Applying the diagonalization, however, spherical waves are converted into plane waves and the Green's function is represented by an angular integration involving diagonal shifts and a diagonal translation operator. Unfortunately, in this form, the translation operator becomes unstable for small distances. Specifically, this operator is a summation of the spherical Hankel functions multiplied with the Legendre polynomials. As depicted in Figure 1.27, the spherical Hankel

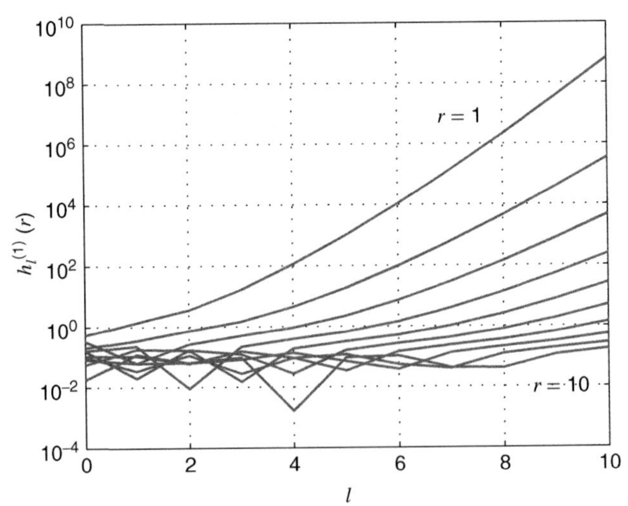

Figure 1.27 Magnitude of the imaginary part of the spherical Hankel function of the first kind $h_l^{(1)}(r)$ with respect to l for different values of r from 1 to 10.

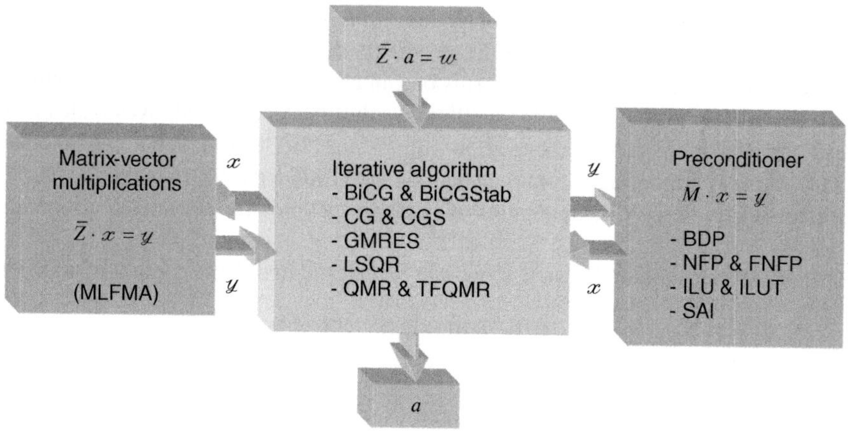

Figure 1.28 Iterative solutions of dense matrix equations derived from integral equations.

functions grow rapidly for small arguments. Hence, the Green's function is to be obtained by adding and subtracting large numbers, which can be prone to numerical problems, depending on the precision used. As a trivial but rigorous solution, avoiding diagonalization for small boxes in MLFMA effectively prevents the low-frequency breakdown, as also detailed in Chapter 3, Section 3.7.

1.12 Iterative Algorithms

Figure 1.28 depicts how MLFMA is used to solve large-scale electromagnetics problems. Specifically, an iterative algorithm performs iterative solutions by employing MLFMA, which provides an efficient means of executing the required MVMs. Note that the iterative algorithm also interacts with a preconditioner, as detailed in this chapter, below, Section 1.13. In the literature, various Krylov-subspace methods are available to solve electromagnetics problems formulated with integral equations [186]–[196]. Some of these methods are listed below:

- BiCG: Biconjugate gradient
- BiCGStab: Biconjugate gradient stabilized [188]
- CG: Conjugate gradient
- CGS: Conjugate gradient squared [189]
- GMRES: Generalized minimal residual [190]
- LSQR: Least-squares QR [191]–[193]
- MINRES: Minimal residual
- QMR: Quasi-minimal residual [194]
- TFQMR: Transpose-free quasi-minimal residual [195]

In all algorithms, the solution is (implicitly) expanded in a Krylov subspace, i.e.,

$$a \in \text{Span}\{w, \overline{Z} \cdot w, \overline{Z}^2 \cdot w, \ldots, \overline{Z}^j \cdot w\}. \tag{1.146}$$

For a given problem, the performance of each algorithm (the number of iterations j) may vary significantly, depending on the shape of the geometry, discretization, and the type of formulation; there is no one algorithm that performs best in all cases. Since reducing the number of iterations is extremely important to obtain efficient solutions, one should investigate and compare the iterative solutions provided by various algorithms [197],[198]. Sequential and parallel implementations of iterative algorithms can be found in [186] and [199]. Chapter 3, Section 3.6 includes many examples and comparisons of iterative methods for solving electromagnetics problems.

The Krylov-subspace methods listed above can be investigated in the context of matrix diagonalization.

1.12.1 Symmetric Lanczos Process

Consider a matrix equation

$$\overline{\boldsymbol{Z}} \cdot \boldsymbol{a} = \boldsymbol{w}, \tag{1.147}$$

where $\overline{\boldsymbol{Z}}$ is an $N \times N$ symmetric matrix. One can define

$$\beta^{(1)}\boldsymbol{v}^{(1)} = \boldsymbol{w} \tag{1.148}$$

and construct

$$\beta^{(k+1)}\boldsymbol{v}^{(k+1)} = \overline{\boldsymbol{Z}} \cdot \boldsymbol{v}^{(k)} - \alpha^{(k)}\boldsymbol{v}^{(k)} - \beta^{(k)}\boldsymbol{v}^{(k-1)}, \tag{1.149}$$

where

$$\alpha^{(k)} = [\boldsymbol{v}^{(k)}]^H \cdot \overline{\boldsymbol{Z}} \cdot \boldsymbol{v}^{(k)} - [\boldsymbol{v}^{(k)}]^H \cdot \beta^{(k)}\boldsymbol{v}^{(k-1)} \tag{1.150}$$

and $\beta^{(k)} \geq 0$ are chosen appropriately such that $\|\boldsymbol{v}^{(k)}\|_2 = 1$. In (1.150), the superscript 'H' represents the Hermitian operation. The recursion can be rewritten as

$$\begin{aligned}[\overline{\boldsymbol{Z}}]_{N\times N} \cdot [\overline{\boldsymbol{V}}]_{N\times k} &= [\overline{\boldsymbol{V}}]_{N\times k} \cdot [\overline{\boldsymbol{\Delta}}]_{k\times k} \\ &\quad + \beta^{(k+1)}[0 \quad 0 \quad \cdots \quad 0 \quad \boldsymbol{v}^{(k+1)}]_{N\times k},\end{aligned} \tag{1.151}$$

where

$$\overline{\boldsymbol{V}} = [\boldsymbol{v}^{(1)} \quad \boldsymbol{v}^{(2)} \quad \cdots \quad \boldsymbol{v}^{(k)}]_{N\times k} \tag{1.152}$$

is a unitary matrix, i.e.,

$$\overline{\boldsymbol{V}}^H \cdot \overline{\boldsymbol{V}} = \overline{\boldsymbol{I}}. \tag{1.153}$$

This process is called tridiagonalization and

$$\overline{\mathbf{\Delta}} = \begin{bmatrix} \alpha^{(1)} & \beta^{(2)} & & & & & \\ \beta^{(2)} & \alpha^{(2)} & \beta^{(3)} & & & & \\ & \beta^{(3)} & \alpha^{(3)} & \beta^{(4)} & & & \\ & & \beta^{(4)} & \alpha^{(4)} & \ddots & & \\ & & & \ddots & \ddots & \beta^{(k-1)} & \\ & & & & \beta^{(k-1)} & \alpha^{(k-1)} & \beta^{(k)} \\ & & & & & \beta^{(k)} & \alpha^{(k)} \end{bmatrix}_{k\times k} \tag{1.154}$$

is tridiagonal.

If the matrix $\overline{\mathbf{Z}}$ is positive definite, the CG method can be applied to solve the matrix equation in (1.147). In this method, a subproblem

$$[\overline{\mathbf{\Delta}}]_{k\times k} \cdot \mathbf{y}^{(k)} = \begin{bmatrix} \beta^{(1)} \\ 0 \\ \vdots \\ 0 \end{bmatrix}_{k\times 1} \tag{1.155}$$

is solved so that the original solution can be obtained as $\mathbf{a}^{(k)} = \overline{\mathbf{V}} \cdot \mathbf{y}^{(k)}$. For the solution of the subproblem, a Cholesky factorization is used, i.e.,

$$[\overline{\mathbf{\Delta}}]_{k\times k} = [\overline{\mathbf{L}}]_{k\times k} \cdot [\overline{\mathbf{D}}]_{k\times k} \cdot [\overline{\mathbf{L}}^{H}]_{k\times k}, \tag{1.156}$$

where $\overline{\mathbf{L}}$ is lower triangular and $\overline{\mathbf{D}}$ is diagonal.

For a general symmetric $\overline{\mathbf{Z}}$ matrix, which is not positive definite, the MINRES method can be applied by minimizing the norm of

$$\begin{bmatrix} & & \overline{\mathbf{\Delta}} & \\ 0 & \cdots & 0 & \beta^{(k)} \end{bmatrix}_{(k+1)\times k} \cdot \mathbf{y}^{(k)} - \begin{bmatrix} \beta^{(1)} \\ 0 \\ \vdots \\ 0 \end{bmatrix}_{(k+1)\times 1} \tag{1.157}$$

to obtain $\mathbf{y}^{(k)}$ and $\mathbf{a}^{(k)} = \overline{\mathbf{V}} \cdot \mathbf{y}^{(k)}$. For the solution of the subproblem in (1.157), a QR factorization is used as

$$\begin{bmatrix} & & \overline{\mathbf{\Delta}} & \\ 0 & \cdots & 0 & \beta^{(k)} \end{bmatrix}_{(k+1)\times k} = [\overline{\mathbf{Q}}]_{(k+1)\times(k+1)} \cdot \begin{bmatrix} \overline{\mathbf{R}} \\ 0 \end{bmatrix}_{(k+1)\times k}, \tag{1.158}$$

where $\overline{\mathbf{Q}}$ is unitary and $\overline{\mathbf{R}}$ is upper triangular.

1.12.2 Nonsymmetric Lanczos Process

Consider a matrix equation in the form of (1.147), where $\overline{\mathbf{Z}}$ is an $N \times N$ nonsymmetric matrix. One can define

$$\beta^{(1)} \boldsymbol{v}^{(1)} = \boldsymbol{w} \tag{1.159}$$

and $\boldsymbol{u}^{(1)} \cdot \boldsymbol{v}^{(1)} = 1$ to construct

$$\beta^{(k+1)} \boldsymbol{u}^{(k+1)} = \overline{\mathbf{Z}} \cdot \boldsymbol{u}^{(k)} - \alpha^{(k)} \boldsymbol{u}^{(k)} - \delta^{(k)} \boldsymbol{u}^{(k-1)} \tag{1.160}$$

$$\delta^{(k+1)} \boldsymbol{v}^{(k+1)} = \overline{\mathbf{Z}} \cdot \boldsymbol{v}^{(k)} - \alpha^{(k)} \boldsymbol{v}^{(k)} - \beta^{(k)} \boldsymbol{v}^{(k-1)}. \tag{1.161}$$

In (1.160) and (1.161),

$$\alpha^{(k)} = [\boldsymbol{u}^{(k)}]^H \cdot \overline{\mathbf{Z}} \cdot \boldsymbol{v}^{(k)}, \tag{1.162}$$

$\beta^{(k)} \geq 0$, and $\delta^{(k)} \geq 0$ are chosen appropriately such that $\|\boldsymbol{v}^{(k)}\|_2 = 1$ and $\|\boldsymbol{u}^{(k)}\|_2 = 1$. The recursion can be rewritten as

$$\begin{aligned}[\overline{\mathbf{Z}}]_{N\times N} \cdot [\overline{\boldsymbol{V}}]_{N\times k} &= [\overline{\boldsymbol{V}}]_{N\times k} \cdot [\overline{\boldsymbol{\Delta}}]_{k\times k} \\ &\quad + \delta^{(k+1)} [0 \quad 0 \quad \cdots \quad 0 \quad \boldsymbol{v}^{(k+1)}]_{N\times k}\end{aligned} \tag{1.163}$$

$$\begin{aligned}[\overline{\mathbf{Z}}]_{N\times N} \cdot [\overline{\boldsymbol{U}}]_{N\times k} &= [\overline{\boldsymbol{U}}]_{N\times k} \cdot [\overline{\boldsymbol{\Delta}}]_{k\times k} \\ &\quad + \beta^{(k+1)} [0 \quad 0 \quad \cdots \quad 0 \quad \boldsymbol{v}^{(k+1)}]_{N\times k},\end{aligned} \tag{1.164}$$

where

$$\overline{\boldsymbol{V}} = [\boldsymbol{v}^{(1)} \quad \boldsymbol{v}^{(2)} \quad \cdots \quad \boldsymbol{v}^{(k)}]_{N\times k} \tag{1.165}$$

and

$$\overline{\boldsymbol{U}} = [\boldsymbol{u}^{(1)} \quad \boldsymbol{u}^{(2)} \quad \cdots \quad \boldsymbol{u}^{(k)}]_{N\times k} \tag{1.166}$$

satisfy

$$\overline{\boldsymbol{U}}^H \cdot \overline{\boldsymbol{V}} = \overline{\boldsymbol{I}}. \tag{1.167}$$

This process is also a tridiagonalization and

$$\overline{\boldsymbol{\Delta}} = \begin{bmatrix} \alpha^{(1)} & \beta^{(2)} & & & & & \\ \delta^{(2)} & \alpha^{(2)} & \beta^{(3)} & & & & \\ & \delta^{(3)} & \alpha^{(3)} & \beta^{(4)} & & & \\ & & \delta^{(4)} & \alpha^{(4)} & \ddots & & \\ & & & \ddots & \ddots & \beta^{(k-1)} & \\ & & & & \delta^{(k-1)} & \alpha^{(k-1)} & \beta^{(k)} \\ & & & & & \delta^{(k)} & \alpha^{(k)} \end{bmatrix}_{k\times k} \tag{1.168}$$

is tridiagonal.

BiCG and QMR methods, as well as their transpose-free variants, i.e., BiCGStab, CGS, and TFQMR, are based on the nonsymmetric Lanczos process. For example, in BiCG, a subproblem similar to (1.155) is solved via an LU factorization in iterative methods factorization, i.e.,

$$\overline{\boldsymbol{\Delta}}_{k\times k} = \overline{\boldsymbol{L}}_{k\times k} \cdot \overline{\boldsymbol{D}}_{k\times k} \cdot \overline{\boldsymbol{R}}_{k\times k}. \tag{1.169}$$

The QMR method is similar to MINRES, and the norm of

$$\begin{bmatrix} & \overline{\boldsymbol{\Delta}} & & \\ 0 & \cdots & 0 & \delta^{(k)} \end{bmatrix}_{(k+1)\times k} \cdot \boldsymbol{y}^{(k)} - \begin{bmatrix} \delta^{(1)} \\ 0 \\ \vdots \\ 0 \end{bmatrix}_{(k+1)\times 1} \tag{1.170}$$

is minimized by using a QR factorization.

1.12.3 Arnoldi Process

The Arnoldi process involves a recursion in the form of

$$\begin{aligned} [\overline{\boldsymbol{Z}}]_{N\times N} \cdot [\overline{\boldsymbol{V}}]_{N\times k} &= [\overline{\boldsymbol{V}}]_{N\times k} \cdot [\overline{\boldsymbol{H}}]_{k\times k} \\ &\quad + \beta^{(k+1)}[0 \quad 0 \quad \cdots \quad 0 \quad \boldsymbol{v}^{(k+1)}]_{N\times k}, \end{aligned} \tag{1.171}$$

$$\begin{aligned} [\overline{\boldsymbol{Z}}]_{N\times N} \cdot [\overline{\boldsymbol{V}}]_{N\times k} &= [\overline{\boldsymbol{V}}]_{N\times k} \cdot [\overline{\boldsymbol{H}}]_{k\times k} \\ &\quad + \beta^{(k+1)}[0 \quad 0 \quad \cdots \quad 0 \quad \boldsymbol{v}^{(k+1)}]_{N\times k}, \end{aligned} \tag{1.172}$$

where $\overline{\boldsymbol{V}} = [\boldsymbol{v}_n^{(1)} \quad \boldsymbol{v}_n^{(2)} \quad \cdots \quad \boldsymbol{v}_n^{(k)}]$ is again a unitary matrix. As opposed to the matrix obtained by a tridiagonalization, $\overline{\boldsymbol{H}}$ is a Hessian matrix. In general, a subproblem similar to (1.155) is considered for the solution. The efficient GMRES method is based on the Arnoldi process.

1.12.4 Golub-Kahan Process

For a general $N \times N$ matrix equation in the form of (1.147), the Golub-Kahan process starts with

$$\beta^{(1)}\boldsymbol{x}^{(1)} = \boldsymbol{w} \tag{1.173}$$

$$\alpha^{(1)}\boldsymbol{v}^{(1)} = \overline{\boldsymbol{Z}}^{H} \cdot \boldsymbol{x}^{(1)} \tag{1.174}$$

and constructs a recursion as

$$\beta^{(k+1)}\boldsymbol{x}^{(k+1)} = \overline{\boldsymbol{Z}} \cdot \boldsymbol{v}^{(k)} - \alpha^{(k)}\boldsymbol{x}^{(k)} \tag{1.175}$$

$$\alpha^{(k+1)}\boldsymbol{v}^{(k+1)} = \overline{\boldsymbol{Z}}^{H} \cdot \boldsymbol{x}^{(k+1)} - \beta^{(k+1)}\boldsymbol{v}^{(k)}, \tag{1.176}$$

where $\alpha^{(k)} \geq 0$ and $\beta^{(k)} \geq 0$ are chosen appropriately such that $\|\boldsymbol{x}^{(k)}\|_2 = 1$ and $\|\boldsymbol{v}^{(k)}\|_2 = 1$. The recursion can be rewritten as

$$[\overline{\boldsymbol{Z}}]_{N\times N} \cdot [\overline{\boldsymbol{V}}]_{N\times k} = [\overline{\boldsymbol{X}}]_{N\times(k+1)} \cdot [\overline{\boldsymbol{\Lambda}}]_{(k+1)\times k} \tag{1.177}$$

$$\begin{aligned}[\overline{\boldsymbol{Z}}^H]_{N\times N} \cdot [\overline{\boldsymbol{X}}]_{N\times(k+1)} &= [\overline{\boldsymbol{V}}]_{N\times k} \cdot [\overline{\boldsymbol{\Lambda}}^H]_{k\times(k+1)} \\ &\quad + \alpha^{(k+1)}[0 \quad 0 \quad \cdots \quad 0 \quad \boldsymbol{v}^{(k+1)}]_{N\times(k+1)},\end{aligned} \tag{1.178}$$

where $\overline{\boldsymbol{\Lambda}}$ is lower-bidiagonal, i.e.,

$$\overline{\boldsymbol{\Lambda}} = \begin{bmatrix} \alpha^{(1)} & & & & & & \\ \beta^{(2)} & \alpha^{(2)} & & & & & \\ & \beta^{(3)} & \alpha^{(3)} & & & & \\ & & \beta^{(4)} & \alpha^{(4)} & & & \\ & & & \ddots & \ddots & & \\ & & & & \beta^{(k-1)} & \alpha^{(k-1)} & \\ & & & & & \beta^{(k)} & \alpha^{(k)} \\ & & & & & & \beta^{(k+1)} \end{bmatrix}_{(k+1)\times k}. \tag{1.179}$$

The LSQR method is based on the Golub-Kahan process, where a subproblem similar to (1.157) is solved via a QR factorization.

1.13 Preconditioning

MLFMA provides the solution of large problems by reducing the complexity of MVMs required by iterative solvers from $\mathcal{O}(N^2)$ to $\mathcal{O}(N \log N)$. For efficient solutions, however, the number of iterations should also be small. Along this direction, preconditioning refers to transforming a matrix equation into an equivalent form with more favorable conditioning characteristics and that can be solved easier with iterative methods [200]. This is usually achieved by multiplying the matrix equation with an approximate inverse of the matrix.

In MLFMA, there are $\mathcal{O}(N)$ near-field interactions, which are calculated directly and are available for constructing preconditioners. Using the one-box-buffer scheme, these interactions are between the basis and testing functions that are located in the same box or in two touching boxes at the lowest level of the tree structure. During solutions via MLFMA, one can reorder the basis and testing functions according to their positions in the multilevel tree. Let N_1 be the number of boxes at the lowest level and $N(C)$ represent the number of basis or testing functions in box $C = 1, 2, \ldots, N_1$. Then, the basis and testing functions in box C are indexed from $N^+(C) + 1$ to $N^+(C) + N(C)$, where

$$N^+(C) = \sum_{C'=1}^{C-1} N(C). \tag{1.180}$$

This way, the impedance matrix has a block structure, where each block represents the interaction of a pair of boxes at the lowest level. In the sparse near-field matrix, only the blocks corresponding to self-interactions of boxes or interactions of two touching boxes involve nonzero elements. An example of reordering is depicted in Figure 1.29, where the self-interactions of the boxes are illustrated for a small 72-unknown problem.

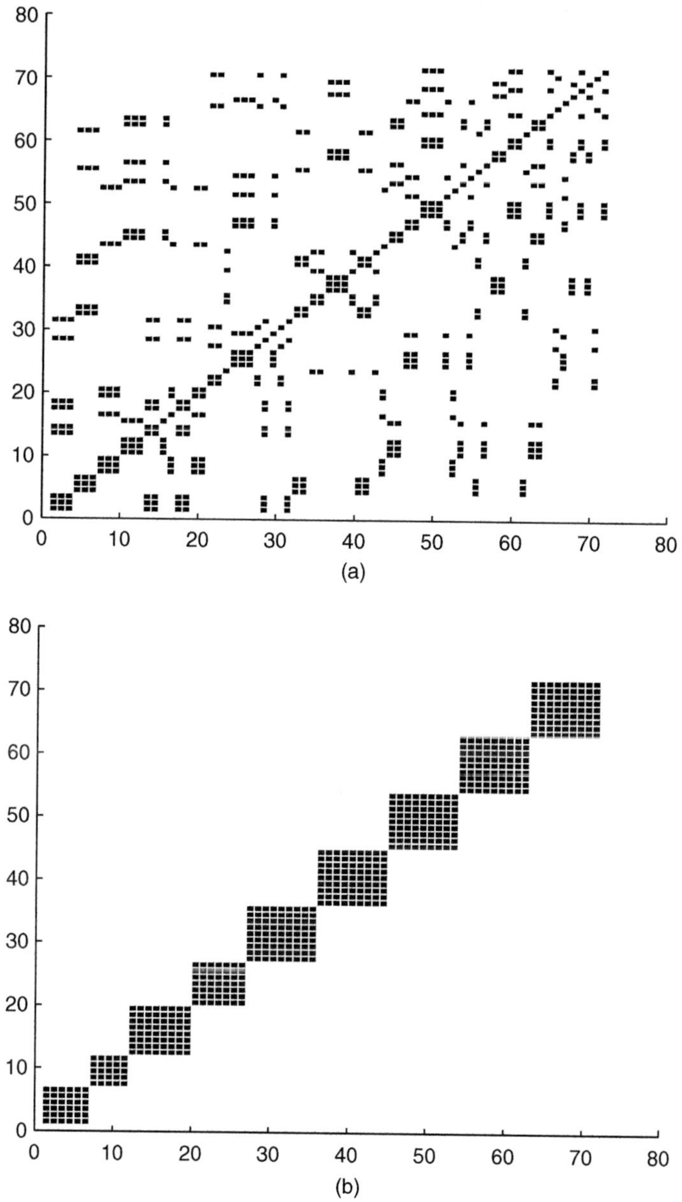

Figure 1.29 Distribution of the matrix elements in the self interactions of the boxes for a 72-unknown problem (a) before reordering and (b) after reordering based on the box indices.

With the direct factorization in preconditioners factorization of the sparse near-field matrix of all near-field interactions, one can obtain the near-field preconditioner (NFP). However, in most cases, near-field interactions are distributed irregularly in the full matrix and cannot be collected in a bandlimited diagonal region, even when using reordering algorithms [201]. Hence, the factorization performed once before iterations and the solution of the preconditioner system applied in each iteration can be expensive in terms of processing time; thus the preconditioner becomes a bottleneck as the problem size grows. As a remedy, a block-diagonal preconditioner (BDP), which is much more efficient, can be obtained by extracting only the self-interactions of the lowest-level boxes [35] (e.g., Figure 1.29). Both the factorization and application of BDP are $\mathcal{O}(N)$; this is appropriate for MLFMA since its complexity is $\mathcal{O}(N \log N)$ and the complexity of the preconditioner should not be higher than that. Due to its simplicity and favorable computing cost, BDP is commonly used for MLFMA solutions of normal and mixed formulations, such as T-N-CFIE. Unfortunately, BDP usually decelerates iterative solutions of tangential formulations, such as T-EFIE, except in a few special cases [202]. In addition, for large problems involving complicated objects, acceleration provided by BDP may not be sufficient, even when normal and mixed formulations are used.

Recently, there have been many efforts to improve iterative solutions of surface integral equations, particularly for metallic objects, using preconditioners [203]–[211]. In the context of MLFMA, one of the most common preconditioning techniques is the incomplete LU (ILU) method [200],[203]–[206]. This is a forward-type preconditioning technique, where the preconditioner matrix $\overline{\boldsymbol{P}}$ approximates the system matrix, and one can solve for

$$\overline{\boldsymbol{P}}^{-1} \cdot \overline{\boldsymbol{Z}} \cdot \boldsymbol{a} = \overline{\boldsymbol{P}}^{-1} \cdot \boldsymbol{w} \tag{1.181}$$

or

$$(\overline{\boldsymbol{Z}} \cdot \overline{\boldsymbol{P}}^{-1}) \cdot (\overline{\boldsymbol{P}} \cdot \boldsymbol{a}) = \boldsymbol{w}, \tag{1.182}$$

instead of the original matrix equation. In (1.181) and (1.182), which correspond to left and right preconditioning, respectively, the solution of $\overline{\boldsymbol{P}} \cdot \boldsymbol{x} = \boldsymbol{y}$ for a given $\boldsymbol{y}$ should be cheaper than the solution of the original matrix equation. During the factorization of the preconditioner matrix, the ILU method sacrifices some of the fill-ins and provides an approximation to the near-field matrix, i.e.,

$$\overline{\boldsymbol{P}} = \overline{\boldsymbol{L}} \cdot \overline{\boldsymbol{U}} \approx \overline{\boldsymbol{Z}}_{NF}. \tag{1.183}$$

It is shown that the ILU preconditioner without a threshold provides inexpensive and good approximations to the near-field matrices for T-N-CFIE, hence it reduces the iteration counts and solution times substantially [206]. For ill-conditioned T-EFIE matrices, however, the ILUT (i.e., threshold-based ILU) preconditioner with pivoting [205] is required to prevent the potential instability. Other successful adoptions of the ILU-type preconditioners are presented in [204].

Despite the remarkable success of the ILU-type preconditioners, they are limited to sequential implementations due to difficulties in parallelizing their factorization algorithms and forward-backward solutions. Hence, the sparse-approximate-inverse (SAI) preconditioner, which is well-suited for parallel implementations, has been more preferable for the solution of large-scale electromagnetics problems [207]–[210]. The SAI preconditioner is based on a backward-type scheme, where the inverse of the system matrix is directly approximated, i.e., $\overline{\boldsymbol{P}} \approx \overline{\boldsymbol{Z}}^{-1}$. In MLFMA, only the near-field matrix is considered, and one can minimize

$$\|\overline{\boldsymbol{I}} - \overline{\boldsymbol{P}} \cdot \overline{\boldsymbol{Z}}_{NF}\|_F, \tag{1.184}$$

where $\| \cdot \|_F$ represents the Frobenius norm. Using the pattern of the near-field matrix for the nonzero pattern of $\overline{\boldsymbol{P}}$ provides some advantages by decreasing the number of QR factorizations required during the minimization in (1.184) [208]. In parallel implementations, one can use row-wise partitioning to distribute the near-field interactions among processors. Therefore, left-preconditioning must be used to accelerate the iterative solutions with the SAI preconditioner. However, right-preconditioning can be used for the symmetric matrix equations, e.g., those derived from T-EFIE [210].

Preconditioners that are based on near-field interactions can be insufficient to accelerate iterative solutions of large-scale problems, especially those formulated with T-EFIE. For more efficient solutions, it is possible to use far-field interactions in addition to near-field interactions and construct more effective preconditioners. As depicted in Figure 1.30, this can be achieved by using flexible solvers and employing approximate and ordinary versions of MLFMA in an inner-outer scheme [211]. Using a reasonable approximation of the inner solutions, the number of outer iterations can be reduced substantially. There are two ways to describe the advantages of this strategy:

- MVMs performed by an ordinary MLFMA are replaced with more efficient MVMs performed by an approximate MLFMA (AMLFMA). Different from the relaxation strategies, however, only a single specific implementation of AMLFMA is sufficient to construct an inner-outer scheme. In addition, reasonable accuracy (without strict limits) is sufficient for the approximation.

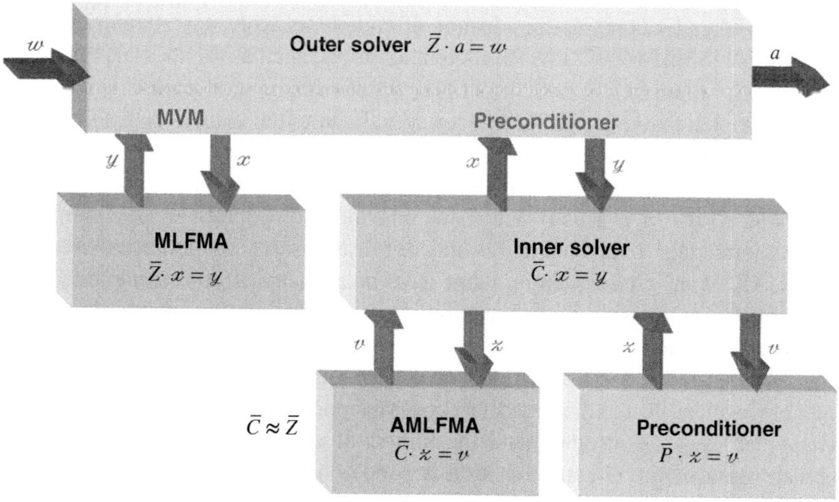

Figure 1.30 An inner-outer scheme involving a flexible iterative solver using an inner solver as a preconditioner.

- Iterative solutions by an ordinary MLFMA are preconditioned with a very strong preconditioner that is constructed by approximating the full matrix instead of the sparse part of the matrix.

Strategies for building a less-accurate MLFMA are discussed in Chapter 3, Section 3.5.

In addition to more efficient solutions, the inner-outer scheme prevents numerical errors that arise because of deviations in the computed residual from the true residual by significantly decreasing the number of outer iterations. This is because the residual gap, i.e., the difference between the true and computed residuals, increases with the number of iterations [212]. Another benefit of the reduction in iteration counts appears when the iterative solutions are performed with the GMRES algorithm, which is usually an optimal method for T-EFIE in terms of the processing time [206],[208]. Even though other variants of GMRES, e.g., the flexible GMRES (FGMRES) [205], require the storage of two vectors per iteration instead of one, nested solutions require significantly less memory than ordinary GMRES solutions because they dramatically reduce the iteration counts.

1.14 Parallelization of MLFMA

For a dense matrix equation involving $\mathcal{O}(N)$ unknowns, MLFMA reduces the complexity of MVMs from $\mathcal{O}(N^2)$ to $\mathcal{O}(N \log N)$, allowing for the solution of large problems with limited computational resources. On the other hand, accurate solutions of many real-life problems require discretizations with millions of elements, leading to matrix equations involving millions of unknowns, which cannot easily be solved with sequential implementations of MLFMA running on a single processor. To solve such large problems, it is helpful to increase computational resources by assembling parallel computing platforms and, at the same time, by parallelizing MLFMA.

Of the various parallelization schemes for MLFMA, the most popular use distributed-memory architectures by constructing clusters of computers with local memories connected via fast networks [158],[214]–[232]. Parallelization tools, such as the message passing interface (MPI) [233], organizing parallel solutions via communication protocols are available. However, the parallelization of MLFMA is not trivial due to the complicated structure of this algorithm [217]. Although other algorithms with higher complexities, such as FMM, have been parallelized very successfully with simple parallelization techniques [216],[226], this is not the case for MLFMA. This is due to the unavoidable occurrences of communications among processors, poor load balancing, and duplications of computations over multiple processors in MLFMA. Consequently, there have been many efforts to improve the parallelization of MLFMA by using novel partitioning strategies, load-balancing algorithms, and optimized communications [217]–[232].

In a typical tree structure of MLFMA, the lowest level involves $\mathcal{O}(N)$ boxes with dimensions in the order of wavelength, and the number of boxes decreases by a factor of four from one level to the next higher level. On the other hand, the number of samples required for the factorization and diagonalization of the Green's function depends on the size of boxes as measured by the wavelength. Consequently, the number of samples increases by a factor of four from one level to the next higher level. It is remarkable that the number of boxes and the number of samples balance each other, and all levels in MLFMA have approximately an equal cost with

$\mathcal{O}(N)$ complexity. Therefore, an efficient parallelization of MLFMA should consider the best possible partitioning at each level. With such a strategy, it would be possible to solve problems with hundreds of millions of unknowns on relatively inexpensive computing platforms with distributed-memory architectures [225]–[232].

In a simple parallelization of MLFMA, boxes are distributed among processors at all levels. Such a technique provides efficient parallelization for small numbers of processors, usually fewer than 16. For more processors, however, the efficiency drops drastically due to poor load balancing. This is because small numbers of boxes in the higher levels of MLFMA cannot be distributed evenly among processors. In addition, dense communications among processors are required. Note that low parallelization efficiency not only increases processing time but also restricts the size of problems that can be solved. As a remedy, a hybrid technique has been proposed to improve the parallelization of MLFMA [217],[221]. This technique is based on using an appropriate partitioning scheme for the higher levels of MLFMA by distributing samples instead of boxes. Due to the improved load balancing in the higher levels, the hybrid technique significantly increases the parallelization efficiency, compared to the simple technique. Nevertheless, the hybrid technique fails to provide efficient solutions when the number of processors is larger than 32 [219],[225].

Recently, a hierarchical parallelization technique has been developed for efficient parallelization of MLFMA [219],[223],[225],[230]–[232]. In this technique, the boxes and their samples are partitioned among processors by employing load-balancing algorithms. Although changing the partitioning between levels bears an additional cost, the hierarchical technique offers a higher parallelization efficiency than previous parallelization techniques for MLFMA. Specifically, this hierarchical partitioning strategy provides two important advantages. First, simultaneous partitioning of the boxes and their samples leads to improved load balancing among processors at each level. Second, communications between processors are reduced, i.e., average package size is enlarged, the number of communication events is reduced, and the communication time is significantly shortened. Details of the hierarchical partitioning strategy and comparisons with previous parallelization techniques can be found in Chapter 4, Section 4.6.

2

Solutions of Electromagnetics Problems with Surface Integral Equations

This chapter presents formulations of electromagnetics problems with surface integral equations. Surface formulations are derived for three-dimensional metallic, dielectric, and composite objects with arbitrary geometries. Various stabilization procedures are discussed for accurate solutions of low-contrast dielectric objects. Accuracy of implementations is extensively investigated for metallic objects and various discretization schemes.

2.1 Homogeneous Dielectric Objects

For dielectric objects with homogeneous material properties, a general procedure for surface formulations can be summarized as follows:

- Apply the surface operators on equivalent surface currents to obtain expressions for secondary fields.
- Enforce the boundary conditions for tangential electric and magnetic fields on the surface of the object either directly or rotationally and derive basic integral equations for the inner and outer media.
- Combine inner and outer integral equations appropriately to obtain a set of equations to solve for equivalent currents.
- Calculate radiated electric and magnetic fields from equivalent currents.

Based on the items above, various surface formulations can be derived by using different combinations of boundary conditions, testing schemes, and scaling operations. Several of them are stable, free of the internal-resonance problem, and commonly used in the literature [51]–[83].

The Multilevel Fast Multipole Algorithm (MLFMA) for Solving Large-Scale Computational Electromagnetics Problems, First Edition. Özgür Ergül and Levent Gürel.

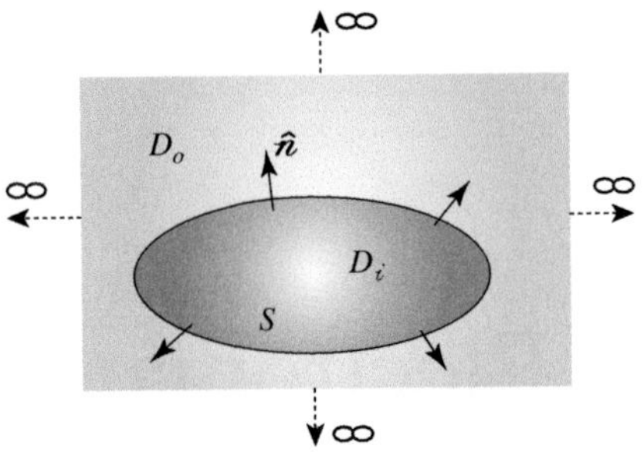

Figure 2.1 A homogeneous object located in a homogeneous space extending to infinity.

2.1.1 Surface Integral Equations

Consider the case in Figure 2.1 involving a homogeneous dielectric object with a closed surface separating two media D_o and D_i with different electrical properties (ϵ_o, μ_o) and (ϵ_i, μ_i). Let incident fields exist outside the object. Using the equivalence principle, the outer and inner problems can be considered separately to derive two sets of integral equations by employing boundary conditions, i.e.,

$$\eta_o \hat{\boldsymbol{n}} \times \mathcal{T}_o\{\boldsymbol{J}\}(\boldsymbol{r}) - \hat{\boldsymbol{n}} \times \mathcal{K}_{PV,o}\{\boldsymbol{M}\}(\boldsymbol{r}) + \frac{\Omega_o}{4\pi}\mathcal{I}\{\boldsymbol{M}\}(\boldsymbol{r}) = -\hat{\boldsymbol{n}} \times \boldsymbol{E}^{inc}(\boldsymbol{r}) \qquad \text{(N-EFIE-O)} \tag{2.1}$$

$$-\eta_o \hat{\boldsymbol{n}} \times \hat{\boldsymbol{n}} \times \mathcal{T}_o\{\boldsymbol{J}\}(\boldsymbol{r}) + \hat{\boldsymbol{n}} \times \hat{\boldsymbol{n}} \times \mathcal{K}_{PV,o}\{\boldsymbol{M}\}(\boldsymbol{r}) - \frac{\Omega_o}{4\pi}\mathcal{I}^{\times n}\{\boldsymbol{M}\}(\boldsymbol{r}) = \hat{\boldsymbol{n}} \times \hat{\boldsymbol{n}} \times \boldsymbol{E}^{inc}(\boldsymbol{r}) \qquad \text{(T-EFIE-O)} \tag{2.2}$$

$$\eta_o^{-1} \hat{\boldsymbol{n}} \times \mathcal{T}_o\{\boldsymbol{M}\}(\boldsymbol{r}) + \hat{\boldsymbol{n}} \times \mathcal{K}_{PV,o}\{\boldsymbol{J}\}(\boldsymbol{r}) - \frac{\Omega_o}{4\pi}\mathcal{I}\{\boldsymbol{J}\}(\boldsymbol{r}) = -\hat{\boldsymbol{n}} \times \boldsymbol{H}^{inc}(\boldsymbol{r}) \qquad \text{(N-MFIE-O)} \tag{2.3}$$

$$-\eta_o^{-1} \hat{\boldsymbol{n}} \times \hat{\boldsymbol{n}} \times \mathcal{T}_o\{\boldsymbol{M}\}(\boldsymbol{r}) - \hat{\boldsymbol{n}} \times \hat{\boldsymbol{n}} \times \mathcal{K}_{PV,o}\{\boldsymbol{J}\}(\boldsymbol{r}) + \frac{\Omega_o}{4\pi}\mathcal{I}^{\times n}\{\boldsymbol{J}\}(\boldsymbol{r}) = \hat{\boldsymbol{n}} \times \hat{\boldsymbol{n}} \times \boldsymbol{H}^{inc}(\boldsymbol{r}) \qquad \text{(T-MFIE-O)} \tag{2.4}$$

for the outer medium D_o and

$$\eta_i \hat{\boldsymbol{n}} \times \mathcal{T}_i\{\boldsymbol{J}\}(\boldsymbol{r}) - \hat{\boldsymbol{n}} \times \mathcal{K}_{PV,i}\{\boldsymbol{M}\}(\boldsymbol{r}) - \frac{\Omega_i}{4\pi}\mathcal{I}\{\boldsymbol{M}\}(\boldsymbol{r}) = 0 \qquad \text{(N-EFIE-I)} \tag{2.5}$$

$$-\eta_i \hat{\boldsymbol{n}} \times \hat{\boldsymbol{n}} \times \mathcal{T}_i\{\boldsymbol{J}\}(\boldsymbol{r}) + \hat{\boldsymbol{n}} \times \hat{\boldsymbol{n}} \times \mathcal{K}_{PV,i}\{\boldsymbol{M}\}(\boldsymbol{r}) + \frac{\Omega_i}{4\pi}\mathcal{I}^{\times n}\{\boldsymbol{M}\}(\boldsymbol{r}) = 0 \qquad \text{(T-EFIE-I)} \tag{2.6}$$

$$\eta_i^{-1} \hat{\boldsymbol{n}} \times \mathcal{T}_i\{\boldsymbol{M}\}(\boldsymbol{r}) + \hat{\boldsymbol{n}} \times \mathcal{K}_{PV,i}\{\boldsymbol{J}\}(\boldsymbol{r}) + \frac{\Omega_i}{4\pi}\mathcal{I}\{\boldsymbol{J}\}(\boldsymbol{r}) = 0 \qquad \text{(N-MFIE-I)} \tag{2.7}$$

$$-\eta_i^{-1} \hat{\boldsymbol{n}} \times \hat{\boldsymbol{n}} \times \mathcal{T}_i\{\boldsymbol{M}\}(\boldsymbol{r}) - \hat{\boldsymbol{n}} \times \hat{\boldsymbol{n}} \times \mathcal{K}_{PV,i}\{\boldsymbol{J}\}(\boldsymbol{r}) - \frac{\Omega_i}{4\pi}\mathcal{I}^{\times n}\{\boldsymbol{J}\}(\boldsymbol{r}) = 0 \qquad \text{(T-MFIE-I)} \tag{2.8}$$

for the inner medium D_i. In (2.1)–(2.8), $\hat{\boldsymbol{n}}$ is the outward normal unit vector, Ω_o and $\Omega_i = 4\pi - \Omega_o$ are the external and internal solid angles, and $\boldsymbol{E}^{inc}$ and $\boldsymbol{H}^{inc}$ are the incident electric and magnetic fields, respectively.

2.1.2 Surface Formulations

For homogeneous dielectric objects, tangential formulations can be obtained by combining tangential integral equations [74] as

$$\begin{bmatrix} a\text{T-EFIE-O} + b\text{T-EFIE-I} \\ c\text{T-MFIE-O} + d\text{T-MFIE-I} \end{bmatrix} \tag{2.9}$$

leading to

$$-\hat{\boldsymbol{n}} \times \hat{\boldsymbol{n}} \times \left[a\eta_o\mathcal{T}_o + b\eta_i\mathcal{T}_i\right]\{\boldsymbol{J}\}(\boldsymbol{r}) + \hat{\boldsymbol{n}} \times \hat{\boldsymbol{n}} \times \left[a\mathcal{K}_{PV,o} + b\mathcal{K}_{PV,i}\right]\{\boldsymbol{M}\}(\boldsymbol{r}) - \frac{1}{4\pi}\left[a\Omega_o - b\Omega_i\right]\mathcal{I}^{\times n}\{\boldsymbol{M}\}(\boldsymbol{r}) = a\hat{\boldsymbol{n}} \times \hat{\boldsymbol{n}} \times \boldsymbol{E}^{inc}(\boldsymbol{r}) \tag{2.10}$$

$$-\hat{\boldsymbol{n}} \times \hat{\boldsymbol{n}} \times \left[c\mathcal{K}_{PV,o} + d\mathcal{K}_{PV,i}\right]\{\boldsymbol{J}\}(\boldsymbol{r}) + \frac{1}{4\pi}\left[c\Omega_o - d\Omega_i\right]\mathcal{I}^{\times n}\{\boldsymbol{J}\}(\boldsymbol{r}) - \hat{\boldsymbol{n}} \times \hat{\boldsymbol{n}} \times \left[c\eta_o^{-1}\mathcal{T}_o + d\eta_i^{-1}\mathcal{T}_i\right]\{\boldsymbol{M}\}(\boldsymbol{r}) = c\hat{\boldsymbol{n}} \times \hat{\boldsymbol{n}} \times \boldsymbol{H}^{inc}(\boldsymbol{r}). \tag{2.11}$$

Among infinitely many possibilities for the scalars $\{a, b, c, d\}$, several choices provide stable formulations, such as

$$a = b = c = d = 1 \tag{2.12}$$

and

$$a = \eta_o^{-1},\ b = \eta_i^{-1},\ c = \eta_o,\ d = \eta_i, \tag{2.13}$$

which lead to the T-PMCHWT formulation [52]–[54] and CTF [74], respectively. Combined tangential formulation CTF can be written as

$$\begin{bmatrix} \mathcal{Z}_{11}^{\text{CTF}} & \mathcal{Z}_{12}^{\text{CTF}} \\ \mathcal{Z}_{21}^{\text{CTF}} & \mathcal{Z}_{22}^{\text{CTF}} \end{bmatrix} \cdot \begin{bmatrix} \boldsymbol{J} \\ \boldsymbol{M} \end{bmatrix}(\boldsymbol{r}) = \hat{\boldsymbol{n}} \times \hat{\boldsymbol{n}} \times \begin{bmatrix} \eta_o^{-1}\boldsymbol{E}^{inc} \\ \eta_o\boldsymbol{H}^{inc} \end{bmatrix}(\boldsymbol{r}), \tag{2.14}$$

where

$$\mathcal{Z}_{11}^{\mathrm{CTF}} = \mathcal{Z}_{22}^{\mathrm{CTF}} = -\hat{\boldsymbol{n}} \times \hat{\boldsymbol{n}} \times \left(\mathcal{T}_o + \mathcal{T}_i\right) \tag{2.15}$$

$$\mathcal{Z}_{12}^{\mathrm{CTF}} = \hat{\boldsymbol{n}} \times \hat{\boldsymbol{n}} \times \left(\eta_o^{-1}\mathcal{K}_{PV,o} + \eta_i^{-1}\mathcal{K}_{PV,i}\right) - \frac{1}{4\pi}\left(\eta_o^{-1}\Omega_o - \eta_i^{-1}\Omega_i\right)\mathcal{I}^{\times n} \tag{2.16}$$

$$\mathcal{Z}_{21}^{\mathrm{CTF}} = -\hat{\boldsymbol{n}} \times \hat{\boldsymbol{n}} \times \left(\eta_o\mathcal{K}_{PV,o} + \eta_i\mathcal{K}_{PV,i}\right) + \frac{1}{4\pi}\left(\eta_o\Omega_o - \eta_i\Omega_i\right)\mathcal{I}^{\times n}. \tag{2.17}$$

Matrix equations obtained from CTF have identical diagonal partitions and they are usually better conditioned than matrix equations obtained from the traditional T-PMCHWT formulation [74].

Different from tangential formulations, normal formulations are derived by using normal integral equations [74], i.e.,

$$\begin{bmatrix} a\text{N-MFIE-O} - b\text{N-MFIE-I} \\ -c\text{N-EFIE-O} + d\text{N-EFIE-I} \end{bmatrix}, \tag{2.18}$$

leading to

$$-\frac{1}{4\pi}\left[a\Omega_o + b\Omega_i\right]\mathcal{I}\{\boldsymbol{J}\}(\boldsymbol{r}) + \hat{\boldsymbol{n}} \times \left[a\mathcal{K}_{PV,o} - b\mathcal{K}_{PV,i}\right]\{\boldsymbol{J}\}(\boldsymbol{r})$$
$$+\hat{\boldsymbol{n}} \times \left[a\eta_o^{-1}\mathcal{T}_o - b\eta_i^{-1}\mathcal{T}_i\right]\{\boldsymbol{M}\}(\boldsymbol{r}) = -a\hat{\boldsymbol{n}} \times \boldsymbol{H}^{inc}(\boldsymbol{r}) \tag{2.19}$$

$$-\hat{\boldsymbol{n}} \times \left[c\eta_o\mathcal{T}_o - d\eta_i\mathcal{T}_i\right]\{\boldsymbol{J}\}(\boldsymbol{r}) - \frac{1}{4\pi}\left[c\Omega_o + d\Omega_i\right]\mathcal{I}\{\boldsymbol{M}\}(\boldsymbol{r})$$
$$+\hat{\boldsymbol{n}} \times \left[c\mathcal{K}_{PV,o} - d\mathcal{K}_{PV,i}\right]\{\boldsymbol{M}\}(\boldsymbol{r}) = c\hat{\boldsymbol{n}} \times \boldsymbol{E}^{inc}(\boldsymbol{r}). \tag{2.20}$$

Choices

$$a = \mu_o,\ b = \mu_i,\ c = \epsilon_o,\ d = \epsilon_i \tag{2.21}$$

and

$$a = b = c = d = 1 \tag{2.22}$$

in (2.18) lead to NMF [51] and CNF [74], respectively. CNF can be written as

$$\begin{bmatrix} \mathcal{Z}_{11}^{\mathrm{CNF}} & \mathcal{Z}_{12}^{\mathrm{CNF}} \\ \mathcal{Z}_{21}^{\mathrm{CNF}} & \mathcal{Z}_{22}^{\mathrm{CNF}} \end{bmatrix} \cdot \begin{bmatrix} \boldsymbol{J} \\ \boldsymbol{M} \end{bmatrix}(\boldsymbol{r}) = -\hat{\boldsymbol{n}} \times \begin{bmatrix} \boldsymbol{H}^{inc} \\ -\boldsymbol{E}^{inc} \end{bmatrix}(\boldsymbol{r}), \tag{2.23}$$

where

$$\mathcal{Z}_{11}^{\mathrm{CNF}} = \mathcal{Z}_{22}^{\mathrm{CNF}} = \hat{\boldsymbol{n}} \times \left(\mathcal{K}_{PV,o} - \mathcal{K}_{PV,i}\right) - \mathcal{I} \tag{2.24}$$

$$\mathcal{Z}_{12}^{\mathrm{CNF}} = \hat{\boldsymbol{n}} \times \left(\eta_o^{-1}\mathcal{T}_o - \eta_i^{-1}\mathcal{T}_i\right) \tag{2.25}$$

$$\mathcal{Z}_{21}^{\mathrm{CNF}} = -\hat{\boldsymbol{n}} \times \left(\eta_o\mathcal{T}_o - \eta_i\mathcal{T}_i\right). \tag{2.26}$$

On the other hand, choosing the scalars as

$$a = \frac{\mu_o}{(\mu_o + \mu_i)},\ b = \frac{\mu_i}{(\mu_o + \mu_i)},\ c = \frac{\epsilon_o}{(\epsilon_o + \epsilon_i)},\ d = \frac{\epsilon_i}{(\epsilon_o + \epsilon_i)} \tag{2.27}$$

in (2.18) leads to MNMF [73], which usually produces better-conditioned matrix equations than NMF and CNF. This formulation can be written as

$$\begin{bmatrix} \mathcal{Z}_{11}^{\text{MNMF}} & \mathcal{Z}_{12}^{\text{MNMF}} \\ \mathcal{Z}_{21}^{\text{MNMF}} & \mathcal{Z}_{22}^{\text{MNMF}} \end{bmatrix} \cdot \begin{bmatrix} \boldsymbol{J} \\ \boldsymbol{M} \end{bmatrix} (\boldsymbol{r}) = -\hat{\boldsymbol{n}} \times \begin{bmatrix} \mu_o(\mu_o + \mu_i)^{-1}\boldsymbol{H}^{inc} \\ -\epsilon_o(\epsilon_o + \epsilon_i)^{-1}\boldsymbol{E}^{inc} \end{bmatrix} (\boldsymbol{r}), \tag{2.28}$$

where

$$\mathcal{Z}_{11}^{\text{MNMF}} = \frac{1}{(\mu_o + \mu_i)}\hat{\boldsymbol{n}} \times \left(\mu_o\mathcal{K}_{PV,o} - \mu_i\mathcal{K}_{PV,i}\right) - \frac{1}{4\pi(\mu_o + \mu_i)}\left(\mu_o\Omega_o + \mu_i\Omega_i\right)\mathcal{I} \tag{2.29}$$

$$\mathcal{Z}_{12}^{\text{MNMF}} = \frac{1}{(\mu_o + \mu_i)}\hat{\boldsymbol{n}} \times \left(\mu_o\eta_o^{-1}\mathcal{T}_o - \mu_i\eta_i^{-1}\mathcal{T}_i\right) \tag{2.30}$$

$$\mathcal{Z}_{21}^{\text{MNMF}} = -\frac{1}{(\epsilon_o + \epsilon_i)}\hat{\boldsymbol{n}} \times \left(\epsilon_o\eta_o\mathcal{T}_o - \epsilon_i\eta_i\mathcal{T}_i\right) \tag{2.31}$$

$$\mathcal{Z}_{22}^{\text{MNMF}} = \frac{1}{(\epsilon_o + \epsilon_i)}\hat{\boldsymbol{n}} \times \left(\epsilon_o\mathcal{K}_{PV,o} - \epsilon_i\mathcal{K}_{PV,i}\right) - \frac{1}{4\pi(\epsilon_o + \epsilon_i)}\left(\epsilon_o\Omega_o + \epsilon_i\Omega_i\right)\mathcal{I}. \tag{2.32}$$

Finally, mixed formulations involve both tangential and normal integral equations. For example, JMCFIE is obtained by using all equations in (2.1)–(2.8) [71], [80], [83], i.e.,

$$\begin{bmatrix} \text{N-MFIE-O} - \text{N-MFIE-I} + \eta_o^{-1}\text{T-EFIE-O} + \eta_i^{-1}\text{T-EFIE-I} \\ -\text{N-EFIE-O} + \text{N-EFIE-I} + \eta_o\text{T-MFIE-O} + \eta_i\text{T-MFIE-I} \end{bmatrix}. \tag{2.33}$$

Combining operators, one can derive JMCFIE as

$$\begin{bmatrix} \mathcal{Z}_{11}^{\text{JMCFIE}} & \mathcal{Z}_{12}^{\text{JMCFIE}} \\ \mathcal{Z}_{21}^{\text{JMCFIE}} & \mathcal{Z}_{22}^{\text{JMCFIE}} \end{bmatrix} \cdot \begin{bmatrix} \boldsymbol{J} \\ \boldsymbol{M} \end{bmatrix} (\boldsymbol{r}) = \hat{\boldsymbol{n}} \times \begin{bmatrix} \eta_o^{-1}\hat{\boldsymbol{n}} \times \boldsymbol{E}^{inc} - \boldsymbol{H}^{inc} \\ \eta_o\hat{\boldsymbol{n}} \times \boldsymbol{H}^{inc} + \boldsymbol{E}^{inc} \end{bmatrix} (\boldsymbol{r}), \tag{2.34}$$

where

$$\mathcal{Z}_{11}^{\text{JMCFIE}} = \mathcal{Z}_{22}^{\text{JMCFIE}} = \mathcal{Z}_{11}^{\text{CTF}} + \mathcal{Z}_{11}^{\text{CNF}} = -\hat{\boldsymbol{n}} \times \hat{\boldsymbol{n}} \times (\mathcal{T}_o + \mathcal{T}_i) + \hat{\boldsymbol{n}} \times (\mathcal{K}_{PV,o} - \mathcal{K}_{PV,i}) - \mathcal{I} \tag{2.35}$$

$$\mathcal{Z}_{12}^{\text{JMCFIE}} = \mathcal{Z}_{12}^{\text{CTF}} + \mathcal{Z}_{12}^{\text{CNF}} = \hat{\boldsymbol{n}} \times \hat{\boldsymbol{n}} \times \left(\eta_o^{-1}\mathcal{K}_{PV,o} + \eta_i^{-1}\mathcal{K}_{PV,i}\right) + \hat{\boldsymbol{n}} \times \left(\eta_o^{-1}\mathcal{T}_o - \eta_i^{-1}\mathcal{T}_i\right) - \frac{1}{4\pi}\left(\eta_o^{-1}\Omega_o - \eta_i^{-1}\Omega_i\right)\mathcal{I}^{\times n} \tag{2.36}$$

$$\mathcal{Z}_{21}^{\text{JMCFIE}} = \mathcal{Z}_{21}^{\text{CTF}} + \mathcal{Z}_{21}^{\text{CNF}} = -\hat{\boldsymbol{n}} \times \hat{\boldsymbol{n}} \times \left(\eta_o\mathcal{K}_{PV,o} + \eta_i\mathcal{K}_{PV,i}\right) - \hat{\boldsymbol{n}} \times \left(\eta_o\mathcal{T}_o - \eta_i\mathcal{T}_i\right) + \frac{1}{4\pi}\left(\eta_o\Omega_o - \eta_i\Omega_i\right)\mathcal{I}^{\times n}. \tag{2.37}$$

Note that JMCFIE is different from conventional CFIE formulations for dielectric objects, e.g., TN-N-CFIE [68], where EFIE and MFIE are combined for the inner and outer media separately.

2.1.3 Discretizations of Surface Formulations

For numerical solutions of surface formulations, surfaces can be discretized by using small planar triangles, on which basis functions are employed to expand unknown surface currents, i.e.,

$$\boldsymbol{J}(\boldsymbol{r}) = \hat{\boldsymbol{n}} \times \boldsymbol{H}(\boldsymbol{r}) \approx \sum_{n=1}^{N} \boldsymbol{a}_J[n]\boldsymbol{b}_n(\boldsymbol{r}) \tag{2.38}$$

$$\boldsymbol{M}(\boldsymbol{r}) = -\hat{\boldsymbol{n}} \times \boldsymbol{E}(\boldsymbol{r}) \approx \sum_{n=1}^{N} \boldsymbol{a}_M[n]\boldsymbol{b}_n(\boldsymbol{r}), \tag{2.39}$$

where $\boldsymbol{b}_n$ represents the nth basis function associated with the nth edge. Testing integral equations, $2N \times 2N$ dense matrix equations are constructed as

$$\begin{bmatrix} \overline{\boldsymbol{Z}}_{11} & \overline{\boldsymbol{Z}}_{12} \\ \overline{\boldsymbol{Z}}_{21} & \overline{\boldsymbol{Z}}_{22} \end{bmatrix} \cdot \begin{bmatrix} \boldsymbol{a}_J \\ \boldsymbol{a}_M \end{bmatrix} = \begin{bmatrix} \boldsymbol{w}_1 \\ \boldsymbol{w}_2 \end{bmatrix}, \tag{2.40}$$

where matrix elements correspond to interactions of basis and testing functions, and RHS vectors are obtained by testing incident fields. Specifically, matrix elements involve combinations of discretized operators, depending on the formulation. Considering the mth testing function $\boldsymbol{t}_m$ and the nth basis function $\boldsymbol{b}_n$, the directly and rotationally tested $\mathcal{K}$, $\mathcal{T}$, and $\mathcal{I}$ operators are discretized as

$$-\hat{\boldsymbol{n}} \times \hat{\boldsymbol{n}} \times \mathcal{K}_{PV} \rightarrow \overline{\boldsymbol{K}}^{T}_{PV}[m,n] = \langle \boldsymbol{t}_m(\boldsymbol{r}), \mathcal{K}_{PV}\{\boldsymbol{b}_n\}(\boldsymbol{r})\rangle \tag{2.41}$$

$$\hat{\boldsymbol{n}} \times \mathcal{K}_{PV} \rightarrow \overline{\boldsymbol{K}}^{N}_{PV}[m,n] = \langle \boldsymbol{t}_m(\boldsymbol{r}), \hat{\boldsymbol{n}} \times \mathcal{K}_{PV}\{\boldsymbol{b}_n\}(\boldsymbol{r})\rangle \tag{2.42}$$

$$-\hat{\boldsymbol{n}} \times \hat{\boldsymbol{n}} \times \mathcal{T} \rightarrow \overline{\boldsymbol{T}}^{T}[m,n] = \langle \boldsymbol{t}_m(\boldsymbol{r}), \mathcal{T}\{\boldsymbol{b}_n\}(\boldsymbol{r})\rangle \tag{2.43}$$

$$\hat{\boldsymbol{n}} \times \mathcal{T} \rightarrow \overline{\boldsymbol{T}}^{N}[m,n] = \langle \boldsymbol{t}_m(\boldsymbol{r}), \hat{\boldsymbol{n}} \times \mathcal{T}\{\boldsymbol{b}_n\}(\boldsymbol{r})\rangle \tag{2.44}$$

$$\begin{aligned} \mathcal{I} \rightarrow \overline{\boldsymbol{I}}^{T}[m,n] &= \langle \boldsymbol{t}_m(\boldsymbol{r}), \mathcal{I}\{\boldsymbol{b}_n\}(\boldsymbol{r})\rangle \\ &= \langle \boldsymbol{t}_m(\boldsymbol{r}), \boldsymbol{b}_n(\boldsymbol{r})\rangle \end{aligned} \tag{2.45}$$

$$\begin{aligned} \mathcal{I}^{\times n} \rightarrow \overline{\boldsymbol{I}}^{N}[m,n] &= \left\langle \boldsymbol{t}_m(\boldsymbol{r}), \mathcal{I}^{\times n}\{\boldsymbol{b}_n\}(\boldsymbol{r})\right\rangle \\ &= \langle \boldsymbol{t}_m(\boldsymbol{r}), \hat{\boldsymbol{n}} \times \boldsymbol{b}_n(\boldsymbol{r})\rangle, \end{aligned} \tag{2.46}$$

where

$$\langle \boldsymbol{t}_m(\boldsymbol{r}), \boldsymbol{f}(\boldsymbol{r})\rangle = \int_{S_m} d\boldsymbol{r} \boldsymbol{t}_m(\boldsymbol{r}) \cdot \boldsymbol{f}(\boldsymbol{r}) \tag{2.47}$$

is a surface integral on the support of the testing function (S_m). Then, the matrix partitions in (2.40) can be derived as

$$\overline{\boldsymbol{Z}}_{11}^{\mathrm{CTF}} = \overline{\boldsymbol{Z}}_{22}^{\mathrm{CTF}} = \overline{\boldsymbol{T}}_o^T + \overline{\boldsymbol{T}}_i^T \tag{2.48}$$

$$\overline{\boldsymbol{Z}}_{12}^{\mathrm{CTF}} = -\eta_o^{-1}\overline{\boldsymbol{K}}_{PV,o}^T - \eta_i^{-1}\overline{\boldsymbol{K}}_{PV,i}^T - \frac{1}{2}\left(\eta_o^{-1} - \eta_i^{-1}\right)\overline{\boldsymbol{I}}^N \tag{2.49}$$

$$\overline{\boldsymbol{Z}}_{21}^{\mathrm{CTF}} = \eta_o\overline{\boldsymbol{K}}_{PV,o}^T + \eta_i\overline{\boldsymbol{K}}_{PV,i}^T + \frac{1}{2}\left(\eta_o - \eta_i\right)\overline{\boldsymbol{I}}^N \tag{2.50}$$

for CTF,

$$\overline{\boldsymbol{Z}}_{11}^{\mathrm{CNF}} = \overline{\boldsymbol{Z}}_{22}^{\mathrm{CNF}} = \overline{\boldsymbol{K}}_{PV,o}^N - \overline{\boldsymbol{K}}_{PV,i}^N - \overline{\boldsymbol{I}}^T \tag{2.51}$$

$$\overline{\boldsymbol{Z}}_{12}^{\mathrm{CNF}} = \eta_o^{-1}\overline{\boldsymbol{T}}_o^N - \eta_i^{-1}\overline{\boldsymbol{T}}_i^N \tag{2.52}$$

$$\overline{\boldsymbol{Z}}_{21}^{\mathrm{CNF}} = -\eta_o\overline{\boldsymbol{T}}_o^N + \eta_i\overline{\boldsymbol{T}}_i^N \tag{2.53}$$

for CNF,

$$\overline{\boldsymbol{Z}}_{11}^{\mathrm{MNMF}} = \frac{1}{(\mu_o + \mu_i)}\left(\mu_o\overline{\boldsymbol{K}}_{PV,o}^N - \mu_i\overline{\boldsymbol{K}}_{PV,i}^N\right) - \frac{1}{2}\overline{\boldsymbol{I}}^T \tag{2.54}$$

$$\overline{\boldsymbol{Z}}_{12}^{\mathrm{MNMF}} = \frac{1}{(\mu_o + \mu_i)}\left(\mu_o\eta_o^{-1}\overline{\boldsymbol{T}}_o^N - \mu_i\eta_i^{-1}\overline{\boldsymbol{T}}_i^N\right) \tag{2.55}$$

$$\overline{\boldsymbol{Z}}_{21}^{\mathrm{MNMF}} = -\frac{1}{(\epsilon_o + \epsilon_i)}\left(\epsilon_o\eta_o\overline{\boldsymbol{T}}_o^N - \epsilon_i\eta_i\overline{\boldsymbol{T}}_i^N\right) \tag{2.56}$$

$$\overline{\boldsymbol{Z}}_{22}^{\mathrm{MNMF}} = \frac{1}{(\epsilon_o + \epsilon_i)}\left(\epsilon_o\overline{\boldsymbol{K}}_{PV,o}^N - \epsilon_i\overline{\boldsymbol{K}}_{PV,i}^N\right) - \frac{1}{2}\overline{\boldsymbol{I}}^T \tag{2.57}$$

for MNMF, and

$$\overline{\boldsymbol{Z}}_{11}^{\mathrm{JMCFIE}} = \overline{\boldsymbol{Z}}_{22}^{\mathrm{JMCFIE}} = \overline{\boldsymbol{Z}}_{11}^{\mathrm{CTF}} + \boldsymbol{Z}_{11}^{\mathrm{CNF}} \tag{2.58}$$

$$= \overline{\boldsymbol{T}}_o^T + \overline{\boldsymbol{T}}_i^T + \overline{\boldsymbol{K}}_{PV,o}^N - \overline{\boldsymbol{K}}_{PV,i}^N - \overline{\boldsymbol{I}}^T \tag{2.59}$$

$$\overline{\boldsymbol{Z}}_{12}^{\mathrm{JMCFIE}} = \overline{\boldsymbol{Z}}_{12}^{\mathrm{CTF}} + \overline{\boldsymbol{Z}}_{12}^{\mathrm{CNF}} \tag{2.60}$$

$$\begin{aligned} &= -\eta_o^{-1}\overline{\boldsymbol{K}}_{PV,o}^T - \eta_i^{-1}\overline{\boldsymbol{K}}_{PV,i}^T \\ &\quad + \eta_o^{-1}\overline{\boldsymbol{T}}_o^N - \eta_i^{-1}\overline{\boldsymbol{T}}_i^N - \frac{1}{2}\left(\eta_o^{-1} - \eta_i^{-1}\right)\overline{\boldsymbol{I}}^N \end{aligned} \tag{2.61}$$

$$\overline{\boldsymbol{Z}}_{21}^{\mathrm{JMCFIE}} = \overline{\boldsymbol{Z}}_{21}^{\mathrm{CTF}} + \overline{\boldsymbol{Z}}_{21}^{\mathrm{CNF}} \tag{2.62}$$

$$= \eta_o\overline{\boldsymbol{K}}_{PV,o}^T + \eta_i\overline{\boldsymbol{K}}_{PV,i}^T - \eta_o\overline{\boldsymbol{T}}_o^N + \eta_i\overline{\boldsymbol{T}}_i^N + \frac{1}{2}\left(\eta_o - \eta_i\right)\overline{\boldsymbol{I}}^N \tag{2.63}$$

for JMCFIE. It should be noted that $\Omega_o = \Omega_i = 2\pi$ on planar triangles. The RHS vectors in (2.40) also depend on the formulation and they involve directly and rotationally tested incident electric and magnetic fields, i.e.,

$$v_E^T[m] = \langle t_m(r), E^{inc}(r) \rangle = \int_{S_m} dr t_m(r) \cdot E^{inc}(r) \tag{2.64}$$

$$v_E^N[m] = \langle t_m(r), \hat{n} \times E^{inc}(r) \rangle = \int_{S_m} dr t_m(r) \cdot \hat{n} \times E^{inc}(r) \tag{2.65}$$

$$v_H^T[m] = \langle t_m(r), H^{inc}(r) \rangle = \int_{S_m} dr t_m(r) \cdot H^{inc}(r) \tag{2.66}$$

$$v_H^N[m] = \langle t_m(r), \hat{n} \times H^{inc}(r) \rangle = \int_{S_m} dr t_m(r) \cdot \hat{n} \times H^{inc}(r). \tag{2.67}$$

Using (2.64)–(2.67), the RHS vectors can be written as

$$\{w_1^{\text{CTF}}, w_2^{\text{CTF}}\} = \{-\eta_o^{-1} v_E^T, -\eta_o v_H^T\} \tag{2.68}$$

$$\{w_1^{\text{CNF}}, w_2^{\text{CNF}}\} = \{-v_H^N, v_E^N\} \tag{2.69}$$

$$\{w_1^{\text{MNMF}}, w_2^{\text{MNMF}}\} = \left\{ -\frac{\mu_o}{(\mu_o + \mu_i)} v_H^N, \frac{\epsilon_o}{(\epsilon_o + \epsilon_i)} v_E^N \right\} \tag{2.70}$$

and

$$\{w_1^{\text{JMCFIE}}, w_2^{\text{JMCFIE}}\} = \{w_1^{\text{CTF}} + w_1^{\text{CNF}}, w_2^{\text{CTF}} + w_2^{\text{CNF}}\} \tag{2.71}$$

$$= \{-\eta_o^{-1} v_E^T - v_H^N, -\eta_o v_H^T + v_E^N\} \tag{2.72}$$

for CTF, CNF, MNMF, and JMCFIE, respectively.

2.1.4 Direct Calculations of Interactions

The electromagnetic interaction between the mth testing function t_m and the nth basis function b_n can be derived for different operators ($\mathcal{T}$ and $\mathcal{K}$) and testing types as

$$\begin{aligned}\overline{T}^T[m,n] = ik \int_{S_m} dr t_m(r) \cdot \int_{S_n} dr' g(r,r') b_n(r') \\ + \frac{i}{k} \int_{S_m} dr t_m(r) \cdot \int_{S_n} dr' \nabla g(r,r') \nabla' \cdot b_n(r')\end{aligned} \tag{2.73}$$

$$\begin{aligned}\overline{T}^N[m,n] = ik \int_{S_m} dr t_m(r) \cdot \hat{n} \times \int_{S_n} dr' g(r,r') b_n(r') \\ + \frac{i}{k} \int_{S_m} dr t_m(r) \cdot \hat{n} \times \int_{S_n} dr' \nabla g(r,r') \nabla' \cdot b_n(r')\end{aligned} \tag{2.74}$$

$$\overline{K}_{PV}^T[m,n] = \int_{S_m} dr t_m(r) \cdot \int_{PV,S_n} dr' b_n(r') \times \nabla' g(r,r') \tag{2.75}$$

$$\overline{K}_{PV}^N[m,n] = \int_{S_m} dr t_m(r) \cdot \hat{n} \times \int_{PV,S_n} dr' b_n(r') \times \nabla' g(r,r'), \tag{2.76}$$

where S_m and S_n represent spatial supports of $\boldsymbol{t}_m$ and $\boldsymbol{b}_n$, respectively. Dielectric formulations can be discretized by employing the RWG functions [57] having a spatial distribution depicted in Figure 1.8. Direct calculations of the interactions in (2.73)–(2.76) can be summarized as follows:

- Using the RWG functions, the interactions in (2.75) and (2.76) can be modified as [168]

$$\overline{\boldsymbol{K}}_{PV}^{T}[m,n] = \int_{S_m} d\boldsymbol{r}\boldsymbol{t}_m(\boldsymbol{r}) \cdot \boldsymbol{b}_n(\boldsymbol{r}) \times \int_{PV,S_n} d\boldsymbol{r}'\nabla' g(\boldsymbol{r},\boldsymbol{r}') \tag{2.77}$$

$$\overline{\boldsymbol{K}}_{PV}^{N}[m,n] = \int_{S_m} d\boldsymbol{r}\left(\boldsymbol{t}_m(\boldsymbol{r}) \times \hat{\boldsymbol{n}}\right) \cdot \boldsymbol{b}_n(\boldsymbol{r}) \times \int_{PV,S_n} d\boldsymbol{r}'\nabla' g(\boldsymbol{r},\boldsymbol{r}'), \tag{2.78}$$

  where the outer integrals are evaluated numerically. Only the principal value of the inner integral is required so that (2.77) and (2.78) are not evaluated for self-interactions of triangles. However, the value of the inner integral is infinite when the testing point is at the edge of the source triangle. Since the singularity is logarithmic and it is quite mild, the interaction of two near-neighboring (touching) triangles can be calculated accurately by sampling observation points strictly inside the testing triangle. In addition, accuracy and efficiency of calculations can be improved by extracting the singularity [124].
- Using divergence-conforming functions, such as the RWG functions, the interaction in (2.73) is modified as

$$\begin{aligned}\overline{\boldsymbol{T}}^{T}[m,n] = {} & ik\int_{S_m} d\boldsymbol{r}\boldsymbol{t}_m(\boldsymbol{r}) \cdot \int_{S_n} d\boldsymbol{r}' g(\boldsymbol{r},\boldsymbol{r}')\boldsymbol{b}_n(\boldsymbol{r}') \\ & + \frac{i}{k}\int_{S_m} d\boldsymbol{r}\boldsymbol{t}_m(\boldsymbol{r}) \cdot \nabla \int_{S_n} d\boldsymbol{r}' g(\boldsymbol{r},\boldsymbol{r}')\nabla' \cdot \boldsymbol{b}_n(\boldsymbol{r}')\end{aligned} \tag{2.79}$$

$$\begin{aligned}= {} & ik\int_{S_m} d\boldsymbol{r}\boldsymbol{t}_m(\boldsymbol{r}) \cdot \int_{S_n} d\boldsymbol{r}' g(\boldsymbol{r},\boldsymbol{r}')\boldsymbol{b}_n(\boldsymbol{r}') \\ & + \frac{i}{k}\int_{S_m} d\boldsymbol{r}\nabla \cdot \left\{\boldsymbol{t}_m(\boldsymbol{r})\int_{S_n} d\boldsymbol{r}' g(\boldsymbol{r},\boldsymbol{r}')\nabla' \cdot \boldsymbol{b}_n(\boldsymbol{r}')\right\} \\ & - \frac{i}{k}\int_{S_m} d\boldsymbol{r}\nabla \cdot \boldsymbol{t}_m(\boldsymbol{r})\int_{S_n} d\boldsymbol{r}' g(\boldsymbol{r},\boldsymbol{r}')\nabla' \cdot \boldsymbol{b}_n(\boldsymbol{r}')\end{aligned} \tag{2.80}$$

$$\begin{aligned}= {} & ik\int_{S_m} d\boldsymbol{r}\boldsymbol{t}_m(\boldsymbol{r}) \cdot \int_{S_n} d\boldsymbol{r}' g(\boldsymbol{r},\boldsymbol{r}')\boldsymbol{b}_n(\boldsymbol{r}') \\ & - \frac{i}{k}\int_{S_m} d\boldsymbol{r}\nabla \cdot \boldsymbol{t}_m(\boldsymbol{r})\int_{S_n} d\boldsymbol{r}' g(\boldsymbol{r},\boldsymbol{r}')\nabla' \cdot \boldsymbol{b}_n(\boldsymbol{r}')\end{aligned} \tag{2.81}$$

  by moving the differential operator onto the testing function in the second term [57]. This way, the hyper-singularity of the $\mathcal{T}$ operator is eliminated. For the second term of (2.74), however, the differential operator must be kept on the Green's function. Then, this term is calculated similarly to the interaction in (2.76). But, as opposed to (2.76), the discretized operator in (2.74) should also be calculated for self-interactions of triangles.

- During the calculation of interactions, computational loops are constructed over triangles instead of basis and testing functions. To facilitate this, integrals over basis and testing functions are divided into many basic (double) integrals that are independent from the alignment of the functions.
- Basic integrals are evaluated in two steps; first, inner integrals are calculated, and second, they are used in forming integrands of outer integrals. Inner integrals are commonly shared by more than one basic integral.
- Inner integrals are performed via a decomposition into numerical and analytical parts [57], [144], [169]. Singularities arising due to the Green's function are extracted this way in order to avoid numerical difficulties and inaccuracies.
- Numerical integrations can be performed by using adaptive methods (such as those described in Section 1.8.3) employing low-order Gaussian quadratures [142]. Such adaptive methods achieve the preset required error bounds by sampling integration points efficiently.

The processing time for the calculation of the interactions in (2.73)–(2.76) depends on medium parameters. When the relative permittivity or permeability of a medium increases, it becomes difficult to evaluate interactions since integrands become more oscillatory. In such cases, accurate calculations of interactions may require extracting more terms to smooth integrands or increasing the number of sampling points for numerical integrations.

Consider the interaction of two half RWG functions $\boldsymbol{t}_{ma}^{\mathrm{RWG}}$ and $\boldsymbol{b}_{nb}^{\mathrm{RWG}}$ associated with the ath triangle of the mth edge and the bth triangle of the nth edge, respectively. Then,

$$\overline{\boldsymbol{I}}^{T,\mathrm{RWG}}[m,n,a,b] = \mathcal{A}_{ma,nb}\int_{S_{ma}} d\boldsymbol{r}(\boldsymbol{r}-\boldsymbol{r}_{ma})\cdot(\boldsymbol{r}-\boldsymbol{r}_{nb}) \tag{2.82}$$

$$\overline{\boldsymbol{I}}^{N,\mathrm{RWG}}[m,n,a,b] = \mathcal{A}_{ma,nb}\int_{S_{ma}} d\boldsymbol{r}(\boldsymbol{r}-\boldsymbol{r}_{ma})\cdot\hat{\boldsymbol{n}}\times(\boldsymbol{r}-\boldsymbol{r}_{nb}) \tag{2.83}$$

$$\begin{aligned}\overline{\boldsymbol{T}}^{T,\mathrm{RWG}}[m,n,a,b] &= ik\mathcal{A}_{ma,nb}\int_{S_{ma}} d\boldsymbol{r}(\boldsymbol{r}-\boldsymbol{r}_{ma})\cdot\boldsymbol{I}_{in}^{(\boldsymbol{r}'-\boldsymbol{r}_{nb})}\\ &\quad-\frac{4i}{k}\mathcal{A}_{ma,nb}\int_{S_{ma}} d\boldsymbol{r}I_{in}^{1}\end{aligned} \tag{2.84}$$

$$\begin{aligned}\overline{\boldsymbol{T}}^{N,\mathrm{RWG}}[m,n,a,b] &= ik\mathcal{A}_{ma,nb}\int_{S_{ma}} d\boldsymbol{r}(\boldsymbol{r}-\boldsymbol{r}_{ma})\cdot\hat{\boldsymbol{n}}\times\boldsymbol{I}_{in}^{(\boldsymbol{r}'-\boldsymbol{r}_{nb})}\\ &\quad-\frac{2i}{k}\mathcal{A}_{ma,nb}\int_{S_{ma}} d\boldsymbol{r}(\boldsymbol{r}-\boldsymbol{r}_{ma})\cdot\hat{\boldsymbol{n}}\times\boldsymbol{I}_{in}^{\nabla'}\end{aligned} \tag{2.85}$$

$$\overline{\boldsymbol{K}}_{PV}^{T,\mathrm{RWG}}[m,n,a,b] = \mathcal{A}_{ma,nb}\int_{S_{ma}} d\boldsymbol{r}(\boldsymbol{r}-\boldsymbol{r}_{ma})\cdot(\boldsymbol{r}-\boldsymbol{r}_{nb})\times\boldsymbol{I}_{in}^{\nabla'} \tag{2.86}$$

$$\overline{\boldsymbol{K}}_{PV}^{N,\mathrm{RWG}}[m,n,a,b] = \mathcal{A}_{ma,nb}\int_{S_{ma}} d\boldsymbol{r}(\boldsymbol{r}-\boldsymbol{r}_{ma})\times\hat{\boldsymbol{n}}\cdot(\boldsymbol{r}-\boldsymbol{r}_{nb})\times\boldsymbol{I}_{in}^{\nabla'}, \tag{2.87}$$

where

$$\mathcal{A}_{ma,nb} = \frac{l_m l_n}{4A_{ma}A_{nb}}\gamma_{ma}\gamma_{nb} \tag{2.88}$$

and

$$I_{in}^{1} = \int_{S_{nb}} d\mathbf{r}' g(\mathbf{r},\mathbf{r}') \tag{2.89}$$

$$\boldsymbol{I}_{in}^{(\mathbf{r}'-\mathbf{r}_{nb})} = \int_{S_{nb}} d\mathbf{r}' g(\mathbf{r},\mathbf{r}')(\mathbf{r}' - \mathbf{r}_{nb}) \tag{2.90}$$

$$\boldsymbol{I}_{in}^{\nabla'} = \int_{S_{nb}} d\mathbf{r}' \nabla' g(\mathbf{r},\mathbf{r}'). \tag{2.91}$$

In (2.88), $\gamma_{nb} = \pm 1$ and $\gamma_{ma} = \pm 1$ depend on the direction of the basis and testing functions on triangles. Interactions of the full basis and testing functions can be obtained as

$$\begin{Bmatrix} \overline{\boldsymbol{I}}^{T,\text{RWG}} \\ \overline{\boldsymbol{I}}^{N,\text{RWG}} \\ \overline{\boldsymbol{T}}^{T,\text{RWG}} \\ \overline{\boldsymbol{T}}^{N,\text{RWG}} \\ \overline{\boldsymbol{K}}_{PV}^{T,\text{RWG}} \\ \overline{\boldsymbol{K}}_{PV}^{N,\text{RWG}} \end{Bmatrix} [m,n] = \sum_{a=1}^{2}\sum_{b=1}^{2} \begin{Bmatrix} \overline{\boldsymbol{I}}^{T,\text{RWG}} \\ \overline{\boldsymbol{I}}^{N,\text{RWG}} \\ \overline{\boldsymbol{T}}^{T,\text{RWG}} \\ \overline{\boldsymbol{T}}^{N,\text{RWG}} \\ \overline{\boldsymbol{K}}_{PV}^{T,\text{RWG}} \\ \overline{\boldsymbol{K}}_{PV}^{N,\text{RWG}} \end{Bmatrix} [m,n,a,b]. \tag{2.92}$$

In order to calculate interactions, a coordinate transformation can be performed such that the basis triangle lies on the x-y plane with one of its edges on the x axis, as depicted in Figure 1.12. Such a transformation is essential in order to easily evaluate analytic integrals appearing in singularity extractions. Then, the interactions become

$$\overline{\boldsymbol{I}}^{T,\text{RWG}}[m,n,a,b] = \mathcal{A}_{ma,nb} \int_{S_{ma}} d\mathbf{r} (\boldsymbol{\rho} - \boldsymbol{\rho}_{ma}) \cdot (\boldsymbol{\rho} - \boldsymbol{\rho}_{nb}) \tag{2.93}$$

$$\overline{\boldsymbol{I}}^{N,\text{RWG}}[m,n,a,b] = \mathcal{A}_{ma,nb} \int_{S_{ma}} d\mathbf{r} (\boldsymbol{\rho} - \boldsymbol{\rho}_{ma}) \cdot \hat{\mathbf{z}} \times (\boldsymbol{\rho} - \boldsymbol{\rho}_{nb}) \tag{2.94}$$

$$\begin{aligned} \overline{\boldsymbol{T}}^{T,\text{RWG}}[m,n,a,b] &= ik\mathcal{A}_{ma,nb} \int_{S_{ma}} d\mathbf{r} (\boldsymbol{\rho} - \boldsymbol{\rho}_{ma}) \cdot \boldsymbol{I}_{in}^{(\boldsymbol{\rho}'-\boldsymbol{\rho}_{nb})} \\ &\quad - \frac{4i}{k}\mathcal{A}_{ma,nb} \int_{S_{ma}} d\mathbf{r} I_{in}^{1} \end{aligned} \tag{2.95}$$

$$\begin{aligned} \overline{\boldsymbol{T}}^{N,\text{RWG}}[m,n,a,b] &= ik\mathcal{A}_{ma,nb} \int_{S_{ma}} d\mathbf{r} (\mathbf{r} - \mathbf{r}_{ma}) \cdot \hat{\boldsymbol{n}} \times \boldsymbol{I}_{in}^{(\boldsymbol{\rho}'-\boldsymbol{\rho}_{nb})} \\ &\quad - \frac{2i}{k}\mathcal{A}_{ma,nb} \int_{S_{ma}} d\mathbf{r} (\mathbf{r} - \mathbf{r}_{ma}) \cdot \hat{\boldsymbol{n}} \times \boldsymbol{I}_{in}^{\nabla'} \end{aligned} \tag{2.96}$$

$$\overline{\boldsymbol{K}}_{PV}^{T,\text{RWG}}[m,n,a,b] = \mathcal{A}_{ma,nb} \int_{S_{ma}} d\mathbf{r} (\mathbf{r} - \mathbf{r}_{ma}) \cdot (\mathbf{r} - \boldsymbol{\rho}_{nb}) \times \boldsymbol{I}_{in}^{\nabla'} \tag{2.97}$$

$$\overline{\boldsymbol{K}}_{PV}^{N,\text{RWG}}[m,n,a,b] = \mathcal{A}_{ma,nb} \int_{S_{ma}} d\mathbf{r} (\mathbf{r} - \mathbf{r}_{ma}) \times \hat{\boldsymbol{n}} \cdot (\mathbf{r} - \boldsymbol{\rho}_{nb}) \times \boldsymbol{I}_{in}^{\nabla'}, \tag{2.98}$$

where $\boldsymbol{\rho} = \hat{\boldsymbol{x}}x + \hat{\boldsymbol{y}}y$ and

$$\boldsymbol{I}_{in}^{(\boldsymbol{\rho}'-\boldsymbol{\rho}_{nb})} = \int_{S_{nb}} d\boldsymbol{r}' g(\boldsymbol{r},\boldsymbol{r}')(\boldsymbol{\rho}' - \boldsymbol{\rho}_{nb}). \tag{2.99}$$

For the integro-differential operators, the interactions are divided into basic integrals as follows:

- The directly tested $\mathcal{T}$ operator involves 7 basic integrals, i.e.,

$$\overline{\boldsymbol{T}}^{T,\mathrm{RWG}}[m,n,a,b] = ikA_{ma,nb}\left\{(x_{ma}x_{nb} + y_{ma}y_{nb})I_1\right.$$
$$\left. -x_{ma}I_2 - y_{ma}I_3 - x_{nb}I_4 - y_{nb}I_5 + I_6 + I_7 - \frac{4}{k^2}I_1\right\}, \tag{2.100}$$

where

$$I_{\{1,4,5\}} = \int_{S_{ma}} d\boldsymbol{r}\,\{1,x,y\}\,I_{in}^1 \tag{2.101}$$

$$I_{\{2,6\}} = \int_{S_{ma}} d\boldsymbol{r}\,\{1,x\}\,I_{in}^{x'} \tag{2.102}$$

$$I_{\{3,7\}} = \int_{S_{ma}} d\boldsymbol{r}\,\{1,y\}\,I_{in}^{y'} \tag{2.103}$$

and

$$I_{in}^{x'} = \int_{S_{nb}} d\boldsymbol{r}' x' g(\boldsymbol{r},\boldsymbol{r}') \tag{2.104}$$

$$I_{in}^{y'} = \int_{S_{nb}} d\boldsymbol{r}' y' g(\boldsymbol{r},\boldsymbol{r}'). \tag{2.105}$$

- The rotationally tested $\mathcal{T}$ operator involves 19 basic integrals, i.e.,

$$\overline{\boldsymbol{T}}^{N,\mathrm{RWG}}[m,n,a,b] = ikA_{ma,nb}\left\{(n_y z_{ma} - n_z y_{ma})I_2 + (n_z x_{ma} - n_x z_{ma})I_3\right.$$
$$+ (n_x z_{ma} y_{nb} - n_y z_{ma} x_{nb} + n_z y_{ma} x_{nb} - n_z x_{ma} y_{nb})I_1$$
$$+ n_z y_{nb} I_4 - n_z x_{nb} I_5 + (n_y x_{nb} - n_x y_{nb})I_6$$
$$+ n_z I_7 - n_z I_8 - n_y I_9 + n_x I_{10}$$
$$- \frac{2}{k^2}\left[(z_{ma}n_y - y_{ma}n_z)I_{11} + (x_{ma}n_z - z_{ma}n_x)I_{14}\right.$$
$$+ n_z I_{12} - n_y I_{13} - n_z I_{15} + n_x I_{16} + (y_{ma}n_x - n_y x_{ma})I_{17}$$
$$\left.\left. + n_y I_{18} - n_x I_{19}\right]\right\}, \tag{2.106}$$

where the normal of the testing triangle is decomposed as

$$\hat{\boldsymbol{n}} = \hat{\boldsymbol{x}}n_x + \hat{\boldsymbol{y}}n_y + \hat{\boldsymbol{z}}n_z, \tag{2.107}$$

and

$$I_{\{1,4,5,6\}} = \int_{S_{ma}} d\boldsymbol{r} \, \{1, x, y, z\} \, I_{in}^{1} \tag{2.108}$$

$$I_{\{2,7,9\}} = \int_{S_{ma}} d\boldsymbol{r} \, \{1, y, z\} \, I_{in}^{x'} \tag{2.109}$$

$$I_{\{3,8,10\}} = \int_{S_{ma}} d\boldsymbol{r} \, \{1, x, z\} \, I_{in}^{y'} \tag{2.110}$$

$$I_{\{11,12,13\}} = \int_{S_{ma}} d\boldsymbol{r} \, \{1, y, z\} \, \hat{\boldsymbol{x}} \cdot \boldsymbol{I}_{in}^{\nabla'} \tag{2.111}$$

$$I_{\{14,15,16\}} = \int_{S_{ma}} d\boldsymbol{r} \, \{1, x, z\} \, \hat{y} \cdot \boldsymbol{I}_{in}^{\nabla'} \tag{2.112}$$

$$I_{\{17,18,19\}} = \int_{S_{ma}} d\boldsymbol{r} \, \{1, x, y\} \, \hat{z} \cdot \boldsymbol{I}_{in}^{\nabla'}. \tag{2.113}$$

- The directly tested $\mathcal{K}$ operator involves 9 basic integrals, i.e.,

$$\begin{aligned}\overline{\boldsymbol{K}}_{PV}^{T,\mathrm{RWG}}[m,n,a,b] = \mathcal{A}_{ma,nb} \{&-z_{ma}y_{nb}I_1 + z_{ma}I_2 + (y_{nb} - y_{ma})I_3 \\ &+ z_{ma}x_{nb}I_4 - z_{ma}I_5 + (x_{ma} - x_{nb})I_6 \\ &+ (x_{ma}y_{nb} - y_{ma}x_{nb})I_7 \\ &+ (y_{ma} - y_{nb})I_8 + (x_{nb} - x_{ma})I_9\},\end{aligned} \tag{2.114}$$

where

$$I_{\{1,2,3\}} = \int_{S_{ma}} d\boldsymbol{r} \, \{1, y, z\} \, \hat{\boldsymbol{x}} \cdot \boldsymbol{I}_{in}^{\nabla'} \tag{2.115}$$

$$I_{\{4,5,6\}} = \int_{S_{ma}} d\boldsymbol{r} \, \{1, x, z\} \, \hat{y} \cdot \boldsymbol{I}_{in}^{\nabla'} \tag{2.116}$$

$$I_{\{7,8,9\}} = \int_{S_{ma}} d\boldsymbol{r} \, \{1, x, y\} \, \hat{z} \cdot \boldsymbol{I}_{in}^{\nabla'}. \tag{2.117}$$

- The rotationally tested $\mathcal{K}$ operator involves 24 basic integrals, i.e.,

$$\begin{aligned}\overline{\boldsymbol{K}}_{PV}^{N,\mathrm{RWG}}[m,n,a,b] = \mathcal{A}_{ma,nb} \{&[y_{nb}(n_x y_{ma} - n_y x_{ma})] I_1 \\ &+ [n_y x_{ma} - n_x(y_{ma} + y_{nb})] I_2 \\ &+ [n_z x_{ma} - n_x z_{ma}] I_3 + [x_{nb}(n_y x_{ma} - n_x y_{ma})] I_4 \\ &+ [n_x y_{ma} - n_y(x_{ma} + x_{nb})] I_5 + [n_z y_{ma} - n_y z_{ma}] I_6 \\ &+ [x_{nb}(n_z x_{ma} - n_x z_{ma}) + y_{nb}(n_z y_{ma} - n_y z_{ma})] I_7 \\ &+ [z_{ma}n_x - n_z(x_{ma} + x_{nb})] I_8 \\ &+ [z_{ma}n_y - n_z(y_{ma} + y_{nb})] I_9\end{aligned}$$

(continued)

$$+ y_{nb}n_y I_{10} - n_y I_{11} - n_z I_{12} + n_x I_{13} + n_x I_{14}$$
$$+ x_{nb}n_x I_{15} - n_x I_{16} - n_z I_{17} + n_y I_{18} + n_y I_{19}$$
$$- n_x I_{20} - n_y I_{21} + \left[n_x x_{nb} + n_y y_{nb} \right] I_{22}$$
$$+ n_z I_{23} + n_z I_{24} \Big\}, \tag{2.118}$$

where

$$I_{\{1,10,2,3,11,12,13,14\}} = \int_{S_{ma}} d\boldsymbol{r} \left\{ 1, x, y, z, xy, xz, y^2, z^2 \right\} \hat{\boldsymbol{x}} \cdot \boldsymbol{I}_{in}^{\nabla'} \tag{2.119}$$

$$I_{\{4,5,15,6,16,17,18,19\}} = \int_{S_{ma}} d\boldsymbol{r} \left\{ 1, x, y, z, xy, yz, x^2, z^2 \right\} \hat{\boldsymbol{y}} \cdot \boldsymbol{I}_{in}^{\nabla'} \tag{2.120}$$

$$I_{\{7,8,9,22,20,21,23,24\}} = \int_{S_{ma}} d\boldsymbol{r} \left\{ 1, x, y, z, xz, yz, x^2, y^2 \right\} \hat{\boldsymbol{z}} \cdot \boldsymbol{I}_{in}^{\nabla'}. \tag{2.121}$$

Finally, the inner integrals in (2.89), (2.91), (2.104), and (2.105) are evaluated as

$$I_{in}^{\{1,x',y'\}} = \frac{1}{4\pi} \int_{S_{nb}} d\boldsymbol{r}' \left\{ 1, x', y' \right\} \frac{\exp(ikR) - 1}{R}$$
$$+ \frac{1}{4\pi} \int_{S_{nb}} d\boldsymbol{r}' \left\{ 1, (x' - x), (y' - y) \right\} \frac{1}{R} + \frac{1}{4\pi} \left\{ 0, x, y \right\} \int_{S_{nb}} d\boldsymbol{r}' \frac{1}{R} \tag{2.122}$$
$$= \frac{1}{4\pi} I_{in,c}^{\{1,x',y'\}} + \frac{1}{4\pi} I_{in,a}^{\{1,(x'-x),(y'-y)\}} + \frac{1}{4\pi} \left\{ 0, x, y \right\} I_{in,a}^{1} \tag{2.123}$$

and

$$\boldsymbol{I}_{in}^{\nabla'} = \frac{1}{4\pi} \int_{S_{nb}} d\boldsymbol{r}' \nabla' \left(\frac{\exp(ikR) - 1 + 0.5k^2R^2}{R} \right)$$
$$+ \frac{1}{4\pi} \int_{S_{nb}} d\boldsymbol{r}' \nabla' \left(\frac{1}{R} \right) - \frac{k^2}{8\pi} \int_{S_{nb}} d\boldsymbol{r}' \nabla' R \tag{2.124}$$
$$= \frac{1}{4\pi} \boldsymbol{I}_{in,c}^{\nabla'} + \frac{1}{4\pi} \boldsymbol{I}_{in,a1}^{\nabla'} - \frac{k^2}{8\pi} \boldsymbol{I}_{in,a2}^{\nabla'} \tag{2.125}$$

by extracting singularities. Note that

$$\lim_{R \to 0} \frac{\exp(ikR) - 1}{R} = ik \tag{2.126}$$

$$\lim_{R \to 0} \nabla' \left(\frac{\exp(ikR) - 1 + 0.5k^2R^2}{R} \right) = 0. \tag{2.127}$$

In (2.123) and (2.125), $I_{in,c}^{\{1,x',y'\}}$ and $\boldsymbol{I}_{in,c}^{\nabla'}$ are calculated numerically using adaptive integration methods (see Chapter 1, Section 1.8.3) or Gaussian quadrature rules [142]. The remaining terms, i.e., $I_{in,a}^{\{1,(x'-x),(y'-y)\}}$, $\boldsymbol{I}_{in,a1}^{\nabla'}$, and $\boldsymbol{I}_{in,a2}^{\nabla'}$, are evaluated analytically (see Chapter 1, Section 1.8.1).

2.1.5 General Properties of Surface Formulations

Using a Galerkin method for the discretization of surface formulations, i.e., using the same set of functions to expand equivalent currents and to test boundary conditions, $\mathcal{T}$ and $\mathcal{I}$ operators are well-tested with $\boldsymbol{t}_m$. On the other hand, $\mathcal{K}$ operator is well-tested with $\hat{\boldsymbol{n}} \times \boldsymbol{t}_m$ [74]. Consequently, tangential formulations, such as CTF, contain well-tested $\mathcal{T}$ operators and/or weakly tested $\mathcal{K}$ and $\mathcal{I}$ operators, i.e.,

$$\bar{\boldsymbol{T}}^{T}[m,n] = \int_{S_m} d\boldsymbol{r}\boldsymbol{t}_m(\boldsymbol{r}) \cdot \mathcal{T}\{\boldsymbol{b}_n\}(\boldsymbol{r}) \tag{2.128}$$

$$\bar{\boldsymbol{K}}_{PV}^{T}[m,n] = \int_{S_m} d\boldsymbol{r}\boldsymbol{t}_m(\boldsymbol{r}) \cdot \mathcal{K}_{PV}\{\boldsymbol{b}_n\}(\boldsymbol{r}) \tag{2.129}$$

$$\bar{\boldsymbol{I}}^{N}[m,n] = \int_{S_m} d\boldsymbol{r}\boldsymbol{t}_m(\boldsymbol{r}) \cdot \hat{\boldsymbol{n}} \times \boldsymbol{b}_n(\boldsymbol{r}). \tag{2.130}$$

On the other hand, normal formulations, such as MNMF and CNF, contain well-tested $\mathcal{K}$ and $\mathcal{I}$ operators and/or weakly tested $\mathcal{T}$ operators, i.e.,

$$\bar{\boldsymbol{K}}_{PV}^{N}[m,n] = \int_{S_m} d\boldsymbol{r}\boldsymbol{t}_m(\boldsymbol{r}) \cdot \hat{\boldsymbol{n}} \times \mathcal{K}_{PV}\{\boldsymbol{b}_n\}(\boldsymbol{r}) \tag{2.131}$$

$$\bar{\boldsymbol{I}}^{T}[m,n] = \int_{S_m} d\boldsymbol{r}\boldsymbol{t}_m(\boldsymbol{r}) \cdot \boldsymbol{b}_n(\boldsymbol{r}) \tag{2.132}$$

$$\bar{\boldsymbol{T}}^{N}[m,n] = \int_{S_m} d\boldsymbol{r}\boldsymbol{t}_m(\boldsymbol{r}) \cdot \hat{\boldsymbol{n}} \times \mathcal{T}\{\boldsymbol{b}_n\}(\boldsymbol{r}). \tag{2.133}$$

In mixed formulations, such as JMCFIE, all three operators can be well-tested. The value of $\bar{\boldsymbol{I}}^{T}[m,n]$ in (2.132) is nonzero for overlapping testing and basis functions, and it typically has a large value for $m = n$. Therefore, matrix equations obtained from normal and mixed formulations are diagonally dominant. In addition, for all stable formulations, such as CTF, CNF, MNMF, and JMCFIE, well-tested operators appear on diagonal partitions of matrix equations. This is a required property for stability.

In general, conditioning and accuracy properties of tangential formulations are determined by well-tested $\mathcal{T}$ operators. Using a Galerkin scheme and divergence-conforming functions, the directly tested $\mathcal{T}$ operator has a weakly singular kernel. This leads to ill-conditioned matrix equations, but very accurate results, provided that the condition number of the matrix is not too large to deteriorate the accuracy of solutions. On the other hand, normal and mixed formulations are usually dominated by well-tested identity operators. This is preferable in terms of efficiency since diagonally dominant matrix equations are well-conditioned and easy to solve iteratively. However, discretization of the identity operator usually involves a large error, which significantly contaminates the accuracy of results.

Numerical Results

In the examples below, accuracy of numerical solutions is investigated when scattering problems are formulated with CTF, CNF, MNMF, and JMCFIE. For all solutions, near-field interactions are calculated with at most 1% error, and far-field interactions are computed by MLFMA

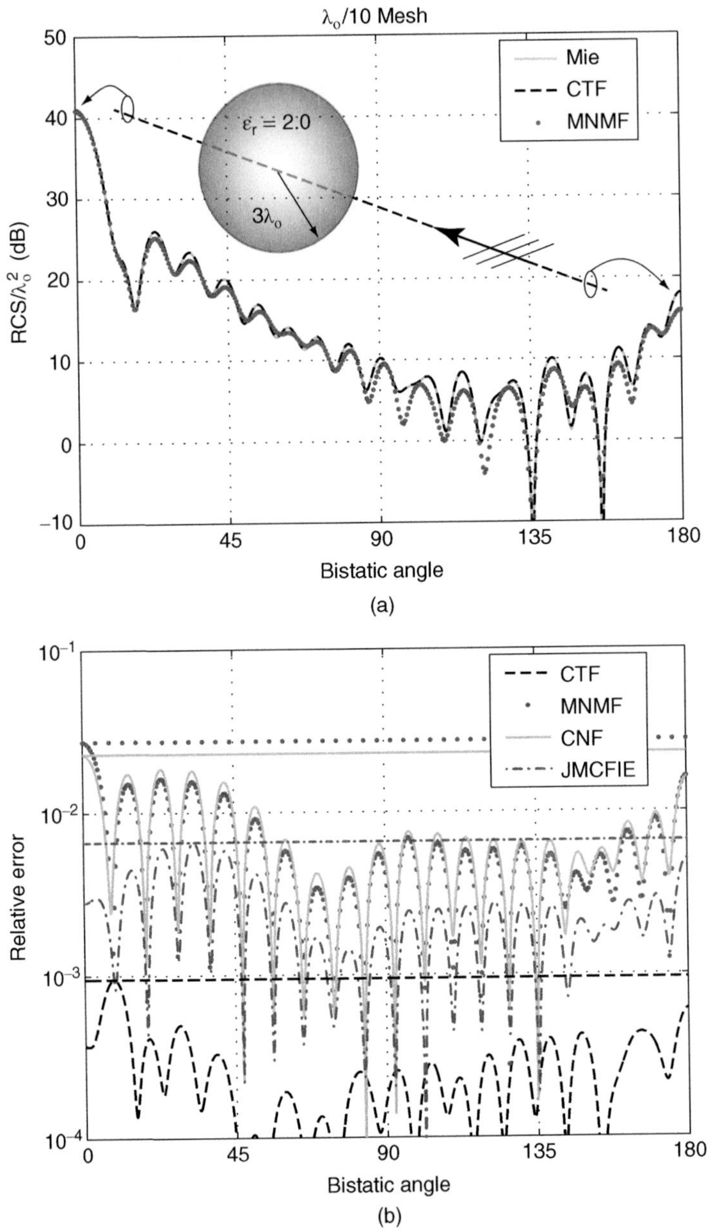

Figure 2.2 (a) The normalized bistatic RCS (RCS/λ_o^2) of a sphere with a radius of $3\lambda_o$ and with a relative permittivity of 2.0 located in free space. (b) The relative error defined in (2.134) for different formulations as a function of the bistatic angle. *Source*: Ergül and Gürel 2009 [80]. Reproduced with permission of IEEE.

with three digits of accuracy. Tree structures are constructed by fixing the box size at the lowest level to $0.25\lambda_o$ and using a bottom-up scheme.

Figure 2.2(a) depicts bistatic RCS values for a sphere with a radius of $3\lambda_o$ and $\epsilon_r = 2.0$. The normalized RCS (RCS/λ_o^2 in dB) is plotted as a function of the observation angle from 0° to 180° on the z-x plane, where 0° corresponds to the forward-scattering direction. It can be observed that the computational values calculated by using CTF are in agreement with the analytical curve obtained by a Mie-series solution. Although the values obtained with MNMF are also close to the analytical curve, they are significantly inaccurate compared to the values obtained with CTF. For a more quantitative information, Figure 2.2(b) presents the relative error in the computational results with respect to the reference analytical solution. All four formulations, i.e., CTF, CNF, MNMF, and JMCFIE, are considered in this plot. The relative error as a function of the bistatic angle θ is defined as

$$\Delta_R(\theta) = \frac{|E^{\infty}_{C,\theta}(\theta) - E^{\infty}_{A,\theta}(\theta)|}{\max\limits_{\theta} |E^{\infty}_{A,\theta}(\theta)|}, \tag{2.134}$$

where $E^{\infty}_{C,\theta}$ and $E^{\infty}_{A,\theta}$ are the computational and analytical values of the far-zone co-polar electric field, i.e.,

$$E^{\infty}_{\theta}(\theta) = \lim_{r\to\infty} \left\{ rE_{\theta}(r, \theta, \phi = 0) \right\}. \tag{2.135}$$

The maximum value of the relative error is also indicated by a horizontal line in the figure for each formulation. Figure 2.2(b) shows that CTF provides the most accurate results, while the normal formulations (MNMF and CNF) are significantly inaccurate compared to CTF. JMCFIE is also worse than CTF, but it is more accurate than the normal formulations. In Figure 2.3, bistatic RCS values and relative errors are presented for a sphere with a radius of $6\lambda_o$ and $\epsilon_r = 4.0$. The results are very similar to the previous case. Figure 2.4 presents similar results for a sphere with a radius of $7.5\lambda_o$ and $\epsilon_r = 2.0$.

Figure 2.5 depicts the normalized bistatic RCS (RCS/λ_o^2 in dB) of a cube with edges of λ_o and $\epsilon_r = 4.0$. The cube is located in free space and illuminated by a plane wave propagating in the z direction with the electric field polarized in the x direction. The bistatic RCS is plotted on the z-x plane, where 0° corresponds to the forward-scattering direction. The object is discretized by using various mesh sizes from $\lambda_o/5$ to $\lambda_o/17$. Convergence to correct RCS values is investigated as the discretization is refined. Figure 2.5 shows that RCS values obtained by using MNMF converge slower than RCS values obtained by using CTF. Specifically, MNMF discretized with $\lambda_o/5$ and $\lambda_o/7$ triangles is unacceptably inaccurate. Using MNMF, there are also significant discrepancies between the results obtained with $\lambda_o/10$ and finer discretizations. Using CTF, however, RCS values convergence rapidly as the discretization is refined.

In general, numerical results on canonical objects verify that the normal formulations MNMF and CNF are consistently inaccurate, compared to CTF and JMCFIE. As discussed later in this chapter, Section 2.3.11, this excessive error in normal formulations is caused by well-tested identity operators discretized with low-order functions, such as the RWG functions. As demonstrated in Figures 2.2–2.4, the discrepancy between the accuracy of results obtained with normal, mixed, and tangential formulations can be significant even for large dielectric objects with smooth surfaces.

Being a combination of CTF and CNF, solutions with JMCFIE are contaminated with the inaccuracy of CNF. Therefore, CTF is preferable to JMCFIE in terms of accuracy. However,

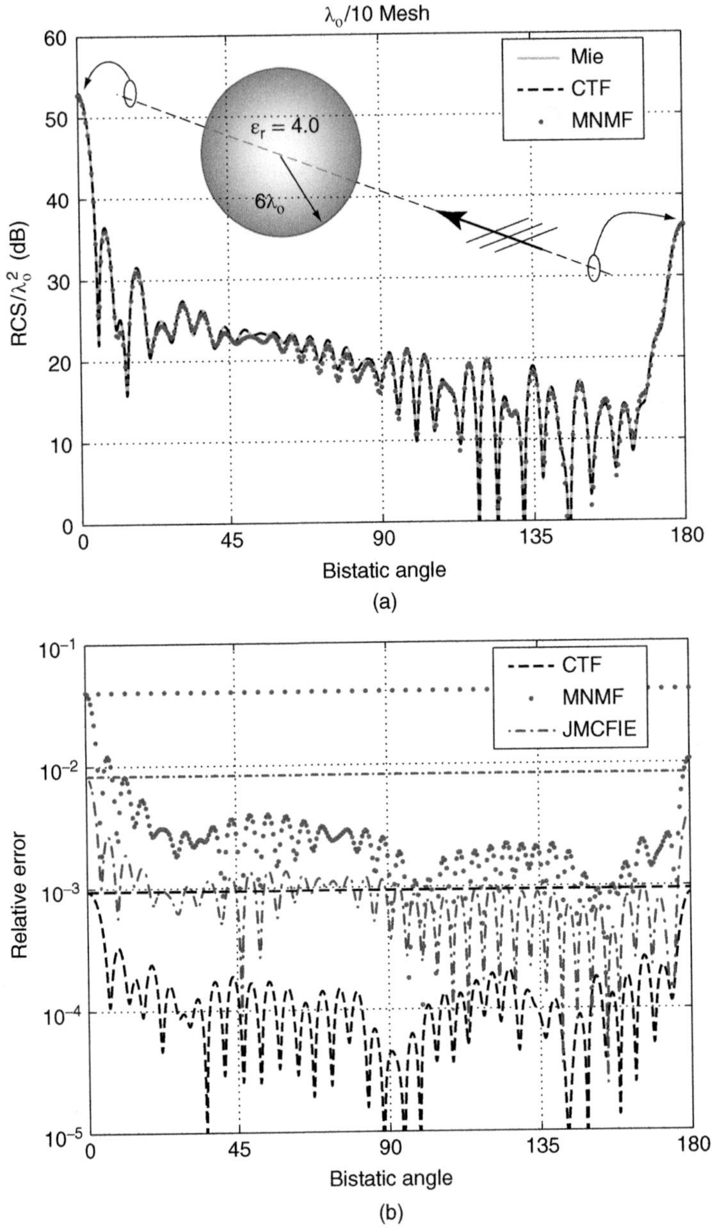

Figure 2.3 (a) The normalized bistatic RCS (RCS/λ_o^2) of a sphere with a radius of $6\lambda_o$ and a relative permittivity of 4.0 located in free space. (b) The relative error defined in (2.134) for different formulations as a function of the bistatic angle. CNF is omitted in this figure since its accuracy is very close to that of MNMF, as depicted in Figure 2.2(b). *Source*: Ergül and Gürel 2009 [80]. Reproduced with permission of IEEE.

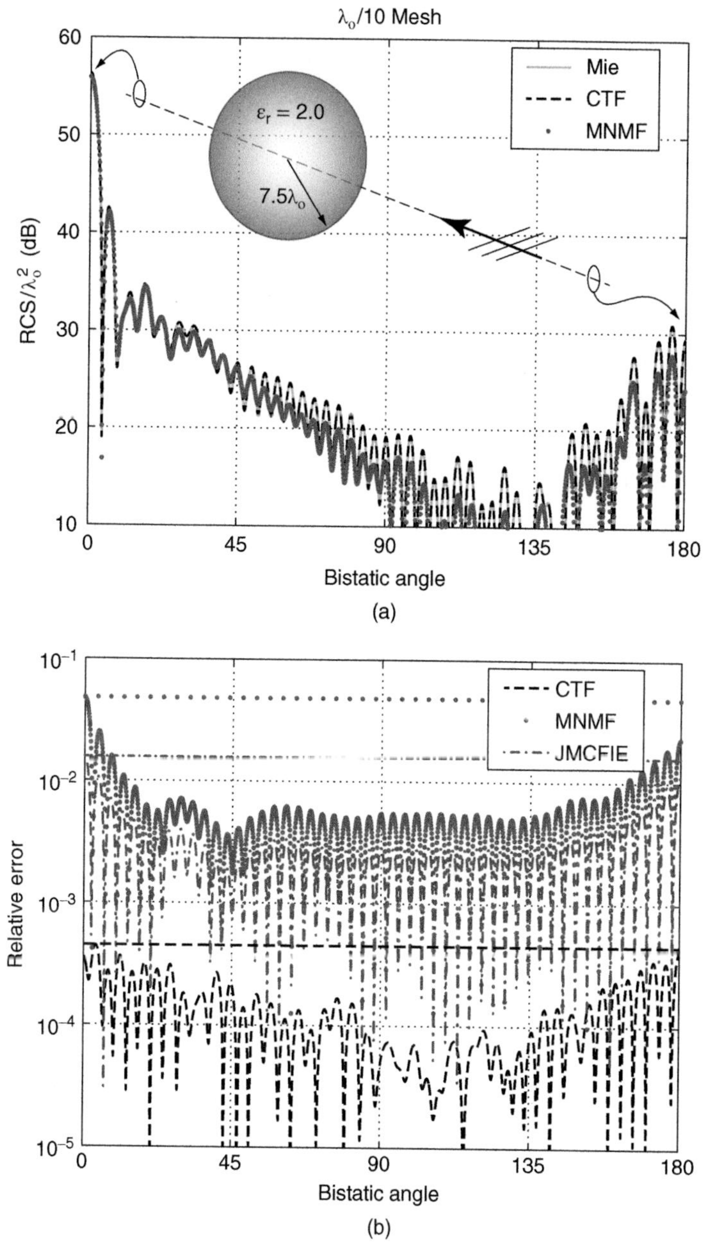

Figure 2.4 (a) The normalized bistatic RCS (RCS/λ_o^2) of a sphere with a radius of 7.5λ_o and a relative permittivity of 2.0 located in free space. (b) The relative error defined in (2.134) for different formulations as a function of the bistatic angle. CNF is omitted in this figure since its accuracy is very close to that of MNMF, as depicted in Figure 2.2(b). *Source*: Ergül and Gürel 2007 [81]. Reproduced with permission of IEEE.

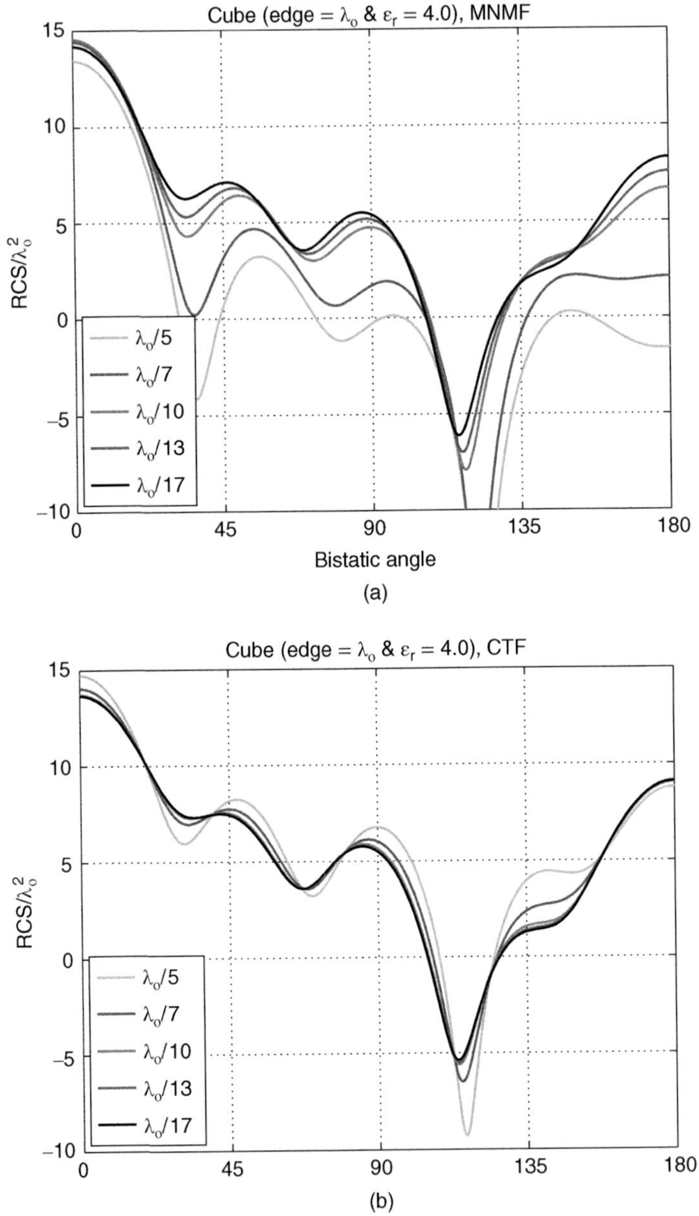

Figure 2.5 The normalized bistatic RCS (RCS/λ_o^2) of a cube with edges of λ_o and a relative permittivity of 4.0 located in free space. RCS values are obtained for different discretizations and by using (a) MNMF and (b) CTF.

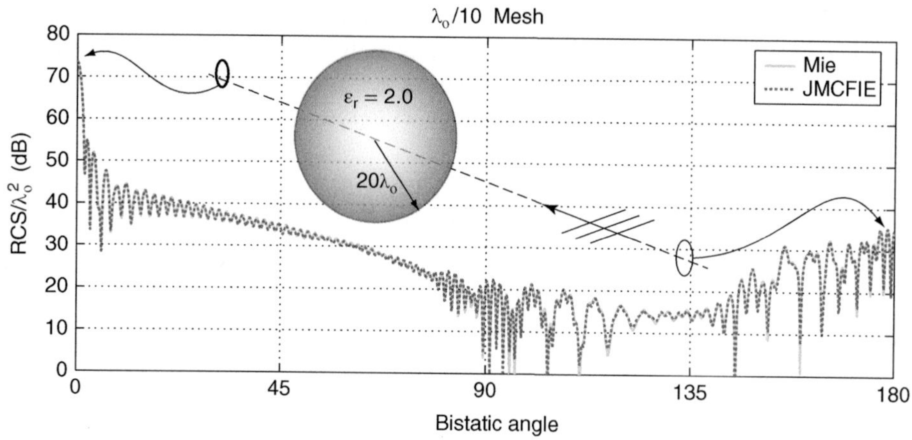

Figure 2.6 The normalized bistatic RCS (RCS/λ_o^2) of a sphere with a radius of $20\lambda_o$ and a relative permittivity of 2.0 located in free space. *Source*: Ergül and Gürel 2009 [80]. Reproduced with permission of IEEE.

JMCFIE can be more suitable for the solution of large problems, which cannot easily be obtained with CTF. For example, Figure 2.6 presents bistatic RCS values for a sphere with a radius of $20\lambda_o$ and $\epsilon_r = 2.0$ discretized with 2 925 708 unknowns. Computational values obtained with JMCFIE are close to the analytical Mie-series solution, and the maximum relative error is 2.4%. This problem cannot be solved in a reasonable number of iterations when it is formulated with CTF, CNF, or MNMF.

2.1.6 *Decoupling for Perfectly Conducting Surfaces*

When an object is a PEC, operators associated with the inner medium as well as the magnetic current disappear in formulations. Then, T-EFIE and T-MFIE in tangential formulations are decoupled, and they can be solved independently to obtain the induced electric current on the surface of the object. For example, CTF in (2.14) reduces to two independent equations, i.e.,

$$-\hat{\boldsymbol{n}} \times \hat{\boldsymbol{n}} \times \mathcal{T}_o\{\boldsymbol{J}\}(\boldsymbol{r}) = \eta_o^{-1}\hat{\boldsymbol{n}} \times \hat{\boldsymbol{n}} \times \boldsymbol{E}^{inc}(\boldsymbol{r}) \tag{2.136}$$

$$-\hat{\boldsymbol{n}} \times \hat{\boldsymbol{n}} \times \mathcal{K}_{PV,o}\{\boldsymbol{J}\}(\boldsymbol{r}) + \frac{\Omega_o}{4\pi}\mathcal{I}^{\times n}\{\boldsymbol{J}\}(\boldsymbol{r}) = \hat{\boldsymbol{n}} \times \hat{\boldsymbol{n}} \times \boldsymbol{H}^{inc}(\boldsymbol{r}). \tag{2.137}$$

Using a Galerkin discretization, T-MFIE in (2.137) is extremely unstable since both $\mathcal{K}$ and identity operators are weakly tested. On the other hand, T-EFIE in (2.136) is stable and is commonly used in the literature [57]. Similarly, in normal formulations, N-MFIE and N-EFIE are decoupled for PEC objects. For example, MNMF in (2.28) reduces to two independent equations, i.e.,

$$\hat{\boldsymbol{n}} \times \mathcal{K}_{PV,o}\{\boldsymbol{J}\}(\boldsymbol{r}) - \frac{\Omega_o}{4\pi}\mathcal{I}\{\boldsymbol{J}\}(\boldsymbol{r}) = -\hat{\boldsymbol{n}} \times \boldsymbol{H}^{inc}(\boldsymbol{r}) \tag{2.138}$$

$$\hat{\boldsymbol{n}} \times \mathcal{T}_o\{\boldsymbol{J}\}(\boldsymbol{r}) = -\eta_o^{-1}\hat{\boldsymbol{n}} \times \boldsymbol{E}^{inc}(\boldsymbol{r}). \tag{2.139}$$

Using a Galerkin discretization, N-EFIE in (2.139) is unstable since the $\mathcal{T}$ operator is weakly tested, whereas N-MFIE in (2.138) is stable and commonly used in the literature [168].

Finally, mixed formulations are also decomposed into two independent formulations for PEC objects. For example, JMCFIE in (2.34) reduces to two mixed formulations, i.e., stable T-N-CFIE

$$-\hat{\boldsymbol{n}} \times \hat{\boldsymbol{n}} \times \mathcal{T}_o\{\boldsymbol{J}\}(\boldsymbol{r}) + \hat{\boldsymbol{n}} \times \mathcal{K}_{PV,o}\{\boldsymbol{J}\}(\boldsymbol{r}) - \frac{\Omega_o}{4\pi}\mathcal{I}\{\boldsymbol{J}\}(\boldsymbol{r})$$
$$= \eta_o^{-1}\hat{\boldsymbol{n}} \times \hat{\boldsymbol{n}} \times \boldsymbol{E}^{inc}(\boldsymbol{r}) - \hat{\boldsymbol{n}} \times \boldsymbol{H}^{inc}(\boldsymbol{r}) \tag{2.140}$$

and unstable N-T-CFIE

$$-\hat{\boldsymbol{n}} \times \mathcal{T}_o\{\boldsymbol{J}\}(\boldsymbol{r}) - \hat{\boldsymbol{n}} \times \hat{\boldsymbol{n}} \times \mathcal{K}_{PV,o}\{\boldsymbol{J}\}(\boldsymbol{r}) + \frac{\Omega_o}{4\pi}\mathcal{I}^{\times n}\{\boldsymbol{J}\}(\boldsymbol{r})$$
$$= \eta_o^{-1}\hat{\boldsymbol{n}} \times \boldsymbol{E}^{inc}(\boldsymbol{r}) + \hat{\boldsymbol{n}} \times \hat{\boldsymbol{n}} \times \boldsymbol{H}^{inc}(\boldsymbol{r}). \tag{2.141}$$

T-N-CFIE in (2.140) is free of the internal-resonance problem and is a suitable formulation for closed conductors [55].

2.1.7 Accuracy with Respect to Contrast

Figure 2.7 presents the backscattered and forward-scattered RCS of a dielectric sphere with a radius of $0.5\lambda_o$ located in free space. The normalized RCS (RCS/λ_o^2) is plotted as a function the contrast ($\epsilon_r - 1$) when the relative permittivity of the sphere changes from $1.0 + 10^{-3}$ to 100.0. For numerical solutions, discretization of the sphere with $\lambda_o/10$ triangles leads to 1860 unknowns. Computational results obtained by using CTF, MNMF, and JMCFIE are compared with analytical results obtained by a Mie-series solution. Figure 2.7 shows that the computational values are consistent with the analytical values for moderate contrasts from 0.1 to 10.0. However, they become inaccurate for very low and high contrasts.

In general, conventional surface formulations are unstable when the contrast of the object is very low. To obtain accurate results, those formulations should be stabilized, as discussed later in this chapter, Section 2.2. On the other hand, the inaccuracy of surface formulations for relatively high contrasts is due to the approximation of the actual geometry in numerical solutions. For example, when planar triangles are used for the discretization, the computational model and the actual geometry of a smooth object are different. If the permittivity or permeability of the object is high, small discrepancies between the model and actual geometry may create large errors. The error is not due to the insufficient discretization of equivalent currents since the discretized model is solved with a desired level of accuracy, e.g., using $\lambda_o/10$ triangles. The error is due to the geometric difference between the actual object and the computational model.

Figure 2.8 presents the backscattered and forward-scattered RCS of a lossy sphere with a radius of $0.5\lambda_o$ located in free space. The relative complex permittivity of the sphere changes from $1.0 + i10^{-3}$ to $1.0 + i100.0$. Computational results obtained by using CTF, MNMF, and JMCFIE are again compared with analytical results. As opposed to the previous case, the surface formulations are accurate when the conductivity is high. In fact, it is well known that $\lambda_o/10$ discretizations can provide accurate results even for smooth PEC objects with infinite conductivity. However, similar to the low-permittivity case, conventional surface formulations

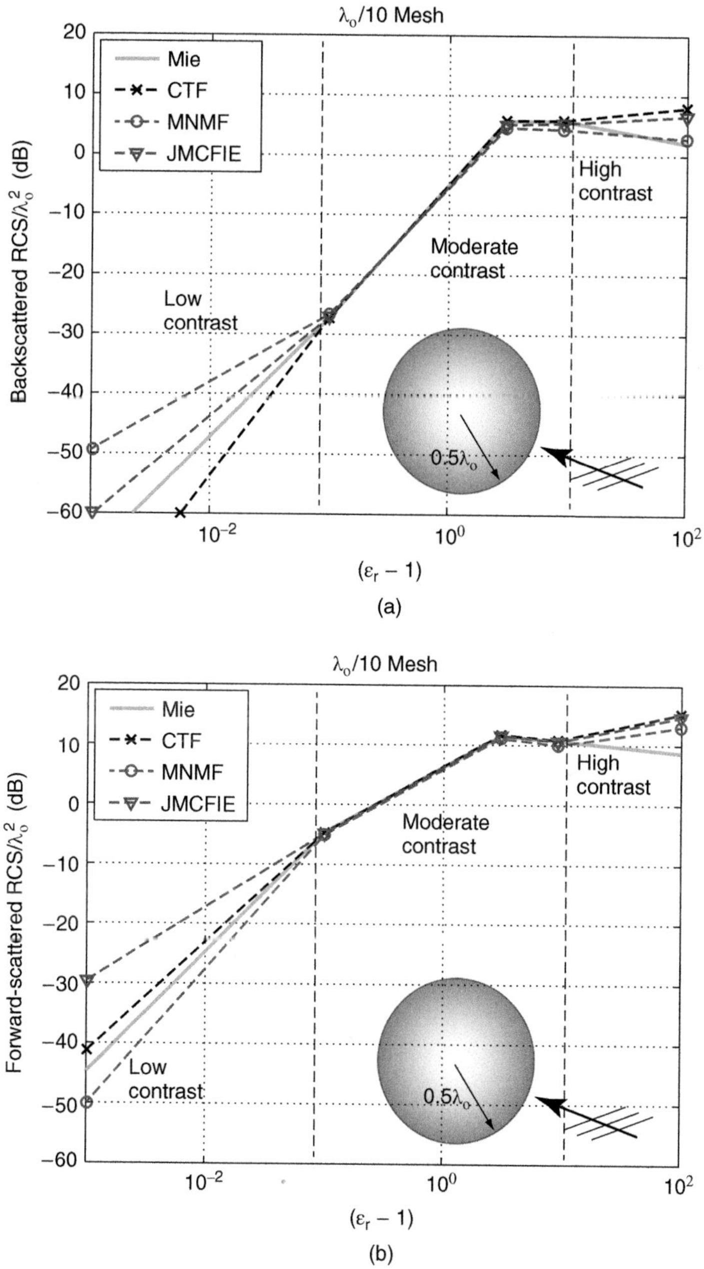

Figure 2.7 The normalized (a) backscattered and (b) forward-scattered RCS (RCS/λ_o^2) of a dielectric sphere with a radius of $0.5\lambda_o$ located in free space. The relative permittivity of the sphere changes from $1.0 + 10^{-3}$ to 100.0.

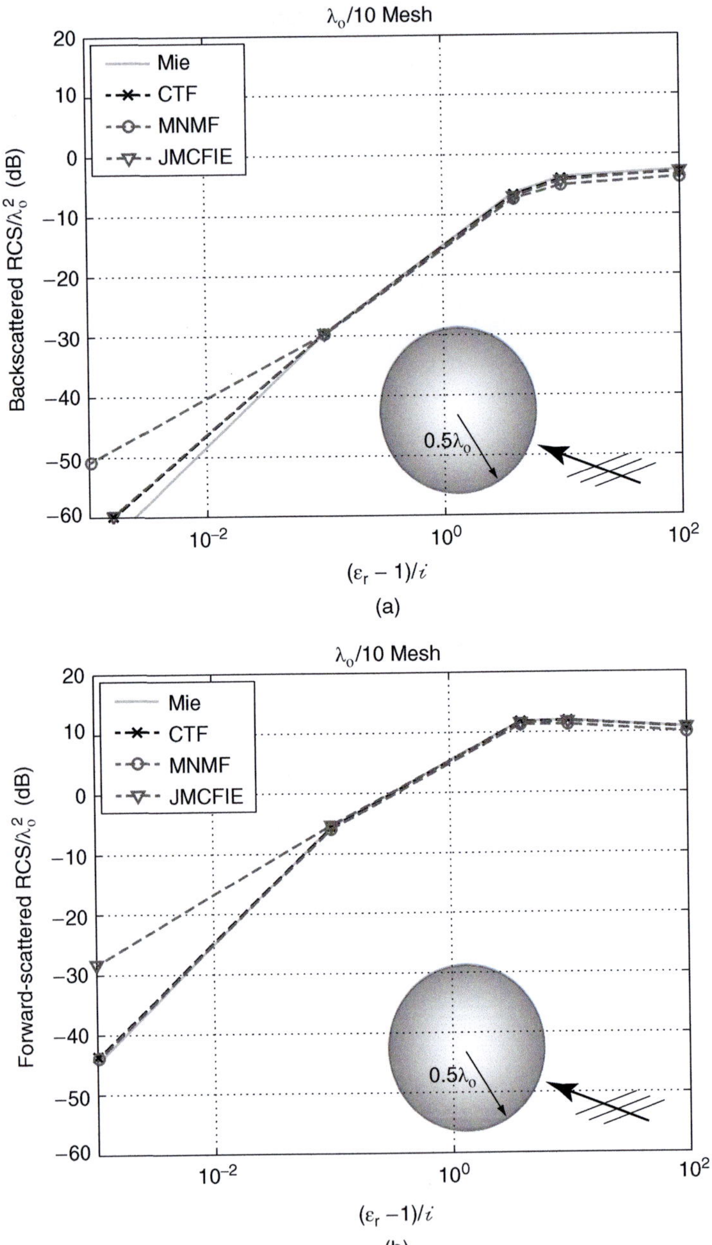

Figure 2.8 The normalized (a) backscattered and (b) forward-scattered RCS (RCS/λ_o^2) of a lossy sphere with a radius of $0.5\lambda_o$ located in free space. The relative complex permittivity of the sphere changes from $1.0 + i10^{-3}$ to $1.0 + i100.0$.

become unstable when the conductivity is low and they provide inaccurate results, as also depicted in Figure 2.8.

In order to show that the geometric discretization is critical for high contrasts, scattering problems involving a sphere with a radius of $0.5\lambda_o$ are investigated. The sphere is located in free space. Two different values for the relative permittivity of the sphere, i.e., $\epsilon_r = 4.0$ (moderate contrast) and $\epsilon_r = 100.0$ (high contrast), are considered. The sphere is illuminated by a plane wave propagating in the z direction with the electric field polarized in the x direction. Figure 2.9 presents the normalized bistatic RCS (RCS/λ_o^2) on the z-x plane. When the contrast is moderate, computational results obtained by using $\lambda_o/10$ discretization are in agreement with Mie-series results. However, as depicted Figure 2.9(b), computational and analytical results become inconsistent for the high contrast case. Figure 2.10 presents the normalized bistatic RCS, when the discretization of the sphere is refined by using $\lambda_o/20$ triangles. It can be observed that the accuracy of computational results is improved for both moderate and high contrasts. However, the improvement is more significant for the high-contrast case. In addition to the better discretization of the equivalent currents, the geometric discretization of the sphere is enhanced by using smaller triangles; this improvement is more visible when the contrast is relatively high.

2.2 Low-Contrast Breakdown and Its Solution

Conventional surface formulations are stable and they provide accurate solutions (with various levels of accuracy, depending on the existence of well-tested identity operators, types of basis and testing functions, discretization method, geometry of the object, etc.), for problems involving objects with moderate contrasts. However, those formulations become inaccurate to compute scattered fields as the contrast of the object decreases, i.e., when electromagnetic material properties of the object and host medium become close to each other. This breakdown, which limits the applicability of surface formulations, does not arise in volume integral equations. Therefore, volume formulations can be used to solve such problems accurately. However, it is also desirable to extend the applicability of surface integral equations to low-contrast problems in order to use the advantages of surface formulations, which may require fewer unknowns compared to volume formulations in many cases.

2.2.1 *A Combined Tangential Formulation*

Stabilization procedures described in this section to eliminate the low-contrast breakdown can be applied to a tangential formulation derived by using

$$a = b = 1, \quad c = d = \eta_o \eta_i \tag{2.142}$$

in (2.9)–(2.11), which is slightly different from the original CTF in (2.14)–(2.17). With these coefficients, the modified CTF (M-CTF) is derived as

$$\begin{bmatrix} \mathcal{Z}_{11}^{\text{M-CTF}} & \mathcal{Z}_{12}^{\text{M-CTF}} \\ \mathcal{Z}_{21}^{\text{M-CTF}} & \mathcal{Z}_{22}^{\text{M-CTF}} \end{bmatrix} \cdot \begin{bmatrix} \boldsymbol{J} \\ \boldsymbol{M} \end{bmatrix} (\boldsymbol{r}) = \hat{\boldsymbol{n}} \times \hat{\boldsymbol{n}} \times \begin{bmatrix} \boldsymbol{E}^{inc} \\ \eta_o \eta_i \boldsymbol{H}^{inc} \end{bmatrix} (\boldsymbol{r}), \tag{2.143}$$

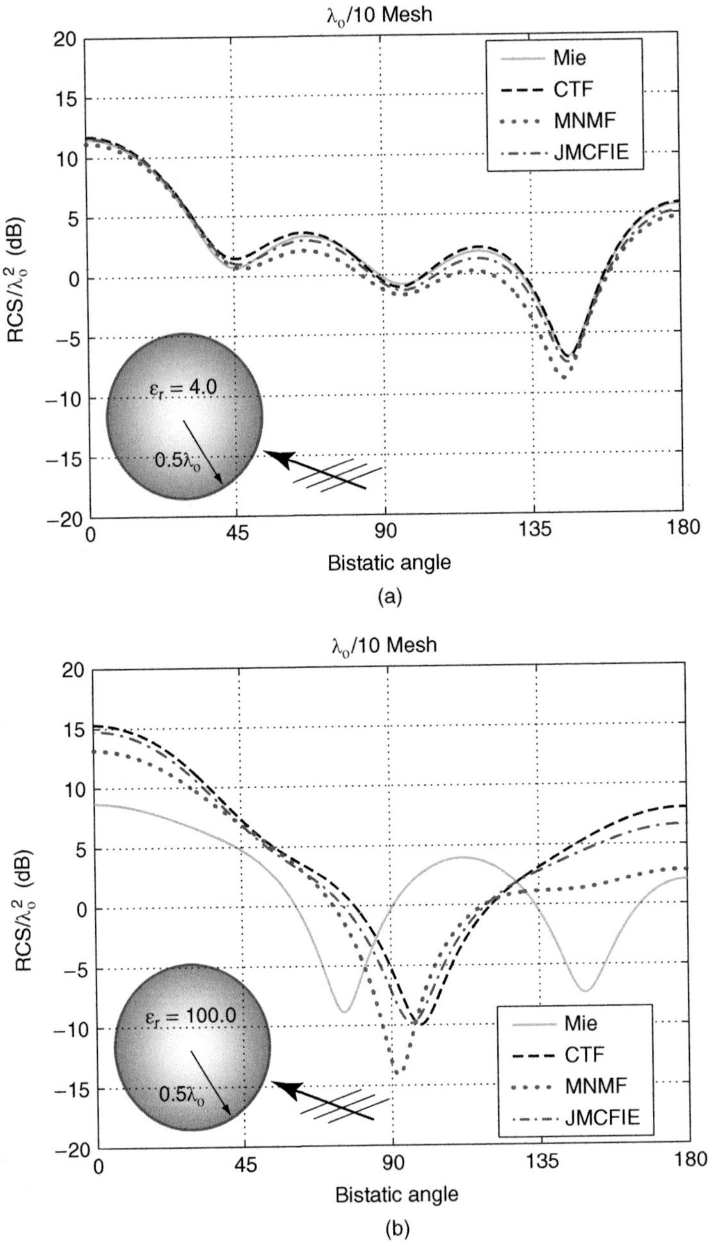

Figure 2.9 The normalized bistatic RCS (RCS$/\lambda_o^2$) of a sphere with a radius of $0.5\lambda_o$ located in free space, when the relative permittivity of the sphere is (a) 4.0 and (b) 100.0. The sphere is discretized with $\lambda_o/10$ triangles.

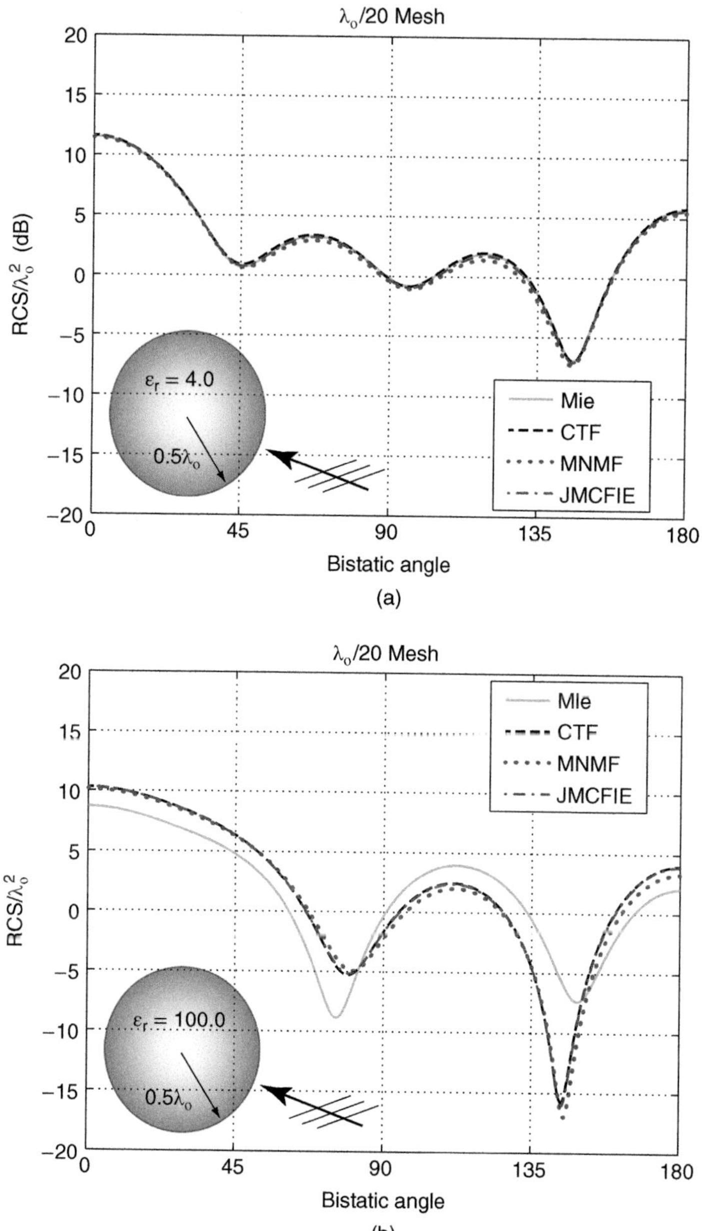

Figure 2.10 The normalized bistatic RCS (RCS/λ_o^2) of a sphere with a radius of $0.5\lambda_o$ located in free space, when the relative permittivity of the sphere is (a) 4.0 and (b) 100.0. The sphere is discretized with $\lambda_o/20$ triangles.

where

$$\mathcal{Z}_{11}^{\text{M-CTF}} = -\hat{\boldsymbol{n}} \times \hat{\boldsymbol{n}} \times \left(\eta_o \mathcal{T}_o + \eta_i \mathcal{T}_i\right) \tag{2.144}$$

$$\mathcal{Z}_{12}^{\text{M-CTF}} = \hat{\boldsymbol{n}} \times \hat{\boldsymbol{n}} \times \left(\mathcal{K}_{PV,o} + \mathcal{K}_{PV,i}\right) - \frac{1}{4\pi}\left(\Omega_o - \Omega_i\right)\mathcal{I}^{\times n} \tag{2.145}$$

$$= \hat{\boldsymbol{n}} \times \hat{\boldsymbol{n}} \times \left(\mathcal{K}_{PV,o} + \mathcal{K}_{PV,i}\right) - \left(1 - \frac{\Omega_i}{2\pi}\right)\mathcal{I}^{\times n} \tag{2.146}$$

$$\mathcal{Z}_{21}^{\text{M-CTF}} = -\eta_o\eta_i \hat{\boldsymbol{n}} \times \hat{\boldsymbol{n}} \times \left(\mathcal{K}_{PV,o} + \mathcal{K}_{PV,i}\right) + \frac{1}{4\pi}\left(\Omega_o - \Omega_i\right)\eta_o\eta_i\mathcal{I}^{\times n} \tag{2.147}$$

$$= -\eta_o\eta_i \hat{\boldsymbol{n}} \times \hat{\boldsymbol{n}} \times \left(\mathcal{K}_{PV,o} + \mathcal{K}_{PV,i}\right) + \eta_o\eta_i\left(1 - \frac{\Omega_i}{2\pi}\right)\mathcal{I}^{\times n} \tag{2.148}$$

$$\mathcal{Z}_{22}^{\text{M-CTF}} = -\hat{\boldsymbol{n}} \times \hat{\boldsymbol{n}} \times \left(\eta_i \mathcal{T}_o + \eta_o \mathcal{T}_i\right). \tag{2.149}$$

Then, the matrix partitions for M-CTF are derived as

$$\overline{\boldsymbol{Z}}_{11}^{\text{M-CTF}} = \eta_o \overline{\boldsymbol{T}}_o^T + \eta_i \overline{\boldsymbol{T}}_i^T \tag{2.150}$$

$$\overline{\boldsymbol{Z}}_{12}^{\text{M-CTF}} = -\overline{\boldsymbol{K}}_{PV,o}^T - \overline{\boldsymbol{K}}_{PV,i}^T \tag{2.151}$$

$$\overline{\boldsymbol{Z}}_{21}^{\text{M-CTF}} = \eta_o\eta_i\left(\overline{\boldsymbol{K}}_{PV,o}^T + \overline{\boldsymbol{K}}_{PV,i}^T\right) \tag{2.152}$$

$$\overline{\boldsymbol{Z}}_{22}^{\text{M-CTF}} = \eta_i \overline{\boldsymbol{T}}_o^T + \eta_o \overline{\boldsymbol{T}}_i^T. \tag{2.153}$$

When discretized, M-CTF is free of identity operators (like the PMCHWT formulation) and involves well-balanced diagonal partitions (like the original CTF) for low contrasts. In addition to M-CTF, CNF in (2.23)–(2.26) will be considered in the context of stabilization.

2.2.2 *Nonradiating Currents*

In surface formulations of a closed object, equivalent surface currents can be decomposed as

$$\boldsymbol{J}(\boldsymbol{r}) = \hat{\boldsymbol{n}} \times \boldsymbol{H}(\boldsymbol{r}) = \hat{\boldsymbol{n}} \times \boldsymbol{H}^{inc}(\boldsymbol{r}) + \hat{\boldsymbol{n}} \times \boldsymbol{H}^{rad}(\boldsymbol{r}) \tag{2.154}$$

$$\boldsymbol{M}(\boldsymbol{r}) = -\hat{\boldsymbol{n}} \times \boldsymbol{E}(\boldsymbol{r}) = -\hat{\boldsymbol{n}} \times \boldsymbol{E}^{inc}(\boldsymbol{r}) - \hat{\boldsymbol{n}} \times \boldsymbol{E}^{rad}(\boldsymbol{r}), \tag{2.155}$$

where $\{\boldsymbol{J}^{inc}, \boldsymbol{M}^{inc}\} = \{\hat{\boldsymbol{n}} \times \boldsymbol{H}^{inc}, -\hat{\boldsymbol{n}} \times \boldsymbol{E}^{inc}\}$ do not radiate outside [66], [96], [110], i.e.,

$$\eta_o \mathcal{T}_o\{\boldsymbol{J}^{inc}\}(\boldsymbol{r}) - \mathcal{K}_{PV,o}\{\boldsymbol{M}^{inc}\}(\boldsymbol{r}) + \frac{\Omega_i}{4\pi}\mathcal{I}^{\times n}\{\boldsymbol{M}^{inc}\}(\boldsymbol{r}) = 0 \tag{2.156}$$

$$\eta_o^{-1} \mathcal{T}_o\{\boldsymbol{M}^{inc}\}(\boldsymbol{r}) + \mathcal{K}_{PV,o}\{\boldsymbol{J}^{inc}\}(\boldsymbol{r}) - \frac{\Omega_i}{4\pi}\mathcal{I}^{\times n}\{\boldsymbol{J}^{inc}\}(\boldsymbol{r}) = 0, \tag{2.157}$$

for $\boldsymbol{r} \in D_o$. Equations (2.156) and (2.157) can be projected onto the surface as

$$\begin{Bmatrix} -\hat{\boldsymbol{n}} \times \hat{\boldsymbol{n}} \times \\ \hat{\boldsymbol{n}} \times \end{Bmatrix} \begin{bmatrix} \eta_o \mathcal{T}_o & -\mathcal{K}_{PV,o} \\ \mathcal{K}_{PV,o} & \eta_o^{-1}\mathcal{T}_o \end{bmatrix} \cdot \begin{bmatrix} \hat{\boldsymbol{n}} \times \boldsymbol{H}^{inc} \\ -\hat{\boldsymbol{n}} \times \boldsymbol{E}^{inc} \end{bmatrix} (\boldsymbol{r}) - \frac{\Omega_i}{4\pi} \begin{Bmatrix} -\hat{\boldsymbol{n}} \times \hat{\boldsymbol{n}} \times \\ \hat{\boldsymbol{n}} \times \end{Bmatrix} \begin{bmatrix} \hat{\boldsymbol{n}} \times \hat{\boldsymbol{n}} \times \boldsymbol{E}^{inc} \\ \hat{\boldsymbol{n}} \times \hat{\boldsymbol{n}} \times \boldsymbol{H}^{inc} \end{bmatrix} (\boldsymbol{r}) = 0 \tag{2.158}$$

or

$$\begin{Bmatrix} -\hat{\boldsymbol{n}} \times \hat{\boldsymbol{n}} \times \\ \hat{\boldsymbol{n}} \times \end{Bmatrix} \begin{bmatrix} \eta_o \mathcal{T}_o & -\mathcal{K}_{PV,o} \\ \mathcal{K}_{PV,o} & \eta_o^{-1}\mathcal{T}_o \end{bmatrix} \cdot \begin{bmatrix} \hat{\boldsymbol{n}} \times \boldsymbol{H}^{inc} \\ -\hat{\boldsymbol{n}} \times \boldsymbol{E}^{inc} \end{bmatrix} (\boldsymbol{r}) = -\frac{\Omega_i}{4\pi} \begin{Bmatrix} -\hat{\boldsymbol{n}} \times \hat{\boldsymbol{n}} \times \\ \hat{\boldsymbol{n}} \times \end{Bmatrix} \begin{bmatrix} \boldsymbol{E}^{inc} \\ \boldsymbol{H}^{inc} \end{bmatrix} (\boldsymbol{r}) \tag{2.159}$$

for $\boldsymbol{r} \in S$. Nonradiating currents (and their extraction) play a major role in the stabilization of surface formulations for low-contrast objects.

2.2.3 *Conventional Formulations in the Limit Case*

As the contrast goes to zero, i.e., $\epsilon_i \to \epsilon_o$ and $\mu_i \to \mu_o$, M-CTF in (2.143)–(2.149) becomes

$$\begin{bmatrix} 0 & -\left(1 - \Omega_i/2\pi\right) \mathcal{I}^{\times n} \\ \eta_o^2 \left(1 - \Omega_i/2\pi\right) \mathcal{I}^{\times n} & 0 \end{bmatrix} \cdot \begin{bmatrix} \boldsymbol{J} \\ \boldsymbol{M} \end{bmatrix} (\boldsymbol{r}) - \hat{\boldsymbol{n}} \times \hat{\boldsymbol{n}} \times \begin{bmatrix} 2\eta_o \mathcal{T}_o & -2\mathcal{K}_{PV,o} \\ 2\eta_o^2 \mathcal{K}_{PV,o} & 2\eta_o \mathcal{T}_o \end{bmatrix} \cdot \begin{bmatrix} \boldsymbol{J} \\ \boldsymbol{M} \end{bmatrix} (\boldsymbol{r}) = \hat{\boldsymbol{n}} \times \hat{\boldsymbol{n}} \times \begin{bmatrix} \boldsymbol{E}^{inc} \\ \eta_o^2 \boldsymbol{H}^{inc} \end{bmatrix} (\boldsymbol{r}) \tag{2.160}$$

or

$$-\hat{\boldsymbol{n}} \times \hat{\boldsymbol{n}} \times \begin{bmatrix} \eta_o \mathcal{T}_o & -\mathcal{K}_{PV,o} \\ \eta_o^2 \mathcal{K}_{PV,o} & \eta_o \mathcal{T}_o \end{bmatrix} \cdot \begin{bmatrix} \boldsymbol{J} \\ \boldsymbol{M} \end{bmatrix} (\boldsymbol{r}) = \frac{\Omega_i}{4\pi} \hat{\boldsymbol{n}} \times \hat{\boldsymbol{n}} \times \begin{bmatrix} \boldsymbol{E}^{inc} \\ \eta_o^2 \boldsymbol{H}^{inc} \end{bmatrix} (\boldsymbol{r}). \tag{2.161}$$

Considering the direct form of (2.159), the solution of (2.161) can be obtained as $\boldsymbol{J} = \hat{\boldsymbol{n}} \times \boldsymbol{H}^{inc}$ and $\boldsymbol{M} = -\hat{\boldsymbol{n}} \times \boldsymbol{E}^{inc}$. When the contrast is zero, CNF in (2.23)–(2.26) reduces to a simpler form, i.e.,

$$-\begin{bmatrix} \mathcal{I} & 0 \\ 0 & \mathcal{I} \end{bmatrix} \cdot \begin{bmatrix} \boldsymbol{J} \\ \boldsymbol{M} \end{bmatrix} (\boldsymbol{r}) = -\hat{\boldsymbol{n}} \times \begin{bmatrix} \boldsymbol{H}^{inc} \\ -\boldsymbol{E}^{inc} \end{bmatrix} (\boldsymbol{r}), \tag{2.162}$$

where the same solution can be obtained trivially. It should be noted that the 'incident currents' $\hat{\boldsymbol{n}} \times \boldsymbol{H}^{inc}, -\hat{\boldsymbol{n}} \times \boldsymbol{E}^{inc}$ do not radiate and conventional forms of surface formulations satisfy the limit case (zero contrast) mathematically. On the other hand, when they are discretized, these formulations fail to provide accurate results for scattered fields from low-contrast objects.

2.2.4 Low-Contrast Breakdown

As the contrast goes to zero, nonradiating currents dominate total currents, while radiating currents, i.e., $\hat{\boldsymbol{n}} \times \boldsymbol{H}^{rad}, -\hat{\boldsymbol{n}} \times \boldsymbol{E}^{rad}$, tend to vanish. As an example, Figure 2.11 presents the equivalent electric and magnetic currents on the surface of a dielectric sphere with a radius of 0.3 m illuminated by a plane wave at 6 GHz. The sphere is in free space and it has a relative dielectric constant of $\epsilon_r = 1.0 + 10^{-4}$. The radiating parts of the currents are also depicted in Figure 2.12. Comparing Figure 2.12 to Figure 2.11, it can be observed that radiating currents form very small portions of total currents. Therefore, when total currents are solved by employing conventional surface formulations, it becomes difficult to perform calculations accurately enough to capture small radiating currents properly. In other words, even though the surface currents $\boldsymbol{J}$ and $\boldsymbol{M}$ are computed with relatively small errors, radiated fields may not be obtained accurately from them, since radiating currents are numerically insignificant compared to nonradiating currents.

2.2.5 Stabilization by Extraction

In order to accurately compute scattered fields from low-contrast objects, nonradiating parts of currents can be extracted to solve only radiating currents as unknowns [95], [96]. Applying such a stabilization procedure to the tangential formulation in (2.143)–(2.149), the stabilized CTF (S-CTF) is derived as

$$\begin{bmatrix} \mathcal{Z}_{11}^{\text{M-CTF}} & \mathcal{Z}_{12}^{\text{M-CTF}} \\ \mathcal{Z}_{21}^{\text{M-CTF}} & \mathcal{Z}_{22}^{\text{M-CTF}} \end{bmatrix} \cdot \begin{bmatrix} \boldsymbol{J}^{rad} \\ \boldsymbol{M}^{rad} \end{bmatrix} (\boldsymbol{r}) = \hat{\boldsymbol{n}} \times \hat{\boldsymbol{n}} \times \begin{bmatrix} \boldsymbol{E}^{inc} \\ \eta_o \eta_i \boldsymbol{H}^{inc} \end{bmatrix} (\boldsymbol{r}) - \begin{bmatrix} \mathcal{Z}_{11}^{\text{M-CTF}} & \mathcal{Z}_{12}^{\text{M-CTF}} \\ \mathcal{Z}_{21}^{\text{M-CTF}} & \mathcal{Z}_{22}^{\text{M-CTF}} \end{bmatrix} \cdot \begin{bmatrix} \boldsymbol{J}^{inc} \\ \boldsymbol{M}^{inc} \end{bmatrix} (\boldsymbol{r}). \tag{2.163}$$

Using (2.144)–(2.149) and the direct form of (2.159),

$$\begin{bmatrix} \mathcal{Z}_{11}^{\text{M-CTF}} & \mathcal{Z}_{12}^{\text{M-CTF}} \\ \mathcal{Z}_{21}^{\text{M-CTF}} & \mathcal{Z}_{22}^{\text{M-CTF}} \end{bmatrix} \cdot \begin{bmatrix} \boldsymbol{J}^{rad} \\ \boldsymbol{M}^{rad} \end{bmatrix} (\boldsymbol{r}) = \begin{bmatrix} \mathcal{Y}_{11} & \mathcal{Y}_{12} \\ \mathcal{Y}_{21} & \mathcal{Y}_{22} \end{bmatrix} \cdot \begin{bmatrix} \boldsymbol{J}^{inc} \\ \boldsymbol{M}^{inc} \end{bmatrix} (\boldsymbol{r}), \tag{2.164}$$

where

$$\mathcal{Y}_{11} = -\hat{\boldsymbol{n}} \times \hat{\boldsymbol{n}} \times \left(\frac{\Omega_o}{\Omega_i} \eta_o \mathcal{T}_o - \eta_i \mathcal{T}_i \right) \tag{2.165}$$

$$\mathcal{Y}_{12} = \hat{\boldsymbol{n}} \times \hat{\boldsymbol{n}} \times \left(\frac{\Omega_o}{\Omega_i} \mathcal{K}_{PV,o} - \mathcal{K}_{PV,i} \right) + \left(1 - \frac{\Omega_i}{2\pi} \right) \mathcal{I}^{\times n} \tag{2.166}$$

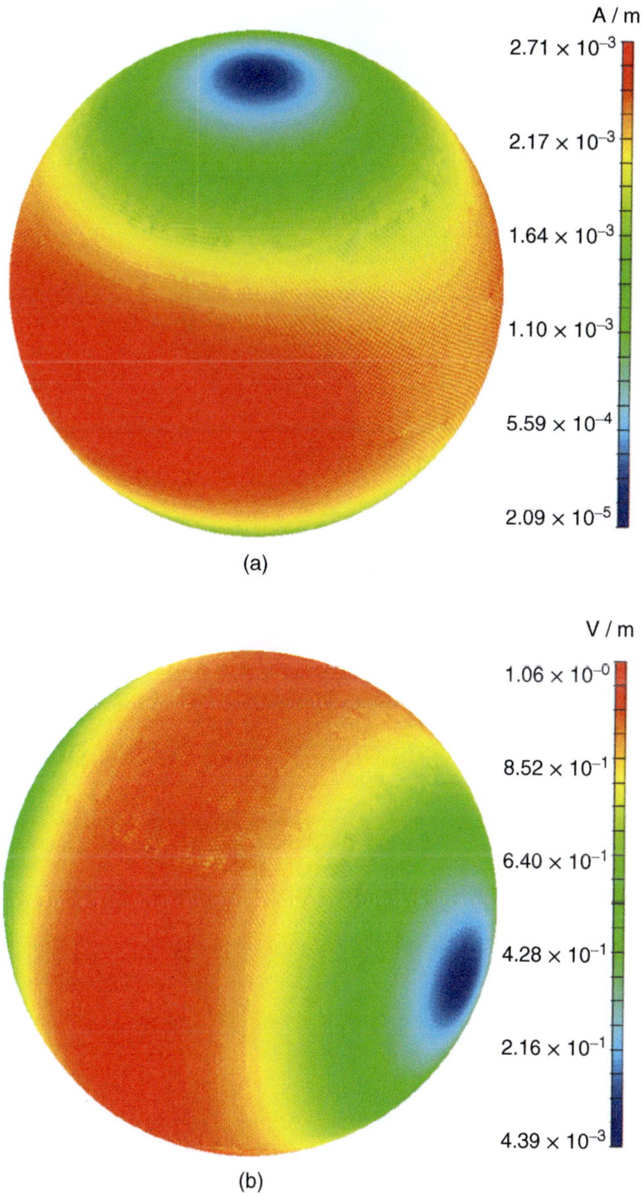

Figure 2.11 The equivalent (a) electric and (b) magnetic currents on a sphere with a radius of 0.3 m illuminated by a plane wave at 6 GHz. The sphere is located in free space and has a relative permittivity of $\epsilon_r = 1.0 + 10^{-4}$. *Source*: Ergül and Gürel 2008 [97]. Reproduced with permission of Elsevier.

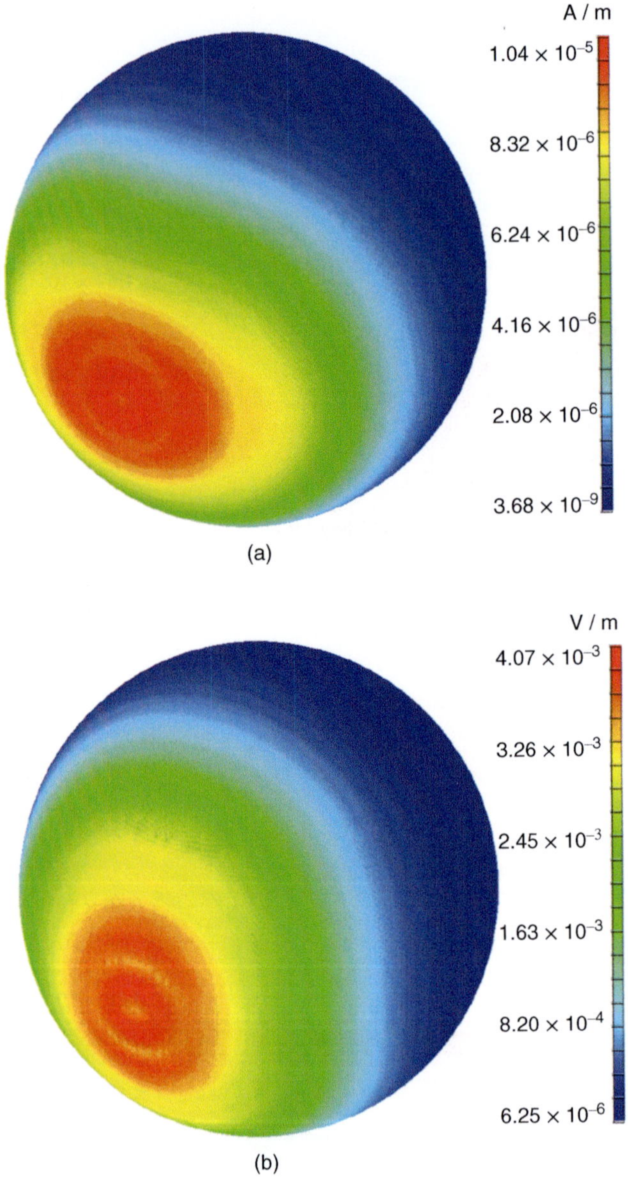

Figure 2.12 Radiating parts of the (a) electric and (b) magnetic currents depicted in Figure 2.11. *Source*: Ergül and Gürel 2008 [97]. Reproduced with permission of Elsevier.

$$\mathcal{Y}_{21} = -\eta_o\eta_i\hat{\boldsymbol{n}} \times \hat{\boldsymbol{n}} \times \left(\frac{\Omega_o}{\Omega_i}\mathcal{K}_{PV,o} - \mathcal{K}_{PV,i}\right) - \eta_o\eta_i\left(1 - \frac{\Omega_i}{2\pi}\right)\mathcal{I}^{\times n} \tag{2.167}$$

$$\mathcal{Y}_{22} = -\hat{\boldsymbol{n}} \times \hat{\boldsymbol{n}} \times \left(\frac{\Omega_o}{\Omega_i}\eta_i\mathcal{T}_o - \eta_o\mathcal{T}_i\right). \tag{2.168}$$

When the contrast goes to zero, the RHS of S-CTF reduces to

$$\begin{bmatrix} \boldsymbol{w}_1^{\text{S-CTF(0)}} \\ \boldsymbol{w}_2^{\text{S-CTF(0)}} \end{bmatrix} = \begin{bmatrix} 0 & (1-\Omega_i/2\pi)\,\mathcal{I}^{\times n} \\ -\eta_o^2(1-\Omega_i/2\pi)\,\mathcal{I}^{\times n} & 0 \end{bmatrix} \cdot \begin{bmatrix} \boldsymbol{J}^{inc} \\ \boldsymbol{M}^{inc} \end{bmatrix}(\boldsymbol{r})$$
$$-\frac{(\Omega_o - \Omega_i)}{\Omega_i}\hat{\boldsymbol{n}} \times \hat{\boldsymbol{n}} \times \begin{bmatrix} \eta_o\mathcal{T}_o & -\mathcal{K}_{PV,o} \\ \eta_o^2\mathcal{K}_{PV,o} & \eta_o\mathcal{T}_o \end{bmatrix} \cdot \begin{bmatrix} \boldsymbol{J}^{inc} \\ \boldsymbol{M}^{inc} \end{bmatrix}(\boldsymbol{r}), \tag{2.169}$$

which is identically zero considering the direct form of (2.159). Hence, the solution is zero, as expected.

Discretization of S-CTF in (2.164) leads to

$$\begin{bmatrix} \overline{\boldsymbol{Z}}_{11}^{\text{M-CTF}} & \overline{\boldsymbol{Z}}_{12}^{\text{M-CTF}} \\ \overline{\boldsymbol{Z}}_{21}^{\text{M-CTF}} & \overline{\boldsymbol{Z}}_{22}^{\text{M-CTF}} \end{bmatrix} \cdot \begin{bmatrix} \boldsymbol{a}_J^{rad} \\ \boldsymbol{a}_M^{rad} \end{bmatrix} = \begin{bmatrix} \overline{\boldsymbol{Y}}_{11} & \overline{\boldsymbol{Y}}_{12} \\ \overline{\boldsymbol{Y}}_{21} & \overline{\boldsymbol{Y}}_{22} \end{bmatrix} \cdot \begin{bmatrix} \boldsymbol{a}_J^{inc} \\ \boldsymbol{a}_M^{inc} \end{bmatrix}, \tag{2.170}$$

where

$$\overline{\boldsymbol{Y}}_{11} = \eta_o\overline{\boldsymbol{T}}_o^T - \eta_i\overline{\boldsymbol{T}}_i^T \tag{2.171}$$

$$\overline{\boldsymbol{Y}}_{12} = -\overline{\boldsymbol{K}}_{PV,o}^T + \overline{\boldsymbol{K}}_{PV,i}^T \tag{2.172}$$

$$\overline{\boldsymbol{Y}}_{21} = \eta_o\eta_i\overline{\boldsymbol{K}}_{PV,o}^T - \eta_o\eta_i\overline{\boldsymbol{K}}_{PV,i}^T \tag{2.173}$$

$$\overline{\boldsymbol{Y}}_{22} = \eta_i\overline{\boldsymbol{T}}_o^T - \eta_o\overline{\boldsymbol{T}}_i^T. \tag{2.174}$$

In (2.170), $\{\boldsymbol{a}_J^{rad}, \boldsymbol{a}_M^{rad}\}$ and $\{\boldsymbol{a}_J^{inc}, \boldsymbol{a}_M^{inc}\}$ are two sets of vectors involving the coefficients to expand radiating and nonradiating currents, respectively, i.e.,

$$\{\hat{\boldsymbol{n}} \times \boldsymbol{H}^{rad}(\boldsymbol{r}), -\hat{\boldsymbol{n}} \times \boldsymbol{E}^{rad}(\boldsymbol{r})\} = \sum_{n=1}^{N} \{\boldsymbol{a}_J^{rad}[n], \boldsymbol{a}_M^{rad}[n]\}\,\boldsymbol{b}_n(\boldsymbol{r}) \tag{2.175}$$

$$\{\hat{\boldsymbol{n}} \times \boldsymbol{H}^{inc}(\boldsymbol{r}), -\hat{\boldsymbol{n}} \times \boldsymbol{E}^{inc}(\boldsymbol{r})\} = \sum_{n=1}^{N} \{\boldsymbol{a}_J^{inc}[n], \boldsymbol{a}_M^{inc}[n]\}\,\boldsymbol{b}_n(\boldsymbol{r}), \tag{2.176}$$

and

$$\boldsymbol{a}_J = \boldsymbol{a}_J^{rad} + \boldsymbol{a}_J^{inc} \tag{2.177}$$

$$\boldsymbol{a}_M = \boldsymbol{a}_M^{rad} + \boldsymbol{a}_M^{inc}. \tag{2.178}$$

Note that only $\boldsymbol{a}_J^{rad}, \boldsymbol{a}_M^{rad}$ contribute to radiated fields.

Extracting nonradiating currents in CNF, one can obtain the stabilized CNF (S-CNF) as

$$\hat{\boldsymbol{n}} \times \begin{bmatrix} \mathcal{Z}_{11}^{\mathrm{CNF}} & \mathcal{Z}_{12}^{\mathrm{CNF}} \\ \mathcal{Z}_{21}^{\mathrm{CNF}} & \mathcal{Z}_{22}^{\mathrm{CNF}} \end{bmatrix} \cdot \begin{bmatrix} \boldsymbol{J}^{rad} \\ \boldsymbol{M}^{rad} \end{bmatrix} (\boldsymbol{r}) = -\hat{\boldsymbol{n}} \times \begin{bmatrix} \boldsymbol{H}^{inc} \\ -\boldsymbol{E}^{inc} \end{bmatrix} (\boldsymbol{r})$$

$$-\hat{\boldsymbol{n}} \times \begin{bmatrix} \mathcal{Z}_{11}^{\mathrm{CNF}} & \mathcal{Z}_{12}^{\mathrm{CNF}} \\ \mathcal{Z}_{21}^{\mathrm{CNF}} & \mathcal{Z}_{22}^{\mathrm{CNF}} \end{bmatrix} \cdot \begin{bmatrix} \boldsymbol{J}^{inc} \\ \boldsymbol{M}^{inc} \end{bmatrix} (\boldsymbol{r}). \tag{2.179}$$

Using (2.23)–(2.26),

$$\hat{\boldsymbol{n}} \times \begin{bmatrix} \mathcal{Z}_{11}^{\mathrm{CNF}} & \mathcal{Z}_{12}^{\mathrm{CNF}} \\ \mathcal{Z}_{21}^{\mathrm{CNF}} & \mathcal{Z}_{22}^{\mathrm{CNF}} \end{bmatrix} \cdot \begin{bmatrix} \boldsymbol{J}^{rad} \\ \boldsymbol{M}^{rad} \end{bmatrix} (\boldsymbol{r})$$

$$= -\hat{\boldsymbol{n}} \times \begin{bmatrix} \mathcal{V}_{11} & \mathcal{V}_{12} \\ \mathcal{V}_{21} & \mathcal{V}_{22} \end{bmatrix} \cdot \begin{bmatrix} \boldsymbol{J}^{inc} \\ \boldsymbol{M}^{inc} \end{bmatrix} (\boldsymbol{r}), \tag{2.180}$$

where

$$\mathcal{V}_{11} = \mathcal{V}_{22} = \mathcal{K}_{PV,o} - \mathcal{K}_{PV,i} \tag{2.181}$$

$$\mathcal{V}_{12} = \mathcal{Z}_{12}^{\mathrm{CNF}} = \eta_o^{-1}\mathcal{T}_o - \eta_i^{-1}\mathcal{T}_i \tag{2.182}$$

$$\mathcal{V}_{21} = \mathcal{Z}_{21}^{\mathrm{CNF}} = -\eta_o\mathcal{T}_o + \eta_i\mathcal{T}_i. \tag{2.183}$$

When the contrast goes to zero, S-CNF reduces to

$$\begin{bmatrix} \mathcal{I} & 0 \\ 0 & \mathcal{I} \end{bmatrix} \cdot \begin{bmatrix} \boldsymbol{J}^{rad} \\ \boldsymbol{M}^{rad} \end{bmatrix} (\boldsymbol{r}) = 0 \tag{2.184}$$

and the solution is again zero. S-CNF is discretized as

$$\begin{bmatrix} \overline{\boldsymbol{Z}}_{11}^{\mathrm{CNF}} & \overline{\boldsymbol{Z}}_{12}^{\mathrm{CNF}} \\ \overline{\boldsymbol{Z}}_{21}^{\mathrm{CNF}} & \overline{\boldsymbol{Z}}_{22}^{\mathrm{CNF}} \end{bmatrix} \cdot \begin{bmatrix} \boldsymbol{a}_J^{rad} \\ \boldsymbol{a}_M^{rad} \end{bmatrix} = -\begin{bmatrix} \overline{\boldsymbol{V}}_{11} & \overline{\boldsymbol{V}}_{12} \\ \overline{\boldsymbol{V}}_{21} & \overline{\boldsymbol{V}}_{22} \end{bmatrix} \cdot \begin{bmatrix} \boldsymbol{a}_J^{inc} \\ \boldsymbol{a}_M^{inc} \end{bmatrix}, \tag{2.185}$$

where

$$\overline{\boldsymbol{V}}_{11} = \overline{\boldsymbol{V}}_{22} = \overline{\boldsymbol{K}}_{PV,o}^{N} - \overline{\boldsymbol{K}}_{PV,i}^{N} \tag{2.186}$$

$$\overline{\boldsymbol{V}}_{12} = \overline{\boldsymbol{Z}}_{12}^{\mathrm{CNF}} = \eta_o^{-1}\overline{\boldsymbol{T}}_o^{N} - \eta_i^{-1}\overline{\boldsymbol{T}}_i^{N} \tag{2.187}$$

$$\overline{\boldsymbol{V}}_{21} = \overline{\boldsymbol{Z}}_{21}^{\mathrm{CNF}} = -\eta_o\overline{\boldsymbol{T}}_o^{N} + \eta_i\overline{\boldsymbol{T}}_i^{N}. \tag{2.188}$$

The left-hand sides (LHSs) of the stable formulations are the same as those of their conventional forms. In other words, the stabilization procedure only requires a modification on the RHSs of the formulations. On the RHSs, both S-CTF and S-CNF involve operators applied on incident fields. Since these operators are already available, the stabilization does not require a significant cost in terms of the processing time and memory usage. The extra cost is only due to the calculation of the modified RHSs before iterative solutions; this can be done in negligible time compared to other parts of implementations.

In their discrete forms, a direct approach to apply operators on incident fields is to expand these fields in a series of basis functions and perform MVMs. Such an expansion can be achieved by using identity operators and solving the equation

$$\begin{bmatrix} \mathcal{I} & 0 \\ 0 & \mathcal{I} \end{bmatrix} \cdot \begin{bmatrix} \hat{\boldsymbol{n}} \times \boldsymbol{H}^{inc} \\ -\hat{\boldsymbol{n}} \times \boldsymbol{E}^{inc} \end{bmatrix} (\boldsymbol{r}) = \begin{bmatrix} \hat{\boldsymbol{n}} \times \boldsymbol{H}^{inc} \\ -\hat{\boldsymbol{n}} \times \boldsymbol{E}^{inc} \end{bmatrix} (\boldsymbol{r}) \tag{2.189}$$

with MOM. Using local basis functions, such as the RWG functions, identity operators lead to sparse matrices, and the discrete form of (2.189) can be written as

$$\begin{bmatrix} \overline{\boldsymbol{I}}^T & 0 \\ 0 & \overline{\boldsymbol{I}}^T \end{bmatrix} \cdot \begin{bmatrix} \boldsymbol{a}_J^{inc} \\ \boldsymbol{a}_M^{inc} \end{bmatrix} = \begin{bmatrix} \boldsymbol{v}_H^N \\ -\boldsymbol{v}_E^N \end{bmatrix}. \tag{2.190}$$

The solution of (2.190) usually requires negligible time; however, using discretized identity operators may deteriorate the accuracy of results, as discussed in the following sections.

Using (2.190) in (2.170) and (2.185), the discretized S-CTF and S-CNF can be written as

$$\begin{bmatrix} \overline{\boldsymbol{Z}}_{11}^{\text{M-CTF}} & \overline{\boldsymbol{Z}}_{12}^{\text{M-CTF}} \\ \overline{\boldsymbol{Z}}_{21}^{\text{M-CTF}} & \overline{\boldsymbol{Z}}_{22}^{\text{M-CTF}} \end{bmatrix} \cdot \begin{bmatrix} \boldsymbol{a}_J^{rad} \\ \boldsymbol{a}_M^{rad} \end{bmatrix}$$

$$= \begin{bmatrix} \overline{\boldsymbol{Y}}_{11} & \overline{\boldsymbol{Y}}_{12} \\ \overline{\boldsymbol{Y}}_{21} & \overline{\boldsymbol{Y}}_{22} \end{bmatrix} \cdot \begin{bmatrix} \overline{\boldsymbol{I}}^T & 0 \\ 0 & \overline{\boldsymbol{I}}^T \end{bmatrix}^{-1} \cdot \begin{bmatrix} \boldsymbol{v}_H^N \\ -\boldsymbol{v}_E^N \end{bmatrix} \tag{2.191}$$

and

$$\begin{bmatrix} \overline{\boldsymbol{Z}}_{11}^{\text{CNF}} & \overline{\boldsymbol{Z}}_{12}^{\text{CNF}} \\ \overline{\boldsymbol{Z}}_{21}^{\text{CNF}} & \overline{\boldsymbol{Z}}_{22}^{\text{CNF}} \end{bmatrix} \cdot \begin{bmatrix} \boldsymbol{a}_J^{rad} \\ \boldsymbol{a}_M^{rad} \end{bmatrix}$$

$$= - \begin{bmatrix} \overline{\boldsymbol{V}}_{11} & \overline{\boldsymbol{V}}_{12} \\ \overline{\boldsymbol{V}}_{21} & \overline{\boldsymbol{V}}_{22} \end{bmatrix} \cdot \begin{bmatrix} \overline{\boldsymbol{I}}^T & 0 \\ 0 & \overline{\boldsymbol{I}}^T \end{bmatrix}^{-1} \cdot \begin{bmatrix} \boldsymbol{v}_H^N \\ -\boldsymbol{v}_E^N \end{bmatrix}, \tag{2.192}$$

respectively.

2.2.6 *Double-Stabilized Combined Tangential Formulation*

The presence of discretized identity operators in (2.190) may deteriorate the accuracy of results obtained with S-CTF and S-CNF. Although this is not critical for S-CNF, which already involves identity operators on the LHS, the accuracy of S-CTF can be affected significantly. Therefore, to obtain more accurate results, one can derive and use the double-stabilized CTF (DS-CTF), which based on calculating coefficients of incident fields by discretizing and solving the direct form of (2.159), i.e.,

$$\begin{bmatrix} \eta_o \overline{\boldsymbol{T}}_o^T & -\overline{\boldsymbol{K}}_{PV,o}^T \\ \overline{\boldsymbol{K}}_{PV,o}^T & \eta_o^{-1} \overline{\boldsymbol{T}}_o^T \end{bmatrix} \cdot \begin{bmatrix} \boldsymbol{a}_J^{inc} \\ \boldsymbol{a}_M^{inc} \end{bmatrix} = -0.5 \begin{bmatrix} \boldsymbol{v}_E^T \\ \boldsymbol{v}_H^T \end{bmatrix}. \tag{2.193}$$

This formulation is completely free of identity operators. On the other hand, the improved accuracy comes at the cost of reduced efficiency since it is necessary to solve an additional dense matrix equation rather than the sparse equation in (2.190). Using (2.193) in (2.170), the discretized DS-CTF can be written as

$$\begin{bmatrix} \overline{Z}_{11}^{\text{M-CTF}} & \overline{Z}_{12}^{\text{M-CTF}} \\ \overline{Z}_{21}^{\text{M-CTF}} & \overline{Z}_{22}^{\text{M-CTF}} \end{bmatrix} \cdot \begin{bmatrix} a_J^{rad} \\ a_M^{rad} \end{bmatrix} =$$

$$-0.5 \begin{bmatrix} \overline{Y}_{11} & \overline{Y}_{12} \\ \overline{Y}_{21} & \overline{Y}_{22} \end{bmatrix} \cdot \begin{bmatrix} \eta_o \overline{T}_o^T & -\overline{K}_{PV,o}^T \\ \overline{K}_{PV,o}^T & \eta_o^{-1} \overline{T}_o^T \end{bmatrix}^{-1} \cdot \begin{bmatrix} v_E^T \\ v_H^T \end{bmatrix}, \tag{2.194}$$

where the extra solution on the RHS is represented by a matrix inversion.

2.2.7 *Numerical Results for Low Contrasts*

In Figure 2.13, the improved accuracy of the stable formulations is demonstrated on scattering problems involving a small sphere with a radius of $0.5\lambda_o$. The sphere is located in free space and illuminated by a plane wave propagating in the z direction with the electric field polarized in the x direction. Discretization of the surface with $\lambda_o/10$ mesh size produces 620 triangles, and 2×930 RWG functions are employed to expand unknown surface currents. Matrix equations involving 1860 unknowns that are obtained by using the conventional and stable formulations are solved iteratively. Matrix elements are computed directly with 10^{-2} relative error. Figure 2.13(a) shows the relative error in the computed values of the far-zone electric field on the z-x plane with respect to different contrasts $(\epsilon_r - 1)$ from 10^{-4} to 10^{-1}. The far-zone co-polar electric field is sampled at 360 points as

$$E_\theta^\infty[n] = \lim_{r\to\infty} \left\{ rE_\theta(r, \theta_n, \phi = 0) \right\}, \tag{2.195}$$

where $\theta_n = (n-1)\pi/360$ for $n = 1, 2, \ldots, 360$. The expression in (2.195) is computed by using different integral-equation formulations and by a Mie-series method as a reference. Then, the relative error is defined as

$$\Delta_R = \frac{\|E_{C,\theta}^\infty - E_{A,\theta}^\infty\|_2}{\|E_{A,\theta}^\infty\|_2}, \tag{2.196}$$

where $E_{C,\theta}^\infty$ and $E_{A,\theta}^\infty$ are computational and analytical values, respectively, and $\|.\|_2$ represents the 2-norm of the complex arrays involving 360 elements. Figure 2.13(a) shows that the M-CTF and CNF formulations encounter stability problems as the contrast decreases and they fail to provide accurate results for low-contrast levels. With the stable formulations, however, the accuracy does not depend on the contrast and the results are accurate even for $\epsilon_r = 1.0 + 10^{-4}$. Note that S-CTF is more accurate than S-CNF. In addition, DS-CTF provides the most accurate results for low-contrast problems, since its implementation is completely free of identity operators. Figure 2.13(b) presents the RCS of the sphere when the contrast is 10^{-3}. The normalized RCS (RCS $/\lambda_o^2$) is plotted on the z-x plane as a function of the observation angle, where 0° corresponds to the forward-scattering direction. It can be observed that the traditional formulations are inaccurate, while the values obtained by using DS-CTF are in agreement with the reference Mie-series solution.

Figure 2.14 shows the solution of a scattering problem involving a sphere with a radius of radius $6\lambda_o$ discretized with 264 006 unknowns. In Figure 2.14(a), the relative error is again

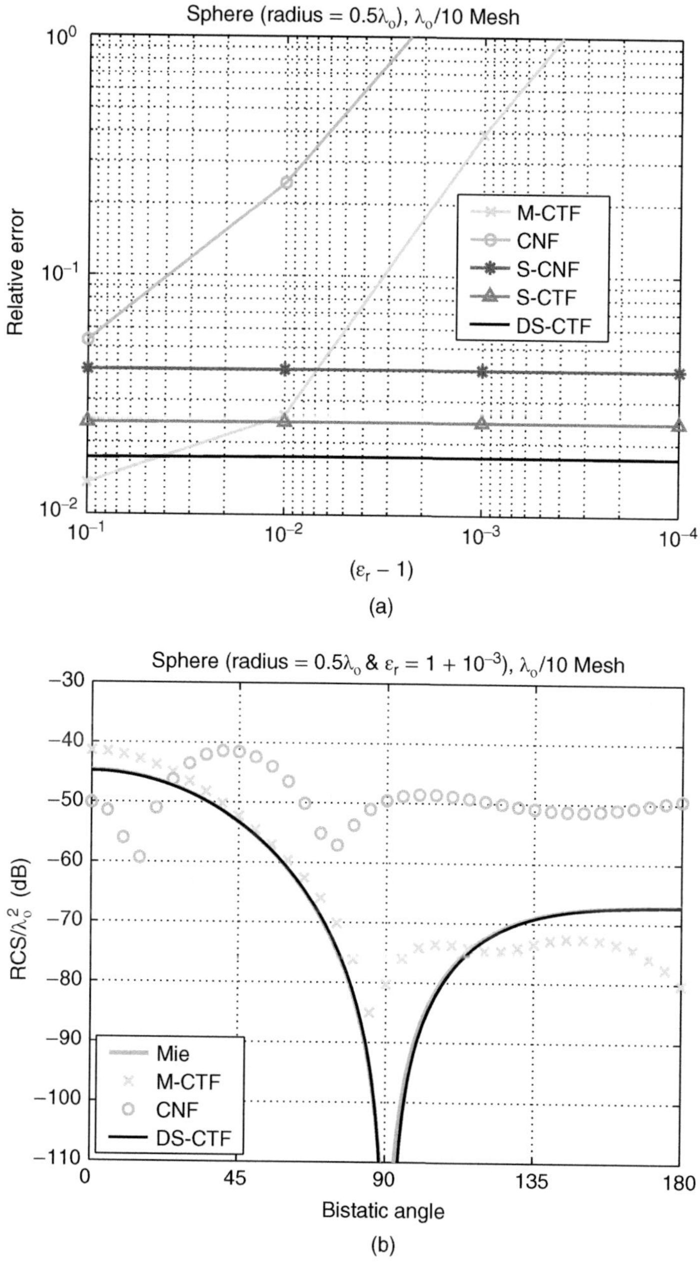

Figure 2.13 (a) The relative error in the solution of scattering problems involving a sphere with a radius of $0.5\lambda_o$ located in free space. (b) The normalized bistatic RCS (RCS/λ_o^2) of a sphere with a radius of $0.5\lambda_o$ and with a relative permittivity of $1.0 + 10^{-3}$ located in free space. *Source*: Ergül and Gürel 2008 [96]. Reproduced with permission of IEEE.

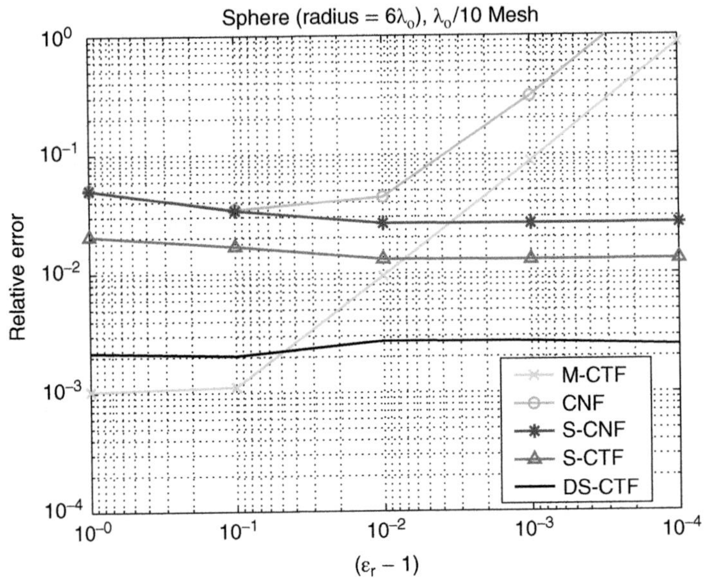

(a)

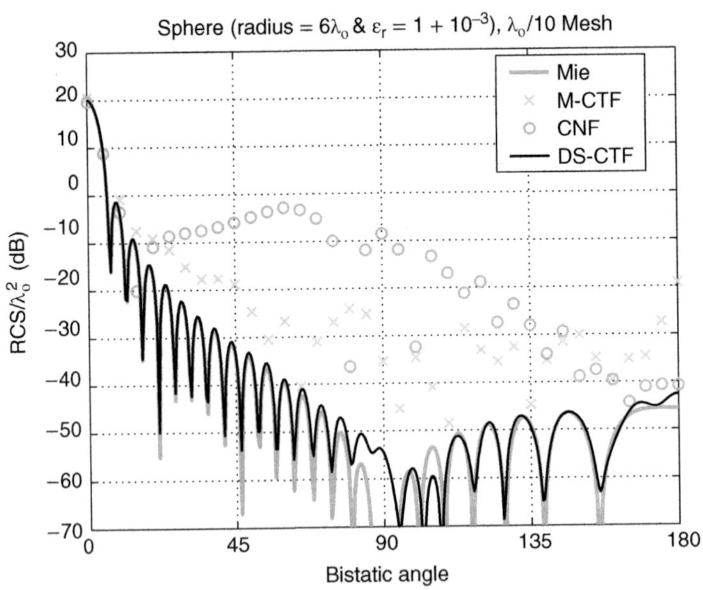

(b)

Figure 2.14 (a) The relative error in the solution of scattering problems involving a sphere with a radius of $6\lambda_o$ located in free space. (b) The normalized bistatic RCS (RCS $/\lambda_o^2$) of a sphere with a radius of $6\lambda_o$ and with a relative permittivity of $1.0 + 10^{-3}$ located in free space. *Source*: Ergül and Gürel 2008 [96]. Reproduced with permission of IEEE.

Table 2.1 Number of CGS iterations (10^{-6} residual error) for the solution of a sphere problem discretized with 264 006 unknowns

	CNF	M-CTF	S-CNF	S-CTF	DS-CTF
Full system	35	638	33	655	616+670
Sparse system	–	–	8	9	–

considered for different formulations with respect to the contrast of the sphere. In MLFMA solutions, near-field interactions are calculated with at most 1% error, and far-field interactions are computed with three digits of accuracy. Similar to the previous example, the stable formulations do not break down as the contrast decreases. In Figure 2.14(b), bistatic RCS values are plotted with respect to the observation angle when the contrast is 10^{-3}. The conventional formulations are accurate only in the forward-scattering direction (0°), while the accuracy is improved significantly by using DS-CTF.

For the solution of the scattering problem involving the sphere of radius $6\lambda_o$ with 0.1 contrast, Table 2.1 lists the number of iterations of the CGS method to reduce the residual error to less than 10^{-6}. Extra costs of S-CTF and S-CNF are negligible compared to M-CTF and CNF, respectively. These stable formulations require only 8–9 iterations to solve the extremely sparse matrix equation in (2.190). On the other hand, using DS-CTF doubles the processing time compared to S-CTF since the solution of the dense matrix equation in (2.193) requires 670 iterations, in addition to 616 iterations for the solution of the main equation. Table 2.1 also shows that both the conventional and stable tangential formulations (M-CTF, S-CTF, and DS-CTF) require more iterations than the normal formulations (CNF and S-CNF). On the other hand, the tangential formulations are generally more accurate to calculate the scattered fields compared to the normal formulations, as shown above.

Figure 2.15 presents the results of a scattering problem involving a cube with edges of 1 m illuminated by a plane wave at 300 MHz. The cube has a relative permittivity of $1.0 + 10^{-4}$ and it is located in free space. Figure 2.15 shows forward-scattered RCS values (m^2) for different discretizations. Triangulation of the geometry with $\lambda_o/5$ and $\lambda_o/30$ mesh sizes leads to 1008 and 36,756 unknowns, respectively. In the MLFMA implementation, near-field interactions are again calculated with 1% error and far-field interactions are computed with three digits of accuracy. Figure 2.15 indicates that all stable formulations tend to converge to a limit (correct) value as the discretization becomes denser. However, the convergence characteristics depend on the formulation type. Specifically, S-CNF and DS-CTF have the slowest and fastest convergence rates, respectively. Another interpretation of Figure 2.15 is that the most accurate results are provided by DS-CTF for a given mesh size. For example, for the conventional $\lambda_o/10$ triangulation, the error of S-CNF is more than 3%; this may not be acceptable since matrix elements are calculated with at most 1% error.

2.2.8 *Breakdown for Extremely Low Contrasts*

In the three stable formulations described above, namely, S-CNF, S-CTF, and DS-CTF, RHSs are obtained by applying the inner and outer operators on incident fields. In addition, operators are subtracted from each other so that RHSs go to zero as the contrast decreases. Hence, these

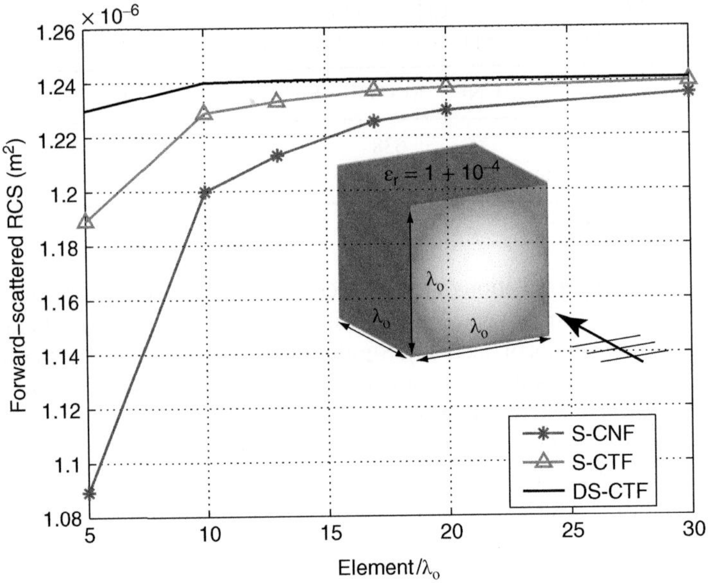

Figure 2.15 The forward-scattered RCS (m^2) of a cube with edges of λ_o and with a relative permittivity of $1.0 + 10^{-4}$ located in free space with respect to various discretizations of the scattering problem. *Source*: Ergül and Gürel 2008 [96]. Reproduced with permission of IEEE.

formulations are called the operator-based-stabilized formulations (OBSFs). Using OBSFs, radiating currents can be computed accurately for low-contrast objects, i.e., when these currents are numerically insignificant compared to nonradiating currents. This is because relatively large nonradiating currents are extracted, and only radiating currents are solved for. Consequently, OBSFs provide accurate results for low-contrast problems that cannot be solved accurately with conventional formulations. However, even OBSFs break down and fail to provide accurate results for extremely low contrasts. The reason is the numerical errors arising during the computation of RHSs of OBSFs, which may become significant when the contrast decreases to very low values and renders the RHS vanishingly small.

To understand numerical problems occurring in OBSFs, consider S-CTF in (2.164)–(2.168) and its discretization in (2.170)–(2.174). The RHS of S-CTF is obtained by applying the integro-differential operators on nonradiating currents, i.e.,

$$\begin{bmatrix} \boldsymbol{w}_1^{\text{S-CTF}} \\ \boldsymbol{w}_2^{\text{S-CTF}} \end{bmatrix} = \begin{bmatrix} \eta_o \overline{\boldsymbol{T}}_o^T & -\overline{\boldsymbol{K}}_{PV,o}^T \\ \eta_o \eta_i \overline{\boldsymbol{K}}_{PV,o}^T & \eta_i \overline{\boldsymbol{T}}_o^T \end{bmatrix} \cdot \begin{bmatrix} \boldsymbol{a}_J^{inc} \\ \boldsymbol{a}_M^{inc} \end{bmatrix} - \begin{bmatrix} \eta_i \overline{\boldsymbol{T}}_i^T & -\overline{\boldsymbol{K}}_{PV,i}^T \\ \eta_o \eta_i \overline{\boldsymbol{K}}_{PV,i}^T & \eta_o \overline{\boldsymbol{T}}_i^T \end{bmatrix} \cdot \begin{bmatrix} \boldsymbol{a}_J^{inc} \\ \boldsymbol{a}_M^{inc} \end{bmatrix}. \tag{2.197}$$

Operators related to the outer and inner media are applied on tangential incident fields (nonradiating currents) in the first and second terms of (2.197), respectively. Then, the second term is subtracted from the first term to compute the RHS. When the contrast is low, the RHS is small, but it is obtained by the subtraction of two terms that are relatively large. Then, depending on the discretization and the accuracy of MVMs, the RHS of S-CTF may not be calculated accurately when the contrast is extremely low, i.e., when the result of the subtraction is extremely small.

In general, OBSFs fail to provide accurate results when the contrast is decreased to extremely low values. Therefore, it is desirable to obtain a robust formulation that is valid for arbitrarily low contrasts. This can be achieved by introducing fictitious fields and forming RHSs based on the difference of fields [97]. This method leads to accurate calculations of RHSs, even when the contrast is extremely low.

2.2.9 Field-Based-Stabilized Formulations

When incident fields are extracted from the LHS, S-CTF in (2.164) can be rewritten as

$$\begin{bmatrix} \mathcal{Z}_{11}^{\text{M-CTF}} & \mathcal{Z}_{12}^{\text{M-CTF}} \\ \mathcal{Z}_{21}^{\text{M-CTF}} & \mathcal{Z}_{22}^{\text{M-CTF}} \end{bmatrix} \cdot \begin{bmatrix} \boldsymbol{J}^{rad} \\ \boldsymbol{M}^{rad} \end{bmatrix}(\boldsymbol{r}) = \frac{\Omega_i}{4\pi}\hat{\boldsymbol{n}}\times\hat{\boldsymbol{n}}\times \begin{bmatrix} \boldsymbol{E}^{inc} \\ \eta_o\eta_i\boldsymbol{H}^{inc} \end{bmatrix}(\boldsymbol{r})$$

$$+\hat{\boldsymbol{n}}\times\hat{\boldsymbol{n}}\times \begin{bmatrix} \eta_i\mathcal{T}_i & -\mathcal{K}_{PV,i} \\ \eta_o\eta_i\mathcal{K}_{PV,i} & \eta_o\mathcal{T}_i \end{bmatrix} \cdot \begin{bmatrix} \boldsymbol{J}^{inc} \\ \boldsymbol{M}^{inc} \end{bmatrix}(\boldsymbol{r}) \tag{2.198}$$

using the direct form of (2.159). At this stage, consider incident fields as functions of medium parameters, i.e.,

$$\boldsymbol{E}^{inc}(\boldsymbol{r}) = \boldsymbol{e}(\boldsymbol{r}, \epsilon_o, \mu_o) \tag{2.199}$$

$$\boldsymbol{H}^{inc}(\boldsymbol{r}) = \boldsymbol{h}(\boldsymbol{r}, \epsilon_o, \mu_o). \tag{2.200}$$

Then, fictitious incident fields can be defined as

$$\boldsymbol{E}_f^{inc}(\boldsymbol{r}) = \boldsymbol{e}(\boldsymbol{r}, \epsilon_i, \mu_i) \tag{2.201}$$

$$\boldsymbol{H}_f^{inc}(\boldsymbol{r}) = \boldsymbol{h}(\boldsymbol{r}, \epsilon_i, \mu_i) \tag{2.202}$$

by using the parameters of the inner medium for the outside. This way, similar to the identities in (2.159), one can derive

$$\begin{Bmatrix} -\hat{\boldsymbol{n}}\times\hat{\boldsymbol{n}}\times \\ \hat{\boldsymbol{n}}\times \end{Bmatrix} \begin{bmatrix} \eta_i\mathcal{T}_i & -\mathcal{K}_{PV,i} \\ \mathcal{K}_{PV,i} & \eta_i^{-1}\mathcal{T}_i \end{bmatrix} \cdot \begin{bmatrix} \hat{\boldsymbol{n}}\times\boldsymbol{H}_f^{inc} \\ -\hat{\boldsymbol{n}}\times\boldsymbol{E}_f^{inc} \end{bmatrix}(\boldsymbol{r})$$

$$= -\frac{\Omega_i}{4\pi}\begin{Bmatrix} -\hat{\boldsymbol{n}}\times\hat{\boldsymbol{n}}\times \\ \hat{\boldsymbol{n}}\times \end{Bmatrix} \begin{bmatrix} \boldsymbol{E}_f^{inc} \\ \boldsymbol{H}_f^{inc} \end{bmatrix}(\boldsymbol{r}). \tag{2.203}$$

By adding and subtracting terms of (2.203) in (2.198), one can obtain

$$\begin{bmatrix} \mathcal{Z}_{11}^{\text{M-CTF}} & \mathcal{Z}_{12}^{\text{M-CTF}} \\ \mathcal{Z}_{21}^{\text{M-CTF}} & \mathcal{Z}_{22}^{\text{M-CTF}} \end{bmatrix} \cdot \begin{bmatrix} \boldsymbol{J}^{rad} \\ \boldsymbol{M}^{rad} \end{bmatrix}(\boldsymbol{r}) = \frac{\Omega_i}{4\pi}\hat{\boldsymbol{n}} \times \hat{\boldsymbol{n}} \times \begin{bmatrix} \boldsymbol{E}^{inc} - \boldsymbol{E}_f^{inc} \\ \eta_o\eta_i\left(\boldsymbol{H}^{inc} - \boldsymbol{H}_f^{inc}\right) \end{bmatrix}(\boldsymbol{r})$$

$$+\hat{\boldsymbol{n}} \times \hat{\boldsymbol{n}} \times \begin{bmatrix} \eta_i\mathcal{T}_i & -\mathcal{K}_{PV,i} \\ \eta_o\eta_i\mathcal{K}_{PV,i} & \eta_o\mathcal{T}_i \end{bmatrix} \cdot \begin{bmatrix} \hat{\boldsymbol{n}} \times \left(\boldsymbol{H}^{inc} - \boldsymbol{H}_f^{inc}\right) \\ -\hat{\boldsymbol{n}} \times \left(\boldsymbol{E}^{inc} - \boldsymbol{E}_f^{inc}\right) \end{bmatrix}(\boldsymbol{r}), \tag{2.204}$$

which is called the field-based-stabilized CTF (FBS-CTF). In contrast to OBSFs, FBS-CTF has a RHS obtained by subtracting real and fictitious incident fields from each other. These subtractions can be performed analytically in the continuous space before the discretization. Then, the operators related to the inner medium can be applied to compute the second term of the RHS in (2.204). It should be noted that the RHS of FBS-CTF is obtained as the sum of two terms, i.e.,

$$\begin{bmatrix} \boldsymbol{w}_{1,1}^{\text{FBS-CTF}} \\ \boldsymbol{w}_{2,1}^{\text{FBS-CTF}} \end{bmatrix} = \frac{\Omega_i}{4\pi}\hat{\boldsymbol{n}} \times \hat{\boldsymbol{n}} \times \begin{bmatrix} \boldsymbol{E}^{inc} - \boldsymbol{E}_f^{inc} \\ \eta_o\eta_i\left(\boldsymbol{H}^{inc} - \boldsymbol{H}_f^{inc}\right) \end{bmatrix}(\boldsymbol{r}) \tag{2.205}$$

and

$$\begin{bmatrix} \boldsymbol{w}_{1,2}^{\text{FBS-CTF}} \\ \boldsymbol{w}_{2,2}^{\text{FBS-CTF}} \end{bmatrix} = \hat{\boldsymbol{n}} \times \hat{\boldsymbol{n}} \times \begin{bmatrix} \eta_i\mathcal{T}_i & -\mathcal{K}_{PV,i} \\ \eta_o\eta_i\mathcal{K}_{PV,i} & \eta_o\mathcal{T}_i \end{bmatrix} \cdot \begin{bmatrix} \hat{\boldsymbol{n}} \times \left(\boldsymbol{H}^{inc} - \boldsymbol{H}_f^{inc}\right) \\ -\hat{\boldsymbol{n}} \times \left(\boldsymbol{E}^{inc} - \boldsymbol{E}_f^{inc}\right) \end{bmatrix}(\boldsymbol{r}), \tag{2.206}$$

which are both small when the contrast is low. Consequently, the RHS can be calculated accurately for arbitrarily low contrasts, and it is sensitive to neither discretization errors nor numerical errors arising during MVMs. FBS-CTF can easily be obtained from the conventional CTF implementation and, similar to S-CTF and S-CNF, its extra cost is negligible.

Using the expansion in (2.189), the discretization of FBS-CTF leads to

$$\begin{bmatrix} \overline{\boldsymbol{Z}}_{11}^{\text{M-CTF}} & \overline{\boldsymbol{Z}}_{12}^{\text{M-CTF}} \\ \overline{\boldsymbol{Z}}_{21}^{\text{M-CTF}} & \overline{\boldsymbol{Z}}_{22}^{\text{M-CTF}} \end{bmatrix} \cdot \begin{bmatrix} \boldsymbol{a}_J^{rad} \\ \boldsymbol{a}_M^{rad} \end{bmatrix} = -0.5 \begin{bmatrix} \boldsymbol{v}_{E,f}^T \\ \eta_o\eta_i\boldsymbol{v}_{H,f}^T \end{bmatrix}$$

$$- \begin{bmatrix} \eta_i\overline{\boldsymbol{T}}_i^T & -\overline{\boldsymbol{K}}_{PV,i}^T \\ \eta_o\eta_i\overline{\boldsymbol{K}}_{PV,i}^T & \eta_o\overline{\boldsymbol{T}}_i^T \end{bmatrix} \cdot \begin{bmatrix} \overline{\boldsymbol{I}}^T & 0 \\ 0 & \overline{\boldsymbol{I}}^T \end{bmatrix}^{-1} \cdot \begin{bmatrix} \boldsymbol{v}_{H,f}^N \\ -\boldsymbol{v}_{E,f}^N \end{bmatrix}, \tag{2.207}$$

where

$$\boldsymbol{v}_{E,f}^T[m] = \int_{S_m} d\boldsymbol{r}\boldsymbol{t}_m(\boldsymbol{r}) \cdot \left[\boldsymbol{E}^{inc}(\boldsymbol{r}) - \boldsymbol{E}_f^{inc}(\boldsymbol{r})\right] \tag{2.208}$$

$$\boldsymbol{v}_{H,f}^T[m] = \int_{S_m} d\boldsymbol{r}\boldsymbol{t}_m(\boldsymbol{r}) \cdot \left[\boldsymbol{H}^{inc}(\boldsymbol{r}) - \boldsymbol{H}_f^{inc}(\boldsymbol{r})\right] \tag{2.209}$$

$$\boldsymbol{v}_{E,f}^N[m] = \int_{S_m} d\boldsymbol{r}\boldsymbol{t}_m(\boldsymbol{r}) \cdot \hat{\boldsymbol{n}} \times \left[\boldsymbol{E}^{inc}(\boldsymbol{r}) - \boldsymbol{E}_f^{inc}(\boldsymbol{r})\right] \tag{2.210}$$

$$\boldsymbol{v}_{H,f}^N[m] = \int_{S_m} d\boldsymbol{r}\boldsymbol{t}_m(\boldsymbol{r}) \cdot \hat{\boldsymbol{n}} \times \left[\boldsymbol{H}^{inc}(\boldsymbol{r}) - \boldsymbol{H}_f^{inc}(\boldsymbol{r})\right] \tag{2.211}$$

are obtained by testing the difference between real and fictitious incident fields.

2.2.10 Numerical Results for Extremely Low Contrasts

Figure 2.16 presents bistatic RCS results for a sphere with a radius of $0.5\lambda_o$. The sphere is located in free space and illuminated by a plane wave propagating in the z direction with the electric field polarized in the x direction. Discretization of the surface with $\lambda_o/10$ mesh size leads to matrix equations involving 1860 unknowns. Matrix elements are computed directly with 10^{-2} relative error. In Figure 2.16, the normalized RCS (RCS/λ_o^2) is plotted as a function of the observation angle on the z-x plane, where 0° corresponds to the forward-scattering direction. The contrast of the sphere is 10^{-5} and 10^{-9} in Figure 2.16(a) and Figure 2.16(b), respectively. For reference, RCS values are also computed analytically by Mie-series solutions. When the contrast is 10^{-5}, the conventional formulations, i.e., M-CTF and CNF, are inaccurate, while the stable formulations provide accurate results that are close to the analytical curve. When the contrast is further reduced to 10^{-9}, OBSFs also fail to agree with the analytical curve. For this extremely low contrast, the only formulation that provides accurate results is FBS-CTF. To further compare the formulations, Figure 2.17 presents the relative error defined in (2.196) as a function of the contrast. The figure shows that the errors of M-CTF and CNF increase sharply when the contrast decreases to below 10^{-1}–10^{-2}, while OBSFs break down when the contrast is about 10^{-5}–10^{-6}. On the other hand, the error of FBS-CTF is almost constant with respect to the contrast. Figure 2.18 presents the relative error, when the computational values are obtained with FMM, instead of MOM. Comparing Figure 2.18 with Figure 2.17, one can observe that the extra error introduced by FMM affects the stability of OBSFs significantly. Specifically, using FMM, those formulations break down earlier, i.e., when the contrast is about 10^{-4}. This is because OBSFs are very sensitive to numerical errors arising during MVMs. On the other hand, FBS-CTF provides stable results even with FMM, as depicted in Figure 2.18.

Figure 2.19 presents the results of scattering problems involving a sphere with a radius of $6\lambda_o$ discretized with $\lambda_o/10$ triangles leading to 264 006 unknowns. The problems are solved by MLFMA, where near-field interactions are calculated with 1% error and far-field interactions are computed with three digits of accuracy. Figure 2.19(a) depicts bistatic RCS values on the z-x plane when the sphere is illuminated by a plane wave propagating in the z direction with the electric field polarized in the x direction and the contrast of the sphere is 10^{-9}. All formulations except for the FBS-CTF fail to provide accurate results compared to the Mie-series solution. RCS values provided by FBS-CTF deviate from analytical results only around 90°, where the scattering is very low. As presented in Figure 2.19(b), the accuracy of FBS-CTF is stable for all values of the contrast from 10^{-1} to 10^{-9}. OBSFs are also stable in the 10^{-1}–10^{-5} range, while they offer different levels of accuracy depending on the use of the identity operator. However, they break down eventually when the contrast decreases to below 10^{-5}. As in the previous examples, conventional CTF and CNF break down immediately below 10^{-1} contrast, testifying to the need for stabilized formulations.

Next, scattering problems involving a cube with edges of λ_o discretized with $\lambda_o/30$ triangles are considered. The cube is in free space and it is illuminated by a plane wave propagating in the z direction with the electric field polarized in the x direction. Three different relative permittivities, i.e., 1.1, $1.0 + 10^{-3}$, and $1.0 + 10^{-6}$, corresponding to 10^{-1}, 10^{-3}, and 10^{-6} contrasts, respectively, are considered. The scattering problems are solved by FMM, where near-field interactions are calculated with 1% error and far-field interactions are computed with three digits of accuracy. Figure 2.20 presents the normalized RCS (RCS/λ_o^2) as a function of the observation angle on the z-x plane, where 0° corresponds to the forward-scattering

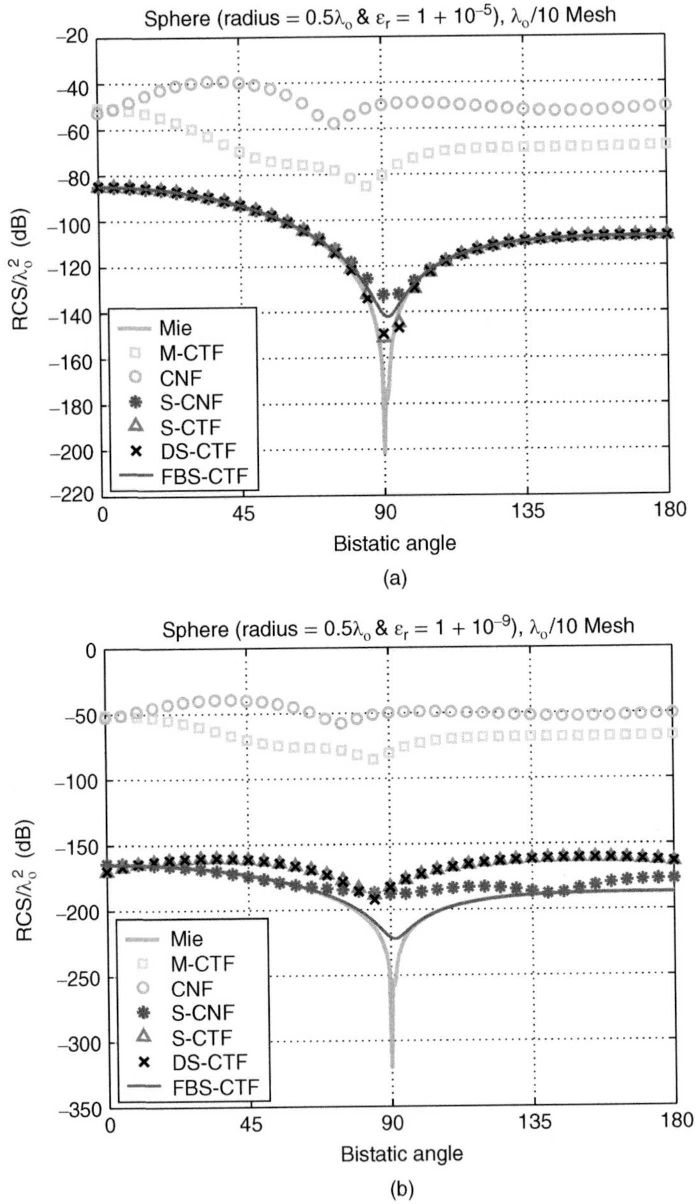

Figure 2.16 The normalized bistatic RCS (RCS$/\lambda_o^2$) of a sphere with a radius of $0.5\lambda_o$ located in free space, when the relative permittivity of the sphere is (a) $1.0 + 10^{-5}$ and (b) $1.0 + 10^{-9}$. *Source*: Ergül and Gürel 2008 [97]. Reproduced with permission of Elsevier.

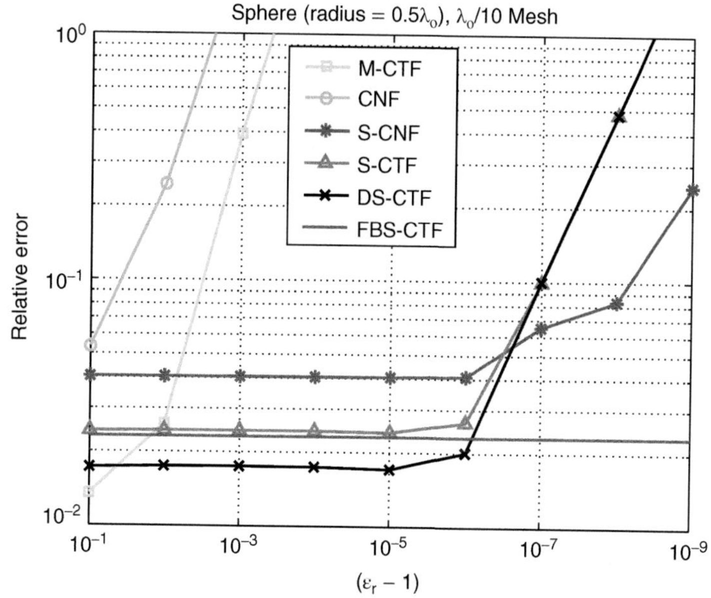

Figure 2.17 The relative error in the solution of scattering problems involving a sphere with a radius of $0.5\lambda_o$ located in free space. *Source*: Ergül and Gürel 2008 [97]. Reproduced with permission of Elsevier.

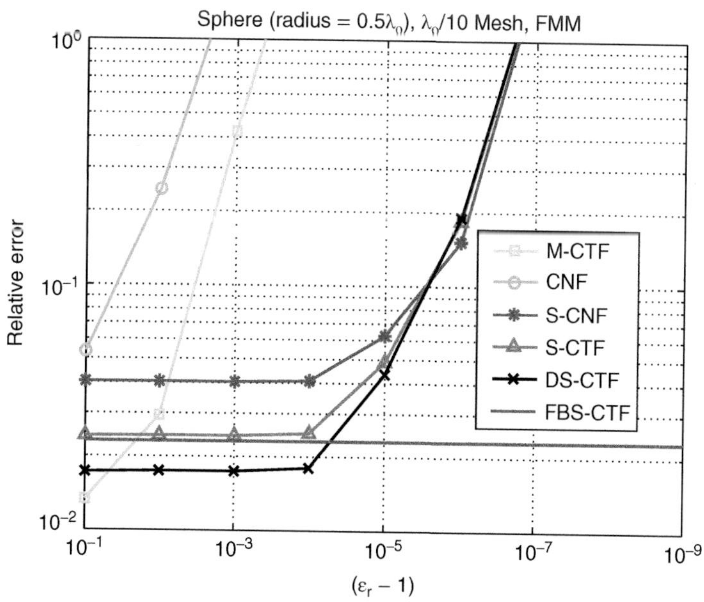

Figure 2.18 The relative error in the solution of scattering problems involving a sphere with a radius of $0.5\lambda_o$ located in free space. The scattering problems are solved by using FMM.

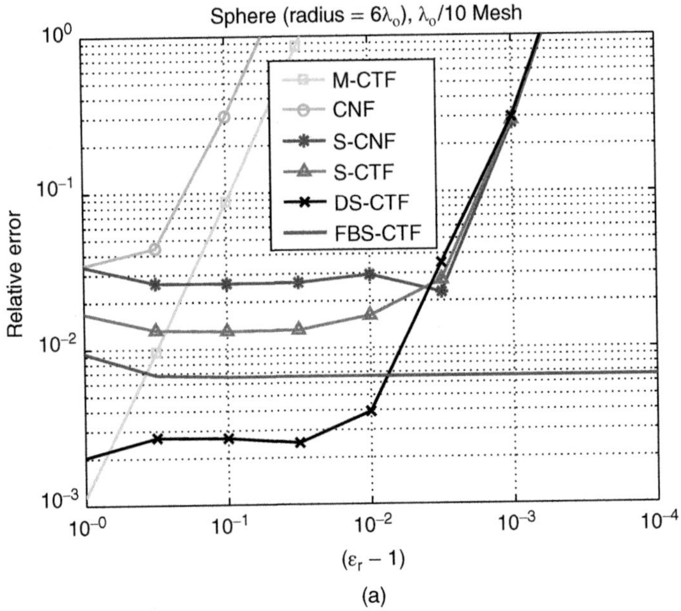

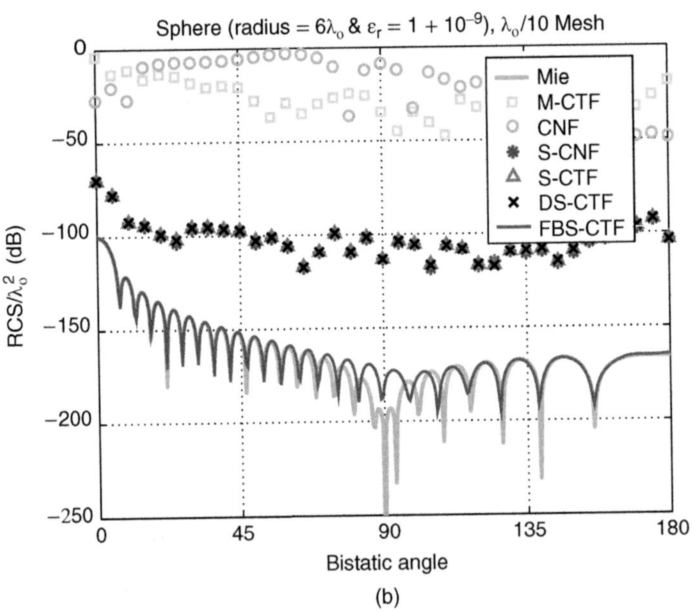

Figure 2.19 (a) The relative error in the solution of scattering problems involving a sphere with a radius of $6\lambda_o$ located in free space. (b) The normalized bistatic RCS (RCS$/\lambda_o^2$) of a sphere with a radius of $6\lambda_o$ and with a relative permittivity of $1.0 + 10^{-9}$ located in free space. *Source*: Ergül and Gürel 2008 [97]. Reproduced with permission of Elsevier.

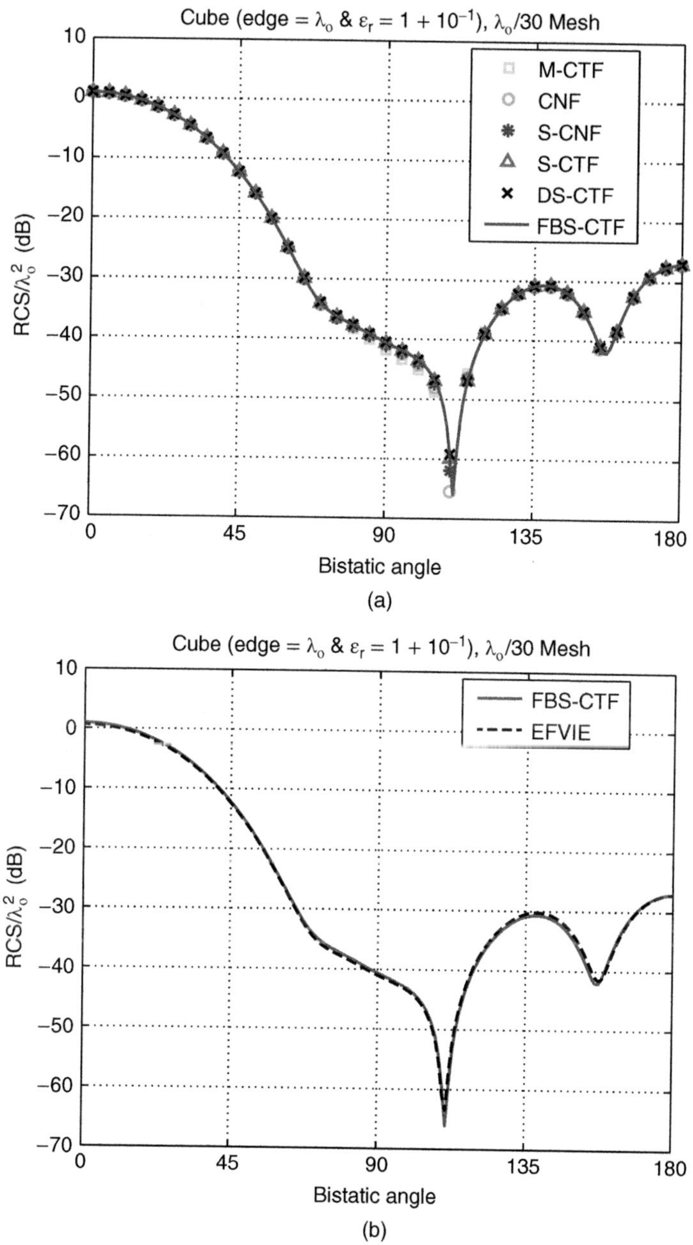

Figure 2.20 (a) The normalized bistatic RCS (RCS$/\lambda_o^2$) of a cube with edges of λ_o located in free space. The relative permittivity of the cube is $1.0 + 10^{-1}$. RCS values are obtained by using surface formulations when the mesh size is $\lambda_o/30$. (b) RCS values obtained with FBS-CTF and EFVIE agree well with each other. *Source*: Ergül and Gürel 2008 [97]. Reproduced with permission of Elsevier.

direction. The contrast of the cube is 10^{-1}. As depicted in Figure 2.20(a), there are relatively small discrepancies among results obtained by using different formulations. In Figure 2.20(b), RCS values obtained by using FBS-CTF are compared with those obtained by using the electric-field volume integral equation (EFVIE) [36], which is immune to low-contrast problems. Figure 2.20(b) confirms that FBS-CTF and other surface formulations are accurate when the contrast of the cube is 10^{-1}. Figure 2.21 compares bistatic RCS values obtained by using various formulations as the contrast of the cube is decreased to 10^{-3}. In this case, RCS values obtained by using the conventional formulations, namely, M-CTF and CNF, are inconsistent with values obtained by using the stable formulations, i.e., S-CNF, S-CTF, DS-CTF, and FBS-CTF. As demonstrated in Figure 2.21(b), FBS-CTF (and other stable formulations) are consistent with VIE. Hence, one can conclude that the stable formulations are accurate, while M-CTF and CNF break down when the contrast of the cube is 10^{-3}. Figure 2.22 presents bistatic RCS results, when the contrast of the cube is very low, i.e., 10^{-6}. As opposed to the previous examples, RCS values obtained with OBSFs (S-CNF, S-CTF, and DS-CTF) and FBS-CTF are not consistent. In Figure 2.22(b), FBS-CTF is again compared with VIE, where the two formulations agree well with each other. This proves that only FBS-CTF provides accurate results, while the other surface formulations break down when the contrast is 10^{-6}.

Figure 2.23 presents the results of scattering problems involving a $\lambda_o \times \lambda_o \times \lambda_o/10$ dielectric slab, where $\lambda_o \approx 1$ m is the wavelength outside the object (free space). The slab is located at the origin as depicted in the insets of the figure, and it is illuminated by a plane wave propagating in the $-z$ direction with the electric field polarized in the x direction. Three different relative permittivities, i.e., 1.1, $1.0 + 10^{-3}$, and $1.0 + 10^{-6}$, corresponding to 0.1, 10^{-3}, and 10^{-6} contrasts, respectively, are considered. The slab is discretized with $\lambda_o/20$ mesh size leading to matrix equations involving 11 424 unknowns. The scattering problems are solved by using FMM, where near-field interactions are calculated with 1% error and far-field interactions are computed with three digits of accuracy. The bistatic RCS is plotted in dBm^2 as a function of the observation angle on the z-x plane. RCS values obtained by using M-CTF and FBS-CTF are compared with those obtained by using EFVIE. As depicted in Figure 2.23(a), M-CTF and FBS-CTF are consistent with EFVIE when the contrast of the slab is relatively large, i.e., 0.1. As the contrast decreases to 10^{-3} and 10^{-6}, however, RCS values obtained with M-CTF become inconsistent with those values obtained with FBS-CTF and EFVIE. Similar to the previous examples, FBS-CTF is accurate and agrees well with the reference EFVIE for all contrasts.

Figure 2.24 presents the solution of scattering problems involving a dielectric ellipsoid in free space. Similar to the slab problems, three different relative permittivities, i.e., 1.1, $1.0 + 10^{-3}$, and $1.0 + 10^{-6}$, corresponding to 0.1, 10^{-3}, and 10^{-6}, contrasts are considered. The object is discretized with 6480 unknowns and the scattering problems are solved by using FMM, where near-field interactions are calculated with 1% error and far-field interactions are computed with three digits of accuracy. Figure 2.24 depicts the bistatic RCS (dBm^2) as a function of the observation angle on the z-x plane, when the ellipsoid is illuminated by a plane wave propagating in the $-z$ direction. It can be observed that both M-CTF and FBS-CTF are consistent with EFVIE when the contrast is relatively large. However, similar to the previous cases, M-CTF becomes inaccurate as the contrast decreases, whereas FBS-CTF is accurate and agrees well with the reference EFVIE, even when the contrast is 10^{-6}.

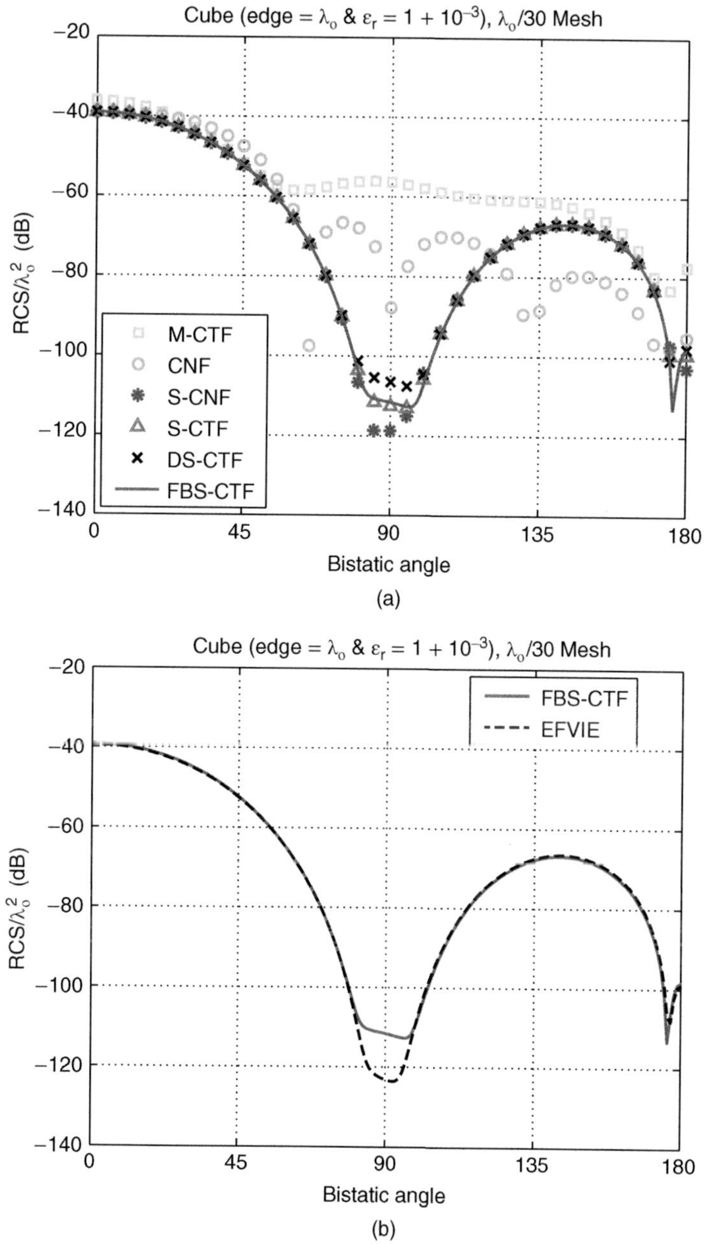

Figure 2.21 (a) The normalized bistatic RCS (RCS $/\lambda_o^2$) of a cube with edges of λ_o located in free space. The relative permittivity of the cube is $1.0 + 10^{-3}$. RCS values are obtained by using surface formulations when the mesh size is $\lambda_o/30$. (b) RCS values obtained with FBS-CTF and EFVIE agree with each other. *Source*: Ergül and Gürel 2008 [97]. Reproduced with permission of Elsevier.

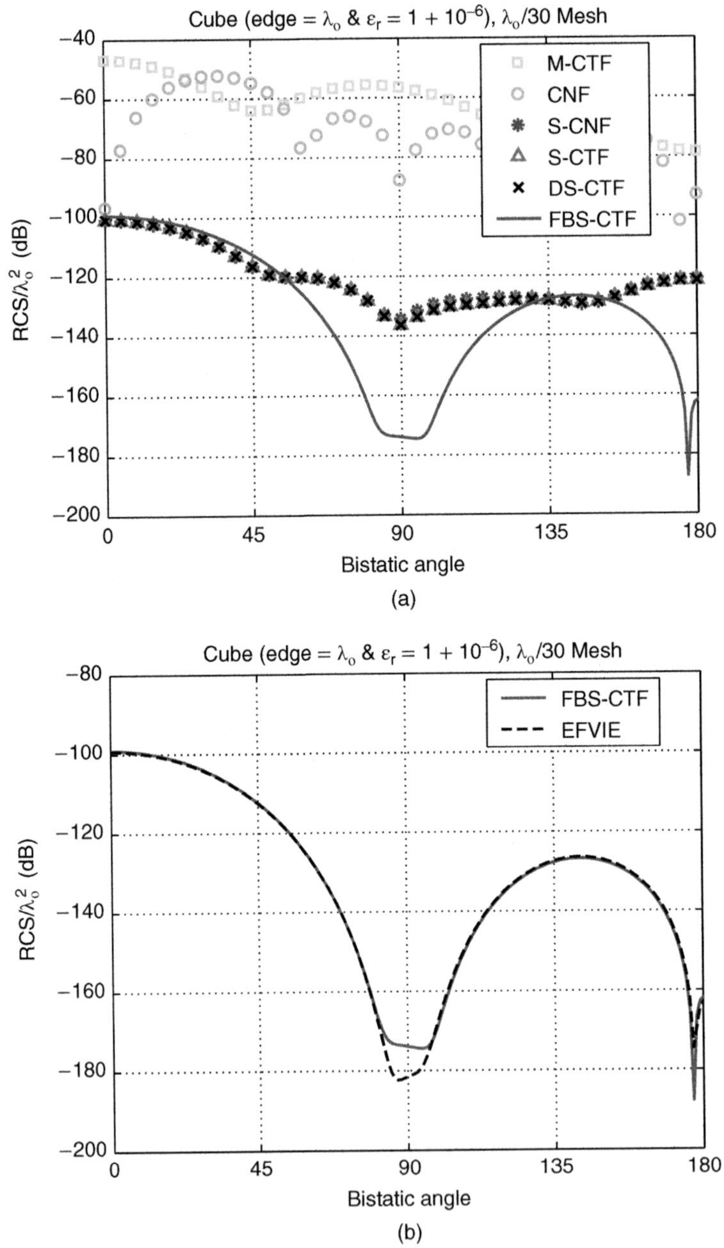

Figure 2.22 (a) The normalized bistatic RCS (RCS/λ_o^2) of a cube with edges of λ_o located in free space. The relative permittivity of the cube is $1.0 + 10^{-6}$. RCS values are obtained by using surface formulations when the mesh size is $\lambda_o/30$. (b) RCS values obtained with FBS-CTF and EFVIE agree with each other. *Source*: Ergül and Gürel 2008 [97]. Reproduced with permission of Elsevier.

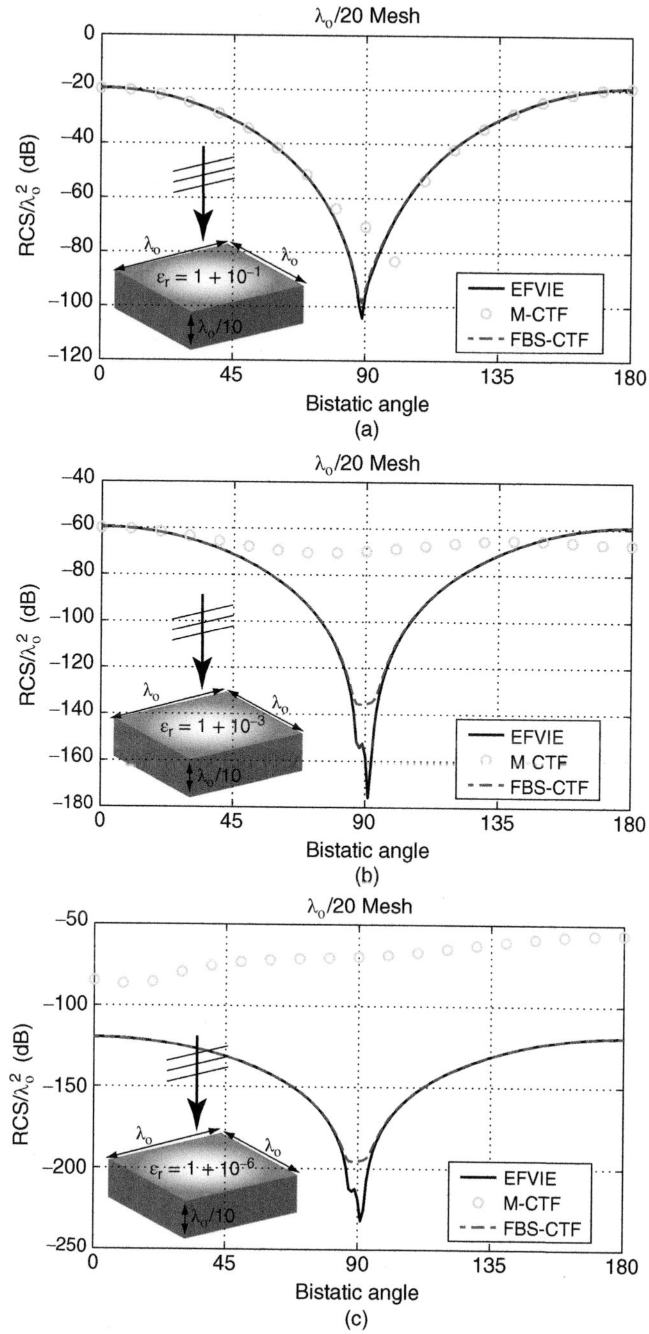

Figure 2.23 The bistatic RCS (dBm^2) of a $\lambda_o \times \lambda_o \times \lambda_o/10$ slab located in free space, when the relative permittivity of the slab is (a) 1.1, (b) $1.0 + 10^{-3}$, and (c) $1.0 + 10^{-6}$.

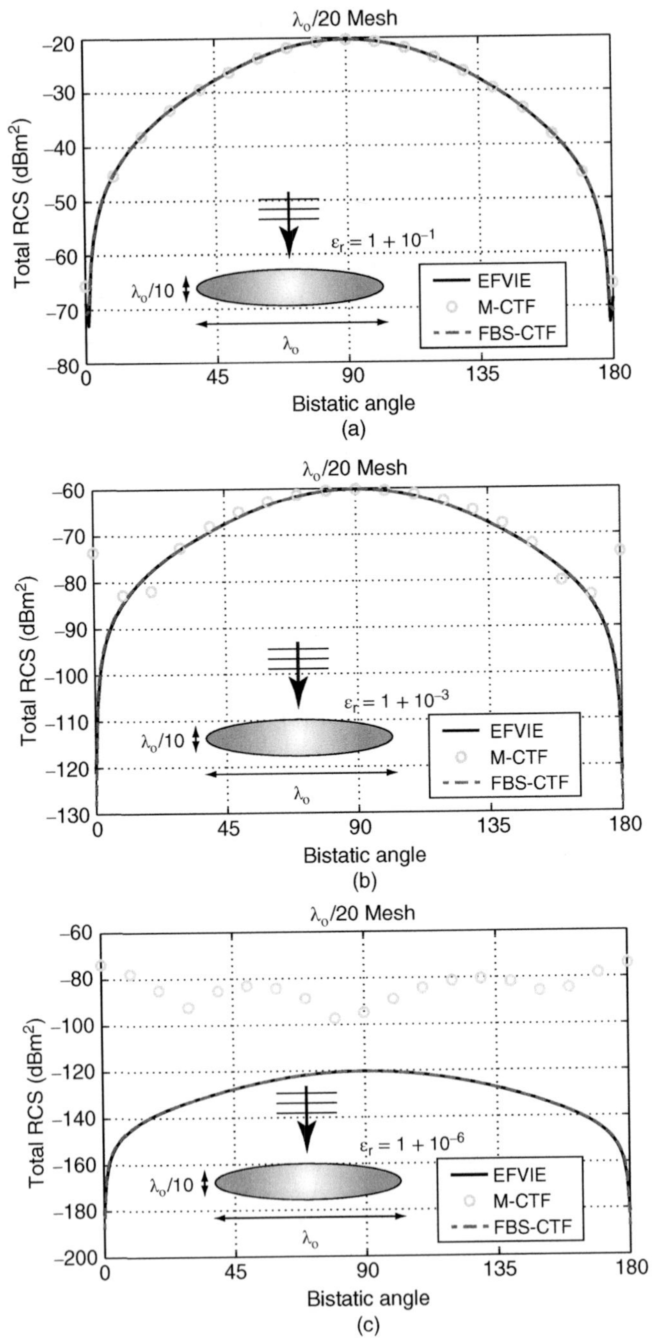

Figure 2.24 The bistatic RCS (dBm^2) of an ellipsoid located in free space, when the relative permittivity of the ellipsoid is (a) 1.1, (b) $1.0 + 10^{-3}$, and (c) $1.0 + 10^{-6}$.

2.3 Perfectly Conducting Objects

For PEC objects, the boundary condition for the tangential electric field can be used to derive EFIE as

$$-\hat{\boldsymbol{n}} \times \hat{\boldsymbol{n}} \times ik \int_S d\boldsymbol{r}' \left[\boldsymbol{J}(\boldsymbol{r}') + \frac{1}{k^2} \nabla' \cdot \boldsymbol{J}(\boldsymbol{r}') \nabla \right] g(\boldsymbol{r}, \boldsymbol{r}') = \hat{\boldsymbol{n}} \times \hat{\boldsymbol{n}} \times \eta^{-1} \boldsymbol{E}^{inc}(\boldsymbol{r}), \tag{2.212}$$

where the scattered or radiated electric field is expressed in terms of the induced (unknown) surface current $\boldsymbol{J}$. Similar to EFIE, MFIE is derived by using the boundary condition for the tangential magnetic field on the surface as

$$-\frac{\Omega_o}{4\pi} \boldsymbol{J}(\boldsymbol{r}) + \hat{\boldsymbol{n}} \times \int_{PV,S} d\boldsymbol{r}' \boldsymbol{J}(\boldsymbol{r}') \times \nabla' g(\boldsymbol{r}, \boldsymbol{r}') = -\hat{\boldsymbol{n}} \times \boldsymbol{H}^{inc}(\boldsymbol{r}), \tag{2.213}$$

where the observation point approaches the closed surface S from the outside. Finally, the conventional CFIE is a convex combination of EFIE and MFIE [55], i.e.,

$$\text{CFIE} = \alpha \text{EFIE} + (1 - \alpha)\text{MFIE}, \tag{2.214}$$

where α is a parameter between 0 and 1.

Discretizations of EFIE and MFIE lead to $N \times N$ matrix equations as

$$\overline{\boldsymbol{Z}}^{\text{EFIE}} \cdot \boldsymbol{a} = \boldsymbol{w}^{\text{EFIE}} \tag{2.215}$$

$$\overline{\boldsymbol{Z}}^{\text{MFIE}} \cdot \boldsymbol{a} = \boldsymbol{w}^{\text{MFIE}}. \tag{2.216}$$

In (2.215) and (2.216),

$$\overline{\boldsymbol{Z}}^{\text{EFIE}}[m,n] = \overline{\boldsymbol{T}}^{T}[m,n] \tag{2.217}$$

$$\overline{\boldsymbol{Z}}^{\text{MFIE}}[m,n] = -\frac{1}{2}\overline{\boldsymbol{I}}^{T}[m,n] + \overline{\boldsymbol{K}}_{PV}^{N}[m,n] \tag{2.218}$$

$$\boldsymbol{w}^{\text{EFIE}}[m] = -\eta^{-1}\boldsymbol{v}_E^T[m] \tag{2.219}$$

$$\boldsymbol{w}^{\text{MFIE}}[m] = -\boldsymbol{v}_H^N[m] \tag{2.220}$$

for $m, n = 1, 2, \ldots, N$, where $\overline{\boldsymbol{K}}_{PV}^{N}$, $\overline{\boldsymbol{T}}^{T}$, $\overline{\boldsymbol{I}}^{T}$, $\boldsymbol{v}_E^T$, and $\boldsymbol{v}_H^N$ are defined in (2.42), (2.43), (2.45), (2.64), and (2.67), respectively. Using the definition in (2.214), matrix equations obtained from CFIE can be written as

$$\overline{\boldsymbol{Z}}^{\text{CFIE}} \cdot \boldsymbol{a} = \boldsymbol{w}^{\text{CFIE}}, \tag{2.221}$$

where

$$\overline{\boldsymbol{Z}}^{\text{CFIE}}[m,n] = \alpha \overline{\boldsymbol{Z}}^{\text{EFIE}}[m,n] + (1-\alpha)\overline{\boldsymbol{Z}}^{\text{MFIE}}[m,n] \tag{2.222}$$

$$\boldsymbol{w}^{\text{CFIE}}[m] = \alpha \boldsymbol{w}^{\text{EFIE}}[m] + (1-\alpha)\boldsymbol{w}^{\text{MFIE}}[m]. \tag{2.223}$$

Hence, the matrix equations for CFIE can easily be obtained from the corresponding matrix equations for EFIE and MFIE.

2.3.1 Comments on the Integral Equations

In surface formulations of electromagnetics problems involving metallic objects, EFIE, MFIE, and CFIE can be employed either at our discretion or due to limitations imposed by the problem. Advantages and disadvantages of those formulations can be listed as follows:

- EFIE can be used to formulate both open and closed surfaces, as detailed later in this chapter, in Section 2.3.3. An open surface can be thought of as the limit case of squeezing a closed surface into zero thickness so that the open surface is formed by merging the two opposite surfaces of an infinitely thin geometry. Then, current distributions on the two sides of the thin structure becomes indistinguishable, and they are simply combined into a single equivalent current to be solved as the unknown of the problem [67], [110]. For open surfaces, EFIE in (2.212) is still applicable since the implied boundary condition remains valid in this limit case. Although the current term $\boldsymbol{J}$ in (2.212) and (2.213) are the same, MFIE is not applicable to open surfaces. Since it contains MFIE, CFIE in (2.214) is also not applicable to open surfaces.
- For closed objects, EFIE and MFIE suffer from internal resonances, as shown below in Section 2.3.2. In vicinities of discrete frequencies, matrix equations obtained from EFIE and MFIE have null-space solutions, i.e.,

$$\overline{Z}^{\mathrm{EFIE}} \cdot \boldsymbol{a} = 0 \quad \text{and} \quad \overline{Z}^{\mathrm{MFIE}} \cdot \boldsymbol{a} = 0, \qquad \boldsymbol{a} \neq 0. \tag{2.224}$$

  As a major advantage, CFIE is free of internal resonances experienced by EFIE and MFIE [55].
- EFIE is a tangential formulation and its discretization usually produces ill-conditioned matrix equations that are difficult to solve iteratively. Being a normal formulation, MFIE involves well-tested identity operators, which lead to well-conditioned matrix equations that are crucial for iterative solvers, such as MLFMA. Choosing a proper coupling parameter α, CFIE leads to better-conditioned matrix equations, compared to both EFIE and MFIE [119], as also shown in the examples in Figure 2.25. Scattering problems involving spheres of various radii, i.e., 0.6λ, 0.75λ, 1.0λ, and 1.5λ, are formulated with EFIE, MFIE, and CFIE using $\alpha = 0.2$. Discretizations with the RWG functions lead to matrix equations involving 1302–8364 unknowns. The problems are solved with CGS accelerated with FMM and BDP. Figure 2.25 shows that iterative solutions of CFIE are faster than those of both EFIE and MFIE, whereas EFIE solutions do not converge in reasonable number of iterations.
- EFIE suffers from the low-frequency breakdown problem [98], [99]. Matrix equations obtained from EFIE discretized with ordinary functions, such as the RWG functions, become increasingly ill-conditioned as the discretization is refined (see below, Section 2.3.4). Since it contains EFIE, CFIE is also affected by the low-frequency breakdown [100]. However, MFIE is stable unless the frequency is extremely low [102].
- MFIE is suitable for hybridization with PO method since both are based on the same magnetic-field boundary condition [65].
- Using a Galerkin scheme, EFIE produces symmetric matrix equations, whereas MFIE and CFIE do not.
- As detailed in Section 2.3.3, MFIE and CFIE involving well-tested identity operators are significantly inaccurate compared to EFIE, especially when they are discretized with low-order functions, such as the RWG functions.

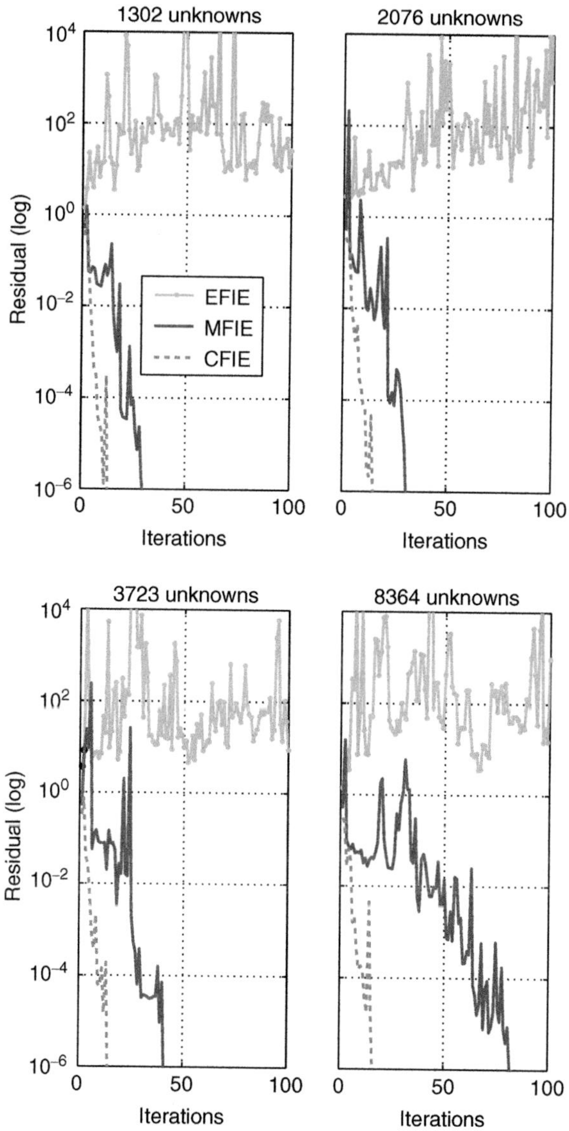

Figure 2.25 Solutions of scattering problems involving spheres of various radii (0.6λ, 0.75λ, λ, and 1.5λ) discretized with 1302–8364 RWG functions. Spheres are illuminated by a plane wave and problems are formulated with EFIE, MFIE, and CFIE ($\alpha = 0.2$). Iterative solutions are performed by CGS accelerated with FMM and BDP. *Source*: Gürel and Ergül 2003 [104]. Reproduced with permission of IEEE.

For closed objects, CFIE is usually preferred since it is free of internal resonances and it leads to better-conditioned matrix equations, compared to EFIE and MFIE. For open surfaces, however, EFIE becomes an inevitable choice. When preconditioned, EFIE can also be more suitable than CFIE for very accurate computations of the far-zone electromagnetic fields scattered from closed surfaces.

2.3.2 *Internal-Resonance Problem*

In order to demonstrate internal resonances encountered in surface formulations of PEC objects, Figure 2.26(a) presents solutions of scattering problems involving a sphere with a radius of 0.3 m. The sphere is illuminated by a plane wave and discretized with 3 cm triangles, on which the RWG functions are defined. Figure 2.26(a) depicts normalized backscattered RCS values (RCS $/\pi a^2$, where a is the radius of the sphere) obtained with EFIE, MFIE, and CFIE, in addition to analytical Mie-series solutions, when a changes from 0.5λ to 1.2λ. It can be observed that RCS values obtained with MFIE are significantly inaccurate in the vicinities of discrete frequencies due to the internal resonances. Although EFIE also suffers from internal-resonance problems, it produces accurate RCS results since the null-space solutions of EFIE do not radiate.

Figure 2.26(b) presents the number of CGS iterations to reduce the residual error to below 10^{-6} for EFIE, MFIE, and CFIE. Iterative solutions of EFIE are accelerated with NFP, whereas MFIE and CFIE solutions are accelerated with BDP. In Figure 2.26(b), peaks at resonance frequencies are observed for both EFIE and MFIE. However, CFIE is free of internal resonances and its iterative solution is stable in the entire frequency range.

2.3.3 *Formulations of Open Surfaces*

Consider a PEC open surface, which is obtained by merging two opposite surfaces of an infinitely thin geometry. Using the same discretization for the opposite surfaces, the electric current is expanded in a series of N basis functions, i.e.,

$$\boldsymbol{J}(\boldsymbol{r}) \approx \sum_{n=1}^{N} \boldsymbol{a}_+[n]\boldsymbol{b}_n(\boldsymbol{r}) + \sum_{n=1}^{N} \boldsymbol{a}_-[n]\boldsymbol{b}_n(\boldsymbol{r}), \tag{2.225}$$

where $\boldsymbol{a}_+[n] = \boldsymbol{a}[n]$ and $\boldsymbol{a}_-[n] = \boldsymbol{a}[n+N]$ represent coefficients (one for each side) of the nth basis function $\boldsymbol{b}_n$. Applying a Galerkin scheme, the same set of functions is employed as testing functions. Then, the discretization of EFIE leads to a $2N \times 2N$ matrix equation in the form of

$$\begin{bmatrix} \overline{\boldsymbol{T}}^T & \overline{\boldsymbol{T}}^T \\ \overline{\boldsymbol{T}}^T & \overline{\boldsymbol{T}}^T \end{bmatrix} \cdot \begin{bmatrix} \boldsymbol{a}_+ \\ \boldsymbol{a}_- \end{bmatrix} = -\eta^{-1} \begin{bmatrix} \boldsymbol{v}_E^T \\ \boldsymbol{v}_E^T \end{bmatrix}, \tag{2.226}$$

which can be reduced to

$$\begin{bmatrix} \overline{\boldsymbol{T}}^T & \overline{\boldsymbol{T}}^T \end{bmatrix} \cdot \begin{bmatrix} \boldsymbol{a}_+ \\ \boldsymbol{a}_- \end{bmatrix} = -\eta^{-1} \boldsymbol{v}_E^T. \tag{2.227}$$

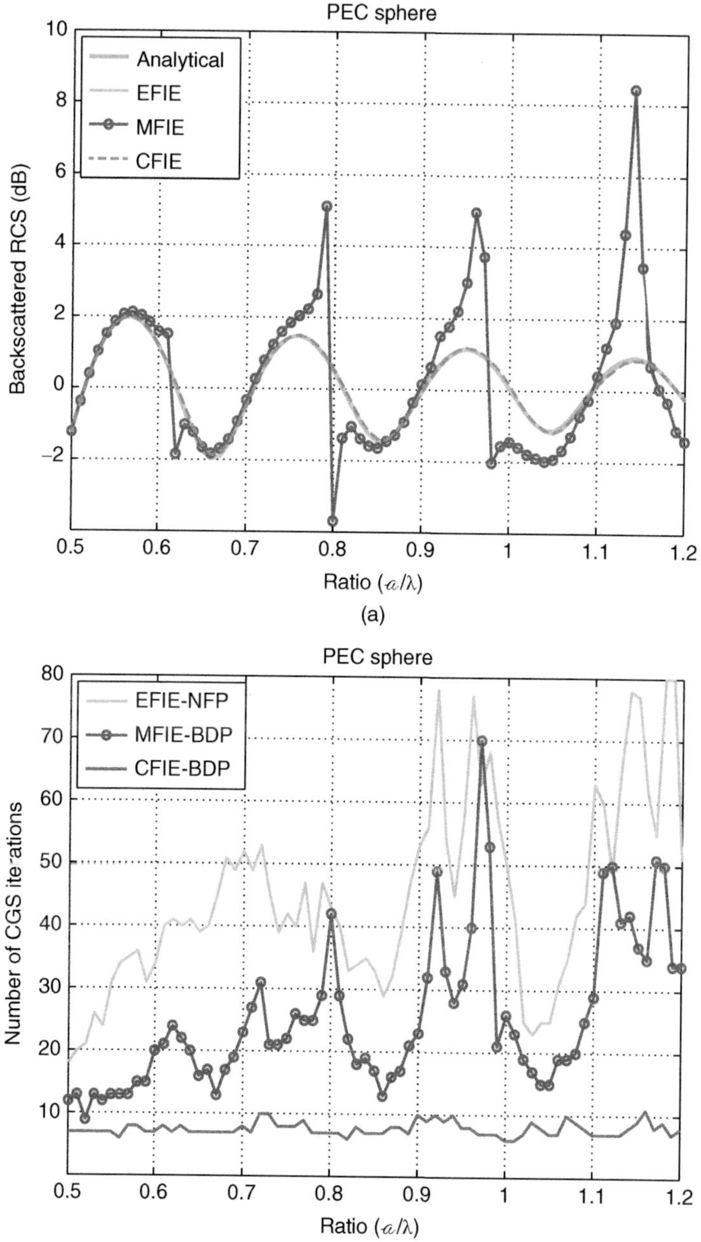

Figure 2.26 Solutions of scattering problems involving a conducting sphere with a radius of 0.3 m. (a) The normalized backscattered RCS (RCS $/\pi a^2$) and (b) the number of CGS iterations (10^{-6} residual error) when the radius of the sphere (a) changes from 0.5λ to 1.2λ.

Instead of solving the under-determined system in (2.231), one can combine the coefficients and obtain an $N \times N$ matrix equation, i.e.,

$$\overline{\boldsymbol{T}}^{T} \cdot (\boldsymbol{a}_{+} + \boldsymbol{a}_{-}) = -\eta^{-1} \boldsymbol{v}_{E}^{T}. \tag{2.228}$$

Then $\boldsymbol{a} = \boldsymbol{a}_{+} + \boldsymbol{a}_{-}$ can be used to compute electromagnetic fields everywhere. On the other hand, a similar discretization of MFIE for the same problem leads to

$$\begin{bmatrix} -0.5\overline{\boldsymbol{I}}^{T} + \overline{\boldsymbol{K}}_{PV}^{N} & 0.5\overline{\boldsymbol{I}}^{T} + \overline{\boldsymbol{K}}_{PV}^{N} \\ -0.5\overline{\boldsymbol{I}}^{T} + \overline{\boldsymbol{K}}_{PV}^{N} & 0.5\overline{\boldsymbol{I}}^{T} + \overline{\boldsymbol{K}}_{PV}^{N} \end{bmatrix} \cdot \begin{bmatrix} \boldsymbol{a}_{+} \\ \boldsymbol{a}_{-} \end{bmatrix} = - \begin{bmatrix} \boldsymbol{v}_{H}^{N} \\ \boldsymbol{v}_{H}^{N} \end{bmatrix}, \tag{2.229}$$

which can be reduced to

$$\begin{bmatrix} -0.5\overline{\boldsymbol{I}}^{T} + \overline{\boldsymbol{K}}_{PV}^{N} & 0.5\overline{\boldsymbol{I}}^{T} + \overline{\boldsymbol{K}}_{PV}^{N} \end{bmatrix} \cdot \begin{bmatrix} \boldsymbol{a}_{+} \\ \boldsymbol{a}_{-} \end{bmatrix} = -\boldsymbol{v}_{H}^{N} \tag{2.230}$$

or

$$\overline{\boldsymbol{K}}_{PV}^{N} \cdot (\boldsymbol{a}_{+} + \boldsymbol{a}_{-}) - 0.5\overline{\boldsymbol{I}}^{T} \cdot (\boldsymbol{a}_{+} - \boldsymbol{a}_{-}) = -\boldsymbol{v}_{H}^{N}. \tag{2.231}$$

Obviously the under-determined system obtained with MFIE in (2.231) cannot be solved to obtain $(\boldsymbol{a}_{+} + \boldsymbol{a}_{-})$.

Similar to MFIE, the conventional CFIE, which is obtained by the convex combination of EFIE and MFIE, is not applicable to open surfaces. Nevertheless, one can derive various combined formulations based on simultaneous solutions of EFIE and MFIE for open surfaces, i.e.,

$$\begin{bmatrix} \alpha_1 \overline{\boldsymbol{T}}^{T} & 0 \\ \alpha_2 \overline{\boldsymbol{K}}_{PV}^{N} & -0.5\alpha_2 \overline{\boldsymbol{I}}^{T} \end{bmatrix} \cdot \begin{bmatrix} (\boldsymbol{a}_{+} + \boldsymbol{a}_{-}) \\ (\boldsymbol{a}_{+} - \boldsymbol{a}_{-}) \end{bmatrix} = - \begin{bmatrix} \alpha_1 \eta^{-1} \boldsymbol{v}_{E}^{T} \\ \alpha_2 \boldsymbol{v}_{H}^{N} \end{bmatrix} \tag{2.232}$$

or

$$\begin{bmatrix} \alpha_1 \overline{\boldsymbol{T}}^{T} & \alpha_1 \overline{\boldsymbol{T}}^{T} \\ \alpha_2 \left(\overline{\boldsymbol{K}}_{PV}^{N} - 0.5\overline{\boldsymbol{I}}^{T} \right) & \alpha_2 \left(\overline{\boldsymbol{K}}_{PV}^{N} + 0.5\overline{\boldsymbol{I}}^{T} \right) \end{bmatrix} \cdot \begin{bmatrix} \boldsymbol{a}_{+} \\ \boldsymbol{a}_{-} \end{bmatrix} = - \begin{bmatrix} \alpha_1 \eta^{-1} \boldsymbol{v}_{E}^{T} \\ \alpha_2 \boldsymbol{v}_{H}^{N} \end{bmatrix}, \tag{2.233}$$

where α_1 and α_2 are combination parameters. However, using alternative combined formulations, such as (2.232) and (2.233), instead of EFIE may not provide any advantage in terms of iterative solutions.

As an example, solutions of a scattering problem involving a PEC half sphere are considered. As depicted in Figure 2.27, the object is illuminated by a plane wave propagating in the $-z$ direction with the electric field polarized in the x direction (Excitation 1) at 2 GHz. Discretization of the object with $\lambda/10$ triangles involves 9991 RWG functions. The problem is formulated with EFIE and with the combined formulation in (2.232) using various values for $\alpha = \alpha_1 = (1 - \alpha_2)$. All solutions are performed iteratively with MLFMA. Table 2.2 lists the number of CGS iterations to reach 10^{-6} residual error. Using the combined formulation instead of EFIE does not accelerate iterative solutions. In fact, EFIE is almost the best formulation in Table 2.2.

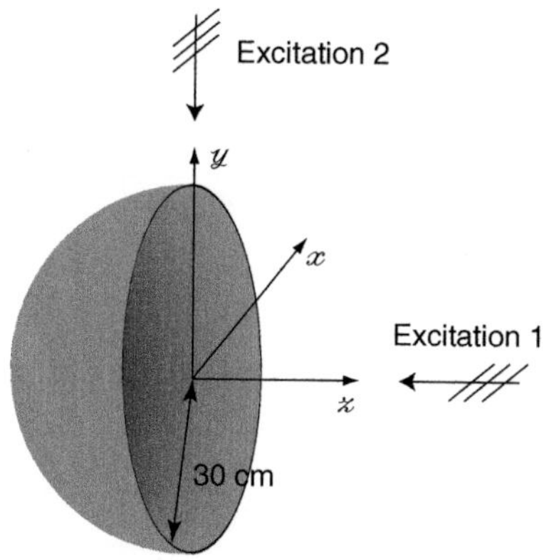

Figure 2.27 Scattering problems involving a half sphere.

Table 2.2 Number of CGS iterations (10^{-6} residual error) for the solution of a scattering problem involving a half sphere depicted in Figure 2.27 at 2 GHz

	Combined formulation (α)								
EFIE	0.1	0.2	0.3	0.4	0.5	0.6	0.7	0.8	0.9
399	549	409	438	413	396	478	461	479	604

2.3.4 Low-Frequency Breakdown

In general, the electric field due to an arbitrary electric current $\boldsymbol{J}$ involves two parts, namely, the vector-potential part

$$\boldsymbol{E}_V(\boldsymbol{r}) = i\omega\mu \int_S d\boldsymbol{r}' \boldsymbol{J}(\boldsymbol{r}') g(\boldsymbol{r}, \boldsymbol{r}') \tag{2.234}$$

and the scalar-potential part

$$\boldsymbol{E}_S(\boldsymbol{r}) = \frac{i}{\omega\epsilon} \int_S d\boldsymbol{r}' \nabla' \cdot \boldsymbol{J}(\boldsymbol{r}') \nabla g(\boldsymbol{r}, \boldsymbol{r}'), \tag{2.235}$$

where $\boldsymbol{E}(\boldsymbol{r}) = \boldsymbol{E}_V(\boldsymbol{r}) + \boldsymbol{E}_S(\boldsymbol{r})$. As the frequency decreases, the scalar-potential part dominates the vector-potential part. In fact, using finite-precision arithmetics, the vector-potential part can be lost during numerical solutions. On the other hand, (2.235) is not a one-to-one map between the electric current and electric field. Specifically, for a given $\boldsymbol{E}_S$, there are infinitely many $\boldsymbol{J}$. This is because, for a nonzero frequency, $\boldsymbol{J}$ may have a solenoidal part $\boldsymbol{J}^{sol}$ such that

$$\nabla' \cdot \boldsymbol{J}^{sol}(\boldsymbol{r}') = 0, \tag{2.236}$$

and this solenoidal part contributes to the electric field only through the vector potential in (2.234). This explains why EFIE breaks down at low frequencies.

A major problem is that the low-frequency breakdown of EFIE can be encountered at an arbitrary frequency as the discretization is refined. Any matrix equation obtained with EFIE can be decomposed as

$$\overline{\boldsymbol{Z}}_V^{\text{EFIE}} \cdot \boldsymbol{a} + \overline{\boldsymbol{Z}}_S^{\text{EFIE}} \cdot \boldsymbol{a} = \boldsymbol{w}^{\text{EFIE}}, \tag{2.237}$$

where

$$\overline{\boldsymbol{Z}}_V^{\text{EFIE}}[m,n] = ik \int_{S_m} d\boldsymbol{r} \boldsymbol{t}_m(\boldsymbol{r}) \cdot \int_{S_n} d\boldsymbol{r}' g(\boldsymbol{r},\boldsymbol{r}') \boldsymbol{b}_n(\boldsymbol{r}') \tag{2.238}$$

$$\overline{\boldsymbol{Z}}_S^{\text{EFIE}}[m,n] = \frac{i}{k} \int_{S_m} d\boldsymbol{r} \boldsymbol{t}_m(\boldsymbol{r}) \cdot \int_{S_n} d\boldsymbol{r}' \nabla g(\boldsymbol{r},\boldsymbol{r}') \nabla' \cdot \boldsymbol{b}_n(\boldsymbol{r}') \tag{2.239}$$

for $m,n = 1,2,\ldots,N$. It can be shown that $\overline{\boldsymbol{Z}}_S^{\text{EFIE}}$ dominates $\overline{\boldsymbol{Z}}_V^{\text{EFIE}}$ when the mesh size decreases. On the other hand, using ordinary basis and testing functions, such as the RWG functions, $\overline{\boldsymbol{Z}}_S^{\text{EFIE}}$ is a rank-deficient matrix. Hence, the overall matrix equation becomes increasingly ill-conditioned as the mesh size is reduced.

A rigorous solution of the low-frequency breakdown of EFIE is using loop-star or loop-tree decomposition methods [103]–[107]. In this section, a simple approach based on separate discretizations of the current and charge distributions is presented for stable solutions of EFIE. Similar formulations are extensively discussed in [108] and [109].

Consider an alternative discretization of EFIE, where the electric current is expanded in a series of the RWG functions and the electric charge density is expanded separately in a series of pulse functions on triangles, i.e.,

$$\frac{i\rho_e(\boldsymbol{r})}{\sqrt{\mu\epsilon}} = \frac{\nabla' \cdot \boldsymbol{J}(\boldsymbol{r}')}{k} \approx \sum_{n=1}^{N_\triangle} \boldsymbol{a}_\rho[n] b_n^{\text{PLS}}(\boldsymbol{r}), \tag{2.240}$$

where $N_\triangle$ is the number of triangles,

$$b_n^{\text{PLS}}(\boldsymbol{r}) = \begin{cases} 1, & \boldsymbol{r} \in S_n^\triangle \\ 0, & \boldsymbol{r} \notin S_n^\triangle \end{cases}, \tag{2.241}$$

and $S_n^\triangle$ is the surface of the nth triangle. Testing EFIE with the RWG functions, an $N \times (N + N_\triangle)$ matrix equation is obtained as

$$\begin{bmatrix} \overline{\boldsymbol{Z}}_V^{\text{EFIE-RWG}} & \overline{\boldsymbol{Z}}_S^{\text{EFIE-R/P}} \end{bmatrix} \cdot \begin{bmatrix} \boldsymbol{a} \\ \boldsymbol{a}_\rho \end{bmatrix} = \frac{i}{k\eta} \boldsymbol{v}_E^T, \tag{2.242}$$

where

$$\overline{\boldsymbol{Z}}_V^{\text{EFIE-RWG}}[m,n] = \int_{S_m} d\boldsymbol{r} \boldsymbol{t}_m^{\text{RWG}}(\boldsymbol{r}) \cdot \int_{S_n} d\boldsymbol{r}' g(\boldsymbol{r},\boldsymbol{r}') \boldsymbol{b}_n^{\text{RWG}}(\boldsymbol{r}') \tag{2.243}$$

$$\overline{\boldsymbol{Z}}_S^{\text{EFIE-R/P}}[m,n'] = \int_{S_m} d\boldsymbol{r} \frac{\nabla \cdot \boldsymbol{t}_m^{\text{RWG}}(\boldsymbol{r})}{k} \int_{S_{n'}^\triangle} d\boldsymbol{r}' g(\boldsymbol{r},\boldsymbol{r}') b_{n'}^{\text{PLS}}(\boldsymbol{r}') \tag{2.244}$$

for $m, n = 1, 2, \ldots, N$ and $n' = 1, 2, \ldots, N_{\triangle}$. To solve the under-determined system in (2.242), consider the Lorentz gauge

$$\nabla \cdot \int_S d\boldsymbol{r}' g(\boldsymbol{r}, \boldsymbol{r}') \boldsymbol{J}(\boldsymbol{r}') = i\omega \int_S d\boldsymbol{r}' g(\boldsymbol{r}, \boldsymbol{r}') \rho_e(\boldsymbol{r}'), \tag{2.245}$$

which describes the relation between the vector and scalar potentials. Discretization of (2.245) leads to an $N_{\triangle} \times (N + N_{\triangle})$ matrix equation

$$\begin{bmatrix} \overline{\boldsymbol{Z}}_V^{\text{EFIE-P/R}} & \overline{\boldsymbol{Z}}_S^{\text{EFIE-PLS}} \end{bmatrix} \cdot \begin{bmatrix} \boldsymbol{a} \\ \boldsymbol{a}_\rho \end{bmatrix} = 0, \tag{2.246}$$

where

$$\overline{\boldsymbol{Z}}_V^{\text{EFIE-P/R}} = \int_{S_{m'}^{\triangle}} d\boldsymbol{r} t_{m'}^{\text{PLS}}(\boldsymbol{r}) \int_{S_n} d\boldsymbol{r}' g(\boldsymbol{r}, \boldsymbol{r}') \frac{\nabla' \cdot \boldsymbol{b}_n^{\text{RWG}}(\boldsymbol{r}')}{k} \tag{2.247}$$

$$\overline{\boldsymbol{Z}}_S^{\text{EFIE-PLS}} = \int_{S_{m'}^{\triangle}} d\boldsymbol{r} t_{m'}^{\text{PLS}}(\boldsymbol{r}) \int_{S_{n'}^{\triangle}} d\boldsymbol{r}' g(\boldsymbol{r}, \boldsymbol{r}') b_{n'}^{\text{PLS}}(\boldsymbol{r}') \tag{2.248}$$

for $m', n' = 1, 2, \ldots, N_{\triangle}$ and $n = 1, 2, \ldots, N$. Finally, (2.242) and (2.246) can be combined to obtain an $(N + N_{\triangle}) \times (N + N_{\triangle})$ matrix equation as

$$\begin{bmatrix} \overline{\boldsymbol{Z}}_V^{\text{EFIE-RWG}} & \overline{\boldsymbol{Z}}_S^{\text{EFIE-R/P}} \\ \overline{\boldsymbol{Z}}_V^{\text{EFIE-P/R}} & \overline{\boldsymbol{Z}}_S^{\text{EFIE-PLS}} \end{bmatrix} \cdot \begin{bmatrix} \boldsymbol{a} \\ \boldsymbol{a}_\rho \end{bmatrix} = \frac{i}{k\eta} \begin{bmatrix} \boldsymbol{v}_E^T \\ 0 \end{bmatrix}. \tag{2.249}$$

The regularized EFIE (REFIE) in (2.249) does not completely eliminate the low-frequency breakdown. However, it is more stable than the conventional EFIE with respect to the mesh size.

As an example, Figure 2.28 presents solutions of scattering problems involving a square patch with edges of 30 cm discretized with 1.5 cm triangles. The patch is located on the x-y plane and illuminated by a plane wave propagating in the $-z$ direction. The frequency is in the range from 7.8125 MHz to 2000 MHz. Figure 2.28(a) depicts the 1-norm condition number for the matrices obtained from EFIE and REFIE with respect to the number of triangles per wavelength. The condition number grows as the mesh size decreases for both EFIE and REFIE. However, REFIE provides better-conditioned matrix equations, especially for dense discretizations. Figure 2.28(b) presents the number of iterations (for 10^{-6} residual error) for the solution of the patch problems. Using BiCGStab for EFIE, convergence cannot be achieved within 2000 iterations when the mesh size is smaller than $\lambda/100$. However, the number of iterations is less than 400 for all problems formulated with REFIE. Figure 2.28(b) also shows that the CGS algorithm is quite successful to solve ill-conditioned matrix equations obtained from EFIE, and the large gap between the iteration counts for EFIE and REFIE disappears when using CGS. In general, the number of iterations in an iterative solution does not only depend on the condition number of the matrix. This is confirmed in this example, where the number of CGS iterations for EFIE and REFIE are close to each other, even though REFIE produces better conditioned matrix equations than EFIE.

Recently, there has been many efforts to eliminate the low-frequency breakdown of EFIE using Calderon identities [101], [110]–[116]. These identities can be derived in various ways.

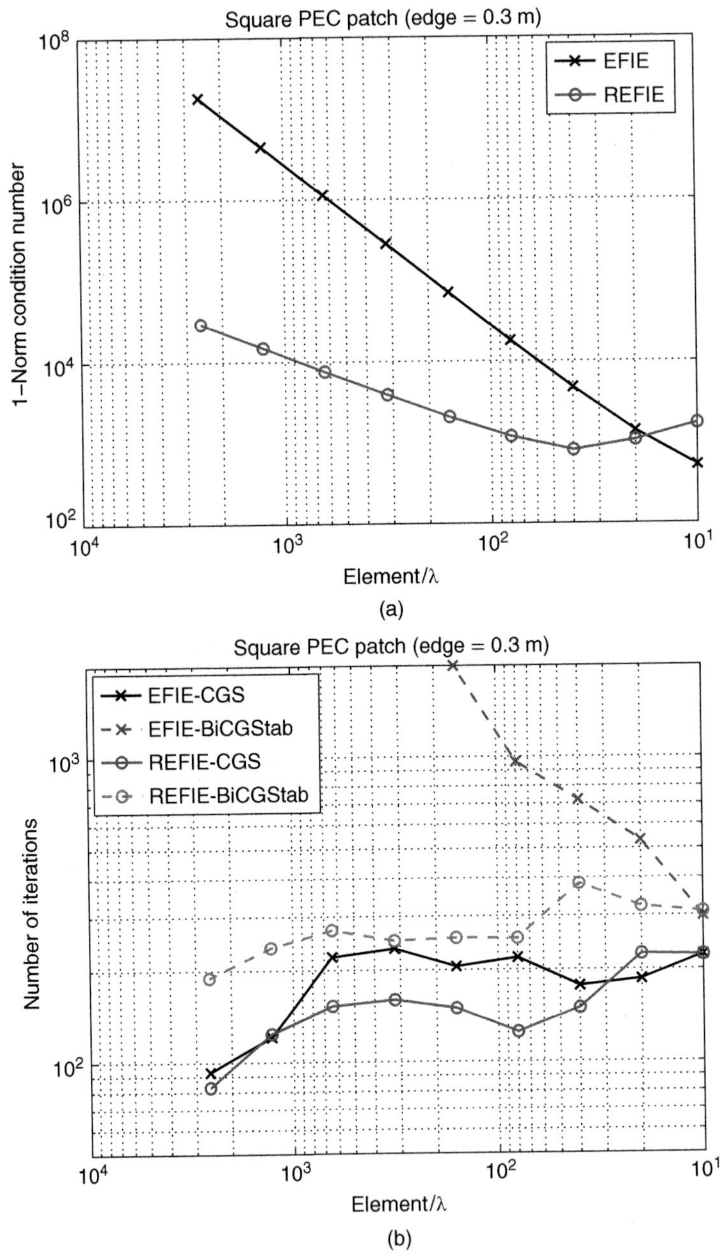

Figure 2.28 Solutions of scattering problems involving a square patch discretized with 1.5 cm triangles. (a) The condition number (1-norm) and (b) iteration counts (10^{-6} residual error) with respect to the number of triangles per wavelength.

For example, consider (2.159) showing the nonradiating property of the incident electromagnetic fields on an arbitrary closed surface S. The rotational form of (2.159) can be rewritten as

$$\begin{bmatrix} \hat{\boldsymbol{n}} \times \mathcal{K}_{PV} & \eta^{-1}\hat{\boldsymbol{n}} \times \mathcal{T} \\ -\eta\hat{\boldsymbol{n}} \times \mathcal{T} & \hat{\boldsymbol{n}} \times \mathcal{K}_{PV} \end{bmatrix} \cdot \begin{bmatrix} \hat{\boldsymbol{n}} \times \boldsymbol{H}^{inc} \\ -\hat{\boldsymbol{n}} \times \boldsymbol{E}^{inc} \end{bmatrix} (\boldsymbol{r}) = -\frac{\Omega_i}{4\pi} \begin{bmatrix} \hat{\boldsymbol{n}} \times \boldsymbol{H}^{inc} \\ -\hat{\boldsymbol{n}} \times \boldsymbol{E}^{inc} \end{bmatrix} (\boldsymbol{r}) \tag{2.250}$$

for $\boldsymbol{r} \in S$, which is valid for arbitrary $\boldsymbol{E}^{inc}$ and $\boldsymbol{H}^{inc}$ satisfying the Maxwell's equations. Multiplying (2.250) with itself and rearranging the terms, one can obtain

$$\left[\left(\hat{\boldsymbol{n}} \times \mathcal{K}_{PV}\right)^2 - (\hat{\boldsymbol{n}} \times \mathcal{T})^2\right](\boldsymbol{r}) = \frac{\Omega_i^2}{16\pi^2}\mathcal{I}(\boldsymbol{r}) \tag{2.251}$$

$$\left[\left(\hat{\boldsymbol{n}} \times \mathcal{K}_{PV}\right)(\hat{\boldsymbol{n}} \times \mathcal{T})\right](\boldsymbol{r}) = -\left[(\hat{\boldsymbol{n}} \times \mathcal{T})\left(\hat{\boldsymbol{n}} \times \mathcal{K}_{PV}\right)\right](\boldsymbol{r}) \tag{2.252}$$

for $r \in S$. The equation (2.251) shows that the $\mathcal{T}$ operator stabilizes itself. Specifically, for smooth surfaces, $\Omega_i(\boldsymbol{r}) = 2\pi$ and

$$(\hat{\boldsymbol{n}} \times \mathcal{T})^2(\boldsymbol{r}) = \frac{1}{4}\mathcal{I}(\boldsymbol{r}) - \left(\hat{\boldsymbol{n}} \times \mathcal{K}_{PV}\right)^2(\boldsymbol{r}). \tag{2.253}$$

The self-stabilization property of the $\mathcal{T}$ operator is valid for closed surfaces, but interestingly, it can be applied locally for open surfaces. For stable solutions, however, (2.253) should be discretized via a mixed scheme involving ordinary and dual basis functions, as detailed in [112].

2.3.5 Accuracy with the RWG Functions

Normal and mixed formulations, such as MFIE and CFIE, provide inaccurate results, compared to tangential formulations, such as EFIE. The discrepancy between results obtained with different formulations is evident especially when they are discretized with low-order basis functions, such as the RWG functions. Although the same physical current distribution is expanded by an identical set of the RWG functions, the same level of accuracy cannot be obtained with EFIE, MFIE, and CFIE. Independent of the shape and size of the geometry, computational results obtained with MFIE and CFIE employing the RWG functions are plagued with a persistent inaccuracy problem.

In order to demonstrate the inaccuracy of MFIE and CFIE compared to EFIE discretized with the RWG functions, Figures 2.29 and 2.30 present solutions of scattering problems involving a PEC sphere with a radius of a. The sphere is illuminated by a plane wave at two different frequencies corresponding to $a = \lambda$ and $a = 2\lambda$. In addition to analytical Mie-series solutions, the problems are solved with EFIE, MFIE, and CFIE ($\alpha = 0.2$). For this purpose, the surface of the sphere is discretized with various triangulations of coarse to fine densities, corresponding to approximately 1000 to 30 000 unknowns. Figures 2.29 and 2.30 depict the backscattered and forward-scattered RCS, respectively, obtained with various mesh sizes. The dots in the plots correspond to the $\lambda/10$ triangulation density at each frequency. Clearly, EFIE results readily converge to the reference analytical results for mesh densities even smaller than $\lambda/10$, whereas MFIE results manifest significant discrepancies with the analytical results not only for $\lambda/10$ triangulation, but also for much higher mesh densities. CFIE solutions in Figures 2.29 and 2.30

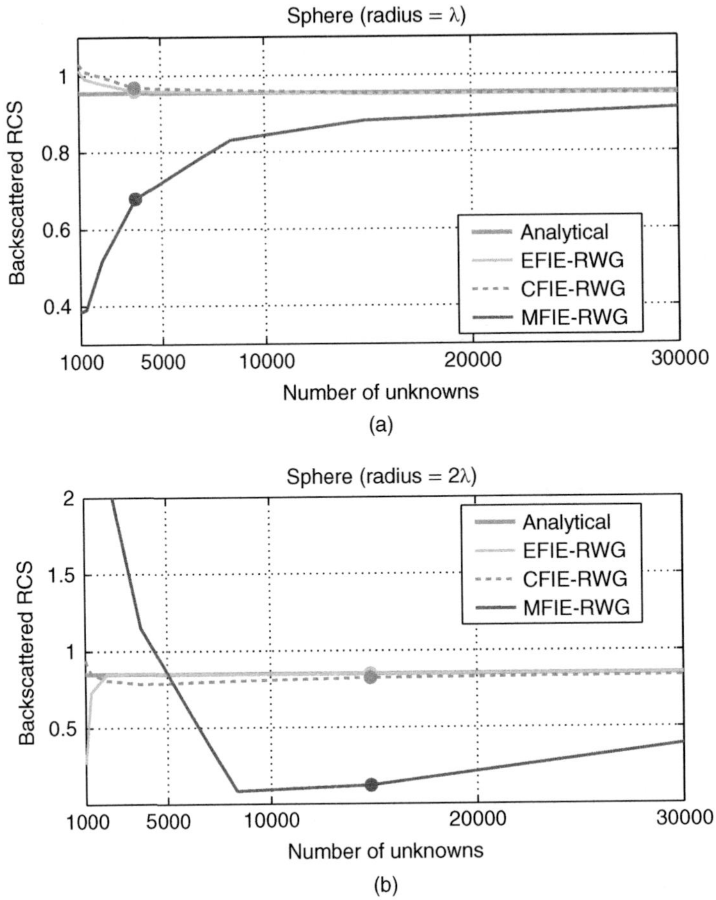

Figure 2.29 The normalized backscattered RCS (RCS $/\pi a^2$) of a PEC sphere with a radius of (a) λ and (b) 2λ as a function of the number of unknowns. The dots on the curves correspond to the $\lambda/10$ discretization. *Source*: Ergül and Gürel 2006 [127]. Reproduced with permission of IEEE.

are obtained by combining EFIE and MFIE with $\alpha = 0.2$. Even though backscattered RCS values presented in Figure 2.29 do not display a significant inaccuracy for the CFIE solutions, it should be emphasized that the backscattered and forward-scattered RCS in Figures 2.29 and 2.30 are obtained from the same CFIE solutions. Therefore, based on the comparison of Mie series, EFIE, and CFIE solutions presented in Figure 2.30, one can conclude that CFIE solutions bear inaccuracies caused by MFIE.

Figure 2.31 presents solutions of scattering problems involving a PEC sphere with a radius of a, when a changes from 0.5λ to 1.5λ. The sphere is discretized with the RWG functions on 2 cm triangles. Relative errors in bistatic RCS values obtained with the EFIE, MFIE, and CFIE ($\alpha = 0.2$) formulations (with respect to the analytical Mie-series solutions) are investigated in two-dimensional plots as functions of the bistatic angle and frequency. In the plots, 0° and 180° correspond to the backscattering and forward-scattering directions, respectively. The relative

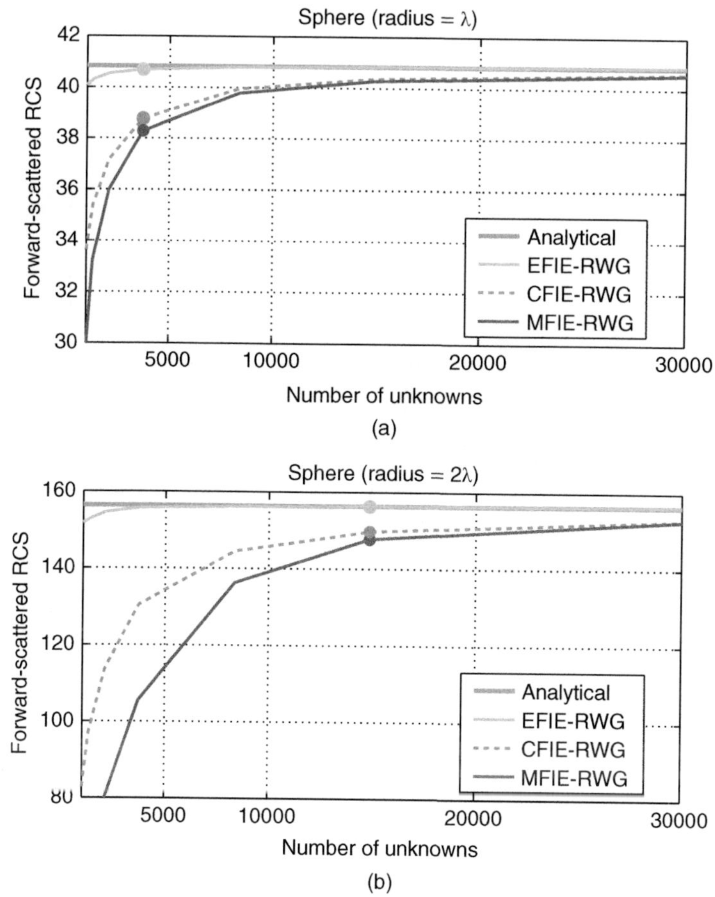

Figure 2.30 The normalized forward-scattered RCS (RCS $/\pi a^2$) of a PEC sphere with a radius of (a) λ and (b) 2λ as a function of the number of unknowns. The dots on the curves correspond to the $\lambda/10$ discretization. *Source*: Ergül and Gürel 2006 [127]. Reproduced with permission of IEEE.

error in the forward-scattering direction as a function of the frequency is investigated separately in Figure 2.31. In the two-dimensional plot, the relative error of EFIE is rarely above 10^{-2}, which occurs due to zero-crossings of the bistatic RCS. In the forward-scattering direction, the relative error of the EFIE solution is below 10^{-3}. On the other hand, the MFIE and CFIE results obtained by the same discretization of the geometry have considerably larger errors. The excessive error exists both for MFIE and CFIE, but MFIE also suffers from the internal-resonance problems, since the null-space solutions of MFIE can radiate to far zone. Those resonances of MFIE appear in the last plot as peaks above a base formed by the CFIE curve.

Next, scattering problems involving a PEC cube with edges of 1 m at 300 MHz and 600 MHz are considered. The cube is located at the origin and it is illuminated by a plane wave propagating in the $-x$ direction with the electric field polarized in the y direction. Various discretizations with 20 cm to 3 cm triangles lead to matrix equations involving 500 to 18 000 unknowns. Figures 2.32 and 2.33 show the magnitude of the y component of the normalized surface

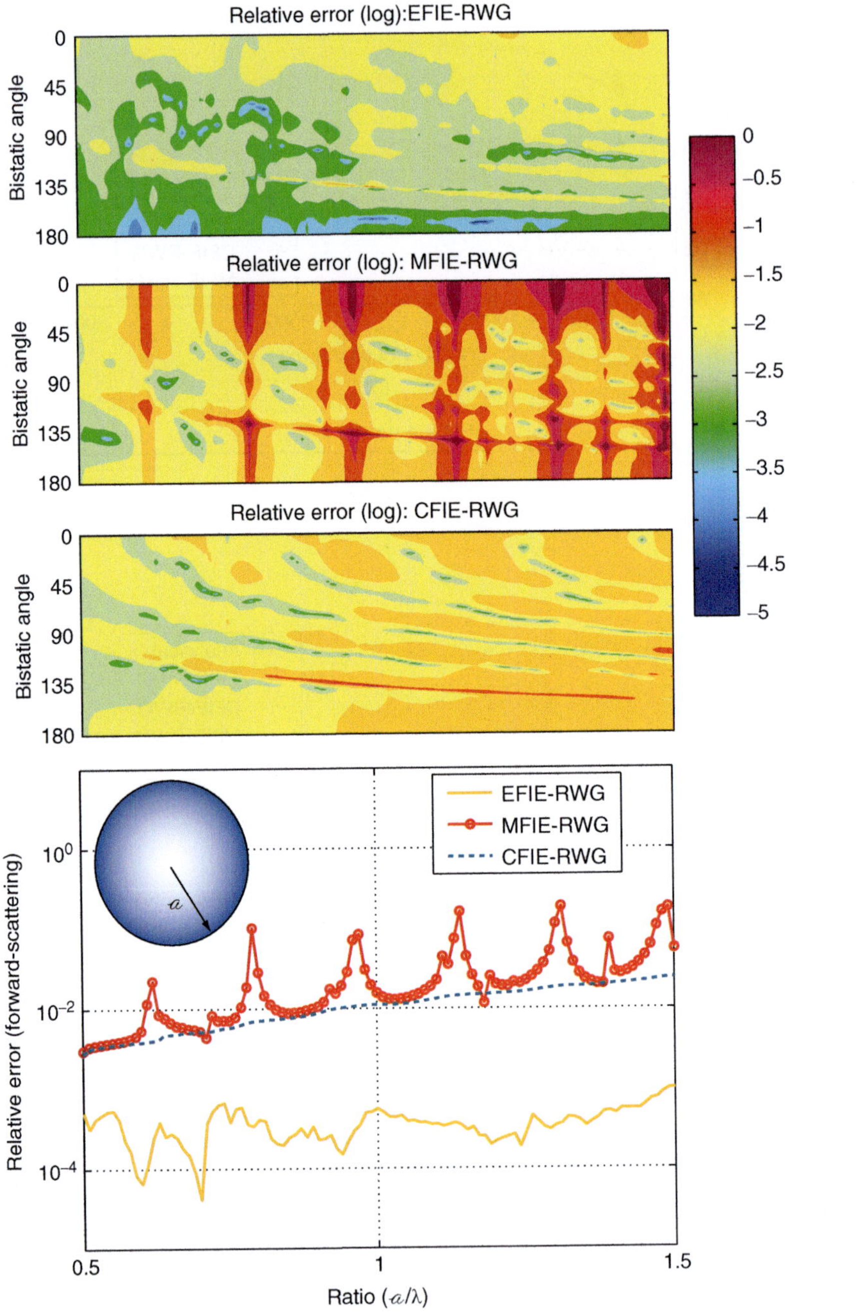

Figure 2.31 The relative error in the bistatic RCS of a PEC sphere with a radius of a, when a changes from 0.5λ to 1.5λ.

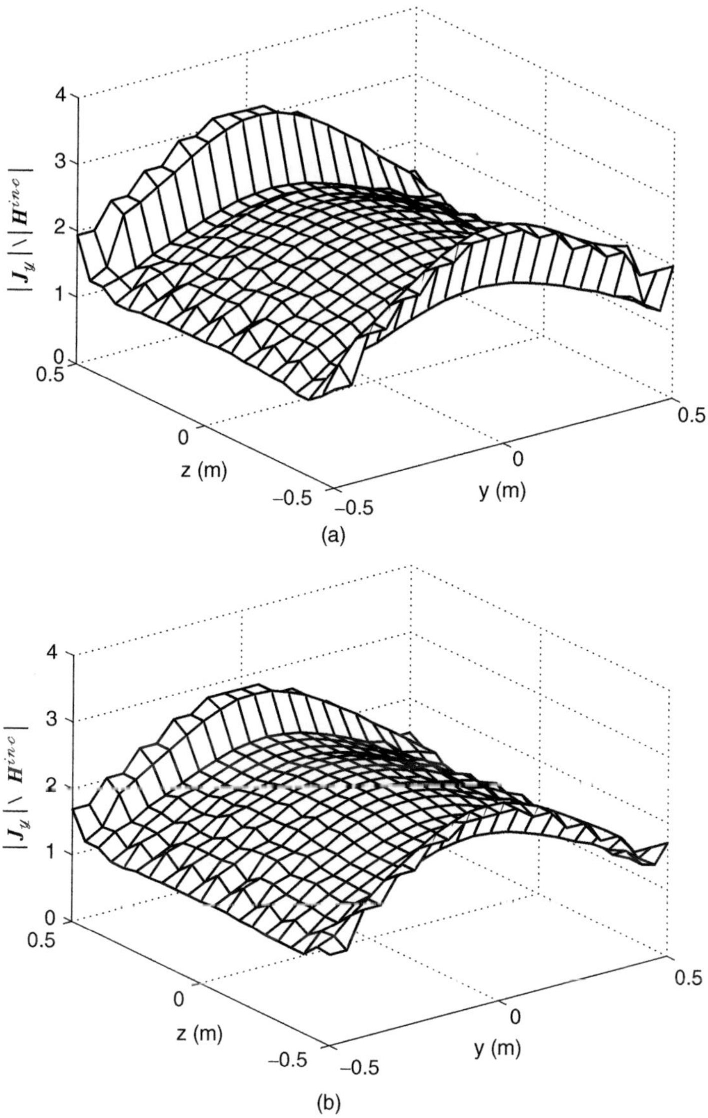

Figure 2.32 The magnitude of the y component of the normalized surface current $\left(|\boldsymbol{J}_y|/|\boldsymbol{H}^{inc}|\right)$ induced on the front surface (at $x = 0.5$ m) of a PEC cube with edges of λ, obtained by using (a) EFIE with RWG and (b) MFIE with RWG. *Source*: Ergül and Gürel 2006 [127]. Reproduced with permission of IEEE.

current $\left(|\boldsymbol{J}_y|/|\boldsymbol{H}^{inc}|\right)$ induced on the front surface of the cube (at $x = 0.5$ m). In Figure 2.32, where $d = \lambda$ and the triangulation size is about $\lambda/10$, and in Figure 2.33, where $d = 2\lambda$ and the triangulation size is about $\lambda/12.5$, one can observe that the modeling of the surface current is different for the EFIE and MFIE implementations using the RWG functions. Even though plots look similar, significant differences can be noticed by focusing on the calculated values at $z = \pm 0.5$ m, where the induced current is expected to be singular. The combination

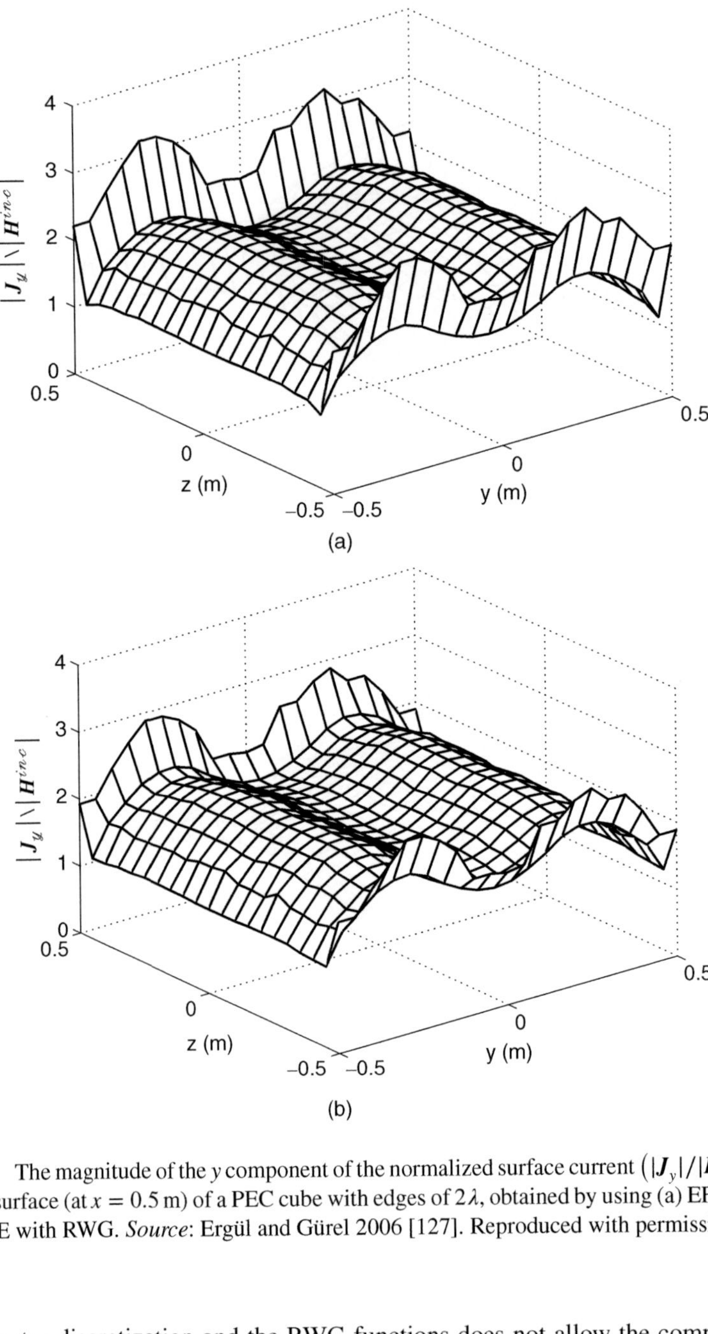

Figure 2.33 The magnitude of the y component of the normalized surface current $\left(|\boldsymbol{J}_y|/|\boldsymbol{H}^{inc}|\right)$ induced on the front surface (at $x = 0.5$ m) of a PEC cube with edges of 2λ, obtained by using (a) EFIE with RWG and (b) MFIE with RWG. *Source*: Ergül and Gürel 2006 [127]. Reproduced with permission of IEEE.

of the geometry discretization and the RWG functions does not allow the computed current values to become singular at those two edges of the cube. Since the physically correct solution is singular, the more accurate computed solutions get, the higher values they would assume at those edges in order to match the singularity as closely as possible. Comparing Figure 2.32(a) to Figure 2.32(b) and Figure 2.33(a) to Figure 2.33(b), EFIE solutions are seen

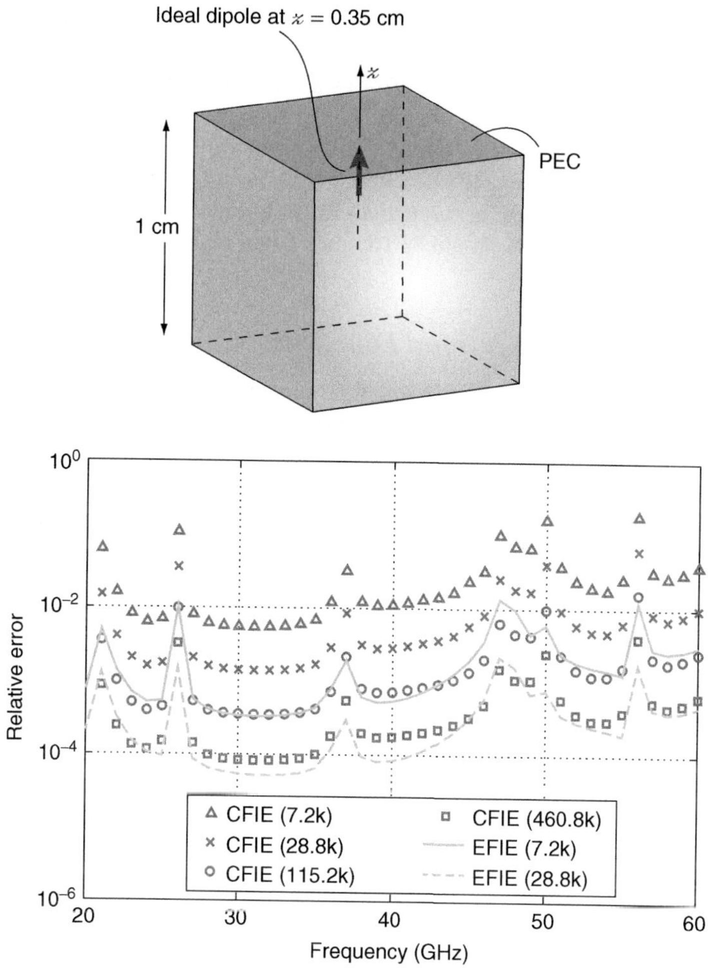

Figure 2.34 Solutions of a radiation problem involving a 1 cm × 1 cm × 1 cm PEC box excited by a Hertzian dipole located inside the box at $z = 0.35$ cm. The relative error is plotted as a function of the frequency from 20 GHz to 60 GHz.

to have consistently higher values than MFIE solutions in spite of employing exactly the same discretization and the RWG functions. Since the only difference is in the integral-equation formulation, this comparison demonstrates the inaccuracy of MFIE relative to EFIE.

To further demonstrate the inaccuracy of CFIE compared to EFIE, Figure 2.34 presents the results of a radiation problem involving a 1 cm × 1 cm × 1 cm PEC box located at the origin. As depicted in Figure 2.34, the box is excited with a Hertzian dipole oriented in the z direction and located inside the box at $z = 0.35$ cm. Ideally, the radiated field outside the box should be zero due to the shielding provided by the closed PEC surface. The radiated field is calculated in the far zone on the x-y plane at $(r = 3\text{ m}, \theta = \pi/2, \phi_n)$, where $\phi_n = (n-1)\pi/180$ for $n = 1, 2, \ldots, 360$. The relative error is defined as the 2-norm of the total electric field divided by the

2-norm of the incident electric field. The total electric field is obtained by adding the incident field due to the Hertzian dipole and the secondary field due to the induced electric current on the cube. Figure 2.34 presents the relative error as a function of the frequency from 20 GHz to 60 GHz. In this range of frequency, the size of the box varies from approximately 0.67λ to approximately 2λ. The radiation problem is discretized with 7200, 28 800, 115 200, and 460 800 unknowns, and solved by a low-frequency MLFMA (LF-MLFMA). We observe that CFIE ($\alpha = 0.2$) is significantly less accurate than EFIE. In order to obtain the same accuracy, the number of unknowns of CFIE should be 16 times larger than that of EFIE.

2.3.6 *Compatibility of the Integral Equations*

Due to the excessive error in MFIE, matrix equations obtained with EFIE and MFIE are incompatible, i.e., a solution vector calculated by using MFIE does not satisfy EFIE with the desired level of accuracy, and vice versa. In fact, for a given problem, coefficient vectors obtained with EFIE and MFIE can be significantly different, i.e.,

$$\left(\overline{\boldsymbol{Z}}^{\mathrm{EFIE}}\right)^{-1} \cdot \boldsymbol{w}^{\mathrm{EFIE}} = \boldsymbol{a}^{\mathrm{EFIE}} \neq \boldsymbol{a}^{\mathrm{MFIE}} = \left(\overline{\boldsymbol{Z}}^{\mathrm{MFIE}}\right)^{-1} \cdot \boldsymbol{w}^{\mathrm{MFIE}}, \tag{2.254}$$

even when the EFIE and MFIE solutions are not contaminated with the internal resonances. Being a convex combination of EFIE and MFIE, CFIE is contaminated with the MFIE inaccuracy. Therefore, coefficients obtained with CFIE are also incompatible with the EFIE and MFIE systems.

Since EFIE is more accurate than MFIE, one can write

$$\boldsymbol{a}^{\mathrm{MFIE}} = \boldsymbol{a}^{\mathrm{EFIE}} + \Delta\boldsymbol{a}^{ME}, \tag{2.255}$$

where the discrepancy between the coefficients obtained with EFIE and MFIE is interpreted as the error in MFIE. The coefficient vector obtained with CFIE is related to those obtained with EFIE and MFIE as

$$\boldsymbol{a}^{\mathrm{CFIE}} = \alpha\left(\overline{\boldsymbol{Z}}^{\mathrm{CFIE}}\right)^{-1} \cdot \overline{\boldsymbol{Z}}^{\mathrm{EFIE}} \cdot \boldsymbol{a}^{\mathrm{EFIE}} + (1-\alpha)\left(\overline{\boldsymbol{Z}}^{\mathrm{CFIE}}\right)^{-1} \cdot \overline{\boldsymbol{Z}}^{\mathrm{MFIE}} \cdot \boldsymbol{a}^{\mathrm{MFIE}}. \tag{2.256}$$

Using (2.255) in (2.256), one can arrive at

$$\boldsymbol{a}^{\mathrm{CFIE}} = \boldsymbol{a}^{\mathrm{EFIE}} + \Delta\boldsymbol{a}^{CE}, \tag{2.257}$$

where

$$\Delta\boldsymbol{a}^{CE} = (1-\alpha)\left(\overline{\boldsymbol{Z}}^{\mathrm{CFIE}}\right)^{-1} \cdot \overline{\boldsymbol{Z}}^{\mathrm{MFIE}} \cdot \Delta\boldsymbol{a}^{ME}, \tag{2.258}$$

which can be interpreted as the CFIE solution being contaminated with the inaccuracy of MFIE.

When the solution of CFIE is obtained by using an iterative algorithm, the residual vector

$$\boldsymbol{r}^{\mathrm{CFIE}} = \boldsymbol{w}^{\mathrm{CFIE}} - \overline{\boldsymbol{Z}}^{\mathrm{CFIE}} \cdot \boldsymbol{a}^{\mathrm{CFIE}} \tag{2.259}$$

is minimized. Using the definition of CFIE in (2.214), the residual vector for CFIE can be written as

$$\boldsymbol{r}^{\mathrm{CFIE}} = \alpha \boldsymbol{r}^{C\to E} + (1-\alpha)\boldsymbol{r}^{C\to M}, \tag{2.260}$$

where

$$\boldsymbol{r}^{C\to E} = \boldsymbol{w}^{\mathrm{EFIE}} - \overline{\boldsymbol{Z}}^{\mathrm{EFIE}} \cdot \boldsymbol{a}^{\mathrm{CFIE}} \tag{2.261}$$

$$\boldsymbol{r}^{C\to M} = \boldsymbol{w}^{\mathrm{MFIE}} - \overline{\boldsymbol{Z}}^{\mathrm{MFIE}} \cdot \boldsymbol{a}^{\mathrm{CFIE}} \tag{2.262}$$

are residual vectors obtained by testing the CFIE solution in the EFIE and MFIE systems, respectively.

An important observation is that the minimization of the norm of $\boldsymbol{r}^{\mathrm{CFIE}}$ does not require the minimization of the norms of $\boldsymbol{r}^{C\to E}$ and $\boldsymbol{r}^{C\to M}$. As an example, consider solutions of radiation and scattering problems involving a $\lambda \times \lambda \times \lambda$ conducting box, when the problems are formulated with CFIE. In the radiation problem, the box is excited with a Hertzian dipole located inside the box at $x = 0.1\lambda$. In the scattering problem, the box is illuminated by a plane wave propagating in the $-x$ direction with the electric field polarized in the y direction. Discretizations with the RWG functions on $\lambda/10$ triangles lead to 2052×2052 matrix equations. Matrix elements are calculated with a maximum of 1% error and iterative solutions are performed by using the BiCGStab algorithm. Iterations are stopped when the residual error

$$\Delta_{res}^{\mathrm{CFIE}} = \frac{\|\boldsymbol{r}^{\mathrm{CFIE}}\|_2}{\|\boldsymbol{w}^{\mathrm{CFIE}}\|_2} \tag{2.263}$$

is reduced to below 10^{-6}.

Figure 2.35 presents the first 30 elements of the residual vectors $\boldsymbol{r}^{\mathrm{CFIE}}$, $\boldsymbol{r}^{C\to E}$, and $\boldsymbol{r}^{C\to M}$, denoted by 'CFIE', 'CFIE in EFIE', and 'CFIE in MFIE', respectively, when $\alpha = 0.2$ in the CFIE solution of the radiation problem involving a $\lambda \times \lambda \times \lambda$ conducting box. It can be observed that $\boldsymbol{r}^{C\to E}$ and $\boldsymbol{r}^{C\to M}$ involve significantly larger elements than $\boldsymbol{r}^{\mathrm{CFIE}}$. Furthermore, the elements of $\boldsymbol{r}^{C\to E}$ are exactly four times those of $\boldsymbol{r}^{C\to M}$. This is because, as the norm of $\boldsymbol{r}^{\mathrm{CFIE}}$ is minimized, the elements of $\boldsymbol{r}^{C\to E}$ and $\boldsymbol{r}^{C\to M}$ are scaled with respect to each other, i.e.,

$$\boldsymbol{r}^{C\to E} \approx -\frac{(1-\alpha)}{\alpha}\boldsymbol{r}^{C\to M}, \tag{2.264}$$

instead of converging to zero. For $\alpha = 0.2$, the ratio of $-(1-\alpha)/\alpha = -4$ is manifested in Figure 2.35. More precisely, the error criteria for the iterative solution, i.e.,

$$\Delta_{res}^{\mathrm{CFIE}} = \frac{\|\boldsymbol{r}^{\mathrm{CFIE}}\|_2}{\|\boldsymbol{w}^{\mathrm{CFIE}}\|_2} = \frac{\|\alpha \boldsymbol{r}^{C\to E} + (1-\alpha)\boldsymbol{r}^{C\to M}\|_2}{\|\boldsymbol{w}^{\mathrm{CFIE}}\|_2} \leq 10^{-6}, \tag{2.265}$$

is achieved without the minimization of $\|\boldsymbol{r}^{C\to E}\|_2$ and $\|\boldsymbol{r}^{C\to M}\|_2$.

Figure 2.36 presents the first 30 elements of the residual vectors $\boldsymbol{r}^{\mathrm{CFIE}}$, $\boldsymbol{r}^{C\to E}$, and $\boldsymbol{r}^{C\to M}$, when $\alpha = 0.5$ in the CFIE solution of the scattering problem involving a $\lambda \times \lambda \times \lambda$ conducting box. Similar to the previous example, $\boldsymbol{r}^{C\to E}$ and $\boldsymbol{r}^{C\to M}$, denoted by 'CFIE in EFIE' and 'CFIE in MFIE', respectively, involve significantly larger elements than $\boldsymbol{r}^{\mathrm{CFIE}}$, denoted by 'CFIE'. As opposed to the previous solution, however, $\boldsymbol{r}^{C\to E} = -\boldsymbol{r}^{C\to M}$ since $-(1-\alpha)/\alpha = -1$. Depending on the value of α, the compatibility of the CFIE solution with the EFIE and MFIE systems

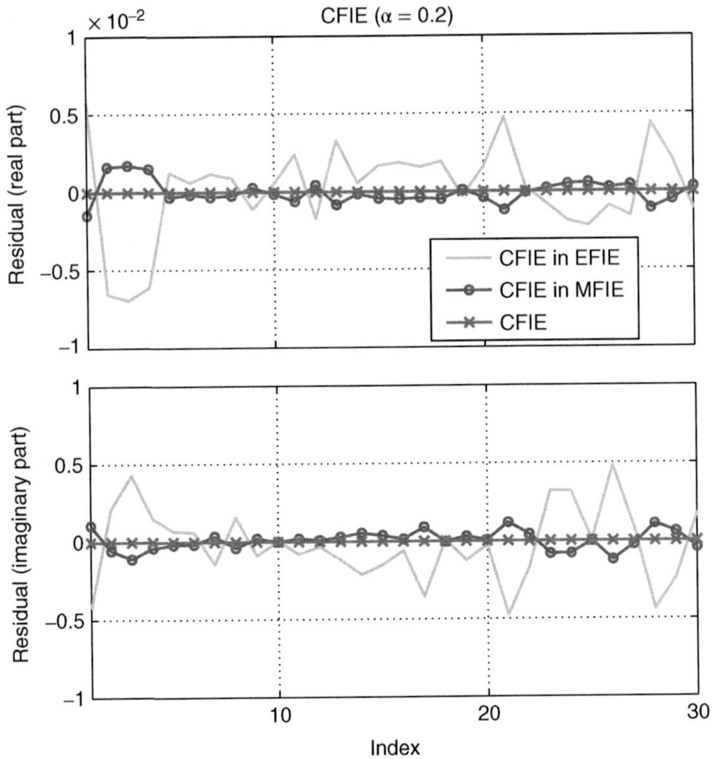

Figure 2.35 The first 30 elements of the residual vectors $\boldsymbol{r}^{\text{CFIE}}$ (CFIE), $\boldsymbol{r}^{C\rightarrow E}$ (CFIE in EFIE), and $\boldsymbol{r}^{C\rightarrow M}$ (CFIE in MFIE) for the iterative CFIE ($\alpha = 0.2$) solution of a radiation problem involving a $\lambda \times \lambda \times \lambda$ conducting box. *Source*: Gürel and Ergül 2009 [122]. Reproduced with permission of IEEE.

varies, but the CFIE solution cannot satisfy both systems at the same time with the desired level of accuracy.

2.3.7 Convergence to Minimum Achievable Error

In an iterative solution of a matrix equation obtained from the discretization of an integral-equation formulation, the accuracy of the solution is saturated and the solution error cannot be reduced any further after a number of iterations. This is because there are various error sources, such as numerical calculations of matrix elements, and the overall error cannot be minimized by only reducing the residual error. Then, in an iterative solution, a further reduction of the residual error becomes meaningless after the minimum solution error is achieved.

A CFIE solution can be significantly incompatible with the corresponding EFIE and MFIE systems, even when the residual error is small. In fact, an iterative solution of CFIE involves a breakpoint, where a further reduction of the residual error does not improve the solution in terms of compatibility with EFIE and MFIE. Since EFIE is more accurate than CFIE, this

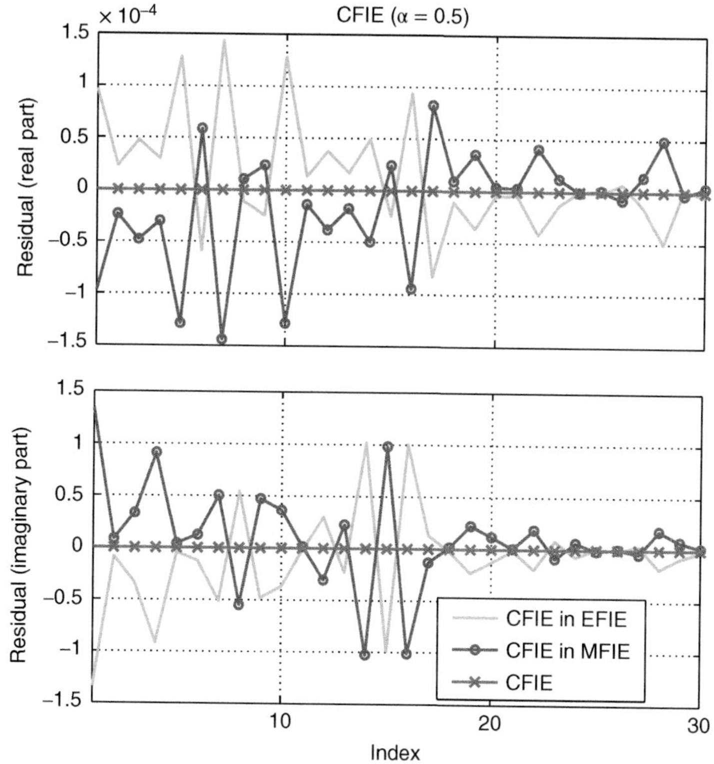

Figure 2.36 The first 30 elements of the residual vectors $\boldsymbol{r}^{\mathrm{CFIE}}$ (CFIE), $\boldsymbol{r}^{C\rightarrow E}$ (CFIE in EFIE), and $\boldsymbol{r}^{C\rightarrow M}$ (CFIE in MFIE) for the iterative CFIE ($\alpha = 0.5$) solution of a scattering problem involving a $\lambda \times \lambda \times \lambda$ conducting box. *Source*: Gürel and Ergül 2009 [122]. Reproduced with permission of IEEE.

breakpoint, where the compatibility of the CFIE solution with the EFIE system is maximized, corresponds to the last useful iteration to obtain the highest possible accuracy with CFIE. It should be noted that the minimum achievable error in an EFIE solution depends on various error sources, and not only on the reduction of the residual error. In the case of CFIE, however, the excessive discretization error in the MFIE part can become a major error source such that the minimum achievable error becomes closely related to the compatibility of the CFIE solution with the EFIE system.

As an example, Figure 2.37 presents iterative solutions of a radiation problem involving a $\lambda \times \lambda \times \lambda$ conducting box excited with a Hertzian dipole located inside the box at $x = 0.1\lambda$. The radiation problem is formulated with CFIE when $\alpha = 0.2$, 0.5, and 0.8. The residual error in (2.263) is plotted with respect to BiCGStab iterations, in addition to

$$\Delta_{res}^{C\rightarrow E} = \frac{\|\boldsymbol{r}^{C\rightarrow E}\|_2}{\|\boldsymbol{w}^{\mathrm{EFIE}}\|_2} \quad \text{and} \quad \Delta_{res}^{C\rightarrow M} = \frac{\|\boldsymbol{r}^{C\rightarrow M}\|_2}{\|\boldsymbol{w}^{\mathrm{MFIE}}\|_2} \tag{2.266}$$

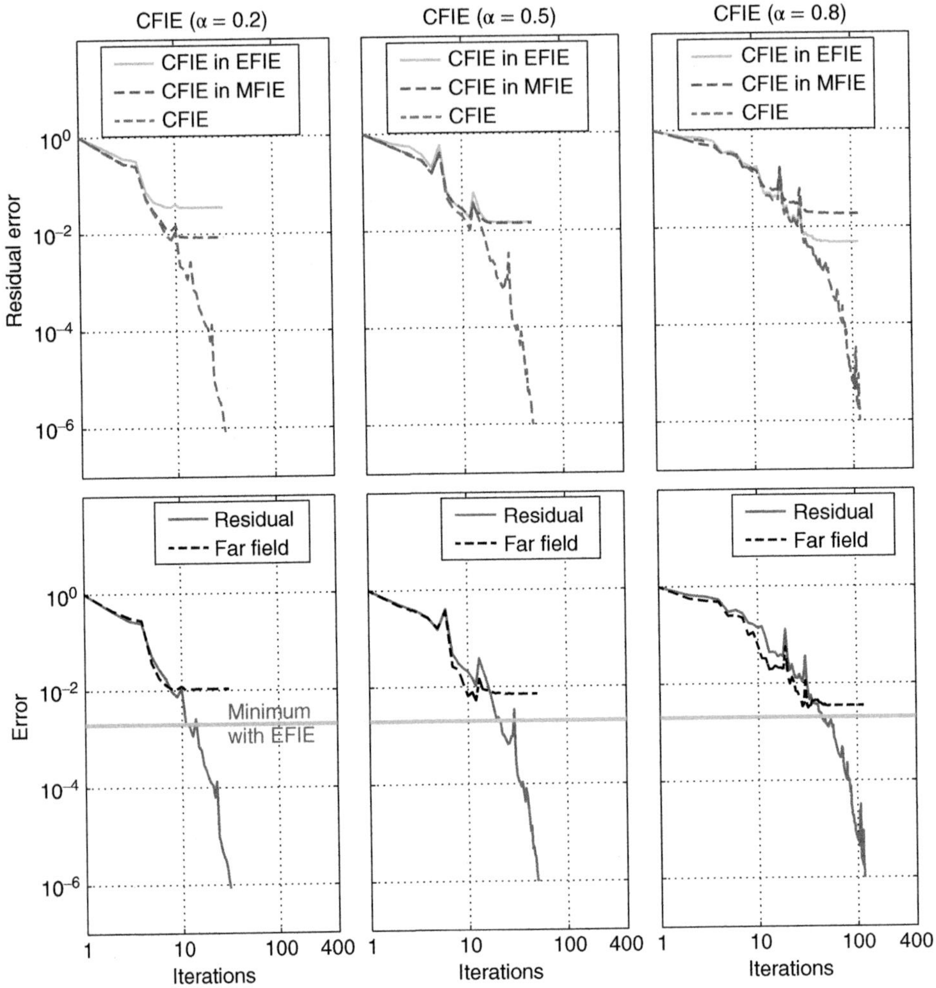

Figure 2.37 Iterative solutions of a radiation problem involving a $\lambda \times \lambda \times \lambda$ conducting box excited with a Hertzian dipole located inside the box at $x = 0.1\lambda$. The residual and far-field errors are plotted with respect to BiCGStab iterations for CFIE when $\alpha = 0.2$, 0.5, and 0.8. *Source*: Gürel and Ergül 2009 [122]. Reproduced with permission of IEEE.

denoted by 'CFIE in EFIE' and 'CFIE in MFIE', respectively, in the first row of Figure 2.37. The following observations can be made:

- The number of iterations is 31, 50, and 117 when $\alpha = 0.2$, 0.5, and 0.8, respectively.
- Compatibility of the CFIE solution with the EFIE and MFIE systems is saturated at about the 10th, 20th, and 50th iterations, respectively. The values of $\Delta_{res}^{C\rightarrow E}$ and $\Delta_{res}^{C\rightarrow M}$, which are obtained by testing the coefficient vector in the EFIE and MFIE systems, respectively, become steady after these breakpoints, and they do not drop further as iterations continue.
- The maximum compatibility of the CFIE solution with the EFIE or MFIE system depends on the value of α.

For the radiation problem, one can also calculate the error in the far-zone electric field at each iteration, in addition to the residual error. The incident and total electric fields are sampled in the far-zone on the x-y plane at $(r = \infty, \theta = \pi/2, \phi_n)$, where $\phi_n = (n-1)\pi/180$ for $n = 1, 2, \ldots, 360$. The total electric field is obtained by combining the incident field radiated by the Hertzian dipole and the secondary field due to the induced electric current on the surface of the cube. Coefficients for the discretized induced current are provided by the iterative algorithm at each iteration. Theoretically, the total electric field should be zero outside the cube, because the source is shielded by a PEC closed surface. Then, the far-field error is defined as

$$\Delta_{FF} = \frac{\|\boldsymbol{E}_{tot}^{\infty}\|_2}{\|\boldsymbol{E}_{inc}^{\infty}\|_2}, \tag{2.267}$$

where $\boldsymbol{E}_{tot}^{\infty}$ and $\boldsymbol{E}_{inc}^{\infty}$ are arrays of 360 elements containing the total and incident electric fields in the far zone. In the second row of Figure 2.37, the far-field error (2.267) in the CFIE solutions is plotted with respect to iterations. The minimum achievable error provided by EFIE is also indicated in each plot. The following observations can be made:

- The accuracy of the CFIE solutions is saturated and cannot be improved any further after some numbers of iterations.
- The last useful iteration and the minimum achievable accuracy in a CFIE solution depends on the value of α.
- For each value of α, the last useful iteration in terms of the solution accuracy corresponds to the breakpoint, where the compatibility of the CFIE solution with the EFIE system is saturated.

Numerical results show that the compatibility of the CFIE solution with the EFIE system is an important indicator in determining the last useful iteration for the highest possible accuracy with CFIE.

Figure 2.38 presents iterative solutions of a scattering problem involving a $\lambda \times \lambda \times \lambda$ conducting box illuminated by a plane wave propagating in the $-x$ direction with the electric field polarized in the y direction. Similar to the previous example, the residual error is plotted as a function of the BiCGStab iterations for CFIE with $\alpha = 0.2$, 0.5, and 0.8. In the first row of Figure 2.38, $\Delta_{res}^{C\to E}$ and $\Delta_{res}^{C\to M}$ are again plotted to demonstrate the compatibility of the CFIE solutions with the EFIE and MFIE systems. For this problem, the saturation occurs at the 7th, 10th, and 30th iterations, respectively, when $\alpha = 0.2$, 0.5, and 0.8. During the solutions, one can also calculate the error in the near-zone electric field at each iteration. The total electric field is sampled inside the cube at $19 \times 19 = 361$ regularly spaced points, i.e.,

$$E_{tot}[n] = \sqrt{|\,E_{tot,x}(x_n, y_n, z)|^2 + |\,E_{tot,y}(x_n, y_n, z)|^2 + |\,E_{tot,z}(x_n, y_n, z)|^2}, \tag{2.268}$$

where $E_{tot,x}$, $E_{tot,y}$, and $E_{tot,z}$ are x, y, and z components of the total electric field, $z = 0$, and $-\lambda/2 \le x_n, y_n \le \lambda/2$ for $n = 1, 2, \ldots, 361$. Because of the shielding effect of the PEC closed surface, the total electric field, which is obtained by combining the incident plane wave and the secondary field due to the induced electric current, should be zero inside the cube. Then, the near-field error is defined as

$$\Delta_{NF} = \frac{\|\boldsymbol{E}_{tot}\|_2}{\|\boldsymbol{E}_{inc}\|_2}, \tag{2.269}$$

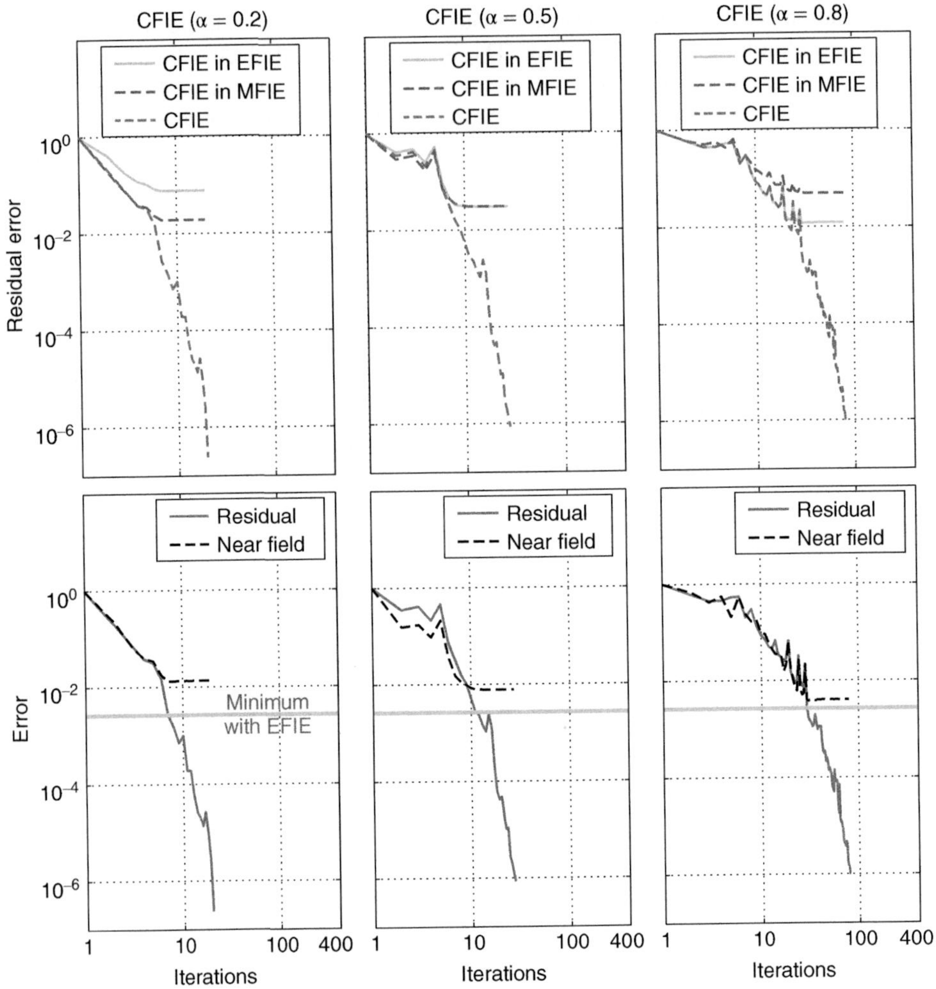

Figure 2.38 Iterative solutions of a scattering problem involving a $\lambda \times \lambda \times \lambda$ conducting box illuminated by a plane wave propagating in the $-x$ direction with the electric field polarized in the y direction. The residual and near-field errors are plotted with respect to BiCGStab iterations for CFIE when $\alpha =$ 0.2, 0.5, and 0.8. *Source*: Gürel and Ergül 2009 [122]. Reproduced with permission of IEEE.

where $\boldsymbol{E}_{tot}$ and $\boldsymbol{E}_{inc}$ are arrays of 361 elements containing the total and incident electric fields at the sample points. As depicted in the second row of Figure 2.38, the last useful iteration for the minimum possible error with CFIE coincides with the breakpoint for the compatibility of the solution with the EFIE system.

Figure 2.39 presents iterative solutions of a scattering problem involving a conducting sphere with a radius of 1.5λ. The sphere is illuminated by a plane wave propagating in the $-z$ direction with the electric field polarized in the x direction. Discretization of the problem with the RWG functions on $\lambda/10$ triangles lead to 8364×8364 matrix equations. Iterative solutions are again performed by using the BiCGStab algorithm. As depicted in the first row of Figure 2.39,

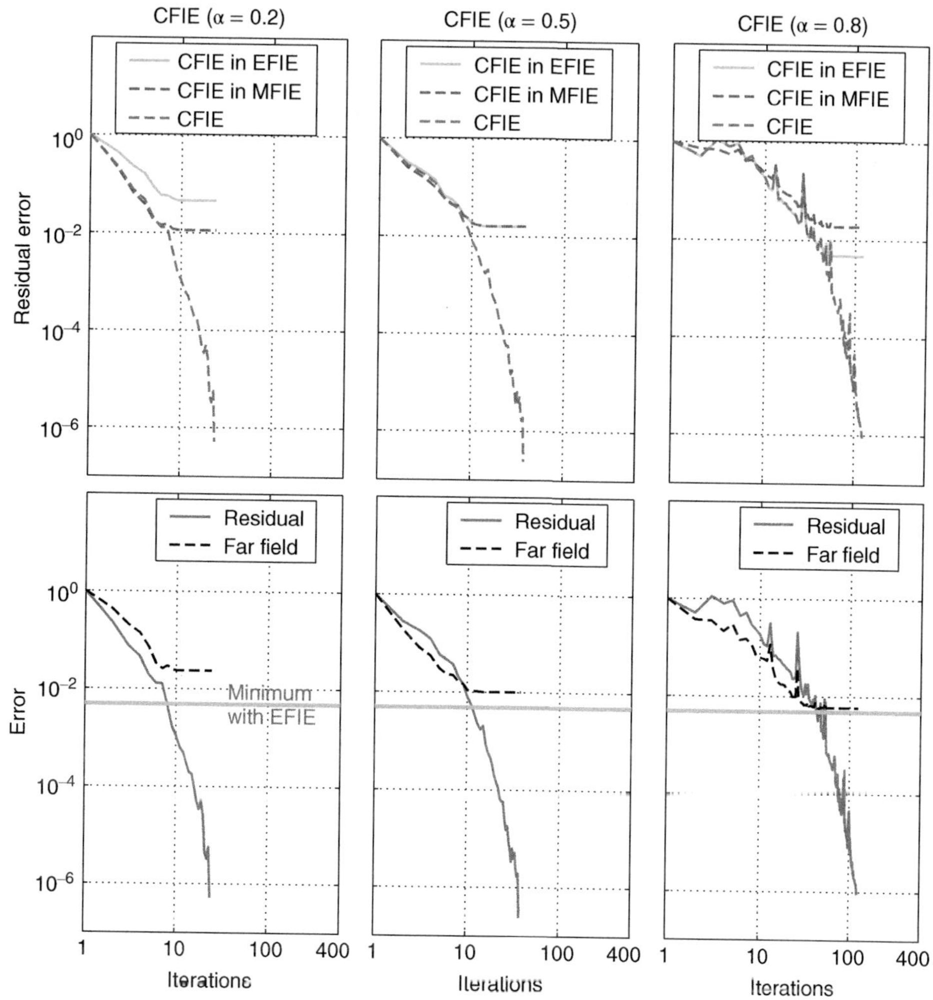

Figure 2.39 Iterative solutions of a scattering problem involving a conducting sphere with a radius of 1.5λ illuminated by a plane wave propagating in the $-x$ direction with the electric field polarized in the y direction. The residual and far-field errors are plotted with respect to BiCGStab iterations for CFIE when $\alpha = 0.2$, 0.5, and 0.8. *Source*: Gürel and Ergül 2009 [122]. Reproduced with permission of IEEE.

the compatibility of the CFIE solutions with the EFIE system saturates at the 9th, 13th, and 56th iterations, respectively, when $\alpha = 0.2$, 0.5, and 0.8. During the iterative solutions of this problem, the far-field error is calculated as

$$\Delta_R = \frac{\|\boldsymbol{E}^{\infty}_{C,\theta} - \boldsymbol{E}^{\infty}_{A,\theta}\|_2}{\|\boldsymbol{E}^{\infty}_{A,\theta}\|_2}, \tag{2.270}$$

where $\boldsymbol{E}^{\infty}_{C,\theta}$ and $\boldsymbol{E}^{\infty}_{A,\theta}$ are arrays of 360 elements containing the computational and analytical values for the co-polar electric field on the x-z plane. Figure 2.39 confirms the previous results that the minimum achievable error and the last useful iteration in CFIE solutions are closely related to the compatibility of solutions with the corresponding EFIE systems.

2.3.8 *Alternative Implementations of MFIE*

Since EFIE in (2.212) contains the electric charge density, i.e.,

$$\nabla \cdot \boldsymbol{J}(\boldsymbol{r}) = i\omega\rho_e(\boldsymbol{r}), \tag{2.271}$$

divergence-conforming basis functions must be used to expand the current density in EFIE. In other words, the basis functions that are employed for the current modeling should also be capable of properly modeling the inherent charge distribution. If an attempt was made to use basis functions that are not divergence conforming, the divergence in (2.212) would create singular blade functions on the edges of the entire discretization that would mathematically undermine computations and that could physically be interpreted as artificial line-charge accumulation and gaps in the conducting surface. In addition, the hyper-singularity in EFIE is usually eliminated by moving the differential operator onto the testing function, as shown in (2.79)–(2.81). Then, testing functions are also restricted to be divergence conforming in EFIE.

In the conventional discretization of the rotationally tested $\mathcal{K}$ operator in (2.76), the differential operator is kept on the Green's function. Therefore, there is no any restriction on basis and testing functions in MFIE. On the other hand, one can move the differential operator from the Green's function onto either the basis function or the testing function. By using standard vector identities and the

$$\nabla' = \nabla'_S + \hat{\boldsymbol{n}}'\frac{\partial}{\partial n'} \tag{2.272}$$

decomposition, one can manipulate (2.76) to obtain

$$\begin{aligned}
\overline{\boldsymbol{K}}^N_{PV}[m,n] &= \int_{S_m} d\boldsymbol{r}\boldsymbol{t}_m(\boldsymbol{r}) \cdot \hat{\boldsymbol{n}} \times \int_{PV,S_n} d\boldsymbol{r}' g(\boldsymbol{r},\boldsymbol{r}')\nabla'_S \times \boldsymbol{b}_n(\boldsymbol{r}') \\
&+ \int_{S_m} d\boldsymbol{r}\boldsymbol{t}_m(\boldsymbol{r}) \cdot \hat{\boldsymbol{n}} \times \int_{PV,S_n} d\boldsymbol{r}'\boldsymbol{b}_n(\boldsymbol{r}') \times \hat{\boldsymbol{n}}'\frac{\partial g(\boldsymbol{r},\boldsymbol{r}')}{\partial n'} \\
&- \int_{S_m} d\boldsymbol{r}\boldsymbol{t}_m(\boldsymbol{r}) \cdot \hat{\boldsymbol{n}} \times \int_{PV,S_n} d\boldsymbol{r}'\nabla'_S \times \left[g(\boldsymbol{r},\boldsymbol{r}')\boldsymbol{b}_n(\boldsymbol{r}')\right],
\end{aligned} \tag{2.273}$$

where the differential operator is partially transferred onto the basis function. In (2.273), $\hat{\boldsymbol{n}}'$ is the outward normal unit vector at the source point $\boldsymbol{r}'$. In this discretization of MFIE, the basis function is restricted to be curl conforming, however, no restriction exists on the testing function. Alternatively, (2.76) can be manipulated to obtain a third form as

$$\begin{aligned}
\overline{\boldsymbol{K}}^N_{PV}[m,n] &= \int_{S_m} d\boldsymbol{r} \left[\nabla_S \times \left(\boldsymbol{t}_m(\boldsymbol{r}) \times \hat{\boldsymbol{n}}\right)\right] \cdot \int_{PV,S_n} d\boldsymbol{r}'\boldsymbol{b}_n(\boldsymbol{r}')g(\boldsymbol{r},\boldsymbol{r}') \\
&+ \int_{S_m} d\boldsymbol{r}\boldsymbol{t}_m(\boldsymbol{r}) \cdot \hat{\boldsymbol{n}} \times \hat{\boldsymbol{n}} \times \int_{PV,S_n} d\boldsymbol{r}'\boldsymbol{b}_n(\boldsymbol{r}')\frac{\partial g(\boldsymbol{r},\boldsymbol{r}')}{\partial n} \\
&- \int_{S_m} d\boldsymbol{r}\nabla_S \cdot \left[\left(\boldsymbol{t}_m(\boldsymbol{r}) \times \hat{\boldsymbol{n}}\right) \times \int_{PV,S_n} d\boldsymbol{r}'\boldsymbol{b}_n(\boldsymbol{r}')g(\boldsymbol{r},\boldsymbol{r}')\right],
\end{aligned} \tag{2.274}$$

where the differential operator is partially applied to the testing function. In this discretization, there is no restriction on the basis function; however, the testing function is restricted to be a divergence-conforming function since

$$\nabla_S \times (\boldsymbol{t}_m(\boldsymbol{r}) \times \hat{\boldsymbol{n}}) = -\hat{\boldsymbol{n}}\nabla \cdot \boldsymbol{t}_m(\boldsymbol{r}). \tag{2.275}$$

For divergence-conforming functions, such as the RWG functions, the expression in (2.76) is appropriate to evaluate matrix elements. Although the same expression is also valid for curl-conforming functions, the modified expression in (2.273) can be preferred for these functions. In (2.273), the inner integral of the third term can be manipulated into a line integral over the edges around the basis function. In addition, the first term in (2.273) is usually simplified by using an expression for the curl of basis functions.

2.3.9 *Curl-Conforming Basis Functions for MFIE*

Investigations on the inaccuracy of MFIE show that neither the numerical evaluation of the singular MFIE integrals, nor the incorrect use of the solid-angle factor, is the fundamental cause of the inaccuracy [124]–[126]. This finding stimulates further studies, especially for a better understanding of how the discretization of MFIE and the modeling of the surface current affects the accuracy of results. This subsection presents implementations of MFIE with the curl-conforming $\hat{\boldsymbol{n}}\times$RWG basis functions. Numerical results show that the accuracy of MFIE can significantly be improved by employing the $\hat{\boldsymbol{n}}\times$RWG functions, instead of the conventional RWG functions.

Spatial distributions of the $\hat{\boldsymbol{n}}\times$RWG functions are shown in Figure 1.9, and can be written as

$$\boldsymbol{b}_n^{\text{nRWG}}(\boldsymbol{r}) = \hat{\boldsymbol{n}} \times \boldsymbol{b}_n^{\text{RWG}}(\boldsymbol{r}), \tag{2.276}$$

where $\boldsymbol{b}_n^{\text{RWG}}(\boldsymbol{r})$ is the associated RWG function. Using (2.276), matrix elements are derived as

$$\bar{\boldsymbol{Z}}^{\text{MFIE-nRWG}}[m,n] = -\frac{1}{2}\bar{\boldsymbol{I}}^{T,\text{nRWG}}[m,n] + \bar{\boldsymbol{K}}_{PV}^{N,\text{nRWG}}[m,n] \tag{2.277}$$

$$= -\frac{1}{2}\bar{\boldsymbol{I}}^{T,\text{RWG}}[m,n] - \bar{\boldsymbol{K}}_{PV}^{N,\text{RWG}}[n,m] \tag{2.278}$$

for $m,n = 1,2,\ldots,N$, where $\bar{\boldsymbol{K}}_{PV}^{N,\text{RWG}}$ and $\bar{\boldsymbol{I}}^{T,\text{RWG}}$ are rotationally tested $\mathcal{K}$ and directly tested $\mathcal{I}$ operators discretized with the RWG functions. Hence, interactions calculated for the $\hat{\boldsymbol{n}}\times$RWG and RWG functions are closely related. When the basis and testing functions associated with the mth and the nth edges do not overlap in space, only the principal-value term remains in (2.277) and (2.278) such that

$$\bar{\boldsymbol{Z}}^{\text{MFIE-nRWG}}[m,n] = \bar{\boldsymbol{K}}_{PV}^{N,\text{nRWG}}[m,n] = -\bar{\boldsymbol{K}}_{PV}^{N,\text{RWG}}[n,m] = -\bar{\boldsymbol{Z}}^{\text{MFIE-RWG}}[n,m], \tag{2.279}$$

where $\bar{\boldsymbol{Z}}^{\text{MFIE-RWG}}$ represents matrix elements for the RWG functions.

It is possible to derive an alternative implementation for the $\hat{\boldsymbol{n}}\times$RWG functions by considering (2.273), i.e.,

$$\begin{aligned}\overline{\boldsymbol{Z}}^{\text{MFIE-nRWG}}[m,n] &= \int_{S_m} d\boldsymbol{r}\boldsymbol{t}_m^{\text{nRWG}}(\boldsymbol{r}) \cdot \hat{\boldsymbol{n}} \times \int_{PV,S_n} d\boldsymbol{r}' g(\boldsymbol{r},\boldsymbol{r}')\nabla'_S \times \boldsymbol{b}_n^{\text{nRWG}}(\boldsymbol{r}') \\ &+ \int_{S_m} d\boldsymbol{r}\boldsymbol{t}_m^{\text{nRWG}}(\boldsymbol{r}) \cdot \hat{\boldsymbol{n}} \times \int_{PV,S_n} d\boldsymbol{r}' \boldsymbol{b}_n^{\text{nRWG}}(\boldsymbol{r}') \times \hat{\boldsymbol{n}}' \frac{\partial g(\boldsymbol{r},\boldsymbol{r}')}{\partial n'} \\ &- \int_{S_m} d\boldsymbol{r}\boldsymbol{t}_m^{\text{nRWG}}(\boldsymbol{r}) \cdot \hat{\boldsymbol{n}} \times \int_{PV,S_n} d\boldsymbol{r}' \nabla'_S \times \left[g(\boldsymbol{r},\boldsymbol{r}')\boldsymbol{b}_n^{\text{nRWG}}(\boldsymbol{r}')\right] \\ &- \frac{1}{2}\overline{\boldsymbol{I}}^{T,\text{nRWG}}[m,n]. \end{aligned} \tag{2.280}$$

Using (2.276) in (2.280), one can obtain

$$\begin{aligned}\overline{\boldsymbol{Z}}^{\text{MFIE-nRWG}}[m,n] &= \int_{S_m} d\boldsymbol{r}\boldsymbol{t}_m^{\text{RWG}}(\boldsymbol{r}) \cdot \int_{PV,S_n} d\boldsymbol{r}' \hat{\boldsymbol{n}}' g(\boldsymbol{r},\boldsymbol{r}')\nabla' \cdot \boldsymbol{b}_n^{\text{RWG}}(\boldsymbol{r}') \\ &+ \int_{S_m} d\boldsymbol{r}\boldsymbol{t}_m^{\text{RWG}}(\boldsymbol{r}) \cdot \int_{PV,S_n} d\boldsymbol{r}' \boldsymbol{b}_n^{\text{RWG}}(\boldsymbol{r}') \frac{\partial g(\boldsymbol{r},\boldsymbol{r}')}{\partial n'} \\ &- \int_{S_m} d\boldsymbol{r}\boldsymbol{t}_m^{\text{RWG}}(\boldsymbol{r}) \cdot \int_{PV,S_n} d\boldsymbol{r}' \hat{\boldsymbol{n}}' \nabla' \cdot \left[g(\boldsymbol{r},\boldsymbol{r}')\boldsymbol{b}_n^{\text{RWG}}(\boldsymbol{r}')\right] \\ &- \frac{1}{2}\overline{\boldsymbol{I}}^{T,\text{RWG}}[m,n]. \end{aligned} \tag{2.281}$$

For the direct calculation of interactions between the $\hat{\boldsymbol{n}} \times$ RWG functions, the alternative implementation in (2.281) is preferred. Consider the interaction of two half $\hat{\boldsymbol{n}} \times$ RWG functions $\boldsymbol{t}_{ma}^{\text{nRWG}}$ and $\boldsymbol{b}_{nb}^{\text{nRWG}}$ associated with the ath triangle of the mth edge and the bth triangle of the nth edge, respectively. Then,

$$\begin{aligned}\overline{\boldsymbol{K}}_{PV}^{N,\text{nRWG}}[m,n,a,b] &= \mathcal{A}_{ma,nb}\left\{2\int_{S_{ma}} d\boldsymbol{r}(\boldsymbol{r}-\boldsymbol{r}_{ma}) \cdot \int_{PV,S_{nb}} d\boldsymbol{r}' \hat{\boldsymbol{n}}' g(\boldsymbol{r},\boldsymbol{r}') \right. \\ &+ \int_{S_{ma}} d\boldsymbol{r}(\boldsymbol{r}-\boldsymbol{r}_{ma}) \cdot \int_{PV,S_{nb}} d\boldsymbol{r}'(\boldsymbol{r}'-\boldsymbol{r}_{nb}) \frac{\partial g(\boldsymbol{r},\boldsymbol{r}')}{\partial n'} \\ &\left. - \int_{S_{ma}} d\boldsymbol{r}(\boldsymbol{r}-\boldsymbol{r}_{ma}) \cdot \int_{PV,S_{nb}} d\boldsymbol{r}' \hat{\boldsymbol{n}}' \nabla' \cdot \left[g(\boldsymbol{r},\boldsymbol{r}')(\boldsymbol{r}'-\boldsymbol{r}_{nb})\right]\right\}, \end{aligned} \tag{2.282}$$

where $\mathcal{A}_{ma,nb}$ is defined in (2.88). After a coordinate transformation,

$$\begin{aligned}\overline{\boldsymbol{K}}_{PV}^{N,\text{nRWG}}[m,n,a,b] &= \mathcal{A}_{ma,nb}\left\{2\int_{S_{ma}} d\boldsymbol{r}(z-z_{ma}) \int_{PV,S_{nb}} d\boldsymbol{r}' g(\boldsymbol{r},\boldsymbol{r}') \right. \\ &+ \int_{S_{ma}} d\boldsymbol{r}(\boldsymbol{\rho}-\boldsymbol{\rho}_{ma}) \cdot \int_{PV,S_{nb}} d\boldsymbol{r}'(\boldsymbol{\rho}'-\boldsymbol{\rho}_{nb}) \frac{\partial g(\boldsymbol{r},\boldsymbol{r}')}{\partial z} \\ &\left. - \int_{S_{ma}} d\boldsymbol{r}(z-z_{ma}) \int_{PV,S_{nb}} d\boldsymbol{r}' \nabla' \cdot \left[g(\boldsymbol{r},\boldsymbol{r}')(\boldsymbol{\rho}'-\boldsymbol{\rho}_{nb})\right]\right\}. \end{aligned} \tag{2.283}$$

Using the surface divergence theorem, the inner integral of the third term in (2.283) can be converted into a line integral, i.e.,

$$\int_{S_{ma}} d\boldsymbol{r}(z - z_{ma}) \int_{PV,S_{nb}} d\boldsymbol{r}'\nabla' \cdot \left[g(\boldsymbol{r},\boldsymbol{r}')(\boldsymbol{\rho}' - \boldsymbol{\rho}_{nb})\right]$$

$$= \int_{S_{ma}} d\boldsymbol{r}(z - z_{ma}) \int_{PV,\partial S_{nb}} d\boldsymbol{r}'\hat{\boldsymbol{u}}' \cdot (\boldsymbol{\rho}' - \boldsymbol{\rho}_{nb}) g(\boldsymbol{r},\boldsymbol{r}'), \tag{2.284}$$

where ∂S_{nb} represents the contour around the basis triangle and $\hat{\boldsymbol{u}}'$ is the unit vector directed perpendicularly outward the triangle. Note that

$$\hat{\boldsymbol{u}}' \cdot (\boldsymbol{\rho}' - \boldsymbol{\rho}_{nb}) = \frac{2A_n}{l_n} \tag{2.285}$$

on the nth edge, but it evaluates to zero on the two other edges of the basis triangle. Then, (2.283) becomes

$$\overline{\boldsymbol{K}}_{PV}^{N,\text{nRWG}}[m,n,a,b] = \mathcal{A}_{ma,nb} \left\{ 2\int_{S_{ma}} d\boldsymbol{r}(z - z_{ma}) \int_{PV,S_{nb}} d\boldsymbol{r}' g(\boldsymbol{r},\boldsymbol{r}') \right.$$

$$+ \int_{S_{ma}} d\boldsymbol{r}(\boldsymbol{\rho} - \boldsymbol{\rho}_{ma}) \cdot \int_{PV,S_{nb}} d\boldsymbol{r}'(\boldsymbol{\rho}' - \boldsymbol{\rho}_{nb}) \frac{\partial g(\boldsymbol{r},\boldsymbol{r}')}{\partial z}$$

$$\left. -\frac{2A_n}{l_n} \int_{S_{ma}} d\boldsymbol{r}(z - z_{ma}) \int_{PV,e_n} d\boldsymbol{r}' g(\boldsymbol{r},\boldsymbol{r}') \right\}. \tag{2.286}$$

The line integral of the Green's function along the nth edge e_n is singular when the observation point approaches this edge. Similar to (2.77) and (2.78), the singularity is logarithmic and does not create numerical problems when the observation points are sampled inside the testing triangle for the near-neighboring interactions. Nevertheless, a singularity-extraction method can be used to improve the evaluation of these interactions [124]. In addition, the order of the integrals can be exchanged as [169]

$$\int_{S_{ma}} d\boldsymbol{r}(z - z_{ma}) \int_{PV,e_n} d\boldsymbol{r}' g(\boldsymbol{r},\boldsymbol{r}') = \int_{PV,e_n} d\boldsymbol{r}' \int_{S_{ma}} d\boldsymbol{r}(z - z_{ma}) g(\boldsymbol{r},\boldsymbol{r}'), \tag{2.287}$$

where the inner integral is now bounded for every case including the sampling of testing points on the edge of the basis triangle.

The interaction of two half $\hat{\boldsymbol{n}}\times$RWG functions can be divided into basic integrals as

$$\overline{\boldsymbol{K}}_{PV}^{N,\text{nRWG}}[m,n,a,b] = \mathcal{A}_{ma,nb} \left\{ -x_{ma}I_6 - y_{ma}I_7 + I_8 + I_9 \right.$$

$$+ (x_{ma}x_{nb} + y_{ma}y_{nb})I_3 - x_{nb}I_4 - y_{nb}I_5 - 2(z_{ma}I_1 - I_2)$$

$$\left. -\frac{2A_{nb}}{l_n}(I_{11} - z_{ma}I_{10}) \right\}, \tag{2.288}$$

where

$$I_{\{1,2\}} = \int_{S_{ma}} d\boldsymbol{r} \{1, z\} I_{in}^1 \tag{2.289}$$

$$I_{\{3,4,5\}} = \int_{S_{ma}} d\boldsymbol{r}\, \{1, x, y\}\, \hat{\boldsymbol{z}} \cdot \boldsymbol{I}_{in}^{\nabla'} \tag{2.290}$$

$$I_{\{6,8\}} = \int_{S_{ma}} d\boldsymbol{r}\, \{1, x\}\, \hat{\boldsymbol{z}} \cdot \boldsymbol{I}_{in}^{x'\nabla'} \tag{2.291}$$

$$I_{\{7,9\}} = \int_{S_{ma}} d\boldsymbol{r}\, \{1, y\}\, \hat{\boldsymbol{z}} \cdot \boldsymbol{I}_{in}^{y'\nabla'} \tag{2.292}$$

$$I_{\{10,11\}} = \int_{S_{ma}} d\boldsymbol{r}\, \{1, z\}\, I_{in}^{e,1}. \tag{2.293}$$

Inner integrals I_{in}^{1} and $\boldsymbol{I}_{in}^{\nabla'}$ are evaluated as in (2.122)–(2.125). In addition,

$$\boldsymbol{I}_{in}^{x'\nabla'} = \boldsymbol{I}_{in}^{(x'-x)\nabla'} + x\boldsymbol{I}_{in}^{\nabla'} \tag{2.294}$$

$$\boldsymbol{I}_{in}^{y'\nabla'} = \boldsymbol{I}_{in}^{(y'-y)\nabla'} + y\boldsymbol{I}_{in}^{\nabla'}, \tag{2.295}$$

where

$$\begin{aligned} \boldsymbol{I}_{in}^{(x'-x)\nabla'} &= \frac{1}{4\pi}\int_{S_{nb}} d\boldsymbol{r}'(x'-x)\nabla'\left(\frac{\exp(ikR)-1}{R}\right) \\ &+ \frac{1}{4\pi}\int_{S_{nb}} d\boldsymbol{r}'(x'-x)\nabla'\left(\frac{1}{R}\right) \end{aligned} \tag{2.296}$$

and

$$\begin{aligned} \boldsymbol{I}_{in}^{(y'-y)\nabla'} &= \frac{1}{4\pi}\int_{S_{nb}} d\boldsymbol{r}'(y'-y)\nabla'\left(\frac{\exp(ikR)-1}{R}\right) \\ &+ \frac{1}{4\pi}\int_{S_{nb}} d\boldsymbol{r}'(y'-y)\nabla'\left(\frac{1}{R}\right). \end{aligned} \tag{2.297}$$

Note that only z components of $\boldsymbol{I}_{in}^{\nabla'}$, $\boldsymbol{I}_{in}^{(x'-x)\nabla'}$, and $\boldsymbol{I}_{in}^{(y'-y)\nabla'}$ are required in (2.290)–(2.292). The basic integrals I_{10} and I_{11} in (2.293) involve a line integral

$$I_{in}^{e,1} = \frac{1}{4\pi}\int_{e_n} dr' \frac{\exp(ikR)}{R}, \tag{2.298}$$

which depends on the orientation of the basis function, as opposed to the other inner integrals.

Numerical Results

To compare the accuracy of solutions obtained by using the RWG and $\hat{\boldsymbol{n}}\times$RWG functions, scattering problems involving a PEC sphere are considered. The problems are solved at two frequencies, i.e., when the radius of the sphere corresponds to λ and 2λ. Figures 2.40 and 2.41 illustrate the normalized backscattered and forward-scattered RCS (RCS $/\pi a^2$), respectively,

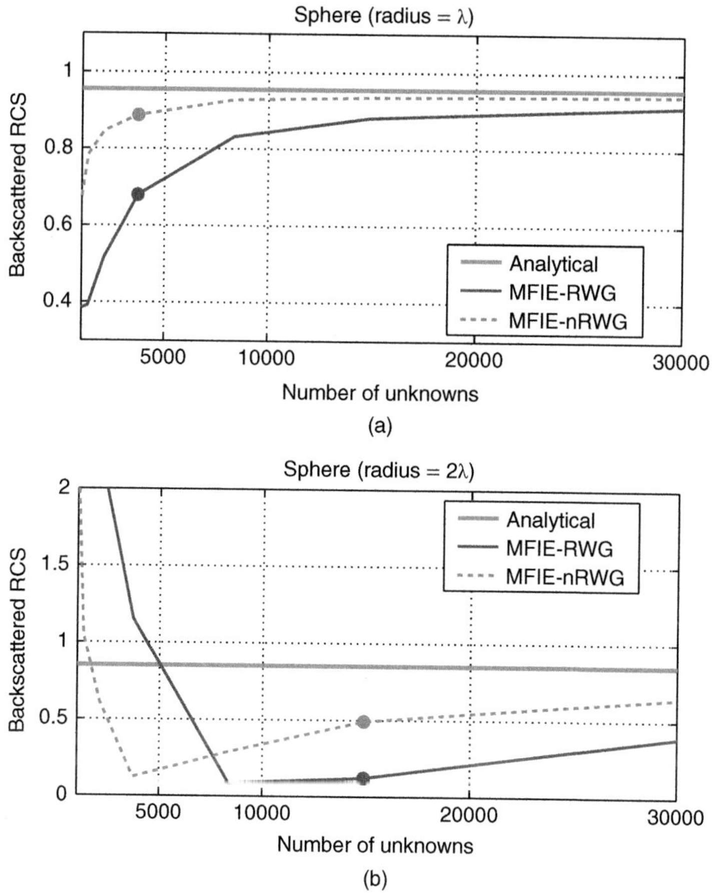

Figure 2.40 The normalized backscattered RCS (RCS $/\pi a^2$) of a PEC sphere with a radius of (a) λ and (b) 2λ as a function of the number of unknowns. The dots on the curves correspond to the $\lambda/10$ discretization. *Source*: Ergül and Gürel 2006 [127]. Reproduced with permission of IEEE.

when the radius of the sphere is $a = \lambda$ and 2λ. Computational values are obtained by FMM implementations employing the RWG and $\hat{\boldsymbol{n}}\times$RWG functions and for various discretizations involving approximately 1000 to 30 000 unknowns. It can be observed that, as N increases, RCS values obtained with the $\hat{\boldsymbol{n}}\times$RWG functions converge to the exact Mie-series results faster than those obtained with the RWG functions, i.e., RCS values obtained with the $\hat{\boldsymbol{n}}\times$RWG functions are more accurate for all discretization densities. The improvement obtained with the $\hat{\boldsymbol{n}}\times$RWG functions over the RWG functions is significant even for the $\lambda/10$ discretization, which is commonly used as a rule of thumb in numerical solutions.

Next, solutions of scattering problems involving a PEC cube with edges of 1 m are considered, when the frequency is 300 MHz and 600 MHz. Figures 2.42(a) and 2.42(b) show that, using the $\hat{\boldsymbol{n}}\times$RWG functions, the modeling of the surface current is visibly improved compared

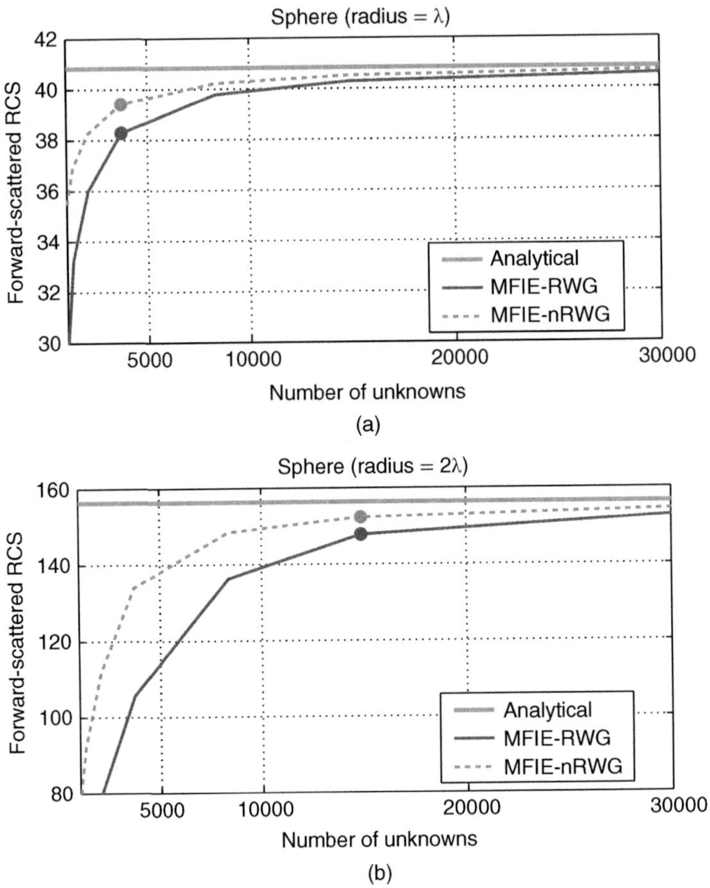

Figure 2.41 The normalized forward-scattered RCS (RCS $/\pi a^2$) of a PEC sphere with a radius of (a) λ and (b) 2λ as a function of the number of unknowns. The dots on the curves correspond to the $\lambda/10$ discretization. *Source*: Ergül and Gürel 2006 [127]. Reproduced with permission of IEEE.

to the RWG functions depicted in Figures 2.32 and 2.33, especially in the representation of singular currents flowing along edges. Specifically, the use of the $\hat{\boldsymbol{n}}\times$RWG functions in the MFIE solution provides higher values at $z = \pm 0.5$ m than the use of the RWG functions with either EFIE or MFIE. It should be remarked that this is merely an indicator for the improved accuracy of MFIE with the use of the $\hat{\boldsymbol{n}}\times$RWG functions, but not an absolute proof since the exact current distribution is not available for comparisons. Nevertheless, RCS results also support the inferences obtained from current solutions. Figures 2.43 and 2.44 demonstrate the backscattered and forward-scattered RCS (in m^2), respectively, when the edges of the cube are λ and 2λ. Comparing the MFIE results, faster convergence (with respect to the improved discretization and the increasing N) of RCS values with the $\hat{\boldsymbol{n}}\times$RWG functions is observed. The improvement is significant especially for relatively coarse discretizations, even for the commonly used $\lambda/10$ triangulation, as marked on the curves.

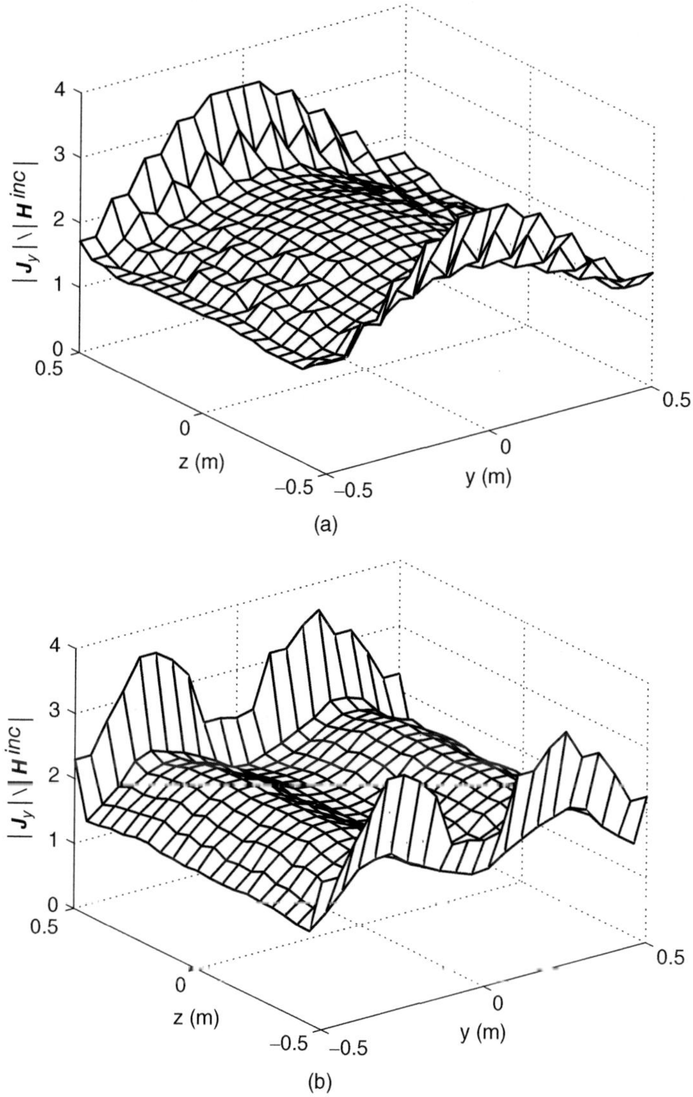

Figure 2.42 The magnitude of the y component of the normalized surface current $(|\boldsymbol{J}_y|/|\boldsymbol{H}^{inc}|)$ induced on the front surface (at $x = 0.5$ m) of a PEC cube with edges of (a) λ and (b) 2λ, obtained by using MFIE with $\hat{\boldsymbol{n}}\times$RWG. *Source*: Ergül and Gürel 2006 [127]. Reproduced with permission of IEEE.

2.3.10 LN-LT Type Basis Functions for MFIE and CFIE

LN-LT type basis functions, i.e., the divergence-conforming LL functions and the curl-conforming $\hat{\boldsymbol{n}}\times$LL functions, can be employed to further improve the accuracy of MFIE and CFIE to the levels of EFIE, without increasing computational requirements. Spatial distributions of the LL functions are depicted in Figure 1.10, where two kinds of

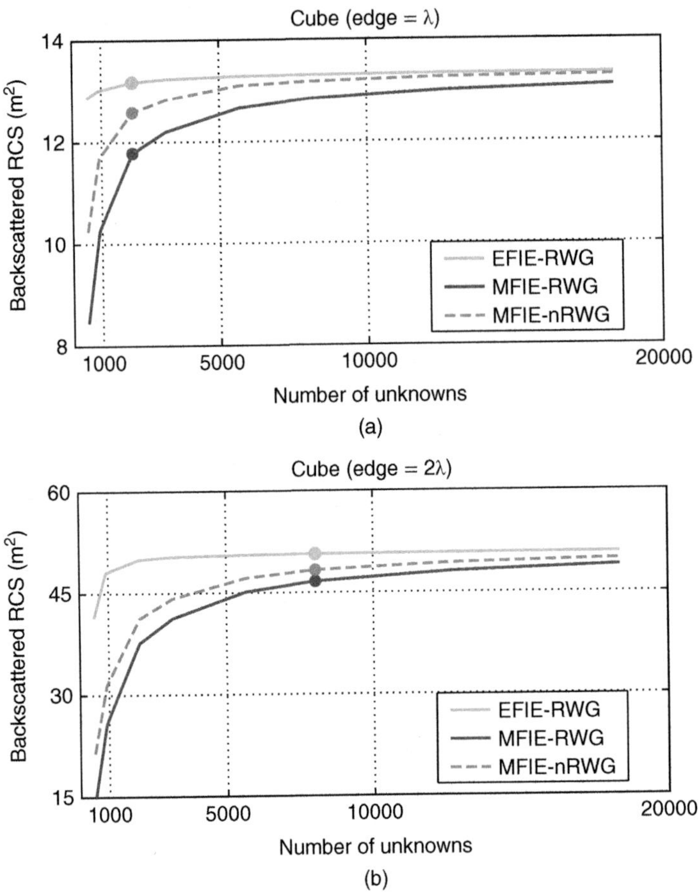

Figure 2.43 The backscattered RCS (m^2) of a PEC cube with edges of (a) λ and (b) 2λ as a function of the number of unknowns. The dots on the curves correspond to the $\lambda/10$ discretization. *Source*: Ergül and Gürel 2006 [127]. Reproduced with permission of IEEE.

the LL functions are defined on the same edge e_n with the expressions given in (1.77) and (1.78). Considering the labeling in Figure 1.10, for both kinds of the LL functions, the spatial distribution

- is parallel to one of the side edges e_p over both of the triangles,
- changes linearly along the edge e_p and becomes maximum at the intersection of that edge with the main edge e_n,
- is exactly zero at the other side edge e_z,
- has a continuous normal component across the main edge e_n, and
- varies linearly at the main edge e_n in both perpendicular and tangential directions.

Because of the final property above, the LL functions are classified as LN-LT type, providing linear variation along the main edge in both directions.

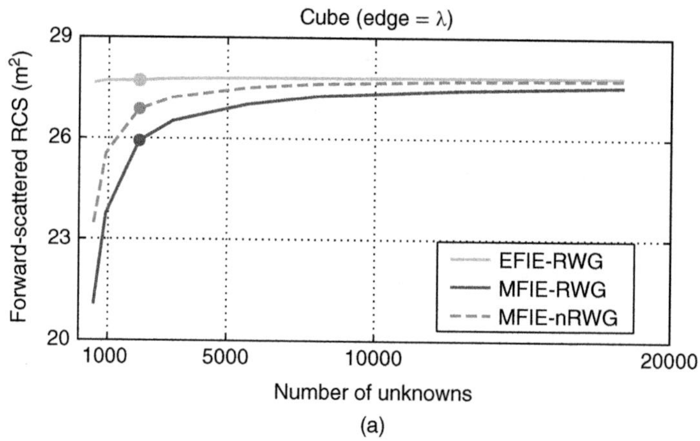

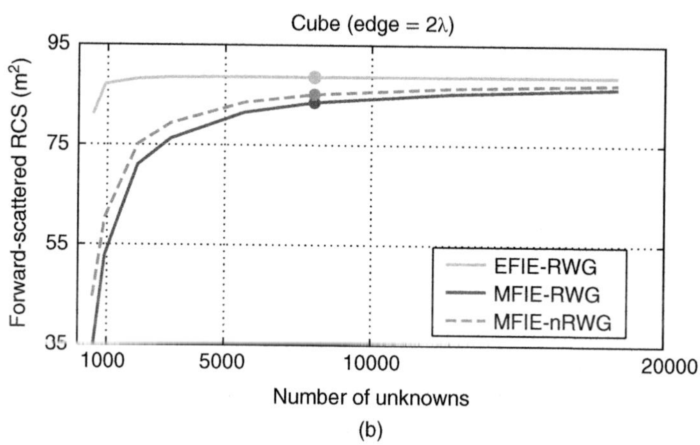

Figure 2.44 The forward-scattered RCS (m^2) of a PEC cube with edges of (a) λ and (b) 2λ as a function of the number of unknowns. The dots on the curves correspond to the $\lambda/10$ discretization. *Source*: Ergül and Gürel 2006 [127]. Reproduced with permission of IEEE.

The LL functions can actually decompose the corresponding RWG function as

$$\boldsymbol{b}_n^{\mathrm{RWG}}(\boldsymbol{r}) = \boldsymbol{b}_n^{\mathrm{LL(1)}}(\boldsymbol{r}) + \boldsymbol{b}_n^{\mathrm{LL(2)}}(\boldsymbol{r}), \tag{2.299}$$

where $\boldsymbol{b}_n^{\mathrm{RWG}}$ represents the RWG function associated with the nth edge. The charge distribution implied by the expressions in (1.77) and (1.78) is

$$\nabla \cdot \boldsymbol{b}_n^{\mathrm{LL(1)}}(\boldsymbol{r}) = \nabla \cdot \boldsymbol{b}_n^{\mathrm{LL(2)}}(\boldsymbol{r}) = \begin{cases} \dfrac{l_n}{2A_{n1}}, & r \in S_{n1} \\ \dfrac{-l_n}{2A_{n2}}, & r \in S_{n2} \\ 0, & \text{otherwise,} \end{cases} \tag{2.300}$$

which is constant and exactly half of that for the RWG functions. Equation (2.300) indicates that the LL functions are divergence-conforming functions since their divergence is finite. Although this property is not crucial for MFIE, which also allows the use of curl-conforming functions depending on the formulation type, the divergence-conforming property of the LL functions becomes essential for EFIE and CFIE implementations.

The LL functions can also be identified as a subset of LN/QT or QN/LT type vector functions discussed in [67]. Contrary to the RWG functions, the LL functions are first-order complete to represent vectors by providing six degrees of freedom to model the linear variation on triangles (see Chapter 1, Section 1.7.1). As depicted in Figure 1.10, there are two kinds of the LL functions defined on the same edge simultaneously to expand the current density better than the RWG functions that are only zeroth-order complete. Improved current modeling with higher-order completeness is achieved at the cost of doubling the number of basis functions compared to the RWG functions for the same triangulation of the geometry. On the other hand, the LL functions provide more accurate RCS results with MFIE and CFIE for the same number of unknowns as the RWG functions. The improvement is present even for curved geometries, where the discretization for the LL functions is coarser and leads to more deviation from the actual geometry to keep the number of unknowns approximately the same as when the RWG functions are used on the same geometry with a finer triangulation.

Although the LL functions are first-order complete to represent vectors, they are only zeroth-order complete to represent the divergence or curl of a vector. The LL functions improve the accuracy of MFIE to levels of EFIE, while they are similar to the RWG functions by providing linear variations for the induced current on triangular domains. In this manner, the LL functions are different from the higher-order techniques in computational electromagnetics appeared in the literature to improve the accuracy of integral equations [130]–[132]. Advantages of the LL functions can be listed as follows:

- The LL functions can be implemented with simple modifications on existing codes employing the RWG functions. Most of the numerical procedures in MLFMA, such as singularity extractions for the near-field interactions and Fourier transforms for the far-field interactions, can be adapted from those of the RWG functions.
- For the same number of unknowns, employing the LL functions in MLFMA implementations does not require extra computational load compared to employing the RWG functions.

The curl-conforming $\hat{\boldsymbol{n}}\times$LL functions are derived from the LL functions as

$$\boldsymbol{b}^{\mathrm{nLL}(1)}(\boldsymbol{r}) = \hat{\boldsymbol{n}} \times \boldsymbol{b}^{\mathrm{LL}(1)}(\boldsymbol{r}) \tag{2.301}$$

$$\boldsymbol{b}^{\mathrm{nLL}(2)}(\boldsymbol{r}) = \hat{\boldsymbol{n}} \times \boldsymbol{b}^{\mathrm{LL}(2)}(\boldsymbol{r}), \tag{2.302}$$

such that

$$\nabla_S \times \boldsymbol{b}^{\mathrm{nLL}(1)}(\boldsymbol{r}) = \hat{\boldsymbol{n}}\nabla \cdot \boldsymbol{b}^{\mathrm{LL}(1)}(\boldsymbol{r}) \tag{2.303}$$

$$\nabla_S \times \boldsymbol{b}^{\mathrm{nLL}(2)}(\boldsymbol{r}) = \hat{\boldsymbol{n}}\nabla \cdot \boldsymbol{b}^{\mathrm{LL}(2)}(\boldsymbol{r}). \tag{2.304}$$

Similar to the LL functions, the $\hat{n}\times$LL functions are also defined in pairs located at the same edge with spatial distributions illustrated in Figure 1.11. For both kinds of the $\hat{n}\times$LL functions, the spatial distribution

- is perpendicular to one of the side edges e_p over both of the triangles,
- changes linearly along the edge e_p and becomes maximum at the intersection of that edge with the main edge e_n,
- is exactly zero at the other side edge e_z,
- has a continuous normal component across the main edge e_n, and
- varies linearly at the main edge e_n in both perpendicular and tangential directions.

The divergence-conforming LL functions can be employed for EFIE, MFIE, and CFIE. As an example, interaction of two half LL functions $\boldsymbol{t}_{ma}^{\mathrm{LL}(1)}$ and $\boldsymbol{b}_{nb}^{\mathrm{LL}(1)}$ associated with the ath triangle of the mth edge and the bth triangle of the nth edge is considered. Both functions are assumed to be the first kind. Using (1.77) and (2.300), the matrix elements are derived as

$$\overline{\boldsymbol{Z}}^{\text{EFIE-LL(1)}}[m,n,a,b] = \overline{\boldsymbol{T}}^{T,\mathrm{LL}(1)}[m,n,a,b] \tag{2.305}$$

$$\overline{\boldsymbol{Z}}^{\text{MFIE-LL(1)}}[m,n,a,b] = \overline{\boldsymbol{K}}_{PV}^{N,\mathrm{LL}(1)}[m,n,a,b] - \frac{1}{2}\overline{\boldsymbol{I}}^{T,\mathrm{LL}(1)}[m,n,a,b], \tag{2.306}$$

where

$$\overline{\boldsymbol{I}}^{T,\mathrm{LL}(1)}[m,n,a,b] = \mathcal{B}_{ma,nb}(\boldsymbol{r}_{m3}-\boldsymbol{r}_{ma})\cdot(\boldsymbol{r}_{n3}-\boldsymbol{r}_{nb})$$
$$\int_{S_{ma}} d\boldsymbol{r}(\boldsymbol{r}-\boldsymbol{r}_{ma})\cdot\left[(\boldsymbol{r}_{m4}-\boldsymbol{r}_{ma})\times\hat{\boldsymbol{n}}\right](\boldsymbol{r}-\boldsymbol{r}_{nb})\cdot\left[(\boldsymbol{r}_{n4}-\boldsymbol{r}_{nb})\times\hat{\boldsymbol{n}}\right] \tag{2.307}$$

$$\overline{\boldsymbol{T}}^{T,\mathrm{LL}(1)}[m,n,a,b] = ik\mathcal{B}_{ma,nb}(\boldsymbol{r}_{m3}-\boldsymbol{r}_{ma})\cdot(\boldsymbol{r}_{n3}-\boldsymbol{r}_{nb})$$
$$\int_{S_m} d\boldsymbol{r}(\boldsymbol{r}-\boldsymbol{r}_{ma})\cdot\left[(\boldsymbol{r}_{m4}-\boldsymbol{r}_{ma})\times\hat{\boldsymbol{n}}\right]$$
$$\int_{S_n} d\boldsymbol{r}'\, g(\boldsymbol{r},\boldsymbol{r}')(\boldsymbol{r}'-\boldsymbol{r}_{nb})\cdot\left[(\boldsymbol{r}_{n4}-\boldsymbol{r}_{nb})\times\hat{\boldsymbol{n}}'\right]$$
$$-\frac{4i}{k}\mathcal{B}_{ma,nb}\int_{S_m} d\boldsymbol{r}\int_{S_n} d\boldsymbol{r}'\, g(\boldsymbol{r},\boldsymbol{r}') \tag{2.308}$$

$$\overline{\boldsymbol{K}}_{PV}^{N,\mathrm{LL}(1)}[m,n,a,b] = \mathcal{B}_{ma,nb}(\boldsymbol{r}_{m3}-\boldsymbol{r}_{ma})\cdot\hat{\boldsymbol{n}}\times(\boldsymbol{r}_{n3}-\boldsymbol{r}_{nb})$$
$$\times\int_{S_{ma}} d\boldsymbol{r}(\boldsymbol{r}-\boldsymbol{r}_{ma})\cdot\left[(\boldsymbol{r}_{m4}-\boldsymbol{r}_{ma})\times\hat{\boldsymbol{n}}\right]$$
$$\int_{PV,S_{nb}} d\boldsymbol{r}'(\boldsymbol{r}'-\boldsymbol{r}_{nb})\cdot\left[(\boldsymbol{r}_{n4}-\boldsymbol{r}_{nb})\times\hat{\boldsymbol{n}}'\right]\nabla' g(\boldsymbol{r},\boldsymbol{r}') \tag{2.309}$$

and

$$\mathcal{B}_{ma,nb} = \frac{l_m l_n}{16(A_{ma})^2(A_{nb})^2}. \tag{2.310}$$

After a coordinate transformation,

$$\overline{\boldsymbol{I}}^{T,\text{LL}(1)}[m,n,a,b] = \mathcal{B}_{ma,nb}(\boldsymbol{\rho}_{m3}-\boldsymbol{\rho}_{ma})\cdot(\boldsymbol{\rho}_{n3}-\boldsymbol{\rho}_{nb})$$
$$\int_{S_{ma}} d\boldsymbol{r}(\boldsymbol{\rho}-\boldsymbol{\rho}_{ma})\cdot\left[(\boldsymbol{\rho}_{m4}-\boldsymbol{\rho}_{ma})\times\hat{z}\right](\boldsymbol{\rho}-\boldsymbol{\rho}_{nb})\cdot\left[(\boldsymbol{\rho}_{n4}-\boldsymbol{\rho}_{nb})\times\hat{z}\right] \tag{2.311}$$

$$\overline{\boldsymbol{T}}^{T,\text{LL}(1)}[m,n,a,b] = ik\mathcal{B}_{ma,nb}(\boldsymbol{\rho}_{m3}-\boldsymbol{\rho}_{ma})\cdot(\boldsymbol{\rho}_{n3}-\boldsymbol{\rho}_{nb})$$
$$\int_{S_m} d\boldsymbol{r}(\boldsymbol{r}-\boldsymbol{r}_{ma})\cdot\left[(\boldsymbol{r}_{m4}-\boldsymbol{r}_{ma})\times\hat{\boldsymbol{n}}\right]$$
$$\int_{S_n} d\boldsymbol{r}' g(\boldsymbol{r},\boldsymbol{r}')(\boldsymbol{\rho}'-\boldsymbol{\rho}_{nb})\cdot\left[(\boldsymbol{\rho}_{n4}-\boldsymbol{\rho}_{nb})\times\hat{z}\right]$$
$$-\frac{4i}{k}\mathcal{B}_{ma,nb}\int_{S_m} d\boldsymbol{r}\int_{S_n} d\boldsymbol{r}' g(\boldsymbol{r},\boldsymbol{r}') \tag{2.312}$$

$$\overline{\boldsymbol{K}}_{PV}^{N,\text{LL}(1)}[m,n,a,b] = \mathcal{B}_{ma,nb}(\boldsymbol{r}_{m3}-\boldsymbol{r}_{ma})\cdot\hat{\boldsymbol{n}}\times(\boldsymbol{\rho}_{n3}-\boldsymbol{\rho}_{nb})$$
$$\times\int_{S_{ma}} d\boldsymbol{r}(\boldsymbol{r}-\boldsymbol{r}_{ma})\cdot\left[(\boldsymbol{r}_{m4}-\boldsymbol{r}_{ma})\times\hat{\boldsymbol{n}}\right]$$
$$\int_{PV,S_{nb}} d\boldsymbol{r}'(\boldsymbol{\rho}'-\boldsymbol{\rho}_{nb})\cdot\left[(\boldsymbol{\rho}_{n4}-\boldsymbol{\rho}_{nb})\times\hat{z}\right]\nabla' g(\boldsymbol{r},\boldsymbol{r}'). \tag{2.313}$$

For the integro-differential operators, the interactions are further divided into basic integrals as

$$\overline{\boldsymbol{T}}^{T,\text{LL}(1)}[m,n,a,b] = ik\mathcal{B}_{ma,nb}(\boldsymbol{\rho}_{m3}-\boldsymbol{\rho}_{ma})\cdot(\boldsymbol{\rho}_{n3}-\boldsymbol{\rho}_{nb})$$
$$\left\{(y_{n4}-y_{nb})\left[(y_{m4}-y_{ma})n_z-(z_{m4}-z_{ma})n_y\right]\left[x_{ma}(x_{nb}I_1-I_2)-x_{nb}I_4+I_5\right]\right.$$
$$+(x_{n4}-x_{nb})\left[(z_{m4}-z_{ma})n_y-(y_{m4}-y_{ma})n_z\right]\left[x_{ma}(y_{nb}I_1-I_3)-y_{nb}I_4+I_6\right]$$
$$+(y_{n4}-y_{nb})\left[(z_{m4}-z_{ma})n_x-(x_{m4}-x_{ma})n_z\right]\left[y_{ma}(x_{nb}I_1-I_2)-x_{nb}I_7+I_8\right]$$
$$+(x_{n4}-x_{nb})\left[(x_{m4}-x_{ma})n_z-(z_{m4}-z_{ma})n_x\right]\left[y_{ma}(y_{nb}I_1-I_3)-y_{nb}I_7+I_9\right]$$
$$+(y_{n4}-y_{nb})\left[(x_{m4}-x_{ma})n_y-(y_{m4}-y_{ma})n_x\right]\left[z_{ma}(x_{nb}I_1-I_2)-x_{nb}I_{10}+I_{11}\right]$$
$$\left.+(x_{n4}-x_{nb})\left[(y_{m4}-y_{ma})n_x-(x_{m4}-x_{ma})n_y\right]\left[z_{ma}(y_{nb}I_1-I_3)-y_{nb}I_{10}+I_{12}\right]\right\}$$
$$-\frac{4i}{k}\mathcal{B}_{ma,nb}I_1 \tag{2.314}$$

and

$$\overline{\boldsymbol{K}}_{PV}^{N,\text{LL}(1)}[m,n,a,b] = \mathcal{B}_{ma,nb}(\boldsymbol{r}_{m3}-\boldsymbol{r}_{ma})\cdot\hat{\boldsymbol{n}}\times(\boldsymbol{\rho}_{n3}-\boldsymbol{\rho}_{nb})\times$$
$$\left\{(y_{n4}-y_{nb})\left[(y_{m4}-y_{ma})n_z-(z_{m4}-z_{ma})n_y\right]\left[x_{ma}(x_{nb}\boldsymbol{I}_1-\boldsymbol{I}_2)-x_{nb}\boldsymbol{I}_4+\boldsymbol{I}_5\right]\right.$$
$$+(x_{n4}-x_{nb})\left[(z_{m4}-z_{ma})n_y-(y_{m4}-y_{ma})n_z\right]\left[x_{ma}(y_{nb}\boldsymbol{I}_1-\boldsymbol{I}_3)-y_{nb}\boldsymbol{I}_4+\boldsymbol{I}_6\right]$$
$$+(y_{n4}-y_{nb})\left[(z_{m4}-z_{ma})n_x-(x_{m4}-x_{ma})n_z\right]\left[y_{ma}(x_{nb}\boldsymbol{I}_1-\boldsymbol{I}_2)-x_{nb}\boldsymbol{I}_7+\boldsymbol{I}_8\right]$$
$$+(x_{n4}-x_{nb})\left[(x_{m4}-x_{ma})n_z-(z_{m4}-z_{ma})n_x\right]\left[y_{ma}(y_{nb}\boldsymbol{I}_1-\boldsymbol{I}_3)-y_{nb}\boldsymbol{I}_7+\boldsymbol{I}_9\right]$$

(continued)

$$+ (y_{n4} - y_{nb}) \left[(x_{m4} - x_{ma})n_y - (y_{m4} - y_{ma})n_x\right] \left[z_{ma}(x_{nb}\boldsymbol{I}_1 - \boldsymbol{I}_2) - x_{nb}\boldsymbol{I}_{10} + \boldsymbol{I}_{11}\right]$$
$$+(x_{n4} - x_{nb}) \left[(y_{m4} - y_{ma})n_x - (x_{m4} - x_{ma})n_y\right] \left[z_{ma}(y_{nb}\boldsymbol{I}_1 - \boldsymbol{I}_3) - y_{nb}\boldsymbol{I}_{10} + \boldsymbol{I}_{12}\right]\Big\}. \quad (2.315)$$

The basic integrals in (2.314) and (2.315) can be listed as

$$I_{\{1,4,7,10\}} = \int_{S_{ma}} d\boldsymbol{r}\, \{1, x, y, z\}\, I^1_{in} \quad (2.316)$$

$$I_{\{2,5,8,11\}} = \int_{S_{ma}} d\boldsymbol{r}\, \{1, x, y, z\}\, I^{x'}_{in} \quad (2.317)$$

$$I_{\{3,6,9,12\}} = \int_{S_{ma}} d\boldsymbol{r}\, \{1, x, y, z\}\, I^{y'}_{in} \quad (2.318)$$

$$\boldsymbol{I}_{\{1,4,7,10\}} = \int_{S_{ma}} d\boldsymbol{r}\, \{1, x, y, z\}\, \boldsymbol{I}^{\nabla'}_{in} \quad (2.319)$$

$$\boldsymbol{I}_{\{2,5,8,11\}} = \int_{S_{ma}} d\boldsymbol{r}\, \{1, x, y, z\}\, \boldsymbol{I}^{x'\nabla'}_{in} \quad (2.320)$$

$$\boldsymbol{I}_{\{3,6,9,12\}} = \int_{S_{ma}} d\boldsymbol{r}\, \{1, x, y, z\}\, \boldsymbol{I}^{y'\nabla'}_{in}, \quad (2.321)$$

where I^1_{in}, $I^{x'}_{in}$, $I^{y'}_{in}$, $\boldsymbol{I}^{\nabla'}_{in}$, $\boldsymbol{I}^{x'\nabla'}_{in}$, and $\boldsymbol{I}^{y'\nabla'}_{in}$ are defined in (2.89), (2.104), (2.105), (2.91), (2.294), and (2.295), respectively.

The curl-conforming $\hat{\boldsymbol{n}}\times$LL functions can be employed only for MFIE. As in the case of the RWG and $\hat{\boldsymbol{n}}\times$RWG functions, matrix elements derived for the LL and $\hat{\boldsymbol{n}}\times$LL functions are closely related. On the other hand, similar to the $\hat{\boldsymbol{n}}\times$RWG functions, one can derive an alternative and more efficient implementation for the $\hat{\boldsymbol{n}}\times$LL functions. Using (2.301) and (2.302) in (2.273), one can obtain

$$\overline{\boldsymbol{K}}^{N,n\mathrm{LL}(1,2)}_{PV}[m,n] = \int_{S_m} d\boldsymbol{r}\boldsymbol{t}^{\mathrm{LL}(1,2)}_m(\boldsymbol{r}) \cdot \int_{PV,S_n} d\boldsymbol{r}'\hat{\boldsymbol{n}}' g(\boldsymbol{r},\boldsymbol{r}')\nabla' \cdot \boldsymbol{b}^{\mathrm{LL}(1,2)}_n(\boldsymbol{r}')$$
$$+ \int_{S_m} d\boldsymbol{r}\boldsymbol{t}^{\mathrm{LL}(1,2)}_m(\boldsymbol{r}) \cdot \int_{PV,S_n} d\boldsymbol{r}'\boldsymbol{b}^{\mathrm{LL}(1,2)}_n(\boldsymbol{r}')\frac{\partial g(\boldsymbol{r},\boldsymbol{r}')}{\partial n'}$$
$$- \int_{S_m} d\boldsymbol{r}\boldsymbol{t}^{\mathrm{LL}(1,2)}_m(\boldsymbol{r}) \cdot \int_{PV,S_n} d\boldsymbol{r}'\hat{\boldsymbol{n}}'\nabla' \cdot \left[g(\boldsymbol{r},\boldsymbol{r}')\boldsymbol{b}^{\mathrm{LL}(1,2)}_n(\boldsymbol{r}')\right]. \quad (2.322)$$

As in the previous cases, the expression in (2.322) should be decomposed into basic integrals for the implementation.

Numerical Results (MFIE)

Figures 2.45 and 2.46 present solutions of scattering problems involving a PEC sphere with a radius of a, when $a = \lambda$, $a = 1.5\lambda$, and $a = 2\lambda$. Discretizations of the sphere with $\lambda/10$ triangles lead to 3723, 8364, and 14,871 edges. The sphere is illuminated by a plane wave and scattering problems are solved with FMM and MLFMA implementations employing the RWG,

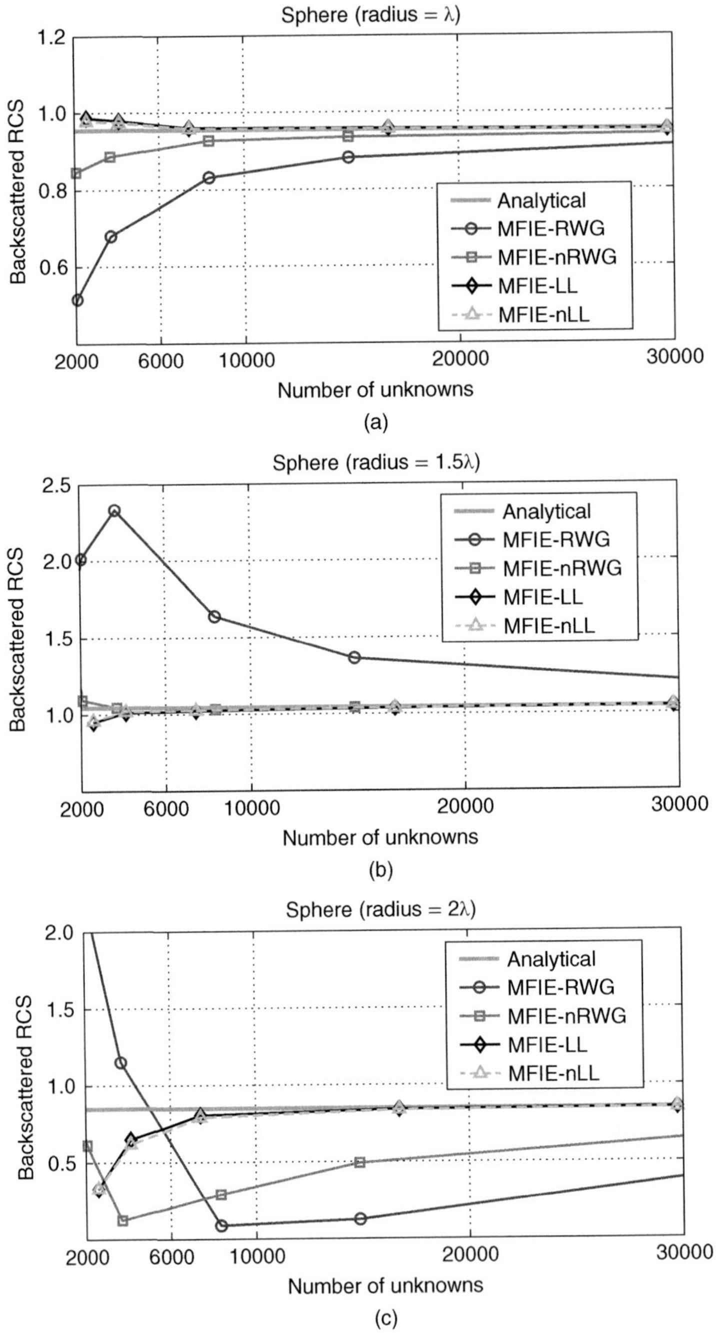

Figure 2.45 The normalized backscattered RCS (RCS $/\pi a^2$) of a PEC sphere with a radius of (a) λ, (b) 1.5λ, and (c) 2λ as a function of the number of unknowns. *Source*: Ergül and Gürel 2006 [128]. Reproduced by permission of Radio Science – American Geophysical Union AGU.

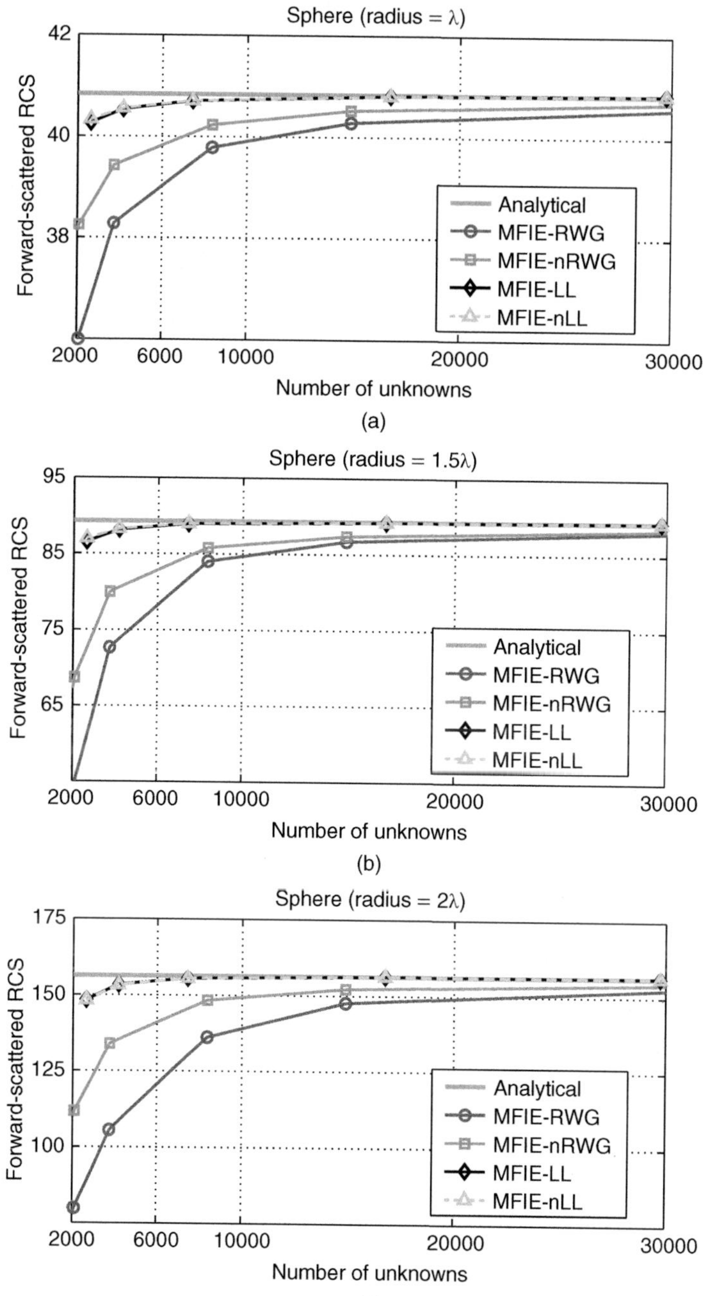

Figure 2.46 The normalized forward-scattered RCS (RCS $/\pi a^2$) of a PEC sphere with a radius of (a) λ, (b) 1.5λ, and (c) 2λ as a function of the number of unknowns. *Source*: Ergül and Gürel 2006 [128]. Reproduced by permission of Radio Science – American Geophysical Union AGU.

$\hat{n}\times$RWG, LL, and $\hat{n}\times$LL functions. Figures 2.45 and 2.46 display the normalized backscattered and forward-scattered RCS (RCS $/\pi a^2$, where a is the radius of the sphere), respectively. RCS values are plotted with respect to the number of unknowns, which corresponds to the number of edges for the RWG and $\hat{n}\times$RWG functions, but twice that for the LL and $\hat{n}\times$LL functions. To compare the performances of the functions, convergence to the analytically calculated value is investigated. It can be observed that the convergence is significantly faster for the LN-LT type functions, namely, the divergence-conforming LL and curl-conforming $\hat{n}\times$LL functions. In other words, for a given number of unknowns, MFIE solutions obtained by these functions are more accurate than those obtained by the RWG and $\hat{n}\times$RWG functions. Also note that the $\hat{n}\times$RWG functions give more accurate RCS values compared to the RWG functions, but the accuracy is relatively poor compared to the LL and $\hat{n}\times$LL functions.

RCS values in Figures 2.45 and 2.46 are plotted with respect to the number of unknowns so that different basis functions can be compared fairly. However, since the LL and $\hat{n}\times$LL functions are defined in pairs at the edges, their triangulation is different from the triangulation of the RWG and $\hat{n}\times$RWG functions to keep the number of unknowns fixed. In other words, a coarser triangulation is employed for the LL and $\hat{n}\times$LL functions. Employing these functions for a given number of unknowns, the number of triangles to model the geometry is half of that for the RWG and $\hat{n}\times$RWG functions. For curved geometries, a coarser triangulation leads to a larger deviation from the actual geometry. Therefore, in the sphere problems, it should be noted that the LL and $\hat{n}\times$LL functions improve the accuracy in spite of the decreased quality of the geometry discretization.

To demonstrate the effect of using different basis functions on the modeling of the induced current, scattering problems involving a PEC cube with edges of 1 m at 300 MHz and 600 MHz are revisited. The cube is located at the origin and illuminated by a plane wave propagating in the $-x$ direction with the electric field polarized in the y direction. Figures 2.47 and 2.48 show the magnitude of the y component of the normalized surface current $\left(|\boldsymbol{J}_y|/|\boldsymbol{H}^{inc}|\right)$ induced on the front surface of the cube when the frequency is 300 MHz and 600 MHz, respectively. The modeled current is obtained for the LN-LT type functions and triangulations with the mesh size of about $\lambda/10$. Comparing Figures 2.47 and 2.48 with Figures 2.32, 2.33, and 2.42, it can be observed that the modeling is significantly improved with the LN-LT type functions compared to the CN-LT type RWG and LN-CT type $\hat{n}\times$RWG functions, especially for the representation of singular currents at the edges. It should be noted that the features and the quality of the current modeling are very similar for both the divergence-conforming LL and curl-conforming $\hat{n}\times$LL functions. Figures 2.49 and 2.50 display the backscattered and forward-scattered RCS (m^2) of the PEC cube, plotted again with respect to the number of unknowns. As in the sphere problems, faster convergence for the LL and $\hat{n}\times$LL functions is observed, while the RWG functions give the worst accuracy for a given number of unknowns in MFIE.

Figure 2.51 presents the boundary-condition error (BCE) in the solution of a scattering problem involving a $\lambda\times\lambda\times\lambda$ conducting box located at the origin. The box is illuminated by a plane wave propagating in the $-x$ direction with the electric field polarized in the y direction. In order to measure the accuracy of the numerical solutions, the magnetic field is sampled to test the boundary condition pointwise on the surface of the box. A relative error

$$\Delta_R(\boldsymbol{r}) = \frac{|\boldsymbol{J}(\boldsymbol{r}) - \hat{\boldsymbol{n}}\times\boldsymbol{H}(\boldsymbol{r})|}{|\hat{\boldsymbol{n}}\times\boldsymbol{H}^{inc}(\boldsymbol{r})|} \tag{2.323}$$

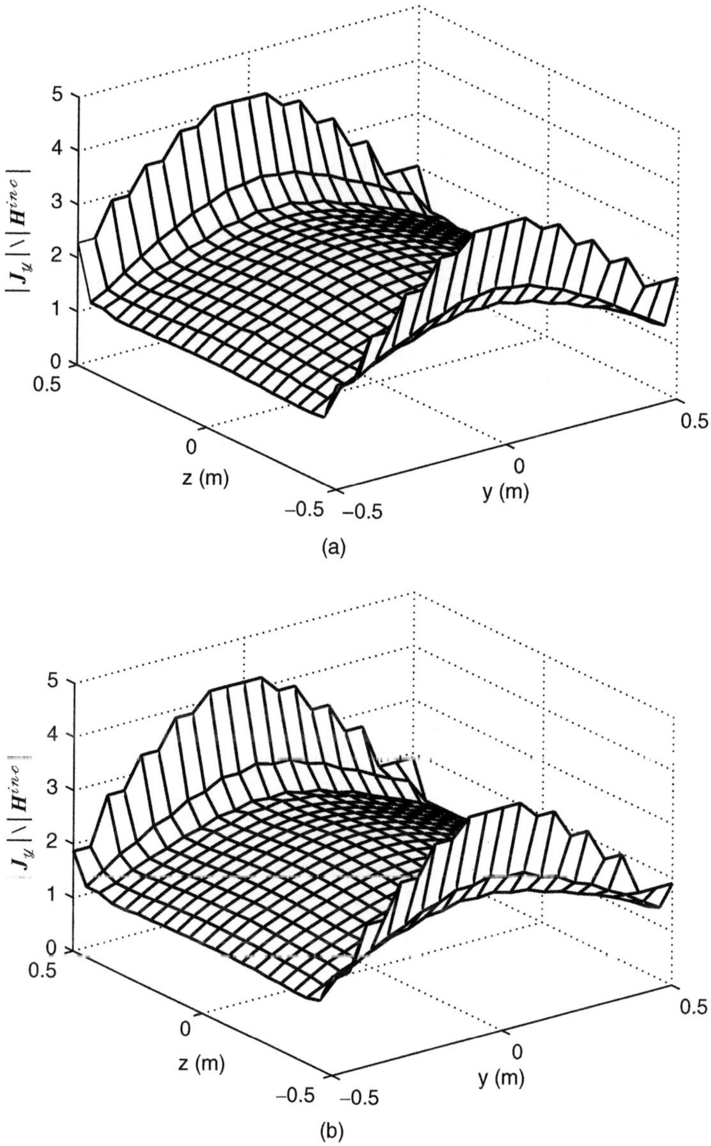

Figure 2.47 The magnitude of the y component of the normalized surface current $\left(|\boldsymbol{J}_y|/|\boldsymbol{H}^{inc}|\right)$ induced on the front surface (at $x = 0.5$ m) of a PEC cube with edges of λ, obtained by using MFIE with (a) LL and (b) $\hat{\boldsymbol{n}}\times$LL. *Source*: Ergül and Gürel 2006 [128]. Reproduced by permission of Radio Science – American Geophysical Union AGU.

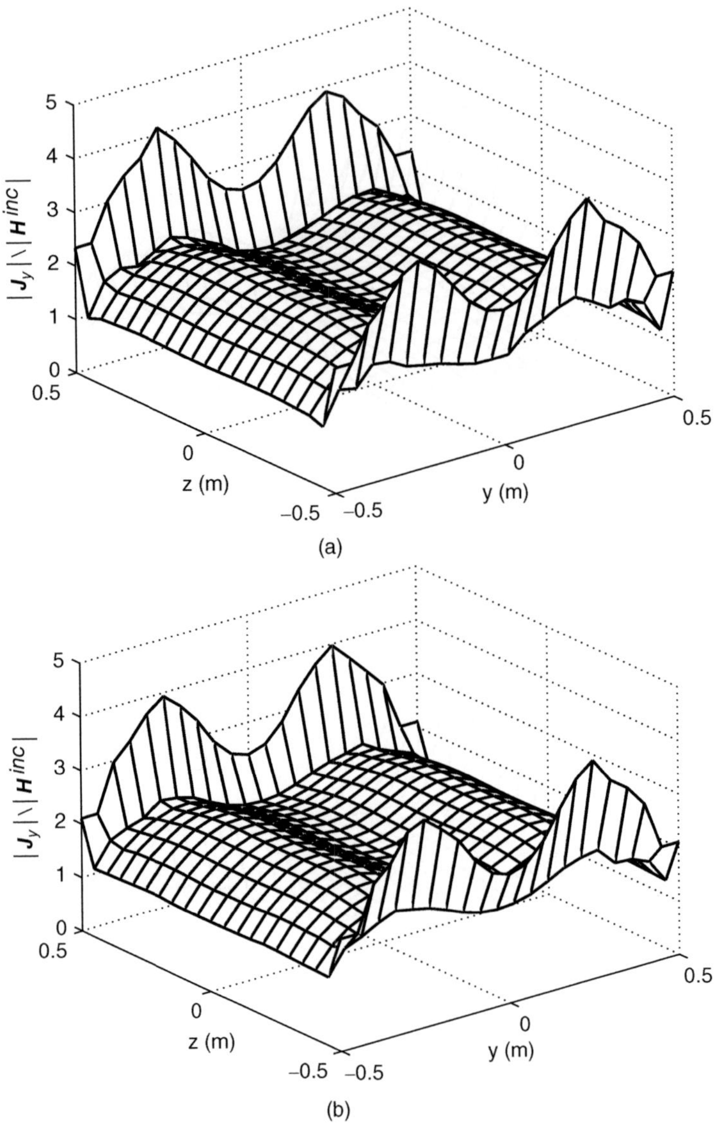

Figure 2.48 The magnitude of the y component of the normalized surface current $(|\boldsymbol{J}_y|/|\boldsymbol{H}^{inc}|)$ induced on the front surface (at $x = 0.5$ m) of a PEC cube with edges of 2λ, obtained by using MFIE with (a) LL and (b) $\hat{\boldsymbol{n}}\times$LL. *Source*: Ergül and Gürel 2006 [128]. Reproduced by permission of Radio Science – American Geophysical Union AGU.

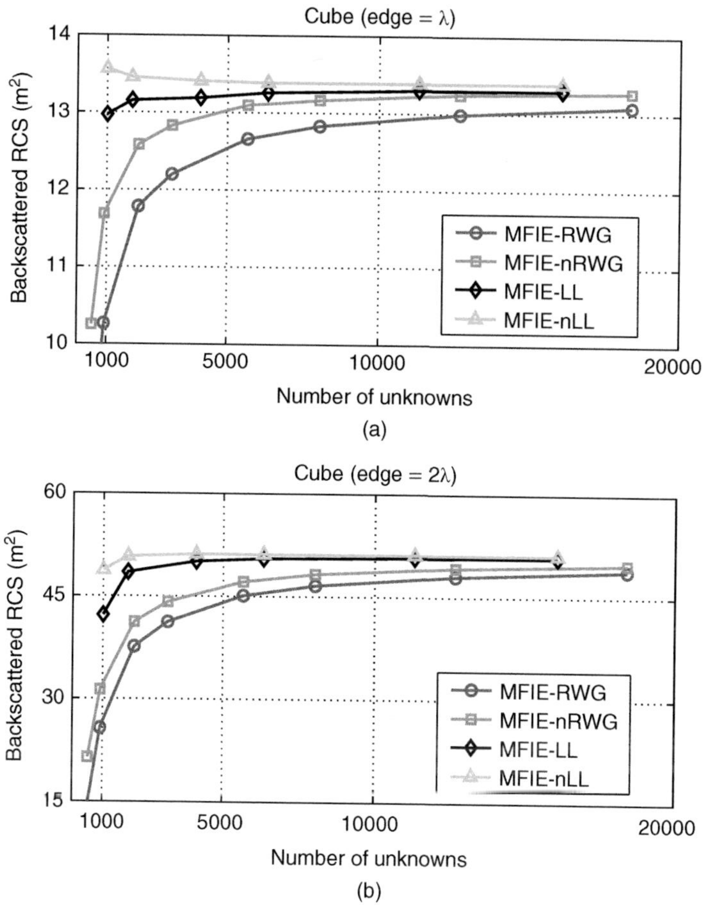

Figure 2.49 The backscattered RCS (m^2) of a PEC cube with edges of (a) λ and (b) 2λ as a function of the number of unknowns. *Source*: Ergül and Gürel 2006 [128]. Reproduced by permission of Radio Science – American Geophysical Union AGU.

is calculated at various observation points $\boldsymbol{r} \in S$. Then, BCE is defined as the integral of the relative error (2.323) over the entire surface, i.e.,

$$\Delta_{BCE}(\boldsymbol{r}) = \int_S d\boldsymbol{r} \Delta_R(\boldsymbol{r}). \tag{2.324}$$

Figure 2.51 depicts BCE as a function of the number of unknowns and for various discretizations of the problem with different basis functions. It can be observed that BCE for the LN-LT type functions is approximately half of that for the RWG and $\hat{\boldsymbol{n}}\times$RWG functions. In other words, the boundary condition for the magnetic field is satisfied better by using the LN-LT type functions. This explains why the LN-LT type functions produce more accurate RCS results than the RWG and $\hat{\boldsymbol{n}}\times$RWG functions.

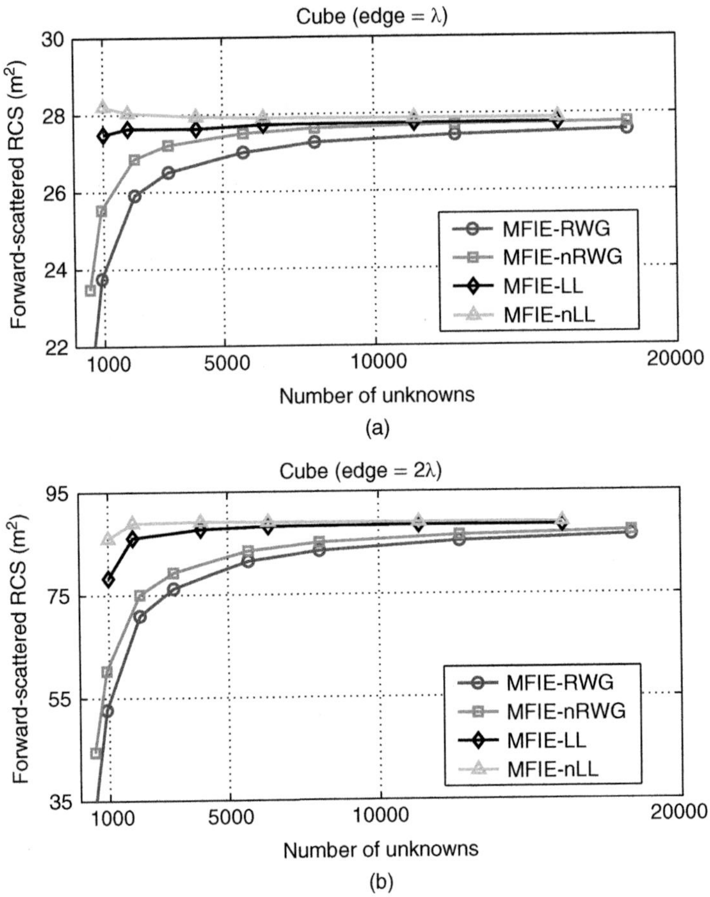

Figure 2.50 The forward-scattered RCS (m^2) of a PEC cube with edges of (a) λ and (b) 2λ as a function of the number of unknowns. *Source*: Ergül and Gürel 2006 [128]. Reproduced by permission of Radio Science – American Geophysical Union AGU.

As an another example, Figure 2.52 presents the results of a scattering problem involving a wing-shaped object with sharp edges and corners. The size of the object is 2λ and it is illuminated by a plane wave propagating in the $-x$ direction with the electric field polarized in the y direction. In Figure 2.52, BCE is plotted as a function of the number of unknowns, when matrix elements are calculated with 1% error. Similar to the cube problem, BCE for the LN-LT functions are significantly less than that for the RWG and $\hat{\boldsymbol{n}} \times$ RWG functions.

Numerical Results (CFIE)

The divergence-conforming LL functions can also be used to improve the accuracy of CFIE, similar to MFIE. As an example, Figure 2.53 presents solutions of scattering problems involving a sphere with a radius of a, when a changes from 0.5λ to 1.5λ. The relative error in bistatic

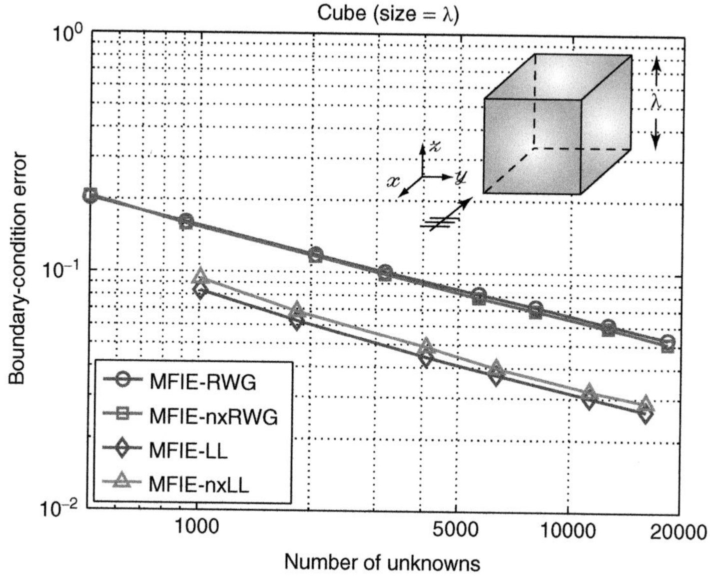

Figure 2.51 BCE in the solution of a scattering problem involving a $\lambda \times \lambda \times \lambda$ conducting box formulated with MFIE and discretized with four different basis functions on various triangulations.

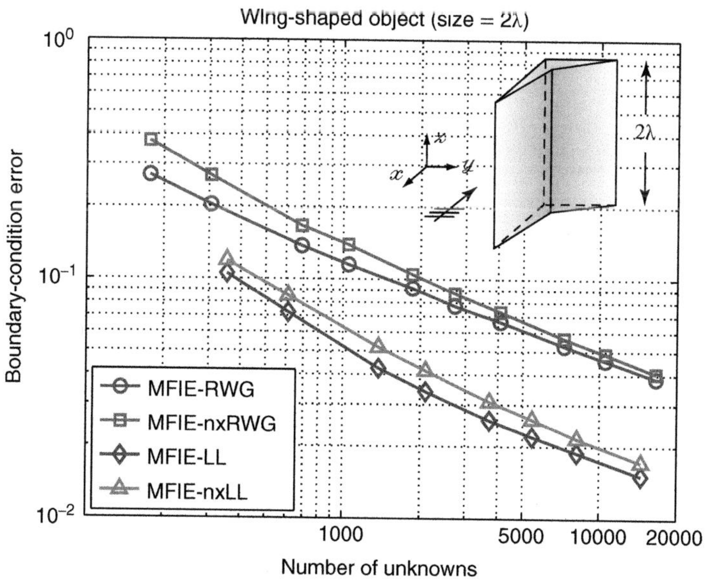

Figure 2.52 BCE in the solution of a scattering problem involving a wing-shaped object formulated with MFIE and discretized with four different basis functions on various triangulations.

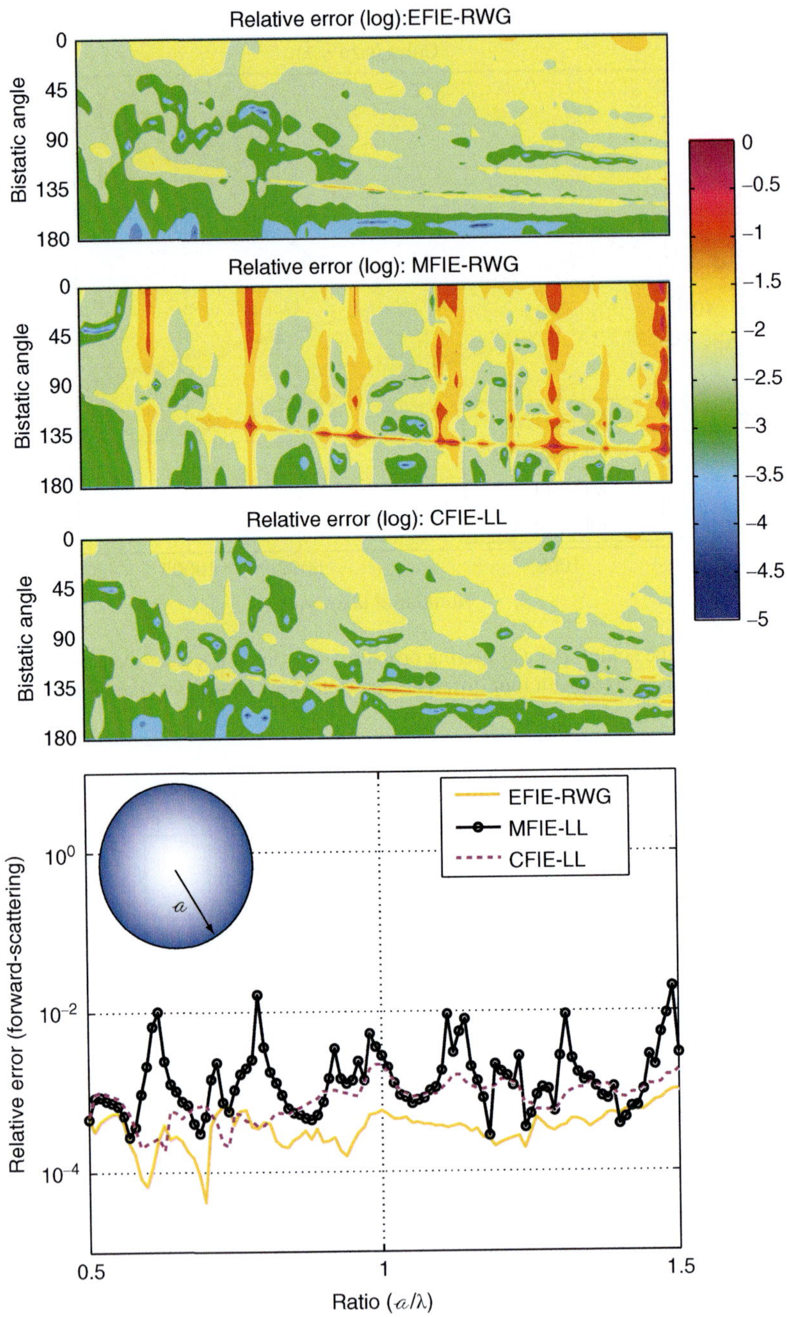

Figure 2.53 The relative error in the bistatic RCS of a PEC sphere with a radius of a, when a changes from 0.5λ to 1.5λ. The computational results are obtained by using EFIE discretized with the RWG functions, as well as MFIE and CFIE discretized with the LL functions.

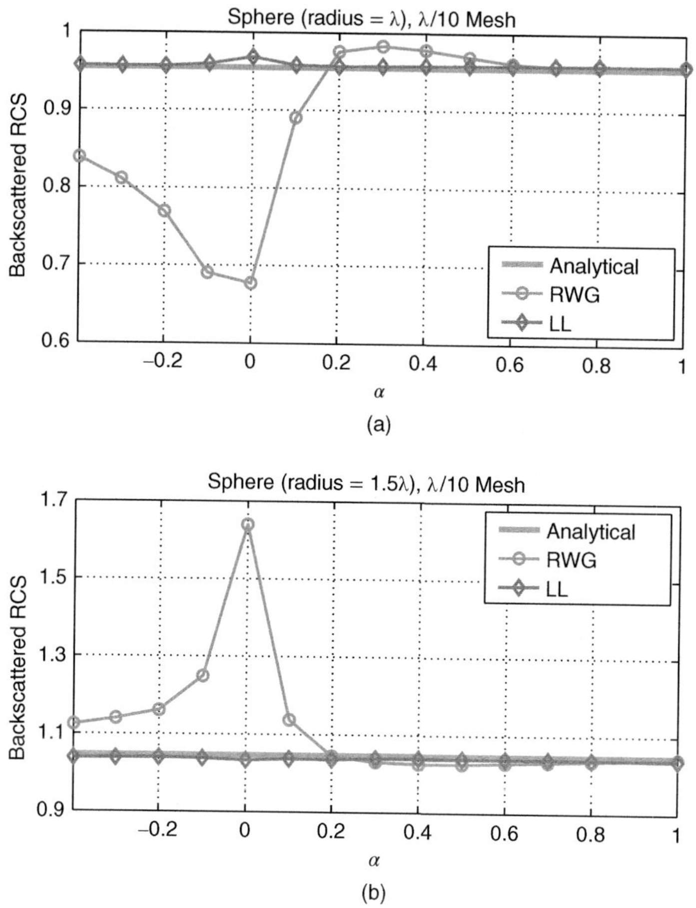

Figure 2.54 The normalized backscattered RCS (RCS $/\pi a^2$) of a PEC sphere with a radius of (a) λ and (b) 2λ as a function of α in CFIE. The number of unknowns is (a) 3723 for RWG and 7446 for LL, (b) 8364 for RWG and 16728 for LL. *Source*: Ergül and Gürel 2006 [128]. Reproduced by permission of Radio Science – American Geophysical Union AGU.

RCS values obtained with the EFIE, MFIE, and CFIE ($\alpha = 0.2$) formulations is plotted as a function of the bistatic angle and frequency. The relative error in the forward-scattering direction is also considered separately as a function of the frequency. EFIE is discretized with the RWG functions, whereas MFIE and CFIE are discretized with the LL functions. Comparing results in Figure 2.53 with the previous results in Figure 2.31, it can be observed that MFIE and CFIE become as accurate as the reference EFIE with the RWG functions. In MFIE solutions, there are two consequences when the RWG functions are replaced by the LL functions. First, the effects of internal resonances are narrower with respect to the frequency for the LL functions. Second, the excessive error of MFIE with the RWG functions disappears and the accuracy is improved for all frequencies. Using the LL functions instead of the RWG functions, accuracy of CFIE is also improved to levels of EFIE.

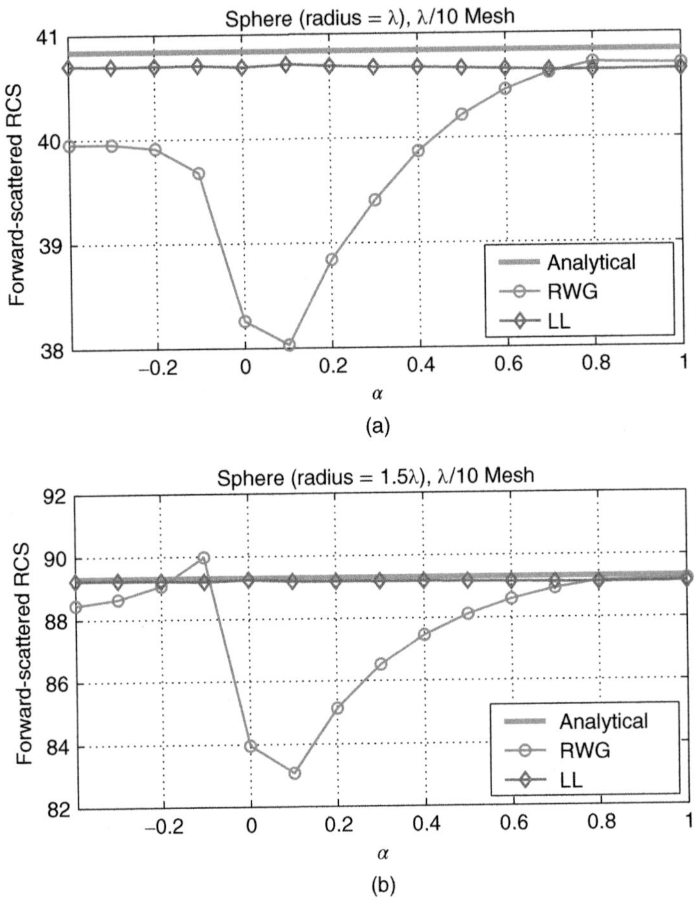

Figure 2.55 The normalized forward-scattered RCS (RCS $/\pi a^2$) of a PEC sphere with a radius of (a) λ and (b) 2λ as a function of α in CFIE. The number of unknowns is (a) 3723 for RWG and 7446 for LL, (b) 8364 for RWG and 16728 for LL. *Source*: Ergül and Gürel 2006 [128]. Reproduced by permission of Radio Science – American Geophysical Union AGU.

To further compare the RWG and LL functions in terms of the accuracy of CFIE implementations, Figures 2.54 and 2.55 present the backscattered and forward-scattered RCS, respectively, of a PEC sphere with a radius of a, when $a = \lambda$ and $a = 1.5\lambda$. For each case, the sphere is discretized with $\lambda/10$ triangles. RCS is calculated for different values of the coupling parameter α. Values at $\alpha = 0$ and $\alpha = 1$ correspond to those obtained by MFIE and EFIE, respectively. Besides the overall improvement in the accuracy by the use of the LL functions, Figures 2.54 and 2.55 demonstrate that the significant and undesired variance in the accuracy with the use of the RWG functions is dramatically reduced with the use of the LL functions. In other words, accuracy is stabilized with respect to the variable α and it becomes weakly dependent on the choice of this variable.

Although there is no optimum α for the accuracy of the LL functions, Figure 2.56 shows that the conditioning of the matrix equation depends strongly on this variable. In this figure,

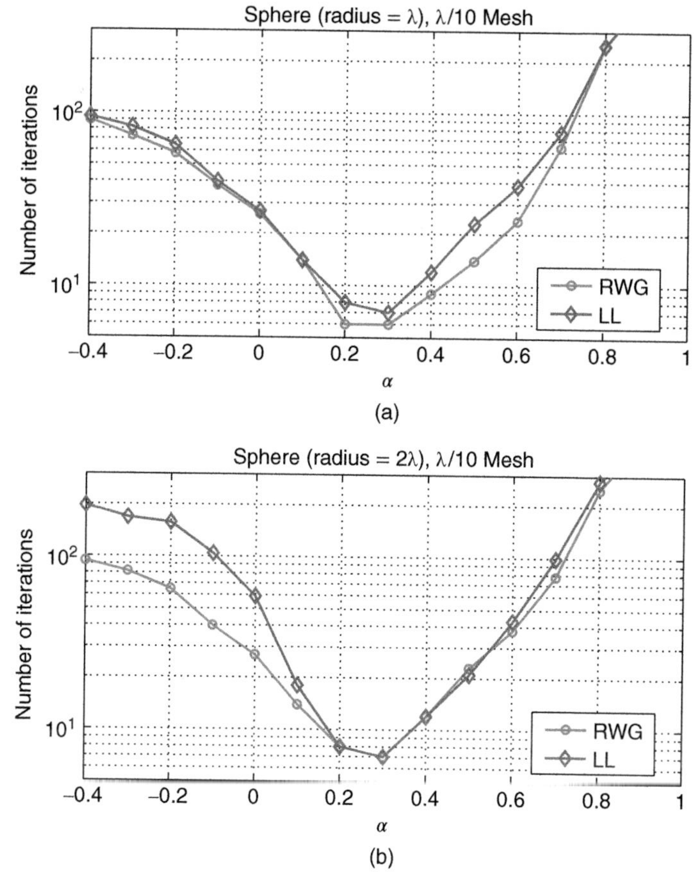

Figure 2.56 Iteration counts (10^{-3} residual error) for the solution of scattering problems involving a sphere of radius (a) λ (3723 edges) and (b) 1.5λ (8364 edges) with respect to α in CFIE. *Source*: Ergül and Gürel 2006 [128]. Reproduced by permission of Radio Science – American Geophysical Union AGU.

CGS iteration counts required to reach 10^{-3} residual error for the FMM solutions of the sphere problems are depicted. FMM employing the CGS solver is accelerated by BDP obtained from the self-interactions of boxes in the FMM structure. It is observed that the approximate range of α from 0.2 to 0.3 is optimum for both the RWG and LL functions.

Figures 2.57 and 2.58 present the backscattered and forward-scattered RCS of a cube with edges of 1 m, when the frequency is 300 MHz and 600 MHz. CFIE formulations ($\alpha = 0.2$) employing the RWG and LL functions are compared. Similar to the previous results, one can observe a significant improvement in the accuracy of CFIE by using the LL functions.

Figure 2.59 presents the solution of a scattering problem involving a stealth airborne target Flamme [149]. The scaled size of the Flamme is 0.6 m corresponding to approximately 6λ at 3 GHz, and it is illuminated by a plane wave propagating in the x-y plane at $\phi = 160°$ angle from the nose. The electric field is polarized in the θ direction (vertical polarization). Figure 2.59 presents the co-polar bistatic RCS (in dBm²) on the x-y plane as a function of

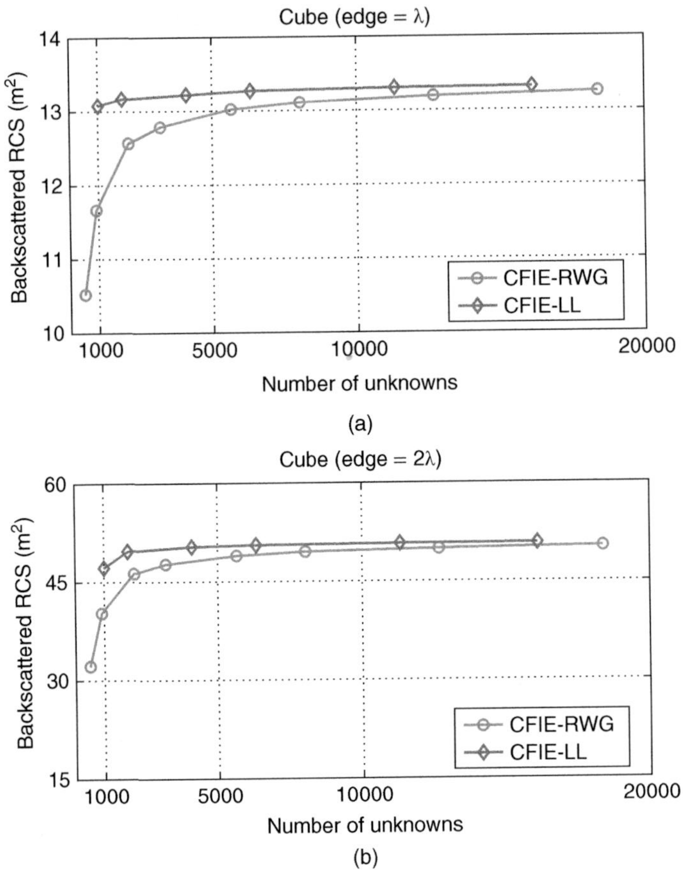

Figure 2.57 The backscattered RCS (m^2) of a PEC cube with edges of (a) λ and (b) 2λ as a function of the number of unknowns. *Source*: Ergül and Gürel 2006 [128]. Reproduced by permission of Radio Science – American Geophysical Union AGU.

the bistatic angle ϕ from 210° to 360°. It can be observed that computational results provided by EFIE employing the RWG functions agree well with measurements. However, the same agreement cannot be obtained when using CFIE discretized with the RWG functions. As also depicted in Figure 2.59, the accuracy of CFIE can be improved to levels of EFIE by employing the LL functions, instead of the RWG functions.

Numerical Results (Large-Scale Problems)

Replacing the RWG functions with the LL functions can significantly improve the accuracy of MFIE and CFIE, even for large and smooth geometries. As an example, scattering problems involving a large sphere with a radius of 6λ are considered. The sphere is illuminated by a plane wave, and scattering problems with different discretizations are solved by using MLFMA implementations employing the MFIE and CFIE ($\alpha = 0.2$) formulations. In Figure 2.60(a), the

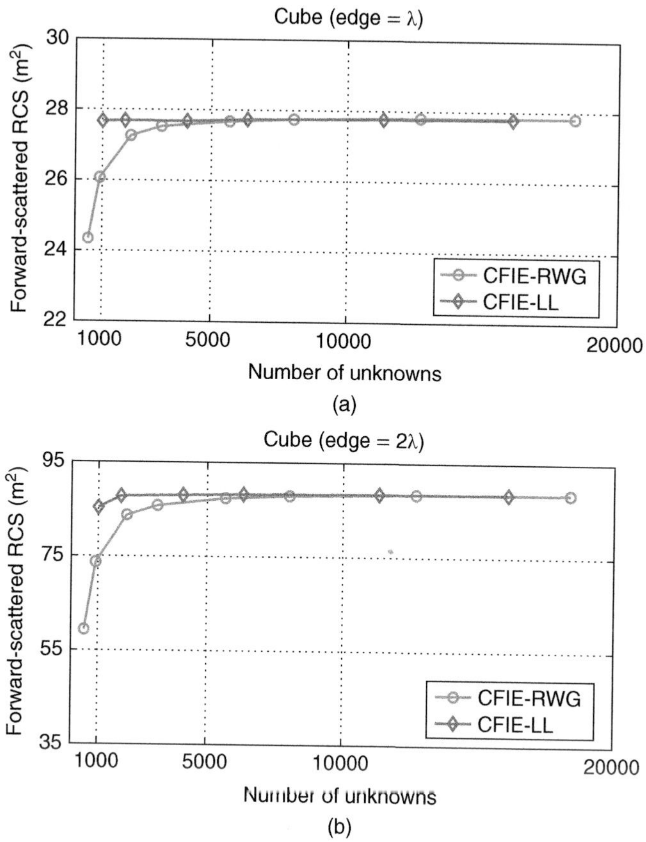

Figure 2.58 The forward-scattered RCS (m^2) of a PEC cube with edges of (a) λ and (b) 2λ as a function of the number of unknowns. *Source*: Ergül and Gürel 2006 [128]. Reproduced by permission of Radio Science – American Geophysical Union AGU.

normalized forward-scattered RCS is plotted with respect to the number of unknowns. Since the size of the scatterer is fixed with respect to the wavelength, the number of unknowns is controlled by changing the fineness of the triangulation. As the number of unknowns increases from about 50 000 to 500 000, all computational curves approach the analytical value obtained by a Mie-series solution. Convergence is significantly faster for the MFIE and CFIE formulations with the LL functions, and MFIE with the RWG functions has the slowest convergence. One can also observe the relatively poor performance of CFIE with the RWG functions. Since the computational resources required for the RWG and LL functions are similar for a fixed number of unknowns, the LL functions are superior to the RWG functions because of the significant boost in the accuracy.

Figure 2.60(b) offers another view for the accuracy of RCS results obtained with the RWG and LL functions. The RCS error, which is defined as the error in the scattered field in reference to an exact Mie-series solution, is plotted on a log-log scale as a function of the number of linear unknowns per wavelength. Then, the slope of the error curve signifies the order of the

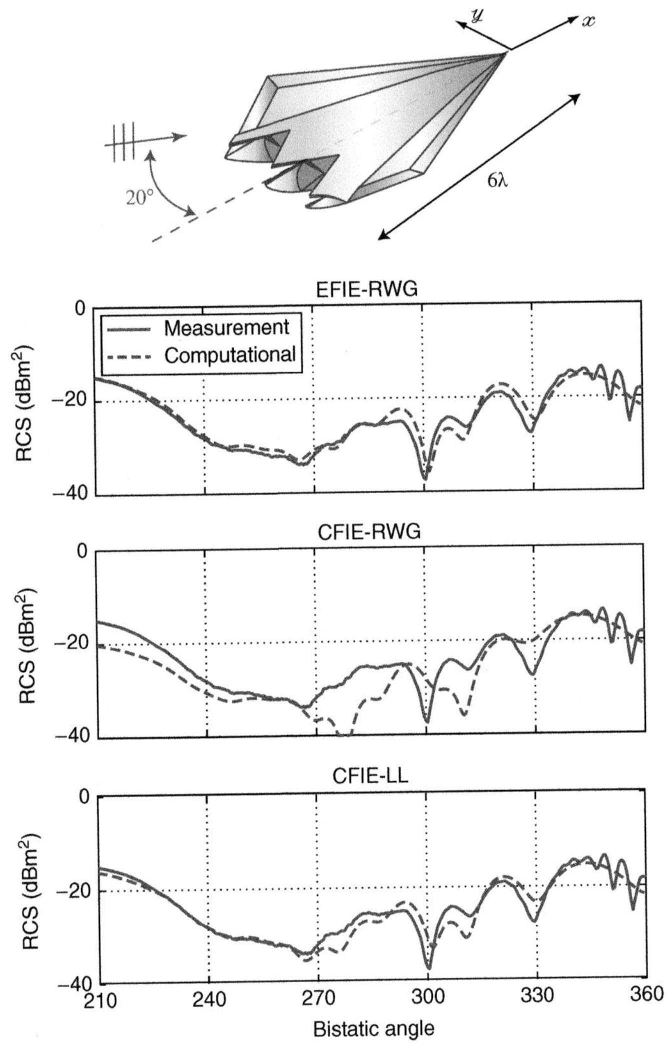

Figure 2.59 The co-polar bistatic RCS (in dBm^2) of the stealth airborne target Flamme at 3 GHz. The maximum dimension of the scaled geometry is 0.6 m, corresponding to approximately 6λ at 3 GHz. The target is illuminated by a plane wave propagating in the x-y plane at 160° angle from the x axis, with the electric field polarized in θ direction (vertical polarization).

convergence of the numerical solution. It is seen in Figure 2.60(b) that the error decreases with higher slopes for the LL functions than for the RWG functions, demonstrating that the LL functions provide higher-order convergence. Note that the slopes of the error curves for the LL functions are not constant, as opposed to the slopes of the error curves for the RWG functions. This is because the LL functions reduce the error so effectively that the residual error is mainly due to the discretization, i.e., the error in the geometrical representation of the sphere.

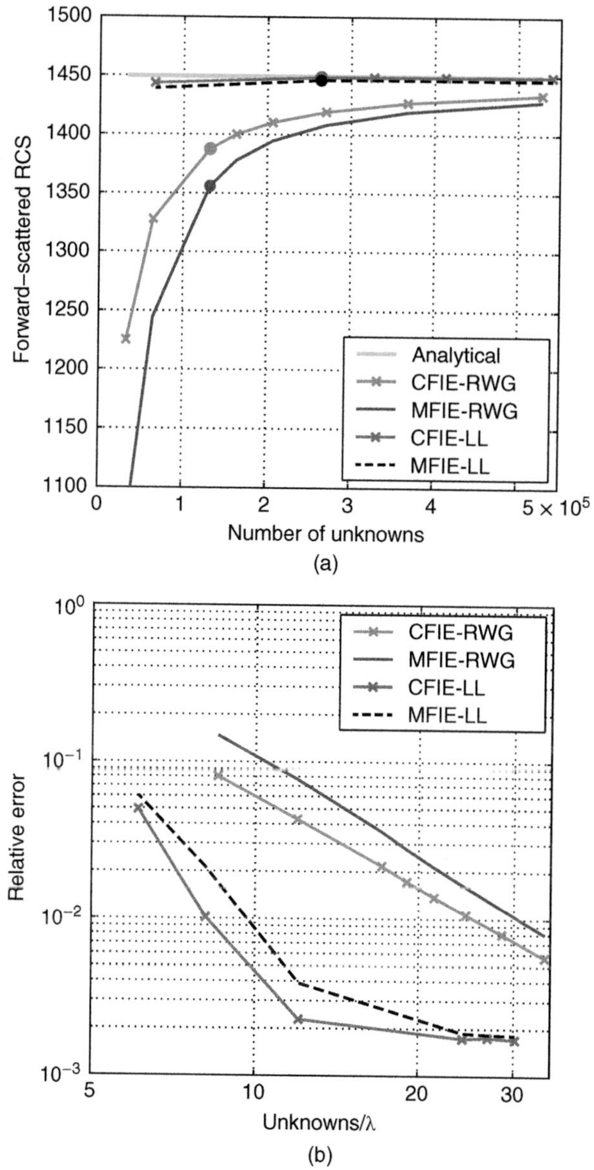

Figure 2.60 (a) The normalized forward-scattered RCS (RCS$/\pi a^2$) of a sphere with a radius of 6λ as a function of the number of unknowns. The dots on the curves correspond to the $\lambda/10$ discretization. (b) The error in the scattered field as a function of the number of linear unknowns per wavelength. *Source*: Ergül and Gürel 2007 [129]. Reproduced with permission of IEEE.

In Figure 2.61(a), the normalized bistatic RCS (RCS $/\pi a^2$) of a sphere with a radius of 6λ is presented. The sphere is discretized with mesh sizes of about $\lambda/8$ for RWG and $\lambda/5$ for LL. Then, the number of unknowns is approximately the same for the RWG and LL cases although the triangulation is coarser for the LL case. In Figure 2.61(a), computational results are obtained with the CFIE ($\alpha = 0.2$) formulation, and analytical results are also displayed. The plot is divided into two parts with different scales of the y axis for clarity of details. For bistatic directions from 0° (backscattering) to 135°, the LL functions unambiguously provide more accurate results. In the range from 135° to 180° (forward-scattering), the improvement is not evident in the dB scale. On the other hand, the relative error is plotted in Figure 2.61(b), where the large error in the forward-scattering direction is visible. The error with the LL functions is consistently lower and under 0.2%. The error with the RWG functions peaks in the forward-scattering direction, where it is above 4%. This error may not acceptable since near-field and far-field interactions of the matrix equation are calculated with a maximum error of 1% and 0.1%, respectively. Furthermore, the relative residual error in the iterative solution is 10^{-6} so that one expects the total computational error to be 1% at most. In the RWG case, the error due to the simultaneous discretization of the geometry and current distribution is significant and dominates the computational error. As Figure 2.61(b) suggests, the excessive error due to the discretization with the RWG functions can be easily prevented by replacing them with the LL functions.

Figure 2.62 presents the results of scattering problems involving a relatively long object with sharp edges and corners, as also described in the figure. The scatterer is composed of planar surfaces only, and therefore, its triangulation does not lead to a deviation from the actual geometry, as opposed to the sphere. The object is illuminated by a plane wave propagating in the $-y$ direction and the forward-scattered RCS is calculated at 4 GHz by employing the MFIE and CFIE ($\alpha = 0.2$) formulations with the RWG and LL functions. Different discretizations with the triangulation size varying from 1.5 cm to 0.375 cm are applied, leading to total of 15,340 to 238,672 triangles, respectively. The $\lambda/10$ triangulation corresponds to 91 782 edges on the object. Figure 2.62(b) presents forward-scattered RCS values with respect to number of unknowns; once again, all four curves tend to converge to each other as the mesh gets denser, the number of unknowns increases, and computations become more accurate. However, the convergence rate depends on the types of the formulation and basis function. MFIE with the RWG functions has the slowest convergence while CFIE with the RWG functions shows relatively poor performance compared to the MFIE and CFIE formulations employing the LL functions. As in the case of the large sphere problem, the LL functions provide significantly improved accuracy for both MFIE and CFIE solutions of the problem.

2.3.11 *Excessive Discretization Error of the Identity Operator*

Recent studies on the accuracy of surface integral equations show that the identity operator is a major error source in numerical solutions. This may explain why normal and mixed formulations involving well-tested identity operators are significantly inaccurate compared to tangential formulations. Discretization of the identity operator is trivial. The integral

$$\bar{\boldsymbol{I}}^T[m,n] = \int_{S_m} d\boldsymbol{r}\boldsymbol{t}_m(\boldsymbol{r}) \cdot \boldsymbol{b}_n(\boldsymbol{r}) \tag{2.325}$$

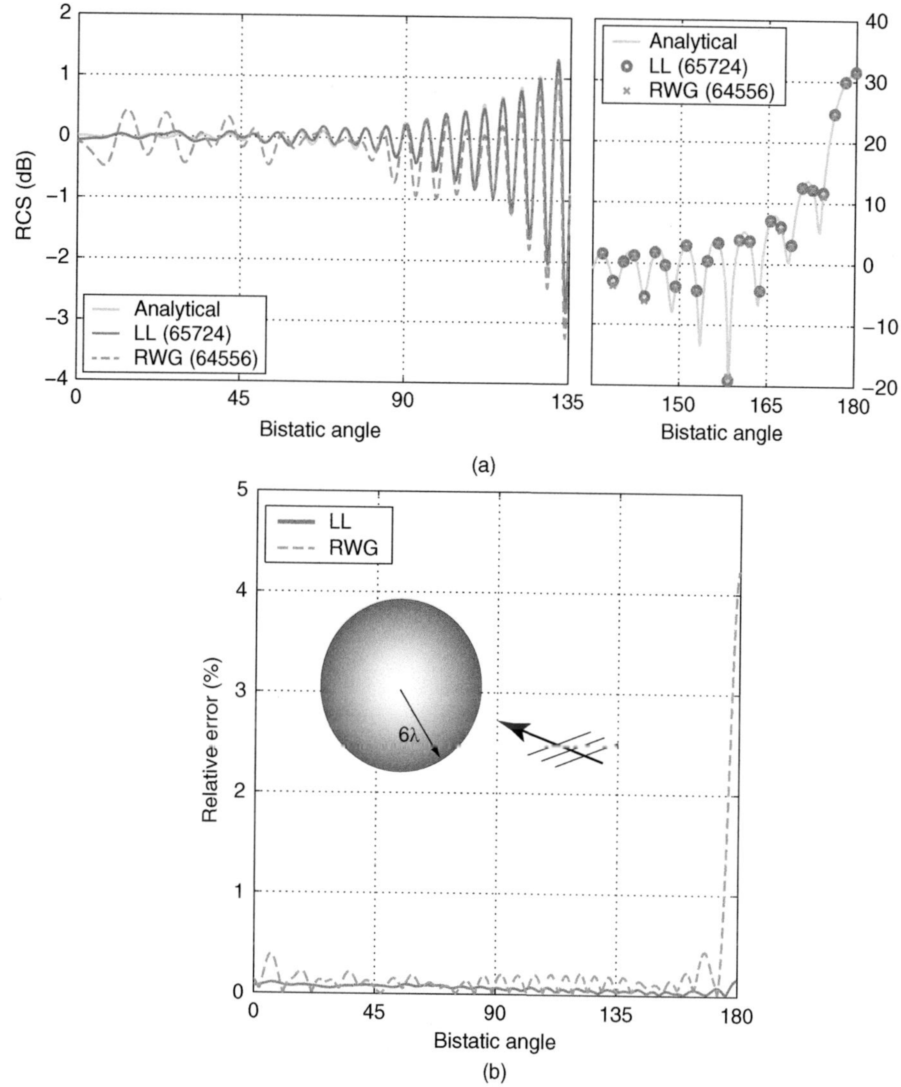

Figure 2.61 (a) The normalized bistatic RCS (RCS $/\pi a^2$) of a sphere with a radius of 6λ as a function of the bistatic angle from 0° (backscattering) to 180° (forward-scattering). (b) The relative error for the bistatic RCS depicted in Figure 2.61(a). The error for the RWG functions is significantly larger than the error for the LL functions, especially in the forward-scattering direction, where the RWG error is above 4%. *Source*: Ergül and Gürel 2007 [129]. Reproduced with permission of IEEE.

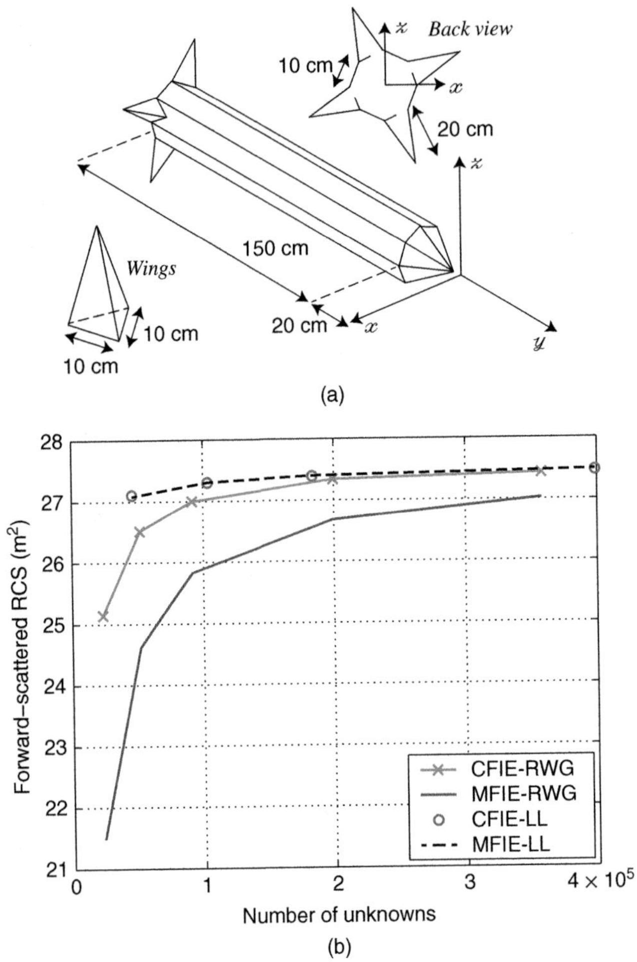

Figure 2.62 (a) A scatterer composed of planar surfaces connected with sharp edges and corners. (b) The forward-scattered RCS (m^2) with respect to the number of unknowns for the scatterer in Figure 2.62(a) illuminated by a plane wave propagating in the $-y$ direction at 4 GHz. The curves are obtained with CFIE and MFIE implementations employing the RWG and the LL functions. *Source*: Ergül and Gürel 2007 [129]. Reproduced with permission of IEEE.

does not contain any singularity and it can be evaluated accurately (even exactly, if desired) by using a low-order Gaussian quadrature rule. On the other hand, the identity operator behaves like an operator with a highly singular kernel [120],[140]. This alternative interpretation can be understood when (2.325) is rewritten as a double integral over basis and testing functions as

$$\bar{\boldsymbol{I}}^T[m,n] = \int_{S_m} d\boldsymbol{r}\boldsymbol{t}_m(\boldsymbol{r}) \cdot \int_{S_n} d\boldsymbol{r}'\,\delta(\boldsymbol{r},\boldsymbol{r}')\boldsymbol{b}_n(\boldsymbol{r}), \tag{2.326}$$

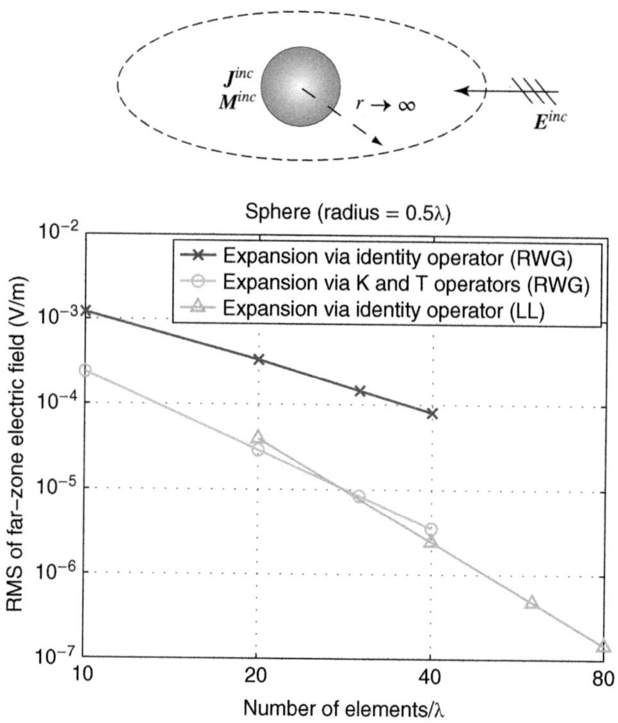

Figure 2.63 RMS of the far-zone electric field due to nonradiating currents $\{\boldsymbol{J}^{inc}(\boldsymbol{r}), \boldsymbol{M}^{inc}(\boldsymbol{r})\}$ expanded on the surface of a sphere with a radius of 0.5λ. The surface is illuminated by a plane wave with unit amplitude propagating in free space.

where $\delta(\boldsymbol{r}, \boldsymbol{r}')$ is a Dirac delta function as the kernel of the operator. Therefore, the discretization of the identity operator may cause an unexpectedly large error, even though the discretization itself involves very small or no error.

To present the excessive error due to the discretization of the identity operator, one can consider the expansion of the nonradiating currents $\{\boldsymbol{J}^{inc}, \boldsymbol{M}^{inc}\}$ in a series of the RWG functions, as in (2.176). For an arbitrary surface, the unknown coefficients $\boldsymbol{a}_J^{inc}$ and $\boldsymbol{a}_M^{inc}$ can be calculated by using two different methods:

- The matrix equation in (2.193) is solved. This method involves only integro-differential operators and does not contain any (well-tested) identity operator.
- The identity equation in (2.190), which is remarkably simpler than the first one, is solved. This method does not involve the discretization of any integro-differential operators.

Note that the two expansion methods are analogous to solutions of electromagnetics problems using tangential and normal/mixed formulations, respectively. In fact, solutions with surface integral equations can be interpreted as the expansion of total currents (instead of nonradiating currents) in a series of basis functions.

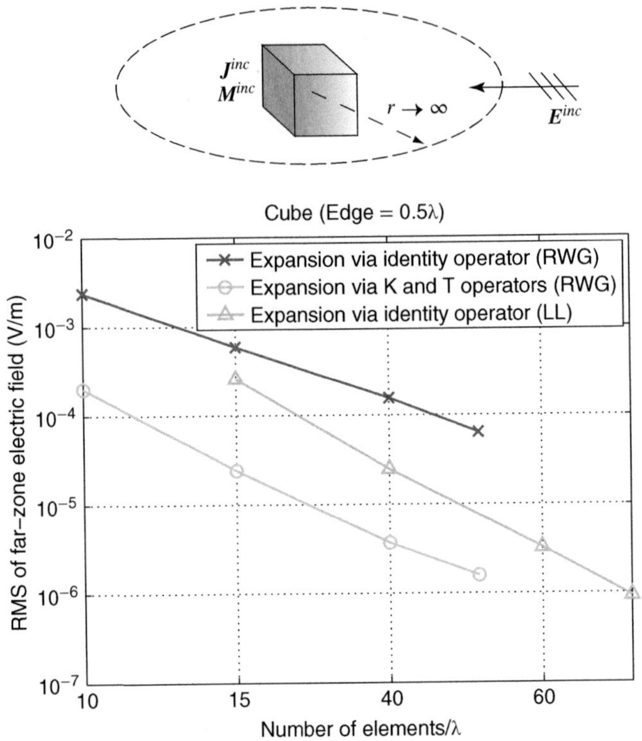

Figure 2.64 RMS of the far-zone electric field due to nonradiating currents $\{\boldsymbol{J}^{inc}(\boldsymbol{r}), \boldsymbol{M}^{inc}(\boldsymbol{r})\}$ expanded on the surface of a cube with 0.5λ edges. The surface is illuminated by a plane wave with unit amplitude propagating in free space.

Figures 2.63 and 2.64 present computational experiments on two different geometries; a sphere with a radius of 0.5λ and a cube with edges of 0.5λ. Both objects are illuminated by a plane wave propagating in the $-x$ direction with the electric field polarized in the y direction. Amplitude of the plane wave is unity. Nonradiating currents $\{\boldsymbol{J}^{inc}, \boldsymbol{M}^{inc}\}$ are expanded in a series of the RWG functions on the objects using the two methods described above, i.e., using directly tested $\mathcal{K}$ and $\mathcal{T}$ operators and using the identity operator. Expansion coefficients are calculated and used to compute radiated fields in the far zone. The far-zone electric field in θ and ϕ directions is sampled at 360 points as

$$\boldsymbol{E}^{\infty}_{\{\theta,\phi\}}[n] = \lim_{r\to\infty} \left\{ r\boldsymbol{E}_{\{\theta,\phi\}}(r, \theta = \pi/2, \phi_n) \right\}, \tag{2.327}$$

where $\phi_n = (n-1)\pi/180$ for $n = 1, 2, \ldots, 360$. Then, the root mean square (RMS) of the total electric field is calculated as

$$f_{\text{RMS}} = \sqrt{\frac{1}{360} \sum_{n=1}^{360} \left(|\, \boldsymbol{E}^{\infty}_{\theta}[n]|^2 + |\, \boldsymbol{E}^{\infty}_{\phi}[n]|^2 \right)}. \tag{2.328}$$

Since the expanded currents $\{\boldsymbol{J}^{inc}, \boldsymbol{M}^{inc}\}$ should not radiate, RMS value in (2.328) directly corresponds to the error in numerical calculations. Using the RWG or LL functions defined on planar triangles, one can compute far-zone fields analytically; hence, the error is only due to the discretization of currents.

Figures 2.63 and 2.64 present the RMS value in (2.328) for the sphere and cube problems, respectively, as a function of the number of elements per wavelength. In both cases, the RMS value decreases as the triangulation becomes finer. However, there exists a significant discrepancy between the results obtained with the two discretization methods. Given a mesh size, the RMS value is consistently smaller with the first method using integro-differential operators compared to the second method using well-tested identity operators. In other words, the discretization of the identity operator generates larger error compared to the discretizations of the $\mathcal{K}$ and $\mathcal{T}$ operators, even when the same set of the RWG functions is used. Figures 2.63 and 2.64 also show that the error can be reduced by using the LL functions for the discretization of the identity operator.

2.4 Composite Objects with Multiple Dielectric and Metallic Regions

Surface formulations can easily be extended to those problems including composite objects with multiple dielectric and PEC regions. The generalized procedure consists of the following main stages:

- Formulate an equivalent problem for each nonmetallic domain by defining equivalent currents, applying operators on currents, and enforcing boundary conditions for tangential fields.
- Perform a discretization for each domain by employing oriented basis and testing functions.
- Combine related unknowns on boundaries and the corresponding equations to form a single matrix equation to solve.
- Calculate radiated electric and magnetic fields from equivalent currents.

This generalized procedure is discussed in [75] in the context of a EFIE-CFIE-PMCHWT formulation, together with various techniques to handle junctions, where three or more domains intersect. In this section, JMCFIE is extended to those problems involving composite structures with multiple dielectric and PEC regions.

Consider a general case involving U regions, namely, $D_0, D_1, \ldots, D_{U-1}$, and D_0 is a region extending to infinity. Each region D_u for $u = 0, 1, \ldots, (U-1)$ is either metallic with perfect conductivity or dielectric with constant electromagnetic parameters, i.e., permittivity ϵ_u and permeability μ_u. Each region D_u has a nonzero volume and is bounded by a closed surface S_u. Then,

$$S_u = \sum_{v=1\ v\neq u}^{U} S_{uv}, \tag{2.329}$$

where $S_{uv} = S_{vu}$ is the interface between the regions D_u and D_v. For an interface S_{uv} for $u < v$, the unit normal vector $\hat{\boldsymbol{n}}$ is directed into the region D_u.

For numerical solutions, surface currents are expanded in a series of the RWG functions. If there is no junction, each basis (RWG) function is located on the interface of two regions,

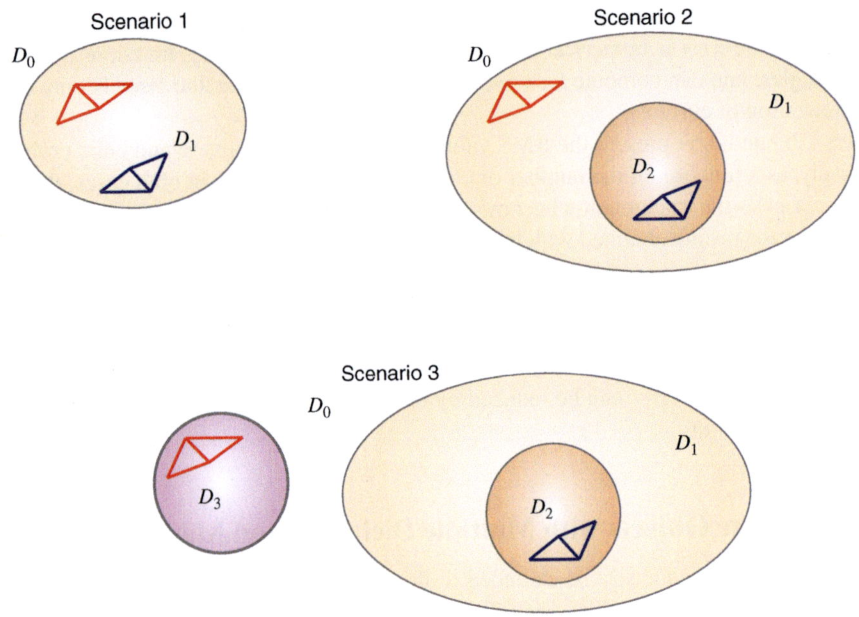

Figure 2.65 Different scenarios for the interactions of the basis and testing functions on composite objects.

such as D_u and D_v. For $u < v$, D_u and D_v are called the first and second regions, respectively, of the basis function. In addition, basis functions are indexed by first considering nonmetallic surfaces, which involve $N_D \leq N$ basis functions. On these surfaces, which separate two dielectric regions, both electric and magnetic currents are expanded in a series of the same set of basis functions $\boldsymbol{b}_n$ for $n = 1, 2, \ldots, N_D$. The remaining $(N - N_D)$ basis functions, if any, are defined on metallic surfaces to expand the electric current. Using a Galerkin scheme for the discretization, the same set of the RWG functions are employed as testing functions, i.e., there are N testing functions to test boundary conditions.

In accordance with the definitions above, Figure 2.65 depicts different scenarios for interactions of basis and testing functions. In the first scenario, the basis (blue) and testing (red) functions are located on the same surface (S_{01}) between D_0 and D_1. Hence, these functions interact through both D_0 and D_1. In the second scenario, the basis function is located on S_{12} between D_1 and D_2, whereas the testing function is located on S_{01}. Hence, in this case, the functions interact only through the region D_1. Finally, in the third scenario, the basis function is located on S_{12}, whereas the testing function is located on S_{03} between D_0 and D_3. Then, these functions do not interact since there is not any common region.

For general composite objects, discretizations of JMCFIE lead to $(N + N_D) \times (N + N_D)$ dense matrix equations in the form of

$$\begin{bmatrix} \left[\overline{\boldsymbol{Z}}_{11}^{\text{JMCFIE}}\right]_{N\times N} & \left[\overline{\boldsymbol{Z}}_{12}^{\text{JMCFIE}}\right]_{N\times N_D} \\ \left[\overline{\boldsymbol{Z}}_{21}^{\text{JMCFIE}}\right]_{N_D\times N} & \left[\overline{\boldsymbol{Z}}_{22}^{\text{JMCFIE}}\right]_{N_D\times N_D} \end{bmatrix} \cdot \begin{bmatrix} \boldsymbol{a}_J \\ \boldsymbol{a}_M \end{bmatrix} = \begin{bmatrix} \boldsymbol{w}_1^{\text{JMCFIE}} \\ \boldsymbol{w}_2^{\text{JMCFIE}} \end{bmatrix}. \tag{2.330}$$

Consider the interaction of the mth testing function $\boldsymbol{t}_m$ and the nth basis function $\boldsymbol{b}_n$, and let a dielectric region D_u be common for the two functions. Then, the corresponding matrix element in the diagonal partition $\overline{\boldsymbol{Z}}_{11}^{\text{JMCFIE}}$ in (2.330) can be written as

$$\overline{\boldsymbol{Z}}_{11}^{\text{JMCFIE}}[m,n] \stackrel{+}{\longleftarrow} -\frac{\xi_n\xi_m}{2}\int_{S_m} d\boldsymbol{r}\boldsymbol{t}_m(\boldsymbol{r})\cdot\boldsymbol{b}_n(\boldsymbol{r})$$
$$+\xi_n\int_{S_m} d\boldsymbol{r}\boldsymbol{t}_m(\boldsymbol{r})\cdot\hat{\boldsymbol{n}}\times\mathcal{K}_{PV,u}\{\boldsymbol{b}_n\}(\boldsymbol{r})$$
$$+\xi_n\xi_m\int_{S_m} d\boldsymbol{r}\boldsymbol{t}_m(\boldsymbol{r})\cdot\mathcal{T}_u\{\boldsymbol{b}_n\}(\boldsymbol{r}), \tag{2.331}$$

where the $\stackrel{+}{\longleftarrow}$ notation is used to indicate the cumulative addition operation, since the value in (2.331) may not be the only contribution to $\overline{\boldsymbol{Z}}_{11}^{\text{JMCFIE}}[m,n]$. Specifically, if $\boldsymbol{t}_m$ and $\boldsymbol{b}_n$ are on the same nonmetallic surface, such as depicted in the first scenario in Figure 2.65, then both regions of these functions are common, and the corresponding matrix element $\overline{\boldsymbol{Z}}_{11}^{\text{JMCFIE}}[m,n]$ involves two sets of contributions, i.e., the interactions of the basis and testing functions through the two regions. In (2.331), signs $\xi_m = \pm 1$ and $\xi_n = \pm 1$ are determined by the orientations of the basis and testing functions. If the common region D_u is the 'first' region for the basis (testing) function, then $\xi_n = +1$ ($\xi_m = +1$); otherwise, $\xi_n = -1$ ($\xi_m = -1$).

When the basis function $\boldsymbol{b}_n$ is not on a metallic surface, i.e., $n \le N_D$, there exists a matrix element $\overline{\boldsymbol{Z}}_{12}^{\text{JMCFIE}}[m,n]$ in (2.330). A contribution to this element due to the interaction of the basis and testing functions through the common region D_u can be written as

$$\overline{\boldsymbol{Z}}_{12}^{\text{JMCFIE}}[m,n] \stackrel{+}{\longleftarrow} -\frac{\xi_n}{2}\eta_u^{-1}\int_{S_m} d\boldsymbol{r}\boldsymbol{t}_m(\boldsymbol{r})\cdot\hat{\boldsymbol{n}}\times\boldsymbol{b}_n(\boldsymbol{r})$$
$$+\xi_n\eta_u^{-1}\int_{S_m} d\boldsymbol{r}\boldsymbol{t}_m(\boldsymbol{r})\cdot\hat{\boldsymbol{n}}\times\mathcal{T}_u\{\boldsymbol{b}_n\}(\boldsymbol{r})$$
$$-\xi_n\xi_m\eta_u^{-1}\int_{S_m} d\boldsymbol{r}\boldsymbol{t}_m(\boldsymbol{r})\cdot\mathcal{K}_{PV,u}\{\boldsymbol{b}_n\}(\boldsymbol{r}). \tag{2.332}$$

When the testing function $\boldsymbol{t}_m$ is not on a metallic surface, i.e., $m \le N_D$, there exists a matrix element $\overline{\boldsymbol{Z}}_{21}^{\text{JMCFIE}}[m,n]$ with a contribution as

$$\overline{\boldsymbol{Z}}_{21}^{\text{JMCFIE}}[m,n] \stackrel{+}{\longleftarrow} \frac{\xi_n}{2}\eta_u\int_{S_m} d\boldsymbol{r}\boldsymbol{t}_m(\boldsymbol{r})\cdot\hat{\boldsymbol{n}}\times\boldsymbol{b}_n(\boldsymbol{r})$$
$$-\xi_n\eta_u\int_{S_m} d\boldsymbol{r}\boldsymbol{t}_m(\boldsymbol{r})\cdot\hat{\boldsymbol{n}}\times\mathcal{T}_u\{\boldsymbol{b}_n\}(\boldsymbol{r})$$
$$+\xi_n\xi_m\eta_u\int_{S_m} d\boldsymbol{r}\boldsymbol{t}_m(\boldsymbol{r})\cdot\mathcal{K}_{PV,u}\{\boldsymbol{b}_n\}(\boldsymbol{r}). \tag{2.333}$$

When both of the basis and testing functions are not on metallic surfaces, there exists a matrix element $\overline{\boldsymbol{Z}}_{22}^{\text{JMCFIE}}[m,n]$, which is equal to the corresponding element of $\overline{\boldsymbol{Z}}_{11}^{\text{JMCFIE}}[m,n]$, i.e.,

$$\overline{\boldsymbol{Z}}_{22}^{\text{JMCFIE}}[m,n] = \overline{\boldsymbol{Z}}_{11}^{\text{JMCFIE}}[m,n], \tag{2.334}$$

for $m \le N_D$ and $n \le N_D$.

In general, each nonmetallic region may host some external sources that produce incident electric and magnetic fields, i.e., $\boldsymbol{E}_u^{inc}$ and $\boldsymbol{H}_u^{inc}$. Consider a testing function $\boldsymbol{t}_m$ on the surface of a nonmetallic region D_u. Incident fields in D_u are tested by $\boldsymbol{t}_m$ and added to the related element in the upper partition of the RHS vector, i.e.,

$$\begin{aligned} \boldsymbol{w}_1^{\text{JMCFIE}}[m] \xleftarrow{+} & -\xi_m \int_{S_m} d\boldsymbol{r} \boldsymbol{t}_m(\boldsymbol{r}) \cdot \hat{\boldsymbol{n}} \times \boldsymbol{H}_u^{inc}(\boldsymbol{r}) \\ & - \xi_m \eta_u^{-1} \int_{S_m} d\boldsymbol{r} \boldsymbol{t}_m(\boldsymbol{r}) \cdot \boldsymbol{E}_u^{inc}(\boldsymbol{r}), \end{aligned} \tag{2.335}$$

where $\xi_m = \pm 1$ is determined by the orientation of $\boldsymbol{t}_m$. In addition, when the testing function $\boldsymbol{t}_m$ is not on a metallic surface, i.e., $m < N_D$, there exists a corresponding element in the second partition of the RHS vector in (2.330). Contribution to this element due to incident fields in region D_u can be written as

$$\begin{aligned} \boldsymbol{w}_2^{\text{JMCFIE}}[m] \xleftarrow{+} & \xi_m \int_{S_m} d\boldsymbol{r} \boldsymbol{t}_m(\boldsymbol{r}) \cdot \hat{\boldsymbol{n}} \times \boldsymbol{E}_u^{inc}(\boldsymbol{r}) \\ & - \xi_m \eta_u \int_{S_m} d\boldsymbol{r} \boldsymbol{t}_m(\boldsymbol{r}) \cdot \boldsymbol{H}_u^{inc}(\boldsymbol{r}). \end{aligned} \tag{2.336}$$

2.4.1 *Special Case: Homogeneous Dielectric Object*

Formulations given in this section can be easily applied to the special cases illustrated in Figure 2.66, i.e., a homogeneous dielectric object and a coated dielectric or metallic object. Assume that incident fields exist only in region D_0, which extends to infinity. In the case of a homogeneous dielectric object, there are two regions, namely, D_0 and D_1. Since there is no metallic surface, $N_D = N$, and the size of the matrix equation obtained from the discretization of JMCFIE is $2N \times 2N$. All RWG functions are located on the same surface, and each matrix element involves two sets of contributions due to the interactions of the basis and testing functions through the regions D_0 and D_1. Using (2.331), (2.332), and (2.333), one can

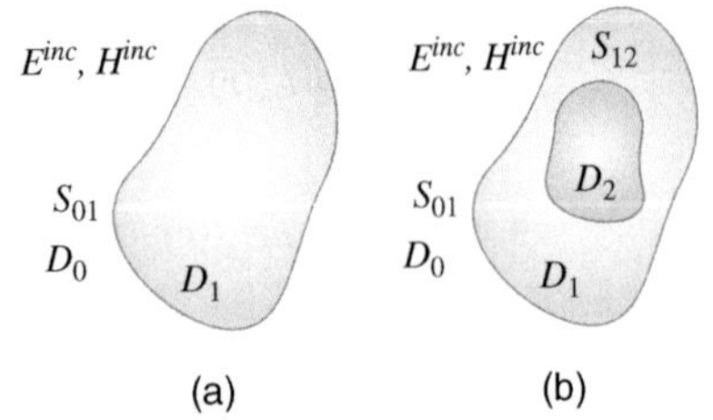

Figure 2.66 Electromagnetics problems involving (a) a single dielectric object and (b) a coated dielectric or metallic object located in homogeneous space.

derive

$$\bar{Z}_{11}^{\text{JMCFIE}} = \bar{Z}_{11}^{\text{JMCFIE}} = -\int_{S_m} d\boldsymbol{r} \boldsymbol{t}_m(\boldsymbol{r}) \cdot \boldsymbol{b}_n(\boldsymbol{r})$$
$$+ \int_{S_m} d\boldsymbol{r} \boldsymbol{t}_m(\boldsymbol{r}) \cdot \hat{\boldsymbol{n}} \times \left[\mathcal{K}_0 - \mathcal{K}_1\right] \{\boldsymbol{b}_n\}(\boldsymbol{r})$$
$$+ \int_{S_m} d\boldsymbol{r} \boldsymbol{t}_m(\boldsymbol{r}) \cdot \left[\mathcal{T}_0 + \mathcal{T}_1\right] \{\boldsymbol{b}_n\}(\boldsymbol{r}) \tag{2.337}$$

$$\bar{Z}_{12}^{\text{JMCFIE}} = \frac{1}{2}(\eta_1^{-1} - \eta_0^{-1}) \int_{S_m} d\boldsymbol{r} \boldsymbol{t}_m(\boldsymbol{r}) \cdot \hat{\boldsymbol{n}} \times \boldsymbol{b}_n(\boldsymbol{r})$$
$$+ \int_{S_m} d\boldsymbol{r} \boldsymbol{t}_m(\boldsymbol{r}) \cdot \hat{\boldsymbol{n}} \times \left[\eta_0^{-1}\mathcal{T}_0 - \eta_1^{-1}\mathcal{T}_1\right] \{\boldsymbol{b}_n\}(\boldsymbol{r})$$
$$- \int_{S_m} d\boldsymbol{r} \boldsymbol{t}_m(\boldsymbol{r}) \cdot \left[\eta_0^{-1}\mathcal{K}_0 + \eta_1^{-1}\mathcal{K}_1\right] \{\boldsymbol{b}_n\}(\boldsymbol{r}) \tag{2.338}$$

$$\bar{Z}_{21}^{\text{JMCFIE}} = \frac{1}{2}(\eta_0 - \eta_1) \int_{S_m} d\boldsymbol{r} \boldsymbol{t}_m(\boldsymbol{r}) \cdot \hat{\boldsymbol{n}} \times \boldsymbol{b}_n(\boldsymbol{r})$$
$$- \int_{S_m} d\boldsymbol{r} \boldsymbol{t}_m(\boldsymbol{r}) \cdot \hat{\boldsymbol{n}} \times \left[\eta_0\mathcal{T}_0 - \eta_1\mathcal{T}_1\right] \{\boldsymbol{b}_n\}(\boldsymbol{r})$$
$$+ \int_{S_m} d\boldsymbol{r} \boldsymbol{t}_m(\boldsymbol{r}) \cdot \left[\eta_0\mathcal{K}_0 + \eta_1\mathcal{K}_1\right] \{\boldsymbol{b}_n\}(\boldsymbol{r}) \tag{2.339}$$

for $m, n = 1, 2, \ldots, N$. In addition, by testing the incident electric and magnetic fields, the elements of the RHS vector can be calculated as

$$\boldsymbol{w}_1^{\text{JMCFIE}} = -\int_{S_m} d\boldsymbol{r} \boldsymbol{t}_m(\boldsymbol{r}) \cdot \hat{\boldsymbol{n}} \times \boldsymbol{H}_0^{inc}(\boldsymbol{r}) - \eta_0^{-1} \int_{S_m} d\boldsymbol{r} \boldsymbol{t}_m(\boldsymbol{r}) \cdot \boldsymbol{E}_0^{inc}(\boldsymbol{r}) \tag{2.340}$$

$$\boldsymbol{w}_2^{\text{JMCFIE}} = \int_{S_m} d\boldsymbol{r} \boldsymbol{t}_m(\boldsymbol{r}) \cdot \hat{\boldsymbol{n}} \times \boldsymbol{E}_0^{inc}(\boldsymbol{r}) - \eta_0 \int_{S_m} d\boldsymbol{r} \boldsymbol{t}_m(\boldsymbol{r}) \cdot \boldsymbol{H}_0^{inc}(\boldsymbol{r}) \tag{2.341}$$

for $m = 1, 2, \ldots, N$. The expressions in (2.337)–(2.341) agree with those in (2.58)–(2.63) and (2.72).

2.4.2 *Special Case: Coated Dielectric Object*

In the case of a coated dielectric object, there are three nonmetallic regions, namely, D_0, D_1, and D_2, while D_1 is between D_0 and D_2, as depicted in Figure 2.66. Since there is no metallic surface, $N_D = N$, and the size of the resulting matrix equation is $2N \times 2N$. Let the first N_{01} basis/testing functions and the remaining $N_{02} = (N - N_{01})$ basis/testing functions be defined on surfaces S_{01} and S_{02}, respectively. Each partition in (2.330) can be divided into four subpartitions, i.e.,

$$\bar{\boldsymbol{Z}}_{ab}^{\text{JMCFIE}} = \begin{bmatrix} \bar{\boldsymbol{Z}}_{ab,11}^{\text{JMCFIE}} & \bar{\boldsymbol{Z}}_{ab,12}^{\text{JMCFIE}} \\ \bar{\boldsymbol{Z}}_{ab,21}^{\text{JMCFIE}} & \bar{\boldsymbol{Z}}_{ab,22}^{\text{JMCFIE}} \end{bmatrix}_{N\times N} \tag{2.342}$$

for $a = 1,2$ and $b = 1,2$. In (2.342), $\bar{\boldsymbol{Z}}_{ab,11}^{\text{JMCFIE}}$ and $\bar{\boldsymbol{Z}}_{ab,22}^{\text{JMCFIE}}$ represent $N_{01} \times N_{01}$ and $N_{02} \times N_{02}$ matrices containing the interactions of the basis and testing functions located on S_{01} and S_{02}, respectively. For $m, n = 1, 2, \ldots, N_{01}$,

$$\begin{aligned} \bar{\boldsymbol{Z}}_{11,11}^{\text{JMCFIE}}[m,n] = \bar{\boldsymbol{Z}}_{22,11}^{\text{JMCFIE}}[m,n] &= -\int_{S_m} d\boldsymbol{r}\boldsymbol{t}_m(\boldsymbol{r}) \cdot \boldsymbol{b}_n(\boldsymbol{r}) \\ &+ \int_{S_m} d\boldsymbol{r}\boldsymbol{t}_m(\boldsymbol{r}) \cdot \hat{\boldsymbol{n}} \times \left[\mathcal{K}_{PV,0} - \mathcal{K}_{PV,1}\right] \{\boldsymbol{b}_n\}(\boldsymbol{r}) \\ &+ \int_{S_m} d\boldsymbol{r}\boldsymbol{t}_m(\boldsymbol{r}) \cdot \left[\mathcal{T}_0 + \mathcal{T}_1\right] \{\boldsymbol{b}_n\}(\boldsymbol{r}) \end{aligned} \tag{2.343}$$

$$\begin{aligned} \bar{\boldsymbol{Z}}_{12,11}^{\text{JMCFIE}}[m,n] &= \frac{1}{2}(\eta_1^{-1} - \eta_0^{-1}) \int_{S_m} d\boldsymbol{r}\boldsymbol{t}_m(\boldsymbol{r}) \cdot \hat{\boldsymbol{n}} \times \boldsymbol{b}_n(\boldsymbol{r}) \\ &+ \int_{S_m} d\boldsymbol{r}\boldsymbol{t}_m(\boldsymbol{r}) \cdot \hat{\boldsymbol{n}} \times \left[\eta_0^{-1}\mathcal{T}_0 - \eta_1^{-1}\mathcal{T}_1\right] \{\boldsymbol{b}_n\}(\boldsymbol{r}) \\ &- \int_{S_m} d\boldsymbol{r}\boldsymbol{t}_m(\boldsymbol{r}) \cdot \left[\eta_0^{-1}\mathcal{K}_{PV,0} + \eta_1^{-1}\mathcal{K}_{PV,1}\right] \{\boldsymbol{b}_n\}(\boldsymbol{r}) \end{aligned} \tag{2.344}$$

$$\begin{aligned} \bar{\boldsymbol{Z}}_{21,11}^{\text{JMCFIE}}[m,n] &= \frac{1}{2}(\eta_0 - \eta_1) \int_{S_m} d\boldsymbol{r}\boldsymbol{t}_m(\boldsymbol{r}) \cdot \hat{\boldsymbol{n}} \times \boldsymbol{b}_n(\boldsymbol{r}) \\ &- \int_{S_m} d\boldsymbol{r}\boldsymbol{t}_m(\boldsymbol{r}) \cdot \hat{\boldsymbol{n}} \times \left[\eta_0\mathcal{T}_0 - \eta_1\mathcal{T}_1\right] \{\boldsymbol{b}_n\}(\boldsymbol{r}) \\ &+ \int_{S_m} d\boldsymbol{r}\boldsymbol{t}_m(\boldsymbol{r}) \cdot \left[\eta_0\mathcal{K}_{PV,0} + \eta_1\mathcal{K}_{PV,1}\right] \{\boldsymbol{b}_n\}(\boldsymbol{r}). \end{aligned} \tag{2.345}$$

For $m, n = (N_{01}+1), (N_{01}+2), \ldots, (N_{01}+N_{02})$, and $(m', n') = (m - N_{01}, n - N_{01})$,

$$\begin{aligned} \bar{\boldsymbol{Z}}_{11,22}^{\text{JMCFIE}}[m',n'] = \bar{\boldsymbol{Z}}_{22,22}^{\text{JMCFIE}}[m',n'] &= -\int_{S_m} d\boldsymbol{r}\boldsymbol{t}_m(\boldsymbol{r}) \cdot \boldsymbol{b}_n(\boldsymbol{r}) \\ &+ \int_{S_m} d\boldsymbol{r}\boldsymbol{t}_m(\boldsymbol{r}) \cdot \hat{\boldsymbol{n}} \times \left[\mathcal{K}_{PV,1} - \mathcal{K}_{PV,2}\right] \{\boldsymbol{b}_n\}(\boldsymbol{r}) \\ &+ \int_{S_m} d\boldsymbol{r}\boldsymbol{t}_m(\boldsymbol{r}) \cdot \left[\mathcal{T}_1 + \mathcal{T}_2\right] \{\boldsymbol{b}_n\}(\boldsymbol{r}) \end{aligned} \tag{2.346}$$

$$\begin{aligned} \bar{\boldsymbol{Z}}_{12,22}^{\text{JMCFIE}}[m',n'] &= \frac{1}{2}(\eta_2^{-1} - \eta_1^{-1}) \int_{S_m} d\boldsymbol{r}\boldsymbol{t}_m(\boldsymbol{r}) \cdot \hat{\boldsymbol{n}} \times \boldsymbol{b}_n(\boldsymbol{r}) \\ &+ \int_{S_m} d\boldsymbol{r}\boldsymbol{t}_m(\boldsymbol{r}) \cdot \hat{\boldsymbol{n}} \times \left[\eta_1^{-1}\mathcal{T}_1 - \eta_2^{-1}\mathcal{T}_2\right] \{\boldsymbol{b}_n\}(\boldsymbol{r}) \end{aligned}$$

(continued)

$$-\int_{S_m} d\boldsymbol{r} \boldsymbol{t}_m(\boldsymbol{r}) \cdot \left[\eta_1^{-1}\mathcal{K}_{PV,1} + \eta_2^{-1}\mathcal{K}_{PV,2}\right]\{\boldsymbol{b}_n\}(\boldsymbol{r}) \tag{2.347}$$

$$\begin{aligned}\overline{\boldsymbol{Z}}_{21,22}^{\mathrm{JMCFIE}}[m',n'] = &\frac{1}{2}(\eta_1 - \eta_2)\int_{S_m} d\boldsymbol{r} \boldsymbol{t}_m(\boldsymbol{r}) \cdot \hat{\boldsymbol{n}} \times \boldsymbol{b}_n(\boldsymbol{r}) \\ &- \int_{S_m} d\boldsymbol{r} \boldsymbol{t}_m(\boldsymbol{r}) \cdot \hat{\boldsymbol{n}} \times \left[\eta_1\mathcal{T}_1 - \eta_2\mathcal{T}_2\right]\{\boldsymbol{b}_n\}(\boldsymbol{r}) \\ &+ \int_{S_m} d\boldsymbol{r} \boldsymbol{t}_m(\boldsymbol{r}) \cdot \left[\eta_1\mathcal{K}_{PV,1} + \eta_2\mathcal{K}_{PV,2}\right]\{\boldsymbol{b}_n\}(\boldsymbol{r}).\end{aligned} \tag{2.348}$$

On the other hand, the off-diagonal subpartitions, i.e., $\overline{\boldsymbol{Z}}_{ab,12}^{\mathrm{JMCFIE}}$ and $\overline{\boldsymbol{Z}}_{ab,21}^{\mathrm{JMCFIE}}$, involve the interactions of the basis and testing functions that are located on different surfaces. These basis and testing functions interact only through the region D_1. For $m = 1, 2, \ldots, N_{01}$, $n = (N_{01} + 1), (N_{01} + 2), \ldots, (N_{01} + N_{02})$, and $n' = n - N_{01}$,

$$\begin{aligned}\overline{\boldsymbol{Z}}_{11,12}^{\mathrm{JMCFIE}}[m,n'] = \overline{\boldsymbol{Z}}_{22,12}[m,n'] = &\frac{1}{2}\int_{S_m} d\boldsymbol{r} \boldsymbol{t}_m(\boldsymbol{r}) \cdot \boldsymbol{b}_n(\boldsymbol{r}) \\ &+ \int_{S_m} d\boldsymbol{r} \boldsymbol{t}_m(\boldsymbol{r}) \cdot \hat{\boldsymbol{n}} \times \mathcal{K}_{PV,1}\{\boldsymbol{b}_n\}(\boldsymbol{r}) \\ &- \int_{S_m} d\boldsymbol{r} \boldsymbol{t}_m(\boldsymbol{r}) \cdot \mathcal{T}_1\{\boldsymbol{b}_n\}(\boldsymbol{r})\end{aligned} \tag{2.349}$$

$$\begin{aligned}\overline{\boldsymbol{Z}}_{12,12}^{\mathrm{JMCFIE}}[m,n'] = &-\frac{\eta_1^{-1}}{2}\int_{S_m} d\boldsymbol{r} \boldsymbol{t}_m(\boldsymbol{r}) \cdot \hat{\boldsymbol{n}} \times \boldsymbol{b}_n(\boldsymbol{r}) \\ &+ \eta_1^{-1}\int_{S_m} d\boldsymbol{r} \boldsymbol{t}_m(\boldsymbol{r}) \cdot \hat{\boldsymbol{n}} \times \mathcal{T}_1\{\boldsymbol{b}_n\}(\boldsymbol{r}) \\ &+ \eta_1^{-1}\int_{S_m} d\boldsymbol{r} \boldsymbol{t}_m(\boldsymbol{r}) \cdot \mathcal{K}_{PV,1}\{\boldsymbol{b}_n\}(\boldsymbol{r})\end{aligned} \tag{2.350}$$

$$\begin{aligned}\overline{\boldsymbol{Z}}_{21,12}^{\mathrm{JMCFIE}}[m,n'] = &\frac{\eta_1}{2}\int_{S_m} d\boldsymbol{r} \boldsymbol{t}_m(\boldsymbol{r}) \cdot \hat{\boldsymbol{n}} \times \boldsymbol{b}_n(\boldsymbol{r}) \\ &- \eta_1\int_{S_m} d\boldsymbol{r} \boldsymbol{t}_m(\boldsymbol{r}) \cdot \hat{\boldsymbol{n}} \times \mathcal{T}_1\{\boldsymbol{b}_n\}(\boldsymbol{r}) \\ &- \eta_1\int_{S_m} d\boldsymbol{r} \boldsymbol{t}_m(\boldsymbol{r}) \cdot \mathcal{K}_{PV,1}\{\boldsymbol{b}_n\}(\boldsymbol{r}).\end{aligned} \tag{2.351}$$

For $m = (N_{01} + 1), (N_{01} + 2), \ldots, (N_{01} + N_{02})$, $n = 1, 2, \ldots, N_{01}$, and $m' = m - N_{01}$,

$$\begin{aligned}\overline{\boldsymbol{Z}}_{11,21}^{\mathrm{JMCFIE}}[m',n] = \overline{\boldsymbol{Z}}_{22,21}[m',n] = &\frac{1}{2}\int_{S_m} d\boldsymbol{r} \boldsymbol{t}_m(\boldsymbol{r}) \cdot \boldsymbol{b}_n(\boldsymbol{r}) \\ &- \int_{S_m} d\boldsymbol{r} \boldsymbol{t}_m(\boldsymbol{r}) \cdot \hat{\boldsymbol{n}} \times \mathcal{K}_{PV,1}\{\boldsymbol{b}_n\}(\boldsymbol{r})\end{aligned}$$

(continued)

$$- \int_{S_m} d\boldsymbol{r} \boldsymbol{t}_m(\boldsymbol{r}) \cdot \mathcal{T}_1\{\boldsymbol{b}_n\}(\boldsymbol{r}) \tag{2.352}$$

$$\begin{aligned}\overline{\boldsymbol{Z}}^{\text{JMCFIE}}_{12,21}[m',n] = {} & \frac{\eta_1^{-1}}{2} \int_{S_m} d\boldsymbol{r} \boldsymbol{t}_m(\boldsymbol{r}) \cdot \hat{\boldsymbol{n}} \times \boldsymbol{b}_n(\boldsymbol{r}) \\ & - \eta_1^{-1} \int_{S_m} d\boldsymbol{r} \boldsymbol{t}_m(\boldsymbol{r}) \cdot \hat{\boldsymbol{n}} \times \mathcal{T}_1\{\boldsymbol{b}_n\}(\boldsymbol{r}) \\ & + \eta_1^{-1} \int_{S_m} d\boldsymbol{r} \boldsymbol{t}_m(\boldsymbol{r}) \cdot \mathcal{K}_{PV,1}\{\boldsymbol{b}_n\}(\boldsymbol{r})\end{aligned} \tag{2.353}$$

$$\begin{aligned}\overline{\boldsymbol{Z}}^{\text{JMCFIE}}_{21,21}[m',n] = {} & -\frac{\eta_1}{2} \int_{S_m} d\boldsymbol{r} \boldsymbol{t}_m(\boldsymbol{r}) \cdot \hat{\boldsymbol{n}} \times \boldsymbol{b}_n(\boldsymbol{r}) \\ & + \eta_1 \int_{S_m} d\boldsymbol{r} \boldsymbol{t}_m(\boldsymbol{r}) \cdot \hat{\boldsymbol{n}} \times \mathcal{T}_1\{\boldsymbol{b}_n\}(\boldsymbol{r}) \\ & - \eta_1 \int_{S_m} d\boldsymbol{r} \boldsymbol{t}_m(\boldsymbol{r}) \cdot \mathcal{K}_{PV,1}\{\boldsymbol{b}_n\}(\boldsymbol{r}).\end{aligned} \tag{2.354}$$

2.4.3 Special Case: Coated Metallic Object

In the case of a coated metallic object, there are two nonmetallic regions, i.e., D_0 and D_1, and the size of the matrix equation is $(N + N_D) \times (N + N_D)$, where $N_D < N$. Let the first N_D basis/testing functions be located on the surface S_{01} and the remaining $N_{02} = (N - N_D)$ basis/testing functions be located on the metallic surface S_{02}. Then, the expressions for the matrix elements in the diagonal partition $\overline{\boldsymbol{Z}}^{\text{JMCFIE}}_{11}$ are the same as the expressions in (2.343), (2.346), (2.349), and (2.352), which are derived for the coated dielectric object. On the other hand, the off-diagonal partition $\overline{\boldsymbol{Z}}^{\text{JMCFIE}}_{12}$ becomes a rectangular matrix with two subpartitions, i.e.,

$$\overline{\boldsymbol{Z}}^{\text{JMCFIE}}_{12} = \begin{bmatrix} \overline{\boldsymbol{Z}}^{\text{JMCFIE}}_{12,11} \\ \overline{\boldsymbol{Z}}^{\text{JMCFIE}}_{12,21} \end{bmatrix}_{N \times N_D}, \tag{2.355}$$

where the matrix elements $\overline{\boldsymbol{Z}}^{\text{JMCFIE}}_{12,11}[m,n]$ and $\overline{\boldsymbol{Z}}^{\text{JMCFIE}}_{12,21}[m',n]$ for $m,n = 1,2,\ldots,N_D$ and $m' = 1,2,\ldots,N_{02}$ are calculated as in (2.344) and (2.353). The off-diagonal partition $\overline{\boldsymbol{Z}}^{\text{JMCFIE}}_{21}$ also becomes a rectangular matrix, i.e.,

$$\overline{\boldsymbol{Z}}^{\text{JMCFIE}}_{21} = \begin{bmatrix} \overline{\boldsymbol{Z}}^{\text{JMCFIE}}_{21,11} & \overline{\boldsymbol{Z}}^{\text{JMCFIE}}_{21,12} \end{bmatrix}_{N_D \times N}, \tag{2.356}$$

where the subpartitions $\overline{\boldsymbol{Z}}^{\text{JMCFIE}}_{21,11}[m,n]$ and $\overline{\boldsymbol{Z}}^{\text{JMCFIE}}_{21,12}[m,n']$ for $m,n = 1,2,\ldots,N_D$, and $n' = 1,2,\ldots,N_{02}$ are calculated as in (2.345) and (2.351). The diagonal partition $\overline{\boldsymbol{Z}}^{\text{JMCFIE}}_{22}$ is an $N_D \times N_D$ matrix with elements $\overline{\boldsymbol{Z}}^{\text{JMCFIE}}_{22}[m,n] = \overline{\boldsymbol{Z}}^{\text{JMCFIE}}_{11}[m,n]$ for $m,n = 1,2,\ldots,N_D$.

Numerical Results

As numerical examples, Figures (2.67) and (2.68) present the solution of large scattering problems discretized with 1,264,128 unknowns. A sphere of radius $5\lambda_o$ is placed inside another sphere of radius $10\lambda_o$. Four different cases are considered:

- The relative permittivities of the core and shell are 4.0 and 2.0, respectively.
- The relative permittivities of the core and shell are 2.0 and 4.0, respectively.
- The core is metallic and the relative permittivity of the shell is 2.0.
- The core is metallic and the relative permittivity of the shell is 4.0.

The scattering problems are solved via 6-level MLFMA, and iteration counts are 101, 283, 75, and 187, respectively, for 10^{-3} residual error. Figures 2.67 and 2.68 presents the

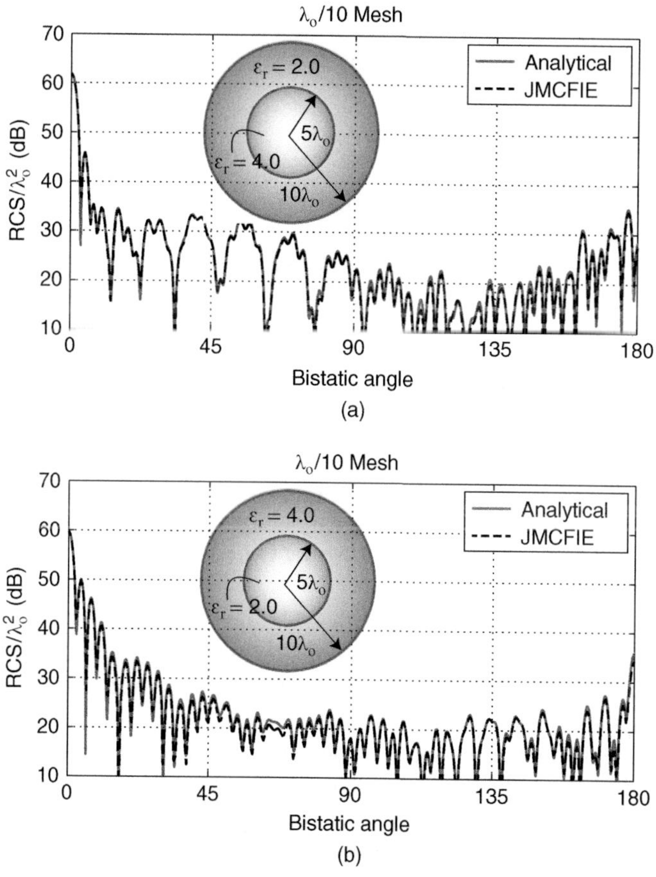

Figure 2.67 The normalized bistatic RCS (RCS$/\lambda_o^2$) of a structure involving spheres of radii $5\lambda_o$ and $10\lambda_o$, when (a) the relative permittivities of the core and shell are 4.0 and 2.0, respectively, and (b) the relative permittivities of the core and shell are 2.0 and 4.0, respectively. *Source*: Ergül and Gürel 2009 [83]. Reproduced by permission of Radio Science – American Geophysical Union AGU.

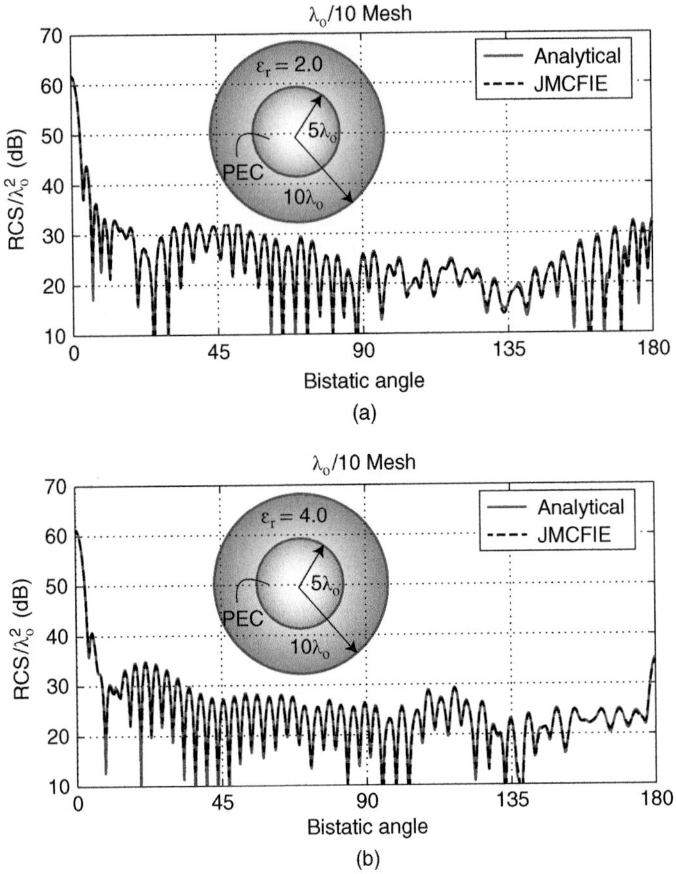

Figure 2.68 The normalized bistatic RCS (RCS $/\lambda_o^2$) of a structure involving spheres of radii $5\lambda_o$ and $10\lambda_o$, when (a) the core is metallic and the relative permittivity of the shell is 2.0 and (b) the core is metallic and the relative permittivity of the shell is 4.0. *Source*: Ergül and Gürel 2009 [83]. Reproduced by permission of Radio Science – American Geophysical Union AGU.

normalized bistatic RCS (RCS $/\lambda_o^2$) values on the *z*-*x* plane, where 0° and 180° correspond to the forward-scattering and backscattering directions, respectively. A good agreement is observed between the computational and analytical results.

Figure 2.69 presents another example involving a spherical object with three dielectric regions. A dielectric sphere of radius $0.3a$ is coated with two dielectric shells of radii $0.5a$ and a, where a changes from $2.0\lambda_o$ to $2.6\lambda_o$. The relative permittivities of the core, inner shell, and outer shell are 1.44, 1.96, and 4.0, respectively. Only one discretization involving 215,304 unknowns is used for the entire frequency range. As depicted in Figure 2.69, solutions are performed efficiently with maximum 101 iterations using 4PBDP. Figure 2.69 also shows the normalized RCS (RCS $/\pi a^2$) in the backscattering direction as a function of a in terms of the wavelength. It can be observed that the computational values are in agreement with the analytical values obtained by Mie-series solutions.

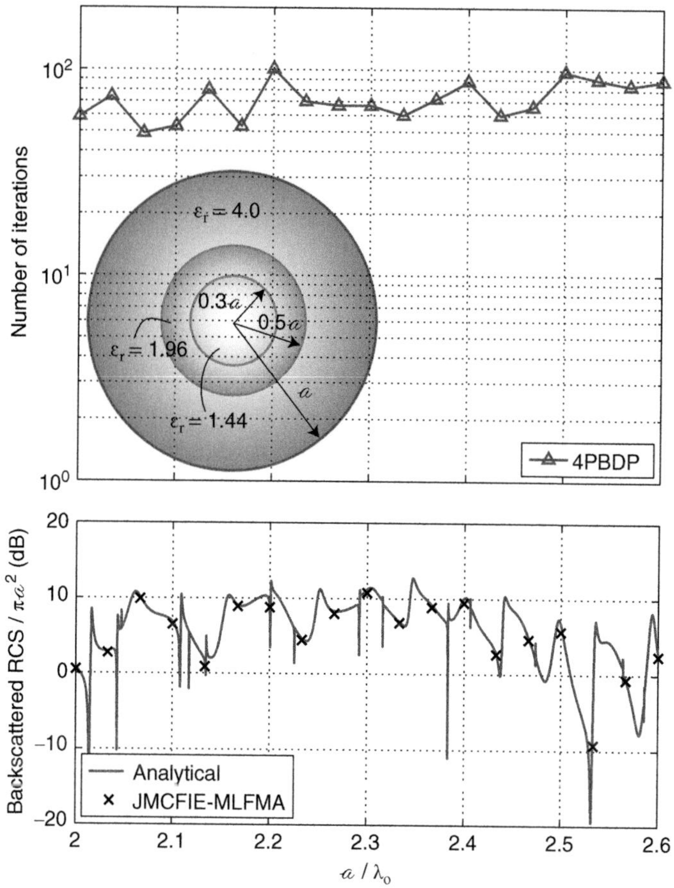

Figure 2.69 Iteration counts for the solution of scattering problems involving a dielectric sphere of radius $0.3a$ coated with two dielectric shells of radii $0.5a$ and a, where a changes from $2\lambda_o$ to $2.6\lambda_o$. The relative permittivities of the core, inner shell, and outer shell are 1.44, 1.96, and 4.0, respectively. The normalized RCS (RCS $/\pi a^2$) of the structure in the backscattering direction is also plotted as a function of a in terms of wavelength. *Source*: Ergül and Gürel 2009 [83]. Reproduced by permission of Radio Science – American Geophysical Union AGU.

2.5 Concluding Remarks

As detailed in this chapter, electromagnetics problems involving three dimensional objects with arbitrary shapes can be formulated rigorously with surface integral equations. Alternative formulations and discretizations are available for metallic, dielectric, and composite objects. Iterative solutions of these problems with sequential and parallel implementations of MLFMA will be focused in the next chapters.

3

Iterative Solutions of Electromagnetics Problems with MLFMA

This chapter is devoted to iterative solutions of electromagnetics problems with MLFMA. Starting from the factorization and diagonalization, MLFMA is discussed in detail. Local interpolations and anterpolations are particularly focused for accurate and efficient implementations. Iterative solutions of metallic and dielectric objects are investigated and compared for different formulations and iterative algorithms. Finally, low-frequency MLFMA (LF-MLFMA) implementations based on multipoles are presented for rigorous analysis of small objects discretized with large numbers of unknowns.

3.1 Factorization and Diagonalization of the Green's Function

FMM and its multilevel version MLFMA are derived from the factorization and diagonalization of the Green's function. As detailed in this section, the factorization of the Green's function is based on the addition theorem.

3.1.1 Addition Theorem

Consider the decomposition of an arbitrary three-dimensional vector $\boldsymbol{r}_{ij} = \boldsymbol{r}_i - \boldsymbol{r}_j = (r_{ij}, \theta_{ij}, \phi_{ij})$ into three vectors as

$$\boldsymbol{r}_{ij} = \boldsymbol{r}_{iI} + \boldsymbol{r}_{IJ} + \boldsymbol{r}_{Jj}, \tag{3.1}$$

where

$$\boldsymbol{r}_{IJ} = \boldsymbol{r}_I - \boldsymbol{r}_J = (r_{IJ}, \theta_{IJ}, \phi_{IJ}) \tag{3.2}$$

$$\boldsymbol{r}_{iI} = \boldsymbol{r}_i - \boldsymbol{r}_I = (r_{iI}, \theta_{iI}, \phi_{iI}) \tag{3.3}$$

$$\boldsymbol{r}_{Jj} = \boldsymbol{r}_J - \boldsymbol{r}_j = (r_{Jj}, \theta_{Jj}, \phi_{Jj}) \tag{3.4}$$

The Multilevel Fast Multipole Algorithm (MLFMA) for Solving Large-Scale Computational Electromagnetics Problems, First Edition. Özgür Ergül and Levent Gürel.

and $|\boldsymbol{r}_{IJ}| > |\boldsymbol{r}_{iI} + \boldsymbol{r}_{Jj}|$. The addition theorem can be written as

$$\Psi_{t,u}(k, \boldsymbol{r}_{ij}) = \sum_{t'=0}^{\infty} \sum_{u'=-t'}^{t'} \Re\{\Psi_{t',u'}(k, \boldsymbol{r}_{iI} + \boldsymbol{r}_{Jj})\}\alpha_{t',u',t,u}(k, \boldsymbol{r}_{IJ}), \tag{3.5}$$

where

$$\Psi_{t,u}(k, \boldsymbol{r}) = h_t^{(1)}(kr)Y_{t,u}(\theta, \phi) \tag{3.6}$$

and

$$\Re\{\Psi_{t,u}(k, \boldsymbol{r})\} = j_t(kr)Y_{t,u}(\theta, \phi) \tag{3.7}$$

represent wave functions (multipole fields). In (3.6) and (3.7) $h_t^{(1)}$ is the spherical Hankel function of the first kind, j_t is the spherical Bessel function of the first kind,

$$Y_{t,u}(\theta, \phi) = \sqrt{\frac{(2t+1)}{4\pi} \frac{(t-u)!}{(t+u)!}} P_t^u(\cos\theta)\exp(iu\phi) \tag{3.8}$$

represents spherical harmonics, and P_t^u is the associated Legendre function. The multipole-to-multipole translation function in (3.5) can be derived as

$$\alpha_{t',u',t,u}(k, \boldsymbol{r}) = \sum_{t''=0}^{\infty} \sum_{u''=-t''}^{t''} 4\pi i^{t'+t''-t}\, \Psi_{t'',u''}(k, \boldsymbol{r}) G_{t,u,t',u',t'',u''}, \tag{3.9}$$

where

$$G_{t,u,t',u',t'',u''} = \int d^2\hat{\boldsymbol{k}} Y_{t,u}(\theta, \phi) Y^*_{t',u'}(\theta, \phi) Y^*_{t'',u''}(\theta, \phi) \tag{3.10}$$

are the Gaunt coefficients. In (3.10), '$*$' represents the complex conjugate operation. Using

$$\int d^2\hat{\boldsymbol{k}} Y^*_{t,u}(\theta, \phi) Y_{t',u'}(\theta, \phi) = \delta_{tt'}\delta_{uu'}, \tag{3.11}$$

$$Y_{t,u}(\theta, \phi) = (-1)^u Y^*_{t,-u}(\theta, \phi), \tag{3.12}$$

and $Y_{0,0}(\theta, \phi) = 1/\sqrt{4\pi}$, one can obtain

$$\begin{aligned} G_{0,0,t',u',t'',u''} &= \int d^2\hat{\boldsymbol{k}} Y_{0,0}(\theta, \phi) Y^*_{t',u'}(\theta, \phi) Y^*_{t'',u''}(\theta, \phi) \\ &= \frac{(-1)^{u''}}{\sqrt{4\pi}} \int d^2\hat{\boldsymbol{k}} Y^*_{t',u'}(\theta, \phi) Y_{t'',-u''}(\theta, \phi) \\ &= \frac{(-1)^{u''}}{\sqrt{4\pi}} \delta_{t't''}\delta_{u'(-u'')} \end{aligned} \tag{3.13}$$

and

$$G_{t,u,0,0,t'',u''} = \frac{1}{\sqrt{4\pi}} \delta_{tt''}\delta_{uu''} \tag{3.14}$$

$$G_{t,u,t',u',0,0} = \frac{1}{\sqrt{4\pi}} \delta_{tt'} \delta_{uu'}. \tag{3.15}$$

Setting $t = u = 0$ in (3.9) and using (3.13), the monopole-to-multipole translation function can be derived as

$$\begin{aligned}
\alpha_{t',u',0,0}(k, \boldsymbol{r}) &= \sum_{t''=0}^{\infty} \sum_{u''=-t''}^{t''} 4\pi i^{t'+t''} \Psi_{t'',u''}(k, \boldsymbol{r}) G_{0,0,t',u',t'',u''} \\
&= \frac{1}{\sqrt{4\pi}} \sum_{t''=0}^{\infty} \sum_{u''=-t''}^{t''} 4\pi i^{t'+t''} \Psi_{t'',u''}(k, \boldsymbol{r}) (-1)^{u''} \delta_{t't''} \delta_{u'(-u'')} \\
&= \sqrt{4\pi} (-1)^{t'+u'} \Psi_{t',-u'}(k, \boldsymbol{r}) \\
&= \sqrt{4\pi} (-1)^{t'+u'} h_{t'}^{(1)}(kr) Y_{t',-u'}(\theta, \phi).
\end{aligned} \tag{3.16}$$

Similarly, setting $t' = u' = 0$ in (3.9) and using (3.14), the multipole-to-monopole translation function is derived as

$$\alpha_{0,0,t,u}(k, \boldsymbol{r}) = \sqrt{4\pi} \Psi_{t,u}(k, \boldsymbol{r}) = \sqrt{4\pi} h_t^{(1)}(kr) Y_{t,u}(\theta, \phi). \tag{3.17}$$

Finally, setting $t' = u' = 0$ in (3.16) or $t = u = 0$ in (3.17), the monopole-to-monopole translation function can be derived as

$$\alpha_{0,0,0,0}(k, \boldsymbol{r}) = \sqrt{4\pi} h_0^{(1)}(kr) Y_{0,0}(\theta, \phi) = h_0^{(1)}(r) = \frac{\exp(ikr)}{ikr}, \tag{3.18}$$

which is simply a scaled version of the homogeneous-space Green's function.

It is remarkable that most of the Gaunt coefficients are identically zero. Considering only nonzero terms, the multipole-to-multipole translation function can be rewritten as

$$\alpha_{t',u',t,u}(k, \boldsymbol{r}) = \sum_{t''=|t-t'|}^{|t+t'|} 4\pi i^{t'+t''-t} \Psi_{t'',u-u'}(k, \boldsymbol{r}) G_{t,u,t',u',t'',u-u'}. \tag{3.19}$$

Using (3.19) and (3.13)–(3.15), it is possible to derive the monopole-to-multipole and multipole-to-monopole translation functions as

$$\begin{aligned}
\alpha_{t',u',0,0}(k, \boldsymbol{r}) &= \sum_{t''=|-t'|}^{|t'|} 4\pi i^{t'+t''} \Psi_{t'',-u'}(k, \boldsymbol{r}) G_{0,0,t',u',t'',-u'} \\
&= 4\pi (-1)^{t'} \Psi_{t',-u'}(k, \boldsymbol{r}) \frac{(-1)^{-u'}}{\sqrt{4\pi}} \delta_{t't'} \delta_{u'u'} \\
&= \sqrt{4\pi} (-1)^{t'+u'} \Psi_{t',-u'}(k, \boldsymbol{r})
\end{aligned} \tag{3.20}$$

and

$$\begin{aligned}
\alpha_{0,0,t,u}(k, \boldsymbol{r}) &= \sum_{t''=|t|}^{|t|} 4\pi i^{t''-t} \Psi_{t,u}(k, \boldsymbol{r}) G_{t,u,0,0,t,u} \\
&= 4\pi \Psi_{t,u}(k, \boldsymbol{r}) \delta_{t't''} \delta_{uu} = \sqrt{4\pi} \Psi_{t,u}(k, \boldsymbol{r}),
\end{aligned} \tag{3.21}$$

respectively.

The addition theorem can be used to factorize the homogeneous-space Green's function, i.e.,

$$g(\boldsymbol{r}_i,\boldsymbol{r}_j) = \frac{ik}{\sqrt{4\pi}}\Psi_{0,0}(k,\boldsymbol{r}_{ij})$$

$$= \frac{ik}{\sqrt{4\pi}}\sum_{t'=0}^{\infty}\sum_{u'=-t'}^{t'}\Re\{\Psi_{t',u'}(k,\boldsymbol{r}_{iI}+\boldsymbol{r}_{Jj})\}\alpha_{t',u',0,0}(k,\boldsymbol{r}_{IJ}). \tag{3.22}$$

Using (3.6), (3.7), (3.12), and (3.20) in (3.22), one can obtain

$$g(\boldsymbol{r}_i,\boldsymbol{r}_j) = ik\sum_{t'=0}^{\infty}\sum_{u'=-t'}^{t'}\Re\{\Psi_{t',u'}(k,\boldsymbol{r}_{iI}+\boldsymbol{r}_{Jj})\}(-1)^{t'+u'}\Psi_{t',-u'}(k,\boldsymbol{r}_{IJ})$$

$$= ik\sum_{t'=0}^{\infty}\sum_{u'=-t'}^{t'}(-1)^{t'}j_{t'}(kr_{iIJj})Y_{t',u'}(\theta_{iIJj},\phi_{iIJj})h_{t'}^{(1)}(kr_{IJ})Y_{t',u'}^{*}(\theta_{IJ},\phi_{IJ})$$

$$= ik\sum_{t'=0}^{\infty}(-1)^{t'}j_{t'}(kr_{iIJj})h_{t'}^{(1)}(kr_{IJ})$$

$$\sum_{u'=-t'}^{t'}Y_{t',u'}(\theta_{iIJj},\phi_{iIJj})Y_{t',u'}^{*}(\theta_{IJ},\phi_{IJ}), \tag{3.23}$$

where

$$\boldsymbol{r}_{iIJj} = (r_{iIJj},\theta_{iIJj},\phi_{iIJj}) = \boldsymbol{r}_{iI}+\boldsymbol{r}_{Jj}. \tag{3.24}$$

The inner summation in (3.23) can be evaluated as

$$\sum_{u'=-t'}^{t'}Y_{t',u'}(\theta_{iIJj},\phi_{iIJj})Y_{t',u'}^{*}(\theta_{IJ},\phi_{IJ}) = \frac{(2t'+1)}{4\pi}P_{t'}(\hat{\mathbf{r}}_{IJ}\cdot\hat{\mathbf{r}}_{iIJj}), \tag{3.25}$$

where $P_{t'}$ is the Legendre polynomial. Inserting (3.25) in (3.23),

$$g(\boldsymbol{r}_i,\boldsymbol{r}_j) = \frac{ik}{4\pi}\sum_{t'=0}^{\infty}(-1)^{t'}(2t'+1)j_{t'}(k\boldsymbol{r}_{iIJj})h_{t'}^{(1)}(k\boldsymbol{r}_{IJ})P_{t'}(\hat{\mathbf{r}}_{IJ}\cdot\hat{\mathbf{r}}_{iIJj}), \tag{3.26}$$

which is called the Gegenbauer's addition theorem.

3.1.2 *Factorization of the Translation Functions*

Consider a multipole-to-multipole translation from $\boldsymbol{r}_i$ to $\boldsymbol{r}_j$, i.e.,

$$\alpha_{t',u',t,u}(k,\boldsymbol{r}_{ij}) = \sum_{t''=0}^{\infty}\sum_{u''=-t''}^{t''}4\pi i^{t'+t''-t}\Psi_{t'',u''}(k,\boldsymbol{r}_{ij})G_{t,u,t',u',t'',u''}. \tag{3.27}$$

Using the addition theorem (3.5) in (3.27) and changing the order of summations,

$$\alpha_{t',u',t,u}(k,\boldsymbol{r}_{ij}) = \sum_{t_1=0}^{\infty}\sum_{u_1=-t_1}^{t_1} 4\pi i^{t'-t}\Re\{\Psi_{t_1,u_1}(k,\boldsymbol{r}_{iIJj})\}$$
$$\sum_{t''=0}^{\infty}\sum_{u''=-t''}^{t''} i^{t''} G_{t,u,t',u',t'',u''}\alpha_{t_1,u_1,t'',u''}(k,\boldsymbol{r}_{IJ}). \tag{3.28}$$

The inner summations can be modified by using (3.9) as

$$\begin{aligned}
&\sum_{t''=0}^{\infty}\sum_{u''=-t''}^{t''} i^{t''} G_{t,u,t',u',t'',u''}\alpha_{t_1,u_1,t'',u''}(k,\boldsymbol{r}_{IJ})\\
&=\sum_{t_2=0}^{\infty}\sum_{u_2=-t_2}^{t_2} 4\pi i^{t_1+t_2}\Psi_{t_2,u_2}(k,\boldsymbol{r}_{IJ})\sum_{t''=0}^{\infty}\sum_{u''=-t''}^{t''} G_{t,u,t',u',t'',u''}G_{t'',u'',t_1,u_1,t_2,u_2}\\
&=\sum_{t_2=0}^{\infty}\sum_{u_2=-t_2}^{t_2} 4\pi i^{t_1+t_2}\Psi_{t_2,u_2}(k,\boldsymbol{r}_{IJ})\sum_{t''=0}^{\infty}\sum_{u''=-t''}^{t''} G_{t,u,t_1,u_1,t'',u''}G_{t'',u'',t',u',t_2,u_2}\\
&=\sum_{t''=0}^{\infty}\sum_{u''=-t''}^{t''} i^{t''+t_1-t'} G_{t,u,t_1,u_1,t'',u''}\alpha_{t',u',t'',u''}(k,\boldsymbol{r}_{IJ})\\
&=\sum_{t''=0}^{\infty}\sum_{u''=-t''}^{t''} i^{t''+t_1-t'} G_{t,u,t'',u'',t_1,u_1}\alpha_{t',u',t'',u''}(k,\boldsymbol{r}_{IJ})
\end{aligned} \tag{3.29}$$

leading to

$$\begin{aligned}
\alpha_{t',u',t,u}(k,\boldsymbol{r}_{ij}) &= \sum_{t_1=0}^{\infty}\sum_{u_1=-t_1}^{t_1} 4\pi i^{t'-t}\Re\{\Psi_{t_1,u_1}(k,\boldsymbol{r}_{iIJj})\}\\
&\quad\sum_{t''=0}^{\infty}\sum_{u''=-t''}^{t''} i^{t''+t_1-t'} G_{t,u,t'',u'',t_1,u_1}\alpha_{t',u',t'',u''}(k,\boldsymbol{r}_{IJ})\\
&=\sum_{t''=0}^{\infty}\sum_{u''=-t''}^{t''} \alpha_{t',u',t'',u''}(k,\boldsymbol{r}_{IJ})\beta_{t'',u'',t,u}(k,\boldsymbol{r}_{iIJj}),
\end{aligned} \tag{3.30}$$

where

$$\beta_{t',u',t,u}(k,\boldsymbol{r}) = \sum_{t''=0}^{\infty}\sum_{u''=-t''}^{t''} 4\pi i^{t'+t''-t}\Re\{\Psi_{t'',u''}(k,\boldsymbol{r})\}G_{t,u,t',u',t'',u''} \tag{3.31}$$

is the multipole-to-multipole shift function. By performing similar derivations, one can also show that

$$\alpha_{t',u',t,u}(k,\boldsymbol{r}_{ij}) = \sum_{t''=0}^{\infty}\sum_{u''=-t''}^{t''} \beta_{t',u',t'',u''}(k,\boldsymbol{r}_{iIJj})\alpha_{t'',u'',t,u}(k,\boldsymbol{r}_{IJ}), \tag{3.32}$$

$$\beta_{t',u',t,u}(k,\boldsymbol{r}_{iIJj}) = \sum_{t''=0}^{\infty} \sum_{u''=-t''}^{t''} \beta_{t',u',t'',u''}(k,\boldsymbol{r}_{iI})\beta_{t'',u'',t,u}(k,\boldsymbol{r}_{Jj}), \tag{3.33}$$

and

$$\alpha_{t',u',t,u}(k,\boldsymbol{r}_{ij}) = \sum_{t_1=0}^{\infty} \sum_{u_1=-t_1}^{t_1} \sum_{t_2=0}^{\infty} \sum_{u_2=-t_2}^{t_2} \beta_{t',u',t_1,u_1}(k,\boldsymbol{r}_{iI}) \alpha_{t_1,u_1,t_2,u_2}(k,\boldsymbol{r}_{IJ})\beta_{t_2,u_2,t,u}(k,\boldsymbol{r}_{Jj}). \tag{3.34}$$

Note that the multipole-to-multipole shift function in (3.31) is the regular part of the multipole-to-multipole translation function in (3.9). Considering only nonzero Gaunt coefficients,

$$\beta_{t',u',t,u}(k,\boldsymbol{r}) = \sum_{t''=|t-t'|}^{|t+t'|} 4\pi i^{t'+t''-t}\Re\{\Psi_{t'',u-u'}(k,\boldsymbol{r})\}G_{t,u,t',u',t'',u-u'}. \tag{3.35}$$

Setting $t = u = 0$ in (3.35) and using (3.13), the monopole-to-multipole shift function is derived as

$$\begin{aligned}
\beta_{t',u',0,0}(k,\boldsymbol{r}) &= \sum_{t''=|-t'|}^{|t'|} 4\pi i^{t'+t''}\Re\{\Psi_{t'',-u'}(k,\boldsymbol{r})\}G_{0,0,t',u',t'',-u'} \\
&= 4\pi(-1)^{t'}\Re\{\Psi_{t',-u'}(k,\boldsymbol{r})\}\frac{(-1)^{-u'}}{\sqrt{4\pi}}\delta_{t't'}\delta_{u'u'} \\
&= \sqrt{4\pi}(-1)^{t'+u'}\Re\{\Psi_{t',-u'}(k,\boldsymbol{r})\} \\
&= \sqrt{4\pi}(-1)^{t'+u'}j_{t'}(kr)Y_{t',-u'}(\theta,\phi).
\end{aligned} \tag{3.36}$$

Similarly, setting $t' = u' = 0$ in (3.35) and using (3.14), the multipole-to-monopole shift function is derived as

$$\begin{aligned}
\beta_{0,0,t,u}(k,\boldsymbol{r}) &= \sum_{t''=|t|}^{|t|} 4\pi i^{t''-t}\Re\{\Psi_{t,u}(k,\boldsymbol{r}_{ij})\}G_{t,u,0,0,t,u} \\
&= 4\pi\Re\{\Psi_{t,u}(k,\boldsymbol{r})\}\frac{1}{\sqrt{4\pi}}\delta_{tt}\delta_{uu} \\
&= \sqrt{4\pi}\Re\{\Psi_{t,u}(k,\boldsymbol{r})\} = \sqrt{4\pi}j_t(kr)Y_{t,u}(\theta,\phi).
\end{aligned} \tag{3.37}$$

Finally, setting $t' = u' = 0$ in (3.36) or $t = u = 0$ in (3.37), the monopole-to-monopole shift function can be derived as

$$\beta_{0,0,0,0}(k,\boldsymbol{r}) = \sqrt{4\pi}j_0(kr)Y_{0,0}(\theta,\phi) = j_0(kr) = \frac{\sin(kr)}{kr}, \tag{3.38}$$

which is the regular part of (3.18).

Consider the application of (3.32) to the monopole-to-monopole translation function, i.e.,

$$\alpha_{0,0,0,0}(k,\boldsymbol{r}_{ij}) = \sum_{t''=0}^{\infty} \sum_{u''=-t''}^{t''} \beta_{0,0,t'',u''}(k,\boldsymbol{r}_{iIJj})\alpha_{t'',u'',0,0}(k,\boldsymbol{r}_{IJ}). \tag{3.39}$$

Using (3.12), (3.16), (3.25), and (3.37) in (3.39), one can obtain the Gegenbauer's addition theorem as

$$\begin{aligned} g(\boldsymbol{r}_i,\boldsymbol{r}_j) &= \frac{ik}{4\pi}\alpha_{0,0,0,0}(k,\boldsymbol{r}_{ij}) \\ &= ik\sum_{t''=0}^{\infty}(-1)^{t''} j_{t''}(kr_{iIJj})h_{t''}^{(1)}(kr_{IJ}) \\ &\quad \sum_{u''=-t''}^{t''}(-1)^{u''}Y_{t'',u''}(\theta_{iIJj},\phi_{iIJj})Y_{t'',-u''}(\theta_{IJ},\phi_{IJ}) \\ &= \frac{ik}{4\pi}\sum_{t''=0}^{\infty}(-1)^{t''}(2t''+1)j_{t''}(kr_{iIJj})h_{t''}^{(1)}(kr_{IJ})P_{t''}(\hat{\boldsymbol{r}}_{IJ}\cdot\hat{\boldsymbol{r}}_{iIJj}), \end{aligned} \tag{3.40}$$

which confirms (3.26).

3.1.3 Expansions

Consider a vector $\boldsymbol{X} = \hat{\boldsymbol{k}}X$, where $X >> r = |\boldsymbol{r}|$ for a given observation point $\boldsymbol{r}$. Then,

$$\frac{\exp(ik|\boldsymbol{X}+\boldsymbol{r}|)}{4\pi|\boldsymbol{X}+\boldsymbol{r}|} \approx \frac{\exp(ikX)}{4\pi X}\exp(i\boldsymbol{k}\cdot\boldsymbol{r}). \tag{3.41}$$

Using the Gegenbauer's addition theorem and the large-argument form of the spherical Hankel function, one can derive

$$\exp(i\boldsymbol{k}\cdot\boldsymbol{r}) = \sum_{t=0}^{\infty} i^t(2t+1)j_t(kr)P_t(\hat{\boldsymbol{k}}\cdot\hat{\boldsymbol{r}}), \tag{3.42}$$

which can be interpreted as the expansion of a plane wave in a series of spherical waves. In order to derive an expression for the expansion of a spherical wave in a series of plane waves, both sides of (3.42) can be multiplied with $P_{t''}(\hat{\boldsymbol{k}}\cdot\hat{\boldsymbol{r}}'')$, where $\hat{\boldsymbol{r}}''$ is an arbitrary direction. Performing angular integrations on both sides,

$$\begin{aligned} &\int d^2\hat{\boldsymbol{k}}\exp(i\boldsymbol{k}\cdot\boldsymbol{r})P_{t''}(\hat{\boldsymbol{k}}\cdot\hat{\boldsymbol{r}}'') \\ &= \int d^2\hat{\boldsymbol{k}}\sum_{t=0}^{\infty} i^t(2t+1)j_t(kr)P_t(\hat{\boldsymbol{k}}\cdot\hat{\boldsymbol{r}})P_{t''}(\hat{\boldsymbol{k}}\cdot\hat{\boldsymbol{r}}'') \\ &= \sum_{t=0}^{\infty} i^t(2t+1)j_t(kr)\int d^2\hat{\boldsymbol{k}}P_t(\hat{\boldsymbol{k}}\cdot\hat{\boldsymbol{r}})P_{t''}(\hat{\boldsymbol{k}}\cdot\hat{\boldsymbol{r}}'') \end{aligned} \tag{3.43}$$

$$= \sum_{t=0}^{\infty} i^t (2t+1) j_t(kr) \frac{4\pi}{(2t''+1)} P_t(\hat{\boldsymbol{r}} \cdot \hat{\boldsymbol{r}}'') \delta_{tt''}$$

$$= 4\pi i^{t''} j_{t''}(kr) P_{t''}(\hat{\boldsymbol{r}} \cdot \hat{\boldsymbol{r}}''). \tag{3.43}$$

Choosing $\hat{\boldsymbol{r}}'' = \hat{\boldsymbol{r}}$, plane-wave expansion of a spherical wave can be derived as

$$j_{t''}(kr) = \frac{(-i)^{t''}}{4\pi} \int d^2\hat{\boldsymbol{k}} \exp(i\boldsymbol{k} \cdot \boldsymbol{r}) P_{t''}(\hat{\boldsymbol{k}} \cdot \hat{\boldsymbol{r}}), \tag{3.44}$$

which is useful for diagonalization.

3.1.4 Diagonalization

Conventional implementations of FMM and MLFMA are based on the diagonalization of the Gegenbauer's addition theorem. Using (3.44) in (3.40) for $\boldsymbol{r} = \boldsymbol{r}_{iIJj}$ and $\boldsymbol{r}'' = \boldsymbol{r}_{IJ}$, one can obtain

$$g(\boldsymbol{r}_i, \boldsymbol{r}_j) = \frac{ik}{(4\pi)^2} \sum_{t''=0}^{\infty} i^{t''} (2t''+1) h_{t''}^{(1)}(kr_{IJ}) \int d^2\hat{\boldsymbol{k}} \exp(i\boldsymbol{k} \cdot \boldsymbol{r}_{iIJj}) P_{t''}(\hat{\boldsymbol{k}} \cdot \hat{\boldsymbol{r}}_{IJ}). \tag{3.45}$$

Changing the order of the summation and angular integration, the homogeneous-space Green's function can be written as

$$g(\boldsymbol{r}_i, \boldsymbol{r}_j) = \frac{ik}{(4\pi)^2} \int d^2\hat{\boldsymbol{k}} \beta(\boldsymbol{k}, \boldsymbol{r}_{iIJj}) \alpha(\boldsymbol{k}, \boldsymbol{r}_{IJ}), \tag{3.46}$$

where

$$\beta(\boldsymbol{k}, \boldsymbol{r}) = \exp(i\boldsymbol{k} \cdot \boldsymbol{r}) \tag{3.47}$$

is the diagonal (plane-wave-to-plane-wave) shift function and

$$\alpha(\boldsymbol{k}, \boldsymbol{r}) = \sum_{t=0}^{\infty} i^t (2t+1) h_t^{(1)}(kr) P_t(\hat{\boldsymbol{k}} \cdot \hat{\boldsymbol{r}}) \tag{3.48}$$

is the diagonal (plane-wave-to-plane-wave) translation function. In practice, the summation in (3.48) is truncated as

$$\alpha_\tau(\boldsymbol{k}, \boldsymbol{r}) = \sum_{t=0}^{\tau} i^t (2t+1) h_t^{(1)}(kr) P_t(\hat{\boldsymbol{k}} \cdot \hat{\boldsymbol{r}}), \tag{3.49}$$

where τ is the truncation number. Using (3.46), the derivatives of the Green's function can also be written in diagonal forms as

$$\{\nabla g\}(\boldsymbol{r}_i, \boldsymbol{r}_j) = \frac{ik}{(4\pi)^2} \int d^2\hat{\boldsymbol{k}} (i\boldsymbol{k}) \exp(i\boldsymbol{k} \cdot \boldsymbol{r}_{iIJj}) \alpha(\boldsymbol{k}, \boldsymbol{r}_{IJ}) \tag{3.50}$$

$$\{\nabla\nabla' g\}(\boldsymbol{r}_i, \boldsymbol{r}_j) = \frac{ik}{(4\pi)^2} \int d^2\hat{\boldsymbol{k}} (\boldsymbol{k}\boldsymbol{k}) \exp(i\boldsymbol{k} \cdot \boldsymbol{r}_{iIJj}) \alpha(\boldsymbol{k}, \boldsymbol{r}_{IJ}) \tag{3.51}$$

to be used for integral equation formulations.

In general, the diagonalization can be applied to all multipole-to-multipole shift and translation functions. For example, consider a multipole-to-multipole shift from $\boldsymbol{r}_i$ to $\boldsymbol{r}_j$, i.e.,

$$\begin{aligned}\beta_{t',u',t,u}(k,\boldsymbol{r}_{ij}) &= \sum_{t''=0}^{\infty}\sum_{u''=-t''}^{t''} 4\pi i^{t'+t''-t}\Re\{\Psi_{t'',u''}(k,\boldsymbol{r}_{ij})\}G_{t,u,t',u',t'',u''}\\ &= \sum_{t''=0}^{\infty}\sum_{u''=-t''}^{t''} 4\pi i^{t'+t''-t} j_{t''}(kr_{ij})Y_{t'',u''}(\theta_{ij},\phi_{ij})\\ &\quad \int d^2\hat{\boldsymbol{k}} Y_{t,u}(\theta,\phi)Y^*_{t',u'}(\theta,\phi)Y^*_{t'',u''}(\theta,\phi).\end{aligned} \tag{3.52}$$

Changing the order of summations and angular integration, and using (3.25) and (3.42), one can obtain

$$\begin{aligned}\beta_{t',u',t,u}(k,\boldsymbol{r}_{ij}) &= 4\pi\int d^2\hat{\boldsymbol{k}}[i^{t'}Y^*_{t',u'}(\theta,\phi)][i^{-t}Y_{t,u}(\theta,\phi)]\\ &\quad \sum_{t''=0}^{\infty} i^{t''} j_{t''}(kr_{ij}) \sum_{u''=-t''}^{t''} Y_{t'',u''}(\theta_{ij},\phi_{ij})Y^*_{t'',u''}(\theta,\phi)\\ &= \int d^2\hat{\boldsymbol{k}}[i^{t'}Y^*_{t',u'}(\theta,\phi)][i^{-t}Y_{t,u}(\theta,\phi)]\\ &\quad \sum_{t''=0}^{\infty} i^{t''}(2t''+1)j_{t''}(kr_{ij})P_{t''}(\hat{\boldsymbol{k}}\cdot\hat{\boldsymbol{r}}_{ij})\\ &= \int d^2\hat{\boldsymbol{k}}[i^{t'}Y^*_{t',u'}(\theta,\phi)]\exp(i\boldsymbol{k}\cdot\boldsymbol{r}_{ij})[i^{-t}Y_{t,u}(\theta,\phi)]\\ &= \int d^2\hat{\boldsymbol{k}} T^{P\to M}_{t',u'}(\theta,\phi)\beta(\boldsymbol{k},\boldsymbol{r}_{ij})T^{M\to P}_{t,u}(\theta,\phi),\end{aligned} \tag{3.53}$$

where

$$T^{M\to P}_{t,u}(\theta,\phi) = i^{-t}Y_{t,u}(\theta,\phi) \tag{3.54}$$

is the multipole-to-plane-wave transformer and

$$T^{P\to M}_{t',u'}(\theta,\phi) = i^{t'}Y^*_{t',u'}(\theta,\phi) \tag{3.55}$$

is the plane-wave-to-multipole transformer. Applying a similar procedure, the multipole-to-multipole translation function can be diagonalized as

$$\alpha_{t',u',t,u}(k,\boldsymbol{r}_{ij}) = \int d^2\hat{\boldsymbol{k}} T^{P\to M}_{t',u'}(\theta,\phi)\beta(\boldsymbol{k},\boldsymbol{r}_{iI})\alpha(\boldsymbol{k},\boldsymbol{r}_{IJ})\beta(\boldsymbol{k},\boldsymbol{r}_{Jj})T^{M\to P}_{t,u}(\theta,\phi). \tag{3.56}$$

Note that (3.56) reduces to (3.46) when $t = u = t' = u' = 0$.

3.2 Multilevel Fast Multipole Algorithm

Matrix equations obtained from the discretization of integral equation formulations can be solved iteratively, where the required MVMs are accelerated by MLFMA. For an $N \times N$ dense matrix equation, MLFMA reduces the complexity of an MVM from $\mathcal{O}(N^2)$ to $\mathcal{O}(N \log N)$. In general, MLFMA splits MVMs as

$$\overline{\boldsymbol{Z}} \cdot \boldsymbol{x} = \overline{\boldsymbol{Z}}_{NF} \cdot \boldsymbol{x} + \overline{\boldsymbol{Z}}_{FF} \cdot \boldsymbol{x}, \tag{3.57}$$

where near-field interactions denoted by $\overline{\boldsymbol{Z}}_{NF}$ are calculated directly and stored in memory to perform the partial multiplications $\overline{\boldsymbol{Z}}_{NF} \cdot \boldsymbol{x}$, while the multiplications involving far-field interactions ($\overline{\boldsymbol{Z}}_{FF} \cdot \boldsymbol{x}$) are performed efficiently using the factorization/diagonalization of the Green's function. Note that, for electromagnetics problems involving homogeneous dielectric objects, the factorization/diagonalization should be applied separately for each medium.

3.2.1 *Recursive Clustering*

Without losing generality, consider a smooth object with an electrical dimension of kD, where $k = 2\pi/\lambda$ is the wavenumber. Discretization (triangulation) of the object with $\lambda/10$ mesh size leads to N unknowns, where $N = \mathcal{O}(k^2D^2)$. The fundamental idea in MLFMA is to replace interactions of basis and testing functions with interactions of groups of functions. Specifically, far-field interactions between distant basis and testing functions are calculated efficiently in a group-by-group manner. In order to calculate far-field interactions (associated with a medium) in a multilevel scheme, a tree structure is constructed by placing the object in a cubic box and recursively dividing the computational domain into subdomains, until the box size is about 0.25λ. A multilevel tree structure with $L = \mathcal{O}(\log(kD)) = \mathcal{O}(\log N)$ levels is obtained by considering nonempty boxes. Although the actual number of levels is $(L + 2)$, the highest two levels are not used directly in MLFMA. At level l from 1 to L, the number of boxes can be approximated as

$$N_l \approx 4^{(1-l)} N_1, \tag{3.58}$$

where $N_1 = \mathcal{O}(N)$. In other words, the number of boxes decreases approximately by a factor of four from a level to the next higher level.

A tree structure in MLFMA can be constructed by using a top-down or a bottom-up scheme. In the top-down scheme, the size of the largest cube enclosing the object is minimized, while the size of the smallest boxes at the lowest level depends on the size of the object and the number of levels. In the bottom-up scheme, however, the size of the smallest boxes is fixed to some value (such as 0.25λ), and the box sizes at the higher levels are recursively doubled until the whole object is enclosed by the largest box. For a given problem, one of the two strategies can be preferable in terms of efficiency and accuracy.

In MLFMA, a multilevel tree structure must be constructed with less than $\mathcal{O}(N \log N)$ complexity. Otherwise, the recursive clustering part can easily become a bottleneck of the implementation. One can apply the following steps to construct a tree structure for a given discretized object.

- Determine the number of levels L and the box size at each level. Identify boxes from the top to the bottom of the tree structure for each unknown. At this stage, boxes are indexed according to their locations in the octree.
- Generate a real array of size N. The value of the nth element of the array is $C + n/N$, where C is the index of the box at the lowest level including nth unknown.
- Sort the real-type array by using a quick-sort algorithm.
- By tracing the sorted array once, find the number of boxes and the number of unknowns in each box.
- Rename each box by considering only nonempty boxes in the tree structure.

These steps work well for extremely large problems discretized with hundreds of millions of unknowns.

3.2.2 Far-Field Interactions

In MLFMA, far-field interactions are calculated in a group-by-group manner using the factorization and diagonalization of the homogeneous-space Green's function [146, 147]. In each MVM, three main stages, i.e., aggregation, translation, and disaggregation, are performed, as described below.

In the aggregation stage, radiated fields of boxes are calculated from the bottom of the tree structure to the highest level ($l = L$). At the lowest level, radiation patterns of basis functions, which are calculated and stored during the setup stage, are multiplied with coefficients provided by the iterative solver and combined to obtain radiated fields of smallest boxes. The radiated field of a box C at a reference point $\boldsymbol{r}_C$ can be calculated as

$$\boldsymbol{S}_C(\boldsymbol{r}_C, \boldsymbol{k}) = \sum_{n \in C} \boldsymbol{x}[n] \boldsymbol{S}_n(\boldsymbol{r}_C, \boldsymbol{k}), \tag{3.59}$$

where $\boldsymbol{S}_n$ represents the radiation pattern of the nth basis function inside the box, and $\boldsymbol{x}[n]$ represents coefficients provided by the iterative solver. In (3.59), the radiation functions $\boldsymbol{S}_C$ and $\boldsymbol{S}_n$ have only θ and ϕ components, and they depend on the angular direction $\hat{\boldsymbol{k}}$. Radiated fields of boxes at higher levels ($l > 1$) are obtained by shifting and combining radiated fields of boxes at lower levels, i.e.,

$$\boldsymbol{S}_C(\boldsymbol{r}_C, \boldsymbol{k}) = \sum_{C' \in C} \beta(\boldsymbol{k}, \boldsymbol{r}_C - \boldsymbol{r}_{C'}) \boldsymbol{S}_{C'}(\boldsymbol{r}_{C'}, \boldsymbol{k}), \tag{3.60}$$

where β is the diagonal shift function defined in (3.47).

In the translation stage of MLFMA, radiated fields of boxes are translated into incoming fields for other boxes. The incoming field at the center of a box C (due to the translations) is calculated as

$$\boldsymbol{G}_C(\boldsymbol{r}_C, \boldsymbol{k}) = \sum_{C' \in \mathcal{F}\{C\}} \alpha_{\tau(l)}(\boldsymbol{k}, \boldsymbol{r}_C - \boldsymbol{r}_{C'}) \boldsymbol{S}_{C'}(\boldsymbol{r}_{C'}, \boldsymbol{k}), \tag{3.61}$$

where $\mathcal{F}\{C\}$ represents boxes in the far-field list of C, and $\alpha_{\tau(l)}$ is the diagonal translation operator defined in (3.49).

In the disaggregation stage, total incoming fields at box centers are calculated from the top of the tree structure to the lowest level. At the highest level, the total incoming field for a box is obtained by the combination of incoming fields due to translations. At the lower levels, however, the incoming field at the center of a box involves a contribution from the incoming field at the center of its parent box. For a box C at a level $l < L$, the total incoming field is calculated as

$$\boldsymbol{G}_C^{+}(\boldsymbol{r}_C, \boldsymbol{k}) = \boldsymbol{G}_C(\boldsymbol{r}_C, \boldsymbol{k}) + \beta(\boldsymbol{k}, \boldsymbol{r}_C - \boldsymbol{r}_{C'})\boldsymbol{G}_{C'}^{+}(\boldsymbol{r}_{C'}, \boldsymbol{k}), \tag{3.62}$$

where $\boldsymbol{G}_C$ is the incoming field due to translations, and C' is the parent box of C, i.e., $C \in C'$. Following disaggregation operations at the lowest level, incoming fields are received by testing functions via an angular integration as

$$\sum_{n=1}^{N} \overline{\boldsymbol{Z}}_{FF}[m,n]\mathbf{x}[n] = \left(\frac{ik}{4\pi}\right)^2 \int d^2\hat{\boldsymbol{k}}\mathbf{F}_m(\boldsymbol{r}_C, \boldsymbol{k}) \cdot \boldsymbol{G}_C^{+}(\boldsymbol{r}_C, \boldsymbol{k}), \tag{3.63}$$

where $\boldsymbol{F}_m$ represents the receiving pattern of the mth testing function inside a box C at the lowest level. Similar to radiation patterns of basis functions, receiving patterns of testing functions are also calculated during the setup of MLFMA.

3.2.3 Radiation and Receiving Patterns

Considering the calculation of far-field interactions as described above, the radiation pattern of a basis function $\boldsymbol{b}_n$ with respect to a reference point $\boldsymbol{r}_{C'}$ can be derived as

$$\boldsymbol{S}_n(\boldsymbol{r}_{C'}, \boldsymbol{k}) = (\overline{\boldsymbol{I}}^{3\times3} - \hat{\boldsymbol{k}}\hat{\boldsymbol{k}}) \cdot \int_{S_n} d\boldsymbol{r}' \exp[-i\boldsymbol{k} \cdot (\boldsymbol{r}' - \boldsymbol{r}_{C'})]\boldsymbol{b}_n(\boldsymbol{r}'), \tag{3.64}$$

where $\overline{\boldsymbol{I}}^{3\times3}$ denotes the 3×3 unit dyad. In contrast to radiation patterns, receiving patterns (of testing functions) depend on the operator and testing type. Specifically, the receiving pattern of a testing function $\boldsymbol{t}_m$ with respect to a reference point $\boldsymbol{r}_C$ can be derived for different operators ($\mathcal{K}$ and $\mathcal{T}$) and testing types (T and N) as

$$\boldsymbol{F}_m^{\mathcal{K},T}(\boldsymbol{r}_C, \boldsymbol{k}) = -\hat{\boldsymbol{k}} \times \int_{S_m} d\boldsymbol{r} \exp[i\boldsymbol{k} \cdot (\boldsymbol{r} - \boldsymbol{r}_C)]\boldsymbol{t}_m(\boldsymbol{r}) \tag{3.65}$$

$$\boldsymbol{F}_m^{\mathcal{T},T}(\boldsymbol{r}_C, \boldsymbol{k}) = (\overline{\boldsymbol{I}}^{3\times3} - \hat{\boldsymbol{k}}\hat{\boldsymbol{k}}) \cdot \int_{S_m} d\boldsymbol{r} \exp[i\boldsymbol{k} \cdot (\boldsymbol{r} - \boldsymbol{r}_C)]\boldsymbol{t}_m(\boldsymbol{r}) \tag{3.66}$$

$$\boldsymbol{F}_m^{\mathcal{K},N}(\boldsymbol{r}_C, \boldsymbol{k}) = -\hat{\boldsymbol{k}} \times \int_{S_m} d\boldsymbol{r} \exp[i\boldsymbol{k} \cdot (\boldsymbol{r} - \boldsymbol{r}_C)]\boldsymbol{t}_m(\boldsymbol{r}) \times \hat{\boldsymbol{n}} \tag{3.67}$$

$$\boldsymbol{F}_m^{\mathcal{T},N}(\boldsymbol{r}_C, \boldsymbol{k}) = (\overline{\boldsymbol{I}}^{3\times3} - \hat{\boldsymbol{k}}\hat{\boldsymbol{k}}) \cdot \int_{S_m} d\boldsymbol{r} \exp[i\boldsymbol{k} \cdot (\boldsymbol{r} - \boldsymbol{r}_C)]\boldsymbol{t}_m(\boldsymbol{r}) \times \hat{\boldsymbol{n}} \tag{3.68}$$

$$= \hat{\boldsymbol{k}} \times \boldsymbol{F}_m^{\mathcal{K},N}(\boldsymbol{r}_C, \boldsymbol{k}). \tag{3.69}$$

For a lossless medium and using a Galerkin scheme,

$$\boldsymbol{F}_m^{\mathcal{T},T}(\boldsymbol{r}_C, \boldsymbol{k}) = \{\boldsymbol{S}_m(\boldsymbol{r}_C, \boldsymbol{k})\}^* \tag{3.70}$$

$$\boldsymbol{F}_m^{\mathcal{K},T}(\boldsymbol{r}_C,\boldsymbol{k}) = -\hat{\boldsymbol{k}} \times \{\boldsymbol{S}_m(\boldsymbol{r}_C,\boldsymbol{k})\}^* \tag{3.71}$$

where '$*$' represents the complex-conjugate operation.

For example, consider a half RWG function associated with the first triangle of the mth edge. Then,

$$\boldsymbol{S}_{m1}^{\mathrm{RWG}}(\boldsymbol{r}_{C'},\boldsymbol{k}) = \frac{l_m}{2A_{m1}}(\overline{\boldsymbol{I}}^{3\times 3} - \hat{\boldsymbol{k}}\hat{\boldsymbol{k}}) \cdot \int_{S_{m1}} d\boldsymbol{r}' \exp[-i\boldsymbol{k}\cdot(\boldsymbol{r}'-\boldsymbol{r}_{C'})](\boldsymbol{r}'-\boldsymbol{r}_{m1}) \tag{3.72}$$

and

$$\boldsymbol{F}_{m1}^{\mathcal{K},N,\,\mathrm{RWG}}(\boldsymbol{r}_C,\boldsymbol{k}) = \frac{l_m}{2A_{m1}}\hat{\boldsymbol{k}} \times \hat{\boldsymbol{n}} \times \int_{S_{m1}} d\boldsymbol{r} \exp[i\boldsymbol{k}\cdot(\boldsymbol{r}-\boldsymbol{r}_C)](\boldsymbol{r}-\boldsymbol{r}_{m1}), \tag{3.73}$$

where the integrals can be calculated analytically. For a first-kind LL function located on the same triangle, the patterns can be written as

$$\boldsymbol{S}_{m1}^{\mathrm{LL}(1)}(\boldsymbol{r}_{C'},\boldsymbol{k}) = \frac{l_m}{4(A_{m1})^2}(\overline{\boldsymbol{I}}^{3\times 3} - \hat{\boldsymbol{k}}\hat{\boldsymbol{k}}) \cdot (\boldsymbol{r}_{m3}-\boldsymbol{r}_{m1})$$

$$[(\boldsymbol{r}_{m4}-\boldsymbol{r}_{m1})\times\hat{\boldsymbol{n}}] \cdot \int_{S_{m1}} d\boldsymbol{r}' \exp[-i\boldsymbol{k}\cdot(\boldsymbol{r}'-\boldsymbol{r}_{C'})](\boldsymbol{r}'-\boldsymbol{r}_{m1}) \tag{3.74}$$

$$\boldsymbol{F}_{m1}^{\mathcal{K},N,\,\mathrm{LL}(1)}(\boldsymbol{r}_C,\boldsymbol{k}) = -\frac{l_m}{4(A_{m1})^2}\hat{\boldsymbol{k}} \times [(\boldsymbol{r}_{m3}-\boldsymbol{r}_{m1})\times\hat{\boldsymbol{n}}]$$

$$[(\boldsymbol{r}_{m4}-\boldsymbol{r}_{m1})\times\hat{\boldsymbol{n}}] \cdot \int_{S_{m1}} d\boldsymbol{r} \exp[i\boldsymbol{k}\cdot(\boldsymbol{r}'-\boldsymbol{r}_C)](\boldsymbol{r}'-\boldsymbol{r}_{m1}) \tag{3.75}$$

where integrals are the same as those in (3.72) and (3.73).

For those formulations involving combinations of operators, a single set of aggregation, translation, and disaggregation stages is sufficient for a MVM associated with a medium. This is because, radiation patterns of basis functions, as well as radiated and incoming fields for boxes do not depend on the operator and testing type. Only receiving patterns of testing functions should be combined according to the formulation. As an example, consider JMCFIE for a homogeneous dielectric object. Using the expressions for the matrix elements in (2.58)–(2.58), the corresponding receiving patterns (of the mth testing functions) can be written as

$$\boldsymbol{F}_{11,m}^{\mathrm{JMCFIE}}(\boldsymbol{r}_C,\boldsymbol{k}_o) = \boldsymbol{F}_{22,m}^{\mathrm{JMCFIE}}(\boldsymbol{r}_C,\boldsymbol{k}_o)$$

$$= \boldsymbol{F}_m^{\mathcal{T},T}(\boldsymbol{r}_C,\boldsymbol{k}_o) + \boldsymbol{F}_m^{\mathcal{K},N}(\boldsymbol{r}_C,\boldsymbol{k}_o) \tag{3.76}$$

$$\boldsymbol{F}_{12,m}^{\mathrm{JMCFIE}}(\boldsymbol{r}_C,\boldsymbol{k}_o) = \eta_o^{-1}\boldsymbol{F}_m^{\mathcal{T},N}(\boldsymbol{r}_C,\boldsymbol{k}_o) - \eta_o^{-1}\boldsymbol{F}_m^{\mathcal{K},T}(\boldsymbol{r}_C,\boldsymbol{k}_o) \tag{3.77}$$

$$\boldsymbol{F}_{21,m}^{\mathrm{JMCFIE}}(\boldsymbol{r}_C,\boldsymbol{k}_o) = \eta_o\boldsymbol{F}_m^{\mathcal{K},T}(\boldsymbol{r}_C,\boldsymbol{k}_o) - \eta_o\boldsymbol{F}_m^{\mathcal{T},N}(\boldsymbol{r}_C,\boldsymbol{k}_o) \tag{3.78}$$

for the outer medium and

$$\boldsymbol{F}_{11,m}^{\mathrm{JMCFIE}}(\boldsymbol{r}_C,\boldsymbol{k}_i) = \boldsymbol{F}_{11,m}^{\mathrm{JMCFIE}}(\boldsymbol{r}_C,\boldsymbol{k}_i)$$

$$= -\boldsymbol{F}_m^{\mathcal{T},T}(\boldsymbol{r}_C,\boldsymbol{k}_i) + \boldsymbol{F}_m^{\mathcal{K},N}(\boldsymbol{r}_C,\boldsymbol{k}_i) \tag{3.79}$$

$$\boldsymbol{F}_{12,m}^{\mathrm{JMCFIE}}(\boldsymbol{r}_C,\boldsymbol{k}_i) = \eta_i^{-1}\boldsymbol{F}_m^{\mathcal{T},N}(\boldsymbol{r}_C,\boldsymbol{k}_i) + \eta_i^{-1}\boldsymbol{F}_m^{\mathcal{K},T}(\boldsymbol{r}_C,\boldsymbol{k}_i) \tag{3.80}$$

$$\boldsymbol{F}_{21,m}^{\text{JMCFIE}}(\boldsymbol{r}_C, \boldsymbol{k}_i) = -\eta_i \boldsymbol{F}_m^{\mathcal{K},T}(\boldsymbol{r}_C, \boldsymbol{k}_i) - \eta_i \boldsymbol{F}_m^{\mathcal{T},N}(\boldsymbol{r}_C, \boldsymbol{k}_i) \tag{3.81}$$

for the inner medium. Note that, for the consistency of orientations, the radiation pattern of the nth basis function is defined as

$$\boldsymbol{S}_{ab,n}^{\text{JMCFIE}}(\boldsymbol{r}_{C'}, \boldsymbol{k}_o) = \boldsymbol{S}_n(\boldsymbol{r}_{C'}, \boldsymbol{k}_o) \tag{3.82}$$

and

$$\boldsymbol{S}_{ab,n}^{\text{JMCFIE}}(\boldsymbol{r}_{C'}, \boldsymbol{k}_i) = -\boldsymbol{S}_n(\boldsymbol{r}_{C'}, \boldsymbol{k}_i) \tag{3.83}$$

for the outer and inner media, respectively, where $a = 1, 2$ and $b = 1, 2$.

3.2.4 *Near-Field Interactions*

In MLFMA, there are also $\mathcal{O}(N^2/N_1) = \mathcal{O}(N)$ near-field interactions, which are calculated directly in the setup stage of the program and stored in memory to be used multiple times during the iterative solution. Using the one-box-buffer scheme, these interactions are between the basis and testing functions that are located in the same box or in two touching boxes at the lowest level. Also note that the near-field matrix $\overline{\boldsymbol{Z}}_{NF}$ has a sparsity of $\mathcal{O}(1/N_1) = \mathcal{O}(1/N)$.

3.2.5 *Sampling*

For each box in the tree structure, radiated and incoming fields are sampled on the unit sphere. It is common to choose samples regularly spaced in the ϕ direction and to use the Gauss-Legendre quadrature in the θ direction [147]. Some examples of this sample scheme are depicted in Figure 3.1. For level $l = 1, 2, \ldots, L$, the number of samples is $S_l^\theta = [\tau(l) + 1]$ and $S_l^\phi = 2[\tau(l) + 1]$ along the θ and ϕ directions, respectively, where $\tau(l)$ is the truncation number for the series in (3.49) to calculate translation operators. For each level l, the truncation number can be determined by the excess bandwidth formula [159] considering the worst-case scenario [170], i.e.,

$$\tau(l) \approx 1.73ka_l + 2.16(d_0)^{2/3}(ka_l)^{1/3}, \tag{3.84}$$

where a_l is the box size and d_0 is the desired digits of accuracy. The sampling rate depends on the box size as measured by the wavelength ($ka_l = 2\pi a_l/\lambda$), and the total number of samples can be approximated as

$$S_l = S_l^\theta S_l^\phi \approx 2^{(l-1)} S_1^\theta \times 2^{(l-1)} S_1^\phi = 4^{(l-1)} S_1, \tag{3.85}$$

where $S_1 = \mathcal{O}(1)$.

Considering the sampling along the angular directions θ and ϕ, aggregations at the lowest level can be rewritten as

$$[\mathbf{s}_C^\theta \ \mathbf{s}_C^\phi]_{S_1\times 2} = \sum_{n\in C} \boldsymbol{x}[n][\boldsymbol{s}_n^\theta \ \boldsymbol{s}_n^\phi]_{S_1\times 2}, \tag{3.86}$$

where $\boldsymbol{s}_n^\theta$ and $\boldsymbol{s}_n^\phi$ are arrays of S_1 elements containing the θ and ϕ components of the radiation pattern of the nth basis function. Similarly, $\boldsymbol{s}_C^\theta$ and $\boldsymbol{s}_C^\phi$ in (3.86) are arrays of S_1 elements

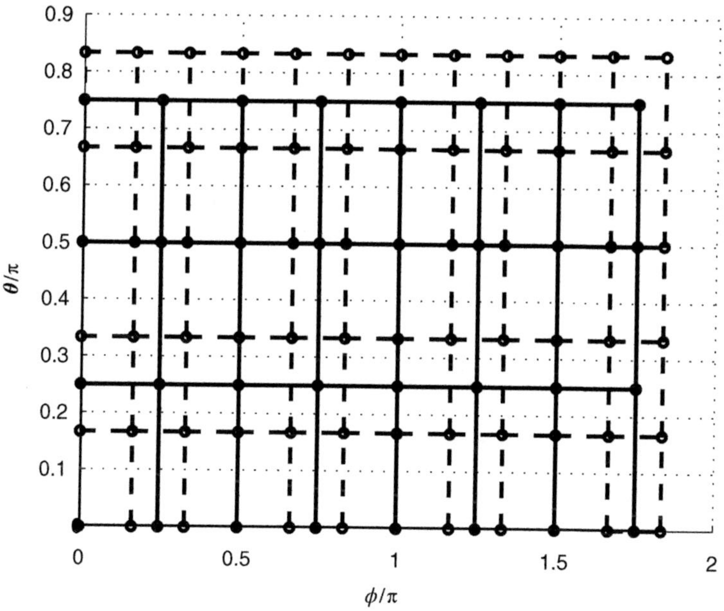

Figure 3.1 Two grids on the unit sphere using regularly spaced points in the ϕ direction and Gauss-Legendre points in the θ direction.

containing the θ and ϕ components of the radiated field of a box C at the lowest level. For a box C at a higher level ($l > 1$),

$$[\boldsymbol{s}_C^\theta \ \boldsymbol{s}_C^\phi]_{S_l\times 2} = \sum_{C'\in C} [\overline{\boldsymbol{\beta}}^{\Diamond}_{C'\to C}]_{S_l\times S_l} \cdot [\overline{\boldsymbol{\Gamma}}_{(l-1)\to l}]_{S_l\times S_{l-1}} \cdot [\boldsymbol{s}_{C'}^\theta \ \boldsymbol{s}_{C'}^\phi]_{S_{l-1}\times 2}, \tag{3.87}$$

where $\overline{\boldsymbol{\beta}}^{\Diamond}_{C'\to C}$ is a diagonal matrix containing the shift functions defined in (3.47) and $\overline{\boldsymbol{\Gamma}}_{(l-1)\to l}$ is a sparse interpolation (upsampling) matrix. The Lagrange interpolation can be used to match different sampling rates of consecutive levels [1].

Similar to aggregations, translation operations can be written in a matrix form as

$$[\boldsymbol{g}_C^\theta \ \boldsymbol{g}_C^\phi]_{S_l\times 2} = \sum_{C'\in\mathcal{F}\{C\}} [\overline{\boldsymbol{\alpha}}^{\Diamond}_{C'\to C}]_{S_l\times S_l} \cdot [\boldsymbol{s}_{C'}^\theta \ \boldsymbol{s}_{C'}^\phi]_{S_l\times 2}, \tag{3.88}$$

where $\boldsymbol{g}_C^\theta$ and $\boldsymbol{g}_C^\phi$ are arrays of S_l elements containing the θ and ϕ components of the incoming field at the center of a box C, and $\overline{\boldsymbol{\alpha}}^{\Diamond}_{C'\to C}$ is a diagonal translation matrix containing the translation functions defined in (3.49). Then, for a box $C \in C'$ at a level $l < L$, the total incoming field is calculated as

$$[\boldsymbol{g}_C^{+,\theta} \boldsymbol{g}_C^{+,\phi}]_{S_l\times 2} = [\boldsymbol{g}_C^\theta \ \boldsymbol{g}_C^\phi]_{S_l\times 2}$$
$$+[\overline{\boldsymbol{\Gamma}}_{(l+1)\to l}]_{S_l\times S_{l+1}} \cdot [\overline{\boldsymbol{\beta}}^{\Diamond}_{C'\to C}]_{S_{l+1}\times S_{l+1}} \cdot [\boldsymbol{g}_{C'}^{+,\theta} \ \boldsymbol{g}_{C'}^{+,\phi}]_{S_{l+1}\times 2}, \tag{3.89}$$

where $\overline{\boldsymbol{\Gamma}}_{(l+1)\rightarrow l}$ is a sparse interpolation (downsampling) matrix. At the lowest level, incoming fields are received by testing functions as

$$\sum_{n=1}^{N} \overline{\boldsymbol{Z}}_{FF}[m,n]\mathbf{a}[n] \approx \left(\frac{ik}{4\pi}\right)^2 \Big\{ [\boldsymbol{f}_m^{\theta}]_{1\times S_1} \cdot [\overline{\boldsymbol{W}}]_{S_1\times S_1} \cdot [\boldsymbol{g}_C^{+,\theta}]_{S_1\times 1} + [\boldsymbol{f}_m^{\phi}]_{1\times S_1} \cdot [\overline{\boldsymbol{W}}]_{S_1\times S_1} \cdot [\boldsymbol{g}_C^{+,\phi}]_{S_1\times 1} \Big\} \tag{3.90}$$

for $m \in C$. In (3.90), $\overline{\boldsymbol{W}}$ is a diagonal matrix involving weights of the angular integration in (3.63), while $\boldsymbol{f}_m^{\theta}$ and $\boldsymbol{f}_m^{\phi}$ are arrays of S_1 elements containing the θ and ϕ components of the receiving pattern of the mth testing function.

3.2.6 Computational Requirements

When MLFMA is used, the memory requirement for a MVM (Θ^{memory}) is proportional to the tree size S_T, i.e.,

$$\Theta^{\text{memory}} \propto S_T = \sum_{l=1}^{L} N_l S_l = \sum_{l=1}^{L} 2N_l[\tau(l)+1]^2. \tag{3.91}$$

The processing time (Θ^{time}) is also related to the tree size as

$$\Theta^{\text{time}} \propto \sum_{l=1}^{L} c_l N_l S_l = \sum_{l=1}^{L} 2c_l N_l[\tau(l)+1]^2, \tag{3.92}$$

where c_l represents relative weights for levels $l = 1, 2, \ldots, L$. Using $N_l \approx 4^{(1-l)}N_1$ and $S_l \approx 4^{(l-1)}S_1$,

$$N_l S_l \approx N_1 S_1 = \mathcal{O}(N) \tag{3.93}$$

and all levels of MLFMA have equal importance with $\mathcal{O}(N)$ complexity in terms of processing time and memory. Then,

$$\Theta^{\text{memory}} = \mathcal{O}(L)\mathcal{O}(N) = \mathcal{O}(N\log N) \tag{3.94}$$

$$\Theta^{\text{time}} = \mathcal{O}(L)\mathcal{O}(N) = \mathcal{O}(N\log N), \tag{3.95}$$

i.e., the complexity of MLFMA is $\mathcal{O}(N\log N)$. Although this is true in general, measurements may present deviations from the ideal case depending on the construction technique for the tree structure, even when N is very large. For example, in Figure 3.2, the tree size (S_T) is plotted as a function of the number of unknowns for the solution of scattering problems involving sphere geometries of various sizes, when the top-down scheme is used to construct the multilevel tree. Far-field interactions are calculated with two digits of accuracy. The radius of the sphere changes from λ to 110λ corresponding to 3723 and 41,883,648 unknowns, respectively, using $\lambda/10$ triangulation. It can be observed that the tree size oscillates around the $O(N\log N)$ curve. Due to such local variations, the processing time and memory requirement for MVMs with respect to N cannot be strictly proportional to $N\log N$. As an example, the tree size grows

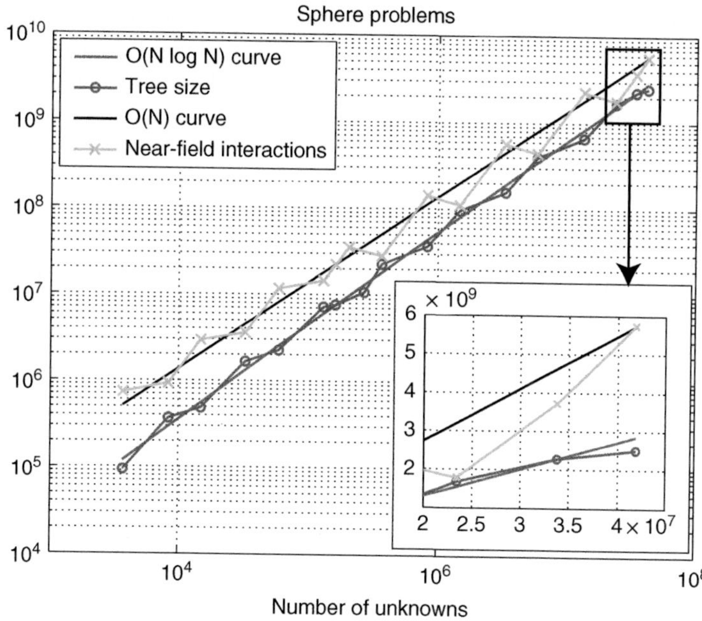

Figure 3.2 Tree size and the number of near-field interactions for the solution of sphere problems using top-down scheme to construct the multilevel tree. *Source*: Ergül and Gürel 2008 [221]. Reproduced with permission of IEEE.

only by 50% when the number of unknowns increases from 23,405,664 to 41,883,648. Then, the memory required for MVMs increases by about 50%, which is below the asymptotical estimation of 85%.

The processing time for the initial setup of MLFMA (prior to the iterative solution) is dominated by calculating near-field interactions, and it is proportional to N^2/N_1. The amount of memory to store near-field interactions can also be significant. Asymptotically, $N_1 = \mathcal{O}(N)$ and near-field interactions has a complexity of $\mathcal{O}(N)$. However, similar to MVMs, local variations in the processing time and memory requirement for near-field interactions may exhibit behavior different than the asymptotical estimation. This is because, as depicted in Figure 3.2, the number of near-field interactions may oscillate around the $\mathcal{O}(N)$ curve when a top-down scheme is used to construct the tree structure. Consequently, variation in the processing time and memory with respect to N can be higher or lower than the asymptotically linear estimate.

Using a Galerkin scheme, the following improvements are possible regarding the near-field interactions:

- For PEC objects, EFIE produces symmetric matrix equations. Hence, only half of near-field interactions are required to to be calculated and stored.
- For homogeneous dielectric objects, CTF, CNF, and JMCFIE leads to identical diagonal blocks. Hence, 3/4 of near-field interactions are required to be calculated and stored. This property is not shared by MNMF, which requires larger memory (4/3 that of others) to store near-field interactions.

Radiation and receiving patterns of basis and testing functions are sampled according to the sampling rate of boxes at the lowest level. These patterns can be calculated and stored in memory before iterative solutions. Since far-field patterns have only θ and ϕ components, they are stored in spherical coordinates. Although the processing time to calculate radiation and receiving patterns is usually negligible, significant amount of memory is required to store them, which can be comparable to the memory used for near-field interactions. If the medium is lossless, the number of samples can be reduced to $S_1 = [\tau(1)/2 + 1] \times [2\tau(1) + 2]$ using the symmetry of patterns. In addition, for a lossless medium and using a Galerkin scheme, the expressions in (3.70) and (3.71) can be used to increase the efficiency. For example, using CTF, only one set of patterns is required for each lossless medium, because both of the receiving patterns, $\boldsymbol{F}_m^{\mathcal{K},T}$ and $\boldsymbol{F}_m^{\mathcal{T},T}$, can be obtained from the related radiation pattern $\boldsymbol{S}_m$. In other words, receiving operations during MVMs can be performed by using radiation patterns (instead of receiving patterns) with small modifications involving a complex conjugation and a cross product with the angular direction $\hat{\boldsymbol{k}}$. However, CNF and MNMF require two sets of patterns, as the receiving patterns $\boldsymbol{F}_m^{\mathcal{K},N}$ and $\boldsymbol{F}_m^{\mathcal{T},N}$ can be derived from each other, but they cannot be obtained directly from the related radiation pattern $\boldsymbol{S}_m$ in spherical coordinates. JMCFIE also requires two sets of patterns, considering all relations between patterns.

Similar to radiation and receiving patterns, translation operators are also calculated and stored in memory before iterations. For each box at any level, there are $\mathcal{O}(1)$ boxes at the same level to translate the radiated field to. In addition, using the symmetry of cubic (identical) boxes, the number of different translation operators is only $7^3 - 3^3 = 316$, independent of the level [158]. Although using the symmetry reduces the number of translation operators significantly, interpolation methods are needed to calculate these operators in $\mathcal{O}(N)$ time during the setup stage [166]. With the optimization of interpolations, as detailed below in Section 3.3.4, both the calculation time and memory for translation operators are insignificant compared to other parts of the implementation, especially when the problem size is large.

As a summary, Table 3.1 lists major parts of MLFMA and their computation requirements for the solution of large problems.

3.2.7 Anterpolation

Interpolations during disaggregations can be replaced with anterpolations (transpose interpolations) to improve the of MLFMA [160]. Figure 3.3 depicts a comparison of interpolation and anterpolation operations in a one-dimensional space. The first plot shows an interpolation from a course grid involving three samples to a fine grid involving four samples. For each target point in the fine grid, two points in the course grid are used. The interpolation weights satisfy $\sum_{i=1}^{3} \gamma_{ij} = 1$ for $j = 1, 2, 3, 4$. The second plot is also an interpolation using two points per target point; but this time, the sampling rate is reduced from four to three using appropriate interpolation weights. On the other hand, the third plot represents an anterpolation operation from three points to four points. Specifically, this operation is derived directly from the interpolation operation (in the first plot) by 'changing the directions of the arrows.' Note that the weights used in the interpolation (γ_{ij}) are not changed and the output function after the anterpolation can be completely different from the input function.

In order to see how anterpolations can be used in MLFMA, consider the interaction of the nth basis function and the mth testing function via a translation at level $l \neq 1$. The corresponding

Table 3.1 Major parts of MLFMA and their computational requirements

Part	Memory	
	Proportional to	Complexity
MVM	$\sum_{l=1}^{L} N_l[\tau(l)+1]^2$	$\mathcal{O}(N\log N)$
Radiation and receiving patterns	$N[\tau(1)+1]^2$	$\mathcal{O}(N)$
Translation operators	$\sum_{l=1}^{L} d_l[\tau(l)+1]^2$	$\mathcal{O}(N)$
Near-field interactions	N^2/N_1	$\mathcal{O}(N)$
	Processing time	
Part	Proportional to	Complexity
MVM	$\sum_{l=1}^{L} c_l N_l[\tau(l)+1]^2$	$\mathcal{O}(N\log N)$
Radiation and receiving patterns	$N[\tau(1)+1]^2$	$\mathcal{O}(N)$
Translation operators	$\sum_{l=1}^{L} d_l[\tau(l)+1]^2$	$\mathcal{O}(N)$
Near-field interactions	N^2/N_1	$\mathcal{O}(N)$

c_l and d_l represent relative weights for levels $l = 1, 2, \ldots, L$

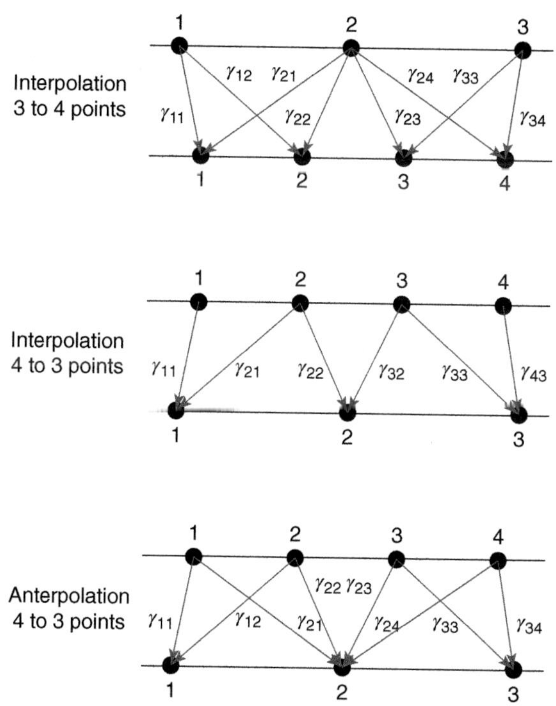

Figure 3.3 Interpolation and anterpolation operators in a one-dimensional space. The plots show an interpolation from the course grid to the fine grid, an interpolation from the fine grid to the course grid, and an anterpolation from the fine grid to the course grid.

matrix element can be written as

$$\overline{Z}[m,n] = \left(\frac{ik}{4\pi}\right)^2 \int d^2\hat{k} \boldsymbol{F}_m(\boldsymbol{r}_C,\boldsymbol{k}) \cdot \boldsymbol{G}_n(\boldsymbol{r}_C,\boldsymbol{k}), \tag{3.96}$$

where $\boldsymbol{G}_n$ is the incoming field due to the nth basis function. Equation (3.96) represents the calculation of a single interaction in MLFMA; but, note that this expression is never used explicitly, because interactions are actually calculated in a group-by-group manner. Since $\tau_l > \tau_1$ for $l > 1$, the sampling rate for $\boldsymbol{G}_n$ is larger than the sampling rate for $\boldsymbol{F}_m$. An accurate way to calculate (3.96) could be to interpolate $\boldsymbol{F}_m$ and to perform the numerical integration in the fine grid. This leads to

$$\begin{aligned}\overline{Z}[m,n] \approx \left(\frac{ik}{4\pi}\right)^2 \Big\{ &[\boldsymbol{g}_n^\theta]_{1\times S_l} \cdot [\overline{\boldsymbol{W}}]_{S_l\times S_l} \cdot [\overline{\boldsymbol{\Gamma}}_{1\to l}]_{S_l\times S_1} \cdot [\boldsymbol{f}_m^\theta]_{S_1\times 1} \\ &+ [\boldsymbol{g}_n^\phi]_{1\times S_l} \cdot [\overline{\boldsymbol{W}}]_{S_l\times S_l} \cdot [\overline{\boldsymbol{\Gamma}}_{1\to l}]_{S_l\times S_1} \cdot [\boldsymbol{f}_m^\phi]_{S_1\times 1} \Big\},\end{aligned} \tag{3.97}$$

where $\boldsymbol{g}_n^\theta$ and $\boldsymbol{g}_n^\phi$ are arrays of S_l elements containing θ and ϕ components of the incoming field due to the nth basis function. On the other hand, the procedure in (3.97) contradicts with the structure of MLFMA, where disaggregations need to be performed downward from top to bottom of the tree. As given in (3.90), incoming fields must be received at the lowest level, i.e.,

$$\begin{aligned}\overline{Z}[m,n] \approx \left(\frac{ik}{4\pi}\right)^2 \Big\{ &[\boldsymbol{f}_m^\theta]_{1\times S_1} \cdot [\overline{\boldsymbol{W}}]_{S_1\times S_1} \cdot [\overline{\boldsymbol{\Gamma}}_{l\to 1}]_{S_1\times S_l} \cdot [\boldsymbol{g}_n^\theta]_{S_l\times 1} \\ &+ [\boldsymbol{f}_m^\phi]_{1\times S_1} \cdot [\overline{\boldsymbol{W}}]_{S_1\times S_1} \cdot [\overline{\boldsymbol{\Gamma}}_{l\to 1}]_{S_1\times S_l} \cdot [\boldsymbol{g}_n^\phi]_{S_l\times 1} \Big\}.\end{aligned} \tag{3.98}$$

The procedure in (3.98) fits into the MLFMA structure, but it is inaccurate since the integration is performed in the coarse grid.

In MLFMA, one can use another procedure, which is obtained by taking the transpose of (3.97), i.e.,

$$\begin{aligned}\overline{Z}[m,n] \approx \left(\frac{ik}{4\pi}\right)^2 \Big\{ &[\boldsymbol{f}_m^\theta]_{1\times S_1} \cdot [(\overline{\boldsymbol{\Gamma}}_{1\to l})^\dagger]_{S_1\times S_l} \cdot [\overline{\boldsymbol{W}}]_{S_l\times S_l} \cdot [\boldsymbol{g}_n^\theta]_{S_l\times 1} \\ &+ [\boldsymbol{f}_m^\phi]_{1\times S_1} \cdot [(\overline{\boldsymbol{\Gamma}}_{1\to l})^\dagger]_{S_1\times S_l} \cdot [\overline{\boldsymbol{W}}]_{S_l\times S_l} \cdot [\boldsymbol{g}_n^\phi]_{S_l\times 1} \Big\},\end{aligned} \tag{3.99}$$

where $(\overline{\boldsymbol{\Gamma}}_{1\to l})^\dagger$ is a transpose interpolation (anterpolation) matrix. This procedure is suitable for MLFMA and its accuracy is exactly the same as the accuracy of the hypothetical procedure given in (3.97). It should be noted that integration weights related to level l also need to be anterpolated; hence, it is appropriate to introduce them during translations when radiated fields are converted into incoming fields.

3.3 Lagrange Interpolation and Anterpolation

In MLFMA implementations, the local Lagrange interpolation is preferable due to its favorable computing cost and controllable error [1, 164, 165]. From the local neighborhood of each target point in the fine grid, where the field is to be obtained by interpolation, $2p \times 2p$ points in the coarse grid are selected. This is illustrated in Figure 3.4 for $p = 2$, where values of the field at 4×4 points (shaded circles) are employed to compute the value at the target point (star).

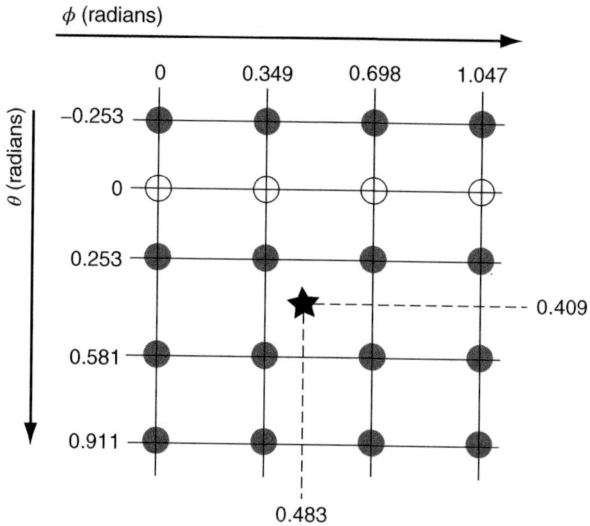

Figure 3.4 Lagrange interpolation employing 4 × 4 points (shaded circles) located in the coarse grid to evaluate the function at a point (star) located in the fine grid. Sampling values of θ and ϕ are specified in radians and selected from a practical case. *Source*: Ergül and Gürel 2006 [164]. Reproduced with permission of IEEE.

Let $f(\theta, \phi)$ be a scalar function representing a radiated or incoming field in MLFMA. Using a two-dimensional Lagrange interpolation, the value of the function at a target point (θ,ϕ) in the fine grid is obtained by using $2p \times 2p$ samples in the coarse grid, i.e.,

$$f(\theta, \phi) \approx \sum_{j=s+1-p}^{s+p} w_j(\phi) \sum_{i=t+1-p}^{t+p} v_i(\theta) f(\theta_i, \phi_j), \tag{3.100}$$

where $w_j(\phi)$ and $v_i(\theta)$ represent interpolation weights derived as

$$w_j(\phi) = \prod_{\substack{m=s+1-p \\ m\neq j}}^{s+p} \frac{\phi - \phi_m}{\phi_j - \phi_m} \tag{3.101}$$

for the ϕ direction, and

$$v_i(\theta) = \prod_{\substack{n=t+1-p \\ n\neq i}}^{t+p} \frac{\theta - \theta_n}{\theta_i - \theta_n} \tag{3.102}$$

for the θ direction, respectively. The reference indices s and t in (3.100)–(3.102) are determined by the location of the target point (θ, ϕ), with respect to samples in the coarse grid.

Since sampling rates for fields are determined by considering their spectral contents, the number of interpolation points $4p^2$ can be fixed for the entire aggregation and disaggregation stages. In other words, the same number of interpolation points can be employed at each level of MLFMA. Although functions being interpolated get richer in terms of harmonic content for the higher levels, the sampling rate is also increased so that a fixed number of interpolation points is sufficient to obtain the same level of accuracy at all levels.

Interpolation of a function at $S_l^\theta \times S_l^\phi$ points requires

$$\Theta^{\text{time}} = 4p^2 S_l^\theta S_l^\phi = 8p^2[\tau(l) + 1]^2 \tag{3.103}$$

operations. If the weights in the θ and ϕ directions are combined, the interpolation in (3.100) can be expressed as a MVM, i.e.,

$$[f]_{S_l \times 1} \approx [\overline{\mathbf{\Gamma}}_{(l-1)\to l}]_{S_l \times S_{l-1}} \cdot [f]_{S_{l-1}\times 1}, \tag{3.104}$$

where f represents one-dimensional arrays involving the samples of f. For interpolations from level $(l-1)$ to l, there are $4p^2$ nonzero elements per row in $\overline{\mathbf{\Gamma}}_{(l-1)\to l}$. The matrix representation in (3.104) is preferred due to its simplicity, and it is very useful for an easy implementation of anterpolations in MLFMA. However, the amount of memory required for the interpolation matrix is proportional to

$$\Theta_{\text{matrix}}^{\text{memory}} = 4p^2 S_l^\theta S_l^\phi = 8p^2[\tau(l) + 1]^2, \tag{3.105}$$

which can be significant for large problems. Considering the original form in (3.100), it is possible to store interpolation weights along the θ and ϕ directions separately, in two arrays of sizes $2p[\tau(l) + 1]$ and $4p[\tau(l) + 1]$, respectively. Then, the total memory used for interpolations from level $(l-1)$ to l becomes

$$\Theta_{\text{array}}^{\text{memory}} = (2p + 4p)[\tau(l) + 1] = 6p[\tau(l) + 1], \tag{3.106}$$

without any change in the number of operations and processing time. The reduction in memory by using the array representation instead of the matrix representation is

$$\frac{\Theta_{\text{array}}^{\text{memory}}}{\Theta_{\text{matrix}}^{\text{memory}}} = \frac{6p[\tau(l) + 1]}{8p^2[\tau(l) + 1]^2} = \frac{3}{4p[\tau(l) + 1]}, \tag{3.107}$$

which is especially significant for higher levels.

3.3.1 Two-Step Method

In general, interpolations and anterpolations constitute a major computational bulk of MLFMA. Therefore, to obtain an efficient solver, it is extremely important to optimize interpolation/anterpolation routines. The number of operations required for the conventional (one-step) Lagrange interpolation from level $(l-1)$ to l is $8p^2[\tau(l) + 1]^2$. This is because there are $[\tau(l) + 1] \times 2[\tau(l) + 1]$ points in the fine grid (samples for level l) and each of these points has $4p^2$ contributions from the coarse grid. On the other hand, locations of sampling points in the θ and ϕ directions are independent of each other. Therefore, interpolations along the two directions can be performed consecutively, as follows:

- Perform an interpolation along the θ direction as

$$f(\theta, \phi_j) \approx \sum_{i=t+1-p}^{t+p} v_i(\theta) f(\theta_i, \phi_j), \tag{3.108}$$

which requires $2pS_l^\theta S_{l-1}^\phi$ operations.

- Perform an interpolation along the ϕ direction using the result of the first step, i.e.,

$$f(\theta,\phi) \approx \sum_{j=s+1-p}^{s+p} w_j(\phi) f(\theta,\phi_j). \tag{3.109}$$

  This step requires $2pS_l^\theta S_l^\phi$ operations.

Consequently, using the two-step method, the processing time required to interpolate the function at $S_l^\theta \times S_l^\phi$ points is

$$\Theta_{\text{two-step}}^{\text{time}} = 2pS_l^\theta(S_{l-1}^\phi + S_l^\phi) = 4p[\tau(l)+1][\tau(l)+\tau(l-1)+2]. \tag{3.110}$$

Comparing processing times required for the one-step and two-step interpolation methods,

$$\begin{aligned}\frac{\Theta_{\text{two-step}}^{\text{time}}}{\Theta_{\text{one-step}}^{\text{time}}} &= \frac{4p[\tau(l)+1][\tau(l)+\tau(l-1)+2]}{8p^2[\tau(l)+1]^2} \\ &< \frac{8p[\tau(l)+1][\tau(l)+1]}{8p^2[\tau(l)+1]^2} = \frac{1}{p}\end{aligned} \tag{3.111}$$

since $\tau(l-1) < \tau(l)$. Therefore, the two-step method is always faster than the one-step method. To store the intermediate array between steps, the two-step method requires a bit more memory than is used in the one-step method. Nevertheless, the speedup in the two-step method more than compensates for the small increase in memory.

The two-step interpolation method is easily implemented by decomposing the conventional one-step method into two successive parts. To demonstrate the acceleration provided by the two-step method, solutions of scattering problems involving a PEC sphere are considered. The radius of the sphere changes from 6λ to 48λ and it is illuminated by a plane wave. The problems are formulated with CFIE and discretized with the RWG functions. Triangulations with $\lambda/10$ mesh size lead to matrix equations involving 132,003 to 8,447,808 unknowns. The problems are solved iteratively, with MVMs performed efficiently by MLFMA. Solutions are parallelized into 16 processes on a cluster of AMD Opteron 870 processors. The hierarchical partitioning strategy is used for the efficient parallelization of MLFMA. Far-field interactions are calculated with two digits of accuracy and interpolation/anterpolation operations are performed using 6×6 stencils ($p = 3$). Table 3.2 lists the processing time required for the aggregation stage, in addition to the speedup offered by the two-step interpolation method. Compared to the conventional one-step method, the two-step method reduces the processing time of the aggregation stage by about 45%. To demonstrate the overall improvement, Table 3.2 also presents the processing time required for MVMs, which are reduced by 25–30% with the two-step interpolation method.

3.3.2 *Virtual Extension of the θ-ϕ Space*

If a target point in the fine grid is close to the boundaries of the sampling space, some interpolation points may fall outside. In other words, the interpolation may need to use sample points at (θ,ϕ), which may have $\theta < 0$, $\theta > \pi$, $\phi < 0$, or $\phi > 2\pi$. To handle these cases, the extended

Table 3.2 Processing time required for an aggregation stage and for an MVM when interpolation/anterpolation Operations are performed by using one-step and two-step interpolation methods

Sphere Problems		Aggregation		
Radius (λ)	Unknowns	One-Step (s)	Two-Step (s)	Reduction
6	132,000	0.646	0.360	44%
7.5	206,499	0.844	0.470	44%
10	367,821	2.18	1.19	45%
15	829,881	3.86	2.14	45%
20	1,462,854	9.47	5.28	44%
30	3,319,524	17.1	9.43	45%
40	5,851,416	40.7	22.3	45%
48	8,447,808	54.7	30.8	44%
Sphere Problems		**MVM**		
Radius (λ)	Unknowns	One-Step (s)	Two-Step (s)	Reduction
6	132,000	1.98	1.47	26%
7.5	206,499	2.52	1.86	26%
10	367,821	6.21	4.46	28%
15	829,881	11.1	8.00	28%
20	1,462,854	27.0	19.6	27%
30	3,319,524	47.0	33.4	29%
40	5,851,416	112	79.2	29%
48	8,447,808	152	109	28%

coverage of the sample space is required. This extended space is periodic in the ϕ direction, i.e.,

$$\begin{aligned} f(\theta,\phi) &= f(\theta,\phi+2\pi), \quad &&\text{if} \quad \phi<0 \\ f(\theta,\phi) &= f(\theta,\phi-2\pi), \quad &&\text{if} \quad \phi>2\pi \end{aligned} \tag{3.112}$$

for an arbitrary function $f(\theta,\phi)$. In the θ direction, however, appropriate shifts (in the ϕ direction) are required. In addition, the value of the function is multiplied by -1 due to reversed directions of θ and ϕ over the poles, i.e.,

$$\begin{aligned} f(\theta,\phi) &= -f(-\theta,\phi+\pi), \quad &&\text{if} \quad \theta<0, \quad 0<\phi<\pi \\ f(\theta,\phi) &= -f(-\theta,\phi-\pi), \quad &&\text{if} \quad \theta<0, \quad \pi<\phi<2\pi \\ f(\theta,\phi) &= -f(2\pi-\theta,\phi+\pi), \quad &&\text{if} \quad \theta>\pi, \quad 0<\phi<\pi \\ f(\theta,\phi) &= -f(2\pi-\theta,\phi-\pi), \quad &&\text{if} \quad \theta>\pi, \quad \pi<\phi<2\pi. \end{aligned} \tag{3.113}$$

3.3.3 Sampling at the Poles

As the problem size grows and more levels are required to construct the tree structure of MLFMA, it becomes critical to minimize the interpolation error. This is because interpolation and anterpolation operations are performed between all consecutive levels and the overall error is accumulated during aggregation and disaggregation stages. In this subsection, a simple method to reduce the error around the two poles of the sphere, i.e., the north pole ($\theta = 0$) and the south pole ($\theta = \pi$), is introduced. This is essential since the error in these regions is usually larger than errors in other regions.

Figure 3.4 demonstrates a practical case, where the value of the field is to be obtained at a point (star) with spherical coordinates $(\theta, \phi) = (0.409, 0.483)$ specified in radians. The figure is related to an aggregation step from a level with a box size of 0.25λ to the next level with a box size of 0.5λ. Using the excess bandwidth formula [159] for three digits of accuracy, the truncation number is 8 and 12 for the lower and higher levels, respectively. Therefore, the number of samples in the θ (ϕ) direction is 9 (18) for the coarse grid and 13 (26) for the fine grid. The interpolation parameter (p) is set to 2 so that 4×4 points (shaded circles) are employed in the interpolation. According to the conventional sampling scheme [147], samples are regularly spaced in the ϕ direction while they are chosen as the Gauss-Legendre points in the θ direction.

In Figure 3.4, there is only one sample in the θ direction on the coarse grid ($\theta = 0.253$) above the target point up to the north pole ($\theta = 0$). Therefore, four points for the interpolation are provided from the region on the other side of the pole. Considering the next sample in the decreasing θ direction, these are points with $\theta = -0.253$. Although this is the best choice, there exists a wide gap in the θ direction from $\theta = -0.253$ to $\theta = 0.253$. These wide gaps created near the poles at all levels lead to larger interpolation errors compared to other regions far from the poles.

To reduce the interpolation error described above, fields can be sampled at the poles, i.e., at $\theta = 0$ and $\theta = \pi$. Although radiated and incoming fields in MLFMA are vectors with two spherical components that can be represented as

$$\boldsymbol{f}(\theta, \phi) = \hat{\boldsymbol{\theta}}\hat{\boldsymbol{\theta}} \cdot \boldsymbol{f}(\theta, \phi) + \hat{\boldsymbol{\phi}}\hat{\boldsymbol{\phi}} \cdot \boldsymbol{f}(\theta, \phi), \tag{3.114}$$

fields are better to be evaluated and stored in x and y directions at the poles. As an example, for the north pole, the θ and ϕ components are extracted as

$$\hat{\boldsymbol{\theta}} \cdot \boldsymbol{f}(0, \phi) = [\cos\phi f_x(\theta = 0) + \sin\phi f_y(\theta = 0)] \tag{3.115}$$

and

$$\hat{\boldsymbol{\phi}} \cdot \boldsymbol{f}(0, \phi) = [\cos\phi f_y(\theta = 0) - \sin\phi f_x(\theta = 0)] \tag{3.116}$$

whenever required for the interpolation. Consequently, independent of the value of ϕ, all interpolations performed near the poles are improved by this technique using (3.115) and (3.116) without having to store values of $\boldsymbol{f}(\theta = 0, \phi)$ for each sample of ϕ. This is illustrated in Figure 3.4, where samples computed with (3.115) and (3.116) are represented by circles located at $\theta = 0$.

To include calculations related to the poles, four more complex numbers are required (two for each pole) for each basis or testing function. The extra memory and the increase in the processing time are negligible compared to base requirements, since the data size for each function is raised from $4[\tau(l)+1]^2$ to $4[\tau(l)+1]^2+4$, where $\tau(l)$ is at least 7 for two digits of accuracy. During aggregation and disaggregation stages, values at the poles are calculated and stored for each box to improve interpolations and anterpolations at all levels. However, the poles do not contribute to angular integrations over unit sphere.

As a numerical example, Figure 3.5 depicts the relative interpolation error related to a basis box C_1 with the size of 2λ at the fourth level from the bottom of the tree structure. For the field of the box, the number of angular samples in the θ and ϕ directions are 33 and 66, respectively, determined by the excess bandwidth formula [159] for three digits of accuracy. By the row-wise arrangement of the θ-ϕ space, the interpolation error is plotted with respect to samples. Only the error in the θ component of the field, which is the dominant component in this example, is plotted. The relative interpolation error is defined as

$$\Delta_R[n] = \frac{|\tilde{f}[n] - f[n]|}{\max |f|}, \tag{3.117}$$

where f is the exact field, $\tilde{f}$ is the perturbed field obtained via interpolations, and $n = 1, 2, \ldots, 2178$ is the sample index. The exact data is calculated by evaluating the field for each basis function inside the cube with the sample rate defined for the fourth level so that it is obtained without any interpolation. However, in the case of the perturbed data, fields of basis functions are sampled according to the smallest box size, which is 0.25λ. Then, three aggregation steps are performed from the lowest (first) level to the fourth level. Consequently, the perturbed data is the practical case, where the interpolation error is introduced at three passages between levels, i.e., from first to second, second to third, and third to fourth.

In Figure 3.5, the interpolation error is plotted when the poles are not employed as in the conventional case (gray) and when they are employed (red). For a clear comparison, maximum errors are also indicated in the plot with horizontal lines. Employing the poles decreases errors, and the maximum relative error is reduced approximately to its half from 1.530×10^{-4} to 7.147×10^{-5}. The reduction in the error is clearly visible at the two ends of the horizontal axis corresponding to points located near the poles. However, the improvement extends beyond these narrow polar regions towards the middle of the horizontal axis corresponding to points located around the equator. This is due to the fact that an interpolation error made in the first steps of the aggregation is propagated towards the equator region in the next steps. Consequently, the use of the poles improves the interpolation accuracy also for the points located far from the poles. In general, improvements obtained by adding the poles in the Lagrange interpolation become more significant as the problem size grows and more levels are required in the MLFMA tree.

Demonstration of the error is not as straightforward as the interpolation error since the anterpolated function is different from the original function. Therefore, to prove that employing poles also increases the accuracy of the anterpolation in MLFMA, Figure 3.6 depicts errors in partial MVMs. After translating the radiation of C_1 into an incoming wave for a testing box C_2, disaggregation steps are performed from the fourth level to the first level. This way, as presented in Figure 3.6(a), one can obtain the error in the partial MVM related to boxes C_1 and C_2 with respect to the index of the testing functions in C_2. Similar to (3.117), the relative

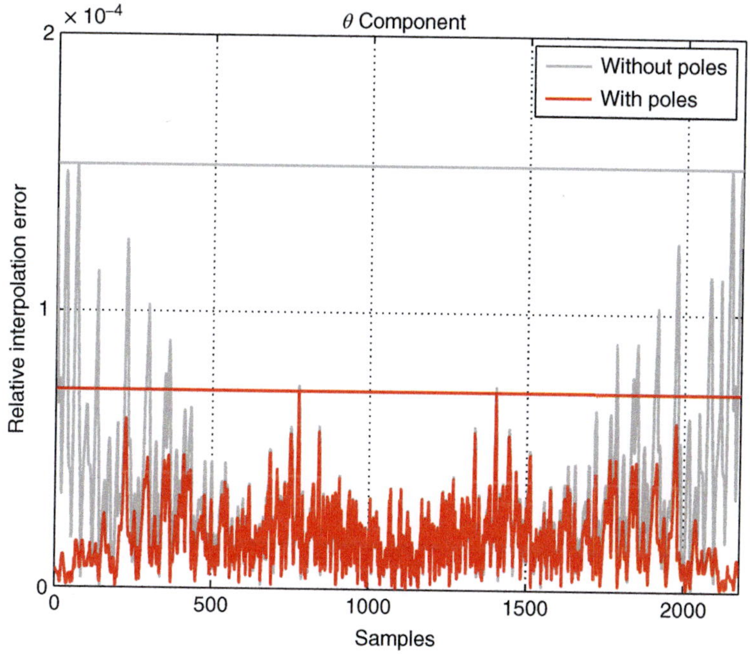

Figure 3.5 Relative interpolation error defined in (3.117) with respect to samples on a 33×66 grid converted into one-dimensional data by a row-wise arrangement of the θ-ϕ space. The reference data is obtained without interpolation. To obtain the interpolated data, aggregation is performed from the lowest (first) level to the fourth level by employing interpolations with (red) and without (gray) poles.

error is defined as

$$\Delta_R[m] = \frac{|\tilde{y}[m] - y[m]|}{\max |y|}, \tag{3.118}$$

where

$$\boldsymbol{y}[m] = \sum_{n=1}^{N(C_1)} \overline{\boldsymbol{Z}}[m,n]\boldsymbol{x}[n] \qquad (m = 1, 2, \ldots, N(C_2)) \tag{3.119}$$

is the result of the partial multiplication without interpolation and anterpolation errors. This reference data is obtained by evaluating radiation and receiving patterns of the basis and testing functions inside the boxes C_1 and C_2, respectively, with the sampling rate defined for the fourth level so that interpolation and anterpolation are not involved. In (3.119), $\boldsymbol{x}[n]$ represents the coefficients of the basis functions inside C_1 for $n = 1, 2, \ldots, N(C_1)$, $\overline{\boldsymbol{Z}}[m,n]$ represents the matrix elements calculated by MLFMA, and m is the testing index from 1 to $N(C_2) = 704$. The perturbed data in (3.117) is obtained by performing usual aggregation and disaggregation steps so that values in Figure 3.6(a) contain both the interpolation and anterpolation errors. Figure 3.6(a) displays a significant improvement in the accuracy obtained by using the poles. Next, the interpolation error is eliminated by computing incoming waves to the center of the

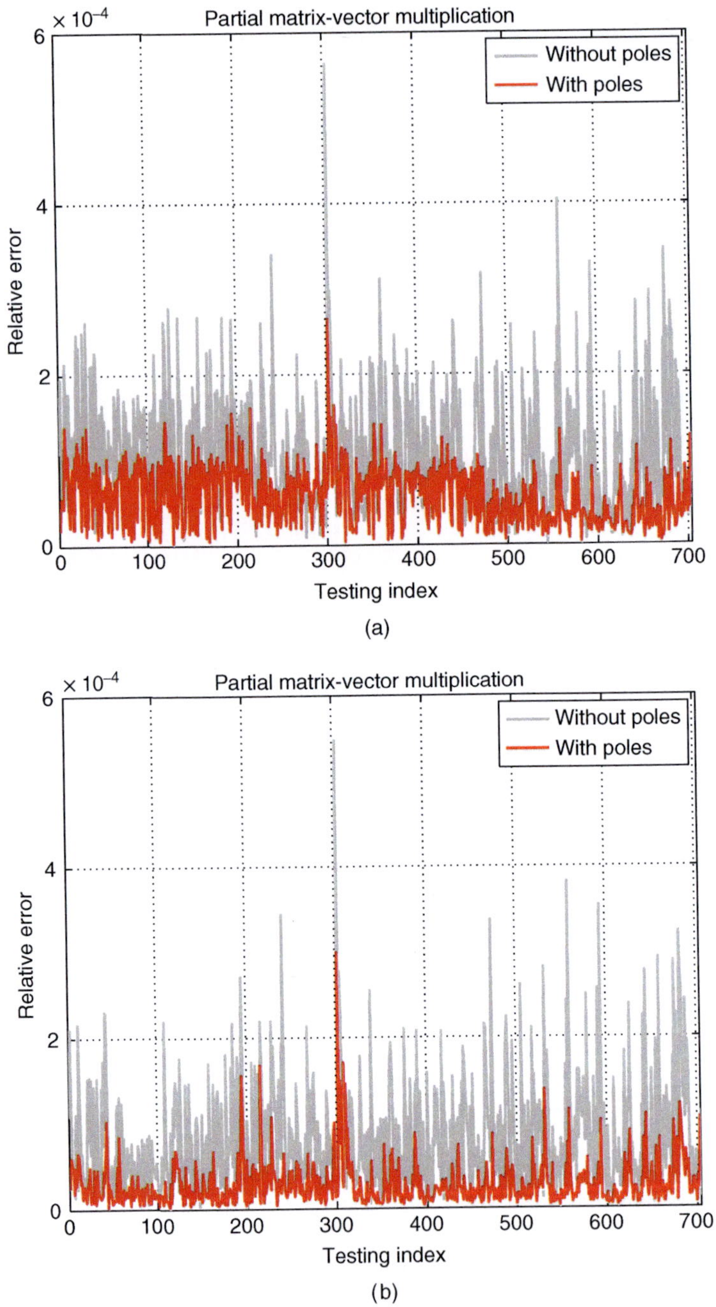

Figure 3.6 Relative errors defined in (3.118) in partial MVMs related to two boxes C_1 and C_2 in the fourth level with respect to the index of the testing function located in C_2. The reference data is obtained without interpolation and anterpolation. (a) The relative error when interpolation and anterpolation are employed. (b) The relative error when interpolation is eliminated and the error is only due to the anterpolation. *Source*: Ergül and Gürel 2006 [164]. Reproduced with permission of IEEE.

testing box C_2 without employing interpolation. The result is depicted in Figure 3.6(b), where the error in the partial MVM is due to only the anterpolation. Therefore, Figure 3.6(b) clearly demonstrates that employing the poles enhances the accuracy of anterpolations, similar to the improvement shown in Figure 3.5 for interpolations.

3.3.4 Interpolation of Translation Operators

In MLFMA, the direct calculation of translation operators requires $\mathcal{O}(N^{3/2})$ operations, and the processing time for their setup becomes substantial as the problem size grows. In [166], the Lagrange interpolation is proposed to efficiently fill translation matrices for large problems. In this section, the Lagrange interpolation of translation operators is revisited and the parameters of the interpolation, namely, the number of interpolation points and the oversampling factor, are optimized to obtain the desired level of the accuracy with the minimum processing time. The optimal parameters are also compared to fixed parameters to demonstrate the improvement obtained with the optimization.

A three-dimensional translation operator between a pair of basis and testing boxes is given in (3.49). For level l, the summation in (3.49) is truncated at $\tau(l)$, which is determined by the excess bandwidth formula in (3.84). In Figure 3.7(a), the truncation number $\tau(l)$ is plotted with respect to the number of accurate digits d_0 and for different values of the box size a_l increasing

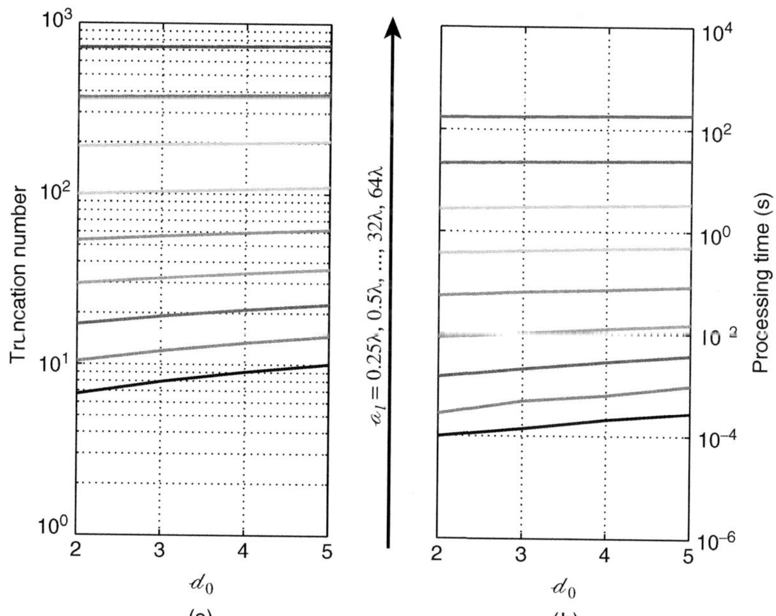

Figure 3.7 (a) Truncation number as a function of d_0 and the box size a_l. (b) Processing time to compute the translation function for a single box-box interaction. In both figures, there are 9 curves for different values of the box size increasing by a factor of two from 0.25λ to 64λ. The lowest and highest curves correspond to 0.25λ and 64λ, respectively. *Source*: Ergül and Gürel 2006 [164]. Reproduced with permission of IEEE.

by factors of two from 0.25λ to 64λ. Figure 3.7(a) demonstrates that L grows rapidly as the box size increases. For a fixed a_l, however, $\tau(l)$ increases gradually with respect to d_0, and the variation is small for large a_l.

In Figure 3.7(b), the processing time to calculate translation operators measured on a 1.8 GHz Opteron 244 processor is plotted with respect to same parameters as in Figure 3.7(a). The values are given for a single interaction between a pair of basis and testing boxes while a typical problem requires the calculation of numerous box-box interactions. Since $\tau(l) = \mathcal{O}(ka_l)$, the processing time to evaluate (3.49) for a fixed $\hat{\boldsymbol{k}}$ is $\mathcal{O}(ka_l)$. In addition, the number of angular directions $\hat{\boldsymbol{k}}$ is $\mathcal{O}(\tau^2(l))$ and the processing time to evaluate (3.49) becomes $O(k^3a_l^3)$ for a box-box interaction. For the lowest level of MLFMA, $\mathcal{O}(k^3a_l^3) = \mathcal{O}(1)$, which is acceptable although the number of boxes at this level is $\mathcal{O}(N)$. However, for the largest boxes of a problem, $\mathcal{O}(ka_l) = \mathcal{O}(N^{1/2})$ and $\mathcal{O}(k^3a_l^3) = \mathcal{O}(N^{3/2})$, which is significant. In fact, as N becomes large, the processing time required to calculate translation operators is dominated by evaluations at the higher levels, although there are a few boxes at those levels. In addition, the setup time for the translation matrix becomes dominant compared to the time required for other parts of MLFMA, even MVMs that can be performed in $\mathcal{O}(N \log N)$ time.

Defining a variable $\varphi = \cos^{-1}(\hat{\boldsymbol{k}} \cdot \hat{\boldsymbol{r}}_{CC'})$, a translation operator can be expressed as a band-limited function of φ [166], i.e.,

$$\alpha_{\tau(l)}(k, R_{CC'}, \varphi) = \sum_{t=0}^{\tau(l)} i^t(2t+1)h_t^{(1)}(kR_{CC'})P_t(\cos\varphi). \tag{3.120}$$

Choosing an oversampling factor q and sampling the operator along φ from 0 to π at $\lfloor q\tau(l) \rfloor$ equally spaced points ($\lfloor \cdot \rfloor$ represents the floor operation), i.e., at $\varphi_i = \pi(i-1)/(\lfloor q\tau(l) \rfloor - 1)$ and $i = 1, \ldots, \lfloor q\tau(l) \rfloor$, the translation operator can be obtained by the Lagrange interpolation at any point as

$$\tilde{\alpha}_{\tau(l)}(k, R_{CC'}, \varphi) = \sum_{i=t+1-p}^{t+p} \alpha_{\tau(l)}(k, R_{CC'}, \varphi_i) w_i(\varphi), \tag{3.121}$$

where $\tilde{\alpha}$ represents the translation function perturbed by the interpolation error,

$$t = \left\lfloor \frac{\varphi(\lfloor q\tau(l) \rfloor - 1)}{\pi} + 1 \right\rfloor, \tag{3.122}$$

and

$$w_i(\varphi) = \prod_{\substack{j=t+1-p \\ j \neq i}}^{t+p} \frac{\varphi - \varphi_j}{\varphi_i - \varphi_j}. \tag{3.123}$$

In (3.121) and (3.123), p is the number of interpolation points employed at each side of the target location φ.

Figures 3.8(a) and 3.8(b) depict the magnitude and phase of the translation operator, respectively, for two boxes separated by $\boldsymbol{r}_{CC'} = \hat{\boldsymbol{x}}2a_l$, where $a_l = 4\lambda$. Using three digits of accuracy, $\tau(l) = 57$. The translation operator is calculated directly, where the function is evaluated at the required points by using (3.120). In the ϕ direction, there are $2[\tau(l) + 1] = 116$ samples that are equally spaced from 0 to 2π. In the θ direction, there are $[\tau(l) + 1] = 58$ samples (zeros of the Legendre polynomial) and they are not equally spaced. Then, there are a total of

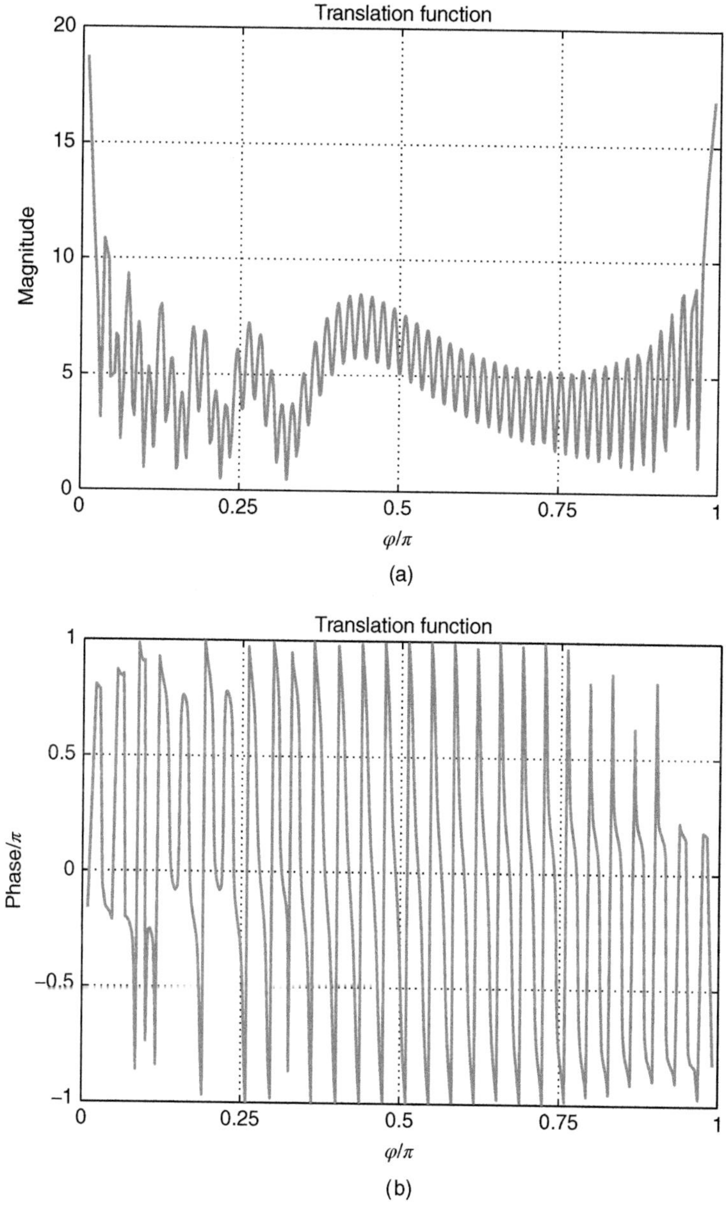

Figure 3.8 (a) Magnitude and (b) phase of the translation function with respect to φ for the case of $a_l = 4\lambda$, $d_0 = 3$, and $\boldsymbol{r}_{CC'} = \hat{\mathbf{x}}2a$. *Source*: Ergül and Gürel 2006 [167]. Reproduced with permission of IEEE.

$2[\tau(l)+1]^2 = 6728$ distinct $\hat{\boldsymbol{k}}$ directions to evaluate the translation operator. It should be noted that the transform from (3.49) to (3.120) not only depends on $\tau(l)$, but also on the relative positions of boxes, i.e., it also depends on $\boldsymbol{r}_{CC'}$.

Before the translation matrix is filled via the Lagrange interpolation, parameters q and p must be determined. For fixed values of d_0 and a_l, a scan over q and p parameters is performed to find their optimal values. Figure 3.9(a) demonstrates the interpolation error with respect to q and p for the case in Figure 3.8. Interpolation error is defined as

$$\Delta_l = \max_n \left\{ \frac{|\tilde{\alpha}_{\tau(l)}(\varphi[n]) - \alpha_{\tau(l)}(\varphi[n])|}{\max\limits_{\varphi}\{|\alpha_{\tau(l)}(\varphi)|\}} \right\}, \tag{3.124}$$

where $n = 1, \ldots, 2[\tau(l)+1]^2$ and $\varphi[n]$ represents sampling points. The interpolation error decreases when q or p is increased. In this case, $d_0 = 3$, which means that MLFMA computes far-field interactions with three digits of accuracy. Thus (p, q) pairs leading to larger than 10^{-3} error are not allowable. In other words, the error introduced by the interpolation of a translation operator should be adjusted according to the desired level of accuracy. This strategy yields a set of (p, q) pairs satisfying the error criterion. Optimization is completed by choosing the (p, q) pair with the minimum processing time. As shown in Figure 3.9(b), the processing time (measured on a 1.8 GHz Opteron 244 processor) to evaluate the translation operator increases as p or q is increased. Then, there exists an optimal (p, q) pair satisfying the desired level of accuracy with the minimum processing time. The parameters p and q can be scanned for various values of a_l and d_0. All possible values of $\boldsymbol{r}_{CC'}$ according to the one-box-buffer scheme [159] should also be checked. In the end, the optimal values listed in Table 3.3 can be obtained with the corresponding speedup compared to the direct calculation. The values presented in Table 3.3 do not depend on the computer platform. The optimal (p, q) pairs are valid for $a_l \geq 4\lambda$ and they are found to be independent of $\boldsymbol{r}_{CC'}$. For smaller boxes such as $a_l = \lambda$ or 2λ, the interpolation does not lead to a significant speedup, and therefore, direct calculations of these translations are preferred. In the case of much smaller boxes such as $a_l = 0.25\lambda$ or 0.5λ, the direct calculation is faster than the interpolation for any (p, q) pair satisfying the desired accuracy.

In Figures 3.10(a) and 3.10(b), the optimal (p, q) pairs are compared to fixed $p = 3, q = 5.0$ values suggested in [166]. In Figure 3.10(a), the interpolation error is plotted with respect to the box size a_l from 4λ to 64λ and for different levels of accuracy, i.e., for $d_0 = 2, 3, 4$, and 5 corresponding to $10^{-2}, 10^{-3}, 10^{-4}$, and 10^{-5} relative errors, respectively. In the optimized case, the error is always below the desired level of accuracy. However, with fixed parameters, error is not controllable and is localized around 10^{-4}. The corresponding speedup is plotted in Figure 3.10(b), where it increases with increasing box size and decreases with increasing

Table 3.3 Speedup obtained by using the optimal (p, q) pair for $a_l \geq 4\lambda$

d_0	(p,q)	$a_l = 4\lambda$	$a_l = 8\lambda$	$a_l = 16\lambda$	$a_l = 32\lambda$	$a_l = 64\lambda$
2	(2,3.5)	14.0	27.5	54.3	108.3	216.0
3	(2,6.5)	10.8	20.2	40.0	77.0	151.9
4	(3,6.0)	7.9	15.0	28.9	56.9	113.7
5	(3,8.5)	7.1	13.0	24.7	48.4	96.6

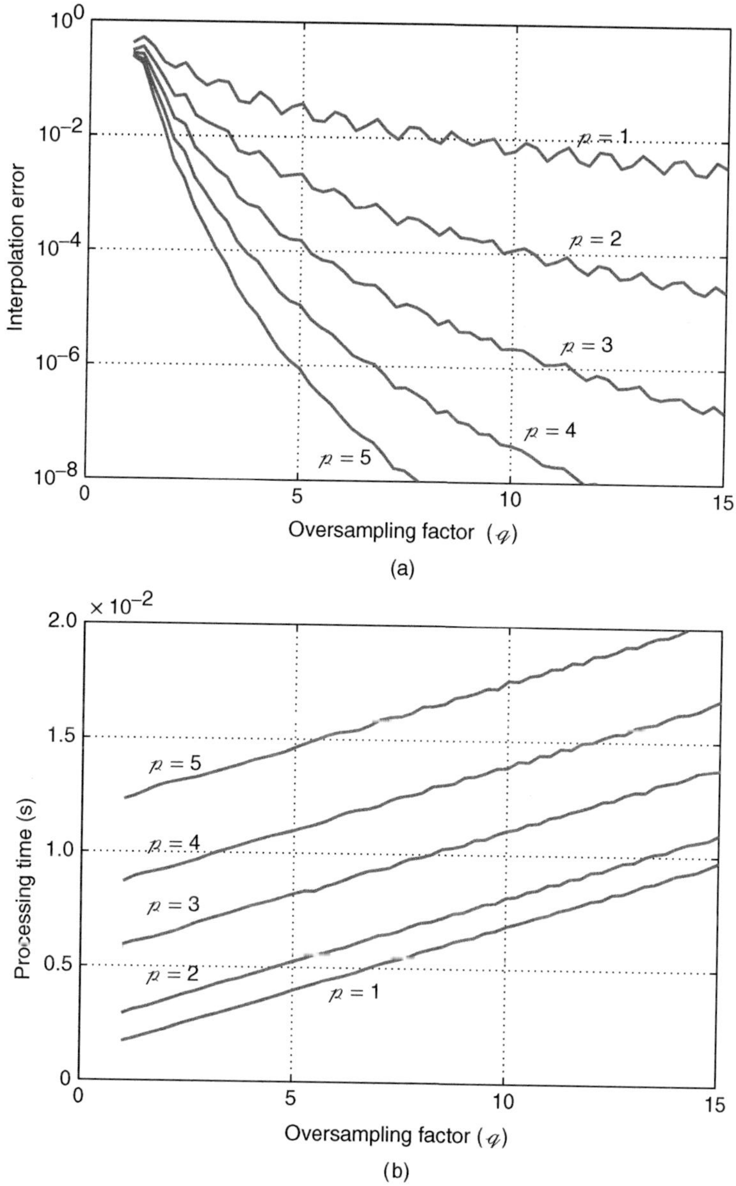

Figure 3.9 (a) Interpolation error and (b) processing time with respect to interpolation parameters p and q for the translation function in Figure 3.8. *Source*: Ergül and Gürel 2006 [167]. Reproduced with permission of IEEE.

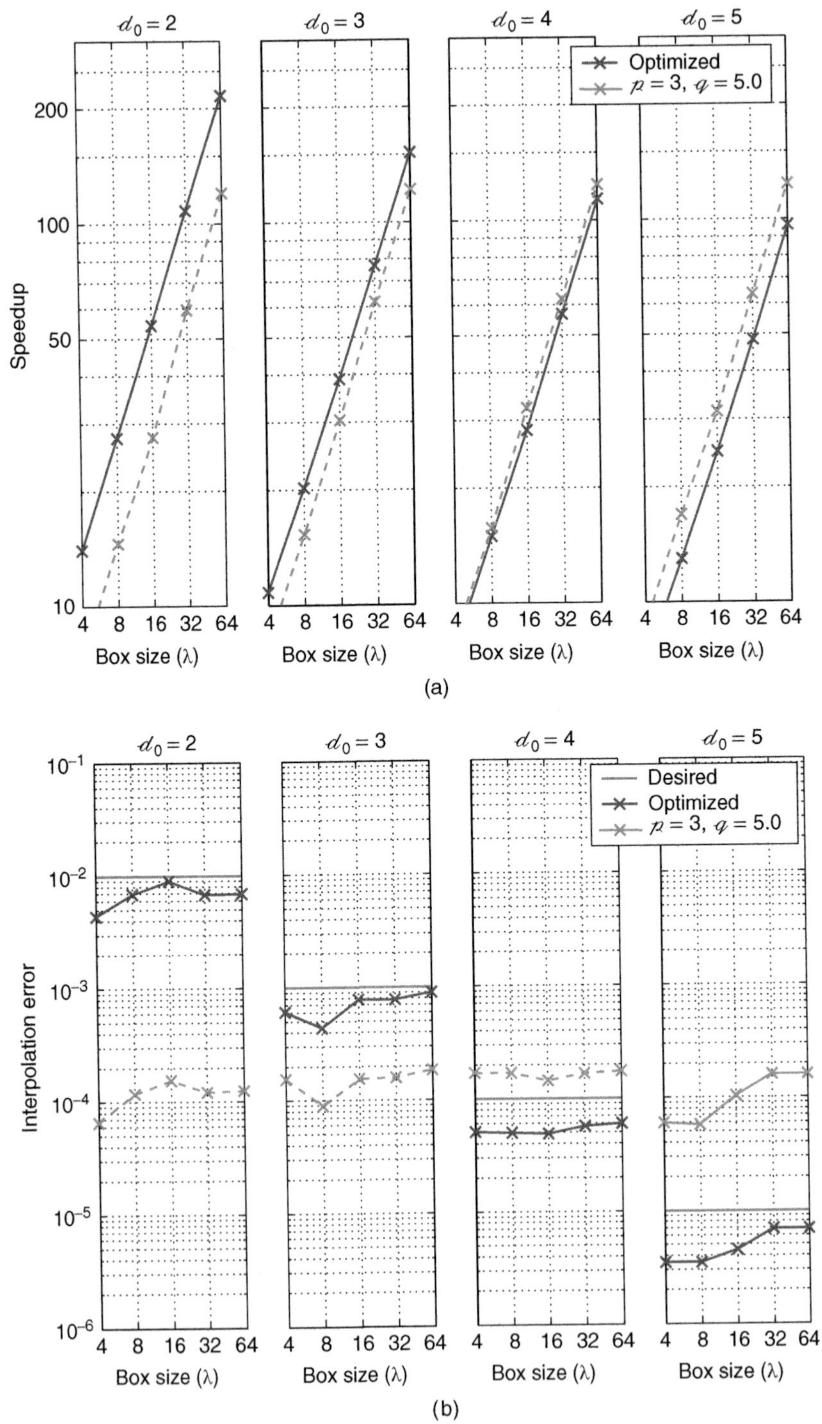

Figure 3.10 (a) Interpolation error and (b) the corresponding speedup for different box sizes from 4λ to 64λ and for $d_0 = 2, 3, 4, 5$. ($\boldsymbol{r}_{CC'} = \hat{\boldsymbol{x}}2a$). *Source*: Ergül and Gürel 2006 [167]. Reproduced with permission of IEEE.

number of accurate digits in the optimized case. This relationship is also evident in Table 3.3. Comparing Figures 3.10(a) and 3.10(b), the following observations can be made:

- For $d_0 = 2$ and 3, fixed $p = 3, q = 5.0$ satisfies the desired level of accuracy but the optimal (p, q) pairs provide higher speedup.
- For $d_0 = 4$ and 5, the fixed $p = 3, q = 5.0$ seems to give higher speedup compared to the optimal (p, q) pairs, however, the accuracy is not satisfied with fixed parameters.

Based on these observations, it can be concluded that optimization is essential to improve the interpolation of translation operators.

To demonstrate the overall improvement obtained with the interpolation, solutions of a scattering problem involving a conducting sphere with a radius of 20λ are considered. This is a 1,462,854-unknown problem solved by a parallel MLFMA implementation with seven levels ($L = 7$). The problem is solved on a cluster of 32 processors (2.6 GHz Pentium 4 Celeron). The box size is 0.25λ for the lowest level and 16λ for the highest level. As an example, if the number of accurate digits d_0 is set to three, then $\tau(l)$ takes values from 8 to 195. The number of different translations is reduced by exploiting the symmetry [158]. During the setup phase of the program, each processor checks all of its box-box interactions to eliminate the unneeded translations. In Figure 3.11(a), the processing time for the calculation of translation operators is plotted with respect to d_0. For both types of calculations (direct and interpolated), the maximum is chosen among the processing times spent by 32 processors. In Figure 3.11(b), the speedup obtained by the interpolation method over direct calculation is plotted as a function of d_0. The speedup is over 14 up to $d_0 = 5$.

3.4 MLFMA for Hermitian Matrix-Vector Multiplications

Some Krylov subspace algorithms, such as LSQR and BiCG, require multiplications with the hermitian of the system matrix, in addition to multiplications with the matrix itself. Those hermitian MVMs can be performed efficiently by reversing the steps of the ordinary FMM or MLFMA. Consider the interaction of the mth testing function with all basis functions using FMM, i.e.,

$$\sum_{n=1}^{N} \overline{Z}[m,n]\boldsymbol{x}[n] = \sum_{n=1}^{N} \overline{Z}_{NF}[m,n]\boldsymbol{x}[n]$$
$$+ \left(\frac{\mathrm{i}k}{4\pi}\right)^2 \int d^2\hat{\boldsymbol{k}}\boldsymbol{F}_m(\boldsymbol{r}_C,\boldsymbol{k}) \cdot \sum_{C'\in\mathcal{F}\{C\}} \alpha_\tau(\boldsymbol{k},\boldsymbol{r}_C - \boldsymbol{r}_{C'}) \sum_{n\in C'} \boldsymbol{S}_n(\boldsymbol{r}_{C'},\boldsymbol{k})\boldsymbol{x}[n], \tag{3.125}$$

where $\boldsymbol{F}_m$ and $\boldsymbol{S}_n$ represent receiving and radiation patterns, respectively. Then, one can derive

$$\sum_{n=1}^{N} \overline{Z}^*[n,m]\boldsymbol{x}[n] = \sum_{n=1}^{N} \overline{Z}^*_{NF}[n,m]\boldsymbol{x}[n]$$
$$+ \left(\frac{ik}{4\pi}\right)^2 \int d^2\hat{\boldsymbol{k}}\boldsymbol{S}^*_m(\boldsymbol{r}_{C'},\boldsymbol{k}) \cdot \sum_{C\in\mathcal{F}\{C'\}} \alpha^*_\tau(\boldsymbol{k},\boldsymbol{r}_C - \boldsymbol{r}_{C'}) \sum_{n\in C} \boldsymbol{F}^*_n(\boldsymbol{r}_C,\boldsymbol{k})\boldsymbol{x}[n]. \tag{3.126}$$

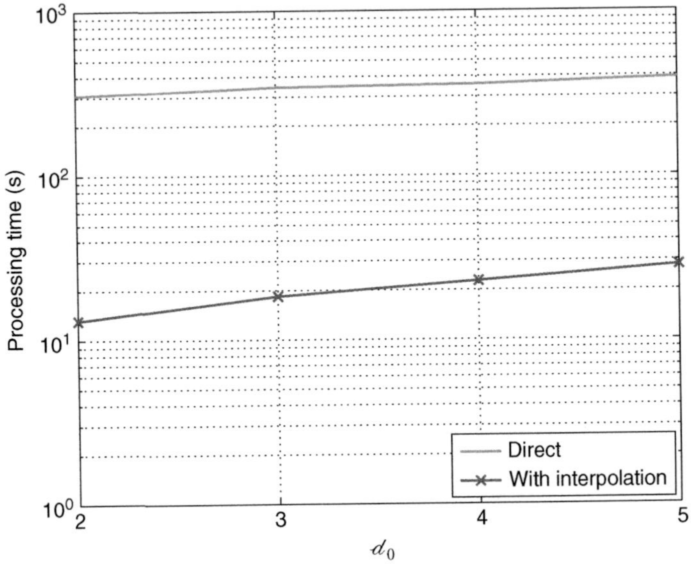

(a)

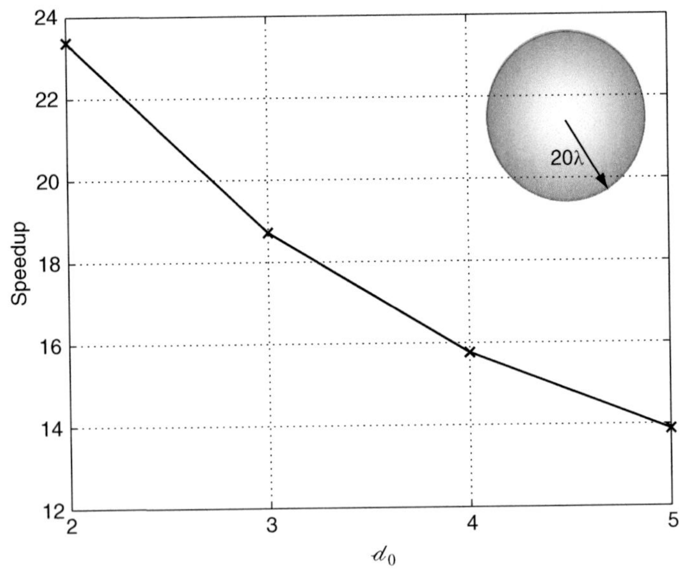

(b)

Figure 3.11 (a) Processing time to compute translation operators for a 1,462,854-unknown sphere problem. (b) Speedup obtained with optimal interpolation compared to the direct calculation of translation operators. *Source*: Ergül and Gürel 2006 [167]. Reproduced with permission of IEEE.

The expression in (3.126) can be interpreted as follows:

- Combine receiving patterns of testing functions in each box.
- Translate receiving patterns of boxes into local receiving patterns for other boxes.
- Receive radiated fields of basis functions locally.

When $\overline{\boldsymbol{Z}}$ is symmetric,

$$\boldsymbol{F}_n^*(\boldsymbol{r}_C, \boldsymbol{k}) = \boldsymbol{S}_n(\boldsymbol{r}_C, \boldsymbol{k}) \tag{3.127}$$

$$\boldsymbol{S}_m^*(\boldsymbol{r}_C, \boldsymbol{k}) = \boldsymbol{F}_m(\boldsymbol{r}_C, \boldsymbol{k}) \tag{3.128}$$

and

$$\sum_{n=1}^{N} \overline{\boldsymbol{Z}}^*[n,m]\boldsymbol{x}[n] = \sum_{n=1}^{N} \overline{\boldsymbol{Z}}_{NF}^*[m,n]\boldsymbol{x}[n]$$
$$+ \left(\frac{ik}{4\pi}\right)^2 \int d^2\hat{\boldsymbol{k}} \boldsymbol{F}_m(\boldsymbol{r}_{C'}, \boldsymbol{k}) \cdot \sum_{C \in \mathcal{F}\{C'\}} \alpha_\tau^*(\boldsymbol{k}, \boldsymbol{r}_C - \boldsymbol{r}_{C'}) \sum_{n \in C} \boldsymbol{S}_n(\boldsymbol{r}_C, \boldsymbol{k})\boldsymbol{x}[n]. \tag{3.129}$$

Using

$$\boldsymbol{F}_m(\boldsymbol{r}_{C'}, \boldsymbol{k}) \cdot \alpha_\tau^*(\boldsymbol{k}, \boldsymbol{r}_C - \boldsymbol{r}_{C'}) \boldsymbol{S}_n(\boldsymbol{r}_C, \boldsymbol{k}) \tag{3.130}$$

$$= \boldsymbol{F}_m^*(\boldsymbol{r}_{C'}, \boldsymbol{k}) \cdot \alpha_\tau^*(\boldsymbol{k}, \boldsymbol{r}_{C'} - \boldsymbol{r}_C) \boldsymbol{S}_n^*(\boldsymbol{r}_C, \boldsymbol{k}), \tag{3.131}$$

one can obtain

$$\sum_{n=1}^{N} \overline{\boldsymbol{Z}}^*[n,m]\boldsymbol{x}[n] = \sum_{n=1}^{N} \overline{\boldsymbol{Z}}^*[m,n]\boldsymbol{x}[n], \tag{3.132}$$

as required.

3.5 Strategies for Building Less-Accurate MLFMA

MLFMA can perform a MVM with a desired level of accuracy, which is controlled by the excess bandwidth formula in (3.84). There are many studies to further improve the reliability of implementations by refining formulas for truncation numbers, especially for small boxes [170]. In most cases, the purpose is to obtain accurate results by suppressing error sources in MLFMA. On the other hand, it is also desirable to build less-accurate forms of MLFMA, which can be more efficient than the original MLFMA. A less-accurate MLFMA can be used to construct a strong preconditioner, where the accuracy is not very critical, but a reasonable approximation is required with high efficiency.

A direct way to construct a less-accurate MLFMA is reducing truncation numbers using (3.84). For example, if the ordinary MLFMA has four digits of accuracy, i.e., $d_0 = 4$, then a less-accurate MLFMA may have one or two digits of accuracy [208]. On the other hand, this strategy has two major disadvantages:

- A less-accurate MLFMA obtained by decreasing d_0 in (3.84) may not be significantly faster than the ordinary MLFMA. Because, truncation numbers loosely depend on d_0 for large boxes at higher levels of MLFMA, such as depicted below in Figure 3.7. For example, let d_0 be reduced from 4 to 1. At the lowest level involving 0.25λ boxes, the truncation number drops significantly by 40% (from 10 to 6). On the other hand, for a higher level with 16λ boxes, the truncation number decreases from 200 to 184, which corresponds to an only 8% reduction. Therefore, reducing the value of d_0 does not provide a significant acceleration, especially when the problem size is large.
- The extra cost of a less-accurate MLFMA obtained by decreasing d_0 can be significant due to radiation and receiving patterns of basis and testing functions calculated during the setup stage. In addition to the ordinary patterns employed by the ordinary MLFMA, a new set of patterns is required for the less-accurate MLFMA with reduced truncation numbers.

As a result, better strategies are required to construct less-accurate and efficient versions of MLFMA.

Another strategy to build a less-accurate MLFMA can be omitting some of far-field interactions. In this case, the number of accurate digits is the same as that for the ordinary MLFMA, but the aggregation, translation, disaggregation stages are omitted for a number of higher levels of the tree structure. The resulting less-accurate MLFMA is called the incomplete MLFMA (IMLFMA), which does not require extra computations during the setup stage. In addition, IMLFMA can easily be obtained via minor modifications to the ordinary MLFMA. On the other hand, this strategy fails to provide an acceptable accuracy for a sufficient speedup. For example, half of the levels must be ignored to obtain a two-fold speedup with IMLFMA compared to the ordinary MLFMA. However, this would lead to a poor approximation since many interactions (much more than 50%) are ignored.

Alternatively, one can use AMLFMA [211], which is based on systematically reducing truncation numbers as

$$\tau_r(l) = \tau_{\min} + a_f(\tau(l) - \tau_{\min}), \tag{3.133}$$

where $\tau_{\min}$ is the minimum truncation number and $\tau(l)$ is the ordinary truncation number for level l. In (3.133), a_f represents the approximation factor in the range from 0.0 to 1.0. As a_f is increased from 0.0 to 1.0, AMLFMA becomes more accurate but less efficient, and it corresponds to ordinary MLFMA when $a_f = 1.0$. Since the truncation number at the lowest level is not modified, AMLFMA does not require extra computations for radiation and receiving patterns of basis and testing functions. Only a new set of translation functions is required, which leads to a negligible extra cost.

As numerical examples, electromagnetics problems involving a sphere of radius 6λ and a $20\lambda \times 20\lambda$ patch are considered. The problems are discretized with 132,003 and 137,792 unknowns, respectively. For both problems, MVMs are performed by AMLFMA with various values of a_f. In addition to AMLFMA, an IMLFMA based on omitting the interactions only at the highest level is tested. The number of levels is four and six for the sphere and patch problems, respectively. The input vector is filled with ones, and the output vectors provided by AMLFMA and IMLFMA are compared with a reference vector provided by the ordinary MLFMA with three digits of accuracy. For each element of the output vector $m = 1, 2, \ldots, N$, the base error is defined as

$$\Delta_b[m] = \lceil log_{10}(\Delta_r[m]) \rceil, \tag{3.134}$$

Table 3.4 Processing time for MVMs via MLFMA and less-accurate implementations of MLFMA

			AMLFMA					
Problem	Unknowns	MLFMA	0.8	0.6	0.4	0.2	0.0	IMLFMA
Sphere	132,003	34.9	28.3	23.2	19.8	16.0	13.0	32.4
Patch	137,792	24.5	19.4	15.2	12.8	10.1	8.1	21.6

where $\lceil\cdot\rceil$ represents the ceil operation and $\Delta_r[m]$ represents the relative error with respect to the reference value provided by the ordinary MLFMA.

Figures 3.12 and 3.13 present the number of elements satisfying various base errors for the sphere and patch problems, respectively. Figure 3.12 shows that the accuracy of the MVM deteriorates slightly when a_f is 0.8 and 0.6 for the sphere problem, i.e., the base error is -3 or below for most of the elements. Those elements are calculated with high accuracy, which is even better than acceptable since the ordinary MLFMA has three digits of accuracy. Reducing the value of a_f to 0.4 and 0.2, accuracy of AMLFMA decreases and numbers of elements satisfying -2 and -1 base errors increase. Finally, when $a_f = 0.0$, i.e., when the truncation number is $\tau_{\min}$ for all levels, there are many elements with 0 base error, which corresponds to minimum 10% relative error. Figure 3.13 presents similar results for a completely different geometry, i.e., $20\lambda \times 20\lambda$ patch. Using AMLFMA, one can adjust the accuracy of MVMs by choosing the approximation factor a_f from 1.0 to 0.0. On the other hand, IMLFMA, which is obtained by omitting only the highest level, provides inaccurate results compared to AMLFMA with $a_f = 0.2 - 0.4$ for both problems.

Figures 3.12 and 3.13 show that one can obtain relatively accurate MVMs by AMLFMA when using $a_f = 0.2 - 0.4$. This cannot be predicted by the excess bandwidth formula in (3.84), which suggests significantly large truncation numbers. This is because the excess bandwidth formula is based on the worst-case scenario for the positions of basis and testing functions inside boxes. In fact, the ordinary MLFMA must use truncation numbers obtained with (3.84) to guarantee the desired level of accuracy. However, for a typical problem, there are many interactions that can be computed accurately using lower truncation numbers. These interactions can be employed in AMLFMA, which can be used to construct powerful preconditioners, where the accuracy is not very critical.

Table 3.4 lists the processing time required for a MVM performed by ordinary MLFMA, AMLFMA, and IMLFMA measured on a cluster of 16 AMD Opteron 870 processors. For both sphere and patch problems, AMLFMA with $a_f = 0.2$ provides a significant acceleration by reducing the processing time more than 50% compared the ordinary MLFMA. Also note that the speedup provided by IMLFMA is less than the speedup provided by AMLFMA, even when $a_f = 0.8$.

3.6 Iterative Solutions of Surface Formulations

In this section, iterative solution of electromagnetics problems involving metallic, dielectric, and composite objects are focused in terms of iterative convergences. Hybrid formulations are

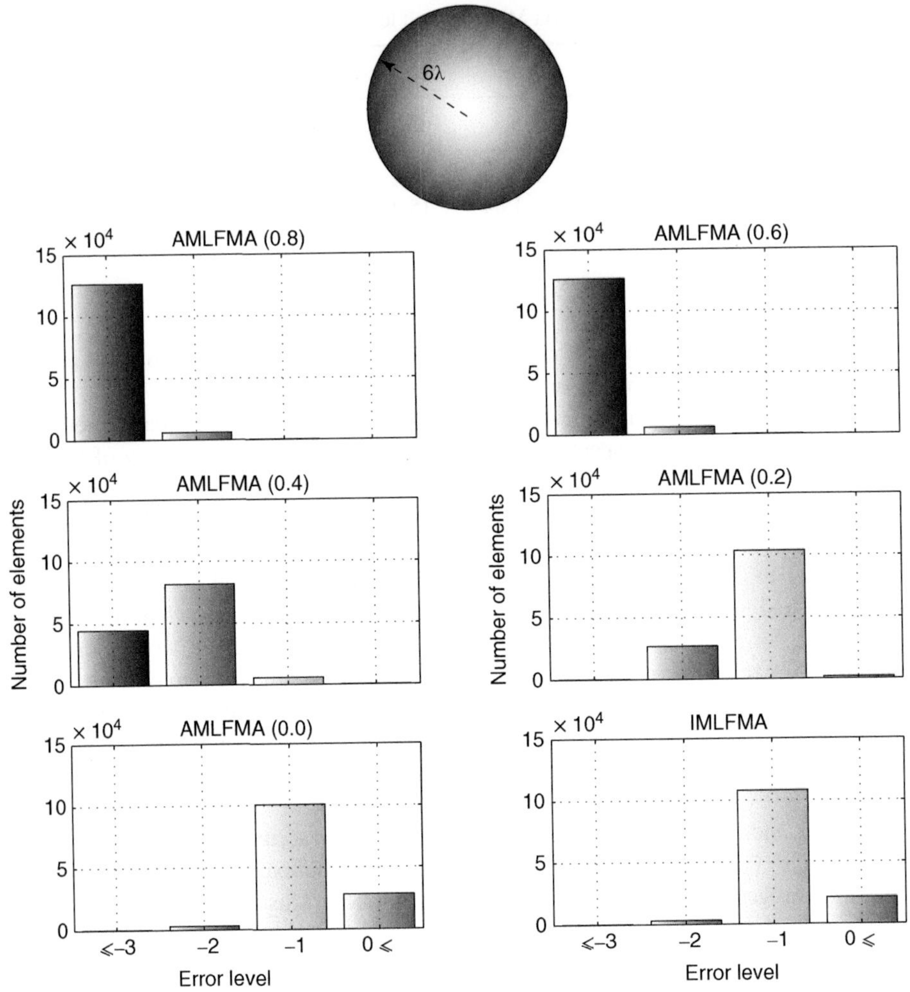

Figure 3.12 Errors in MVMs performed by AMLFMA with $a_f = 0.0–0.8$ and by IMLFMA (omitting the highest level) for a sphere problem discretized with 132,003 unknowns. The reference data is obtained by using an ordinary MLFMA with three digits of accuracy. *Source*: Ergül and Gürel 2006 [167]. Reproduced with permission of IEEE.

derived for improved iterative solutions of metallic objects with coexisting open and closed parts.

3.6.1 *Hybrid Formulations of PEC Objects*

Real-life electromagnetics problems often involve both thin and thick conductors that need to be modeled with open and closed surfaces, respectively. Since MFIE is not applicable to open surfaces, conventional solutions are forced to employ EFIE even when only a small part of the

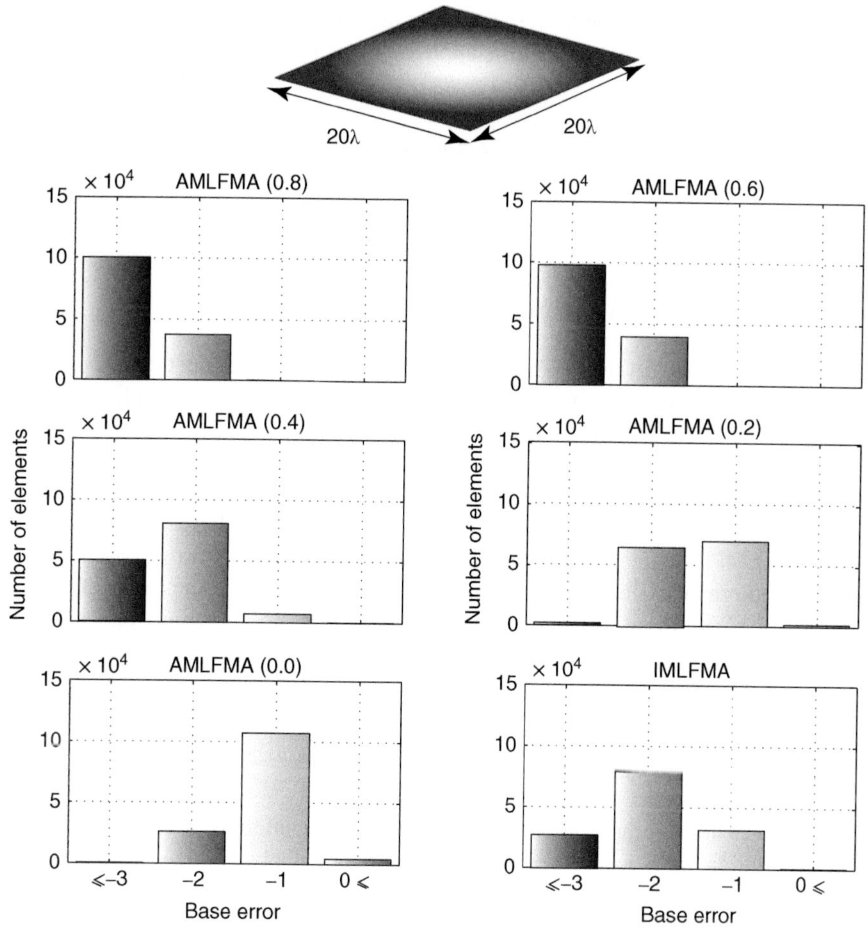

Figure 3.13 Errors in MVMs performed by AMLFMA with $a_f = 0.0$–0.8 and by IMLFMA (omitting the highest level) for a patch problem discretized with 137,792 unknowns. The reference data is obtained by using an ordinary MLFMA with three digits of accuracy.

geometry is an open surface. Unfortunately, EFIE produces ill-conditioned matrix equations that are difficult to solve with an iterative solver, especially when the problem includes some closed parts. Consequently, as the problem size increases, the solution of composite problems with coexisting open and closed surfaces becomes extremely difficult. Although iterations are accelerated via efficient algorithms, such as MLFMA, convergence is usually not reached within a reasonable number of iterations. To remedy this, hybrid formulations can be developed for efficient solutions of composite problems without sacrificing accuracy [59, 64, 82].

Extending the definition of CFIE in (2.214), hybrid integral equations (HIEs) can be written as

$$\text{HIE} = \alpha(\boldsymbol{r})\text{EFIE} + [1 - \alpha(\boldsymbol{r})]\text{MFIE}, \tag{3.135}$$

where the parameter α now depends on the observation point. HIEs employ MFIE or CFIE on closed parts of the geometry, while EFIE is applied on open parts, as usual. Discretization of (3.135) leads to

$$\overline{Z}^{\text{HIE}} \cdot \boldsymbol{a} = \boldsymbol{w}^{\text{HIE}}, \tag{3.136}$$

where

$$\overline{Z}^{\text{HIE}}[m,n] = \alpha_m \overline{Z}^{\text{EFIE}}[m,n] + (1-\alpha_m)\overline{Z}^{\text{MFIE}}[m,n] \tag{3.137}$$

$$\boldsymbol{w}^{\text{HIE}}[m] = \alpha_m \boldsymbol{w}^{\text{EFIE}}[m] + (1-\alpha_m)\boldsymbol{w}^{\text{MFIE}}[m]. \tag{3.138}$$

The matrix equation in (3.136)–(3.138) has a variable α_m depending on the index of the row (m) in the impedance matrix, i.e., the index of the testing function. This provides the freedom to choose different linear combinations for different testing functions, even using pure EFIE ($\alpha_m = 1$) or pure MFIE ($\alpha_m = 0$) for some m. In general, one can obtain better-conditioned matrix equations by including CFIE interactions in the solution of composite problems. Also note that the traditional CFIE can be interpreted as a special case of HIEs for closed surfaces, where EFIE and MFIE are combined with a fixed parameter for each testing case.

MLFMA solutions of HIEs, where α_m depends on the location of the testing function, is similar to MLFMA solutions of EFIE, MFIE, and CFIE, and do not require any special procedures. This is because, in MLFMA, radiated and incoming fields do not depend on the type of the formulation and α_m appears only in receiving patterns of testing functions.

Numerical Results

Figure 3.14 shows three composite problems, where relative amounts of open and closed surfaces are different. In numerical solutions considered below, all objects are located in free space and discretized with the RWG functions. The problems are also summarized in Table 3.5. The first one in Figure 3.14(a) is a radiation problem involving a dipole antenna (open surface) of length 1 m located over a PEC rectangular box (closed surface) with dimensions of 1 m × 5 m × 5 m. This structure is an example of problems with almost completely closed geometries, except for very minor open parts. The problem is solved at 300 MHz, and the discretization of the object leads to 24,186 unknowns on the box and only 19 unknowns on the antenna. The feed of the antenna is simulated by a delta-gap source located at the center of the strip.

Figure 3.14(b) shows an example of problems involving mostly closed parts but with some open parts. A cross-shaped open surface is placed over a closed ellipsoid surface with dimensions shown in the figure. The object is investigated at 500 MHz and illuminated by a plane wave propagating in the $-x$ direction with the electric field polarized in the y direction. Discretization of the object leads to 115,023 unknowns on the ellipsoid (closed surface) and 16,608 unknowns on the cross-shaped part (open surface).

As an example to problems with comparable amounts of open and closed parts, Figure 3.14(c) shows a sphere with radius 30 cm placed over a 60 cm × 60 cm square patch. The structure is illuminated by the same plane wave as in the second problem. The problem is solved at two frequencies; at 1 GHz, the number of unknowns is 3768 on the sphere and 1301 on the patch, whereas at 6 GHz, there are 132,336 unknowns on the sphere and 49,200 on the patch.

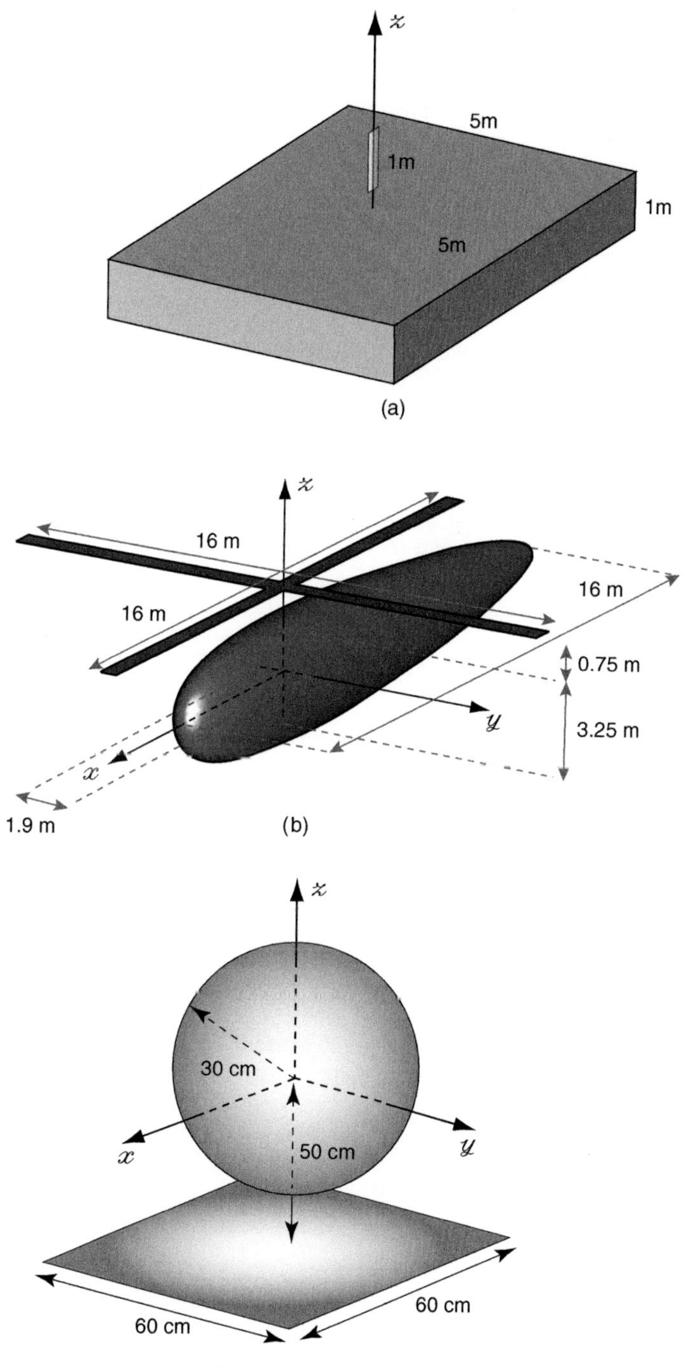

Figure 3.14 Various composite problems (a) with almost completely closed geometries, except for very minor open parts, (b) involving more closed parts than open parts, and (c) with comparable amounts of open and closed parts.

Table 3.5 Composite problems involving open and closed PEC parts

		Number of unknowns	
Problem	Frequency	Closed parts	Open parts
Problem 1: Figure 3.14(a)	300 MHz	24,186	19
Problem 2: Figure 3.14(b)	500 MHz	115,023	16,608
Problem 3: Figure 3.14(c)	1 GHz	3768	1301
Problem 3: Figure 3.14(c)	6 GHz	132,336	49,200

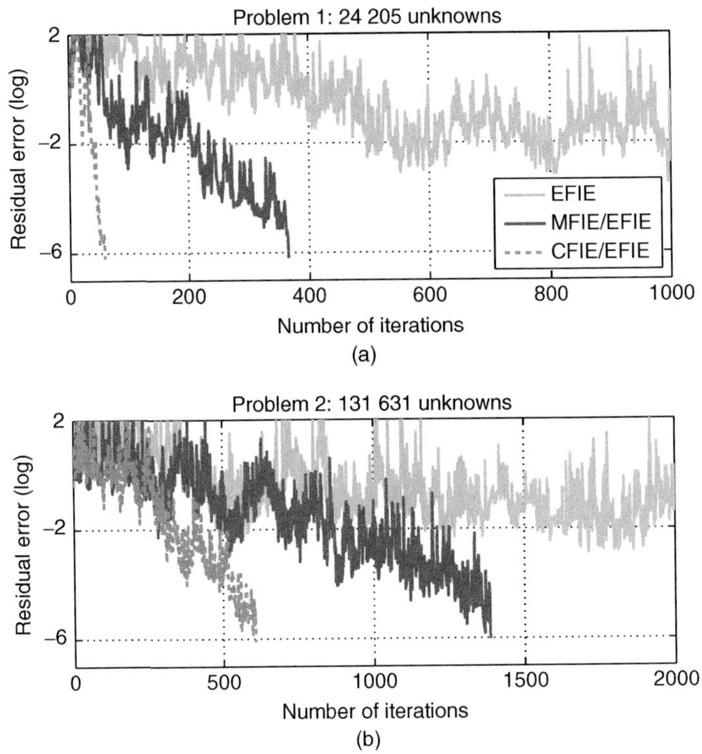

Figure 3.15 Convergence characteristics of the problems in Figure 3.14(a) and Figure 3.14(b) for different formulations, i.e., EFIE, MFIE/EFIE, and CFIE/EFIE, solved with CGS. *Source*: Ergül and Gürel 2009 [82]. Reproduced with permission of IEEE.

In Figures 3.15 and 3.16, convergence characteristics are depicted for the CGS algorithm and when different formulations are applied to the composite problems. In all examples, iterative convergence is the slowest for the conventional EFIE. Especially for the first two problems, EFIE does not reach the 10^{-6} residual error within a reasonable number of iterations. As shown in Figures 3.15(a) and 3.15(b), the hybrid CFIE/EFIE formulation dramatically improves the convergence of these two problems, when CFIE with $\alpha_m = 0.2$ is applied to closed parts. On the

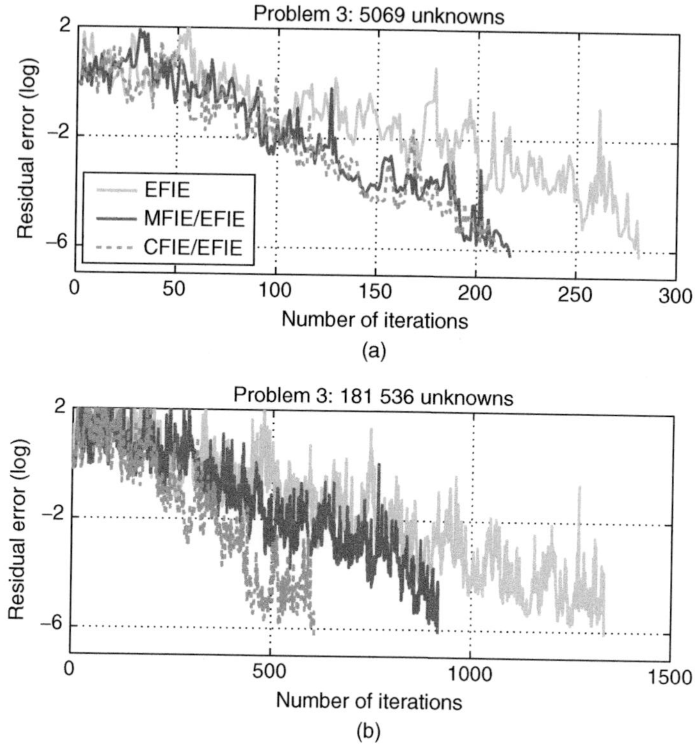

Figure 3.16 Convergence characteristics of the problem in Figure 3.14(c) for different formulations, i.e., EFIE, MFIE/EFIE, and CFIE/EFIE, solved with CGS. *Source*: Ergül and Gürel 2009 [82]. Reproduced with permission of IEEE.

other hand, when MFIE is applied to closed parts, leading to a hybrid MFIE/EFIE formulation, improvement is not sufficient.

For the third problem containing comparable amounts of open and closed surfaces, there is improvement with the hybrid CFIE/EFIE formulation, but it is less than those in the previous two problems. For this problem, EFIE converges better than the first and second problems due to the larger portion of open surface. The CFIE/EFIE formulation is still much better than EFIE for the larger problem in Figure 3.16(b). However, the acceleration provided by CFIE/EFIE is not very significant for the smaller problem in Figure 3.16(a). In fact, iterations in Figure 3.16(a) are at the frequency where EFIE shows a quite good performance. In Figure 3.17, a frequency sweep from 500 MHz to 1200 MHz is demonstrated for the same problem. EFIE suffers from internal resonances, and the number of iterations peaks at various frequencies. On the other hand, the hybrid CFIE/EFIE formulation is free of the internal resonances so that the number of iterations is steady and consistently in the same low range.

Figures 3.18 and 3.19 demonstrate convergence characteristics for the solution of Problem 1 (in Table 3.5) using various iterative methods. In general, the hybrid CFIE/EFIE formulation consistently leads to the fastest convergence. Table 3.6 lists the number of iterations for the same problem to reach 10^{-6} residual error. When the EFIE formulation is applied to the entire

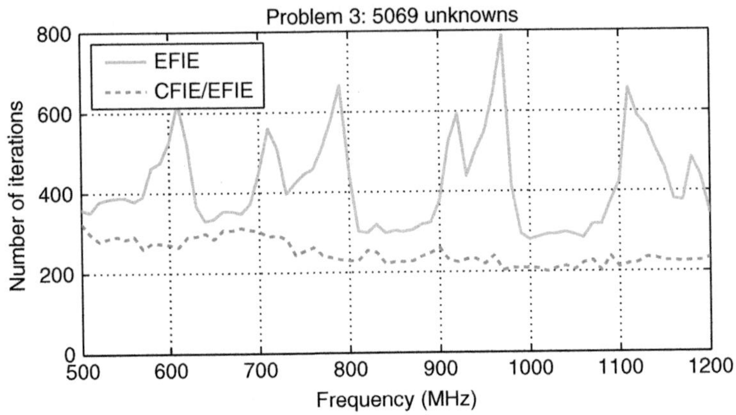

Figure 3.17 Iteration counts to reach 10^{-6} residual error for the problem in Figure 3.14(c) in the frequency range from 500 MHz to 1200 MHz. *Source*: Ergül and Gürel 2009 [82]. Reproduced with permission of IEEE.

Table 3.6 Number of iterations (10^{-6} residual error) for the solution of problem 1 in Table 3.5 without preconditioning

Method	EFIE	MFIE/EFIE	CFIE
BiCG	NC	495	72
BiCGStab	NC	377	47
CGS	1488	372	63
GMRES	NC	388	60
LSQR	1307	568	101
TFQMR	1407	358	52

NC: No convergence.

problem, CGS, LSQR, and TFQMR converge within 2000 iterations, while BiCG, BiCGStab, and GMRES (with a restart parameter of 200 iterations) do not. With CFIE/EFIE, the number of iterations drops for all cases, but GMRES performs the best in terms of the number of MVMs. This is because GMRES requires only one MVM per iteration while others require two. On the other hand, GMRES requires more memory than the other iterative methods in Table 3.6; hence BiCGStab might be preferred as problem size grows and it becomes critical to reduce memory.

Convergence characteristics for preconditioned systems are depicted in Figures 3.20 and 3.21. For the three problems, EFIE and HIEs are solved with CGS preconditioned by BDPs. EFIE solutions with NFP are also presented in the same figures for comparison. To easily check the effect of preconditioning on different problems and formulations, Table 3.7 lists the number of iterations to reach 10^{-6} residual error. Looking at Figure 3.20, Figure 3.21, and Table 3.7, the following observations can be made:

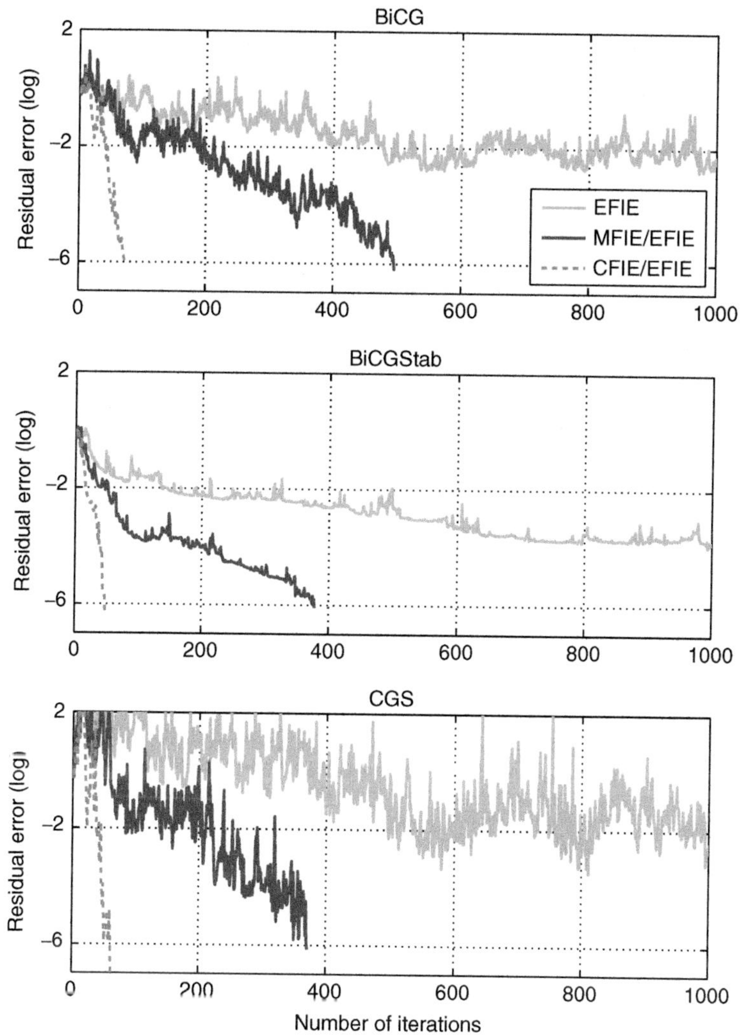

Figure 3.18 Convergence characteristics of the problem in Figure 3.14(a) for various iterative methods and different formulations, i.e., EFIE, MFIE/EFIE, and CFIE/EFIE. *Source*: Ergül and Gürel 2009 [82]. Reproduced with permission of IEEE.

- For the first problem with a completely closed geometry except for a very minor open part, the convergence of HIEs is significantly improved by the efficient BDP. However, for EFIE, BDP does not work and the convergence is worse than the no-preconditioner (NP) case. The strong NFP improves the convergence of EFIE, although the number of iterations is still high, compared to HIEs.
- For the second problem, where there are more closed than open parts, BDP does not work even with HIEs. This can be explained with the worsening of the condition of the system since more open parts need to be formulated with EFIE, compared to the first problem.

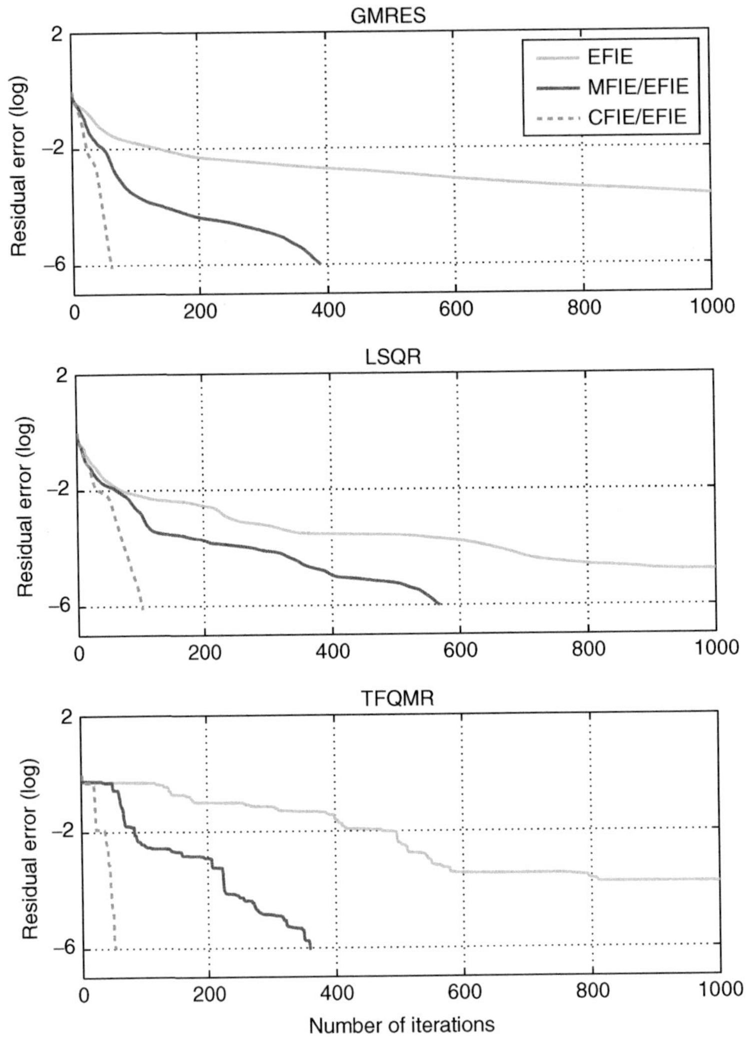

Figure 3.19 Convergence characteristics of the problem in Figure 3.14(a) for various iterative methods and different formulations, i.e., EFIE, MFIE/EFIE, and CFIE/EFIE. *Source*: Ergül and Gürel 2009 [82]. Reproduced with permission of IEEE.

Nevertheless, the hybrid CFIE/EFIE formulation is preferable since its NP case is still better than EFIE with a strong NFP.

- For the third problem, where open and closed parts are comparable, the effect of EFIE becomes significant. As in the second problem, BDP does not work even for HIEs. In addition to this, EFIE with NFP performs better than HIEs with NP. Consequently, NFP provides a trade-off between memory and number of iterations. There is an alternative way using the hybrid CFIE/EFIE formulation with NFP leading to fewer iterations. However, due to the nonsymmetric impedance matrix, this would be more expensive in terms of memory than the EFIE case.

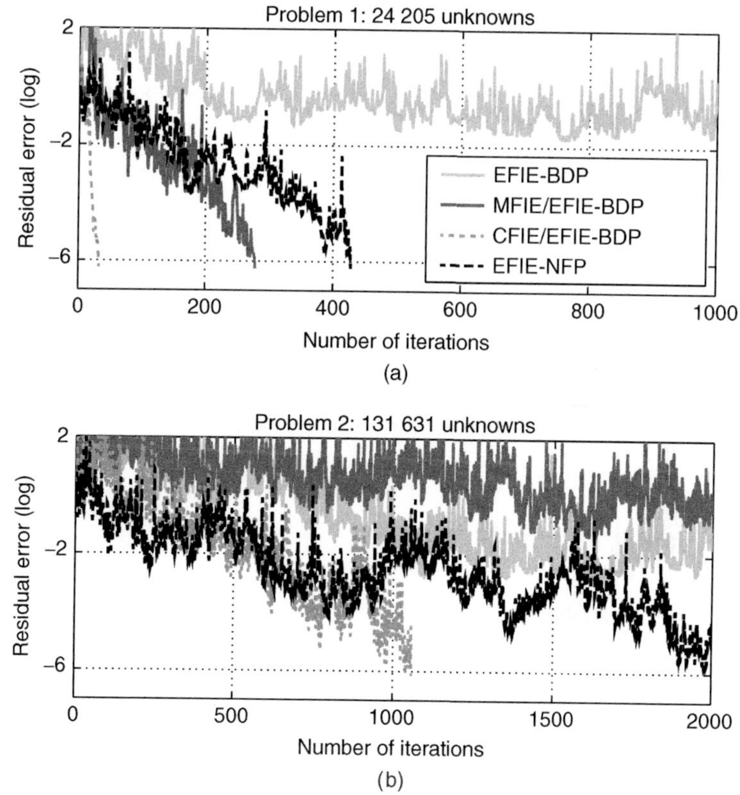

Figure 3.20 Convergence characteristics of the problems in Figures 3.14(a) and 3.14(b) for the preconditioned CGS and different formulations, i.e., EFIE, MFIE/EFIE, and CFIE/EFIE. *Source*: Ergül and Gürel 2009 [82]. Reproduced with permission of IEEE.

In Figure 3.22, the variable α_m employed in the convex combination of EFIE and MFIE in the CFIE part of hybrid formulations is investigated. Two different cases are demonstrated in the figure, where the number of iterations is depicted with respect to the variable α_m applied on closed parts of the problems. The four curves in each figure represent residual errors from 10^{-3} to 10^{-6}. Comparing Figures 3.22(a) and 3.22(b), one can observe that the dependence on α_m is directly based on the type of geometry. In Problem 1, where only a small portion of the geometry is open, there is an optimal α_m of about 0.2–0.4 for the best convergence. In both directions to MFIE/EFIE ($\alpha_m = 0$) and to EFIE ($\alpha_m = 1$), the number of iterations increases. On the other hand, in the case of Problem 3, the open part is relatively larger, and it is difficult to set an optimal α_m for $\alpha_m \neq 1$. It can be concluded that choosing α_m around 0.2 does not guarantee the exact optimal convergence for all cases, but it provides a number of iterations close to the optimal case.

In order to test the accuracy of HIEs, RCS results are considered for the problems in Figures 3.14(a) and 3.14(c). In Figure 3.23, normalized RCS values (RCS/λ^2) are plotted on the z-x plane with respect to θ. Highest RCS values in Figures 3.23(b) and 3.23(c) appear at 270°, which corresponds to the forward-scattering direction. For both radiation and scattering

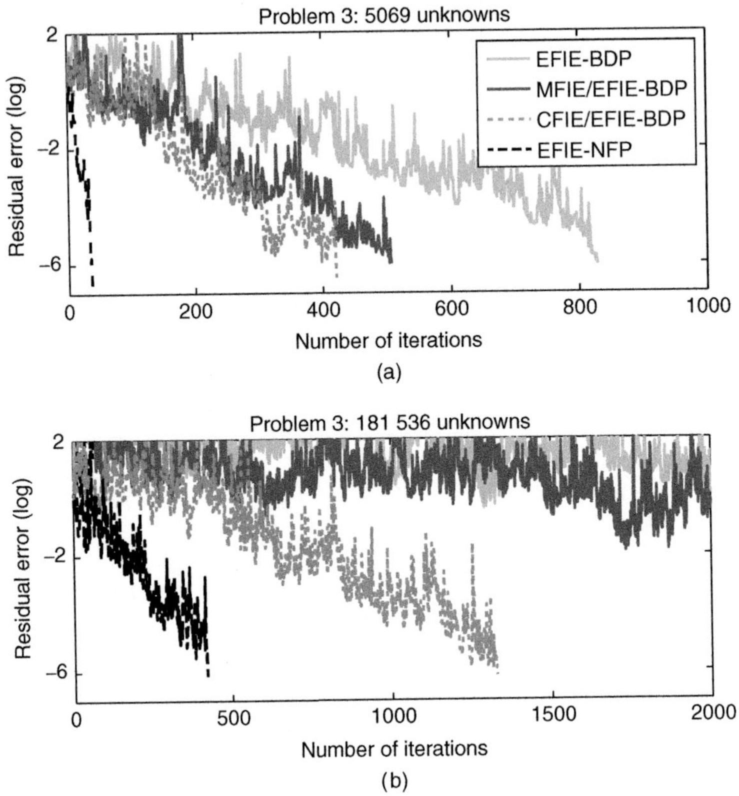

Figure 3.21 Convergence characteristics of the problem in Figure 3.14(c) for the preconditioned CGS and different formulations, i.e., EFIE, MFIE/EFIE, and CFIE/EFIE. *Source*: Ergül and Gürel 2009 [82]. Reproduced with permission of IEEE.

problems, it can be observed that the hybrid CFIE/EFIE formulation provides accurate results that agree well with the reference EFIE solutions.

3.6.2 *Iterative Solutions of Normal Equations*

Transforming the original matrix equations into normal equations may improve the convergence of EFIE significantly [197, 198]. This subsection presents the solution of EFIE by employing the LSQR algorithm [191, 192], which corresponds to a stable application of the CG algorithm on normal equations. Despite the squaring of the condition number due to such a transformation into normal equations, LSQR may improve the convergence rate of iterative solutions of EFIE by performing better than many other iterative algorithms. It is also observed that the improved convergence is peculiar to EFIE, and LSQR does not have a good performance for CFIE solutions. Acceleration due to the transformation into normal equations can also be useful to reduce the memory required for the GMRES algorithm.

Table 3.7 Number of CGS iterations (10^{-6} residual error) for the solution of composite problems

Problem/Unknowns	Preconditioner	EFIE	MFIE/EFIE	CFIE/EFIE
Problem 1 / 24,205	NP	1488	372	63
Problem 1 / 24,205	BDP	NC	280	33
Problem 1 / 24,205	NFP	429	-	-
Problem 2 / 131,631	NP	NC	1387	607
Problem 2 / 131,631	BDP	NC	NC	1058
Problem 2 / 131,631	NFP	NC	-	-
Problem 3 / 5069	NP	280	218	219
Problem 3 / 5069	BDP	828	507	421
Problem 3 / 5069	NFP	37	-	-
Problem 3 / 181,536	NP	1334	919	611
Problem 3 / 181,536	BDP	NC	NC	1327
Problem 3 / 181,536	NFP	420	-	-

NC: no convergence.

Matrix equations in (2.212), (2.213), and (2.214) can be transformed into normal equations as

$$\overline{Z} \cdot a = w \quad \rightarrow \quad \overline{Z}^H \cdot \overline{Z} \cdot a = \overline{Z}^H \cdot w \tag{3.139}$$

with

$$\overline{Z}^H = (\overline{Z}^\dagger)^*, \tag{3.140}$$

where '†' and '∗' represent transpose and complex conjugate operations, respectively. Note that $\overline{Y} = \overline{Z}^H \cdot \overline{Z}$ is a Hermitian (self-adjoint) matrix, i.e., $\overline{Y}^H = \overline{\mathbf{Y}}$. Using a Galerkin scheme, EFIE matrices become symmetric so that only the complex conjugate operation is required in (3.140). The ordinary MLFMA can be employed to perform MVMs related to the complex-conjugate matrix as

$$y = \overline{Z}^* \cdot x = (\overline{Z} \cdot x^*)^*, \tag{3.141}$$

where only the input (x) and output (y) vectors are required to be modified. In the case of non-Galerkin EFIE, however, a transpose operation is required in (3.140), which can be implemented carefully by reversing steps of the ordinary MLFMA, as described above in Section 3.4. The resulting transpose MLFMA is also required for MFIE and CFIE, which lead to non-symmetric matrices, even with the Galerkin method.

Figure 3.24 presents iterative solutions of a scattering problem involving a PEC sphere with a radius of 1.5λ, where the residual error is plotted with respect to the iteration counts. The sphere is illuminated by a plane wave, and matrix equations involving 8364 unknowns are solved by the CGS algorithm using MLFMA. For both CFIE and EFIE formulations, convergences of ordinary and transformed matrix equations are compared. In addition to (3.139), a transformation

$$\overline{Z} \cdot a = w \quad \rightarrow \quad \overline{Z} \cdot \overline{Z} \cdot a = \overline{Z} \cdot w, \tag{3.142}$$

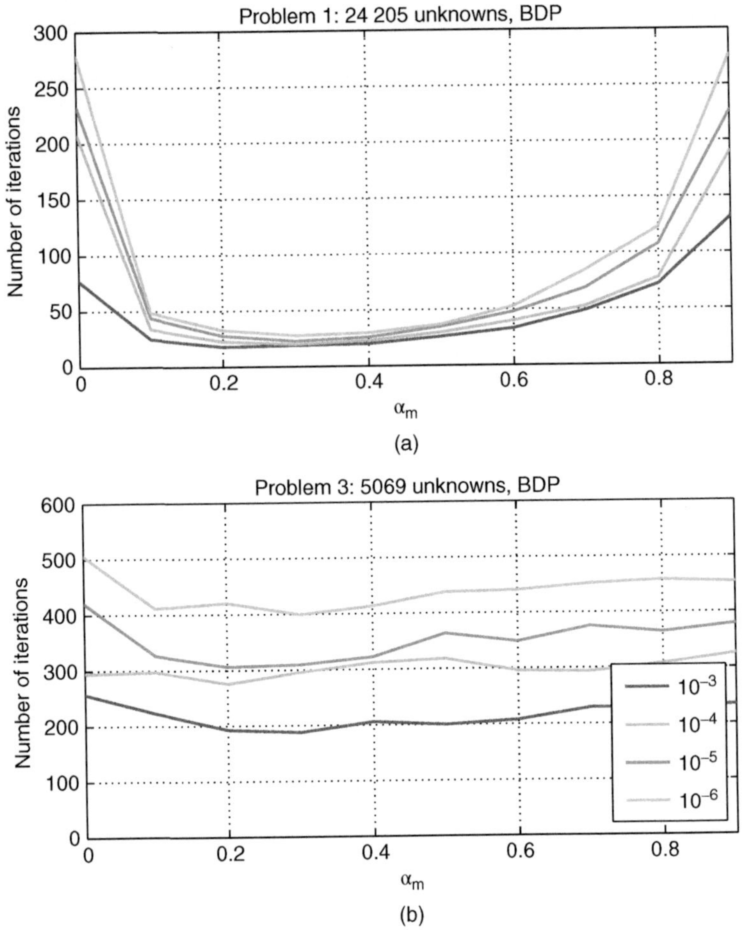

Figure 3.22 Iteration counts with respect to the variable α_m, which is applied to closed parts of the problems in Figures 3.14(a) and 3.14(c). *Source*: Ergül and Gürel 2009 [82]. Reproduced with permission of IEEE.

where both sides of the ordinary matrix equation are multiplied with the impedance matrix, is considered. The new matrices in (3.139) and (3.142) obtained by the transformations, i.e., $\overline{Z}^H \cdot \overline{Z}$ and $\overline{Z} \cdot \overline{Z}$, have condition numbers that are squares of the condition numbers of the impedance matrix $\overline{Z}$. With higher condition numbers, both transformations might decrease the convergence rates. However, Figure 3.24 shows that this is not true and one may obtain different results from the two transformations, especially for EFIE:

- The transformation in (3.142) decreases the convergence rates significantly for both CFIE and EFIE compared to the ordinary solutions. The transformed matrix equations require more iterations, especially in the case of EFIE, where convergence to an error of less than 0.1 cannot be achieved within 1000 iterations.

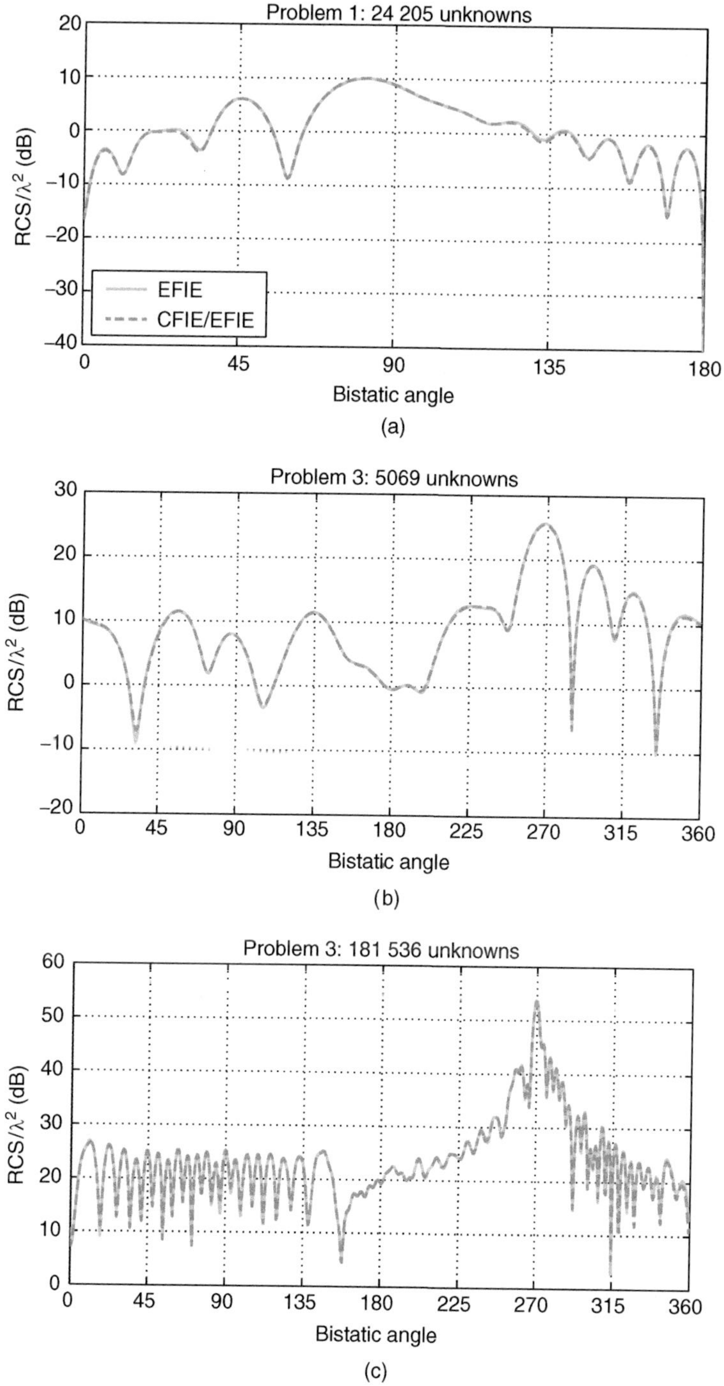

Figure 3.23 Normalized bistatic RCS (RCS/λ^2) on the z-x plane for the structures in Figures 3.14(a) and 3.14(c). *Source*: Ergül and Gürel 2009 [82]. Reproduced with permission of IEEE.

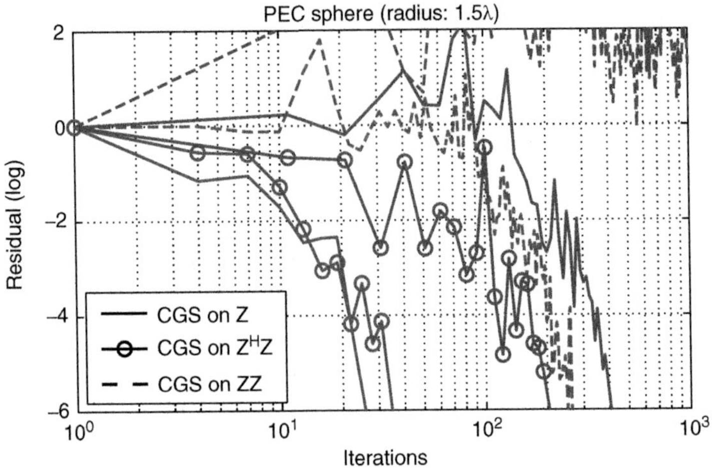

Figure 3.24 Iterative solutions of a scattering problem involving a PEC sphere with a radius of 1.5λ formulated with CFIE (red) and EFIE (blue). Matrix equations have 8364 unknowns and solutions are performed by the CGS algorithm applied on the ordinary equation and transformed equations in (3.140) and (3.142), labeled by '$Z^H Z$' and 'ZZ', respectively. *Source*: Ergül and Gürel 2009 [82]. Reproduced with permission of IEEE.

- With the transformation in (3.139), CFIE has a faster convergence compared to the solution obtained by the transformation in (3.142), but it is still slower than the ordinary solution. However, in the case of EFIE, the transformation in (3.139) leads to an even faster convergence than the ordinary solution.

The results presented in Figure 3.24 can be confirmed by considering various scattering problems involving different geometries with closed and open surfaces. The transformation in (3.139) tends to improve the convergence rate of iterative solutions of EFIE.

In Figure 3.24, the transformed equation in (3.139) offers a faster convergence rate compared to the ordinary equation for EFIE. However, it does not increase the efficiency of the iterative solution since the transformed equation requires four MVMs per CGS iteration while the ordinary solution requires only two. Hence, the transformed equation in (3.139) can be solved using the CG algorithm in order to reduce the number of MVMs to two per iteration, while maintaining benefits of the transformation. Among different types of CG algorithms for normal equations, the LSQR algorithm is a stable method based on the Lanczos bidiagonalization as detailed in [191, 192]. Then, without having to consider the transformation in (3.139) explicitly, one can simply feed the LSQR algorithm with MVMs related to the impedance matrix $\overline{\mathbf{Z}}$ and its transpose complex conjugate $\overline{\mathbf{Z}}^H$. Then, LSQR can perform the CG solutions of the transformed equation in an implicit and stable way.

Numerical Results (LSQR)

Figure 3.25 presents iterative solutions of a scattering problem discretized with 14,871 unknowns. The problem involves a sphere with a radius of 2λ illuminated by a plane wave.

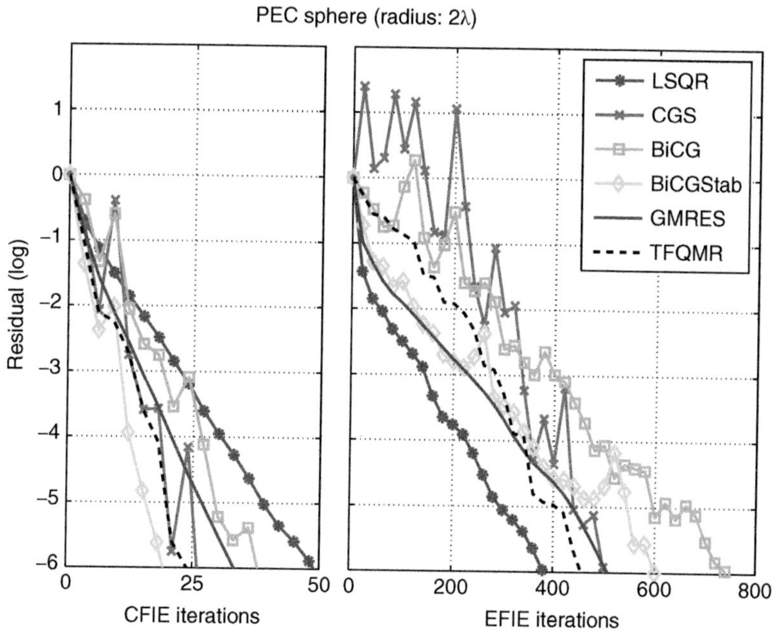

Figure 3.25 Iterative solutions of a scattering problem involving a sphere with a radius of 2λ formulated with (a) CFIE and (b) EFIE. Matrix equations involving 14,871 unknowns are solved by using various iterative algorithms. *Source*: Ergül and Gürel 2008 [197]. Reproduced with permission of IEEE.

Ordinary equations obtained with CFIE and EFIE are solved iteratively by employing various algorithms, including LSQR. Note that the solution of the ordinary equations in (2.212) and (2.214) using LSQR corresponds to a stable solution of the transformed equation in (3.139) using the CG algorithm [193]. All iterative algorithms in Figure 3.25 require two MVMs for each iteration except GMRES, which requires only one MVM per iterations at the cost of increased memory. Figure 3.25(a) shows that most efficient solutions of the CFIE formulation are performed by BiCGStab and the no-restart GMRES. In terms of the processing time, GMRES is better than the other iterative algorithms since it requires only 34 MVMs to reduce the residual error to less than 10^{-6}. However, BiCGStab is preferable since it requires only 20 iterations, or 40 MVMs, without using the extra memory needed by GMRES. On the other hand, LSQR is inefficient compared to all other iterative algorithms by requiring 50 iterations, or 100 MVMs, to reduce the error to same levels. Therefore, LSQR appears to be a bad choice for the CFIE solution of the problem.

In contrast to CFIE, Figure 3.25(b) shows that LSQR has significantly different convergence properties for the EFIE solution of the problem; it is not the slowest converging iterative algorithm as in Figure 3.25(a). Indeed, it requires 383 iterations, or 766 MVMs, to reduce the error to less than 10^{-6} without using the extra memory needed by GMRES, which requires 510 MVMs. In other words, LSQR provides faster convergence compared to BiCG, BiCGStab, CGS, and TFQMR, and it is a good alternative to GMRES, which offers a faster convergence at the cost of extra memory. The improved convergence provided by LSQR is due to favorable properties of the transformation in (3.139) for EFIE. The same improvement is not observed in

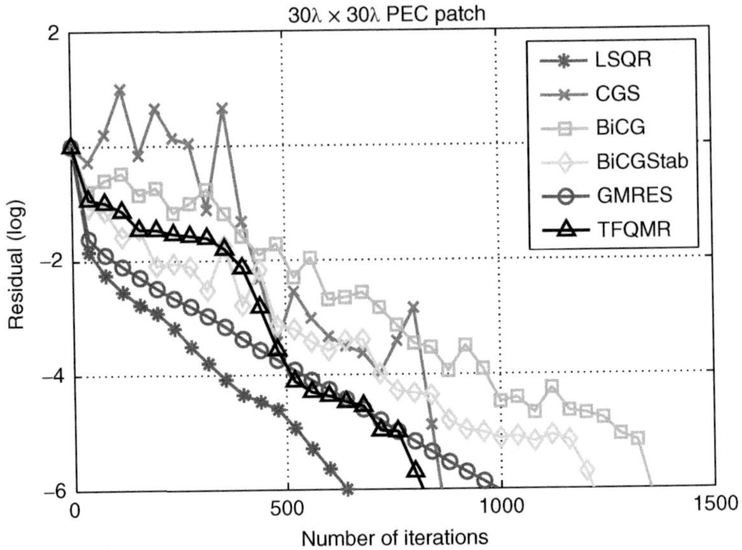

Figure 3.26 Iterative solutions of a scattering problem involving a $30\lambda \times 30\lambda$ patch discretized with 310,383 unknowns. *Source*: Ergül and Gürel 2008 [197]. Reproduced with permission of IEEE.

CFIE solutions since the transformation in (3.139) does not improve the convergence of CFIE, as also depicted in Figure 3.24.

In Figure 3.25, it is clearly observed that LSQR provides an improved convergence for EFIE, but it is still inefficient compared to CFIE solutions. In general, for the solution of problems involving closed surfaces, CFIE is very efficient and may negate the necessity of employing LSQR with EFIE. On the other hand, for problems with open surfaces, EFIE is the inevitable choice and the improved convergence provided by LSQR can become important. As an example, Figures 3.26 and 3.27 present the solution of scattering problems involving a $d \times d$ conducting patch, where $d = 30$ cm and corresponds to various values from 12λ to 30λ. The patch is located on the x-y plane and illuminated by a plane wave propagating in the $-z$ direction. Discretizations of the object for different frequencies lead to relatively large matrix equations with the number of unknowns ranging from 49,200 to 310,383. Scattering problems are solved by MLFMA parallelized into 16 processes on a cluster of AMD Opteron 870 processors. Figure 3.26 presents the number of iterations when $d = 30\lambda$ and the number of unknowns is 310,383. In terms of the iteration counts, LSQR provides the fastest convergence compared to the other iterative algorithms and it requires 713 iterations, or 1426 MVMs, to reduce the error to less than 10^{-6}. On the other hand, since GMRES requires only one MVM per iteration, a fair comparison should be based on the processing time of iterative solutions. In addition, such a comparison is necessary in order to consider different setup times required by iterative algorithms, which become considerable as the number of unknowns increases. In this manner, Figure 3.27 presents solution times with respect to the number of unknowns. The fixed setup time of MLFMA is omitted, considering only the time required to perform the iterative solutions of reduce the residual error to less than 10^{-6}. It can be observed that no-restart GMRES provides the fastest solution, especially as the problem size grows. However, as also

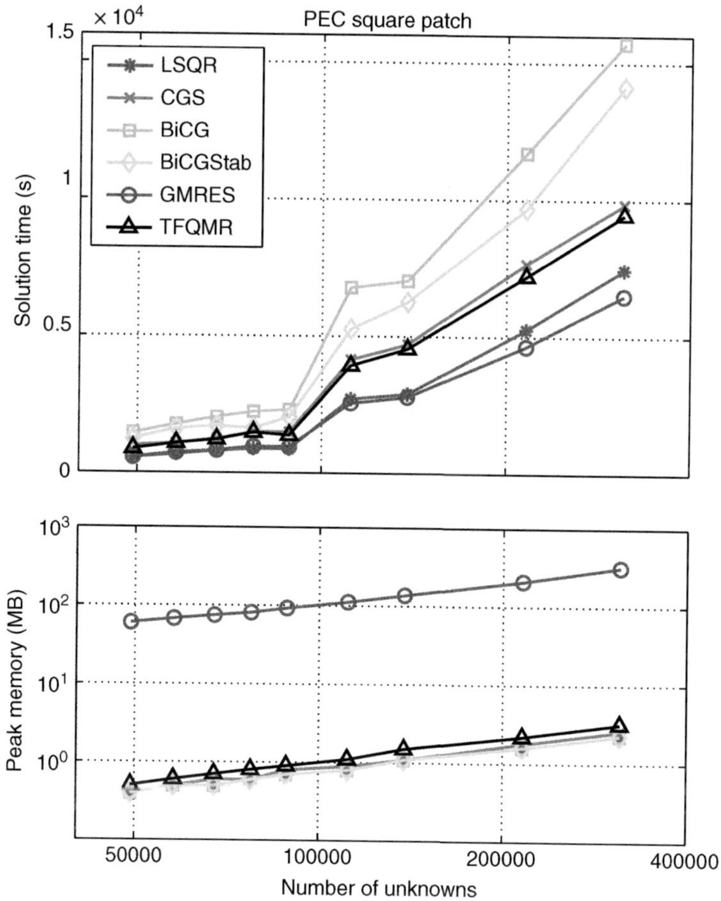

Figure 3.27 Processing time and peak memory per processor with respect to the number of unknowns for the solution of scattering problems involving a patch geometry with various dimensions from $12\lambda \times 12\lambda$ to $30\lambda \times 30\lambda$. Only the time and memory required for iterative solutions are considered. *Source:* Ergül and Gürel 2008 [197]. Reproduced with permission of IEEE.

seen in Figure 3.27, it requires a peak memory that is significantly larger than the other iterative algorithms. As an example, for the solution of the largest problem in Figure 3.27 with 310,383 unknowns, GMRES requires 321 MB memory per processor, while the total memory usage (including the memory required by MLFMA) is only 456 MB per processor. Thus, leaving the memory-hungry GMRES aside, among the memory-efficient solutions provided by the other algorithms, LSQR requires the minimum processing time, as depicted in Figure 3.27. In other words, LSQR performs better than CGS, BiCG, BiCGStab, and TFQMR, and it is a strong alternative to no-restart GMRES, which has considerably larger memory requirements.

To further demonstrate the improved convergence provided by LSQR, two scattering problems involving a half sphere (HS) and an open prism (OP), which are depicted in Figures 2.27 and 3.28, respectively, are considered. The half-sphere problem is solved at 2 GHz and

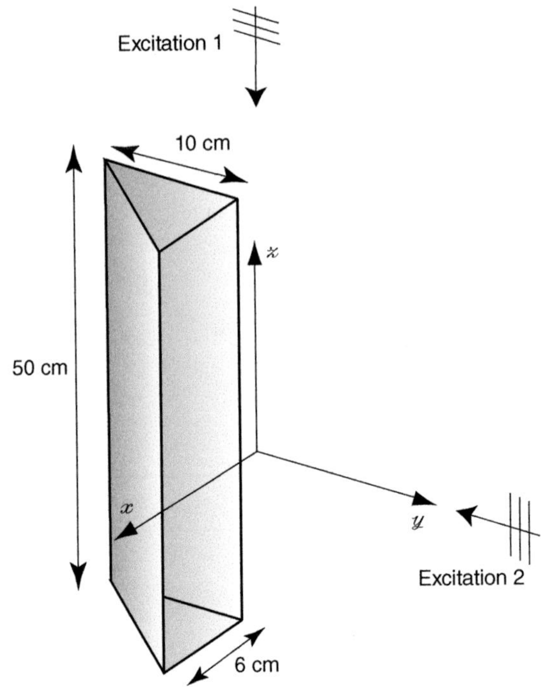

Figure 3.28 Scattering problems involving an open prism. *Source*: Ergül and Gürel 2008 [197]. Reproduced with permission of IEEE.

7.5 GHz, while the open-prism problem is solved at 5 GHz and 16 GHz. For the lower frequencies, discretizations with about $\lambda/10$ triangulation lead to 9911 and 11,351 unknowns for the half sphere and the open prism, respectively (small problems). For the higher frequencies, similar discretizations with $\lambda/10$ mesh size lead to 116,596 and 127,925 unknowns (large problems). Parallel MLFMA is employed to perform MVMs. For each geometry, two scattering problems involving the plane-wave excitations described in Figures 2.27 and 3.28 are solved. As an iterative solver, the LSQR algorithm is employed, in addition to GMRES, CGS, BiCG, BiCGStab, and TFQMR. Figure 3.29 presents the processing time required by various iterative algorithms for solutions. The processing time is measured on a cluster of 16 AMD Opteron 870 processors for the convergence of the residual error to less than 10^{-6}. It can be observed that LSQR outperforms other iterative algorithms, except for GMRES with a restart parameter of 1000. But, once again, GMRES requires considerably larger memory than all other algorithms. As an example, for the solution of the large half-sphere problem, the memory requirement of GMRES is about 140 MB per processor, while it is only 1–1.5 MB for other algorithms, including LSQR.

Numerical Results (Using GMRES for Normal Equations)

The improved convergence provided by LSQR is due to favorable properties of normal equations in (3.139) for EFIE. One can also use the benefits of normal equations by

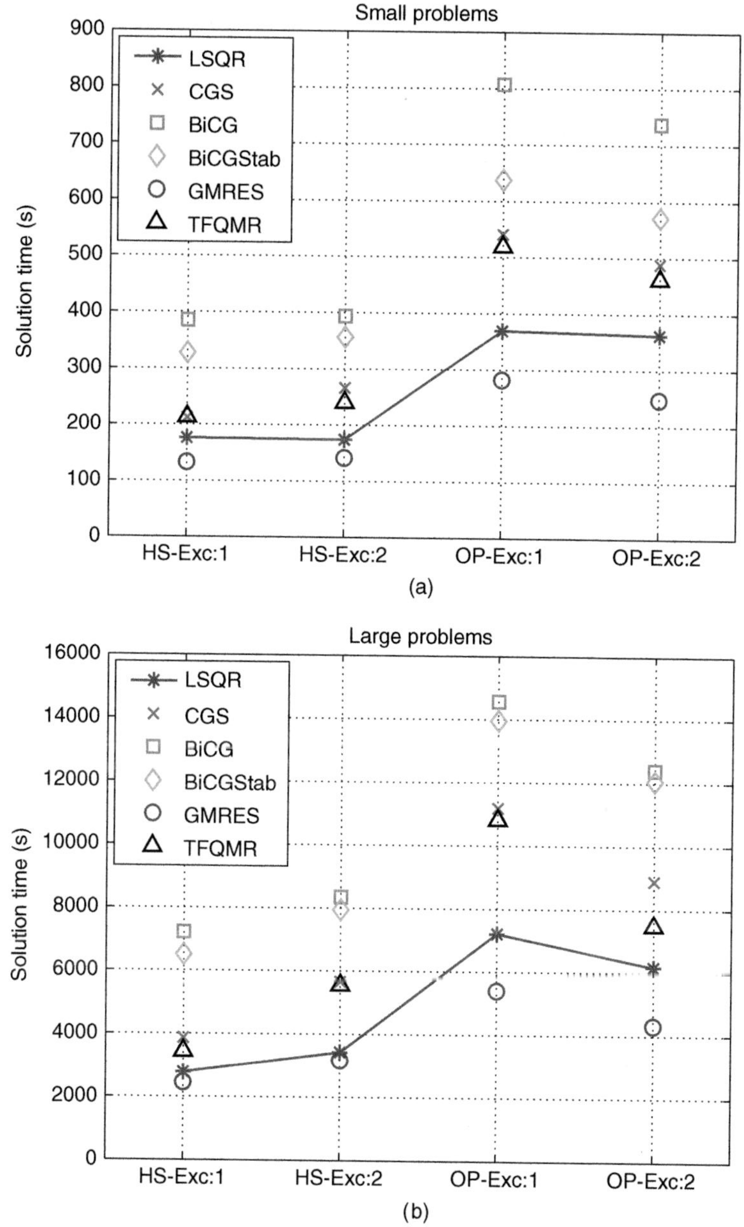

Figure 3.29 The processing time required by various iterative algorithms for MLFMA solutions of scattering problems described in Figures 2.27 and 3.28.

Table 3.8 Solutions of scattering problems involving a patch geometry of various sizes

		GMRES		
Size	Unknowns	Iterations	Time (s)	Memory (MB)
$12\lambda \times 12\lambda$	49,200	620	568	30
$18\lambda \times 18\lambda$	111,405	776	2372	81
$20\lambda \times 20\lambda$	137,792	793	2635	107
$25\lambda \times 25\lambda$	215,338	877	4232	180
$30\lambda \times 30\lambda$	310,383	991	6070	290
		GMRES-NE		
Size	N	Iterations	Time (s)	Memory (MB)
$12\lambda \times 12\lambda$	49,200	254	445	13
$18\lambda \times 18\lambda$	111,405	335	1959	36
$20\lambda \times 20\lambda$	137,792	360	2269	49
$25\lambda \times 25\lambda$	215,338	413	3620	87
$30\lambda \times 30\lambda$	310,383	467	5016	139

employing GMRES. As an example, scattering problems involving a square patch with various sizes from $12\lambda \times 12\lambda$ to $30\lambda \times 30\lambda$ are revisited. The problems are solved on a cluster of 16 AMD Opteron 870 processors. Table 3.8 lists the number of iterations for 10^{-6} residual error, solution time, and memory per processor, when GMRES is used to solve the original and transformed (normal) matrix equations. Only the memory required for GMRES is considered. Compared to conventional solutions, the number of iterations is reduced by 53–59% when GMRES is applied on normal equations (denoted as GMRES-NE). The processing time, however, does not decrease by the same amount, since the number of MVMs per iteration is two for GMRES-NE. Nevertheless, for the largest problem discretized with 310,383 unknowns, the processing time is reduced by 17%, while the number of MVMs decreases only by 6%. This is because the cost of GMRES itself increases quadratically with the number of iterations [190]. Hence, reducing the number of iterations is preferable even when the number of MVMs does not decrease significantly. Table 3.8 shows that the memory required for GMRES is reduced by 52–57%. For the largest problem, MLFMA itself requires only 135 MB memory per processor. Then, the total memory usage (including GMRES and MLFMA) per processor is reduced from 425 MB to 274 MB, just by solving the normal equation instead of the original equation with GMRES.

Figure 3.30 presents the solution of a scattering problem involving a two-dimensional array (wall) that consists of 11×11 spheres. The wall is illuminated by a plane wave with normal incidence (normal to the plane of the array) at 500 GHz. Spheres have a radius of 0.3 mm and they are arranged regularly with 1 mm periodicity in the x and y directions. The overall size of the wall is 5.3×5.3×0.6 mm and is discretized with 112,530 RWG functions. Figure 3.30(b) presents convergence histories when the problem is solved with GMRES applied to original

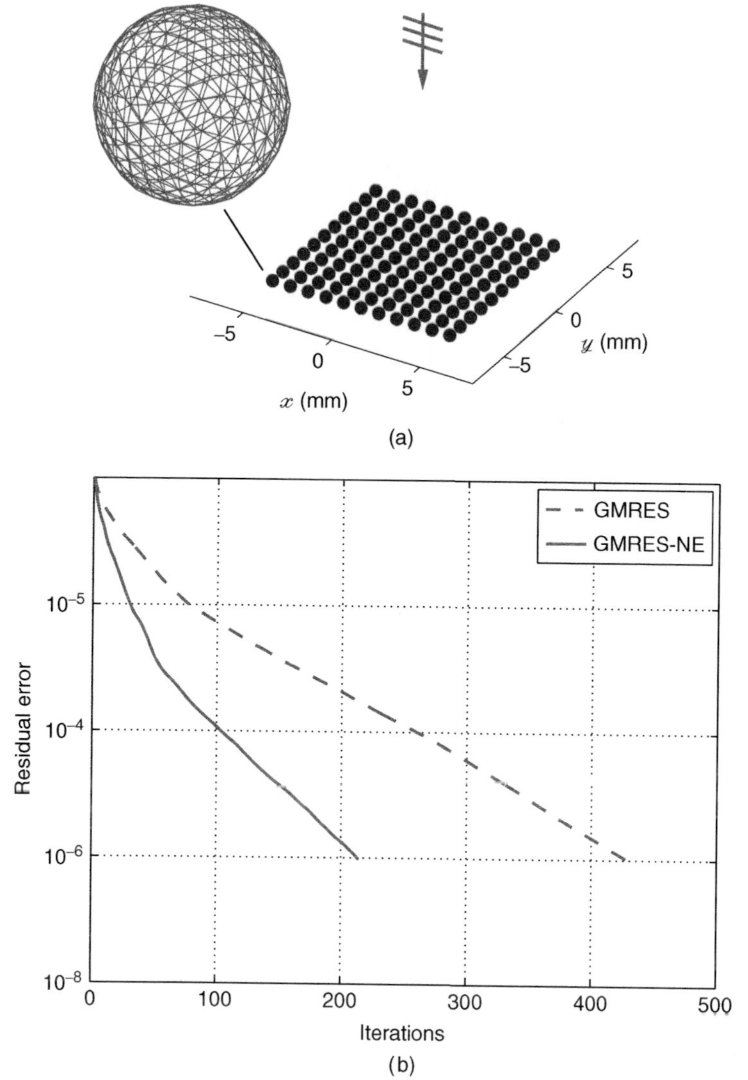

Figure 3.30 (a) A scattering problem involving an 11×11 array of PEC spheres illuminated by a plane wave. (b) Convergence histories when the scattering problem is solved with GMRES applied to the original and transformed matrix equations.

and transformed matrix equations. For the original matrix equation, the number of iterations (for 10^{-6} residual error) is 429. Transforming the matrix equation increases the convergence rate and the number of iterations is reduced to 213. Hence, the memory required for GMRES is reduced from 750 MB to 390 MB, and the total memory (including GMRES and MLFMA) is reduced from 1447 MB to 1087 MB by solving the normal equation instead of the original equation.

Limitations

Transformations into normal equations can improve iterative solutions of EFIE. However, the improvement is usually restricted to ordinary discretizations with $\lambda/10$ triangles. For example, in Figure 3.31, iterative solutions of scattering problems involving a square patch with edges of 30 cm are revisited. In Figure 3.31(a), the patch is discretized with 49,200 unknowns and the frequency changes from 3 GHz to 15 GHz. The mesh size corresponds to $\lambda/8$ and $\lambda/40$ at 15 GHz and 3 GHz, respectively. Due to the low-frequency breakdown of EFIE, the iterative convergence becomes difficult as the mesh size decreases with respect to wavelength. However, as also depicted in Figure 3.31(a), the number of iterations increases faster for LSQR, compared to the other iterative algorithms. In other words, the improved convergence provided by LSQR vanishes for dense discretizations. Figure 3.31(b) presents another experiment, where a $10\lambda \times 10\lambda$ patch is discretized with various mesh sizes from $\lambda/12$ to $\lambda/30$. In this case, the frequency is fixed to 10 GHz, but the number of unknowns changes from 49,200 to 310,383. Similar to the previous example, the rapid convergence of the LSQR algorithm compared to the other algorithms deteriorates as the mesh size decreases.

In addition to low performance for dense discretizations, the LSQR algorithm can be difficult to accelerate with preconditioners. For example, Table 3.9 lists the number of MVMs required for the half-sphere and open-prism problems discretized with 9911 and 11,351 unknowns, respectively. Solutions with CGS, BiCGStab, and LSQR are compared when iterations are accelerated with NFP, in addition to the NP case. Without preconditioning, LSQR performs better than the other algorithms. However, using NFP, both CGS and BiCGStab are accelerated much more than LSQR, which becomes the worst choice among the three algorithms.

Despite many limitations that prevent the use of LSQR, the advantages of this algorithm and the transformation into normal equations for EFIE are interesting. It needs further research to reveal how such a transformation that squares the condition number can improve the solution of EFIE and how it can be generalized to other types of problems.

3.6.3 *Iterative Solutions of Dielectric Objects*

For electromagnetics problems involving dielectric objects, two types of efficient preconditioners, namely, the two-partition block-diagonal preconditioner (2PBDP) and the four-partition block-diagonal preconditioner (4PBDP), can be employed to reduce the number of iterations

Table 3.9 Number of MVMs (10^{-6} residual error) for the solution of scattering problems involving a half sphere and open prism

	CGS		BiCGStab		LSQR	
Problem	NP	NFP	NP	NFP	NP	NFP
Half sphere, excitation 1 at 2 GHz	798	38	1240	36	657	174
Half sphere, excitation 2 at 2 GHz	988	38	1346	38	655	172
Open prism, excitation 1 at 5 GHz	1342	72	1606	60	935	448
Open prism, excitation 2 at 5 GHz	1220	56	1426	54	899	422

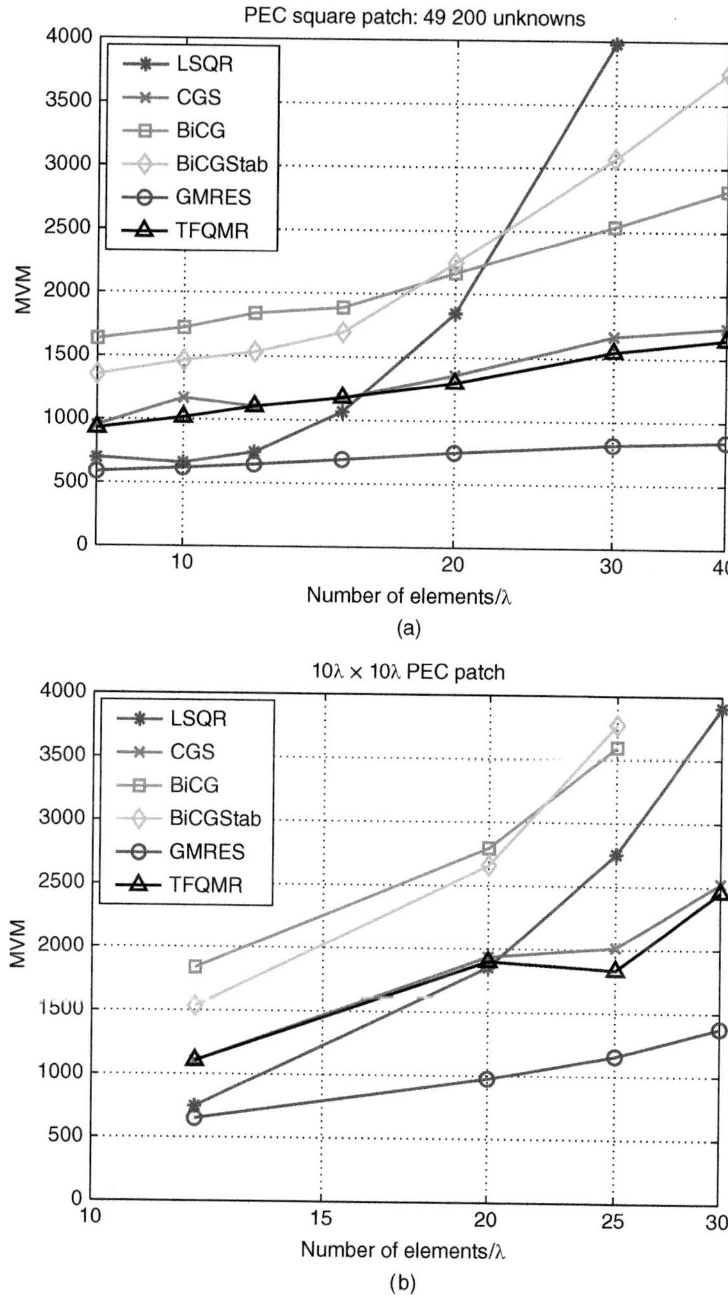

Figure 3.31 Number of MVMs as a function of the number of elements per wavelength for the solution of scattering problems involving a square patch with edges of 30 cm. (a) The patch is discretized with 49,200 unknowns and the frequency changes from 3 GHz to 15 GHz. (b) The frequency is 10 GHz and the patch is discretized with 49,200 to 310,383 unknowns.

required for solutions. Matrix equations in the form of (2.40) can be preconditioned as

$$\overline{\boldsymbol{P}}^{-1} \cdot \begin{bmatrix} \overline{\boldsymbol{Z}}_{11} & \overline{\boldsymbol{Z}}_{12} \\ \overline{\boldsymbol{Z}}_{21} & \overline{\boldsymbol{Z}}_{22} \end{bmatrix} \cdot \begin{bmatrix} \boldsymbol{a}_J \\ \boldsymbol{a}_M \end{bmatrix} = \overline{\boldsymbol{P}}^{-1} \cdot \begin{bmatrix} \boldsymbol{w}_1 \\ \boldsymbol{w}_2 \end{bmatrix}, \tag{3.143}$$

where $\overline{\boldsymbol{P}}$ is a preconditioner matrix. In MLFMA, there are $\mathcal{O}(N)$ near-field interactions, which are calculated directly and are available for constructing preconditioners. BDP, which is based on using self-interactions of boxes at the lowest level, is commonly used to accelerate MLFMA solutions of electromagnetics problems involving PEC objects formulated with MFIE and CFIE [35]. The preconditioner matrix, which has a block-diagonal structure, can be inverted and used efficiently with $\mathcal{O}(N)$ complexity. A direct extension of BDP for dielectric problems, which is called 2PBDP, involves self-interactions of boxes at the lowest level in diagonal partitions, i.e.,

$$\overline{\boldsymbol{P}}_{2P} = \begin{bmatrix} \overline{\boldsymbol{P}}_{11} & 0 \\ 0 & \overline{\boldsymbol{P}}_{22} \end{bmatrix}, \tag{3.144}$$

where $\overline{\boldsymbol{P}}_{11} \approx \overline{\boldsymbol{Z}}_{11}$ and $\overline{\boldsymbol{P}}_{22} \approx \overline{\boldsymbol{Z}}_{22}$ are block-diagonal matrices. Then, a preconditioned matrix equation can be written as

$$\begin{bmatrix} \overline{\boldsymbol{B}}_{11} \cdot \overline{\boldsymbol{Z}}_{11} & \overline{\boldsymbol{B}}_{11} \cdot \overline{\boldsymbol{Z}}_{12} \\ \overline{\boldsymbol{B}}_{22} \cdot \overline{\boldsymbol{Z}}_{21} & \overline{\boldsymbol{B}}_{22} \cdot \overline{\boldsymbol{Z}}_{22} \end{bmatrix} \cdot \begin{bmatrix} \boldsymbol{a}_J \\ \boldsymbol{a}_M \end{bmatrix} = \begin{bmatrix} \overline{\boldsymbol{B}}_{11} \cdot \boldsymbol{w}_1 \\ \overline{\boldsymbol{B}}_{22} \cdot \boldsymbol{w}_2 \end{bmatrix}, \tag{3.145}$$

where $\overline{\boldsymbol{B}}_{11} = \overline{\boldsymbol{P}}_{11}^{-1}$ and $\overline{\boldsymbol{B}}_{22} = \overline{\boldsymbol{P}}_{22}^{-1}$ are also block-diagonal matrices. Similar to BDP, 2PBDP accelerates iterative solutions of normal and mixed formulations, such as CNF, MNMF, and JMCFIE.

To further improve iterative solutions of normal and mixed formulations, one can use 4PBDP, which is based on using diagonal blocks, i.e., self-interactions of boxes at lowest levels, in all four partitions of matrix equations. This way, some of the large elements in off-diagonal matrix partitions ($\overline{\boldsymbol{Z}}_{12}$ and $\overline{\boldsymbol{Z}}_{21}$) are considered in constructing an effective preconditioner. The resulting preconditioner matrices are in the form of

$$\overline{\boldsymbol{P}}_{4P} \approx \begin{bmatrix} \overline{\boldsymbol{P}}_{11} & \overline{\boldsymbol{P}}_{12} \\ \overline{\boldsymbol{P}}_{21} & \overline{\boldsymbol{P}}_{22} \end{bmatrix}. \tag{3.146}$$

Since $\overline{\boldsymbol{P}}_{12}$ and $\overline{\boldsymbol{P}}_{21}$ are also block-diagonal matrices, the inverse of $\overline{\boldsymbol{P}}_{4P}$ can be evaluated efficiently as

$$\overline{\boldsymbol{P}}_{4P}^{-1} = \begin{bmatrix} \overline{\boldsymbol{B}}_{11} & \overline{\boldsymbol{B}}_{12} \\ \overline{\boldsymbol{B}}_{21} & \overline{\boldsymbol{B}}_{22} \end{bmatrix}, \tag{3.147}$$

where

$$\overline{\boldsymbol{B}}_{11} = \overline{\boldsymbol{P}}_{11}^{-1} \cdot \left[\overline{\boldsymbol{I}} + \overline{\boldsymbol{P}}_{12} \cdot \overline{\boldsymbol{S}}^{-1} \cdot \overline{\boldsymbol{P}}_{21} \cdot \overline{\boldsymbol{P}}_{11}^{-1} \right] \tag{3.148}$$

$$\overline{\boldsymbol{B}}_{12} = -\overline{\boldsymbol{P}}_{11}^{-1} \cdot \overline{\boldsymbol{P}}_{12} \cdot \overline{\boldsymbol{S}}^{-1} \tag{3.149}$$

$$\overline{\boldsymbol{B}}_{21} = -\overline{\boldsymbol{S}}^{-1} \cdot \overline{\boldsymbol{P}}_{21} \cdot \overline{\boldsymbol{P}}_{11}^{-1} \tag{3.150}$$

$$\overline{\boldsymbol{B}}_{22} = \overline{\boldsymbol{S}}^{-1}, \tag{3.151}$$

and

$$\overline{\boldsymbol{S}} = \overline{\boldsymbol{P}}_{22} - \overline{\boldsymbol{P}}_{21} \cdot \overline{\boldsymbol{P}}_{11}^{-1} \cdot \overline{\boldsymbol{P}}_{12} \tag{3.152}$$

is the Schur complement of $\overline{\boldsymbol{P}}_{11}$. The matrix operations in (3.148)–(3.152), i.e., matrix-matrix multiplications, inversion of $\overline{\boldsymbol{P}}_{11}$, and inversion of $\overline{\boldsymbol{S}}$ can be performed efficiently in $\mathcal{O}(N)$ time using $\mathcal{O}(N)$ memory. Numerical experiments show that the extra cost of 4PBDP with respect to 2PBDP is always negligible, considering the overall cost of solutions with MLFMA. Nevertheless, 4PBDP can significantly improve the efficiency of solutions by reducing the number of iterations, and it is especially useful when the acceleration provided by 2PBDP is not sufficient.

In general, iterative solutions of CNF and JMCFIE become difficult as the contrast increases, i.e., when the permittivity and permeability change significantly across dielectric interfaces. The main reason is the existence of off-diagonal matrix partitions, which are numerically sensitive to the contrast. The off-diagonal partitions of JMCFIE and CNF are significantly unbalanced due to the multiplications with η^{-1} and η in $\overline{\boldsymbol{Z}}_{12}$ and $\overline{\boldsymbol{Z}}_{21}$, respectively. Although this may not be critical for low contrasts, the off-diagonal partition $\overline{\boldsymbol{Z}}_{21}$ dominates the overall matrix, as the contrast of the object increases. The elements in the off-diagonal partition of CNF can be written as

$$\begin{aligned}\overline{\boldsymbol{Z}}_{21}^{\mathrm{CNF}}[m,n] = i\omega \int_{S_m} d\boldsymbol{r}\boldsymbol{t}_m(\boldsymbol{r}) \cdot \hat{\boldsymbol{n}}(\boldsymbol{r}) \times \\ \left\{ \int_{S_n} d\boldsymbol{r}' \boldsymbol{b}_n(\boldsymbol{r}') \left[\mu_i g_i(\boldsymbol{r},\boldsymbol{r}') - \mu_o g_o(\boldsymbol{r},\boldsymbol{r}')\right] \right. \\ \left. + \frac{1}{\omega^2} \int_{S_n} d\boldsymbol{r}' \nabla' \cdot \boldsymbol{b}_n(\boldsymbol{r}') \nabla \left[\frac{g_i(\boldsymbol{r},\boldsymbol{r}')}{\varepsilon_i} - \frac{g_o(\boldsymbol{r},\boldsymbol{r}')}{\varepsilon_o}\right] \right\}.\end{aligned} \tag{3.153}$$

Note that $\overline{\boldsymbol{Z}}_{21}^{\mathrm{CNF}}$ is also a major contribution in $\overline{\boldsymbol{Z}}_{21}^{\mathrm{JMCFIE}}$ for sufficiently high contrasts. Using a Taylor series expansion for the exponential in the Green's function,

$$\left[\mu_i g_i(\boldsymbol{r},\boldsymbol{r}') - \mu_o g_o(\boldsymbol{r},\boldsymbol{r}')\right] = \frac{1}{4\pi R} \sum_{s=0}^{\infty} \frac{(i\omega R)^s}{s!} \left[\mu_i(\mu_i\varepsilon_i)^{s/2} - \mu_o(\mu_o\varepsilon_o)^{s/2}\right] \tag{3.154}$$

and

$$\begin{aligned}\nabla \left[\frac{g_i(\boldsymbol{r},\boldsymbol{r}')}{\varepsilon_i} - \frac{g_o(\boldsymbol{r},\boldsymbol{r}')}{\varepsilon_o}\right] = \frac{\hat{\boldsymbol{R}}}{4\pi R^2} \sum_{s=0}^{\infty} \frac{(i\omega R)^{s+1}}{s!} \left[\frac{(\mu_i\varepsilon_i)^{s/2+1/2}}{\varepsilon_i} - \frac{(\mu_o\varepsilon_o)^{s/2+1/2}}{\varepsilon_o}\right] \\ - \frac{\hat{\boldsymbol{R}}}{4\pi R^2} \sum_{s=0}^{\infty} \frac{(i\omega R)^s}{s!} \left[\frac{(\mu_i\varepsilon_i)^{s/2}}{\varepsilon_i} - \frac{(\mu_o\varepsilon_o)^{s/2}}{\varepsilon_o}\right],\end{aligned} \tag{3.155}$$

where $\mathbf{R} = (\boldsymbol{r} - \boldsymbol{r}') = \hat{\boldsymbol{R}}R$. Note that $1/R$ and $1/R^2$ singularities in (3.154) and (3.155) exist when $\mu_o \neq \mu_i$ and $\varepsilon_o \neq \varepsilon_i$, respectively. In addition, numerical values of the expressions in

(3.154) and (3.155), thus $\overline{\mathbf{Z}}_{21}^{\mathrm{CNF}}$, grow rapidly with the increasing contrast. As a result, the overall matrix equation becomes significantly unbalanced and difficult to solve iteratively. For those problems formulated with CNF and JMCFIE, 4PBDP, which employs large elements in $\overline{\mathbf{Z}}_{21}$, can be very effective to accelerate iterative solutions.

Numerical Results

In the examples below, the efficiency of numerical solutions is investigated when scattering problems are formulated with CTF, CNF, MNMF, and JMCFIE. For all solutions, near-field interactions are calculated with at most 1% error, and far-field interactions are computed by MLFMA with three digits of accuracy. Tree structures are constructed by fixing the size of boxes at the lowest level to $0.25\lambda_o$ and using a bottom-up scheme.

Figure 3.32 presents the amount of memory required for MLFMA solutions of scattering problems involving a dielectric sphere with a radius of 0.3 m located in free space. The sphere is illuminated by a plane wave propagating in the z direction with the electric field polarized in the x direction. The radius of the sphere changes from $0.75\lambda_o$ to $7.5\lambda_o$, where λ_o is the wavelength in free space. Discretizations with $\lambda_o/10$ triangulations lead to 4142 and 412,998 unknowns, respectively, for radii $0.75\lambda_o$ and $7.5\lambda_o$. The relative permittivity of the sphere is $\epsilon_r = 2.0$. Figure 3.32 presents the peak memory required for MLFMA solutions with respect to the number of unknowns. The peak memory depends on the formulation type mainly because

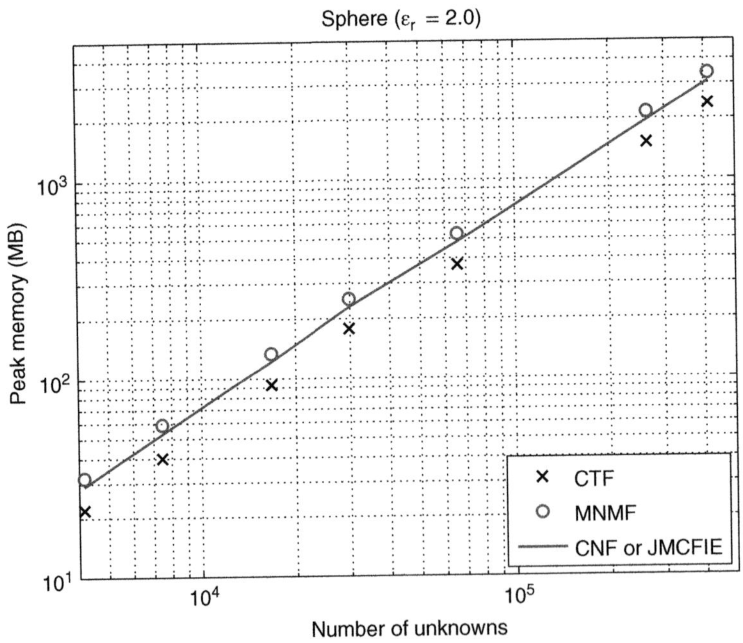

Figure 3.32 Peak memory required for MLFMA solutions of scattering problems involving a sphere with a relative permittivity of 2.0 located in free space. The radius of the sphere is in the range from $0.75\lambda_o$ to $7.5\lambda_o$. *Source*: Ergül and Gürel 2009 [80]. Reproduced with permission of IEEE.

of different storage requirements for near-field interactions (identical or nonidentical diagonal partitions) and far-field patterns of basis and testing functions (one set or two sets of patterns). Having identical diagonal partitions and using one set of far-field patterns, CTF requires less memory than the other formulations. CNF and JMCFIE require two sets of far-field patterns, and their memory usage is larger than CTF. MNMF has nonidentical diagonal partitions and requires two sets of far-field patterns, leading to a larger memory usage than CTF, CNF, and JMCFIE. Although the memory is not critical for small problems, it becomes more important as the problem size grows. For example, when the radius is $7.5\lambda_o$, the peak memory is 2370 MB and 3385 MB for CTF and MNMF, respectively.

Table 3.10 lists the number of CGS iterations to reduce the residual error to below 10^{-3} for scattering problems involving a sphere with a relative permittivity of 4.0. The radius of the sphere changes from $0.75\lambda_o$ to $7.5\lambda_o$. As the problem size gets larger, the setup time becomes negligible, compared with the time required for iterations. Then, the processing time of MLFMA solutions is directly proportional to the iteration counts. For all formulations, two efficient preconditioners, i.e., 2PBDP and 4PBDP, are applied in addition to the no-preconditioner case. The following observations can be made:

- Preconditioning matrix equations with 2PBDP and 4PBDP does not accelerate the iterative convergence for CTF. In fact, these low-cost preconditioners decelerate the convergence and increase the number of iterations for this formulation. Especially for large problems, convergence cannot be achieved within 2000 iterations, when 2PBDP and 4PBDP are used for CTF. A negative effect of efficient preconditioners was also observed for other tangential formulations, such as EFIE for the solution of PEC objects (see above, Section 3.6.1). However, 2PBDP and 4PBDP significantly accelerate the iterative convergence for the normal formulations (CNF and MNMF) and JMCFIE.
- Although they are both normal formulations, iteration counts for CNF and MNMF differ significantly; convergence is consistently faster for MNMF. For large problems, CNF fails to provide quick convergence even when compared to CTF without preconditioning.
- Using 4PBDP provides faster convergence than 2PBDP, especially for JMCFIE. This is due to strong off-diagonal partitions of JMCFIE, in addition to its strong diagonal partitions involving well-tested identity operators.
- For small problems, the iteration counts for MNMF are lower than JMCFIE. This can be observed in Table 3.10, when the radius of the sphere is in the range from $0.75\lambda_o$ to $3\lambda_o$. For larger problems, however, the convergence for JMCFIE becomes faster than the convergence for MNMF. When the radius of the sphere is $7.5\lambda_o$, the number of iterations is consistently lower for JMCFIE with and without preconditioning.

To further compare the dielectric formulations, i.e., CTF, CNF, MNMF, and JMCFIE, in terms of efficiency, scattering problems involving a dielectric cube with edges of $4\lambda_o$ are considered. Similar to the sphere problems, the cube is illuminated by a plane wave propagating in the z direction with the electric field polarized in the x direction. Discretization of the problem with $\lambda_o/10$ triangulation leads to 64,548 unknowns. The relative permittivity of the cube changes from 2.0 to 16.0. Figure 3.33(a) depicts the number of CGS iterations with respect to the contrast of the cube, i.e., $(\epsilon_r - 1)$, to reduce the residual error below 10^{-3}. For small contrasts, CTF has the slowest convergence, while the normal formulations, i.e., CNF and MNMF, offer the fastest convergence. As the contrast increases, however, the convergence of CNF

Table 3.10 Number of CGS iterations (10^{-3} residual error) for the solution of sphere problems with $\varepsilon_r = 4.0$

Sphere problems		CTF			CNF		
Radius	Unknowns	NP	2PBDP	4PBDP	NP	2PBDP	4PBDP
$0.75\lambda_o$	4152	102	344	325	48	47	32
$1.00\lambda_o$	7446	245	623	557	87	52	42
$1.50\lambda_o$	16,728	259	1145	959	184	115	76
$2.00\lambda_o$	29,742	213	1317	1183	272	182	90
$3.00\lambda_o$	65,724	678	NC	NC	305	278	128
$6.00\lambda_o$	264,006	417	NC	NC	NC	1583	755
$7.50\lambda_o$	412,998	369	NC	NC	1153	1005	468
Sphere problems		MNMF			JMCFIE		
Radius	Unknowns	NP	2PBDP	4PBDP	NP	2PBDP	4PBDP
$0.75\lambda_o$	4152	34	26	19	46	34	28
$1.00\lambda_o$	7446	46	30	30	94	48	37
$1.50\lambda_o$	16,728	66	51	32	129	77	63
$2.00\lambda_o$	29,742	66	54	46	122	86	54
$3.00\lambda_o$	65,724	144	111	108	173	189	113
$6.00\lambda_o$	264,006	332	234	174	414	276	148
$7.50\lambda_o$	412,998	380	303	321	269	230	123

NP: no preconditioner, NC: no convergence.

decelerates significantly and this formulation has the poorest convergence when $\varepsilon_r = 16.0$. Figure 3.33(a) also shows that MNMF provides the most efficient solutions and the number of iterations for this formulation is almost constant when the contrast increases from 3.0 to 15.0. Figure 3.33(b) presents the number of iterations when 4PBDP is used to accelerate the convergence. Similar to the sphere problems, CTF solutions are decelerated with 4PBDP and the convergence cannot be achieved within 2000 iterations for contrasts from 3.0 to 11.0. Comparing Figure 3.33(b) with Figure 3.33(a), one can also observe that 4PBDP significantly reduces the number of iterations for CNF and JMCFIE, but it provides less improvement for MNMF. Using 4PBDP, MNMF still has the lowest iteration counts, but the efficiency of JMCFIE becomes close to the efficiency of MNMF.

Table 3.10 and Figure 3.33 show that MNMF and JMCFIE are the most appropriate formulations in terms of efficiency. These formulations are further investigated by considering both various preconditioning schemes and different iterative algorithms to improve the efficiency of solutions. As an example, Figure 3.34 presents the number of iterations for the sphere problems when $\varepsilon_r = 2.0$. In addition to CGS with no preconditioning and with 4PBDP, solutions obtained by employing the BiCGStab algorithm are considered. BiCGStab is known to provide rapid convergence for normal and combined formulations. Figure 3.34 shows that the number of iterations is reduced for both MNMF and JMCFIE, if BiCGStab is employed, instead of CGS.

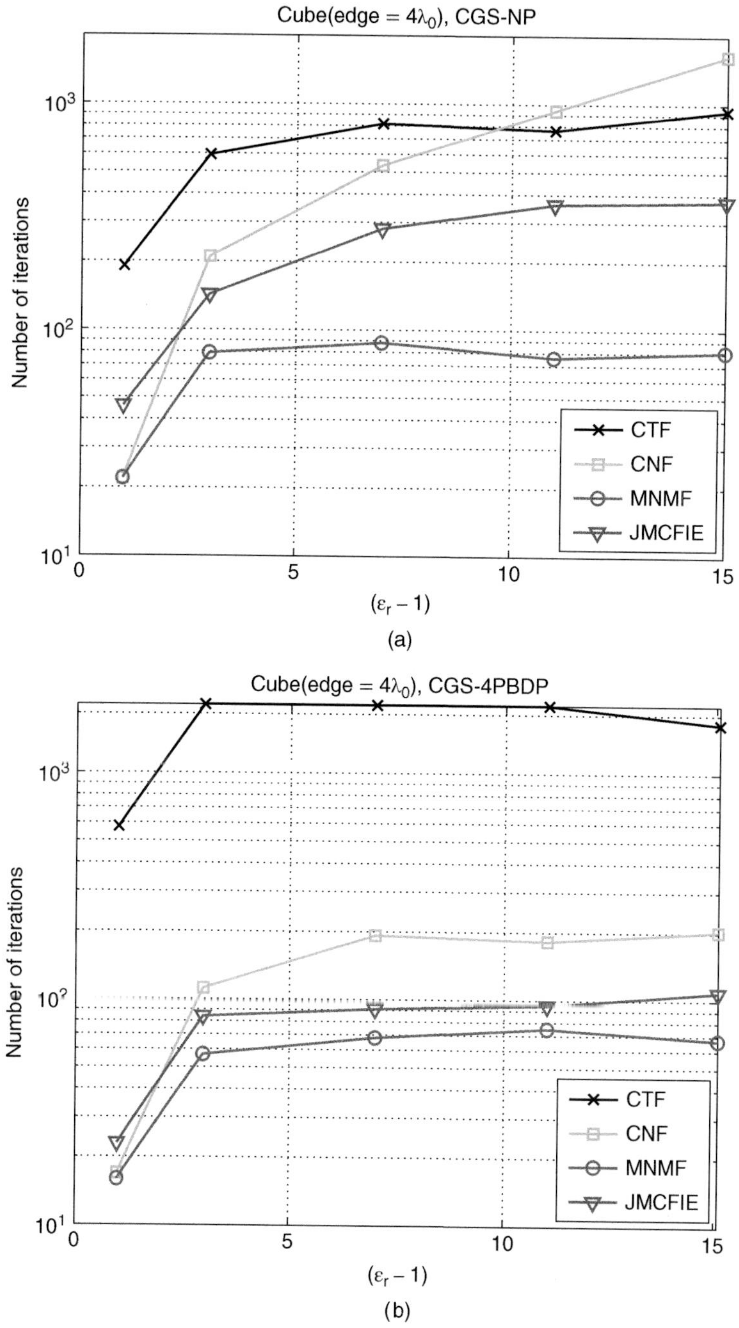

Figure 3.33 Iteration counts for the solution of scattering problems involving a cube with edges of $4\lambda_o$ located in free space. The relative permittivity of the cube changes from 2.0 to 16.0. Iterative solutions are performed by CGS (a) without preconditioning and (b) accelerated with 4PBDP. *Source*: Ergül and Gürel 2009 [80]. Reproduced with permission of IEEE.

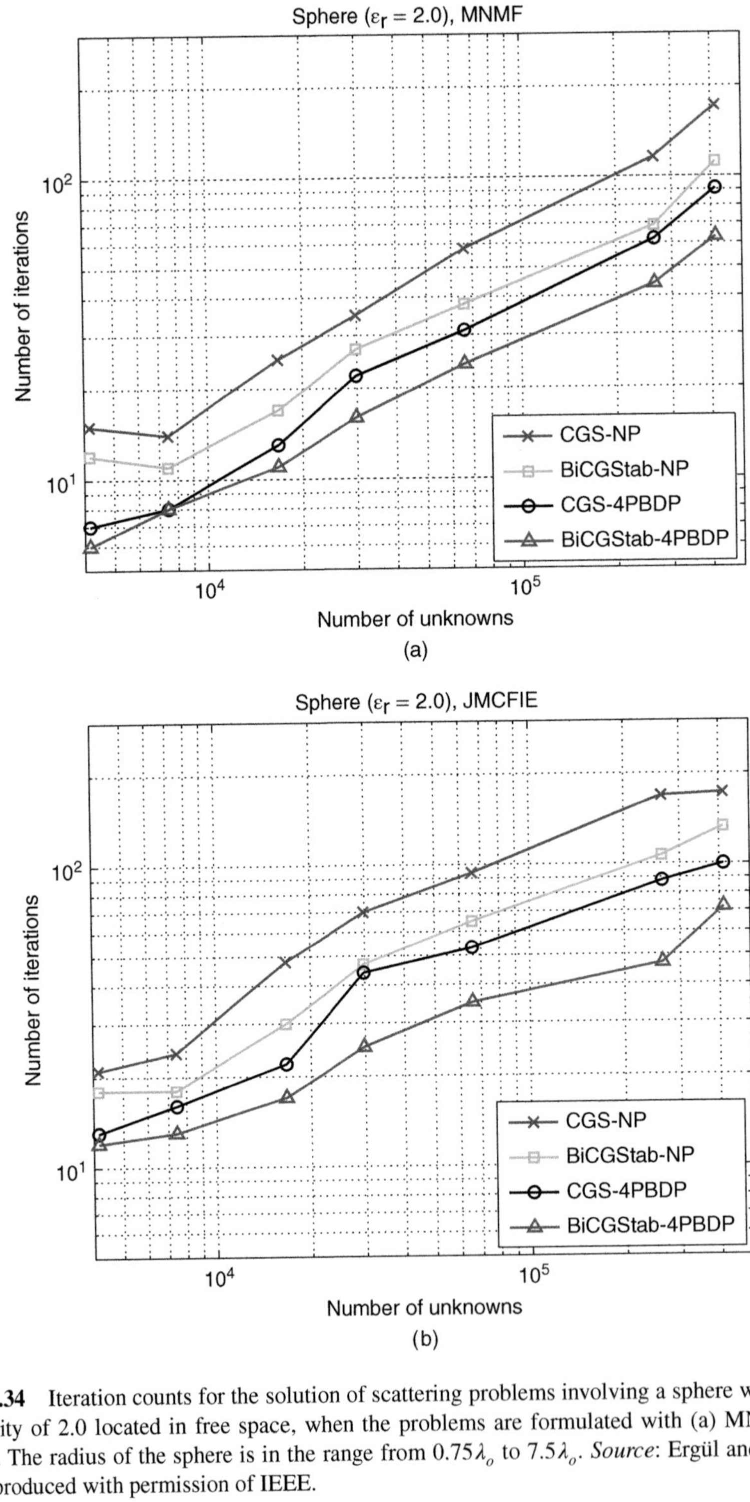

Figure 3.34 Iteration counts for the solution of scattering problems involving a sphere with a relative permittivity of 2.0 located in free space, when the problems are formulated with (a) MNMF and (b) JMCFIE. The radius of the sphere is in the range from $0.75\lambda_o$ to $7.5\lambda_o$. *Source*: Ergül and Gürel 2009 [80]. Reproduced with permission of IEEE.

MNMF and JMCFIE are also compared for the solution of very large dielectric problems. Figure 3.35(a) presents the number of iterations for the sphere problems (with $\varepsilon_r = 2.0$), when the solutions are performed by BiCGStab accelerated with 4PBDP. This time, the frequency is extended to 20 GHz and the radius of the sphere grows up to $20\lambda_o$. At 20 GHz, discretization of the sphere with $\lambda_o/10$ triangulation leads to 2,925,708 unknowns. Figure 3.35(a) shows that solutions with JMCFIE become significantly faster than MNMF for large problems. Figure 3.35(b) presents the results of a similar experiment, where scattering problems involving a dielectric cube ($\varepsilon_r = 4.0$) are solved by using BiCGStab accelerated with 4PBDP. The size of the cube changes from λ_o to $20\lambda_o$, where λ_o is the wavelength in free space. The number of unknowns due to $\lambda_o/10$ triangulations is in the range from 4104 to 1,624,320. Iteration counts required for both 10^{-2} and 10^{-3} residual errors are plotted with respect to the number of unknowns. Similar to the previous case, solutions with JMCFIE become faster than MNMF for large problems.

In general, JMCFIE leads to more efficient solutions than MNMF, when the problem size is sufficiently large. Further investigations show that the better performance of JMCFIE may become more evident when a problem involves complicated targets. As an example, Figure 3.36 presents the results of a problem involving a 5-layer periodic structure in free space excited by a Hertzian dipole. Dimensions of the structure and the position of the source are detailed in Figure 3.36(a). Discretization of the structure with 10 cm triangulation size leads to 38,700 unknowns. For the relative permittivity of the structure, two different values, i.e., $\varepsilon_r = 2.0$ and $\varepsilon_r = 4.0$, are considered. Figure 3.36(b) depicts the iteration counts for 10^{-3} residual error as a function of the frequency from 200 MHz to 300 MHz, when problems are solved by using BiCGStab accelerated with 4PBDP. It can be observed that MNMF offers faster solutions when $\varepsilon_r = 2.0$, and the number of iterations is halved compared to JMCFIE. When $\varepsilon_r = 4.0$, however, the number of iterations for MNMF increases rapidly as the frequency changes from 200 MHz to 300 MHz. At 300 MHz, convergence cannot be achieved within 1000 iterations by using MNMF. On the other hand, JMCFIE is more stable in the same frequency range, and it provides significantly faster solutions when $\varepsilon_r = 4.0$ and the frequency is larger than 200 MHz.

3.6.4 *Iterative Solutions of Composite Objects with Multiple Dielectric and Metallic Regions*

For a general problem involving U regions, MLFMA must be applied for each nonmetallic region separately [154]. Consider the formulation of a composite problem with JMCFIE, as described in Chapter 2, Section 2.4. For each region D_u, one can perform four MVMs with the four partitions of the system matrix, i.e.,

$$y_1 = \overline{Z}_{11,u}^{\text{JMCFIE}} \cdot \boldsymbol{x}_J + \overline{Z}_{12,u}^{\text{JMCFIE}} \cdot \boldsymbol{x}_M \tag{3.156}$$

$$y_2 = \overline{Z}_{21,u}^{\text{JMCFIE}} \cdot \boldsymbol{x}_J + \overline{Z}_{22,u}^{\text{JMCFIE}} \cdot \boldsymbol{x}_M, \tag{3.157}$$

where the coefficients $\boldsymbol{x}_J$ and $\boldsymbol{x}_M$ are again provided by the iterative algorithm. As in the previous cases, a set of aggregation, translation, and disaggregation stages is performed once for a multiplication with a partition, although each partition involves a combination of the integro-differential operators. At the beginning of an aggregation stage, radiation patterns of

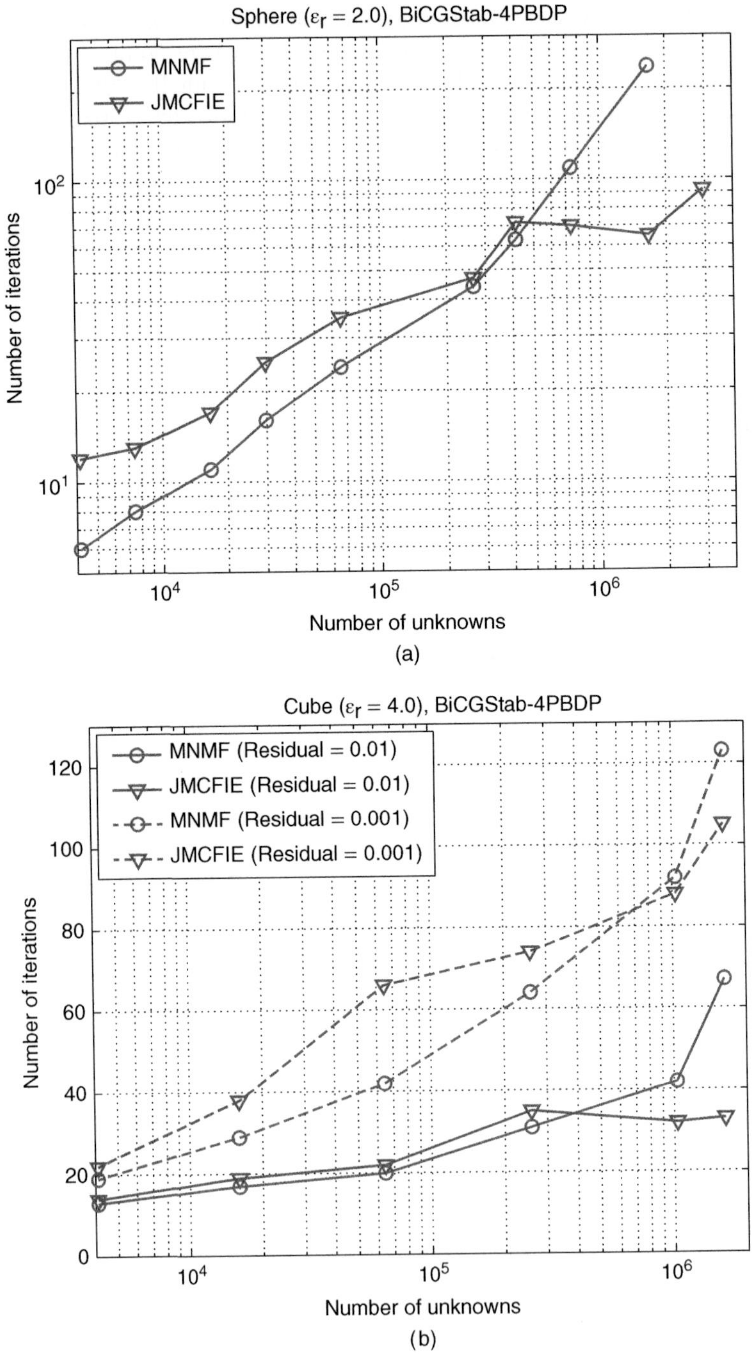

Figure 3.35 Iteration counts for the solution of scattering problems involving (a) a sphere with a relative permittivity of 2.0 and (b) a cube with a relative permittivity of 4.0 located in free space. The radius of the sphere is in the range from $0.75\lambda_o$ to $20\lambda_o$ and the edge length of the cube is in the range from λ_o to $20\lambda_o$. Iterative solutions are performed by employing BiCGStab accelerated with 4PBDP. *Source*: Ergül and Gürel 2009 [80]. Reproduced with permission of IEEE.

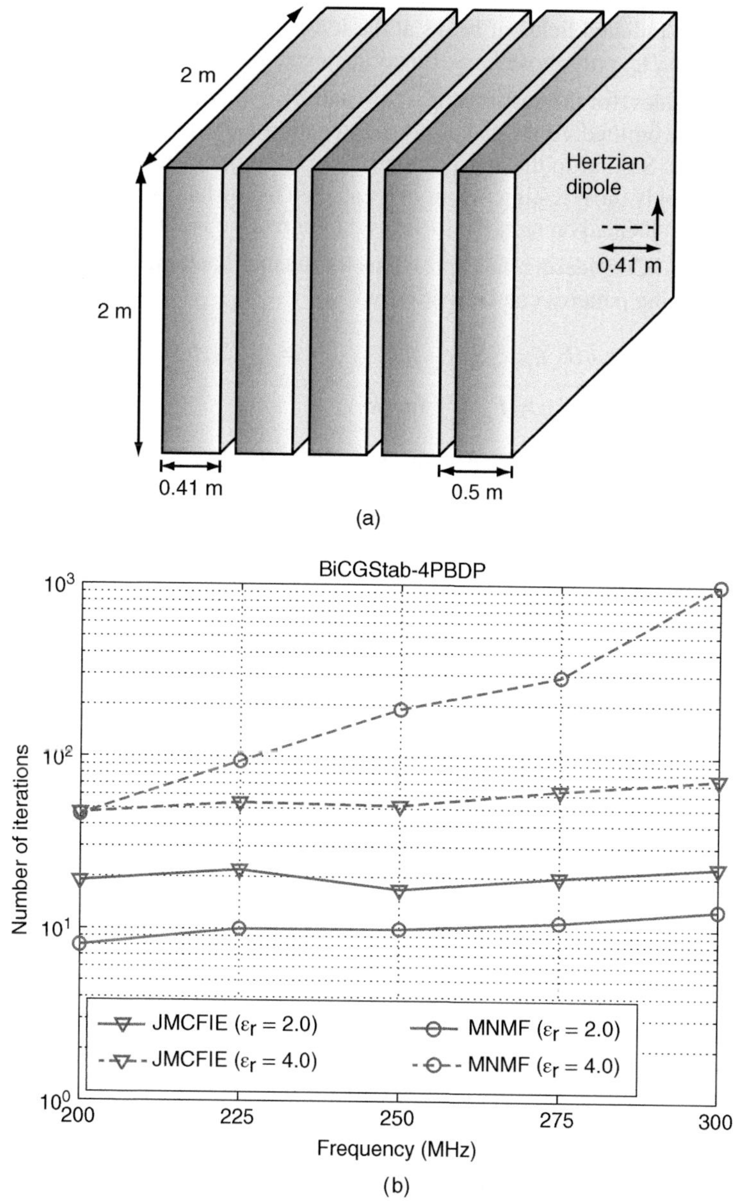

Figure 3.36 (a) A 5-layer periodic dielectric structure illuminated by a Hertzian dipole in free space. (b) Iteration counts (using BiCGStab accelerated with 4PBDP) for the solution of the problem when the frequency changes from 200 MHz to 300 MHz, and the relative permittivity of the structure is 2.0 and 4.0. *Source*: Ergül and Gürel 2009 [80]. Reproduced with permission of IEEE.

basis functions are multiplied with the coefficients provided by the iterative algorithm and combined to obtain radiated fields of boxes at the lowest level. In the aggregation stage performed for a region D_u, only those basis functions located on the surface of the region (S_u) are considered. Besides, for the partitions $\overline{\boldsymbol{Z}}^{\text{JMCFIE}}_{12,u}$ and $\overline{\boldsymbol{Z}}^{\text{JMCFIE}}_{22,u}$, the basis functions located on metallic surfaces are omitted. At the end of a disaggregation stage, incoming fields are received by testing functions. Similar to the aggregation stage, a disaggregation stage performed for a region D_u involves only those testing functions located on the surface of the region. In addition, the testing functions located on metallic surfaces do not receive incoming fields for partitions $\overline{\boldsymbol{Z}}^{\text{JMCFIE}}_{21,u}$ and $\overline{\boldsymbol{Z}}^{\text{JMCFIE}}_{22,u}$. Considering the expressions for the matrix elements in (2.331)–(2.333), radiation and receiving patterns can be written as

$$\boldsymbol{S}^{\text{JMCFIE}}_{ab,n,u}(\boldsymbol{r}_{C'},\boldsymbol{k}_u) = \xi_n \boldsymbol{S}_n(\boldsymbol{r}_{C'},\boldsymbol{k}_u) \qquad (a=\{1,2\}, b=\{1,2\}) \tag{3.158}$$

$$\begin{aligned}\boldsymbol{F}^{\text{JMCFIE}}_{11,m,u}(\boldsymbol{r}_C,\boldsymbol{k}_u) &= \boldsymbol{F}^{\text{JMCFIE}}_{22,m,u}(\boldsymbol{r}_C,\boldsymbol{k}_u)\\ &= \xi_m \boldsymbol{F}^{\mathcal{T},T}_m(\boldsymbol{r}_C,\boldsymbol{k}_o) + \boldsymbol{F}^{\mathcal{K},N}_m(\boldsymbol{r}_C,\boldsymbol{k}_o)\end{aligned} \tag{3.159}$$

$$\boldsymbol{F}^{\text{JMCFIE}}_{12,m,u}(\boldsymbol{r}_C,\boldsymbol{k}_u) = \eta_u^{-1}\boldsymbol{F}^{\mathcal{T},N}_m(\boldsymbol{r}_C,\boldsymbol{k}_u) - \xi_m \eta_u^{-1}\boldsymbol{F}^{\mathcal{K},T}_m(\boldsymbol{r}_C,\boldsymbol{k}_u) \tag{3.160}$$

$$\boldsymbol{F}^{\text{JMCFIE}}_{21,m,u}(\boldsymbol{r}_C,\boldsymbol{k}_u) = \xi_m \eta_u \boldsymbol{F}^{\mathcal{K},T}_m(\boldsymbol{r}_C,\boldsymbol{k}_u) - \eta_u \boldsymbol{F}^{\mathcal{T},N}_m(\boldsymbol{r}_C,\boldsymbol{k}_u), \tag{3.161}$$

where $\boldsymbol{F}^{\mathcal{K},T}_m$, $\boldsymbol{F}^{\mathcal{T},T}_m$, $\boldsymbol{F}^{\mathcal{K},N}_m$, and $\boldsymbol{F}^{\mathcal{T},N}_m$ are defined in (3.65)–(3.69).

Numerical Results

In the following examples, iterative solutions of various scattering problems involving composite structures are investigated when the solutions are accelerated with 2PBDP and 4PBDP, in addition to the NP case. For those problems involving metallic surfaces, the off-diagonal partitions of 4PBDP are rectangular matrices. In addition, the blocks in the off-diagonal partitions are not necessarily square, and some of them can be rectangular, depending on the object and the recursive clustering scheme in MLFMA. In all examples, the scatterers are located in free space and illuminated by a plane wave propagating in the z direction with the electric field polarized in the x direction. Surfaces are discretized with the RWG functions on $\lambda_o/10$ triangulations, where λ_o is the wavelength outside the objects (free space), and iterative solutions are performed using the BiCGStab algorithm. MVMs are accelerated via MLFMA, and the relative residual error for the iterative convergence is set to 10^{-3}.

Figure 3.37 depicts the number of iterations for the solution of scattering problems involving a spherical composite object with multiple dielectric regions. A dielectric sphere of radius a is coated with a dielectric shell of radius $2a$, where a changes from $0.5\lambda_o$ to $2.5\lambda_o$. Discretizations lead to 13,176 and 316,032 unknowns, respectively, when $a = 0.5\lambda_o$ and $a = 2.5\lambda_o$. Figure 3.37(a) presents iteration counts with respect to the number of unknowns when the relative permittivities of the core and shell are 4.0 and 2.0, respectively. In this case, 2PBDP reduces the number of iterations substantially in comparison to the NP case, while 4PBDP does not provide a significant improvement over 2PBDP. On the other hand, when the permittivities of the shell and core are exchanged, one can obtain the number of iterations depicted in Figure 3.37(b), where 4PBDP presents a superior performance in comparison to 2PBDP. Due to the relatively high contrast between the shell and free space, solutions of JMCFIE

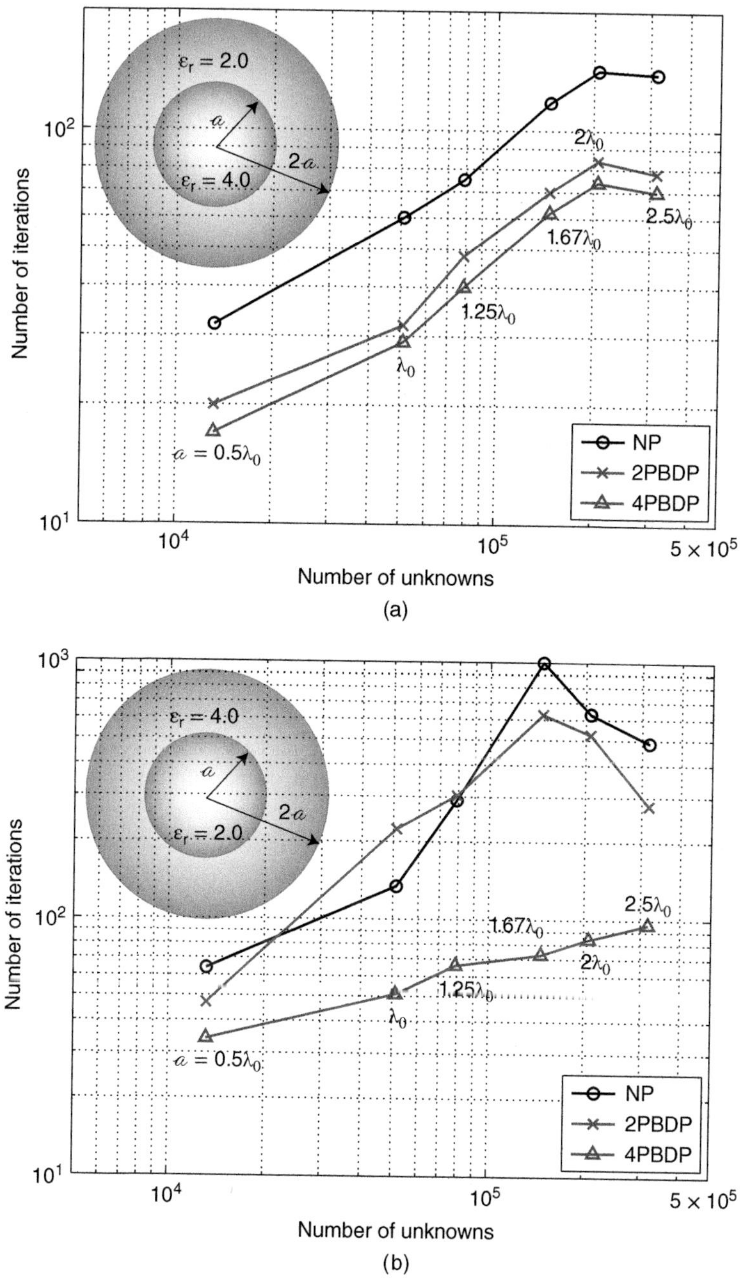

Figure 3.37 Iteration counts for the solution of scattering problems involving a dielectric sphere of radius a coated with a dielectric shell of radius $2a$, where a changes from $0.5\lambda_o$ to $2.5\lambda_o$. (a) Low-contrast case when the relative permittivities of the core and shell are 4.0 and 2.0, respectively. (b) High-contrast case when the relative permittivities of the core and shell are 2.0 and 4.0, respectively. *Source*: Ergül and Gürel 2009 [80]. Reproduced with permission of IEEE.

become difficult without preconditioning. For example, when $a = 1.67\lambda_o$, convergence cannot be achieved in 1000 iterations. 2PBDP accelerates the convergence for large problems, but the improvement is not sufficient. Using 4PBDP, the number of iterations is less than 100 for all solutions in Figure 3.37.

Next, iterative solutions of scattering problems involving a spherical composite object with dielectric and metallic parts are considered. A metallic sphere of radius a is coated with a dielectric shell of radius $2a$, where a changes from $0.5\lambda_o$ to $2.5\lambda_o$. Figure 3.38 presents the number of iterations with respect to the number of unknowns. Similar to the previous example, 2PBDP reduces the number of iterations significantly for the low-contrast case, i.e., when the relative permittivity of the shell is 2.0, as depicted in Figure 3.38(a). In this case, 4PBDP provides some improvement over 2PBDP, as the problem size grows. When the relative permittivity of the shell is 4.0, however, 4PBDP accelerates the iterative solutions significantly, compared to 2PBDP. In fact, 2PBDP decelerates solutions for large problems, and there is a large discrepancy between the performances of 2PBDP and 4PBDP. Using 4PBDP, the number of iterations is again less than 100 for all solutions in Figure 3.38.

In addition to the spherical problems, electromagnetics problems involving composite objects with sharp edges and corners are considered. Figure 3.39 presents iteration counts for the solution of scattering problems involving a coated dielectric cube. The core and shell have edges of a and $2a$, respectively, where a changes from $0.5\lambda_o$ to $2.5\lambda_o$. Faces of the object are parallel to the coordinate axes. Discretizations lead to matrix equations involving 9864 to 228,132 unknowns. Figure 3.39(a) depicts the number of iterations with respect to the number of unknowns, when the relative permittivities of the core and shell are 4.0 and 2.0, respectively. Results are similar to those for the spherical object depicted in Figure 3.37(a), i.e., 2PBDP accelerates the iterative solutions significantly, and 4PBDP further reduces the number of iterations slightly compared to 2PBDP. When the relative permittivities of the core and shell are exchanged, however, 4PBDP performs much better than 2PBDP, as depicted in Figure 3.39(b). On the other hand, unlike the solutions of the spherical object in Figure 3.38(b), 2PBDP is also effective for the high-contrast case in Figure 3.39(b). This is probably due to larger numbers of elements being used in constructing 2PBDP for the cubic object, i.e., boxes at the lowest level of the multilevel tree are more populated for the cubic object than for the spherical object. Nevertheless, 4PBDP is again preferable for all solutions in Figure 3.39.

Figure 3.40 presents the solution of scattering problems involving a coated metallic cube. Sizes of the core and shell are the same as those in the coated dielectric cube. Similar to the previous examples, 4PBDP provides the most efficient results, and it presents improved convergence in comparison to 2PBDP, when the contrast is high.

3.7 MLFMA for Low-Frequency Problems

MLFMA is a powerful method; however, it suffers from low-frequency breakdowns, which limit its applicability to those problems involving small details with respect to wavelength. A major limitation is that the box size in MLFMA cannot be very small compared to wavelength since the plane wave expansion becomes invalid for short distances. Hence, when an ordinary MLFMA is applied to low-frequency problems discretized with very small elements, the lowest-level boxes may involve many discretization elements. As a result, the complexity of

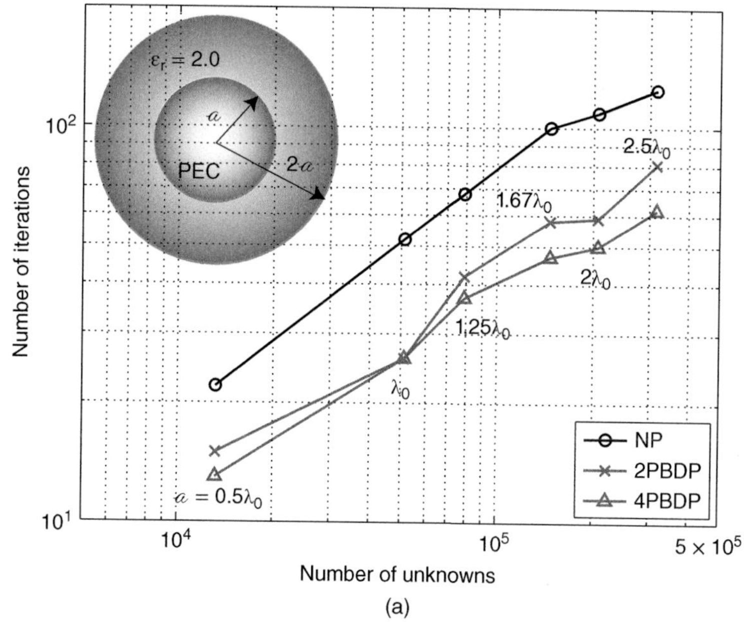

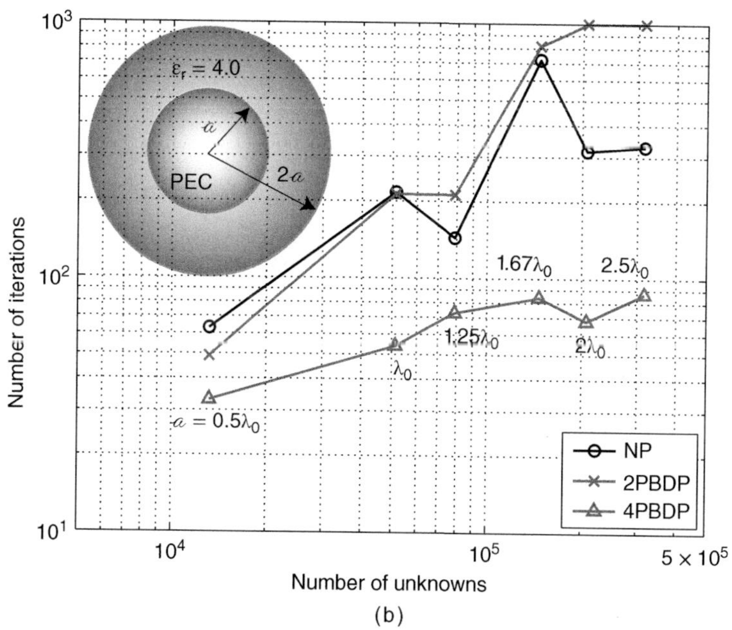

Figure 3.38 Iteration counts for the solution of scattering problems involving a PEC sphere of radius a coated with a dielectric shell of radius $2a$, where a changes from $0.5\lambda_o$ to $2.5\lambda_o$. The relative permittivity of the shell is (a) 2.0 and (b) 4.0. *Source*: Ergül and Gürel 2009 [80]. Reproduced with permission of IEEE.

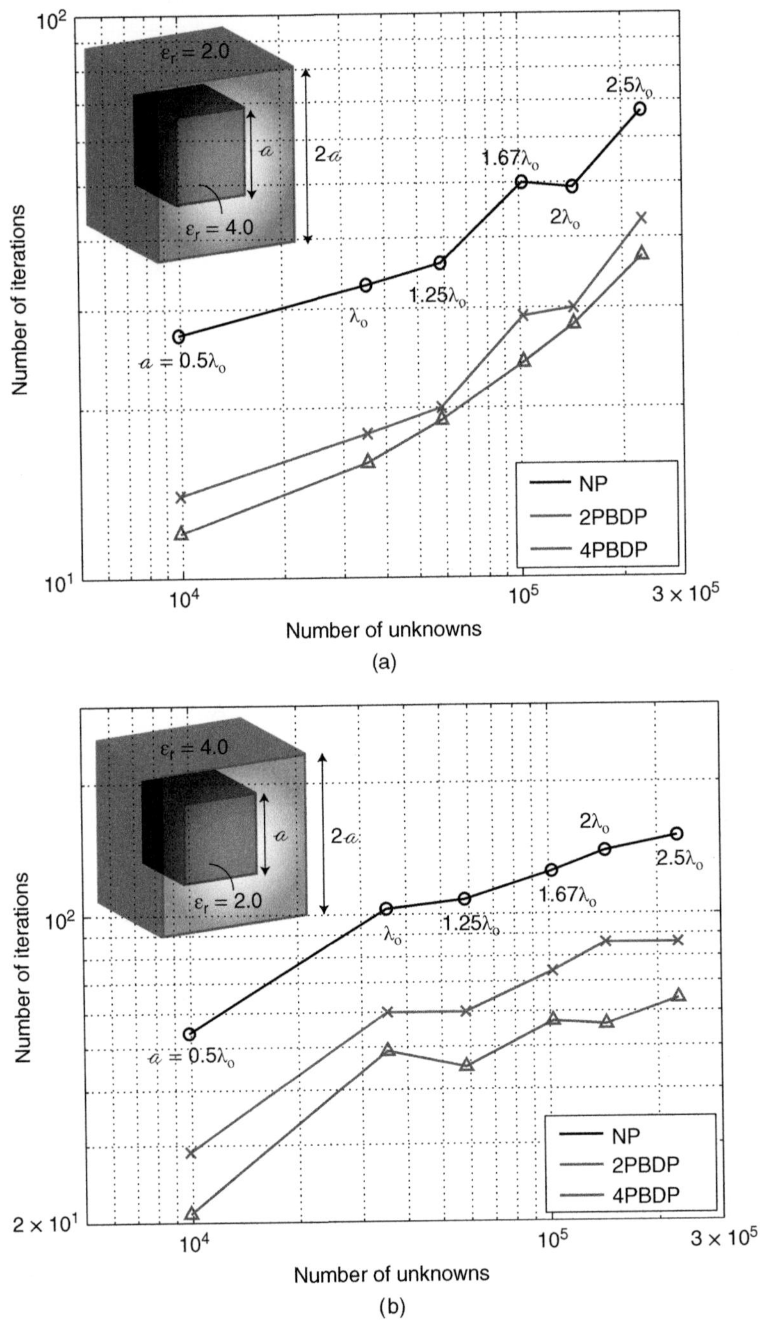

Figure 3.39 Iteration counts for the solution of scattering problems involving a dielectric cube coated with a dielectric shell. The core and shell have edges of a and $2a$, respectively, where a changes from $0.5\lambda_o$ to $2.5\lambda_o$. (a) Low-contrast case when the relative permittivities of the core and shell are 4.0 and 2.0, respectively. (b) High-contrast case when the relative permittivities of the core and shell are 2.0 and 4.0, respectively. *Source*: Ergül and Gürel 2009 [80]. Reproduced with permission of IEEE.

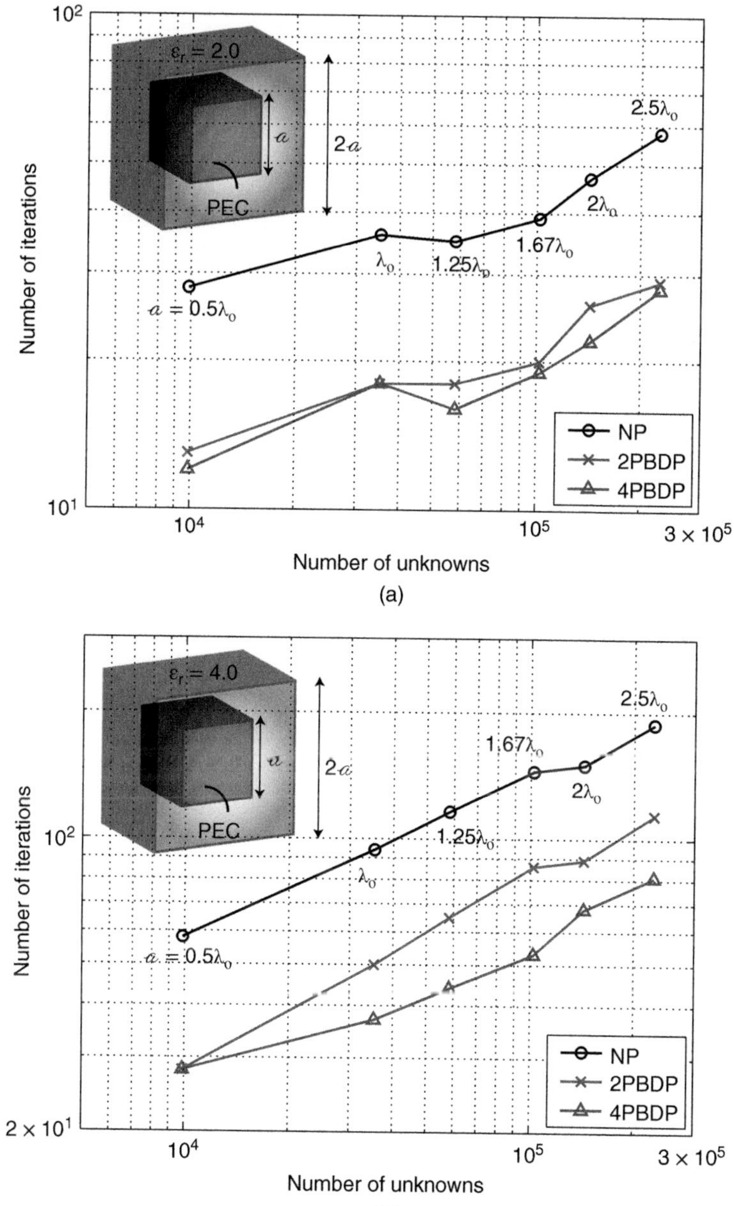

Figure 3.40 Iteration counts for the solution of scattering problems involving a PEC cube coated with a dielectric shell. The core and shell have edges of a and $2a$, respectively, where a changes from $0.5\lambda_o$ to $2.5\lambda_o$. The relative permittivity of the shell is (a) 2.0 and (b) 4.0. *Source*: Ergül and Gürel 2009 [80]. Reproduced with permission of IEEE.

the MLFMA implementation can be more than $\mathcal{O}(N \log N)$ due to excessively large numbers of near-field interactions that must be calculated directly.

For low-frequency problems, an efficient implementation of LF-MLFMA based on multipoles [178]–183 is presented in this section. Fields are expanded in terms of multipoles, but they are not converted into plane waves to avoid low-frequency breakdown. This way, the box size is not restricted, and objects can be partitioned into boxes that can be much smaller than the wavelength. Without the plane wave expansion, translations are not diagonalized and they involve dense MVMs, which may decrease the efficiency. On the other hand, number of multipoles does not depend on the box size for small boxes, and the resulting LF-MLFMA implementation has $\mathcal{O}(N)$ complexity, provided that the object is not very large. Since the multipole expansion becomes inefficient for very large boxes, it is required to combine LF-MLFMA and the ordinary MLFMA to obtain a broadband implementation for dense discretizations of very large objects.

3.7.1 Factorization of the Matrix Elements

Low-frequency problems can be solved efficiently by using the factorization of the Green's function without diagonalization. Consider the solution of electromagnetics problems involving metallic objects discretized with the RWG functions. Matrix elements related to far-zone interactions can be written as

$$\begin{aligned}\overline{\boldsymbol{Z}}_{FF}^{\text{EFIE}}[m,n] &= ik\int_{S_m} d\boldsymbol{r}\boldsymbol{t}_m(\boldsymbol{r})\cdot\int_{S_n} d\boldsymbol{r}' g(\boldsymbol{r},\boldsymbol{r}')\boldsymbol{b}_n(\boldsymbol{r}')\\ &\quad -\frac{i}{k}\int_{S_m} d\boldsymbol{r}\nabla\cdot\boldsymbol{t}_m(\boldsymbol{r})\int_{S_n} d\boldsymbol{r}' g(\boldsymbol{r},\boldsymbol{r}')\nabla'\cdot\boldsymbol{b}_n(\boldsymbol{r}') \qquad (3.162)\\ &= \overline{\boldsymbol{Z}}_{FF,V}^{\text{EFIE}}[m,n] + \overline{\boldsymbol{Z}}_{FF,S}^{\text{EFIE}}[m,n] \qquad (3.163)\end{aligned}$$

and

$$\overline{\boldsymbol{Z}}_{FF}^{\text{MFIE}}[m,n] = \int_{S_m} d\boldsymbol{r}(\boldsymbol{t}_m(\boldsymbol{r})\times\hat{\boldsymbol{n}})\cdot\int_{PV,S_n} d\boldsymbol{r}'\boldsymbol{b}_n(\boldsymbol{r}')\times\nabla' g(\boldsymbol{r},\boldsymbol{r}') \qquad (3.164)$$

for EFIE and MFIE, respectively. Using (3.34), the Green's function can be factorized as

$$\begin{aligned}g(\boldsymbol{r},\boldsymbol{r}') = \frac{ik}{4\pi}\sum_{t_1=0}^{\infty}\sum_{u_1=-t_1}^{t_1}\sum_{t_2=0}^{\infty}\sum_{u_2=-t_2}^{t_2} &\beta_{0,0,t_1,u_1}(k,\boldsymbol{r}-\boldsymbol{r}_C)\\ &\alpha_{t_1,u_1,t_2,u_2}(k,\boldsymbol{r}_{CC'})\beta_{t_2,u_2,0,0}(k,\boldsymbol{r}_{C'}-\boldsymbol{r}'), \qquad (3.165)\end{aligned}$$

where $|\boldsymbol{r}_{CC'}| = |\boldsymbol{r}_C - \boldsymbol{r}_{C'}| > |\boldsymbol{r}-\boldsymbol{r}'-\boldsymbol{r}_C+\boldsymbol{r}_{C'}|$. Using (3.165) in (3.162)–(3.164), one can obtain

$$\begin{aligned}\overline{\boldsymbol{Z}}_{FF,V}^{\text{EFIE}}[m,n] = \left(\frac{ik}{4\pi}\right)^2\sum_{t_1=0}^{\infty}\sum_{u_1=-t_1}^{t_1}\sum_{t_2=0}^{\infty}\sum_{u_2=-t_2}^{t_2} &\boldsymbol{F}_{m,t_1,u_1}^{\text{EFIE}}(\boldsymbol{r}_C,k)\\ &\cdot\alpha_{t_1,u_1,t_2,u_2}(k,\boldsymbol{r}_{CC'})\boldsymbol{S}_{n,t_2,u_2}^{\text{EFIE}}(\boldsymbol{r}_{C'},k) \qquad (3.166)\end{aligned}$$

$$\overline{Z}^{\mathrm{EFIE}}_{FF,S}[m,n] = \left(\frac{ik}{4\pi}\right)^2 \sum_{t_1=0}^{\infty} \sum_{u_1=-t_1}^{t_1} \sum_{t_2=0}^{\infty} \sum_{u_2=-t_2}^{t_2} \boldsymbol{F}^{\mathrm{EFIE}}_{m,t_1,u_1}(\boldsymbol{r}_C, k)$$

$$\alpha_{t_1,u_1,t_2,u_2}(k, \boldsymbol{r}_{CC'}) \boldsymbol{S}^{\mathrm{EFIE}}_{n,t_2,u_2}(\boldsymbol{r}_{C'}, k) \tag{3.167}$$

$$\overline{Z}^{\mathrm{MFIE}}_{FF}[m,n] = \left(\frac{ik}{4\pi}\right)^2 \sum_{t_1=0}^{\infty} \sum_{u_1=-t_1}^{t_1} \sum_{t_2=0}^{\infty} \sum_{u_2=-t_2}^{t_2} \boldsymbol{F}^{\mathrm{MFIE}}_{m,t_1,u_1}(\boldsymbol{r}_C, k)$$

$$\cdot\, \alpha_{t_1,u_1,t_2,u_2}(k, \boldsymbol{r}_{CC'}) \boldsymbol{S}^{\mathrm{MFIE}}_{n,t_2,u_2}(\boldsymbol{r}_{C'}, k), \tag{3.168}$$

where

$$\boldsymbol{S}^{\mathrm{EFIE}}_{n,t,u}(\boldsymbol{r}_{C'}, k) = \boldsymbol{S}^{\mathrm{MFIE}}_{n,t,u}(\boldsymbol{r}_{C'}, k) = \sqrt{4\pi} \int_{S_n} d\boldsymbol{r}' \boldsymbol{b}_n(\boldsymbol{r}') \beta_{t,u,0,0}(k, \boldsymbol{r}_{C'} - \boldsymbol{r}') \tag{3.169}$$

$$S^{\mathrm{EFIE}}_{n,t,u}(\boldsymbol{r}_{C'}, k) = \frac{\sqrt{4\pi}}{ik} \int_{S_n} d\boldsymbol{r}' \nabla' \cdot \boldsymbol{b}_n(\boldsymbol{r}') \beta_{t,u,0,0}(k, \boldsymbol{r}_{C'} - \boldsymbol{r}') \tag{3.170}$$

are radiation patterns of the nth basis function and

$$\boldsymbol{F}^{\mathrm{EFIE}}_{m,t,u}(\boldsymbol{r}_C, k) = \sqrt{4\pi} \int_{S_m} d\boldsymbol{r} \boldsymbol{t}_m(\boldsymbol{r}) \beta_{0,0,t,u}(k, \boldsymbol{r} - \boldsymbol{r}_C) \tag{3.171}$$

$$F^{\mathrm{EFIE}}_{m,t,u}(\boldsymbol{r}_C, k) = \frac{\sqrt{4\pi}}{ik} \int_{S_m} d\boldsymbol{r} \nabla \cdot \boldsymbol{t}_m(\boldsymbol{r}) \beta_{0,0,t,u}(k, \boldsymbol{r} - \boldsymbol{r}_C) \tag{3.172}$$

$$\boldsymbol{F}^{\mathrm{MFIE}}_{m,t,u}(\boldsymbol{r}_C, k) = \frac{\sqrt{4\pi}}{ik} \int_{S_m} d\boldsymbol{r} (\boldsymbol{t}_m(\boldsymbol{r}) \times \hat{\boldsymbol{n}}) \times \nabla \beta_{0,0,t,u}(k, \boldsymbol{r} - \boldsymbol{r}_C) \tag{3.173}$$

are receiving patterns of the mth testing function. Note that, for a lossless medium and using a Galerkin scheme,

$$\boldsymbol{F}^{\mathrm{EFIE}}_{m,t,u}(\boldsymbol{r}_C, k) = \{\boldsymbol{S}^{\mathrm{EFIE}}_{m,t,u}(\boldsymbol{r}_C, \boldsymbol{k})\}^* \tag{3.174}$$

$$F^{\mathrm{EFIE}}_{m,t,u}(\boldsymbol{r}_C, k) = -\{S^{\mathrm{EFIE}}_{m,t,u}(\boldsymbol{r}_C, \boldsymbol{k})\}^* \tag{3.175}$$

since

$$\beta_{t',u',t,u}(k, \boldsymbol{r}) = \beta^*_{t,u,t',u'}(k, -\boldsymbol{r}) \tag{3.176}$$

for a real k.

In (3.166)–(3.168), matrix elements are calculated by a direct translation between the box centers $\boldsymbol{r}_{C'}$ and $\boldsymbol{r}_C$, where radiation and receiving patterns are evaluated. However, as in the ordinary MLFMA, multilevel interactions can be designed by performing shifts before and after translations. For example, let a multipole-to-multipole translation be performed between $\boldsymbol{r}_D$ and $\boldsymbol{r}_{D'}$, which are close to the receiving and radiation centers, i.e., $\boldsymbol{r}_C$ and $\boldsymbol{r}_{C'}$, respectively. Then, the translation function in (3.166)–(3.168) is further factorized as

$$\alpha_{t_1,u_1,t_2,u_2}(k, \boldsymbol{r}_{CC'}) = \sum_{t_3=0}^{\infty} \sum_{u_3=-t_3}^{t_3} \sum_{t_4=0}^{\infty} \sum_{u_4=-t_4}^{t_4} \beta_{t_1,u_1,t_3,u_3}(k, \boldsymbol{r}_{CD})$$

$$\alpha_{t_3,u_3,t_4,u_4}(k, \boldsymbol{r}_{DD'}) \beta_{t_4,u_4,t_2,u_2}(k, \boldsymbol{r}_{D'C'}), \tag{3.177}$$

where $\boldsymbol{r}_{CD} = \boldsymbol{r}_C - \boldsymbol{r}_D$ and $\boldsymbol{r}_{D'C'} = \boldsymbol{r}_{D'} - \boldsymbol{r}_{C'}$.

In practice, the summations in (3.166)–(3.168) and (3.177), as well as those in the translation and shift functions, must be truncated at $t = \tau$. A worst-case analysis can be performed in order to determine truncation numbers required for a desired level of accuracy. Consider two identical boxes with edges of $a_l = 2^{l-2}\lambda$, where l changes from -7 to 4, i.e., $\lambda/512 \leq a_l \leq 4\lambda$. The 'source' box is centered at $(x, y, z) = (0, 0, 0)$ and the 'observation' box is centered at $(x, y, z) = (2a_l, 0, 0)$, according to a one-box-buffer scheme. Then, a total of 64 different cases are considered, when the source and observation points are located at box corners, i.e., $\boldsymbol{r}' = (\pm a_l/2, \pm a_l/2, \pm a_l/2)$ and $\boldsymbol{r} = (2a_l \pm a_l/2, \pm a_l/2, \pm a_l/2)$. The Green's function is calculated by using the factorization in (3.22) with multipole-to-multipole translations between box centers. The relative error of the factorization compared to the exact evaluation of the Green's function is obtained for different values of the truncation number τ. Considering 64 different cases, the maximum error is selected to determine the number of accurate digits for a given truncation number. As a result, the truncation numbers required for various box sizes and different numbers of accurate digits (d_0) from 1 to 5 are obtained, as listed in Table 3.11. One can observe that, for a desired level of accuracy, the truncation number is almost constant when the box size is small. For very large boxes, however, it grows rapidly as the box size increases; this limits the applicability of multipole-to-multipole interactions for large boxes.

The truncation numbers listed in Table 3.11 are also tested for two-level interactions. Once again, two $a_l \times a_l \times a_l$ boxes centered at $(x, y, z) = (0, 0, 0)$ and $(x, y, z) = (2a_l, 0, 0)$ are considered. This time, each box contains a smaller box with edges of $a_l/2$. Source and observation points are located at the corners of the smaller boxes. A total of $8 \times 8 \times 8 \times 8 = 4096$ cases are tested considering eight different locations for the source and observation points, as well as eight different placement of the smaller boxes inside the larger boxes. In each case, the homogeneous-space Green's function is calculated by using a factorization involving two shifts on the source side (from the source point to the center of the smaller box and from the center of the smaller box to the center of the larger box), a multipole-to-multipole translation between

Table 3.11 Truncation numbers obtained by a worst-case analysis of the factorization of the Green's function in (3.22)

		Number of accurate digits				
l	Box Size	$d_0 = 1$	$d_0 = 2$	$d_0 = 3$	$d_0 = 4$	$d_0 = 5$
4	4λ	25	28	31	33	36
3	2λ	14	16	19	21	27
2	λ	8	10	12	19	25
1	$\lambda/2$	5	8	12	19	24
0	$\lambda/4$	4	7	11	17	24
−1	$\lambda/8$	3	7	11	17	24
−2	$\lambda/16$	3	7	11	17	24
−3	$\lambda/32$	3	7	11	17	24
−4	$\lambda/64$	3	7	11	17	24
−5	$\lambda/128$	3	7	11	17	24
−6	$\lambda/256$	3	7	11	17	24
−7	$\lambda/512$	3	7	11	17	24

Table 3.12 Maximum relative error in a two-level calculation of the Green's function using the truncation numbers in Table 3.11

Box Size		Maximum relative error		
Large	Small	$d_0 = 2$	$d_0 = 3$	$d_0 = 4$
4λ	2λ	1.1882E-2	6.2268E-4	8.2273E-5
2λ	λ	5.1073E-3	4.3117E-4	8.3946E-5
λ	$\lambda/2$	3.9394E-3	8.3753E-4	1.3247E-4
$\lambda/2$	$\lambda/4$	4.5135E-3	1.0285E-3	1.0108E-4
$\lambda/4$	$\lambda/8$	7.8453E-3	1.0573E-3	9.7895E-5
$\lambda/8$	$\lambda/16$	7.1366E-3	1.0489E-3	9.6000E-5
$\lambda/16$	$\lambda/32$	6.9732E-3	1.0468E-3	9.5533E-5
$\lambda/32$	$\lambda/64$	6.9331E-3	1.0463E-3	9.5417E-5
$\lambda/64$	$\lambda/128$	6.9232E-3	1.0461E-3	9.5388E-5
$\lambda/128$	$\lambda/256$	6.9207E-3	1.0461E-3	9.5380E-5
$\lambda/256$	$\lambda/512$	6.9201E-3	1.0461E-3	9.5379E-5

the centers of the larger boxes, and two shifts on the observation side (from the center of the larger box to the center of the smaller box and from the center of the smaller box to the observation point). Table 3.12 lists the maximum relative error of the factorization compared to the analytical evaluation of the Green's function for different box sizes and desired accuracy. One can observe that, in some cases, the error is larger than desired. This is because the truncation numbers listed in Table 3.11 are obtained for single-level interactions and extra shifts in multilevel interactions may deteriorate the accuracy of the factorization. Nevertheless, the relative errors listed in Table 3.12 are obtained for the worst case scenarios and the target errors are generally satisfied.

3.7.2 Low-Frequency MLFMA

LF-MLFMA can be implemented by using the factorizations of matrix elements in (3.162)–(3.164). The recursive clustering and construction of the tree structure in LF-MLFMA are very similar to those in the ordinary MLFMA. However, as opposed to MLFMA, there is no limitation for the box size. Without considering the electrical size of the object, a tree structure of $L = \mathcal{O}(\log N)$ levels can be constructed by placing the object in a cubic box and recursively dividing the computational domain into subdomains. Then, far-field interactions are calculated in a group-by-group manner via three main stages, i.e., aggregation, translation, and disaggregation. For level $l = 1, 2, \ldots, L$, the number of multipoles to represent fields is $M_l = [\tau(l) + 1]^2$.

In the aggregation stage, radiated fields of boxes are calculated in terms of multipoles from the bottom of the tree structure to the highest level ($l = L$). Aggregations at the lowest level can be written as

$$[s_C^x \; s_C^y \; s_C^z \; s_C^s]_{M_1 \times 4} = \sum_{n \in C} \boldsymbol{x}[n][s_n^x \; s_n^y \; s_n^z \; s_n^s]_{M_1 \times 4}, \qquad (3.178)$$

where $\boldsymbol{x}[n]$ represent coefficients provided by the iterative solver, $\{\boldsymbol{s}_n^x, \boldsymbol{s}_n^y, \boldsymbol{s}_n^z, \boldsymbol{s}_n^s\}$ are arrays of M_1 elements containing the radiation pattern of the nth basis function, and $\{\boldsymbol{s}_C^x, \boldsymbol{s}_C^y, \boldsymbol{s}_C^z, \boldsymbol{s}_C^s\}$ are arrays of M_1 elements containing the radiated field of a box C at the lowest level. Note that $\boldsymbol{s}_n^s$ and $\boldsymbol{s}_C^s$, which are related to the scalar electric potential, are not required for MFIE. For a box C at a higher level ($l > 1$),

$$[\boldsymbol{s}_C^x \; \boldsymbol{s}_C^y \; \boldsymbol{s}_C^z \; \boldsymbol{s}_C^s]_{M_l \times 4} = \sum_{C' \in C} [\overline{\boldsymbol{\beta}}_{C' \to C}]_{M_l \times M_{l-1}} \cdot [\boldsymbol{s}_{C'}^x \; \boldsymbol{s}_{C'}^y \; \boldsymbol{s}_{C'}^z \; \boldsymbol{s}_{C'}^s]_{M_{l-1} \times 4}, \tag{3.179}$$

where $\overline{\boldsymbol{\beta}}_{C' \to C}$ is an $M_l \times M_{l-1}$ dense matrix containing multipole-to-multipole shift functions $\beta_{t',u',t,u}(k, \boldsymbol{r}_C - \boldsymbol{r}_{C'})$ for $t' = 1, 2, \ldots, \tau(l)$ and $t = 1, 2, \ldots, \tau(l-1)$.

In the translation stage, radiated fields of boxes are translated into incoming fields for other boxes. The incoming field at the center of a box C due to boxes in the far-field of C, i.e., $\mathcal{F}\{C\}$, can be obtained as

$$[\boldsymbol{g}_C^x \; \boldsymbol{g}_C^y \; \boldsymbol{g}_C^z \; \boldsymbol{g}_C^s]_{M_l \times 4} = \sum_{C' \in \mathcal{F}\{C\}} [\overline{\boldsymbol{\alpha}}_{C' \to C}]_{M_l \times M_l} \cdot [\boldsymbol{s}_{C'}^x \; \boldsymbol{s}_{C'}^y \; \boldsymbol{s}_{C'}^z \; \boldsymbol{s}_{C'}^s]_{M_l \times 4}, \tag{3.180}$$

where $\{\boldsymbol{g}_C^x, \boldsymbol{g}_C^y, \boldsymbol{g}_C^z, \boldsymbol{g}_C^s\}$ are arrays of M_l elements containing the incoming field at the center of a box C and $\overline{\boldsymbol{\alpha}}_{C' \to C}$ is an $M_l \times M_l$ dense matrix containing multipole-to-multipole translation functions $\alpha_{t',u',t,u}(k, \boldsymbol{r}_C - \boldsymbol{r}_{C'})$ for $t' = 1, 2, \ldots, \tau(l)$ and $t = 1, 2, \ldots, \tau(l)$.

In the disaggregation stage, total incoming fields at box centers are calculated from the top of the tree structure to the lowest level. For a box C at a level $l < L$, the total incoming field is calculated as

$$\begin{aligned}
[\boldsymbol{g}_C^{+,x} \; \boldsymbol{g}_C^{+,y} \; \boldsymbol{g}_C^{+,z} \; \boldsymbol{g}_C^{+,s}]_{M_l \times 4} &= [\boldsymbol{g}_C^x \; \boldsymbol{g}_C^y \; \boldsymbol{g}_C^z \; \boldsymbol{g}_C^s]_{M_l \times 4} \\
&+ [\overline{\boldsymbol{\beta}}_{C' \to C}]_{M_l \times M_{l+1}} \cdot [\boldsymbol{g}_{C'}^{+,x} \; \boldsymbol{g}_{C'}^{+,y} \; \boldsymbol{g}_{C'}^{+,z} \; \boldsymbol{g}_{C'}^{+,s}]_{M_{l+1} \times 4}
\end{aligned} \tag{3.181}$$

by combining contributions from translations and incoming field at the center of the parent box C'. In (3.181), $\overline{\boldsymbol{\beta}}_{C' \to C}$ is an $M_l \times M_{l+1}$ dense matrix containing multipole-to-multipole shift functions $\beta_{t',u',t,u}(k, \boldsymbol{r}_C - \boldsymbol{r}_{C'})$ for $t' = 1, 2, \ldots, \tau(l-1)$ and $t = 1, 2, \ldots, \tau(l)$. For a lossless medium,

$$[\overline{\boldsymbol{\beta}}_{C' \to C}]_{M_l \times M_{l+1}} = \{[\overline{\boldsymbol{\beta}}_{C \to C'}]_{M_{l+1} \times M_l}\}^H, \tag{3.182}$$

where 'H' represents the Hermitian operation, since $\beta_{t',u',t,u}(k, \boldsymbol{r}_C - \boldsymbol{r}_{C'}) = \beta^*_{t,u,t',u'}(k, \boldsymbol{r}_{C'} - \boldsymbol{r}_C)$. At the lowest level, total incoming fields are received by testing functions as

$$\begin{aligned}
\sum_{n=1}^{N} \overline{\boldsymbol{Z}}_{FF}[m,n]\boldsymbol{a}[n] \approx \left(\frac{ik}{4\pi}\right)^2 \{ &[\boldsymbol{f}_m^x]_{1 \times M_1} \cdot [\boldsymbol{g}_C^{+,x}]_{M_1 \times 1} + [\boldsymbol{f}_m^y]_{1 \times M_1} \cdot [\boldsymbol{g}_C^{+,y}]_{M_1 \times 1} \\
&+ [\boldsymbol{f}_m^z]_{1 \times M_1} \cdot [\boldsymbol{g}_C^{+,z}]_{M_1 \times 1} + [\boldsymbol{f}_m^s]_{1 \times M_1} \cdot [\boldsymbol{g}_C^{+,s}]_{M_1 \times 1} \}
\end{aligned} \tag{3.183}$$

for $m \in C$, where $\{\boldsymbol{f}_m^x, \boldsymbol{f}_m^y, \boldsymbol{f}_m^z, \boldsymbol{f}_m^s\}$ are arrays of M_1 elements containing the receiving pattern of the mth testing function.

Using multipoles, the processing time and memory requirement for the aggregation, translation, and disaggregation stages at level l are proportional to $N_l M_l^4$, where $N_l \approx 4^{(1-l)} N_1$ is the number of boxes. For small objects, $M_l = \mathcal{O}(1)$, which is independent of N, and LF-MLFMA is dominated by computations at the lowest level with $\mathcal{O}(N)$ complexity. On the other hand, LF-MLFMA is not appropriate for large-scale problems since the number of multipoles can be prohibitively large at the higher levels. As a remedy, the ordinary MLFMA should be employed at higher levels, where the multipole representation becomes inefficient. The resulting broadband implementation of MLFMA can provide efficient solutions of large-scale problems involving small details with respect to wavelength [183].

3.7.3 *Broadband MLFMA*

In a broadband MLFMA implementation using multipoles, radiated fields represented by multipoles at level l are converted into plane waves as

$$[\boldsymbol{s}_C^x \ \boldsymbol{s}_C^y \ \boldsymbol{s}_C^z \ \boldsymbol{s}_C^s]_{S_l \times 4} = \overline{\boldsymbol{T}}^{M\to P}_{S_l \times M_l} \cdot [\boldsymbol{s}_C^x \ \boldsymbol{s}_C^y \ \boldsymbol{s}_C^z \ \boldsymbol{s}_C^s]_{M_l \times 4}, \tag{3.184}$$

where $\overline{\boldsymbol{T}}^{M\to P}$ is an $S_l \times M_l$ transformer matrix containing the multipole-to-plane-wave transformers in (3.54). Note that only the θ and ϕ components of radiated fields are required using the plane-wave representation. They can be obtained as

$$\boldsymbol{s}_C^\theta[n] = \cos(\theta[n])\cos(\phi[n])\boldsymbol{s}_C^x[n] + \cos(\theta[n])\sin(\phi[n])\boldsymbol{s}_C^y[n] - \sin(\theta[n])\boldsymbol{s}_C^z[n] \tag{3.185}$$

$$\boldsymbol{s}_C^\phi[n] = -\sin(\theta[n])\sin(\phi[n])\boldsymbol{s}_C^x[n] + \sin(\theta[n])\cos(\phi[n])\boldsymbol{s}_C^y[n] \tag{3.186}$$

for $n = 1, \ldots, S_l$, where $\theta[n]$ and $\phi[n]$ represent the spherical coordinates of samples on the unit sphere. When fields are converted into plane waves, the aggregation, translation, and disaggregation stages at level l and higher levels can be performed as in the ordinary MLFMA. After the disaggregation operations are performed from the highest level to level l, incoming fields represented by plane waves are transformed back to the cartesian system and converted into multipoles as

$$[\boldsymbol{s}_C^x \ \boldsymbol{s}_C^y \ \boldsymbol{s}_C^z \ \boldsymbol{s}_C^s]_{M_l \times 4} = \overline{\boldsymbol{T}}^{P\to M}_{M_l \times S_l} \cdot [\boldsymbol{s}_C^x \ \boldsymbol{s}_C^y \ \boldsymbol{s}_C^z \ \boldsymbol{s}_C^s]_{S_l \times 4}, \tag{3.187}$$

where $\overline{\boldsymbol{T}}^{P\to M}$ is an $M_l \times S_l$ transformer matrix containing the plane-wave-to-multipole transformers in (3.55).

3.7.4 *Numerical Results*

In order to test the accuracy of LF-MLFMA using multipoles, solutions of scattering problems involving a PEC sphere of radius 0.3 m discretized with 930 RWG functions are considered. The sphere is illuminated by a plane wave propagating in the $-z$ direction with the electric field polarized in the x direction in free space. The frequency is in the range from 15.625 MHz to 500 MHz. Hence, the diameter of the sphere changes from approximately $\lambda/32$ to λ, whereas the mesh size changes from approximately $\lambda/320$ to $\lambda/10$. The problems are formulated with EFIE, MFIE, and CFIE, and solutions are performed by using a very accurate MOM implementation and LF-MLFMA with three digits of accuracy. Figure 3.41(a) presents the relative

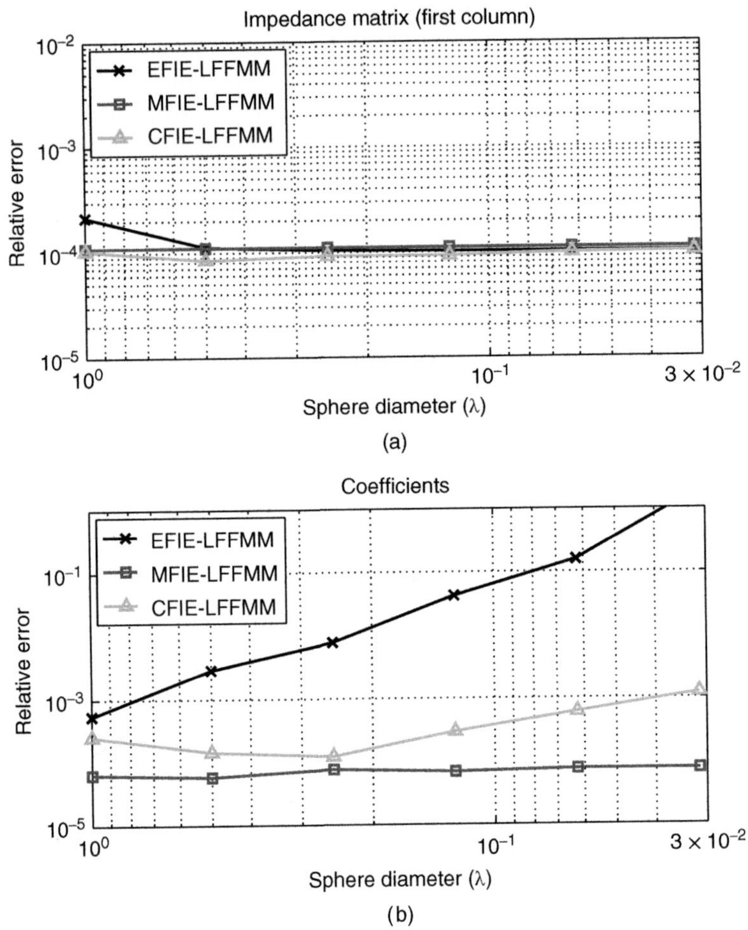

Figure 3.41 Relative error (a) in the first column of the impedance matrix and (b) in expansion coefficients obtained by using LF-MLFMA compared to a very accurate MOM implementation for the solution of scattering problems involving a PEC sphere.

error in the first column of the impedance matrix as a function of sphere size with respect to wavelength. The relative error is defined as

$$\Delta_r = \frac{\max_m\{|\overline{\mathbf{Z}}^{\text{LF-MLFMA}}[m,1] - \overline{\mathbf{Z}}^{\text{MOM}}[m,1]|\}}{\max_m\{|\overline{\mathbf{Z}}^{\text{MOM}}[m,1]|\}}, \tag{3.188}$$

where $\overline{\mathbf{Z}}^{\text{LF-MLFMA}}$ and $\overline{\mathbf{Z}}^{\text{MOM}}$ are matrices obtained by using LF-MLFMA and the reference MOM implementation, respectively. One can observe that matrix elements can be computed accurately with the desired level of accuracy using LF-MLFMA. Specifically, the relative error is less than 10^{-3} for all formulations and at all frequencies. On the other hand, accurate computations of matrix elements do not necessarily lead to accurate solutions. Figure 3.41(b) presents

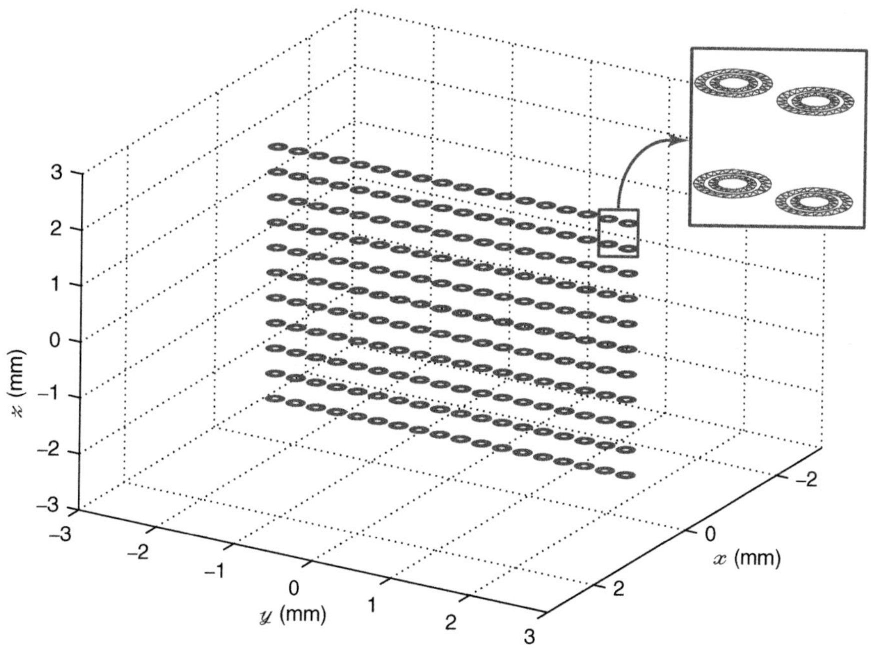

Figure 3.42 One-layer metamaterial wall involving 18 × 11 SRRs.

the relative error (defined with respect to the reference MOM solution as in (3.188)) in coefficients to expand the electric current obtained by using LF-MLFMA. In the case of EFIE, the relative error increases as the mesh size decreases. This is because EFIE itself suffers from a low-frequency breakdown and three digits of accuracy provided by LF-MLFMA becomes insufficient when the condition number is very large. Note that CFIE is also affected by the low-frequency breakdown problem, but MFIE provides stable solutions in the entire frequency range since the frequency is not extremely low.

In order to demonstrate the efficiency of LF-MLFMA, solutions of real-life electromagnetics problems involving metamaterials are considered. Figure 3.42 depicts a 1-layer metamaterial wall, which is constructed by periodically arranging 18 × 11 split-ring resonators (SRRs). A single SRR has dimensions in the order of microns and resonates at about 100 GHz when embedded into a homogeneous host medium with a relative permittivity of 4.8 [22]. An incident field is generated by a Hertzian dipole located at $x = -1.2$ mm. For numerical solutions, surfaces are discretized with $\lambda/100$ triangles, where λ is the wavelength in the host medium. Such a dense discretization is required for accurate modeling of SRRs that involve small details with respect to wavelength. The problems are formulated with EFIE and matrix equations involving 16,236 unknowns are solved iteratively by the GMRES method without restart. MVMs are performed by MLFMA and LF-MLFMA with 2 digits of accuracy. Iterative solutions are also accelerated by the SAI preconditioner.

Figure 3.43 presents the solution of electromagnetics problems involving the 1-layer SRR wall depicted in Figure 3.42. The number of iterations for 10^{-3} residual error, solution time, and total time are plotted with respect to frequency from 90 GHz to 110 GHz. The number

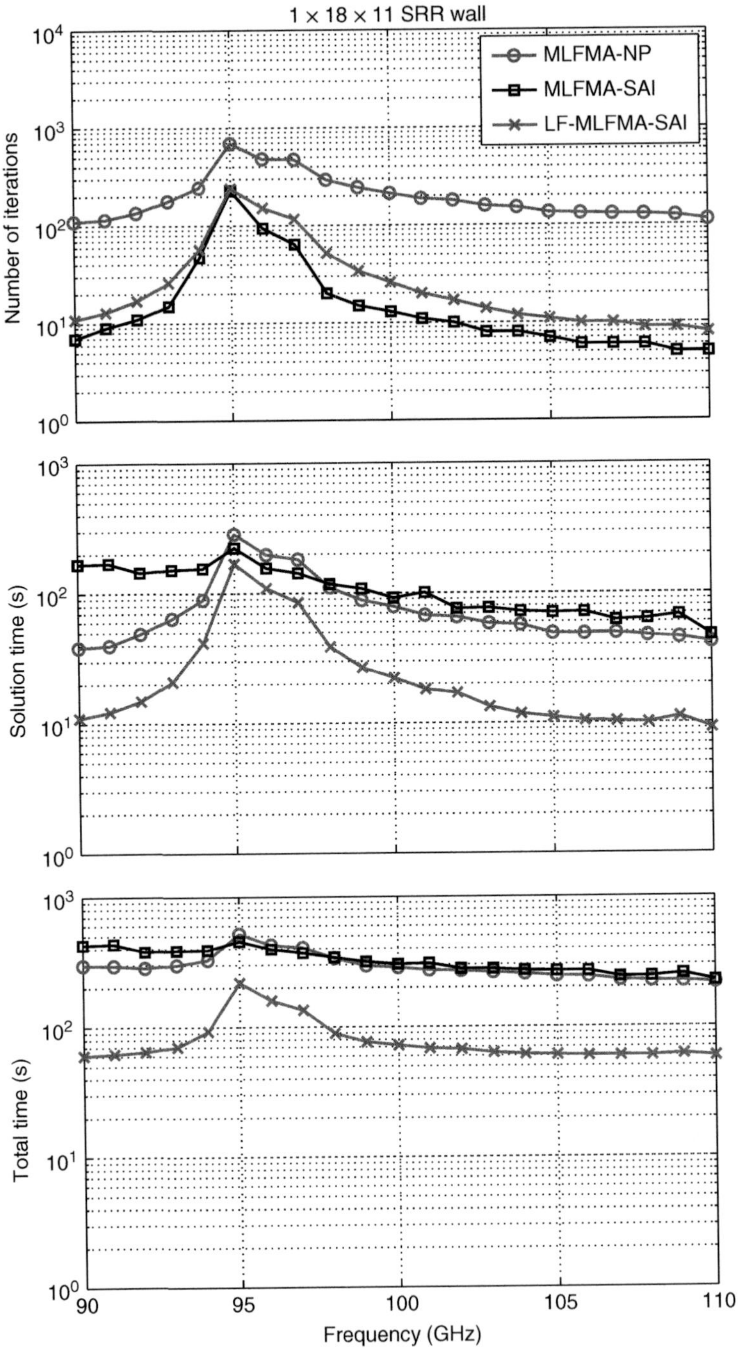

Figure 3.43 Solutions of electromagnetics problems involving a 1-layer SRR wall discretized with 16,236 unknowns depicted in Figure 3.42. The number of GMRES iterations (10^{-3} residual error), solution time, and total time including setup and solution times are plotted with respect to frequency.

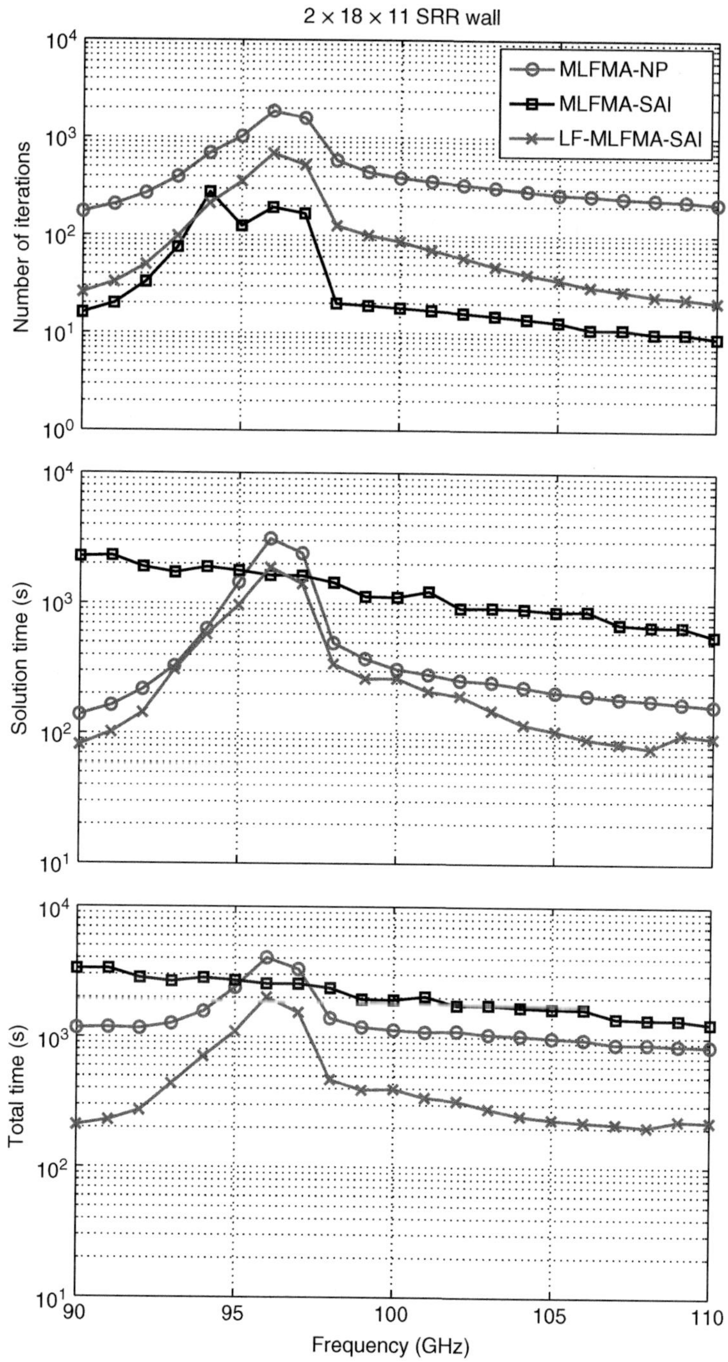

Figure 3.44 Solutions of electromagnetics problems involving a 2-layer SRR wall discretized with 32,472 unknowns. The number of GMRES iterations (10^{-3} residual error), solution time, and total time including setup and solution times are plotted with respect to frequency.

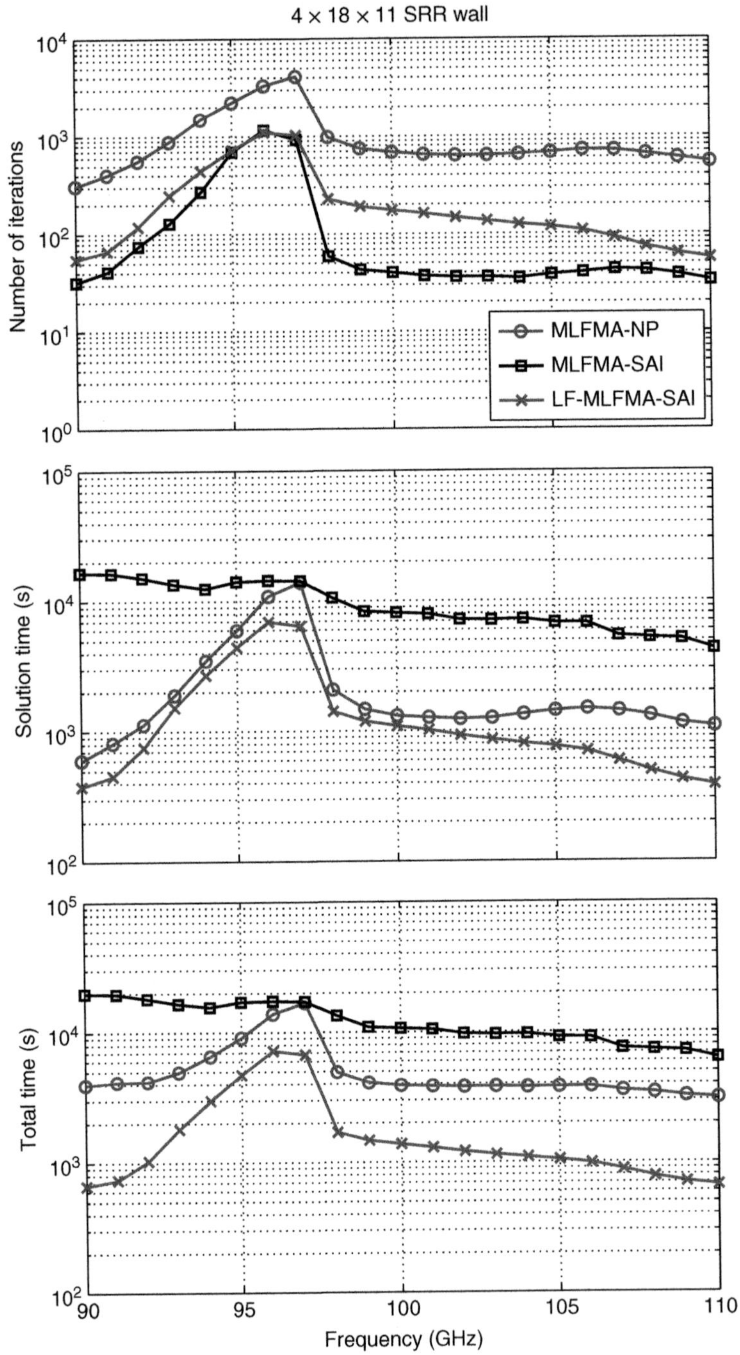

Figure 3.45 Solutions of electromagnetics problems involving a 4-layer SRR wall discretized with 64,944 unknowns. The number of GMRES iterations (10^{-3} residual error), solution time, and total time including setup and solution times are plotted with respect to frequency.

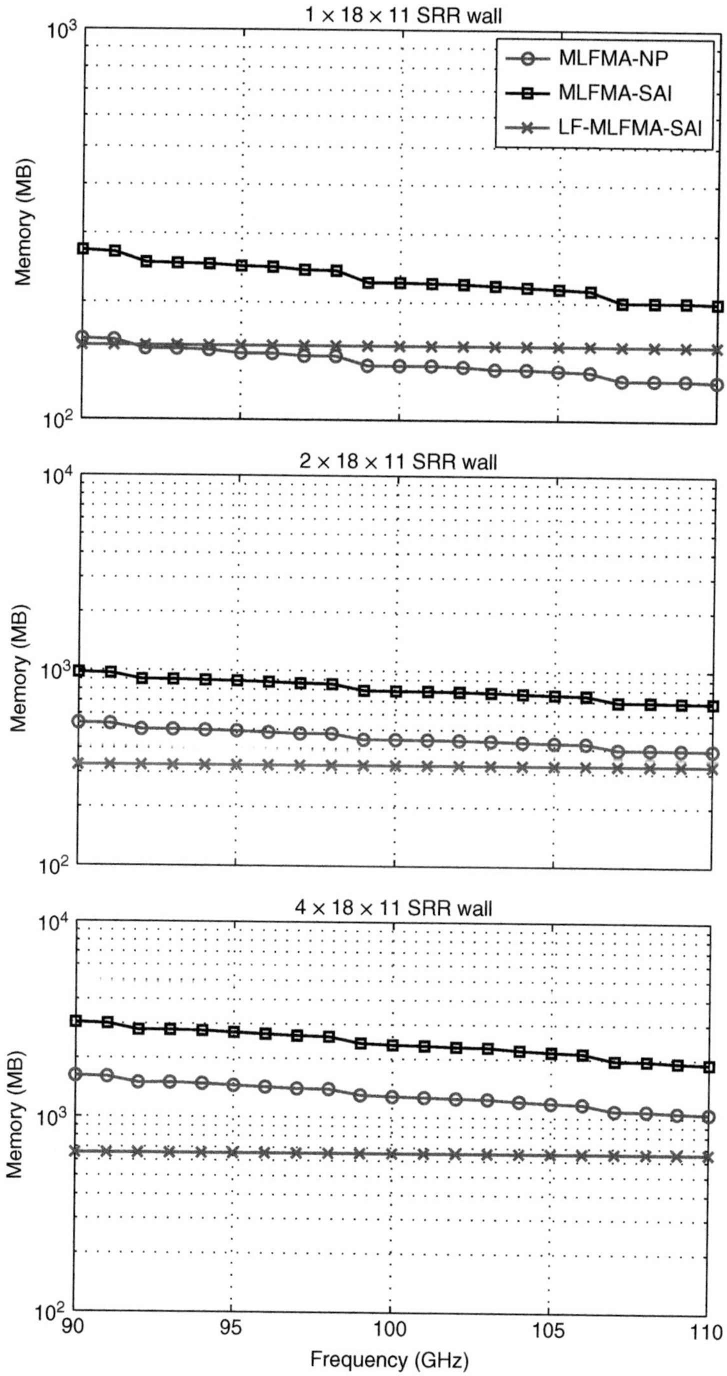

Figure 3.46 Memory required for the solution of electromagnetics problems involving 1-layer, 2-layer, and 4-layer SRR walls discretized with 16,236, 32,472, and 64,944 unknowns, respectively.

of iterations become maximum at 95 GHz due to a numerical resonance. Considering only MLFMA, the SAI preconditioner reduces the number of iterations significantly compared to the NP case. On the other hand, the solution time is not reduced due to the additional factorization cost of the preconditioner. In other words, reducing the number of iterations does not necessarily accelerate solutions. One can also observe that using LF-MLFMA instead of MLFMA increases the number of iterations since the number of near-field interactions used to construct the preconditioner is smaller in LF-MLFMA, compared to MLFMA. On the other hand, the most efficient solutions are provided by LF-MLFMA. Superior performance of LF-MLFMA is more apparent in terms of the total time, which includes the setup time dominated by the calculation of near-field interactions in addition to the solution time.

Figures 3.44 and 3.45 present the number of iterations, solution time, and total time with respect to frequency for 2-layer and 4-layer SRR walls involving $2 \times 18 \times 11$ and $4 \times 18 \times 11$ SRRs, respectively. It can be observed that LF-MLFMA provides very efficient solutions, compared to MLFMA. Figure 3.46 presents the memory required for the solution of 1-layer, 2-layer, and 4-layer SRR walls. In addition to faster solutions, LF-MLFMA requires less memory than MLFMA.

3.8 Concluding Remarks

As shown in this chapter, electromagnetics problems formulated with surface integral equations can be solved accurately and efficiently with MLFMA. This algorithm is a powerful tool to reduce the complexity of matrix-vector multiplications required for iterative solutions without sacrificing the accuracy. Using MLFMA, it becomes possible to solve large-scale problems that cannot be analyzed via direct methods. In the next chapter, parallelization of MLFMA will be considered for the solution of even larger problems discretized with millions of unknowns.

4

Parallelization of MLFMA for the Solution of Large-Scale Electromagnetics Problems

Parallelization of MLFMA is essential for the solution of very large electromagnetics problems discretized with millions of unknowns. This chapter presents the hierarchical partitioning strategy and various high-performance computing techniques for the efficient parallelization of MLFMA on distributed-memory architectures. Due to improved load balancing and reduced communications, the hierarchical strategy offers a higher parallelization efficiency than previous approaches, especially when the number of processors is large. Higher parallelization efficiency translates into the ability to solve larger problems with available resources. Using optimizations and load-balancing algorithms along with the hierarchical strategy allows for accurate analysis of complicated targets that are larger than 1000λ on moderately parallel computers.

4.1 On the Parallelization of MLFMA

Because of the complicated structure of the algorithm, parallelization of MLFMA is not trivial and several issues must be carefully considered.

- Partitioning: The main task in the parallelization of MLFMA is partitioning the data among processors with minimal duplication. As discussed in Section 4.6.7, the hierarchical partitioning strategy allows for efficient distribution of the multilevel tree structure.
- Load balancing: Parallelization cannot be achieved efficiently without distributing tasks equally among processors. Load-balancing algorithms must be used for each part of the implementation to obtain high parallelization efficiency.
- Communications: In parallel MLFMA, processors need to communicate with each other to transfer data. Using appropriate partitioning schemes and load-balancing algorithms

The Multilevel Fast Multipole Algorithm (MLFMA) for Solving Large-Scale Computational Electromagnetics Problems, First Edition. Özgür Ergül and Levent Gürel.

significantly reduces the data traffic. However, remaining communications must be organized carefully. For high efficiency, it is also essential to use high-speed networks to connect processors.

4.2 Parallel Computing Platforms for Numerical Examples

Numerical examples presented in this chapter are performed on the parallel computers listed in Table 4.1. In each of these computers, nodes are identical and connected via an Infiniband network, while processors in a node are connected through high-speed mainboard buses. In addition, memory in a node is available to all cores in the node. Note that, in the context of parallelization, the terms 'processor' and 'core' are used synonymously. For a solution on p processors, solutions are performed on the maximum number of nodes available, i.e., the number of processes per node is minimized. In other words, if a code is parallelized into 2^i processes and if $2^i \leq N_n$, where N_n is the number of nodes, then 2^i nodes are used, each running only one process. This ensures that parallel implementations are tested rigorously by fully using the network between nodes. When $2^i > N_n$, however, the solution is parallelized over N_n nodes, and $2^i/N_n$ processors are employed per node.

Obviously, the clusters in Table 4.1 that are involving multicore nodes are nonuniform in terms of communication rates between cores. As an example, Figure 4.1 presents the time required for one-to-one communications measured on the B-Clovertown cluster. A single data package of 4 kB is communicated between all pairs of the cores of the cluster. One can observe large discrepancies between the time required for the intra-node and inter-node communications. Specifically, the package can be communicated in more than 10 μs when the cores are located in different nodes, whereas it takes less than 6 μs when the cores are located in the same node. The rate of the intra-node communications also depends on the location of the cores in the node, e.g., communications are faster if the cores are in the same processor package. For the generalization of numerical results in this chapter, the nonuniform property of the clusters in Table 4.1 are not explicitly used. However, it is worth mentioning that the hierarchical strategy described below in Section 4.6 is very suitable for such nonuniform platforms by facilitating processor arrangements to minimize inter-node communications.

Table 4.1

Computer	Nodes	Processor per node	Memory per node
B-Clovertown	8	2 Quad-Core E5345 (2.33 GHz)	32 GB
C-Clovertown	128	2 Quad-Core E5345 (2.33 GHz)	16 GB
S-Clovertown	16	2 Quad-Core X5355 (2.66 GHz)	16 GB
B-Harpertown	16	2 Quad-Core X5472 (3.00 GHz)	32 GB
U-Nehalem	64	2 Quad-Core X5550 (2.67 GHz)	24 GB
S-Nehalem	16	2 Quad-Core X5560 (2.80 GHz)	32 GB
N-Nehalem	16	2 Quad-Core X5560 (2.80 GHz)	48 GB
S-Westmere	16	2 Quad-Core X5670 (2.93 GHz)	48 GB
S-Tigerton	16	2 Quad-Core X7350 (2.93 GHz)	32 GB
S-Dunnington	16	4 Six-Core E7450 (2.40 GHz)	48 GB
S-Nehalem-EX	16	4 Eight-Core L7555 (1.87 GHz)	128 GB

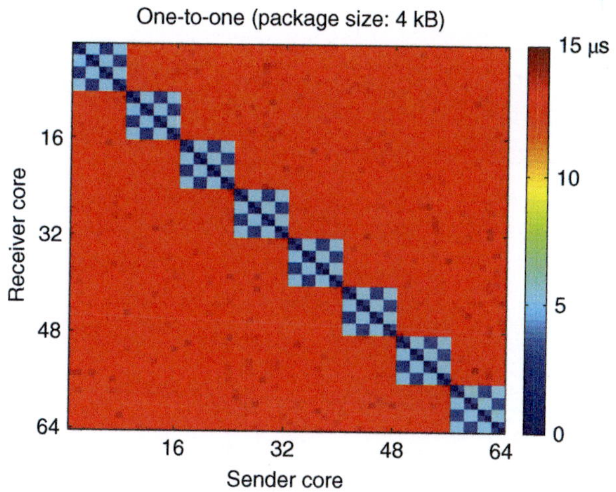

Figure 4.1 Time required for one-to-one communications measured on the B-Clovertown cluster. A data package of size 4 kB is communicated between all processors (cores) of the cluster.

4.3 Electromagnetics Problems for Numerical Examples

Figure 4.2 presents the objects that are employed for testing parallel implementations of MLFMA. In addition to the sphere, two canonical objects, namely a $1 \times 3.5 \times 0.1$ m rectangular box and a wing-shaped object with sharp edges and corners, are considered. These are illustrated in Figures 4.2(a) and 4.2(b), respectively. The NASA Almond [234] in Figure 4.2(c) and the Flamme [149] in Figure 4.2(d) are famous benchmarking objects, whose scattering problems are also investigated in this chapter. Note that all objects in Figure 4.2 involve closed surfaces, which can be formulated with CFIE when these surfaces are PECs.

In all numerical examples, problems are discretized with the RWG functions on $\lambda/10$ mesh size, where λ is the wavelength in the host medium. Both near-field and far-field interactions are calculated with maximum 1% error. All iterative solutions are performed by using BiCGStab accelerated with BDPs. For relatively small problems involving 1–20 millions unknowns, the target residual error is chosen as 10^{-6}, while this value is usually relaxed to 10^{-3} for the larger problems, provided that the accuracy is not sacrificed.

4.4 Simple Parallelizations of MLFMA

Simple parallelizations of MLFMA are usually based on distributing boxes among processors at all levels. One can apply load balancing for both near-field and far-field interactions. Details of a simple parallelization of MLFMA are presented in the following subsections.

4.4.1 Near-Field Interactions

In the parallelization of MLFMA, near-field interactions should be distributed equally among processors using a load-balancing algorithm. Considering the sparse near-field matrix, rows

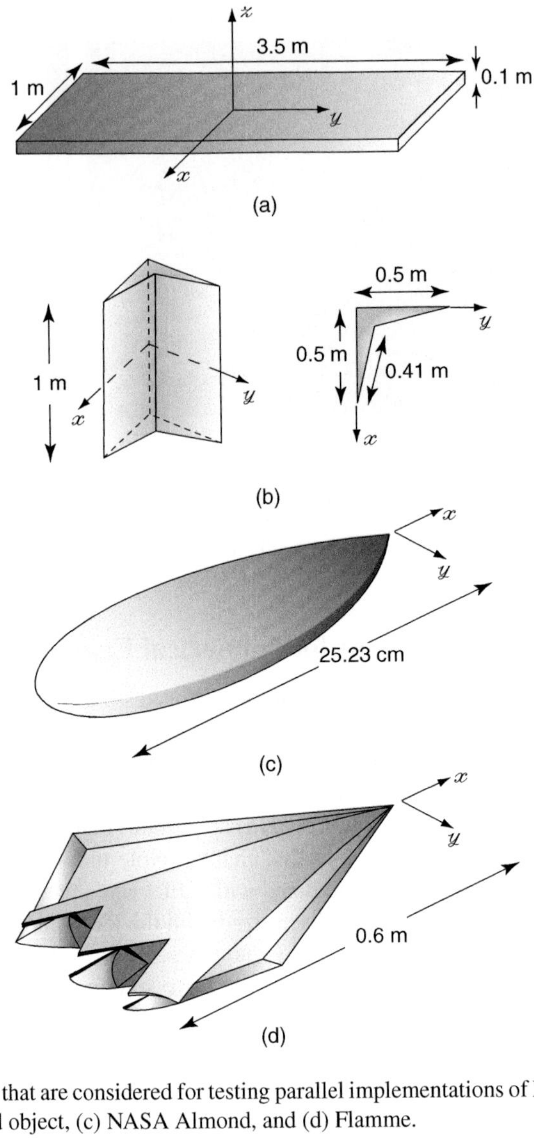

Figure 4.2 Objects that are considered for testing parallel implementations of MLFMA: (a) rectangular box, (b) wing-shaped object, (c) NASA Almond, and (d) Flamme.

$m = 1, 2, \ldots, N$ are assigned to processors in such a way that all processors have approximately equal numbers of near-field interactions (near-field partitioning). Distributing rows equally among processors usually fails to provide good load balancing, even for the solution of those problems involving symmetrical geometries, such as the sphere (see below, Section 4.7.3). After the distribution, near-field interactions are calculated in each processor without any communication.

4.4.2 Far-Field Interactions

Calculation of far-field interactions is organized according to the partitioning of the tree structure (far-field partitioning). One can choose a level of distribution (LOD) to distribute boxes among processors using a load-balancing algorithm that considers the combined load of all descendants (children, grandchildren, etc.) of each box at LOD. The combined load for a box is the size of the subtree attached to the box; all descendants weighted by the number of field samples are counted. The load-balancing algorithm assigns the whole branch of the tree starting at an LOD box to the same processor. This way, aggregation and disaggregation stages up to LOD can be performed without any communication. During the setup stage, each processor calculates and stores radiation and receiving patterns of basis and testing functions included in its local tree according to the far-field partitioning.

Using the simple parallelization strategy, there can be two different schemes for the parallelization of higher levels above LOD. In the first scheme, boxes are still distributed among processors equally as much as possible. Then, a box and its parent box may belong to different processors. As a consequence, aggregation and disaggregation stages involve dense communications for higher levels of the tree structure. Rather than this scheme, one can choose a second one, where boxes at levels above LOD are allowed to be duplicated in different processors, if required. This way, all aggregation and disaggregation operations can be performed without any communication. Note that, in this scheme, when radiated fields are translated into incoming fields, contributions from different processors are combined automatically for duplicated boxes.

Translations in MLFMA constitute a critical part in terms of the efficiency of the simple parallelization strategy. This is because dense one-to-one communications are required between processors. In general, each processor sends some data (radiated fields) to all other processors. These communications can be organized by using a communication map, which consists of interaction layers to match processors. Specifically, for p processors, it can be shown that translation communications can be achieved in $(p-1)$ steps, as depicted in Figure 4.3 for a 8-processor case. After processors are paired, the following operations are performed on the receiver and sender sides of each pair:

- The sender and receiver determine box-box interactions involving basis boxes on the sender side and testing boxes on the receiver side.
- Radiated fields of basis boxes are sent one by one.
- When the radiated field of a basis box is received by the receiver, all translations involving this basis box and testing boxes owned by the receiver are performed. This ensures that the same data is not transferred more than once.

To further improve the efficiency of translations, nonblocking send and receive operations of MPI can be used to transfer the data. In addition to inter-processor translations, there are also intra-processor translations that are related to boxes located in the same processor. These translations can be performed independently in each processor, without any communication.

In the setup of MLFMA, each processor is tasked with calculating a set of translation operators that will be required during MVMs. For a given translation, the operator is calculated

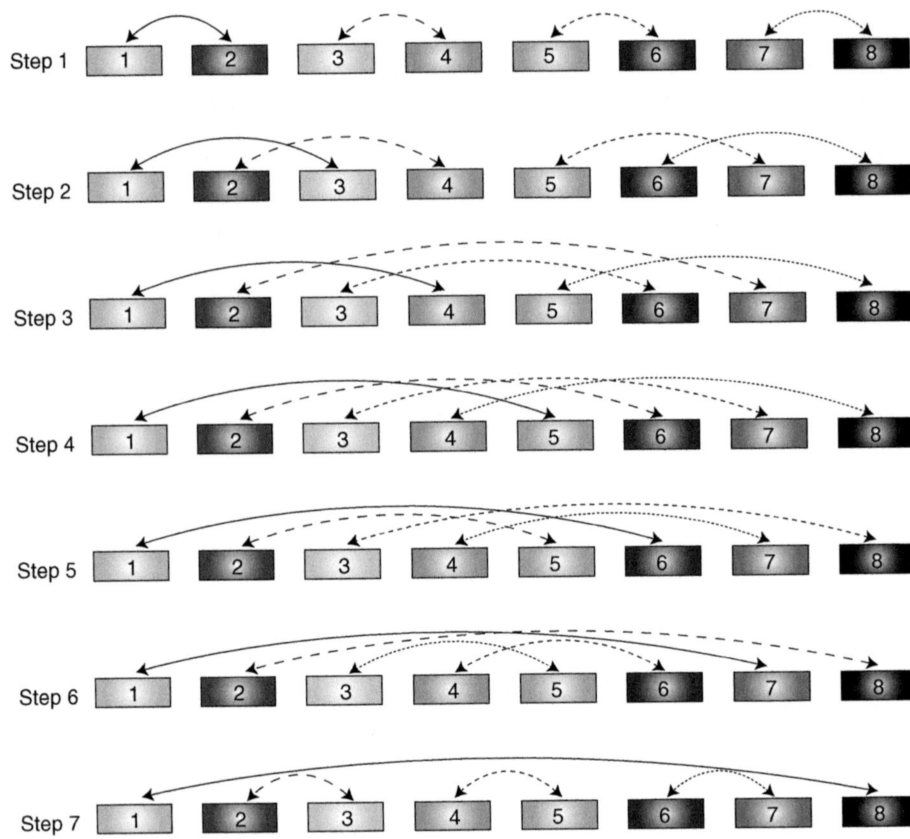

Figure 4.3 Processor pairing for translations.

by the processor working on the testing box. Due to symmetry, a translation operator can be used for many interactions at a level. Therefore, some translation operators are duplicated and included in more than one processor. Those duplications may decrease the efficiency of the parallelization, especially due to increasing cost of these operators at higher levels.

Finally, to match near-field and far-field partitioning schemes during MVMs, all-to-one and one-to-all communications are required, as depicted in Figure 4.4. After near-field computations are performed, the partitioning of the output vector is modified for far-field computations. Note that the iterative solver is also parallelized according to the far-field partitioning. The processing time for these communications is usually negligible. Similar all-to-all communications are also required during solutions of preconditioner systems.

4.5 The Hybrid Parallelization Strategy

The simple parallelization strategy that is based on assigning each box to one processor fails to provide efficient solutions. This is mainly due to dense communications between processors, duplication of computations, and unbalanced distribution of the workload among processors.

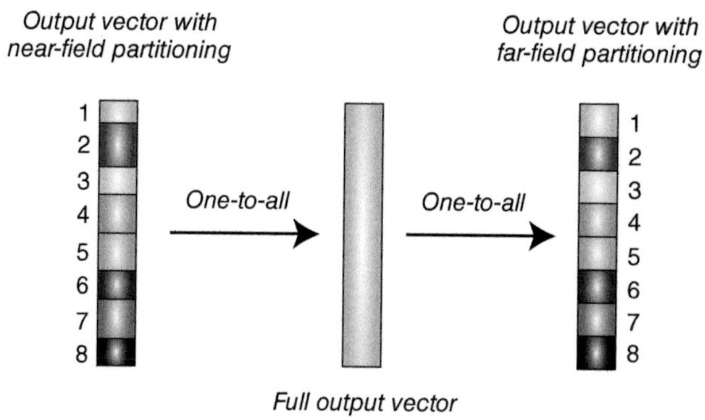

Figure 4.4 A demonstration of communications performed in each MVM to match near-field and far-field partitioning schemes.

Since such problems arise mostly at the higher levels of MLFMA, the hybrid partitioning strategy, which applies different partitioning schemes for lower and higher levels, can be used to improve the efficiency [217], [221]. In this strategy, processor assignments are made on the basis of fields of boxes at higher levels. In other words, each box at higher levels is shared by all processors, while each processor is assigned to the same portion of fields for all boxes.

As an example, Figure 4.5 depicts a 4-level tree structure ($L = 4$), where the levels are represented by two-dimensional rectangles. The horizontal and vertical dimensions of the rectangles correspond to boxes and θ samples of fields, respectively. At the lower levels (distributed levels), there are many boxes with small numbers of samples for radiated and incoming fields. Therefore, it is appropriate to distribute boxes at these levels by assigning each of them to a single processor. At the higher levels (shared levels), however, it is easier to distribute field samples among the processors by assigning each box to all processors, since there are a few boxes at these levels with large numbers of samples. Note that field samples are partitioned only along the θ direction for an easy implementation of interpolation and anterpolation operations.

Distributing field samples instead of boxes at the higher levels of MLFMA has two major consequences:

- Translations can be performed without any communication at higher levels, whereas aggregation and disaggregation stages require one-to-one communications.
- Load balancing is improved since the distribution of samples is easier than the distribution of boxes at higher levels.

In the following subsections, details of the hybrid strategy are presented by emphasizing the major differences compared to the simple strategy.

4.5.1 Aggregation Stage

Using the hybrid strategy, the partitioning scheme is changed during the aggregation stage at LOD by employing an all-to-all communication, as shown in Figure 4.6. For each box, samples

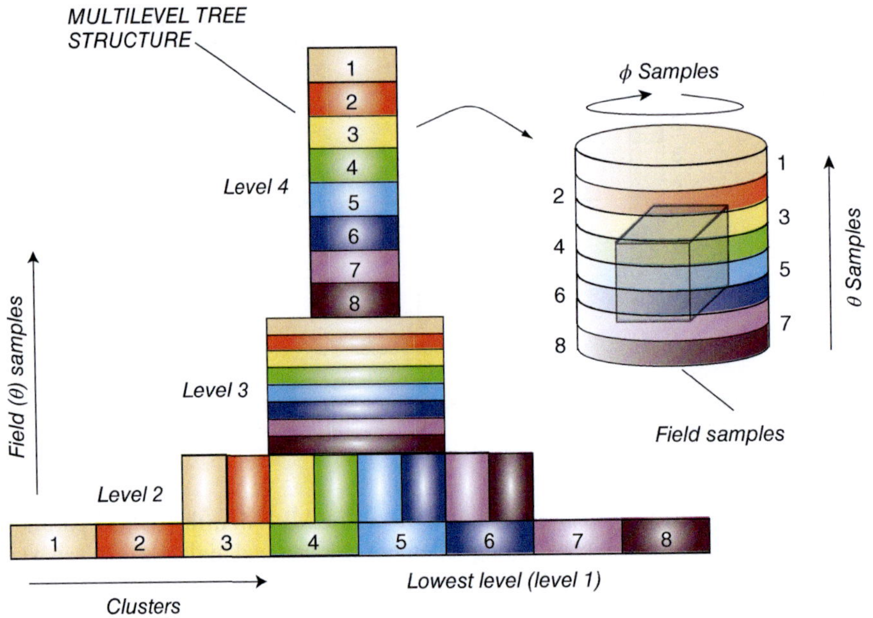

Figure 4.5 Distribution of a 4-level tree structure among eight processors using the hybrid partitioning strategy.

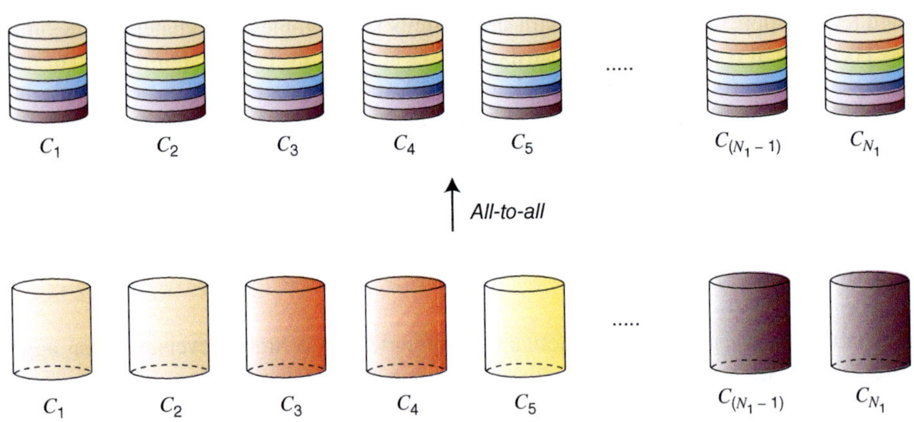

Figure 4.6 All-to-all communications performed at LOD to change the far-field partitioning scheme from distributed levels to shared levels.

of the radiated field stored in a processor is distributed among all processors. From LOD to the highest level L, the aggregation stage involves one-to-one communications that are required for the interpolation of fields. This is illustrated in Figure 4.7, where an interpolation is performed on samples of a box C located at level 3 of the tree structure in Figure 4.5. As an example, only the interpolation in processor 5 is depicted although similar operations are also performed

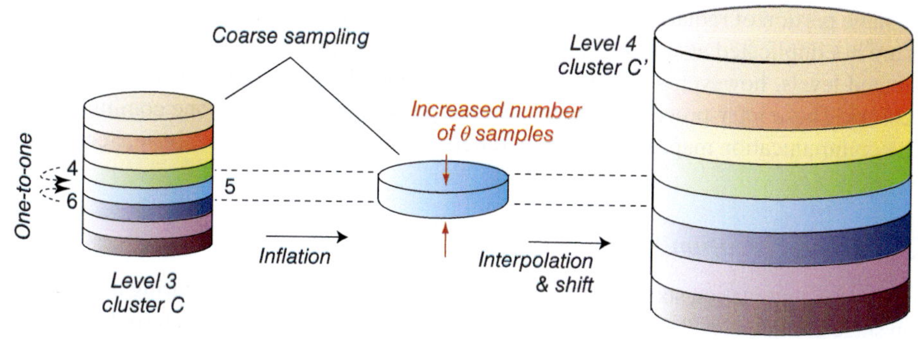

Figure 4.7 Interpolations at shared levels involving one-to-one communications.

in other processors. To compute the data at each sample in the fine grid, a set of samples are used in the coarse grid. Even though a local interpolation method is used, some of those coarse samples may be located in other processors. Therefore, one-to-one communications are performed to provide the required data (inflation). After the data is prepared, interpolation and shifting operations are performed to include the contribution of the box C in the radiated field of its parent box C'.

It should be note that communications at shared levels are mainly required between the processors located close to each other. In other words, a processor with index n requires data from its neighbors, i.e., $(n-1)$ and $(n+1)$. On the other hand, depending on the partitioning and number of interpolation points, more data might be required from other processors next to neighbors. One can apply a load-balancing algorithm to distribute fields appropriately so that the amount of the data transferred among all processors is minimized. However, as the number of processors increases and field samples are distributed over many processors, dense one-to-one communications cannot be avoided; this may reduce the efficiency of the hybrid partitioning strategy.

For each problem, the number of distributed and shared levels can be determined by an optimization. For this purpose, LOD is assigned to a series of possible levels and the distribution of boxes and their fields is monitored. For some levels (higher levels), the distribution of fields is better than the distribution of boxes, i.e., field samples can be partitioned evenly among processors, but not boxes. For others (lower levels), however, boxes can be partitioned easily, while it is difficult to partition field samples among processors. Then, one can choose LOD such that distributing fields (boxes) is more preferable for all levels above (below) LOD. The choice of LOD depends on the tree structure (hence the geometry of the object) as well as the number of processors. However, experiments show that, for a given problem, LOD is insensitive to the latter parameter if only a small number (e.g., 2 to 16) of processors are employed.

4.5.2 *Translation Stage*

Using the hybrid strategy, all translations at shared levels can be performed without any communication since field samples are distributed among processors and a processor is assigned to the same portion of radiated and incoming fields for all boxes. In addition, translation operators are not duplicated at shared levels, since they are also partitioned among processors. At distributed levels, however, some translations are needed between boxes located in different processors. These inter-processor translations are performed via one-to-one communications using a communication map, as in simple parallelizations of MLFMA (see above, Section 4.4).

4.5.3 Disaggregation Stage

Similar to the aggregation stage, one-to-one communications are required during the disaggregation stage at shared levels. Since the anterpolation is the transpose of the interpolation, some of the samples obtained from an anterpolation operation should be sent to other processors. This is because interpolations during the aggregation stage are performed by using inflated data prepared by one-to-one communications among processors. As the reverse of this process, anterpolations produce inflated data, which must be deflated via one-to-one communications. This is illustrated in Figure 4.8, where the processor 5 performs an anterpolation operation on samples of box C' for its subboxes C. Some of the resulting data in the coarse grid is used locally, while the rest are sent to the other processors (deflation), i.e., exactly the reverse of the inflation in interpolations. As the disaggregation proceeds down to LOD, the partitioning is changed via an all-to-all communication. Then, the disaggregation is performed from LOD to the lowest level without any communication. Finally, at the lowest level, each processor performs angular integrations and produces a partition of the output vector to be returned to the iterative solver.

4.5.4 Communications in Hybrid Parallelizations

As a summary, Table 4.2 lists communications required during MVMs using the hybrid parallelization strategy. For distributed and shared levels, communications are required in translation and aggregation/disaggregation stages, respectively. In addition to those one-to-one communications, the hybrid parallelization strategy requires collective all-to-one, one-to-all, and all-to-all communications.

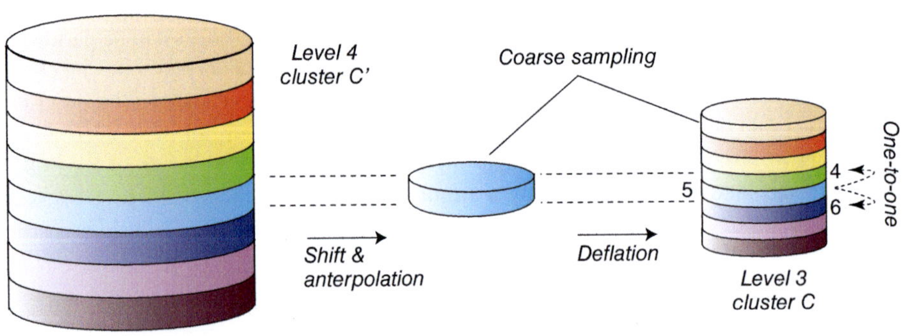

Figure 4.8 Anterpolations (transpose interpolations) at shared levels involving one-to-one communications.

Table 4.2 Communications required during MVMs using the hybrid partitioning strategy

Part	Communication
Near-field interactions	All-to-one and one-to-all
Distributed aggregation/disaggregation	None
Aggregation/disaggregation at LOD	All-to-all
Shared aggregation/disaggregation	One-to-one
Distributed translation	One-to-one
Shared translation	None

4.5.5 *Numerical Results with the Hybrid Parallelization Strategy*

To demonstrate the efficiency provided by the hybrid partitioning strategy, Figure 4.9 presents the solution of a scattering problem involving a PEC sphere with a radius of 24λ. The sphere is illuminated by a plane wave propagating in the $-x$ direction with the electric field polarized in the y direction. The problem is discretized with 2,111,952 unknowns and solved on the S-Clovertown cluster. Figure 4.9 depicts the efficiency (with respect to the solution with a single processor) when the solution is parallelized into 2, 4, 8, 12, and 16 processes. The parallelization efficiency is defined as

$$\varepsilon_p = \frac{T_1}{pT_p}, \tag{4.1}$$

where T_p is the processing time of the solution using p processors. Figure 4.9(a) shows that the overall efficiency (setup and iterative solution) is above 85% when the number of processors is 16. In this case, the efficiency ratios for the setup and solution parts are about 97% and 80%, respectively. One can observe in Figure 4.9(a) that the setup part is parallelized very efficiently, since this part is communication-free and involves computations (especially near-field interactions) that are perfectly distributed among processors using load-balancing algorithms.

Figure 4.9(b) presents the parallelization efficiency for the aggregation, translation, and disaggregation stages, in addition to the overall efficiency for MVMs. It can be observed that the aggregation and disaggregation stages are parallelized with about 87% efficiency, while the efficiency for the translation stage is 59% for the 16-processor case. To further investigate the parallelization, Figure 4.10 presents the processing time and efficiency (with respect to the solution with two processors) for various categorized parts of MVMs. The following observations can be made:

- The aggregation and disaggregation stages at the distributed levels ($l = 1, 2, 3, 4$) constitute the significant part of the processing time of MVM. These stages are perfectly parallelized, thanks to the load-balancing algorithm for the distributed levels.
- The parallelization efficiency of the aggregation and disaggregation stages at the shared levels (from $l = 5$ to $l = L = 7$) is also quite high. However, the efficiency drops to about 80% for the 16-processor case. This is due to the increasing amount of one-to-one communications for interpolations and anterpolations.

- Parallelization efficiency of the communication-free (intra-processor) translations is in the 80%–100% range. All of translations at the shared levels and some of those at the distributed levels are communication-free.
- Translations that are performed with communications (inter-processor translations) and all-to-all communications performed at LOD exhibit reduced efficiency as the number of processors increases. Since they take longer processing time, the inter-processor translations affect the overall efficiency more than the all-to-all communications.

In general, inter-processor translations are bottlenecks of the hybrid parallelization strategy (as of the simple parallelization strategy). Since these translations are performed at distributed levels, their negative contributions can be minimized by increasing the number of shared levels. However, aggregation and disaggregation at lower levels cannot be performed efficiently by

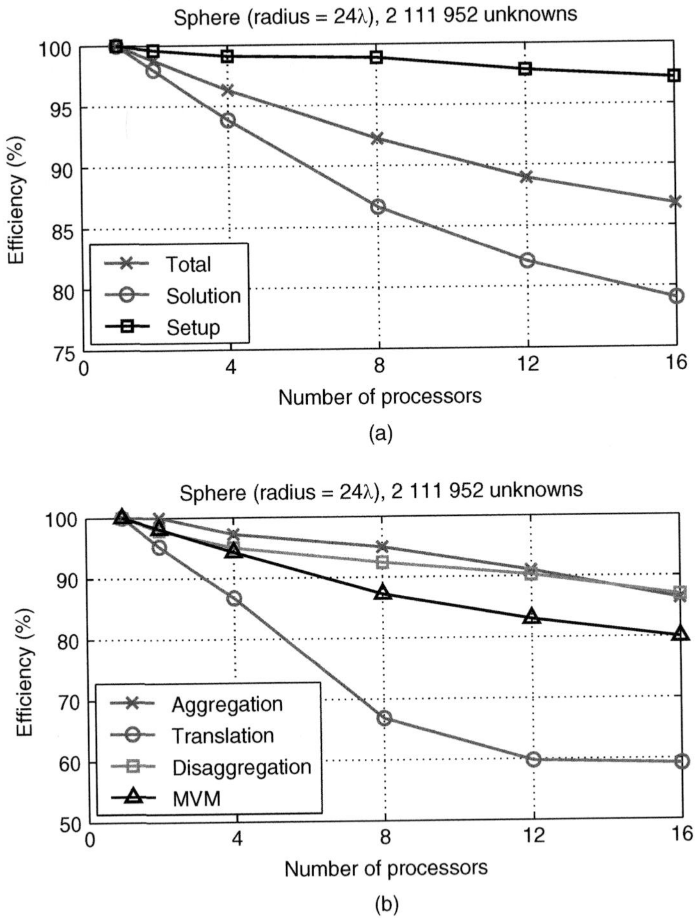

Figure 4.9 Parallelization efficiency for the solution of a scattering problem involving a PEC sphere of radius 24λ discretized with 2,111,952 unknowns. *Source*: Ergül and Gürel 2008 [221]. Reproduced with permission of IEEE.

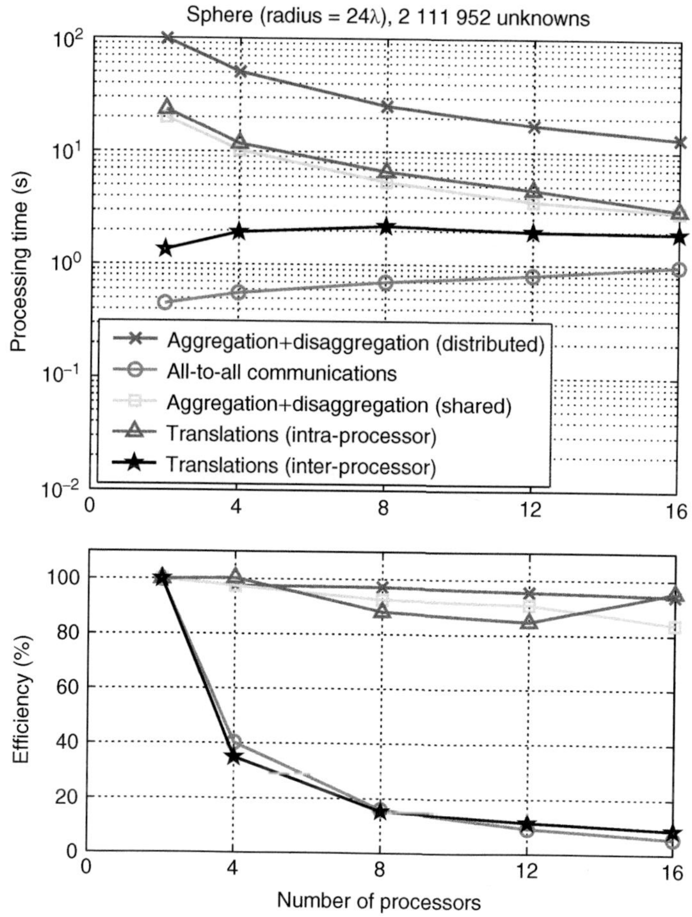

Figure 4.10 Processing time and parallelization efficiency for various categorized parts of MVMs for the solution of a scattering problem involving a PEC sphere of radius 24λ discretized with 2,111,952 unknowns. *Source*: Ergül and Gürel 2008 [221]. Reproduced with permission of IEEE.

partitioning coarsely sampled fields. In fact, numbers of distributed and shared levels should be optimized for the solution of each problem. The numerical results presented in Figures 4.9 and 4.10 are obtained with the best possible choice for the number of distributed/shared levels in terms of the overall parallelization efficiency.

Figure 4.11 presents the solution of a scattering problem involving a sphere of radius 80λ discretized with 23,405,664 unknowns on the S-Clovertown cluster. The sphere is illuminated by a plane wave propagating in the $-x$ direction with the electric field polarized in the y direction. In Figure 4.11(a), the total processing time is depicted for all processors from 1 to 16. After the input and recursive clustering part (1), computations of translation operators (2) and radiation/receiving patterns (4) require negligible time. Calculation of near-field interactions (3) dominates the setup time, which is about 94 minutes. Then the solution part (5), involving a total of 34 MVMs, is performed in about 155 minutes. The processing time for a MVM is

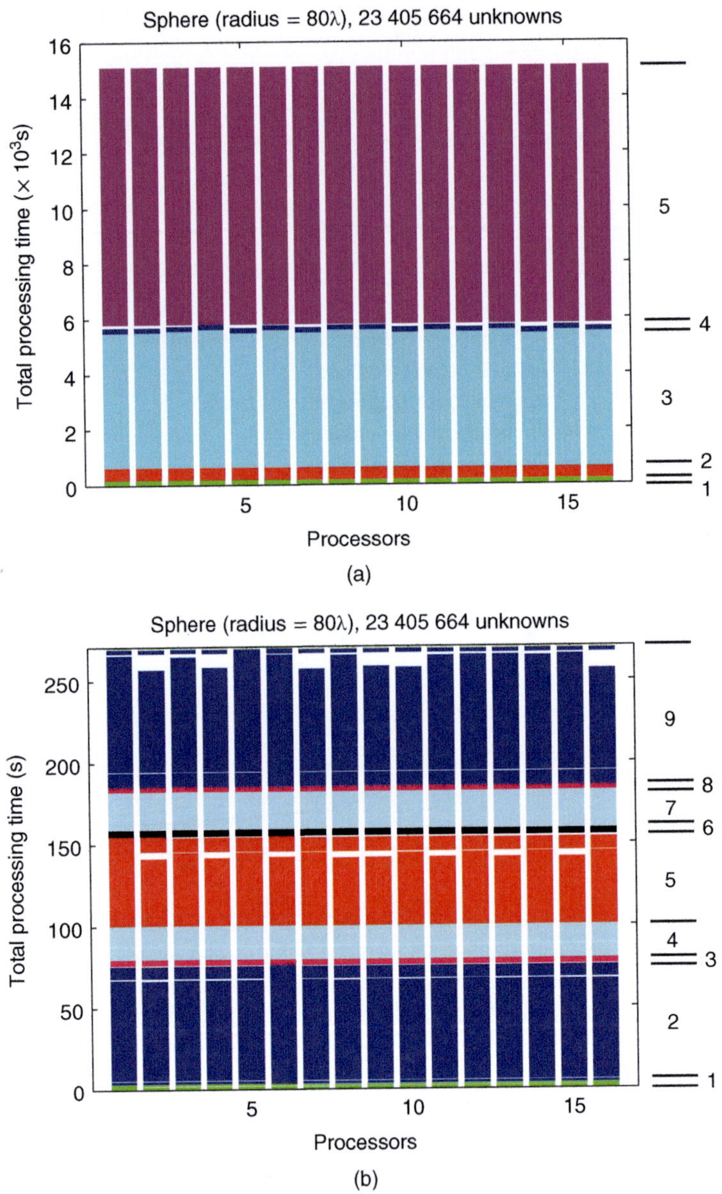

Figure 4.11 Time diagrams for the solution of a scattering problem involving a sphere of radius 80λ discretized with 23,405,664 unknowns using the hybrid parallelization strategy on 16 processors of the S-Clovertown cluster. (a) Overall time and (b) single MVM. In the diagrams, white areas correspond to waits before operations that require synchronization. *Source*: Ergül and Gürel 2008 [221]. Reproduced with permission of IEEE.

depicted in 4.11(b), including the near-field stage (1), aggregation/disaggregation at distributed levels (2,9), all-to-all communications (3,8), aggregation/disaggregation at shared levels (4,7), communication-free (intra-processor) translations (5), and inter-processor translations (6). The most problematic parts in terms of parallelization efficiency, i.e., all-to-all communications and inter-processor translations, require negligible time compared to other parts of the MVM. This is commonly observed with large-sized problems and supports the conclusion that the parallelization efficiency for a fixed number of processors usually increases as the problem size grows.

Even though the hybrid parallelization strategy increases the parallelization efficiency significantly, compared to simple parallelizations, the improvement can be insufficient, especially when the number of processors is large. As a remedy, the hierarchical partitioning strategy that is well suited for the multilevel structure of MLFMA can be used. With the enhanced load balancing offered by such a technique, the parallelization of MLFMA can be improved significantly.

4.6 The Hierarchical Parallelization Strategy

The hierarchical partitioning strategy is based on the simultaneous partitioning of boxes and their fields at all levels. Partitioning in both directions (boxes and samples of fields) is adjusted appropriately by considering the number of boxes and the number of samples at each level. As an example, consider again a 4-level tree structure ($L = 4$), where each level is represented by a rectangular box. A hierarchical partitioning of the tree structure among eight processors is depicted in Figure 4.12. At the lowest level, boxes are distributed among eight processors, and each box is assigned to a single processor, without any partitioning of field samples. Then, at the next level ($l = 2$), field samples are partitioned among two groups of processors, i.e., (1,3,5,7) and (2,4,6,8), while the number of box partitions is reduced to four. At this level, samples of each box are shared by two processors. As proceeding to the higher levels, the number of partitions for boxes and samples of fields are systematically decreased and increased, respectively. In the following subsections, the hierarchical partitioning strategy is detailed by considering the main stages of MLFMA.

4.6.1 *Hierarchical Partitioning of Tree Structures*

Consider the parallelization of MLFMA on a cluster of p processors, where $p = 2^i$ for some integer i. Using the hierarchical partitioning strategy, the number of partitions for boxes at level l is chosen as

$$p_{l,c} = \max\left\{\frac{p}{2^{(l-1)}}, 1\right\} = \max\left\{p\,2^{(1-l)}, 1\right\}. \tag{4.2}$$

Note that boxes are not partitioned for levels $l > \log_2(p)$, if such a level exists. The number of boxes assigned to each processor can be approximated as

$$N_l^p \approx \frac{N_l}{p_{l,c}} \approx \left\{\begin{matrix} 2^{(1-l)}N_1/p, & l \leq \log_2(p) \\ 4^{(1-l)}N_1, & l > \log_2(p) \end{matrix}\right\}. \tag{4.3}$$

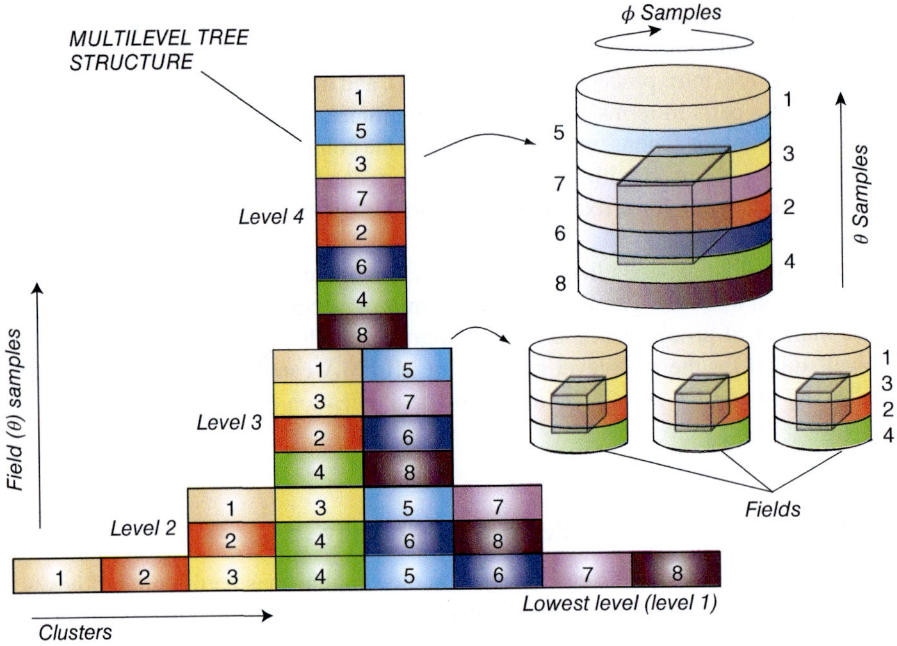

Figure 4.12 Distribution of a 4-level tree structure among eight processors using the hierarchical partitioning strategy. *Source*: Ergül and Gürel 2009 [225]. Reproduced with permission of IEEE.

In addition, samples of fields are divided into

$$p_{l,s} = \frac{p}{p_{l,c}} = \min\left\{2^{(l-1)}, p\right\} \tag{4.4}$$

partitions along the θ direction for level l. The number of θ samples assigned to each processor is

$$S_l^{\theta,p} \approx \frac{S_l^{\theta}}{p_{l,s}} \approx \left\{ \begin{array}{ll} S_1^{\theta}, & l \leq \log_2(p) \\ 2^{(l-1)}S_1^{\theta}/p, & l > \log_2(p) \end{array} \right\}. \tag{4.5}$$

Also considering the sampling in the ϕ direction, the total number of samples per processor can be written as

$$S_l^p = S_l^{\theta,p} S_l^{\phi} \approx \left\{ \begin{array}{ll} 2^{(l-1)}S_1, & l \leq \log_2(p) \\ 4^{(l-1)}S_1/p, & l > \log_2(p) \end{array} \right\}. \tag{4.6}$$

Finally, the size of the local data at each processor is

$$N_l^p S_l^p \approx N_1 S_1/p \tag{4.7}$$

for $l = 1, 2, \ldots, L$.

4.6.2 Aggregation Stage

For the partitioned tree structure in Figure 4.12, the aggregation operations from level 3 to level 4 are depicted in Figure 4.13. Since field samples are partitioned among processors, one-to-one communications are required to inflate the local data in accordance with the interpolation requirements. As shown in Figure 4.13, one-to-one communications occur within two separate groups of processors that are located in the same columns, i.e., $(1,2,3,4)$ and $(5,6,7,8)$. As an important advantage of the hierarchical partitioning strategy, distribution of θ samples into large numbers of partitions is avoided. Therefore, communications are required only between pairs of processors located next to each other. For example, processor 3 communicates with processors 1 and 2, but not with processor 4. Processors 3 and 4 would need to communicate with each other if the number of the θ samples required for the interpolation is larger than the number of the θ samples per processor. However, using the hierarchical strategy, the number of partitions along the θ direction, hence the number of θ samples per processor, can be adjusted such that those secondary communications between distant processors are totally avoided.

When the required data is prepared by one-to-one communications for a box, its radiated field is interpolated and shifted to the center of its parent box. Temporary levels involving parent boxes and field samples after interpolation and shifting operations are denoted as intermediate levels. As an example, for the partitioned tree structure in Figure 4.12, level 3.5 is depicted in Figure 4.13. Following the interpolation, the number of samples along the θ direction assigned to each processor is doubled, i.e.,

$$S_{l+1/2}^{\theta,p} \approx 2S_l^{\theta,p} \approx \begin{Bmatrix} 2S_1^{\theta}, & l \leq \log_2(p) \\ 2^l S_1^{\theta}/p, & l > \log_2(p) \end{Bmatrix}. \tag{4.8}$$

At the same time, the number of boxes in each processor can be written as

$$N_{l+1/2}^{p} \approx \frac{N_l^p}{4} \approx \begin{Bmatrix} 2^{-(1+l)}N_1/p, & l \leq \log_2(p) \\ 4^{(-l)}N_1, & l > \log_2(p) \end{Bmatrix}. \tag{4.9}$$

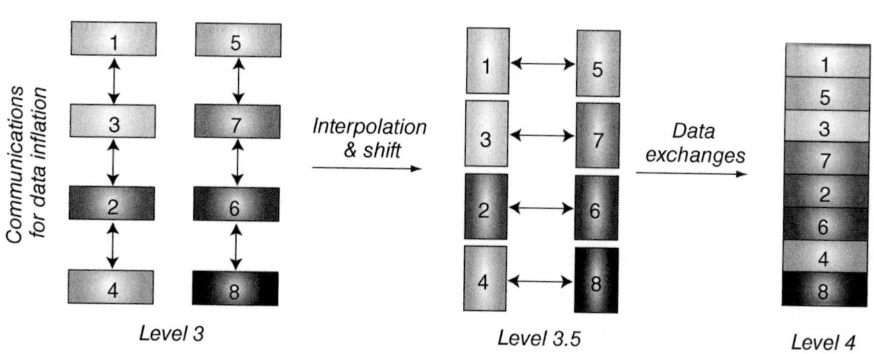

Figure 4.13 Aggregation operations from level 3 to level 4 for the partitioned tree structure in Figure 4.12. *Source*: Ergül and Gürel 2009 [225]. Reproduced with permission of IEEE.

Intermediate levels are defined temporarily and used to arrange the data in each processor, before communications are performed to modify the partitioning according to the employed hierarchical structure.

From an intermediate level $(l + 1/2)$ to the next level $(l + 1)$, data is exchanged among processors, if $l \leq \log_2(p)$. As depicted in Figure 4.13, processors are paired according to their positions in the partitioning scheme. Each processor performs the following communications:

- Send half of field samples of each box at the intermediate level.
- Receive the complementary data, which involves field samples of some boxes, from the associated processor.

With data exchanges, the number of boxes in each processor is doubled with respect to the number of boxes at the intermediate level, while the number of samples along the θ direction is halved. Then,

$$N_{l+1}^{p} \approx \begin{Bmatrix} 2^{(-l)}N_1/p, & (l+1) \leq \log_2(p) \\ 4^{(-l)}N_1, & (l+1) > \log_2(p) \end{Bmatrix} \tag{4.10}$$

and

$$S_{l+1}^{\theta,p} \approx \begin{Bmatrix} S_1^{\theta}, & (l+1) \leq \log_2(p) \\ 2^{l}S_1^{\theta}/p, & (l+1) > \log_2(p) \end{Bmatrix}, \tag{4.11}$$

which agree with the expressions in (4.3) and (4.5), respectively.

4.6.3 Translation Stage

Using the hierarchical partitioning strategy, one-to-one communications are required during the translation stage for levels $l \leq \log_2(p)$, since boxes are partitioned and some translations are needed between boxes located in different processors. As in the simple and hybrid parallelization strategies, those communications are achieved by pairing processors and transferring radiated fields of boxes between pairs. On the other hand, using the hierarchical partitioning strategy, communications are required only among processors that are located in the same row of the partitioning. For example, as depicted in Figure 4.14, communications at level 2 are performed within two separate groups of processors, i.e., $(1, 3, 5, 7)$ and $(2, 4, 6, 8)$. In general, for inter-processor translations at level l, each processor is paired one by one with the other $(p_{l,c} - 1)$ processors. Once a pairing is established, radiated fields of boxes are transferred and translations are performed by the receiver processor.

4.6.4 Disaggregation Stage

As usual, the parallelization of the disaggregation stage is very similar to the parallelization of the aggregation stage using the hierarchical partitioning strategy. When incoming fields are calculated at box centers at level l, partitioning is modified via data exchanges among processors, if $l \leq \log_2(p) + 1$. This way, the partitioning at level $(l - 1/2)$ is generated, as required for anterpolation and shifting operations. Then, incoming fields at box centers are anterpolated and shifted to the centers of their subboxes at level $(l - 1)$. Following an anterpolation operation, some of the resulting data are used locally, while the rest are sent to other processors. Similar

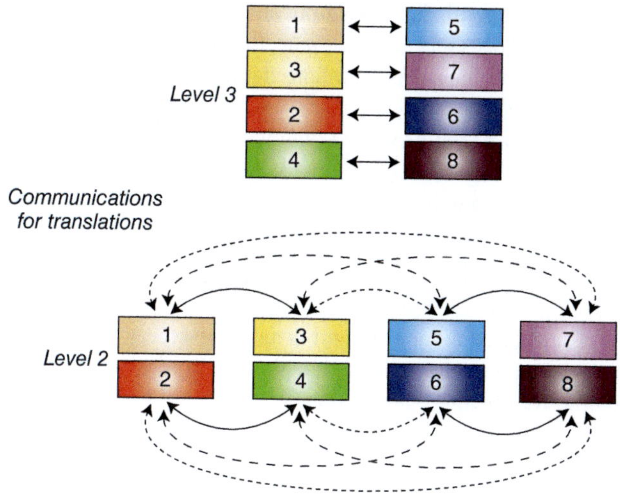

Figure 4.14 One-to-one communications during the translation stage at levels 2 and 3 of the partitioned tree structure in Figure 4.12. *Source*: Ergül and Gürel 2009 [225]. Reproduced with permission of IEEE.

to communications during interpolations, data is transferred among neighboring processors in the same column of the partitioning scheme.

4.6.5 *Communications in Hierarchical Parallelizations*

Using the hierarchical partitioning strategy, computations on the tree structure are distributed among processors with improved load balancing, compared to previous strategies based on partitioning with respect to only boxes or only samples of fields. However, there are still unavoidable communications among processors, which may reduce the efficiency of the parallelization. In this section, these communications are investigated in detail.

Communications in Aggregation and Disaggregation Stages

During an operation in a processor at level $l = 2, 3, \ldots, (L-1)$, the amount of data required from other processors for each box is proportional to the number of samples in the ϕ direction. Considering also the number of boxes per processor, the communication volume for interpolations at level l can be written as

$$v_{int,l} \propto 2^{(l-1)} S_1^{\phi} \begin{Bmatrix} 2^{(1-l)} N_1/p, & l \leq \log_2(p) \\ 4^{(1-l)} N_1, & l > \log_2(p) \end{Bmatrix}$$

$$= \begin{Bmatrix} N_1 S_1^{\phi}/p, & l \leq \log_2(p) \\ N_1 S_1^{\phi}/2^{(l-1)}, & l > \log_2(p) \end{Bmatrix} \leq N_1 S_1^{\phi}/p \qquad (4.12)$$

for $l = 2, 3, \ldots, (L-1)$. In general, the communication volume $v_{int,l}$ tends to decrease with the increasing number of processors p. Also note that aggregations at the lowest level do not involve one-to-one communications to inflate data, since samples are not partitioned at this level.

For levels $l \leq \log_2(p)$, half of the produced data is exchanged between pairs of processors to modify the partitioning in accordance with the hierarchical structure. The volume of these communications can be expressed as

$$v_{exc,l} \propto \left\{ \begin{array}{cc} N_1 S_1/p, & l \leq \log_2(p) \\ 0, & l > \log_2(p) \end{array} \right\} \leq N_1 S_1/p \tag{4.13}$$

for $l = 1, 2, \ldots, (L-1)$, where the upper bound is again inversely proportional to the number of processors p. The communication volume for the disaggregation stage is the same as the communication volume for the aggregation stage.

Communications in Translation Stage

Using the hierarchical partitioning strategy, each processor is paired one by one with $(p_{l,c} - 1) = \max\left\{p\,2^{(1-l)} - 1, 0\right\}$ processors to perform inter-processor translations at level l. For each pair, the number of box-box interactions required to be performed is proportional to the number of boxes per processor. In addition, the size of the data transferred in each interaction is proportional to the number of local samples per box (S_l^p). Therefore, the communication volume for translations can be written as

$$v_{trans,l} \propto \max\left\{p\,2^{(1-l)} - 1, 0\right\} N_l^p S_l^p \leq N_1 S_1 / 2^{(l-1)} \tag{4.14}$$

for $l = 1, 2, \ldots, L$. Note that the upper bound for the communication volume in (4.14) depends on the level. Specifically, communications during translations may become significant for lower levels, i.e., when l is small. In fact, due to these communications, the translation stage is a major bottleneck in hierarchical parallelizations of MLFMA (similar to simple and hybrid parallelizations).

Although it is a rather standard technique in parallelization, it is worth mentioning that the translation stage is very convenient to be improved by overlapping communications and computations. A simple illustration is shown in Figure 4.15, where two processors (p_1 and p_2) are involved in a series of communications, i.e., p_1 prepares and sends data to be processed by p_2. Using nonblocking communications of MPI, it is possible to reduce the wait periods for synchronizations. Specifically, by using nonblocking communications, p_1 does not wait for the completion of the data transfer and immediately starts preparing the next data after the communication is successfully buffered. Various nonblocking communication schemes are available in MPI and can be found in [233]. These schemes are very useful for translations, which involves repetitive communications and data processing.

4.6.6 Irregular Partitioning of Tree Structures

In the previous sections, the partitioning is assumed to change regularly on the tree structure, i.e., the number of partitions of boxes decreases from a level to the next upper level and the number of partitions of field samples is adjusted accordingly. In fact, for a given problem, one

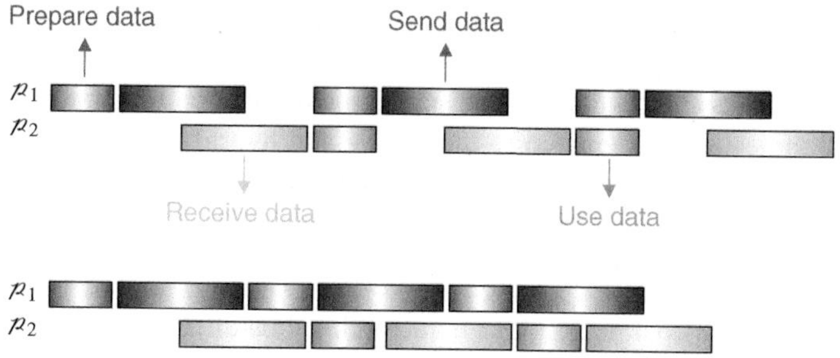

Figure 4.15 Improvement of the parallelization by overlapping communications and computations.

can choose the number of partitions at each level depending on optimizations. For example, in some cases, the number of partitions may not change from a level to the next level. This may be required to avoid excessive communications during aggregation and disaggregation stages. Partitioning at the lowest level is also a parameter and, as opposed to the common practices described above, field samples can be partitioned at this level depending on the number of processors and the number of boxes/samples at this level. In general, optimizations should be applied at each level to ensure that the hierarchical partitioning is optimal.

4.6.7 *Comparisons with Previous Parallelization Strategies*

In this subsection, the hierarchical partitioning strategy is compared with two previous approaches, namely, the simple and hybrid partitioning strategies. As detailed above in Section 4.4, a simple parallelization of MLFMA can be based on the distribution of boxes among processors at all levels. A major disadvantage of the simple parallelization strategy is the difficulty in distributing small numbers of boxes at higher levels. When the number of processors is large, those boxes must be duplicated over multiple processors. Otherwise, a large amount of data is required to be communicated during aggregation and disaggregation stages. In addition, when using the simple parallelization strategy, translations involve dense communications among processors.

The hybrid parallelization strategy has been successfully developed to improve the parallelization of MLFMA [217]. As detailed above in Section 4.5, lower (distributed) levels of MLFMA are partitioned as in the simple strategy, i.e., boxes are distributed among processors. At higher (shared) levels, however, samples of fields are distributed, instead of boxes. Unlike the hierarchical partitioning, samples at a shared level are distributed among all processors, without any partitioning of boxes. Distributing samples provides improved load balancing and communication-free translations for higher levels. On the other hand, problems arise for some levels at the middle of the tree, where it is not efficient to distribute either fields or boxes among processors.

The hierarchical partitioning strategy provide two important advantages, compared to the simple and hybrid strategies.

- Improved Load-Balancing: Partitioning both boxes and samples of fields leads to improved load balancing of workload among processors at each level. Consequently, duplication of computations, which may occur in simple parallelizations, and waits for synchronizations of processors are minimized.

- Reduced Communications: Although the hierarchical partitioning strategy increases the types of communications, compared to the simple and hybrid approaches, the amount of data transferred is decreased. In addition, due to improved load balancing, the average package size is enlarged and the number of communication events is reduced. As a result, the communication time is significantly shortened.

To demonstrate reduced communications in hierarchical parallelizations, Table 4.3 presents the solution of a scattering problem involving a PEC sphere of radius 20λ discretized with

Table 4.3 Communications during a MVM for the solution of a scattering problem involving a sphere discretized with 1,462,854 unknowns on 64 processors

	Hybrid		Hierarchical	
	Number of events	Amount (bytes)	Number of events	Amount (bytes)
Interpolation at $l = 7$	1360	646,272	136	107,712
Interpolation at $l = 6$	6960	1,030,080	292	108,040
Interpolation at $l = 5$	0	0	590	148,680
Interpolation at $l = 4$	0	0	1170	182,520
Data exchange at $l = 5$	0	0	2	55,552
Data exchange at $l = 4$	0	0	2	73,920
Data exchange at $l = 3$	0	0	2	106,560
Data exchange at $l = 2$	0	0	2	122,640
Data exchange at $l = 1$	0	0	2	213,304
Change partitioning	1160	171,680	0	0
Aggregation total	9480	1,848,032	2198	1,118,928
Aggregation ratio	100%	100%	23%	61%
Disaggregation total	9480	1,848,032	2198	1,118,928
Disaggregation ratio	100%	100%	23%	61%
Translation at $l = 7$	0	0	56	111,104
Translation at $l = 6$	0	0	148	139,120
Translation at $l = 5$	0	0	235	156,288
Translation at $l = 4$	425	749,700	427	179,340
Translation at $l = 3$	910	615,160	1005	313,560
Translation at $l = 2$	1812	587,088	1861	602,964
Translation at $l = 1$	3228	464,832	3483	501,552
Translation total	6375	2,416,780	7215	2,003,928
Translation ratio	100%	100%	113%	83%
MVM	25,335	6,112,844	11,611	4,241,784
MVM ratio	100%	100%	46%	69%
Average package size	241 bytes (100%)		365 bytes (151%)	

1,462,854 unknowns. The problem is solved by a 7-level MLFMA parallelized with hybrid and hierarchical strategies on 64 processors. The hybrid strategy involves three shared and four distributed levels. The number of box partitions in the hierarchical strategy is 64, 64, 32, 16, 8, 4, and 2 from the lowest level to the highest level. Note that field samples are not distributed for levels $l = 1$ and $l = 2$. Table 4.3 lists the number of communications and the amount of data communicated during a MVM. It can be observed that the total amount of communications is reduced by 31% using the hierarchical partitioning strategy compared to the hybrid strategy. In addition, the average size of packages is increased by 51% and the number of communication events is reduced by 54%. These results are also confirmed with other experiments involving different numbers of processors and different objects with diverse geometries.

Another important advantage of the hierarchical partitioning strategy appears when MLFMA is employed on a cluster with multiprocessor nodes. Most of the mainboards built recently have multiple processors connected via high-speed buses. Then, parallel computers are constructed by clustering a number of multiprocessor computing units, i.e., nodes, instead of processors. Resulting parallel computers are highly nonuniform, since communications among processors in the same node are significantly faster than those among processors located in different nodes (see Figure 4.1). Using multicore processors further complicates the situation, since communications within nodes also have diverse rates, depending on whether inter-core communications are taking place in the same processor or between two processors in the same node. The hierarchical parallelization strategy are suitable for such parallel platforms.

As an example, let the tree structure in Figure 4.12 be partitioned among two nodes, each having four processors, i.e., processors 1–4 and processors 5–8 are located in two different nodes. Then, all communications during the aggregation and disaggregation stages from level 1 to level 3 are performed inside nodes. Inter-node communications are required only for translations and for data exchanges during aggregation and disaggregation stages between level 3 and level 4. In general, the hierarchical partitioning strategy facilitates processor arrangements in nonuniform platforms to minimize inter-node communications. However, in the numerical examples presented in this chapter, this advantage is not used directly. Hence, the improved efficiency obtained with hierarchical partitioning strategy is general and valid for all types of distributed-memory architectures.

4.6.8 Numerical Results with the Hierarchical Parallelization Strategy

The improved efficiency provided by the hierarchical partitioning strategy, compared to previous parallelization approaches, is demonstrated on scattering problems involving PEC spheres of various sizes discretized with millions of unknowns. Each sphere is illuminated by a plane wave propagating in the $-x$ direction with the electric field polarized in the y direction. Figure 4.16 presents the parallelization efficiency obtained for the solution of a scattering problem involving a sphere of radius 20λ discretized with 1,462,854 unknowns. Figure 4.16(a) depicts the efficiency for the total time (including the setup and iterations), when the solution is parallelized on 2, 4, 8, 16, 32, 64, and 128 processors of the S-Tigerton cluster. It can be observed that the hierarchical strategy improves the parallelization efficiency significantly, compared to the simple and hybrid strategies, especially when the number of processors is large. The hybrid parallelization strategy performs better than the simple parallelization strategy; however, its efficiency drops rapidly for $p > 32$, and it becomes inefficient, compared to the hierarchical parallelization strategy. Using 128 processors, the

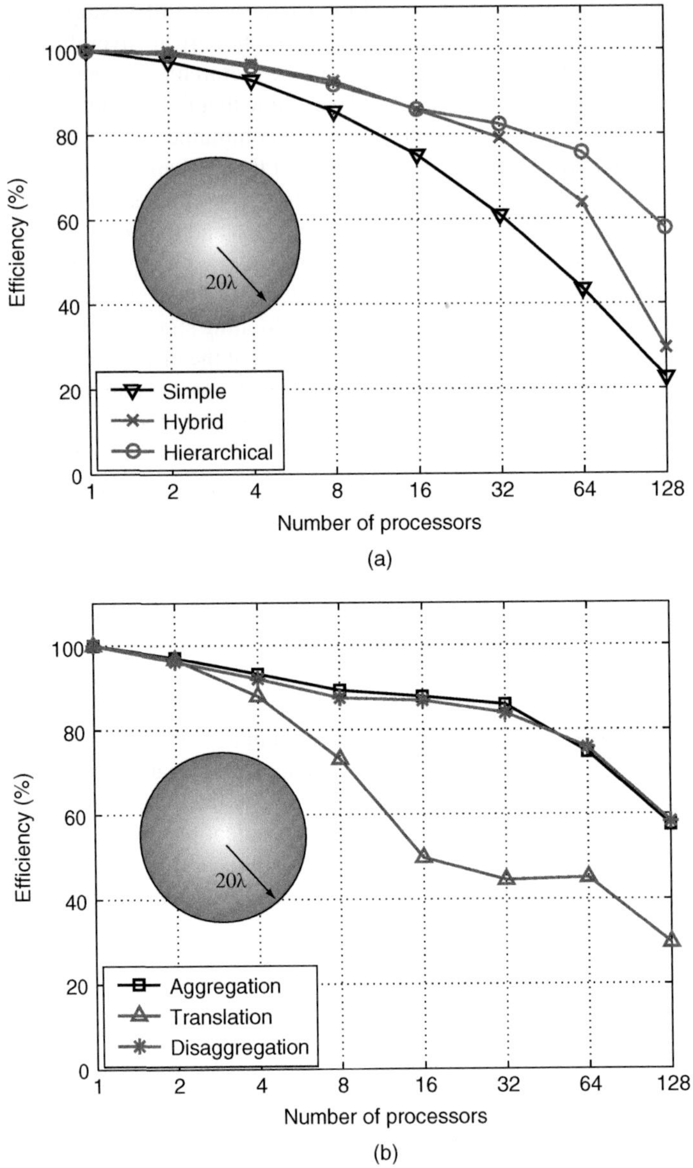

Figure 4.16 Parallelization efficiency with respect to the number of processors for the solution of a scattering problem involving a PEC sphere of radius 20λ discretized with 1,462,854 unknowns. (a) Overall efficiency including setup and iterations, when the solution is parallelized by using the simple, hybrid, and hierarchical strategies. (b) Efficiencies for MLFMA stages, i.e., aggregation, translation, and disaggregation, using the hierarchical strategy. *Source*: Ergül and Gürel 2009 [225]. Reproduced with permission of IEEE.

hierarchical parallelization strategy provides 58% efficiency, corresponding to a 74-fold speedup with respect to the single-processor solution.

Figure 4.16(b) presents the parallelization efficiency for the three main stages of MVMs, i.e., aggregation, translation, and disaggregation, using the hierarchical parallelization strategy. One can observe that (once again) the translation stage becomes a major bottleneck when the number of processors increases. For a solution on 128 processors, the parallelization efficiency of translations drops below 30%. This is because the communication volume for translations, given in (4.14), does not scale with the number of processors p, unlike the communication volume for aggregation and disaggregation stages. In addition, many communications required for inter-processor translations occur among processors located in different nodes. Then, the rate of communications during translations is mostly limited by the speed of the network. Nevertheless, even the parallelization of translations is improved with the hierarchical strategy, and the overall efficiency provided by the hierarchical strategy is consistently higher than those obtained with the simple and hybrid strategies.

Figure 4.17 presents the parallelization efficiency for the solution of scattering problems involving PEC spheres of radii 40λ and 60λ discretized with 5,851,416 and 13,278,096 unknowns, respectively, where the efficiency is defined with respect to solutions with two and four processors of the S-Tigerton cluster. Similar to the previous results, the parallelization efficiency is increased significantly by using the hierarchical partitioning strategy.

Even though Figures 4.16 and 4.17 compare relative performances of different parallelization strategies, it should be emphasized that they do not provide a complete information for the efficiency of solutions. In general, one should also consider the following factors.

- Clock Rate of Processors: Although using faster processors leads to faster solutions, the parallelization efficiency can be degraded as the computation time is reduced. This is because the communication time becomes more significant when the processing time for computations is small.
- Efficient Implementation of the Algorithm: Minimizing the processing time of MLFMA may also have an adverse effect on the parallelization efficiency. For example, radiation and receiving patterns of the basis and testing functions are calculated during the setup of the program and used efficiently during iterations. Calculating the patterns on the fly in each MVM without storing them is also a common practice for low memory implementations. That would increase the processing time, but the parallelization efficiency would also increase, since those calculations can be parallelized very efficiently.
- Accuracy Parameters: Accuracy of solutions also affects the parallelization efficiency. For example, most of communications during aggregation and disaggregation stages could be avoided by reducing the number of interpolation points. This would increase the parallelization efficiency, but the accuracy of solutions would deteriorate.

Note that the efficiency results presented in Figures 4.16 and 4.17 are obtained under strict circumstances, using a fast and accurate implementation of MLFMA on a cluster of processors with a relatively high clock rate. To quantify the efficiency of solutions, Table 4.4 lists the processing time, when the three problems are solved on 128 processors of the S-Tigerton cluster. Using the hierarchical parallelization strategy, the largest problem with 13,278,096 unknowns is solved in less than one hour.

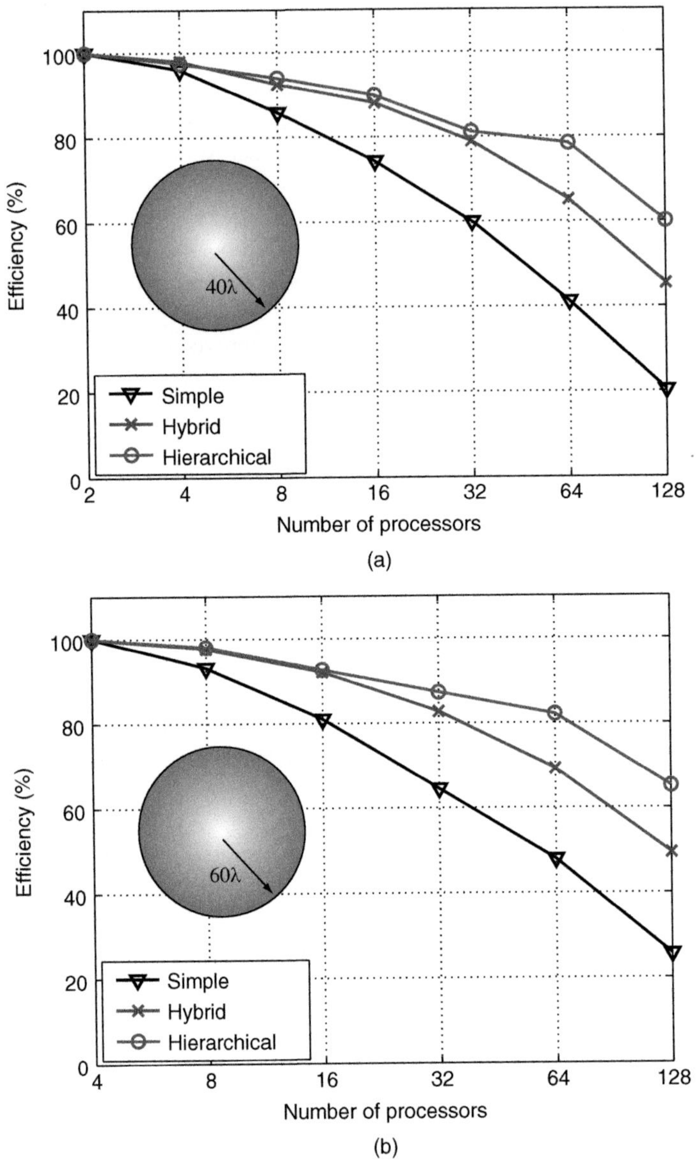

Figure 4.17 Parallelization efficiency for the solution of scattering problems involving (a) a PEC sphere of radius 40λ discretized with 6,851,416 unknowns and (b) a PEC sphere of radius 60λ discretized with 13,278,096 unknowns. Parallel efficiency is defined with respect to two and four processors, respectively. *Source*: Ergül and Gürel 2009 [225]. Reproduced with permission of IEEE.

Table 4.4 Solutions of sphere problems using the hierarchical parallelization strategy

Sphere radius	20λ	40λ	60λ
Unknowns	1,462,854	5,851,416	13,278,096
Number of iterations	27	30	43
	Setup time (s)		
Simple	184	1065	2630
Hybrid	89	356	812
Hierarchical	88	348	797
	Solution time (s)		
Simple	512	2467	5400
Hybrid	442	1162	3122
Hierarchical	184	804	2257

4.7 Efficiency Considerations for Parallel Implementations of MLFMA

The hierarchical partitioning strategy enables the efficient parallelization of MLFMA by improving load balancing, reducing communications, and facilitating processor arrangements for faster communications. On the other hand, efficient solutions require more than efficient partitioning and one needs to consider other factors, such as programming, system hardware and software, load balancing, and optimizations. This section provides an overview of these factors for a basic understanding of efficient parallel implementations of MLFMA.

4.7.1 *Efficient Programming*

It is well-known that programming is as essential as the underlying algorithms to obtain efficient implementations. Specifically, implementations must be optimized to eliminate all bottlenecks and to avoid stagnation when the problem size is large. As an example of impacts of programming style to the efficiency of solutions, Figure 4.18 presents a simple experiment involving multiplications of $N \times N$ dense matrices ($\overline{X}$ and $\overline{Y}$) using the Fortran 90 programming language. The processing time is measured on an AMD Opteron 870 processor as a function of N^3. Theoretically, the complexity of matrix-matrix multiplications is $\mathcal{O}(N^3)$. Two different schemes are considered to perform multiplications in the form of $\overline{V} = \overline{X} \cdot \overline{Y}$.

- Ordinary: Loops are constructed as

$$
\begin{aligned}
&\text{for } m = 1, 2, \ldots, N \\
&\quad \text{for } n = 1, 2, \ldots, N \\
&\qquad \text{for } k = 1, 2, \ldots, N \\
&\qquad\quad \overline{V}[m,n] = \overline{V}[m,n] + \overline{X}[m,k]\overline{Y}[k,n].
\end{aligned}
$$

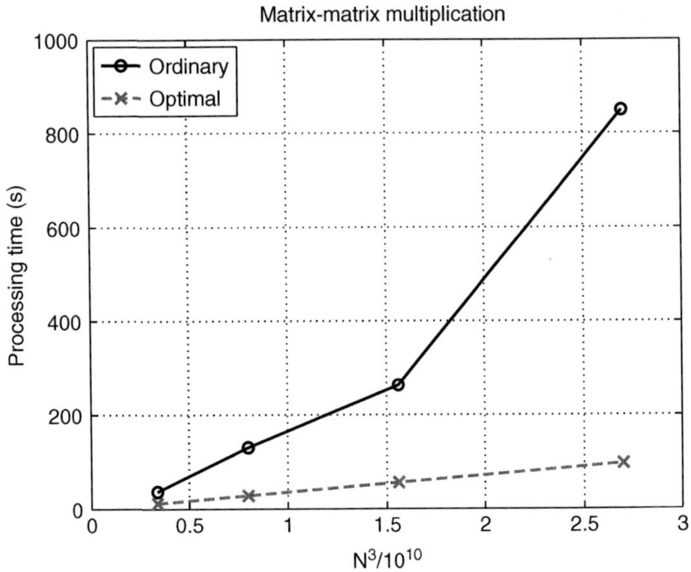

Figure 4.18 Processing time required for the multiplication of $N \times N$ dense matrices on an AMD Opteron 870 processor. *Source*: Ergül and Gürel 2011 [230]. Reproduced with permission of IEEE.

- Improved: Loops are constructed as

$$\text{for } n = 1, 2, \ldots, N$$
$$\text{for } k = 1, 2, \ldots, N$$
$$\text{for } m = 1, 2, \ldots, N$$
$$\overline{\boldsymbol{V}}[m, n] = \overline{\boldsymbol{V}}[m, n] + \overline{\boldsymbol{X}}[m, k]\overline{\boldsymbol{Y}}[k, n].$$

Figure 4.18 shows that, using the first (ordinary) scheme, the processing time of the matrix-matrix multiplications does not change linearly with respect to N^3, i.e., the complexity of matrix-matrix multiplications seems larger than $\mathcal{O}(N^3)$. In addition, the processing time is significantly longer for the first (ordinary) scheme compared to the second (improved) case, especially when N is large. A major difference between these two schemes is the use of the data stored in memory. Using the Fortran programming language, matrices are stored columnwise in memory, and it is always faster to access a local data compared to a distributed data. When the loops are constructed by following the ordinary scheme, $\overline{\boldsymbol{X}}$ and $\overline{\boldsymbol{V}}$ are not used in an efficient way. Specifically, in the ordinary scheme, both $\overline{\boldsymbol{X}}$ and $\overline{\boldsymbol{V}}$ are used row-wise, which leads to significant delays when accessing data in memory. Even the complexity of matrix-matrix multiplications seems larger than the expected due to the inefficient use of memory. As depicted above in Figure 4.18, simply changing the order of loops rectifies the complexity and provides the desired efficiency.

4.7.2 System Software

Software of the parallel computer should be optimized for the best possible performance of parallel implementations of MLFMA. As an example, Figure 4.19 presents an optimization of intra-node communications between eight processors of the B-Clovertown cluster. A simple update of the basic input/output system (BIOS) significantly accelerates intra-node communications. Specifically, the latency decreases and the bandwidth increases as depicted in Figures 4.19(a) and 4.19(b), respectively.

Figure 4.20 presents a similar acceleration of inter-node communications by updating the Infiniband driver. The processing time required for an integer transfer is plotted when processors located in different nodes perform communications with each other. If the number of processors is larger than two, multiple communications are performed at the same time. Hence, the communication time increases as the number of processors increases. Figure 4.20 shows that an update of the Infiniband driver can significantly accelerate inter-node communications.

4.7.3 Load Balancing

Load balancing in MLFMA can be performed with data-binning techniques since the multilevel structure, as well as near-field and far-field interactions, are known at the beginning of the program. The effectiveness of different load-balancing algorithms and resulting parallelization efficiencies can be monitored on various problems.

Using cubic boxes in MLFMA provides many computational advantages thanks to the symmetry. At the same time, since the box size is fixed, basis and testing functions are distributed unequally among boxes at the lowest level. As an example, Figure 4.21 presents the solution of a scattering problem involving a PEC sphere of radius 20λ discretized with 1,462,854 unknowns. The problem is solved by a 7-level MLFMA using a bottom-up scheme. Figure 4.21(a) depicts a histogram of the lowest-level boxes in terms of the number of unknowns. It can be observed that the number of unknowns per box changes in a wide range from 1 to 42. Also note that about 7000 out of 109,790 boxes involve only one unknown. Figure 4.21(b) depicts the number of near-field interactions assigned to each processor when the solution is parallelized on a cluster of 32 processors and the rows of the matrix equation are distributed equally among processors. Since the number of unknowns per box vary significantly, the number of near-field interactions per row also vary significantly, even for a simple sphere geometry. Hence, an equal distribution of rows leads to an unequal distribution of unknowns among processors. As a remedy, one can use a load-balancing algorithm in order to distribute near-field interactions (instead of rows) equally among processors, leading to the near-field partitioning. Note that a different load balancing of far-field interactions leads to the far-field partitioning, which can be significantly different from the near-field partitioning.

Numerical Examples

Table 4.5 presents the solution of large-scale electromagnetics problems discretized with more than 100 million unknowns on the B-Harpertown cluster. A scattering problem involving a PEC

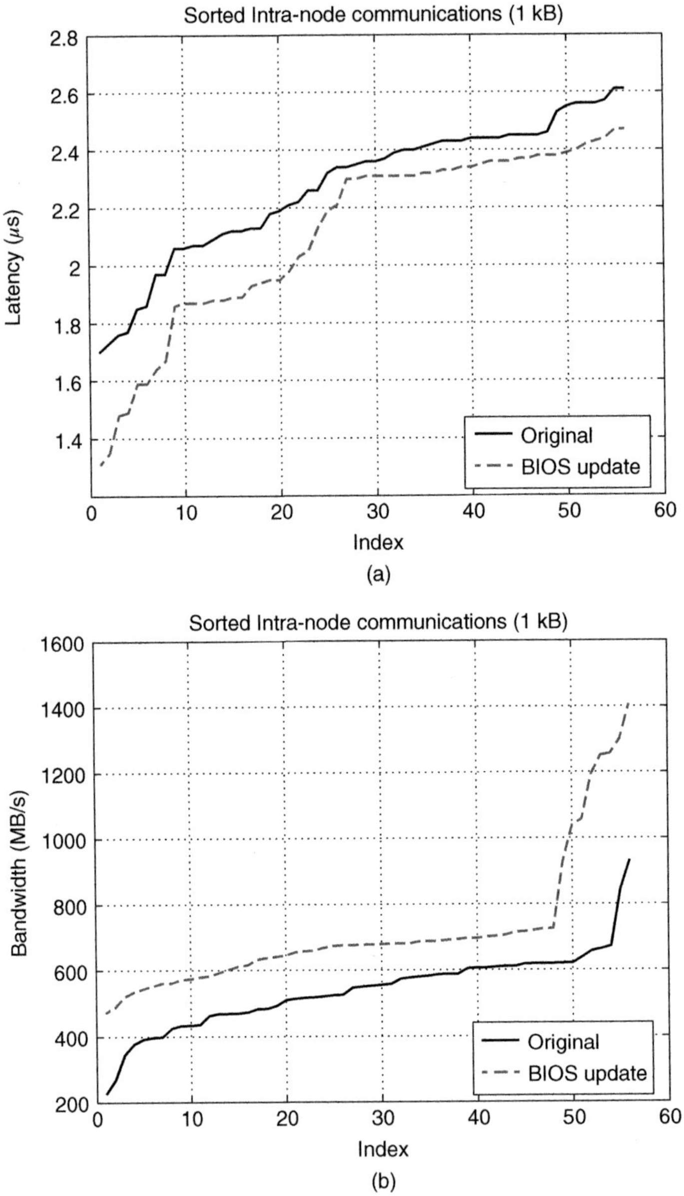

Figure 4.19 Optimization of intra-node communications in the B-Clovertown cluster via a BIOS update. (a) Sorted latency (μs) and (b) sorted bandwidth (MB/s)

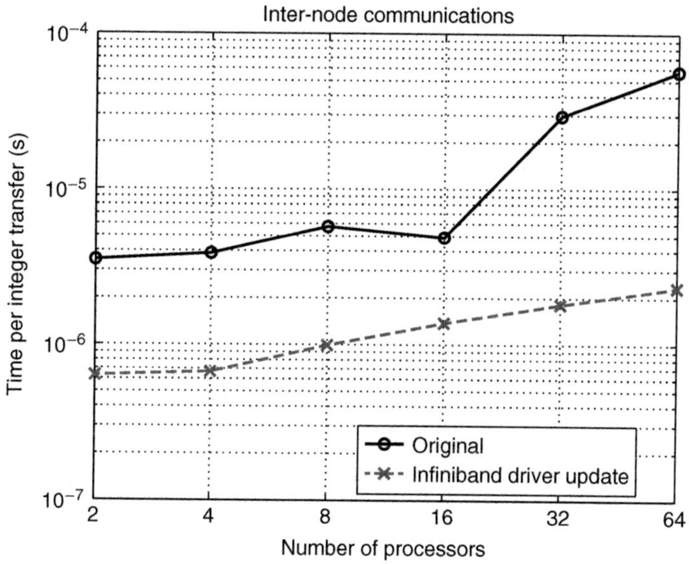

Figure 4.20 Optimization of inter-node communications in the B-Clovertown cluster via an Infiniband driver update.

sphere of radius 180λ discretized with 135,164,928 unknowns is solved by a 10-level MLFMA. In addition, scattering problems involving the Flamme at 360 GHz, the NASA Almond at 850 GHz, and the wing-shaped object at 120 GHz are considered (see Section 4.3). The sphere is illuminated by a plane wave propagating in the $-x$ direction with the electric field polarized in the y direction. Other three objects are illuminated by a plane wave propagating in the x-y plane at a 30° angle from the x axis with the electric field polarized in the z direction (V polarization). Solutions are performed on 16, 32, and 64 processors of the B-Harpertown cluster. Table 4.5 lists the processing time, as well as the parallelization efficiency obtained for 32 and 64 processors with respect to 16 processors. In addition to the setup and solution parts, aggregation, translation, and disaggregation stages are considered separately. Note that 'MVM Rest' accounts for near-field interactions, all-to-all communications to change the partitioning of the output vector before far-field interactions, and some input/output operations. Table 4.3 shows that, using 64 processors, the parallelization efficiency for the total time is more than 80% for all problems.

Table 4.6 lists the total memory required for the solutions listed in Table 4.5. Increasing the number of processors raises the memory requirement since the efficiency of the parallelization is not perfectly 100%. Nevertheless, using 64 processors, the parallelization efficiency in terms of memory usage is also larger than 80% for all problems.

Figure 4.22 presents some details of the processing time for the 64-processor solutions listed in Table 4.5. It can be observed that the calculation of near-field interactions and the iterative solution are dominant parts. The processing time for near-field interactions is perfectly balanced among processors for all cases, except for the wing-shaped object. This is because all near-field interactions are calculated accurately with 1% error and the number of integration points, thus the processing time, for an interaction depends on the relative

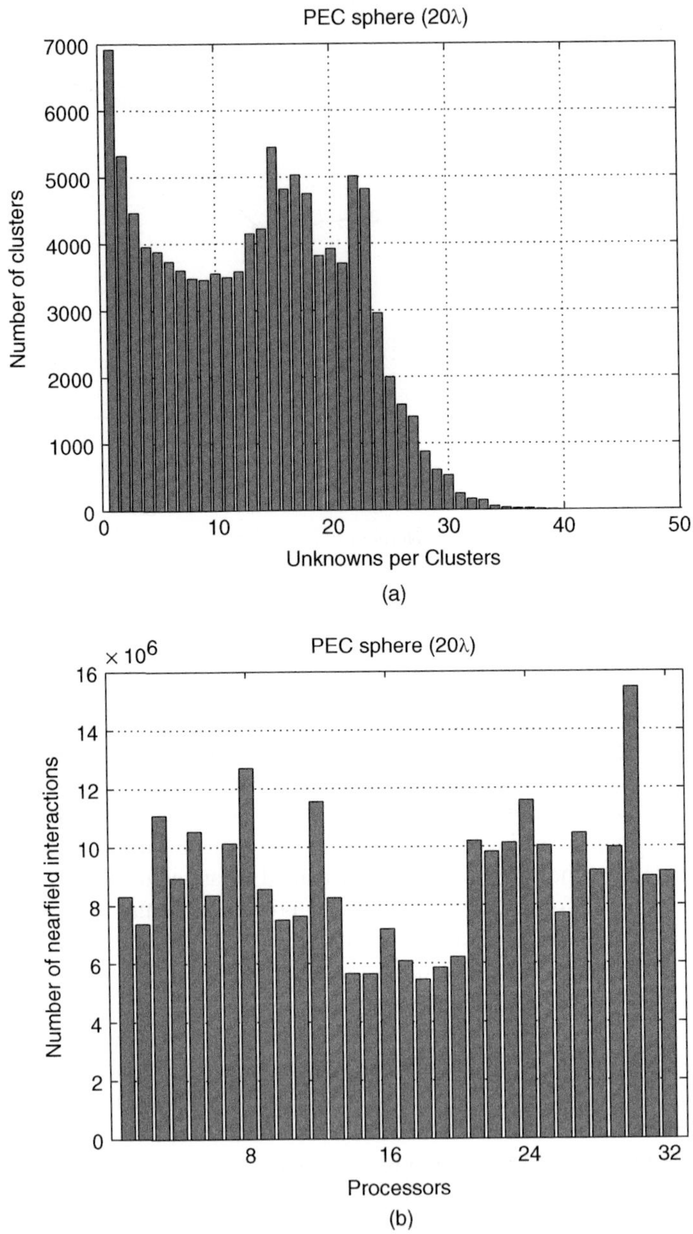

Figure 4.21 Solution of a scattering problem involving a PEC sphere discretized with 1,462,854 unknowns. (a) Histogram of the lowest-level boxes in terms of the number of unknowns. (b) Number of near-field interactions assigned to each processor when the rows of the matrix equation are distributed equally among 32 processors.

Table 4.5 Solutions of large-scale electromagnetics problems involving more than 100 million unknowns with parallel MLFMA on the B-Harpertown cluster

Problem	Sphere	Flamme	Almond	Wing
Size	360λ	720λ	715λ	400λ
Unknowns	135,164,928	134,741,760	125,167,104	121,896,960
Levels	10	11	11	10
Box size	0.176λ	0.176λ	0.175λ	0.195λ
Excitation	H (0°)	V (30°)	V (30°)	V (30°)
Iterations	23	44	20	17
	Processing time for 16 processors			
Setup (minutes)	480	537	435	366
Solution (minutes)	489	644	330	177
Total (minutes)	975	1186	769	546
MVM (s)	618	425	449	285
Aggregation (s)	200	138	154	98
Translation (s)	194	129	128	69
Disaggregation (s)	209	144	155	106
MVM rest	15	14	12	12
	Processing time for 32 processors			
Setup (minutes)	243 (99%)	272 (99%)	219 (99%)	186 (98%)
Solution (minutes)	263 (93%)	369 (87%)	171 (97%)	99 (89%)
Total (minutes)	511 (95%)	646 (92%)	394 (98%)	289 (95%)
MVM (s)	327 (95%)	234 (91%)	239 (94%)	155 (92%)
Aggregation (s)	103 (97%)	72 (96%)	77 (100%)	51 (96%)
Translation (s)	105 (92%)	75 (86%)	72 (89%)	40 (86%)
Disaggregation (s)	106 (99%)	76 (95%)	80 (97%)	54 (98%)
MVM Rest (s)	13	11	10	10
	Processing time for 64 processors			
Setup (minutes)	126 (95%)	136 (99%)	109 (100%)	96 (95%)
Solution (minutes)	161 (76%)	203 (79%)	101 (82%)	62 (71%)
Total (minutes)	292 (84%)	345 (86%)	215 (89%)	163 (84%)
MVM (s)	194 (80%)	128 (83%)	137 (82%)	92 (78%)
Aggregation (s)	58 (86%)	37 (93%)	43 (90%)	28 (88%)
Translation (s)	64 (76%)	41 (79%)	39 (82%)	21 (82%)
Disaggregation (s)	60 (87%)	38 (95%)	45 (86%)	30 (88%)
MVM rest (s)	12	12	10	13

positions of basis and testing functions. On the discretized wing-shaped object, there are difficult interactions that require longer processing time, compared to other interactions. Hence, using a load-balancing algorithm based on the equal distribution of near-field interactions without considering their computation times, those difficult interactions may accumulate in some processors, leading to unequal processing times for the near-field stage.

Table 4.6 Peak memory (GB) required for the solutions in Table 4.5

Problem	Sphere	Flamme	Almond	Wing
Size	360λ	720λ	715λ	400λ
Unknowns	135,164,928	134,741,760	125,167,104	121,896,960
16 processors	424	427	385	431
32 processors	438 (97%)	455 (94%)	414 (93%)	443 (97%)
64 processors	467 (91%)	513 (83%)	471 (82%)	500 (86%)

Nevertheless, distributing near-field interactions equally among processors may still be preferred since this is usually the best choice in terms of memory. For the 64-processor solutions, Figure 4.23 presents details of MVMs, which involve near-field, aggregation, intra-processor translation, inter-processor translation, and disaggregation stages. It can be observed that, using the hierarchical strategy and load-balancing algorithms, the workload is distributed very efficiently.

Figure 4.24 presents some details of the memory usage for the 64-processor solutions listed in Table 4.5. In this figure, the total amount of memory used in each computing node from 1 to 16 is considered. A majority of memory is required to store near-field interactions and radiation and receiving patterns of basis and testing functions. Another significant contribution is due to aggregation and disaggregation arrays, which contain radiated and incoming fields of boxes calculated during MVMs. Memory used in computing nodes may not be the same, mostly due to different amounts of radiation and receiving patterns of basis and testing functions assigned to processors. Those patterns are distributed according to the far-field partitioning of the tree structure. In general, boxes at the lowest level are distributed almost equally among processors. However, box populations, i.e., numbers of basis and testing functions in boxes, may vary significantly. A load-balancing algorithm, which accounts for populations of boxes, can be used to improve the distribution of radiation and receiving patterns. However, this method may deteriorate the load balancing of MVMs in terms of the processing time.

4.7.4 *Memory Recycling and Optimizations*

Although the minimization of the processing time is crucial in parallel solutions, efficient use of the available memory becomes more critical as the problem size grows. Parallel implementations of MLFMA usually have complicated memory footprints. Dynamic memory allocations and deallocations are essential to recycle the used memory as much as possible. An important issue is the optimization of memory during MVMs (usually the peak memory). Applying memory recycling, all unnecessary data structures are deallocated; only essential data structures remain allocated before iterative solutions. During an iterative solution, a majority of memory is used for near-field interactions, radiation/receiving patterns of basis/testing functions, translation operators, and aggregation/disaggregation arrays. For solving large problems on distributed-memory architectures, it is essential to distribute these data structures equally among processors. Otherwise, even though the total amount of memory is sufficient to solve a given problem, memory required by a specific processor can exceed the maximum memory limit available for that processor, and this may prevent the solution of the problem.

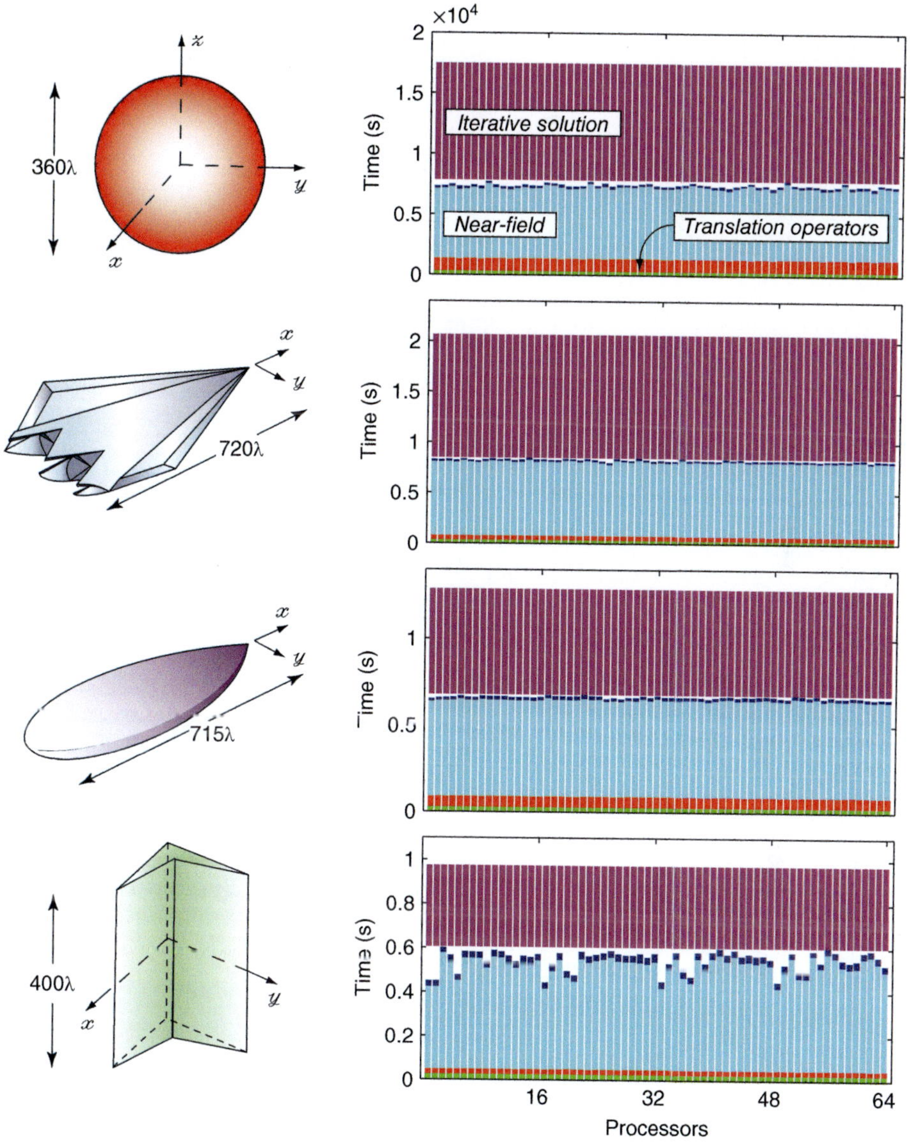

Figure 4.22 Details of the processing time for the solution of large scattering problems on 64 processors of the B-Harpertown cluster.

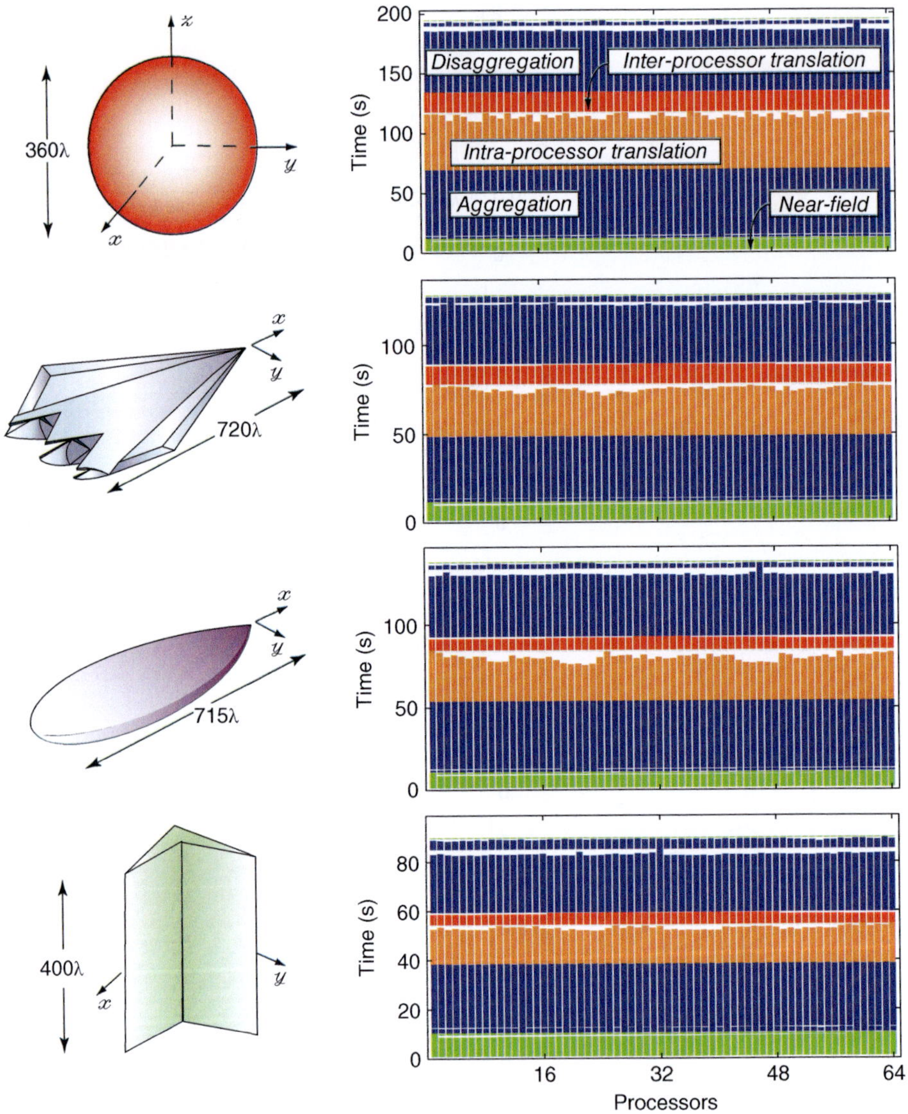

Figure 4.23 Details of MVMs for the solution of large scattering problems on 64 processors of the B-Harpertown cluster.

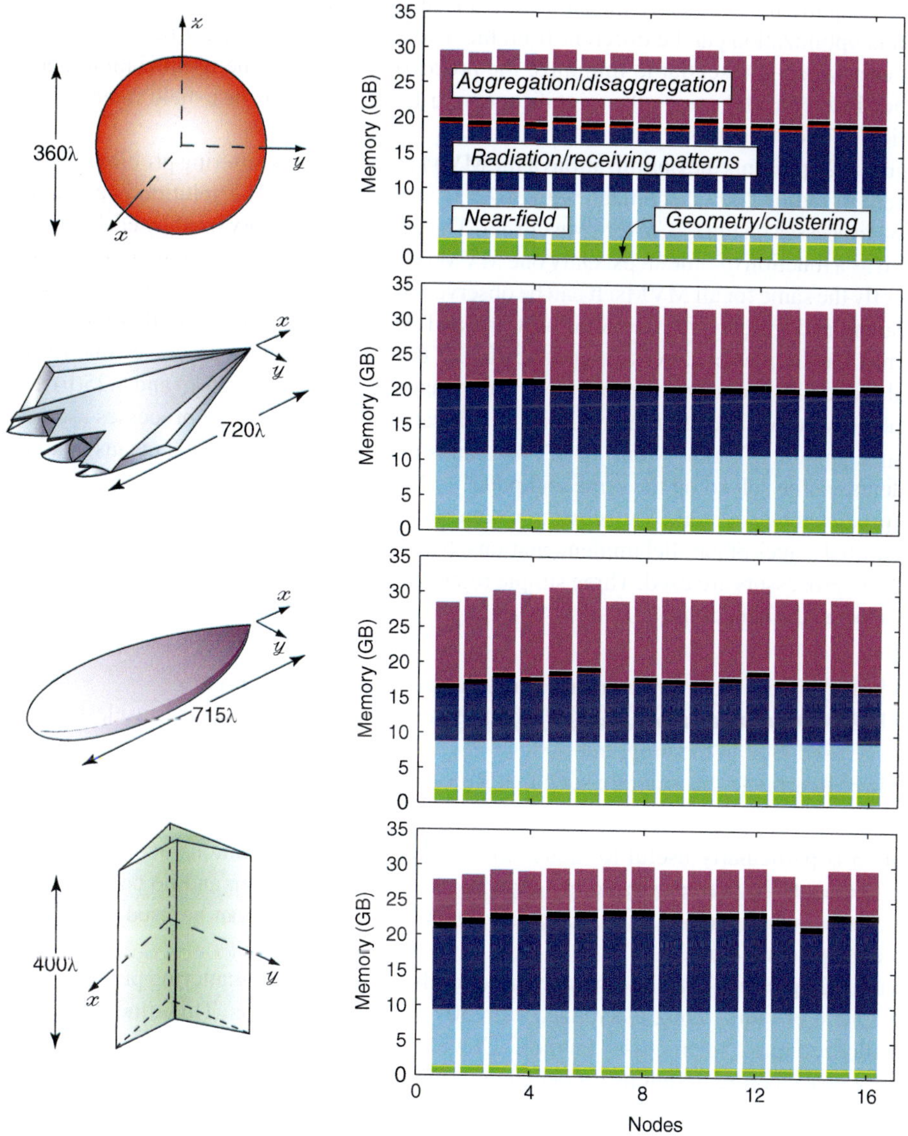

Figure 4.24 Details of memory used for the solution of large scattering problems on 64 processors of the B-Harpertown cluster.

Hence, parallel MLFMA implementations should be carefully optimized such that all processors require approximately the same amount of memory during iterative solutions. Note that this optimization can be different from the load-balancing scheme for the multilevel tree structure, which is essential to minimize the processing time. For the optimization memory, one needs to consider some other significant contributions, such as near-field interactions, in addition to tree structures.

Figure 4.25 presents the optimization of the MVM memory on a scattering problem involving a sphere of radius 260λ discretized with 307,531,008 unknowns. The solution is parallelized into 64 processes on the C-Clovertown cluster. The amount of memory used by each process is plotted as a function of time steps. Only one MVM is considered since the memory requirement is exactly the same for all MVMs. It can be observed that the used memory is not monotonically increasing and it fluctuates due to allocations and deallocations. Also note that the memory required for the master process is quite different than those of other processes due to some initial sequential operations for data input and management. Figures 4.25(a) and 4.25(b) depict the memory before and after the optimization. It can be observed that all processes use nearly the same amount of memory during MVM, thanks to the optimization. This means that larger problems can be solved on the same computer.

As predicted by Amdahls law, sequential data structures, particularly those that are allocated in the initial stages of parallel implementations, become bottlenecks as the problem size grows and more processors are used. Three simple rules are used repetitively to avoid memory peaks and associated stagnations:

- Rule 1: Allocate memory for a data structure just before it is required.
- Rule 2: Deallocate memory used for a data structure as soon as it becomes useless so that it can be used later in the program.
- Rule 3: Rearrange the program such that Rules 1 and 2 can be further applied.

Rule 3 is particularly useful to reduce memory peaks before iterations and MVMs. Code rearrangements for memory recycling depend on the implementation and no common procedure exists. But, Figure 4.26 illustrates how such rearrangements can be used to effectively reduce the peak memory.

Figure 4.27 presents again the memory used for the solution of the scattering problem involving a sphere of radius 260λ discretized with 307,531,008 unknowns. This time, the solution is parallelized into 128 processes on the C-Clovertown cluster. Using 128 processes, there are memory peaks before iterations, as depicted in Figure 4.27(a). These peaks are major bottlenecks before the memory recycling. Using code rearrangements and memory recycling, the peak memory of the implementation can be reduced significantly (from 12.6 GB to 10.4 GB), as depicted in Figure 4.27(b). This reduction translates into the ability to solve larger problems on the same computer.

4.7.5 Parallel Environment

Parallel solutions are naturally affected by the parallel environment. This subsection presents effects of some factors, such as other running problems, distribution of processes, MPI version, and faulty processors.

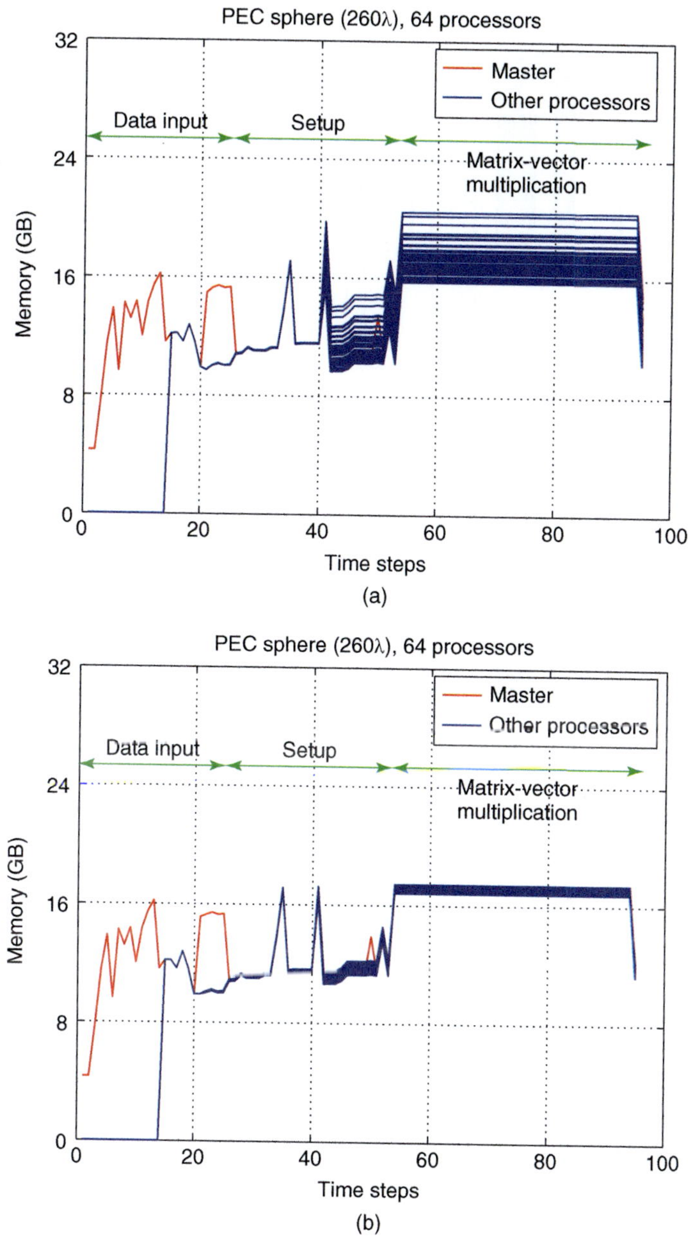

Figure 4.25 Memory used for the solution of a scattering problem involving a sphere of radius 260λ discretized with 307,531,008 unknowns. The solution is parallelized into 128 processes using the hierarchical strategy. Memory for each process is plotted as a function of time steps (a) before and (b) after the optimization of the MVM memory.

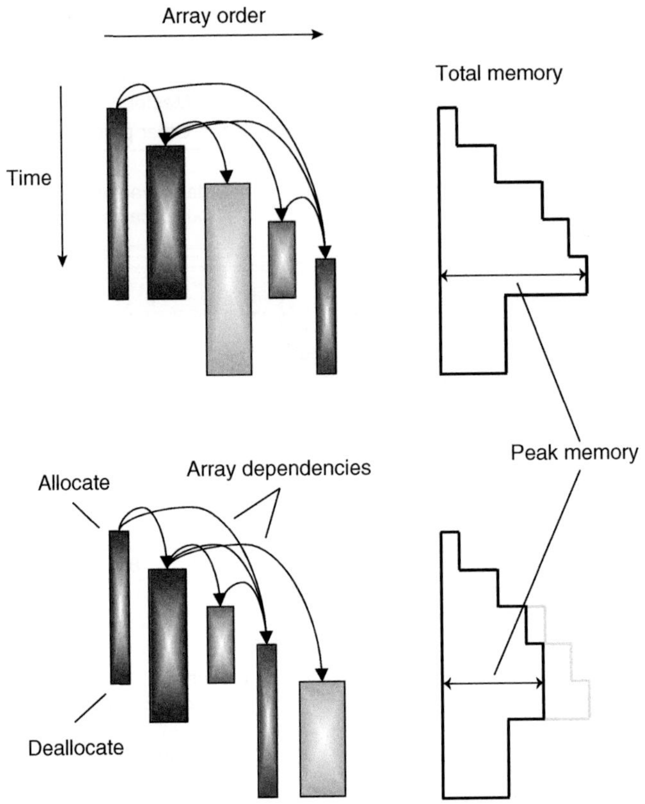

Figure 4.26 Code rearrangements for memory recycling.

Other Running Programs

Figure 4.28 presents the solution of a scattering problem involving a PEC sphere of radius 20λ discretized with 1,462,854 unknowns. The problem is solved by a 7-level MLFMA using the hybrid strategy and a bottom-up scheme on the B-Clovertown cluster. Two different cases are considered:

- A system program requiring a significant processing load is employed on a single processor of a node, in addition to the parallel MLFMA.
- All processors are reserved for the parallel MLFMA.

Figures 4.28(a) and 4.28(b) depict the processing time for the setup and iterative solution parts, respectively, with respect to the number of processors. It can be observed that the processing time decreases regularly as the number of processors increases up to 32 in both cases, i.e., with and without the system program. When the number of processors is further increased to 64, however, the existence of the system program affects the processing time significantly, even though this program is employed on a single processor. In fact, for the solution part, the processing time increases from 32 processors to 64 processors. This experiment shows the

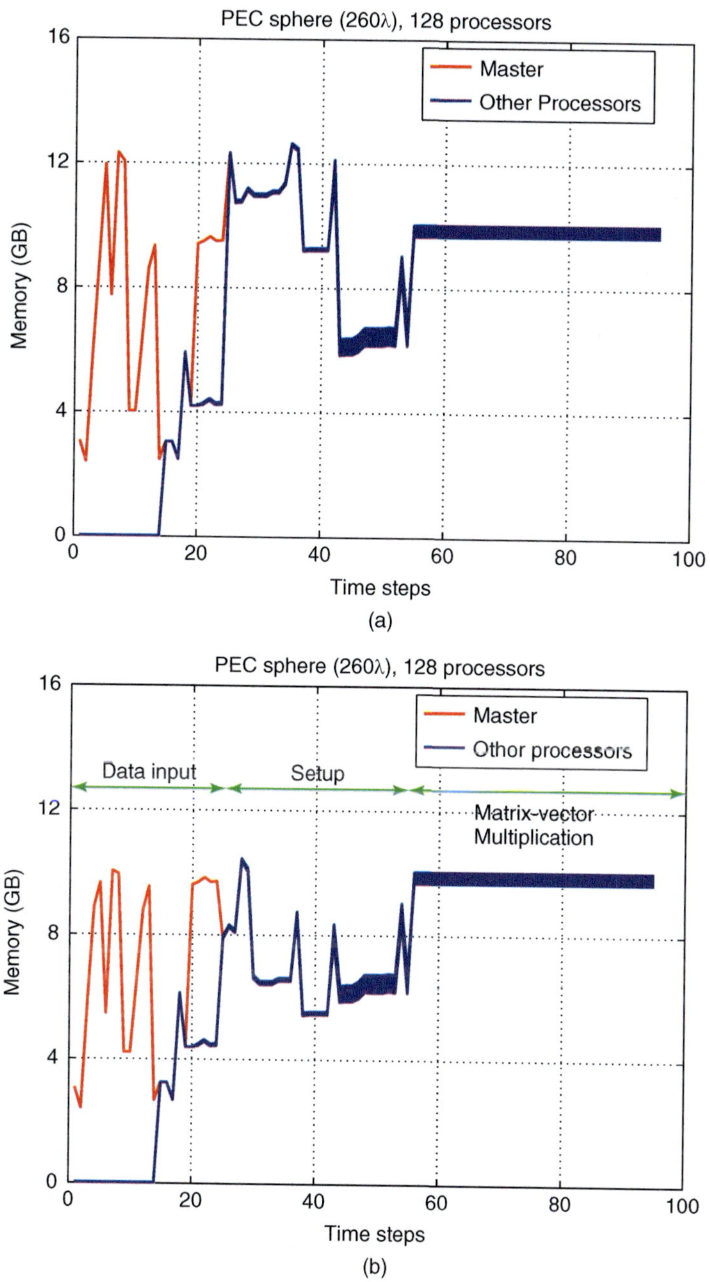

Figure 4.27 Memory used for the solution of a scattering problem involving a sphere of radius 260λ discretized with 307,531,008 unknowns. The solution is parallelized into 128 processes using the hierarchical strategy. Memory for each process is plotted as a function of time steps (a) before and (b) after code rearrangements followed by memory recycling. *Source*: Gürel and Ergül 2013 [235]. Reproduced with permission of IEEE.

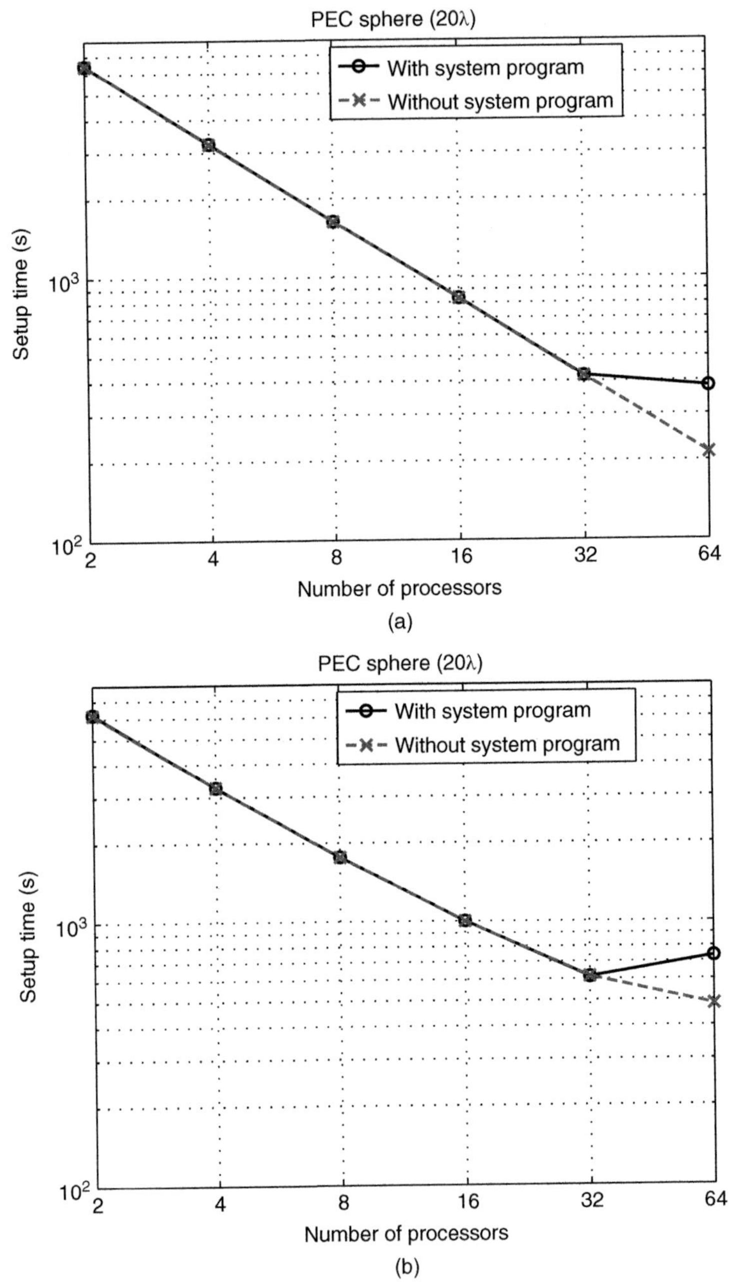

Figure 4.28 Processing time for the solution of a scattering problem involving a PEC sphere of radius 20λ discretized with 1,462,854 unknowns. (a) Setup and (b) solution times are investigated with and without a system program employed on the B-Clovertown cluster.

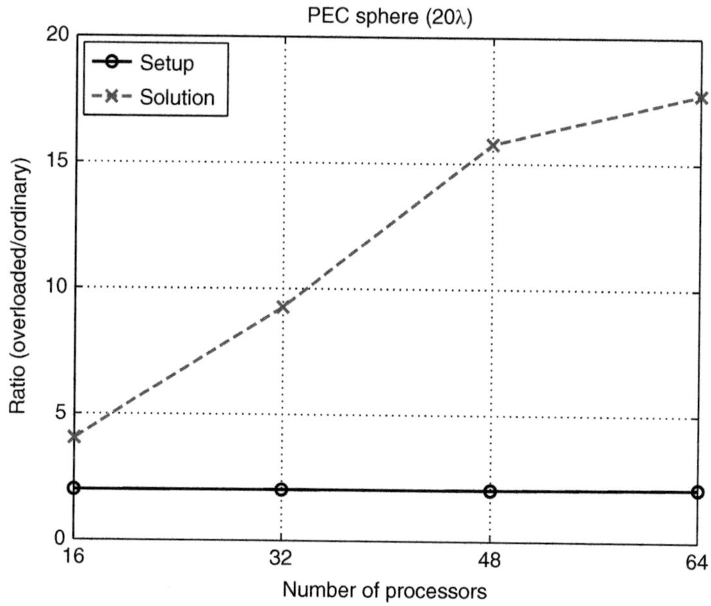

Figure 4.29 Solutions of a scattering problem involving a PEC sphere of radius 20λ discretized with 1,462,854 unknowns on the B-Clovertown cluster. The processing times obtained when the parallel computer is employed as usual and when it is overloaded are compared.

necessity of a 'master node' to host system programs without affecting the processing time of parallel implementations.

Figure 4.29 presents again the solution of the sphere problem involving 1,462,854 unknowns on the B-Clovertown cluster. In addition to ordinary solutions on 16–64 processors, overloaded solutions are considered by employing dummy programs at the same time on all 64 processors of the cluster. In Figure 4.29, the ratio of processing times required for the overloaded and ordinary solutions is plotted as a function of the number of processors. It can be observed that the setup time is doubled since each processor is actually used with 50% efficiency. On the other hand, overloading significantly affects the solution part, which involves many communications between processors. Specifically, using 32 and 64 processors, overloading leads to more than 15-fold deceleration for the solution part.

As discussed above in Section 4.6.7, using slower processors may increase the parallelization efficiency since the computation time tends to increase and dominate the communication time. This observation does not contradict with the results in Figure 4.29. Specifically, overloading in Figure 4.29 does not only slow down processors but also decreases the performance of the whole system, including communications. The setup part, which involves only computations, is affected by 50% due to the reduced effective clock rate. The solution part, however, is affected more than 50% since it requires synchronizations between processors, and hence, delays in different processors tend to accumulate.

Distribution of Processes

Figure 4.30 presents solutions of a scattering problem involving a PEC sphere of radius 20λ discretized with 1,462,854 unknowns on the S-Tigerton cluster. Solutions are parallelized among various numbers of processors. Figure 4.30(a) depicts a comparison of two different cases, i.e., when 8 and 16 nodes of the cluster are employed. In both cases, the number of active nodes is maximized for all solutions. One can observe that the ratio of processing times on 8 and 16 nodes is close to one, except for the 64-processor case. In this case, using 8 nodes instead of 16 nodes increases the solution time by 8%. This experiment shows that using all processors in a node may decrease the performance of the parallel computer, even though a sufficient number of processors is available. Figure 4.30(b) presents the solution of the same problem on 16 nodes of the S-Tigerton cluster when the number of nodes is maximized (distributed) and when the number of nodes is minimized (full). For example, 16 processors are selected from 16 nodes (1 processor per each node) and 2 nodes (8 processors per each node) in the distributed and full schemes, respectively. Figure 4.30(b) confirms that using all processors in a node increases the iterative solution time. Note that, this is not observed for the setup part, which only involves computations and does not require any communications between processors.

MPI Version

The choice of the MPI program is another major parameter, which affects the parallelization of MLFMA. For a given parallel computer, one of the interfaces may perform better than others. As an example, Figure 4.31(a) presents a comparison of different MPI versions for the solution of a scattering problem involving a PEC sphere of radius 20λ discretized with 1,462,854 unknowns on the B-Clovertown cluster. One can observe that MVAPICH leads to faster solutions than LAM-MPI and Open-MPI, especially when the number of processors is large. It is remarkable that the setup time does not depend on the MPI program, but the solution time does. Using 64 processors, Open-MPI and LAM-MPI are 100% and 30% slower than MVAPICH for the solution part.

On the B-Clovertown cluster, MVAPICH is also more stable than other MPI versions for the solution of large-scale problems. As an example, Figure 4.31(b) presents the solution of a scattering problem involving a PEC sphere of radius 110λ discretized with 41,883,648 unknowns. The processing time required for each MVM is plotted. It can be observed that the first two MVMs require significantly more time than the others using LAM-MPI. This undesirable effect is not observed when using MVAPICH that provides more stable solutions than LAM-MPI. It should be emphasized that these findings on the MPI versions are based on their current versions and they are measured on the B-Clovertown cluster, i.e., they may not be generalized to other parallel computers. Nevertheless, the results in Figure 4.31 shows that the choice of the MPI program can be crucial in terms of the efficiency and stability of parallel solutions.

Faulty Processors

As the number of processors of parallel computers increases, it becomes more probable that some processors suffer from technical problems, such as overheating. Such faulty processors directly affect the efficiency of parallel solutions, but unfortunately, their effects may not be

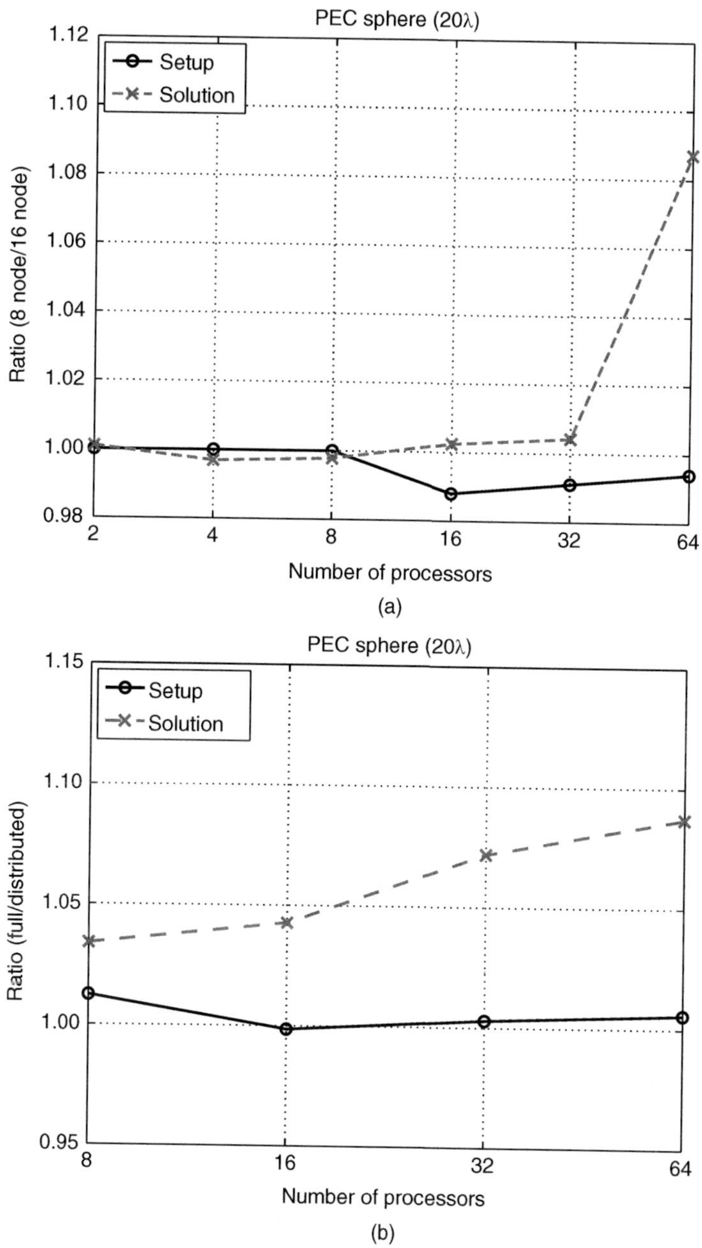

Figure 4.30 Solutions of a scattering problem involving a PEC sphere of radius 20λ discretized with 1,462,854 unknowns on the S-Tigerton cluster. The processing times obtained when (a) different numbers of nodes are employed and (b) processors are distributed among nodes in different (full and distributed) ways are compared.

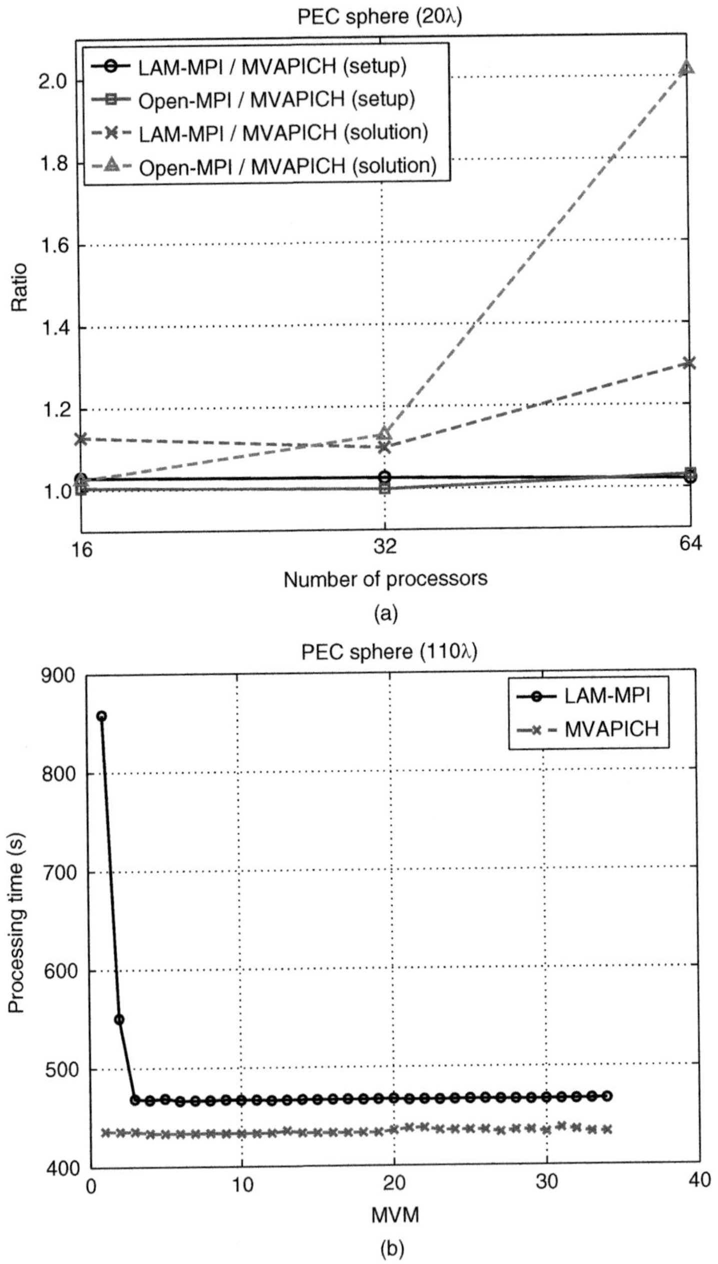

Figure 4.31 Comparisons of MPI versions on scattering problems involving (a) a PEC sphere of radius 20λ discretized with 1,462,654 unknowns and (b) a PEC sphere of radius 110λ discretized with 41,883,648 unknowns.

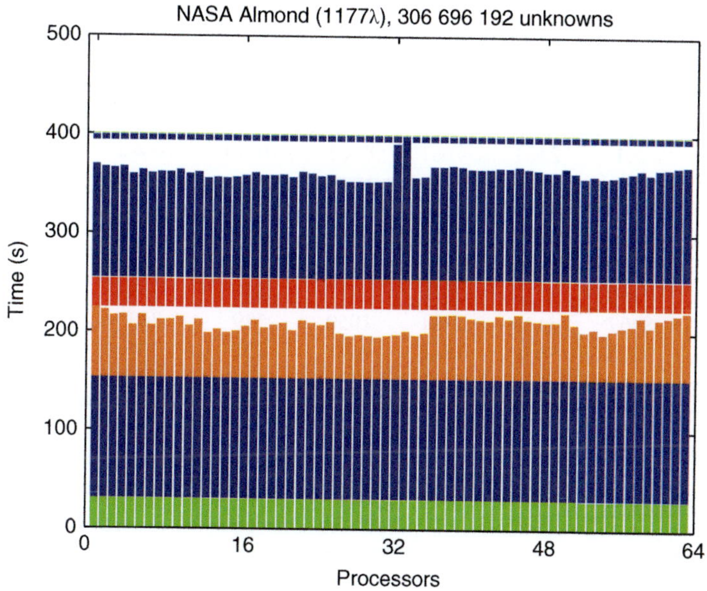

Figure 4.32 Solution of a scattering problem involving the NASA Almond at 1.4 THz. The problem is discretized with 306,696,192 unknowns and the solution is parallelized into 64 processes on the U-Nehalem cluster. Details of a MVM are shown for all processors.

understood easily. As an example, Figure 4.32 presents the solution of a scattering problem involving the NASA Almond at 1.4 THz discretized with 306,696,192. The solution is parallelized into 64 processes on the U-Nehalem cluster. Figure 4.32 depicts the details of a MVM, which involves near-field, aggregation, intra-processor translation, inter-processor translation, and disaggregation stages. White areas correspond to wait periods before synchronizations. It can be observed that two of the processors require significantly longer time during the disaggregation stage. Hence, other processors have to wait before the final synchronization and the parallelization efficiency drops. The results in Figure 4.32 looks like an unbalanced workload among processors. But, in fact, two processors suffer from overheating problems and they significantly affect the overall performance of the parallel implementation on the cluster.

4.7.6 *Parallel Computers*

Obviously, efficiency of parallel implementations may vary significantly depending on the parallel computer. There are many parameters, such as the processor model, clock rate, cache size, memory access rate, and network, and all these parameters make contribution to the overall efficiency of parallel solutions. As an example, Tables 4.7–4.9 present the solution of a scattering problem involving a PEC sphere of radius 20λ discretized with 1,462,854 unknowns. The problem is solved on various parallel computers listed in Table 4.1 after various time and memory optimizations described earlier in this chapter. Tables 4.7–4.9 presents the setup, iterative solution, and total computation times for the solution of the scattering problem on

Table 4.7 Setup time for a scattering problem involving a PEC sphere discretized with 1,462,854 unknowns on different numbers of processors of the computers listed in Table 4.1

	Number of nodes × Number of processes per node							
Computer	1 × 1	2 × 1	4 × 1	8 × 1	16 × 1	16 × 2	16 × 4	16 × 8
C-Clovertown	12,560	6319	3203	1612	817	411	210	109
B-Harpertown	7767	3599	1851	933	467	242	120	63
U-Nehalem	8259	4131	2095	1049	534	269	137	73
S-Nehalem	7950	3954	2058	1011	557	279	141	70
N-Nehalem	8467	4138	2111	1050	529	263	133	65
S-Westmere	7294	3658	1844	930	472	237	120	62
S-Dunnington	9114	4612	2348	1197	591	297	154	80

Table 4.8 Iterative solution time for a scattering problem involving a PEC sphere discretized with 1,462,854 unknowns on different numbers of processors of the computers listed in Table 4.1

	Number of nodes × Number of processes per node							
Computer	1 × 1	2 × 1	4 × 1	8 × 1	16 × 1	16 × 2	16 × 4	16 × 8
C-Clovertown	6907	3564	1892	1029	563	319	201	150
B-Harpertown	5510	2913	1564	833	456	259	168	142
U-Nehalem	5189	2760	1456	774	432	255	167	135
S-Nehalem	4910	2855	1498	829	455	232	140	122
N-Nehalem	5542	2938	1572	847	459	243	147	93
S-Westmere	4388	2396	1311	690	385	208	130	95
S-Dunnington	6634	4240	2604	1479	911	477	363	350

Table 4.9 Total computing time for a scattering problem involving a PEC sphere discretized with 1,462,854 unknowns on different numbers of processors of the computers Listed in Table 4.1

	Number of nodes × Number of processes per node							
Computer	1 × 1	2 × 1	4 × 1	8 × 1	16 × 1	16 × 2	16 × 4	16 × 8
C-Clovertown	19,474	9891	5103	2649	1388	738	420	270
B-Harpertown	12,679	6515	3419	1768	926	505	292	209
U-Nehalem	13,462	6905	3565	1838	980	539	316	219
S-Nehalem	12,862	6819	3566	1852	1023	522	291	203
N-Nehalem	13,011	7079	3686	1899	991	508	282	161
S-Westmere	11,684	6064	3157	1630	867	447	253	162
S-Dunnington	15,751	8855	4956	2681	1507	778	525	440

these computers using different numbers of nodes and processors per node. The setup time includes (and is dominated by) the computation of near-field interactions. The iterative solution time includes 27 iterations (54 MVMs) to reduce the residual error to below 10^{-6}. Finally, the total time includes the setup and iterative solution, as well as data input and management. It can be seen that the total time of the fastest solution is 161 seconds on 128 processors of the N-Nehalem cluster. It is remarkable that, among the parallel computers listed in Table 4.1, N-Nehalem does not have the fastest processors. Figure 4.33 depicts the solution of the scattering problem on three different clusters, i.e., S-Nehalem, N-Nehalem, and S-Westmere. Both the total computing time and the corresponding parallelization efficiency are plotted with respect to the number of processors from 1 to 128. It can be observed that the parallelization efficiency, which is defined with respect to one-processor solutions for each cluster, changes significantly depending on the computer. As also discussed above in Section 4.6.7, the parallelization efficiency can be misleading to assess the actual efficiency of parallel solutions. And, in fact, as demonstrated in Tables 4.7–4.9 and Figure 4.33, a direct comparison of the computing time may provide a better assessment of the efficiency on different computers.

4.8 Accuracy Considerations for Parallel Implementations of MLFMA

MLFMA is a full-wave solver with a controllable accuracy. However, accuracy of a solution and its dependence on various error sources of the algorithm may not be obvious. For a given problem, its discretization (e.g., the number of unknowns) and other accuracy parameters may not be sufficient to assess the actual accuracy of the solution. This uncertainty of accuracy makes it difficult to rigorously compare different implementations, since the accuracy of solutions obtained with these implementations can be different and direct comparisons can be unfair. Increasing the problem size and the number of unknowns (a recent trend in computational electromagnetics) should be supported and accompanied by the investigation of the accuracy of results.

In general, MLFMA implementations involve various error sources that can be categorized as follows:

- Discretization of surfaces using small elements, e.g., planar triangles, with respect to wavelength.
- Discretization of integral-equation formulations using a set of basis and testing functions.
- Direct calculation of the near-field interactions using numerical integration methods, e.g., Gaussian quadratures and adaptive techniques.
- Factorization and diagonalization of the far-field interactions, e.g., truncation, clustering scheme, and translation strategies.
- Multilevel aggregation and disaggregation, e.g., interpolation and anterpolation techniques.
- Iterative solution, e.g., convergence to the target residual error.

All these error sources are important and affect the accuracy of solutions, but some of them may dominate the others depending on the problem and implementation. Recent analyses based on numerical comparisons show that some error sources are more dominant and thus should be suppressed for more accurate solutions. At the same time, identifying less-effective

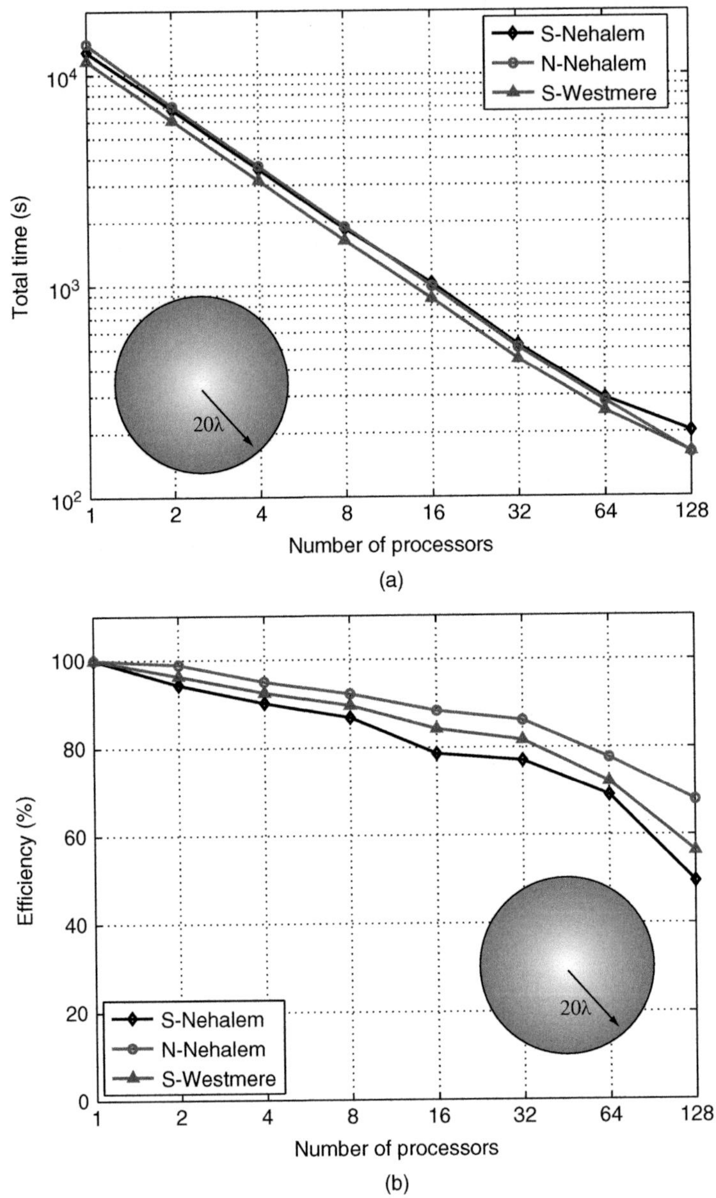

Figure 4.33 Solution of a scattering problem involving a PEC sphere of radius 20λ discretized with 1,462,854 unknowns on three different clusters. (a) Total computing time and (b) the corresponding parallelization efficiency are plotted with respect to the number of processors from 1 to 128.

error sources may allow for the derivation of more efficient implementations by systematically relaxing the corresponding parameters.

As an example, solutions of a scattering problem involving a PEC sphere of radius $a = 96\lambda$ discretized with 33,791,232 unknowns are considered. The sphere is illuminated by a plane wave propagating in the $-x$ direction with the electric field polarized in the y direction. The problem is formulated with CFIE = αEFIE + $(1 - \alpha)$MFIE. In order to demonstrate the effect of the combination factor α on the accuracy, two different solutions with $\alpha = 0.2$ and $\alpha = 0.6$ are considered. Both near-field and far-field interactions are calculated with maximum 1% error. Convergence to 0.001 residual error is achieved in 20 and 66 iterations for $\alpha = 0.2$ and $\alpha = 0.6$, respectively. Figures 4.34 and 4.35 present the normalized RCS (RCS$/\pi a^2$ in dB) on the x-y plane as a function of the bistatic angle ϕ from 0° to 180°, where 0° and 180° correspond to the backscattering and forward-scattering directions, respectively. RCS results around the backscattering and forward-scattering directions are particularly focused in separate plots. In general, the computational results agree well with the analytical results obtained by a Mie-series solution. However, one can find different levels of error in a detailed analysis, such as the focused plot around the backscattering direction. Specifically, the error is more significant in Figure 4.34, i.e., when $\alpha = 0.2$ and the MFIE contribution is large in the CFIE formulation. For a more quantitative analysis, the relative error given in (2.196) is calculated when the computational and analytical values are sampled at 0.1° intervals. For the results depicted in Figures 4.34 and 4.35 (when $\alpha = 0.2$ and $\alpha = 0.6$), the relative error is found to be 2.41% and 0.69%, respectively. These values clearly show that the combination factor α of CFIE is an important parameter for the accuracy, even for very large problems.

Figures 4.36 and 4.37 present solutions of another scattering problem involving a PEC sphere of radius 280λ discretized with 374,490,624 unknowns when $\alpha = 0.2$ and $\alpha = 0.6$, respectively. Similar to the previous example, the accuracy of the solution is significantly affected by the choice of the combination factor α of CFIE. Specifically, for this problem, the relative error defined in (2.196) is 1.95% and 0.58% when $\alpha = 0.2$ and $\alpha = 0.6$, respectively. The number of BiCGStab iterations to reduce the residual error to below 10^{-3} increases from 31 to 89 when α changes from 0.2 to 0.6; but this increase in the iteration counts (hence the computing time) can be acceptable because of the significantly improved accuracy.

Table 4.10 lists the relative error in various solutions of the scattering problem involving the sphere of radius 96λ. For both $\alpha = 0.2$ and $\alpha = 0.6$, accuracy is investigated when the far-field interactions are calculated with zero, one, two, and three digits of accuracy. These values correspond to 100%, 10%, 1%, and 0.1% errors in the worst-case scenario. The following observations can be made:

- When $\alpha = 0.2$, the number of accurate digits in the far-field interactions has a negligible effect on the final result. For those solutions, the error due to the low-order discretization of the identity operator (in the MFIE contribution) dominates the error arising during the calculation of the far-field interactions.
- When $\alpha = 0.6$, the discretization error due to the identity operator is moderately suppressed. Hence, calculating the far-field interactions accurately becomes more critical compared to the $\alpha = 0.2$ case. On the other hand, the accuracy is slightly improved when the far-field interactions are calculated with two or three digits of accuracy instead of one digit of accuracy. In other words, there is no need to calculate the far-field interactions with high accuracy

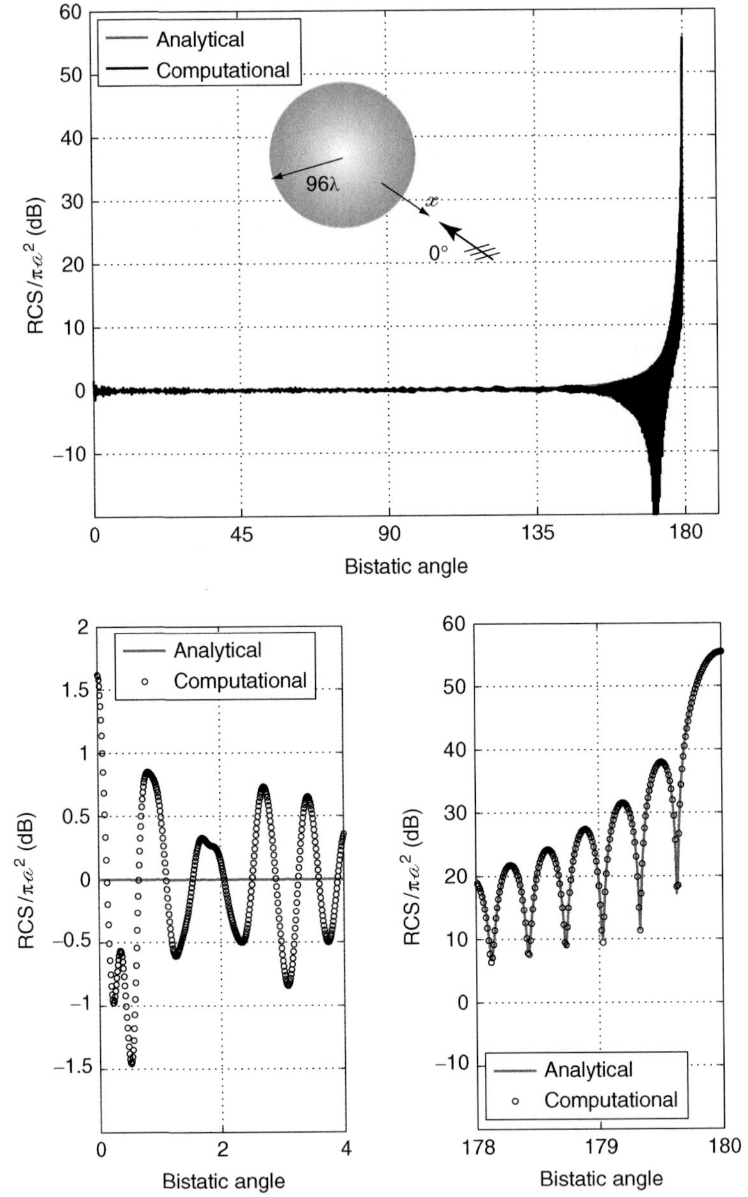

Figure 4.34 Normalized bistatic RCS ($RCS/\pi a^2$) of a sphere with a radius of 96λ (discretized with 33,791,232 unknowns) from 0° to 180°, where 0° and 180° correspond to the backscattering and forward-scattering directions, respectively. RCS values are zoomed around the backscattering and forward-scattering directions in separate plots. Computational values are obtained with a CFIE formulation using $\alpha = 0.2$. *Source*: Ergül and Gürel 2011 [236]. Reproduced with permission of IEEE.

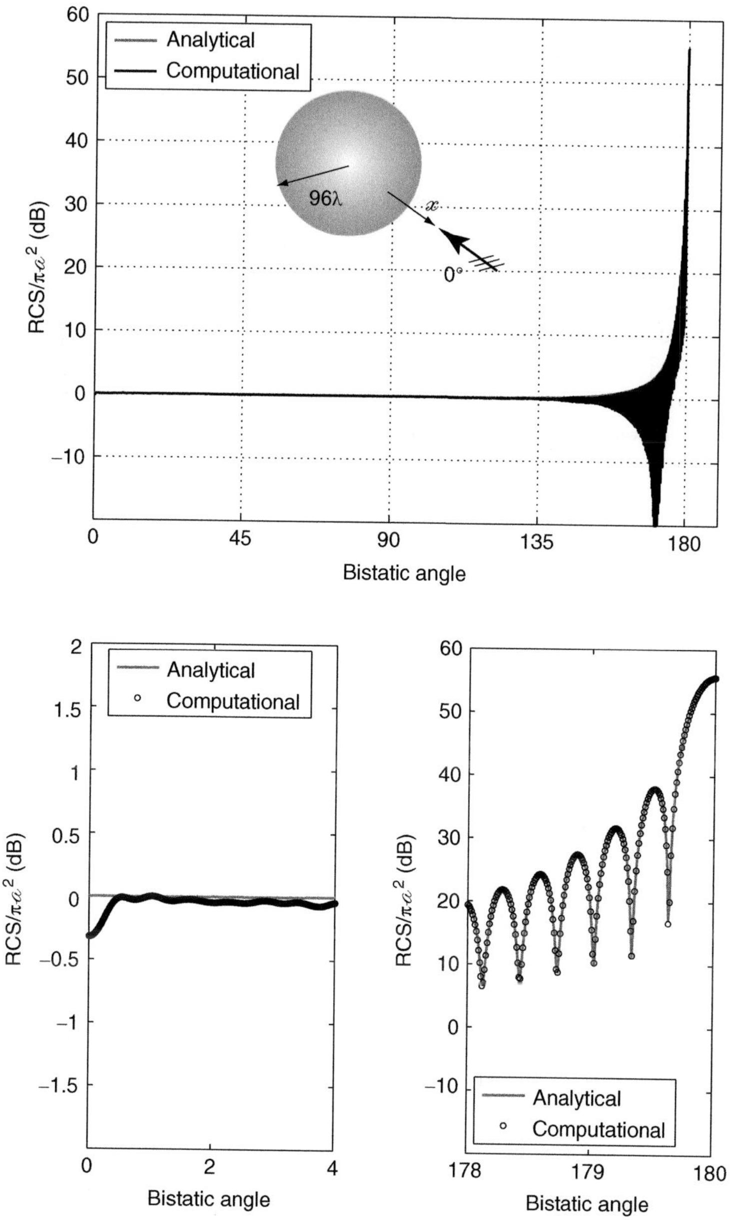

Figure 4.35 Normalized bistatic RCS ($RCS/\pi a^2$) of a sphere with a radius of 96λ (discretized with 33,791,232 unknowns) from 0° to 180°, where 0° and 180° correspond to the backscattering and forward-scattering directions, respectively. RCS values are zoomed around the backscattering and forward-scattering directions in separate plots. Computational values are obtained with a CFIE formulation using $\alpha = 0.6$. *Source*: Ergül and Gürel 2011 [236]. Reproduced with permission of IEEE.

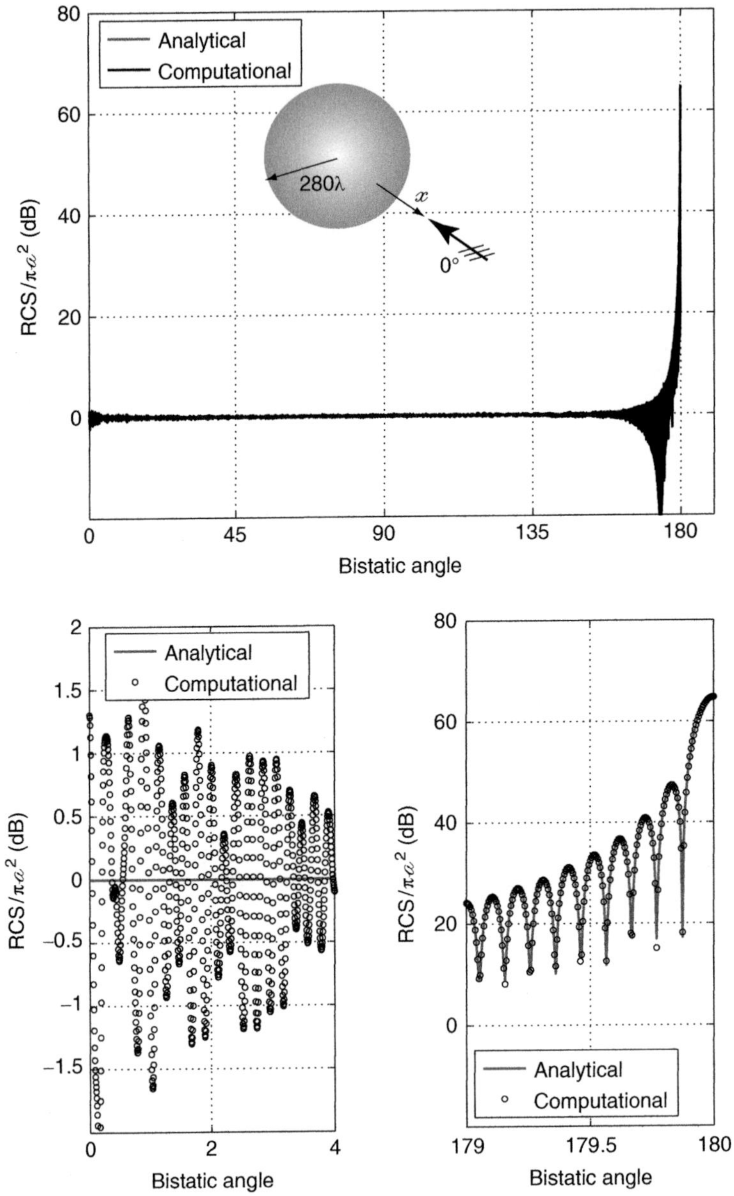

Figure 4.36 Normalized bistatic RCS (RCS/πa^2) of a sphere with a radius of 280λ (discretized with 374,490,624 unknowns) from 0° to 180°, where 0° and 180° correspond to the backscattering and forward-scattering directions, respectively. RCS values are zoomed around the backscattering and forward-scattering directions in separate plots. Computational values are obtained with a CFIE formulation using $\alpha = 0.2$. *Source*: Ergül and Gürel 2011 [236]. Reproduced with permission of IEEE.

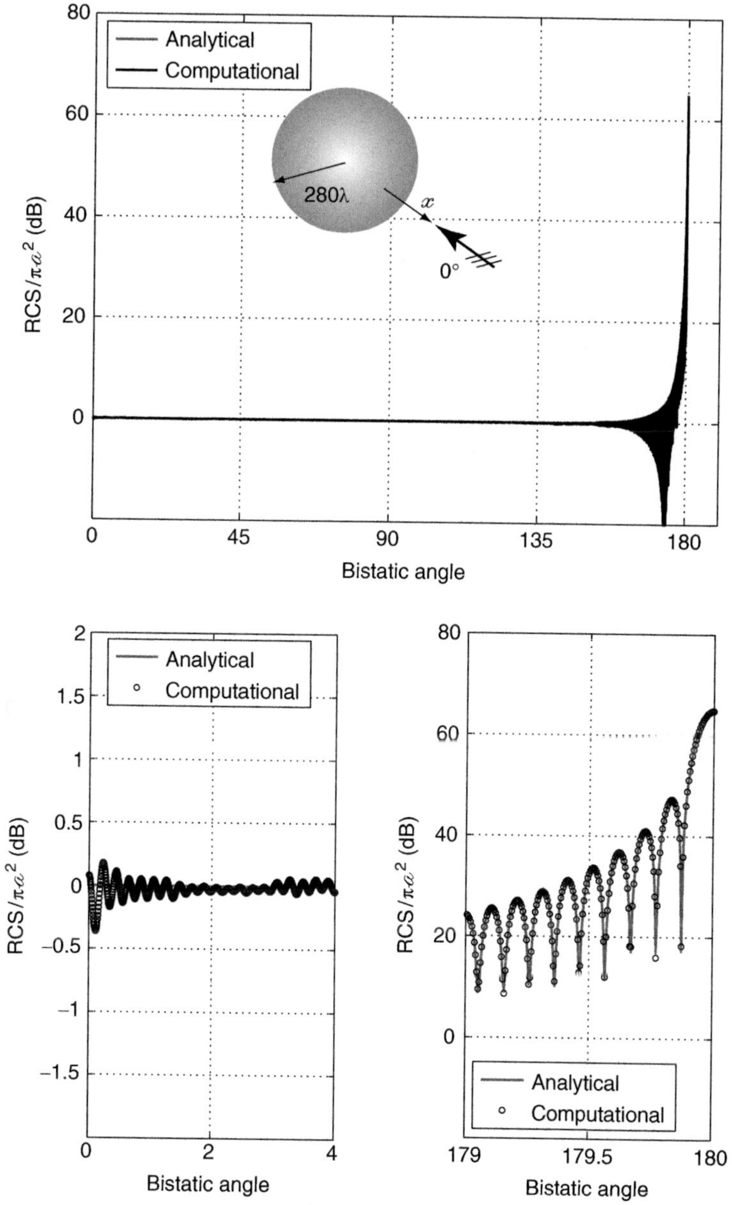

Figure 4.37 Normalized bistatic RCS ($\text{RCS}/\pi a^2$) of a sphere with a radius of 280λ (discretized with 374,490,624 unknowns) from 0° to 180°, where 0° and 180° correspond to the backscattering and forward-scattering directions, respectively. RCS values are zoomed around the backscattering and forward-scattering directions in separate plots. Computational values are obtained with a CFIE formulation using $\alpha = 0.6$. *Source*: Ergül and Gürel 2011 [236]. Reproduced with permission of IEEE.

Table 4.10 Error in the solution of scattering problems involving a PEC sphere of radius 96λ discretized with 33,791,232 unknowns

	α of CFIE = 0.2			
Accurate digits (MLFMA)	0	1	2	3
Relative error (RCS)	3.96%	2.52%	2.41%	2.38%
	α of CFIE = 0.6			
Accurate digits (MLFMA)	0	1	2	3
Relative error (RCS)	2.95%	0.74%	0.69%	0.66%

(requiring high computational requirements) unless the excessive discretization error is further suppressed. In these solutions, note that the near-field interactions are calculated with maximum 1% error. Hence, achieving 0.1% error (three digits of accuracy) in the final results may not be possible, even when EFIE ($\alpha = 1.0$) is used instead of CFIE and the far-field interactions are computed with three digits of accuracy.

Interestingly, numerical examples show that RCS values obtained with the 0.6 CFIE parameter (and only one digit of accuracy in the far-field interactions) can be more accurate than those obtained with the 0.2 CFIE parameter (and three digits of accuracy in the far-field interactions) even though the far-field interactions are calculated 100 times more accurately in the latter.

4.8.1 Mesh Quality

An important parameter, which is often discarded in large-scale simulations, is the mesh quality. In addition to the mesh size, the quality of the discretization is essential to obtain accurate solutions of electromagnetics problems. Nevertheless, it becomes increasingly difficult to the detect errors in the discretization, as the problem size gets larger and the discretization involves many triangles. As an example to those errors encountered during discretizations, Figure 4.38 presents the solution of a scattering problem involving the Flamme. The electric current induced on the surface of the target is investigated. It can be observed that the current values become unusually high at two symmetric locations. By zooming into those locations, extremely narrow triangles with very large aspect ratios are observed. Such contaminations of accuracy of discretizations in solutions due to distorted discretization elements, such as triangles with large aspect ratios, can be difficult to detect without investigating the induced current on each and every portion of the surface.

4.9 Solutions of Large-Scale Electromagnetics Problems Involving PEC Objects

By developing and employing parallel implementations of MLFMA using the hierarchical partitioning strategy, electromagnetics problems discretized with hundreds of millions of

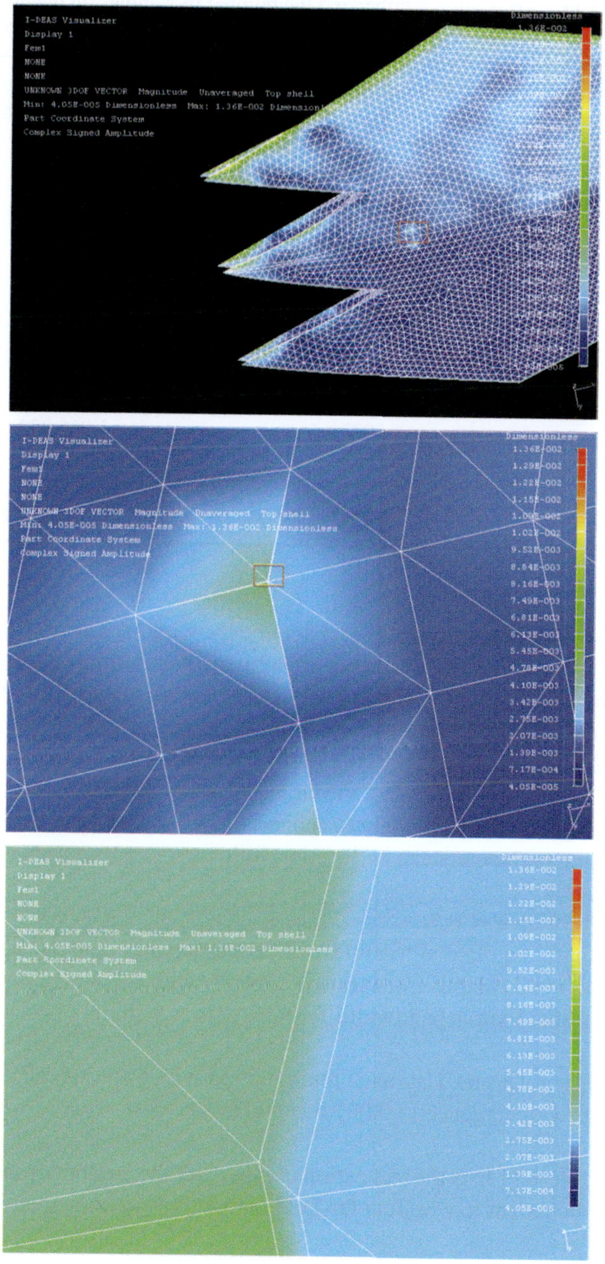

Figure 4.38 Induced electric current on the surface of the Flamme. Unusual currents due to triangles with high aspect ratios are clearly visible.

Table 4.11 Truncation number and the number of samples on the unit sphere to obtain two digits of accuracy for different box sizes in MLFMA

Box Size	Truncation number	Number of samples (on the unit sphere)
0.175λ	5	72
0.35λ	8	162
0.70λ	13	392
1.4λ	22	1058
2.8λ	39	3200
5.6λ	72	10,658
11.2λ	136	37,583
22.4λ	262	138,338
44.8λ	510	522,242
89.6λ	1003	2,016,032
179.2λ	1986	7,896,338
358.4λ	3945	31,141,832
716.8λ	7857	123,496,328

unknowns can be solved both efficiently and accurately without resorting to approximation techniques. This section presents examples of numerical solutions of very large electromagnetics problems involving PEC objects with canonical and complicated geometries (see Figure 4.2). Since they are obtained by using a full-wave solver, numerical results presented in this section can be used for benchmarking purposes.

In all solutions presented in this section, the target accuracy is defined as maximum 1% error in far-zone electromagnetic fields. In order to achieve this error threshold with the best possible efficiency, some of the parameters can be listed as follows:

- Formulation: CFIE with $\alpha = 0.5$.
- Discretization: Nearly equilateral triangles with edges smaller than $\lambda/10$.
- Near-Field Interactions: Adaptive integration methods with 0.5% maximum error.
- Far-Field Interactions: Excess bandwidth formula with two digits of accuracy in the worst case.
- Interpolations/Anterpolations: Lagrange interpolation with 6×6 stencils.
- Iterative Solutions: Convergence to 0.001 residual error.

As an example, Table 4.11 lists the truncation number and the number of samples on the unit sphere for different box sizes to compute far-field interactions in numerical solutions.

4.9.1 PEC Sphere

Figures 4.39–4.49 present solutions of increasingly large scattering problems involving a PEC sphere of radius $a = 0.3$ m. The sphere is illuminated by a plane wave propagating in the $-x$ direction with the electric field polarized in the y direction when the frequency changes from 20 GHz to 340 GHz. In this range, the diameter of the sphere changes from approximately 40λ

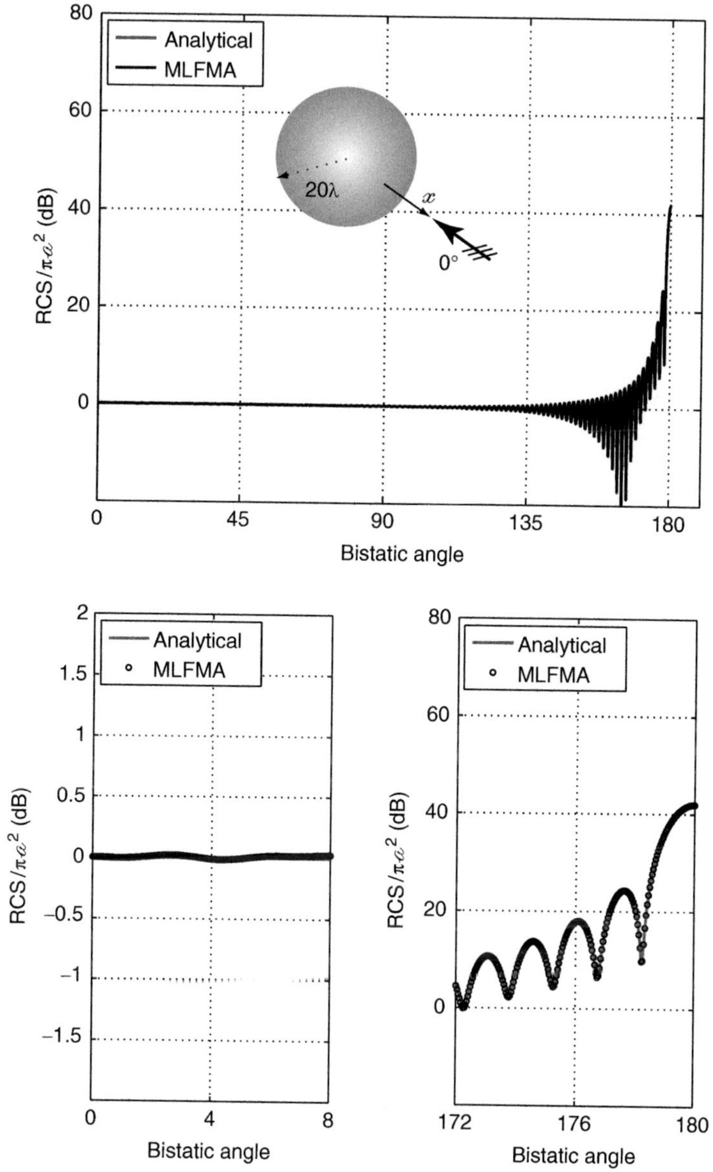

Figure 4.39 Normalized bistatic RCS ($\text{RCS}/\pi a^2$) of a sphere with a radius of 20λ (discretized with 1,462,854 unknowns) from 0° to 180°, where 0° and 180° correspond to the backscattering and forward-scattering directions, respectively. RCS values are zoomed around the backscattering and forward-scattering directions in separate plots. Computational values provided by the parallel MLFMA implementation with maximum 1% error agree well with an analytical Mie-series solution.

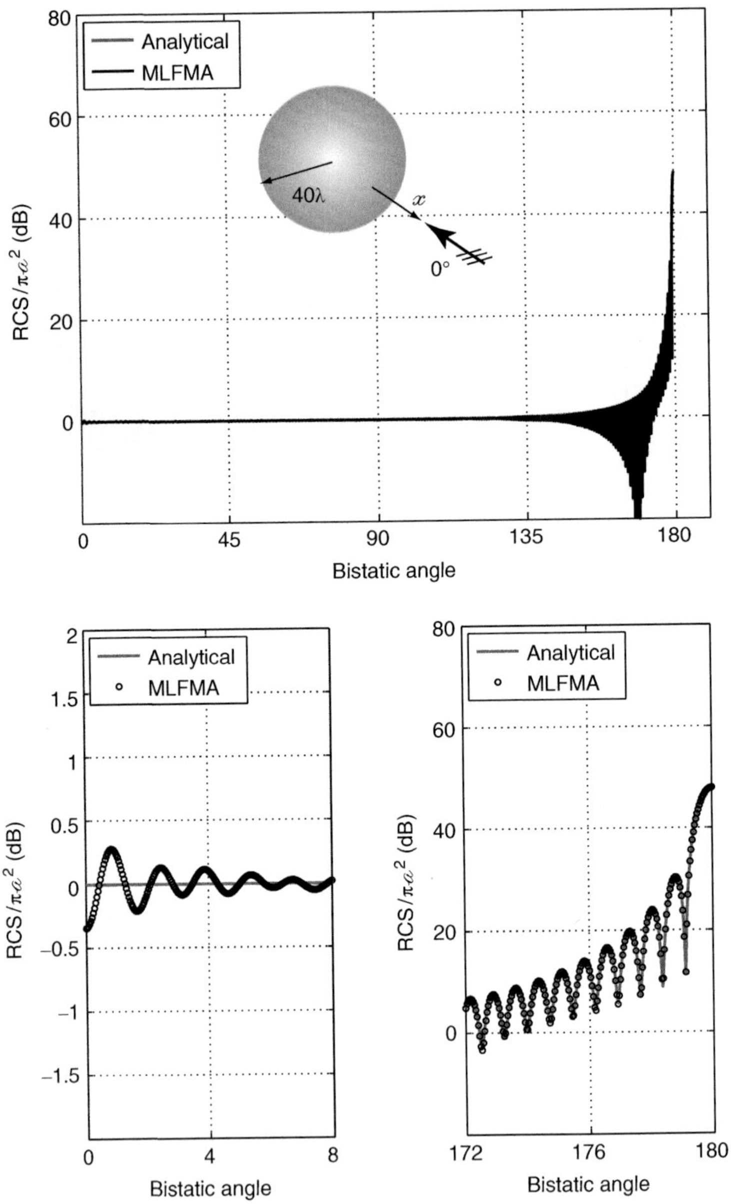

Figure 4.40 Normalized bistatic RCS ($RCS/\pi a^2$) of a sphere with a radius of 40λ (discretized with 5,851,416 unknowns) from 0° to 180°, where 0° and 180° correspond to the backscattering and forward-scattering directions, respectively. RCS values are zoomed around the backscattering and forward-scattering directions in separate plots. Computational values provided by the parallel MLFMA implementation with maximum 1% error agree well with an analytical Mie-series solution.

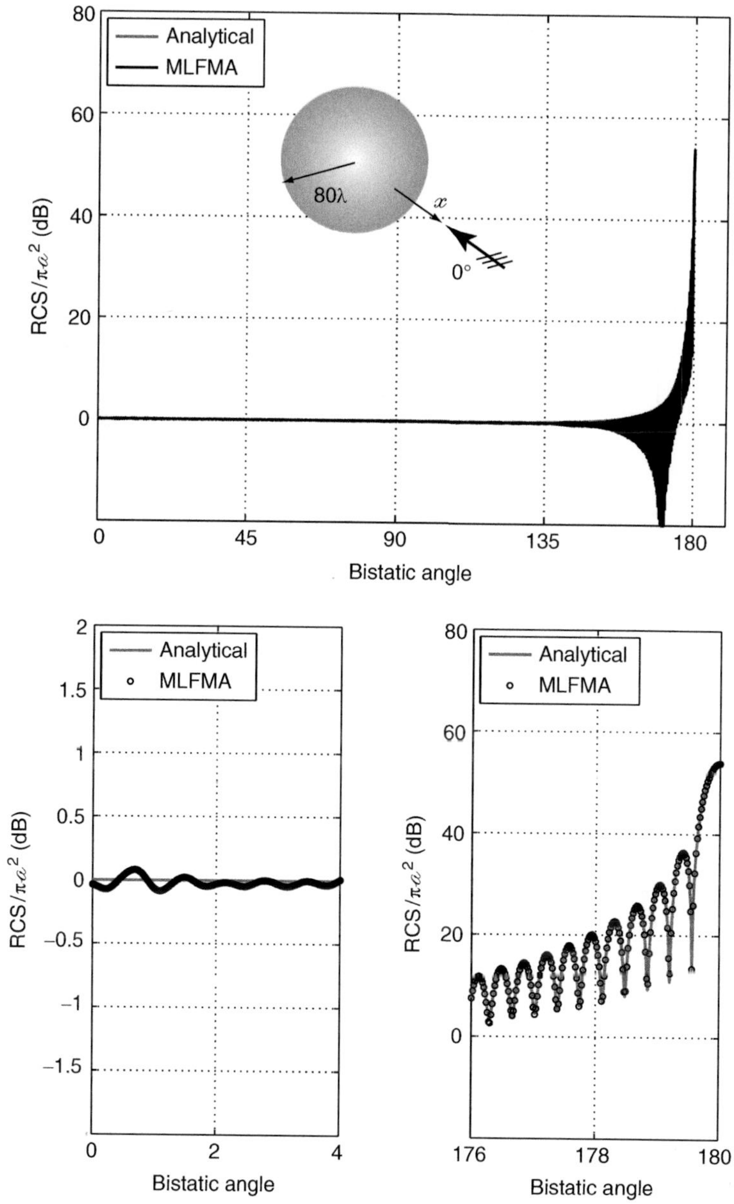

Figure 4.41 Normalized bistatic RCS ($RCS/\pi a^2$) of a sphere with a radius of 80λ (discretized with 23,405,664 unknowns) from 0° to 180°, where 0° and 180° correspond to the backscattering and forward-scattering directions, respectively. RCS values are zoomed around the backscattering and forward-scattering directions in separate plots. Computational values provided by the parallel MLFMA implementation with maximum 1% error agree well with an analytical Mie-series solution.

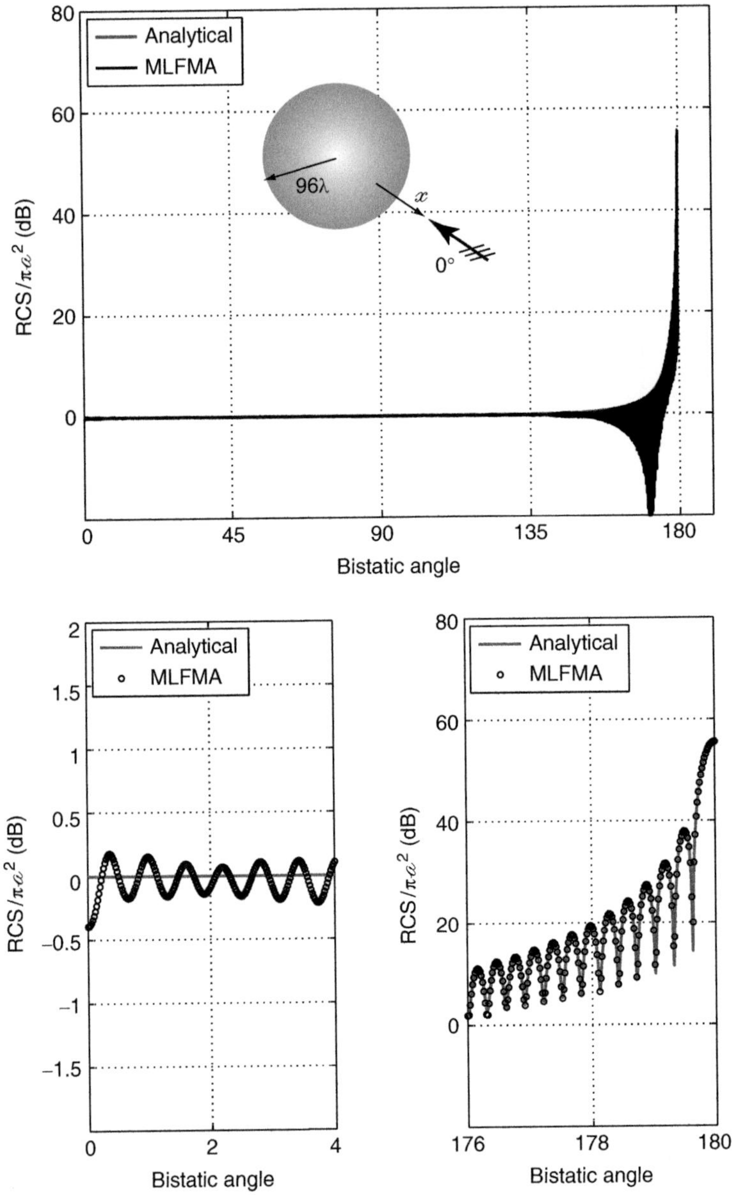

Figure 4.42 Normalized bistatic RCS (RCS/πa^2) of a sphere with a radius of 96λ (discretized with 33,791,232 unknowns) from 0° to 180°, where 0° and 180° correspond to the backscattering and forward-scattering directions, respectively. RCS values are zoomed around the backscattering and forward-scattering directions in separate plots. Computational values provided by the parallel MLFMA implementation with maximum 1% error agree well with an analytical Mie-series solution.

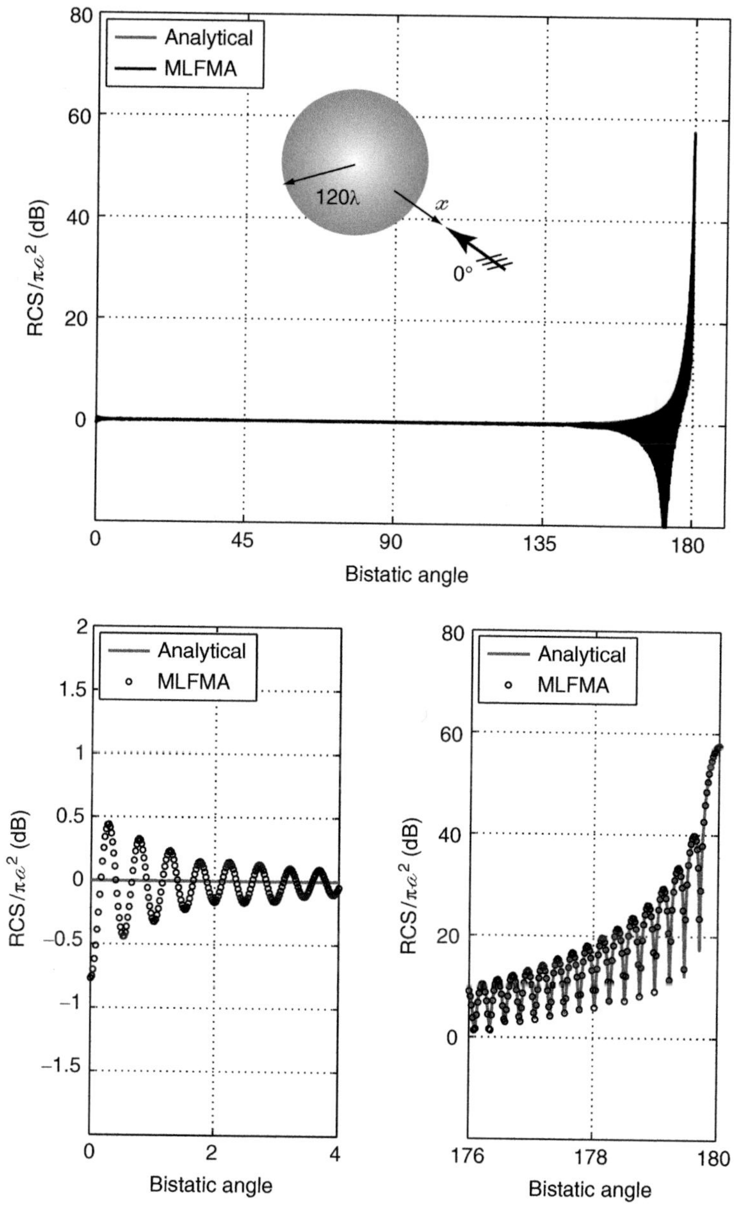

Figure 4.43 Normalized bistatic RCS (RCS/πa^2) of a sphere with a radius of 120λ (discretized with 53,112,384 unknowns) from 0° to 180°, where 0° and 180° correspond to the backscattering and forward-scattering directions, respectively. RCS values are zoomed around the backscattering and forward-scattering directions in separate plots. Computational values provided by the parallel MLFMA implementation with maximum 1% error agree well with an analytical Mie-series solution.

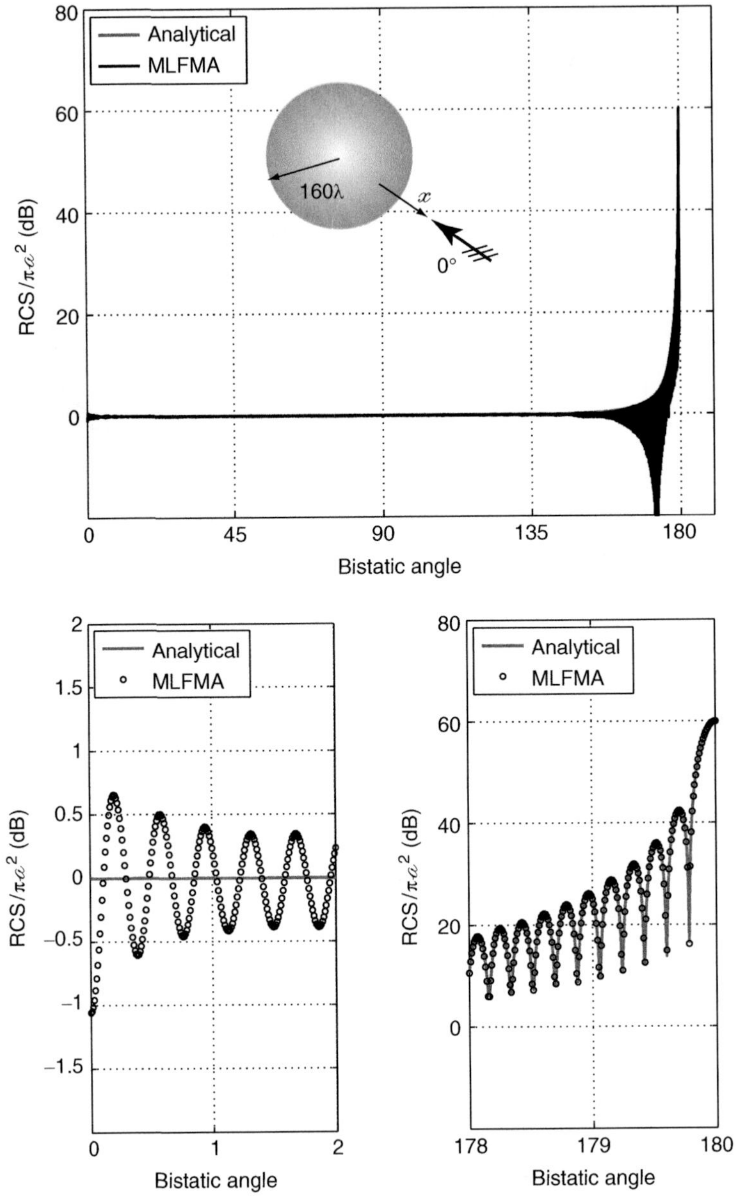

Figure 4.44 Normalized bistatic RCS ($RCS/\pi a^2$) of a sphere with a radius of 160λ (discretized with 93,622,656 unknowns) from 0° to 180°, where 0° and 180° correspond to the backscattering and forward-scattering directions, respectively. RCS values are zoomed around the backscattering and forward-scattering directions in separate plots. Computational values provided by the parallel MLFMA implementation with maximum 1% error agree well with an analytical Mie-series solution.

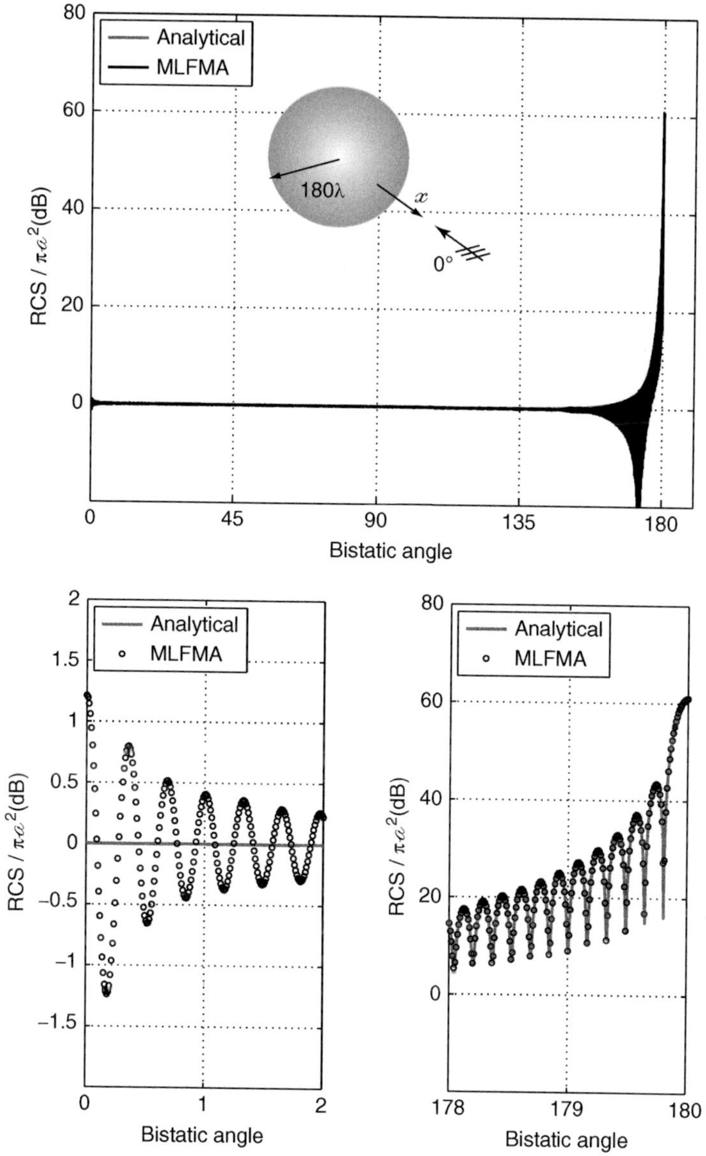

Figure 4.45 Normalized bistatic RCS (RCS/πa^2) of a sphere with a radius of 180λ (discretized with 135,164,928 unknowns) from 0° to 180°, where 0° and 180° correspond to the backscattering and forward-scattering directions, respectively. RCS values are zoomed around the backscattering and forward-scattering directions in separate plots. Computational values provided by the parallel MLFMA implementation with maximum 1% error agree well with an analytical Mie-series solution.

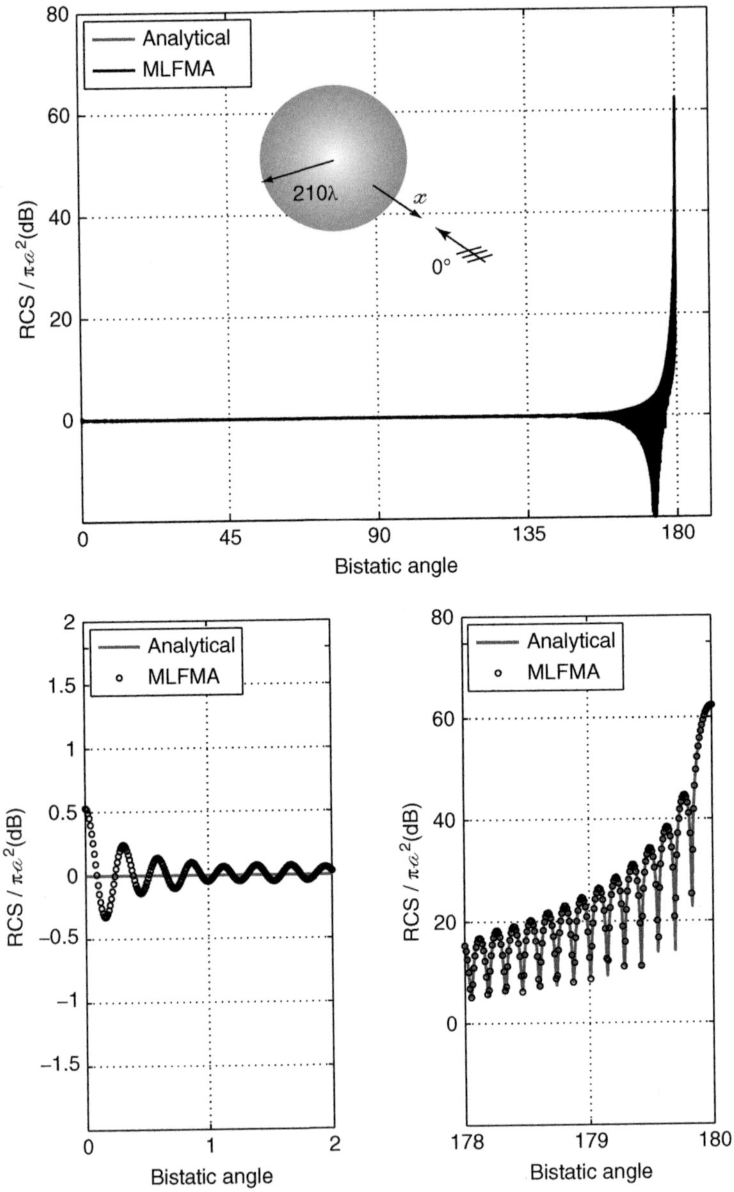

Figure 4.46 Normalized bistatic RCS (RCS/πa^2) of a sphere with a radius of 210λ (discretized with 204,823,296 unknowns) from 0° to 180°, where 0° and 180° correspond to the backscattering and forward-scattering directions, respectively. RCS values are zoomed around the backscattering and forward-scattering directions in separate plots. Computational values provided by the parallel MLFMA implementation with maximum 1% error agree well with an analytical Mie-series solution. *Source*: Gürel and Ergül 2013 [235]. Reproduced with permission of IEEE.

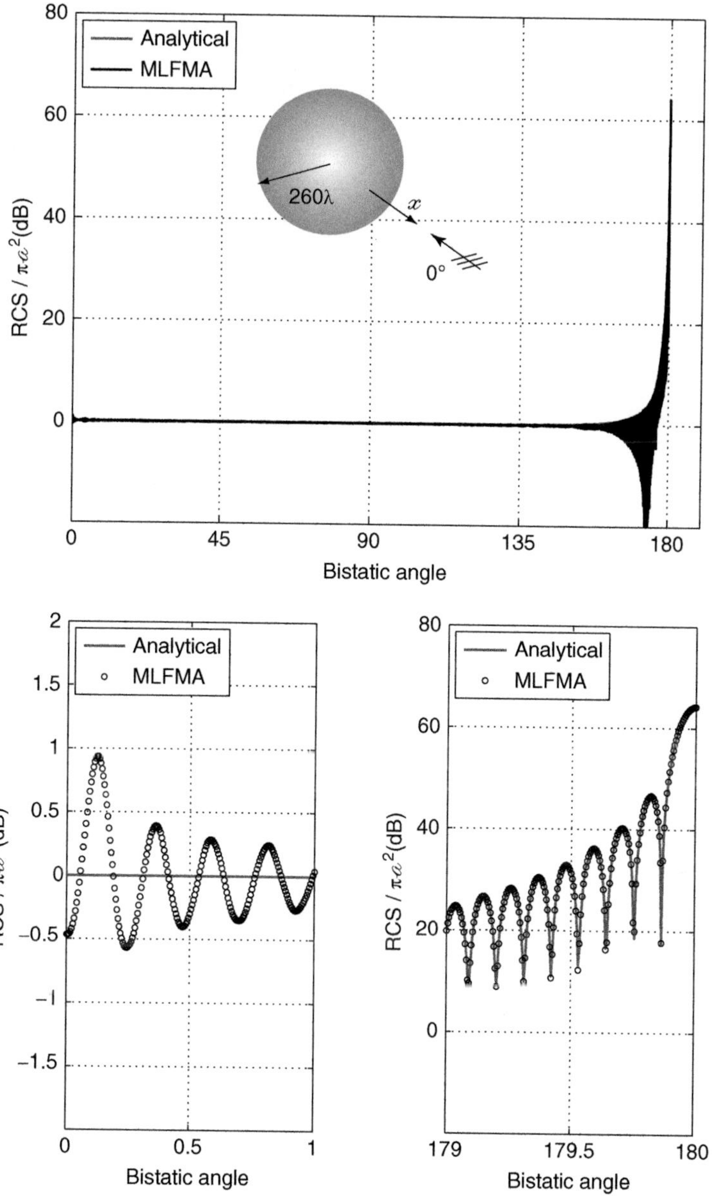

Figure 4.47 Normalized bistatic RCS (RCS/πa^2) of a sphere with a radius of 260λ (discretized with 307,531,008 unknowns) from 0° to 180°, where 0° and 180° correspond to the backscattering and forward-scattering directions, respectively. RCS values are zoomed around the backscattering and forward-scattering directions in separate plots. Computational values provided by the parallel MLFMA implementation with maximum 1% error agree well with an analytical Mie-series solution. *Source*: Gürel and Ergül 2011 [237]. Reproduced with permission of IEEE.

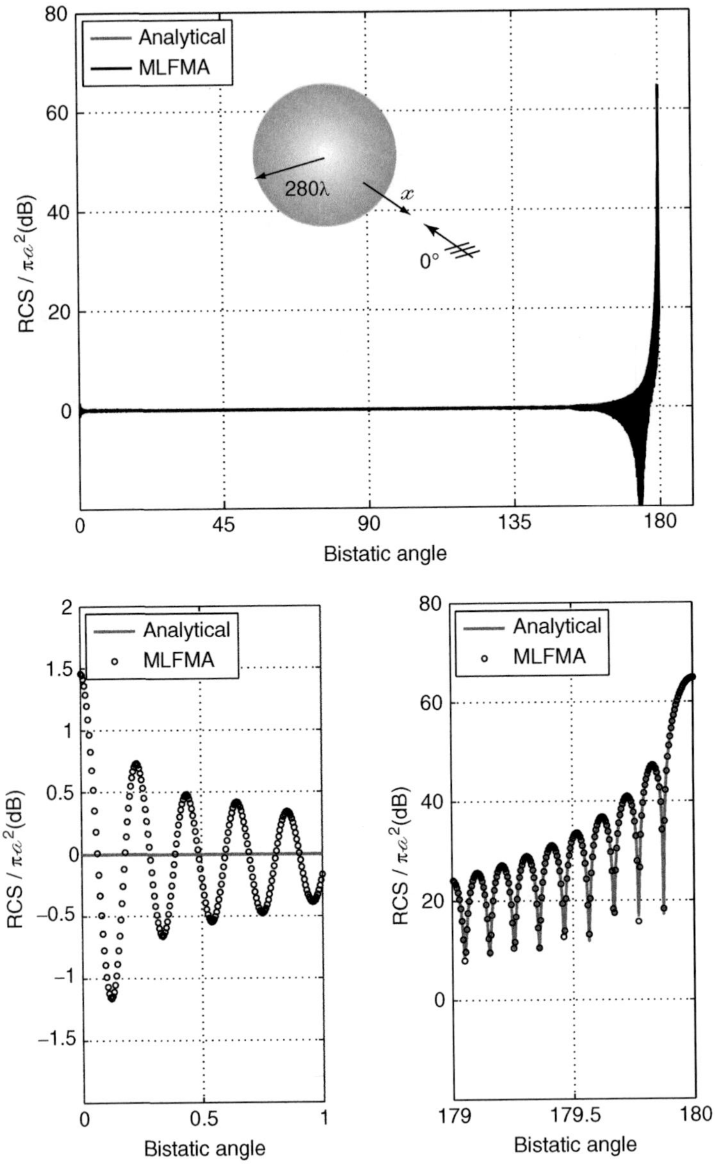

Figure 4.48 Normalized bistatic RCS (RCS/πa^2) of a sphere with a radius of 280λ (discretized with 374,490,624 unknowns) from 0° to 180°, where 0° and 180° correspond to the backscattering and forward-scattering directions, respectively. RCS values are zoomed around the backscattering and forward-scattering directions in separate plots. Computational values provided by the parallel MLFMA implementation with maximum 1% error agree well with an analytical Mie-series solution.

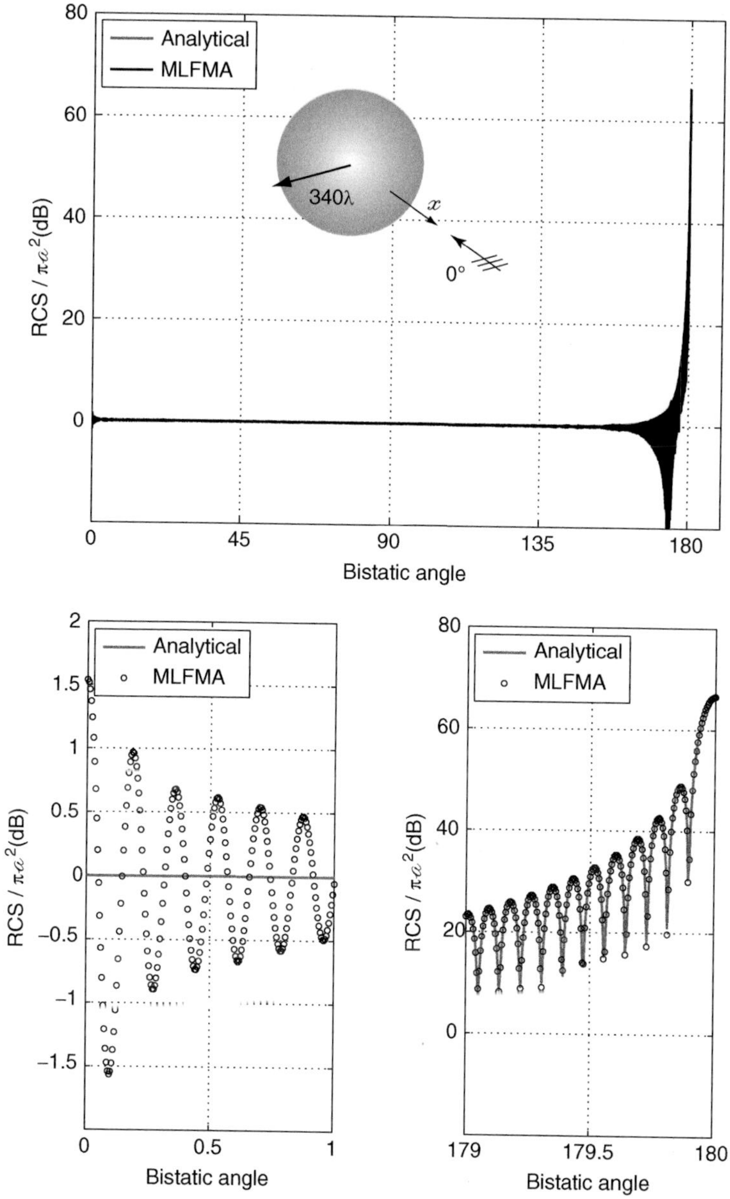

Figure 4.49 Normalized bistatic RCS (RCS/πa^2) of a sphere with a radius of 340λ (discretized with 540,659,712 unknowns) from 0° to 180°, where 0° and 180° correspond to the backscattering and forward-scattering directions, respectively. RCS values are zoomed around the backscattering and forward-scattering directions in separate plots. Computational values provided by the parallel MLFMA implementation with maximum 1% error agree well with an analytical Mie-series solution.

Table 4.12 Solutions of increasingly large electromagnetics problems involving a sphere of radius $a = 0.3$ m on the S-Nehalem-EX cluster

Frequency (GHz)	Unknowns	Iterations (10^{-3} residual)	Total time (minutes)
20	1,462,854	21	4
40	5,851,416	27	16
80	23,405,664	33	61
96	33,791,232	39	107
120	53,112,384	44	183
160	93,622,656	47	333
180	135,164,928	47	471
210	204,823,296	50	647
260	307,531,008	55	1080
280	374,490,624	58	1430
340	540,659,712	65	3632*

to approximately 680λ. Figures 4.39–4.49 depict the normalized RCS (RCS$/\pi a^2$ in dB) on the x-y plane as a function of the bistatic angle ϕ from 0° to 180°, where 0° and 180° correspond to the backscattering and forward-scattering directions, respectively. RCS results around the backscattering and forward-scattering directions are particularly focused in separate plots. It can be observed that the computational results agree well with the analytical results obtained by Mie-series solutions. Table 4.12 lists the number of BiCGStab iterations (for 10^{-3} residual error) and the total computing time measured on the S-Nehalem-EX cluster. All solutions are parallelized into 128 processes except for the largest problem, which is solved on 64 processors. Note that a direct solution of this problem involving 540 million unknowns without using MLFMA would require 2.3 exabytes (1 exabyte $= 10^9$ GB) of memory, and it would take more than 4000 years just to construct the matrix equation on the same S-Nehalem-EX cluster. In addition, such a hypothetical solution would not be able to attain 1% error due to accumulation of rounding errors in finite-precision arithmetics.

Table 4.13 lists the relative error in far-zone electric fields obtained by the computational solutions of the scattering problems in Table 4.12. The relative error defined in (2.196) is calculated in different angular ranges, i.e., 0°–30°, 0°–90°, and 0°–180°, when the computational and analytical values are sampled at 0.1° intervals. Considering the 0°–180° range, it can be observed that the target (maximum 1% error) is generally satisfied. As usual, it is more difficult to attain high accuracy around the backscattering direction due to lower RCS values. But, even in the critical 0°–30° range, the largest error in Table 4.13 is 2.19%.

4.9.2 Other Canonical Problems

Figure 4.50 presents the solution of scattering problems involving a $1 \times 3.5 \times 0.1$ m PEC box (see above, Section 4.3) at 75 GHz. At this frequency, the size of the box corresponds to approximately 875λ, and its discretization leads to matrix equations involving 174,489,600 unknowns. The box is illuminated by two plane waves propagating on the x-z plane at 30° and

Table 4.13 Relative error in far-zone electric fields obtained by the Computational solutions of increasingly large scattering problems listed in Table 4.12

Frequency (GHz)	Unknowns	Relative error		
		0°–30°	0°–90°	0°–180°
20	1,462,854	0.0052	0.0055	0.0092
40	5,851,416	0.0092	0.0079	0.0097
80	23,405,664	0.0099	0.0095	0.0125
96	33,791,232	0.0147	0.0111	0.0099
120	53,112,384	0.0122	0.0092	0.0099
160	93,622,656	0.0170	0.0123	0.0104
180	125,164,928	0.0209	0.0139	0.0093
210	204,823,296	0.0116	0.0088	0.0072
260	307,531,008	0.0219	0.0156	0.0090
280	374,490,624	0.0183	0.0127	0.0087
340	540,659,712	0.0203	0.0136	0.0082

60° angles from the z axis with the electric field polarized in the ϕ direction. The total number of iterations is 32 and two solutions are performed in a total of 10 hours on the S-Dunnington cluster. Figure 4.50 depicts the co-polar bistatic RCS (dBms stands for dBm^2) on the x-z plane as a function of the bistatic angle θ. In addition to the forward-scattering direction, RCS values make peaks at two specular reflection directions, i.e., at 150° and 330° for the 30° illumination and at 120° and 300° for the 60° illumination. Both the directions and magnitudes of these peaks are consistent with reflections from the large top surface and the small side surface.

Figure 4.51 presents the solution of scattering problems involving a wing-shaped object with a size of 1 m (see above, Section 4.3) at 150 GHz. The object is discretized with 172,228,608 unknowns, and it is illuminated by two plane waves propagating on the x-y plane at 30° and 60° angles from the x axis. The electric field is polarized in the θ direction. The total number of iterations is 36 and two solutions are performed in a total of 12 hours on the S-Dunnington cluster. Figure 4.51 depicts the co-polar bistatic RCS (dBms) on the x-y plane as a function of the bistatic angle ϕ. It can be observed that RCS values of the wing-shaped object make peaks at three directions due to specular reflections, in addition to the forward-scattering direction. Two of these peaks are due to specular reflections from the two large flat surfaces of the wing, but the third one, interestingly, corresponds to double reflection from both surfaces. This result can be used to benchmark single-bounce and multi-bounce high-frequency approximations.

Measuring the accuracy of solutions for arbitrary geometries is not trivial, especially for very large problems. But there are special cases that can be investigated to check the accuracy. As an example, Figure 4.52 presents the solution of a radiation problem involving a $200\lambda \times 200\lambda \times 200\lambda$ PEC box discretized with 75,276,288 unknowns. A Hertzian dipole oriented in the z direction is located on the top of the box. Figure 4.52 presents the z component of the normalized electric field in the far zone on the x-y plane. In addition to the incident field radiated by the Hertzian dipole, the secondary field due to the induced electric current on the surface of the cube is considered. It can be observed that the secondary field is 30–40 dB less than the incident field, and the incident field is very close to the total field. In fact, the secondary field can be interpreted as the noise created by the box to the radiation of the Hertzian

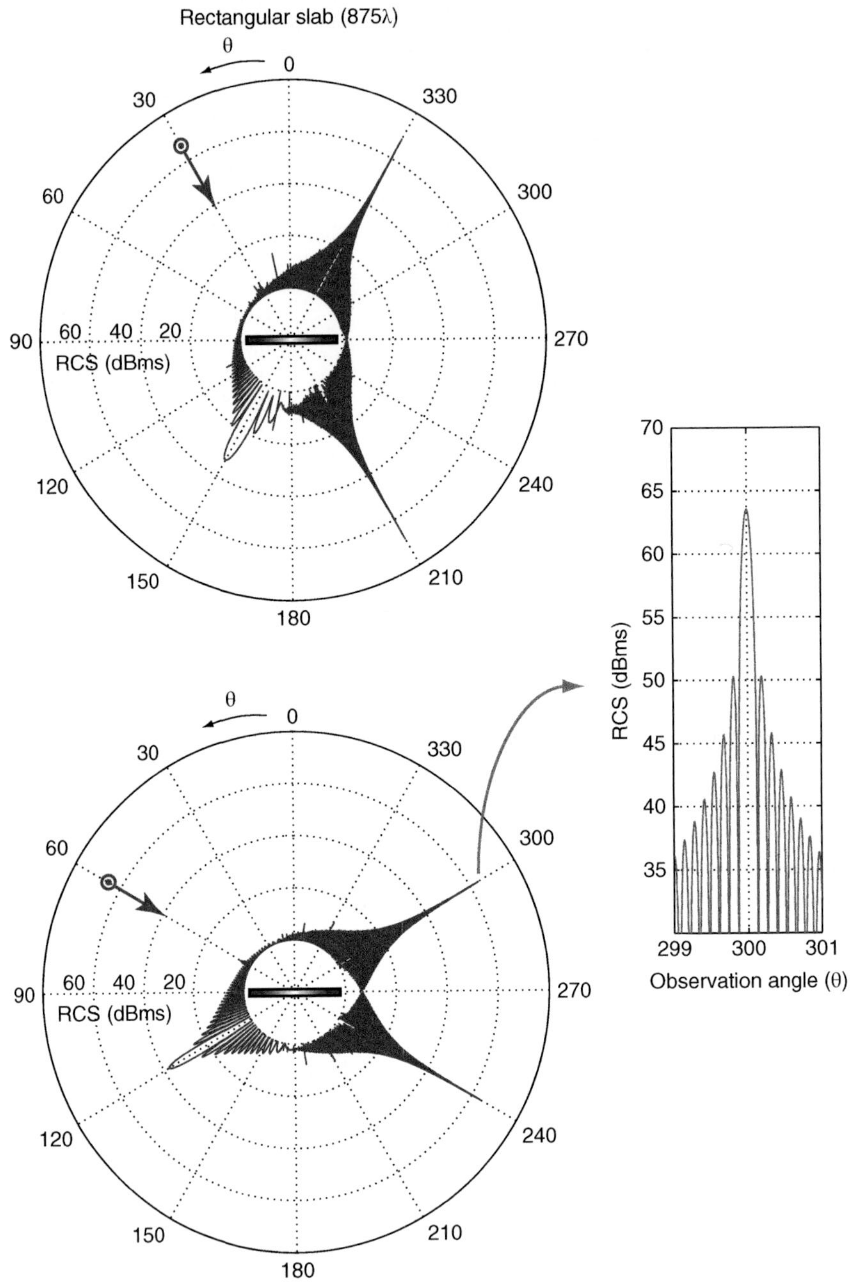

Figure 4.50 Co-polar bistatic RCS (dBms) of the rectangular box in Figure 4.2 at 75 GHz on the *x*-*z* plane. The box is illuminated by plane waves propagating on the *x*-*z* plane at 30° and 60° angles from the *z* axis with the electric field polarized in the ϕ direction. *Source*: Ergül and Gürel 2011 [230]. Reproduced with permission of IEEE.

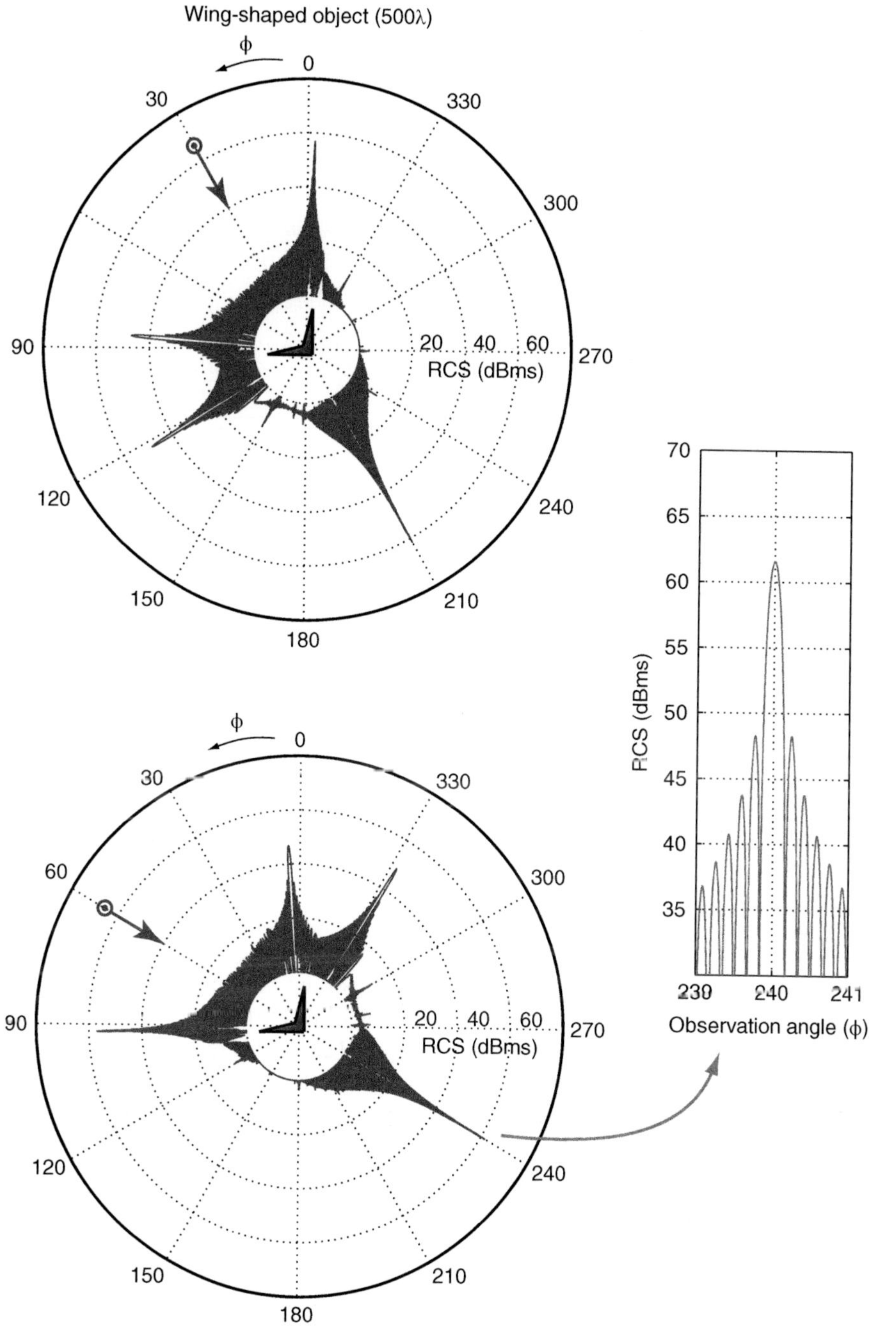

Figure 4.51 Co-polar bistatic RCS (dBms) of the wing-shaped object in Figure 4.2 at 150 GHz on the *x*-*y* plane. The object is illuminated by plane waves propagating on the *x*-*y* plane at 30° and 60° angles from the *x* axis with the electric field polarized in the θ direction. *Source*: Ergül and Gürel 2011 [230]. Reproduced with permission of IEEE.

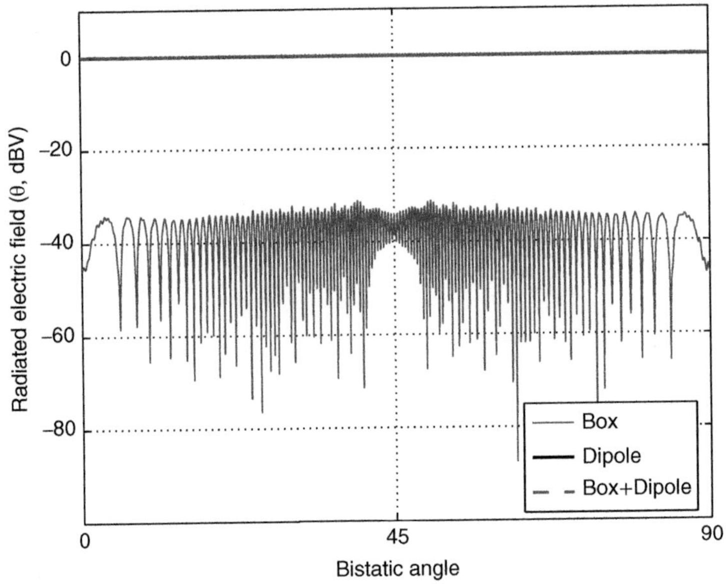

Figure 4.52 Radiated electric field (dBV) in the far zone on the x-y plane due to a Hertzian dipole located on a $200\lambda \times 200\lambda \times 200\lambda$ conducting box.

dipole. Since the box is very large, it behaves like an infinite ground plane and it does not affect the radiation pattern of the dipole. This also shows that the numerical solution performed by MLFMA is accurate and leads to physically expected results.

4.9.3 NASA Almond

The NASA Almond depicted in Figure 4.2(c) is an important benchmark problem, which is commonly used for testing numerical implementations of electromagnetics solvers, particularly because its geometry can be defined mathematically [234]. This section presents solutions of various large-scale electromagnetics problem involving the NASA Almond. All solutions are again performed with a target of maximum 1% error in far-zone fields.

Figures 4.53–4.55 present the solution of scattering problems involving the NASA Almond at 850 GHz, i.e., when the the size of the object corresponds to approximately 715λ. Discretization of the object with $\lambda/10$ mesh size leads to matrix equations involving 126,167,104 unknowns. As also depicted in Figure 4.2(c), the object is located on the x-y plane such that its nose is directed towards the x axis. It is illuminated by various plane waves propagating on the x-y plane with the electric field polarized in the ϕ and θ directions. Figures 4.53 and 4.54 present the number of BiCGStab iterations and processing times (setup, iterative solution, and total) in minutes when the solutions are performed by an 11-level MLFMA on the B-Harpertown cluster. The number of iterations and processing times are plotted with respect to the illumination angle measured from the x axis. Due to the smooth surface of the NASA Almond, the number of iterations and solution time do not change significantly as the

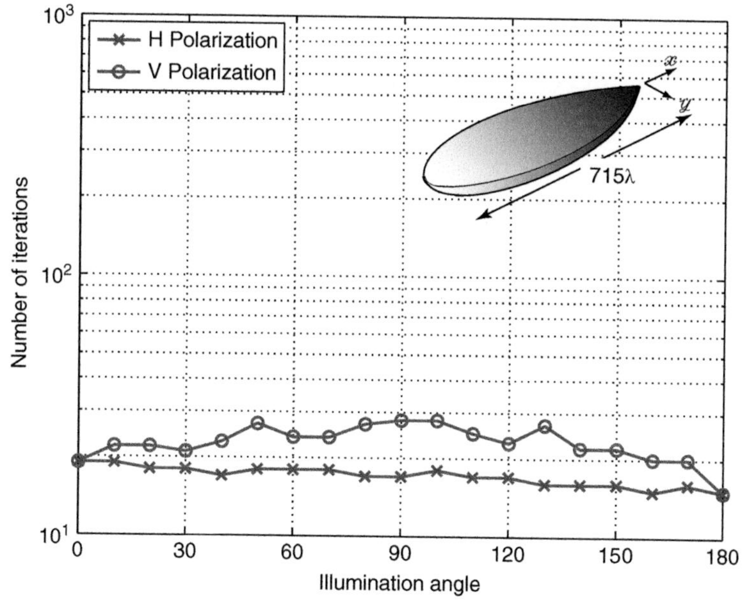

Figure 4.53 Number of BiCGStab iterations (with respect to the illumination angle measured from the x axis) for the solution of scattering problems involving the NASA Almond at 850 GHz discretized with 125,167,104 unknowns. *Source*: Ergül and Gürel 2011 [230]. Reproduced with permission of IEEE.

illumination angle changes. Figure 4.55 presents the co-polar bistatic RCS (dBms) on the x-y plane as a function of observation and illumination angles. It can be observed that, in addition to the forward-scattering direction, RCS of the NASA Almond is quite high in a range of observation angles, depending on the illumination angle. Specifically, for both ϕ and θ polarizations, large RCS values from $(120° - \phi_i)$ to $(240° - \phi_i)$, where ϕ_i is the illumination angle, are observed. On the other hand, there exists no strong specular reflection from the NASA Almond, and its RCS does not exhibit any visible peak except for in the forward-scattering direction.

Figures 4.56–4.58 present numerical experiments on the NASA Almond at 1.1 THz, i.e., when the size of the object is approximately 925λ. At this frequency, discretization of the object with $\lambda/10$ mesh size leads to matrix equations involving 203,476,224 unknowns. Similar to the previous examples, the object is illuminated by various plane waves propagating on the x-y plane. Figures 4.56 and 4.57 present the number of BiCGStab iterations and processing times when solutions are performed by a 12-level MLFMA on the C-Clovertown cluster. The co-polar bistatic RCS (dBms) on the x-y plane is depicted in Figure 4.58 for both ϕ and θ polarizations. It can be observed that the RCS results are very similar to those at the lower frequency, depicted in Figure 4.55.

Figures 4.59–4.63 present the solution of increasingly large scattering problems involving the NASA Almond. The frequency changes from 112.5 GHz to 1.8 THz, i.e., the size of the object changes from approximately 94λ to approximately 1514λ. Discretizations with $\lambda/10$ triangles lead to matrix equations involving 2,157,462 to 552,310,272 unknowns. At each frequency, two different illuminations are considered; a plane wave propagating in the $-x$

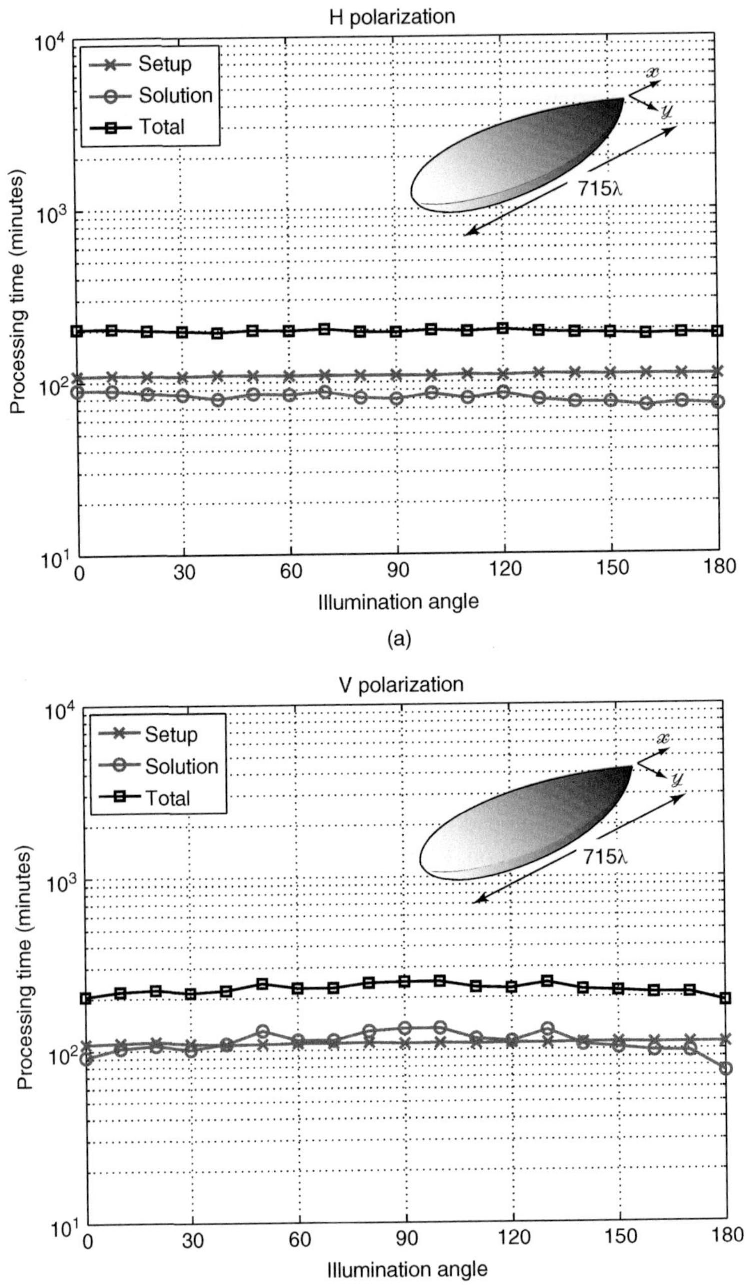

Figure 4.54 Processing times (with respect to the illumination angle measured from the x axis) for the solution of scattering problems involving the NASA Almond at 850 GHz discretized with 125,167,104 unknowns (measured on the B-Harpertown cluster). *Source*: Ergül and Gürel 2011 [230]. Reproduced with permission of IEEE.

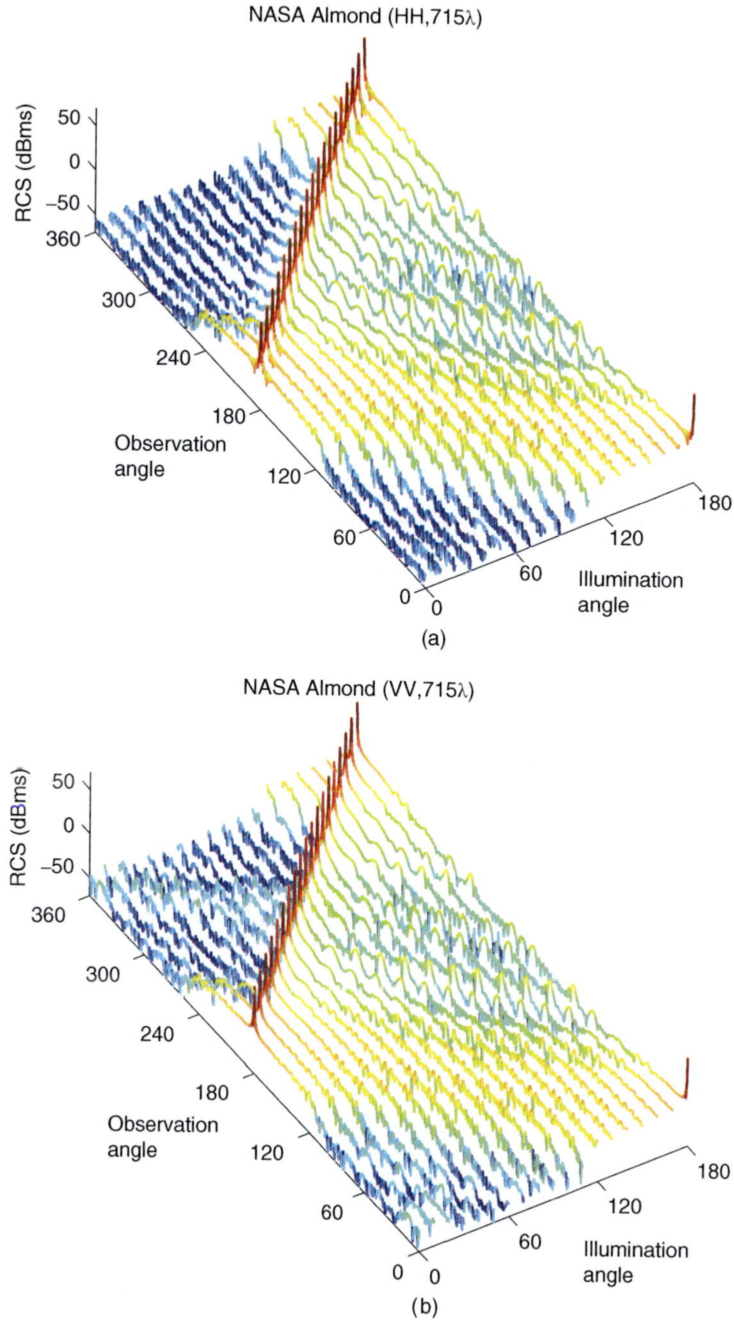

Figure 4.55 Co-polar bistatic RCS (dBms) of the NASA Almond at 850 GHz. RCS is plotted on the x-y plane as a function of the observation and illumination angles when the incident electric field is polarized in (a) ϕ and (b) θ directions. *Source*: Ergül and Gürel 2011 [230]. Reproduced with permission of IEEE.

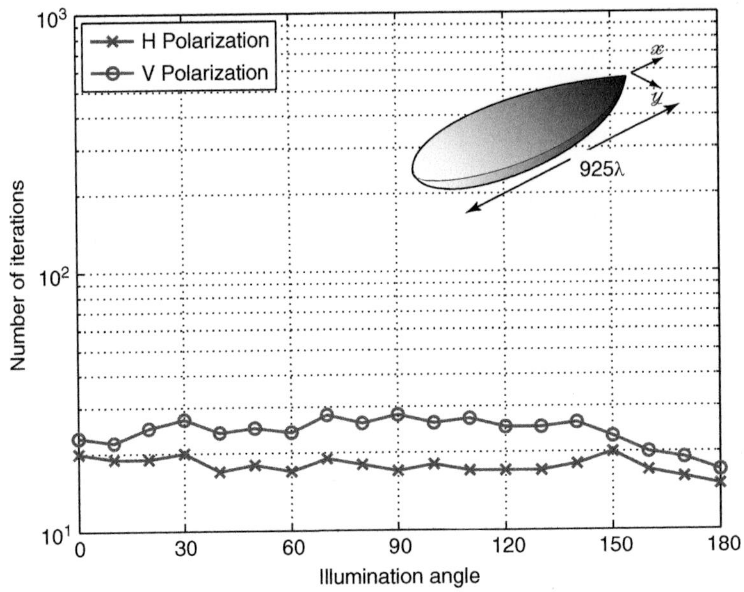

Figure 4.56 Number of BiCGStab iterations (with respect to the illumination angle measured from the x axis) for the solution of scattering problems involving the NASA Almond at 1.1 THz discretized with 203,476,224 unknowns.

direction (0° from the x axis) and a plane wave propagating on the x-y plane at 30° angle from the x axis. For both illuminations, the electric field is polarized in the ϕ direction. Figures 4.59–4.63 depict the co-polar and cross-polar bistatic RCS (dBms) on the x-y plane with respect to the bistatic angle ϕ from 0° to 180°. The following observations can be made for all frequencies:

- RCS is generally higher in the 120°–240° and 90°–210° ranges for the 0° and 30° illuminations, respectively.
- There exists no strong specular reflection from the object.
- The cross-polar RCS is very low compared to the co-polar RCS.
- The co-polar RCS in the backscattering direction is very low, and the NASA Almond exhibits a stealth ability. For example, at the highest frequency, the backscattered co-polar RCS is 90 dB lower than the forward-scattered co-polar RCS.

Table 4.14 lists the number of BiCGStab iterations (for 10^{-3} residual error) and the total computing time measured on the S-Nehalem-EX cluster for the solution of large-scale scattering problems involving the NASA Almond. All solutions are parallelized into 64 processes. Note that the largest problem discretized with more than 550 million unknowns can be solved by using a 13-level MLFMA in 54–56 hours per excitation.

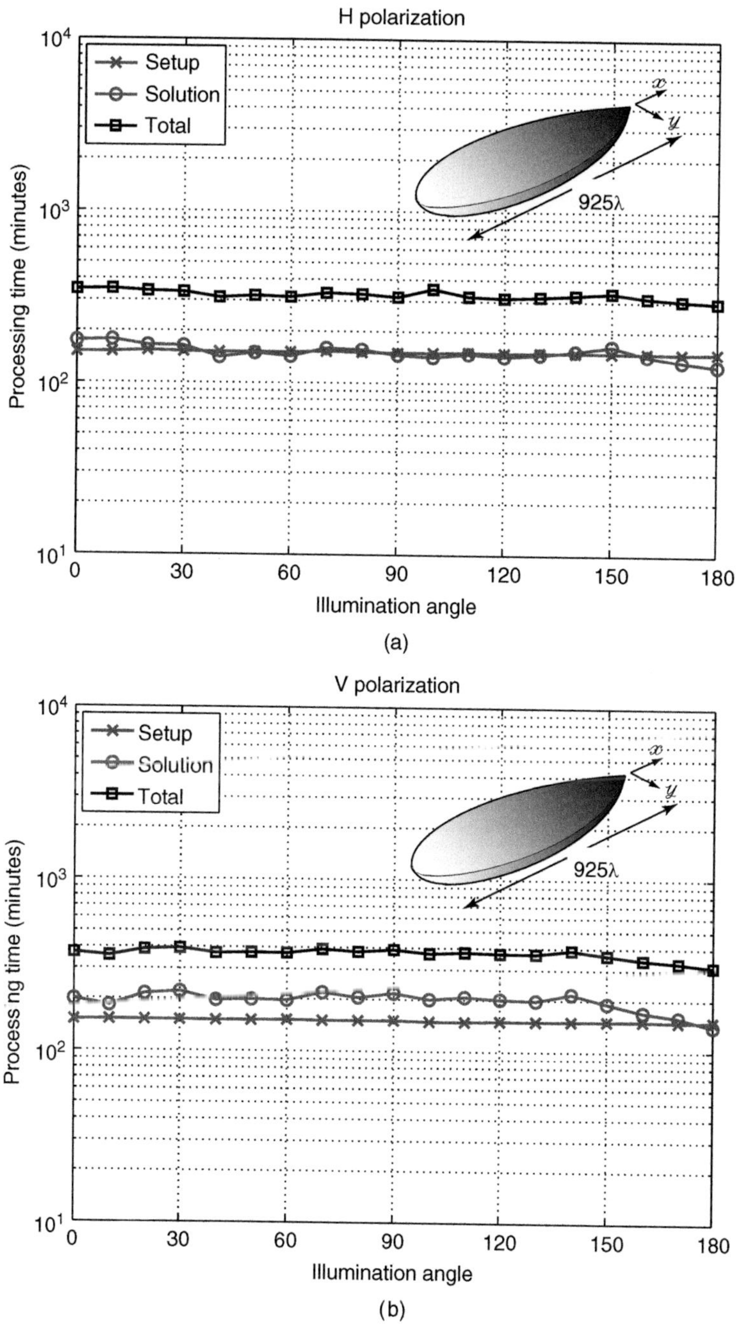

Figure 4.57 Processing times (with respect to the illumination angle measured from the *x* axis) for the solution of scattering problems involving the NASA Almond at 1.1 THz discretized with 203,476,224 unknowns (measured on the C-Clovertown cluster)

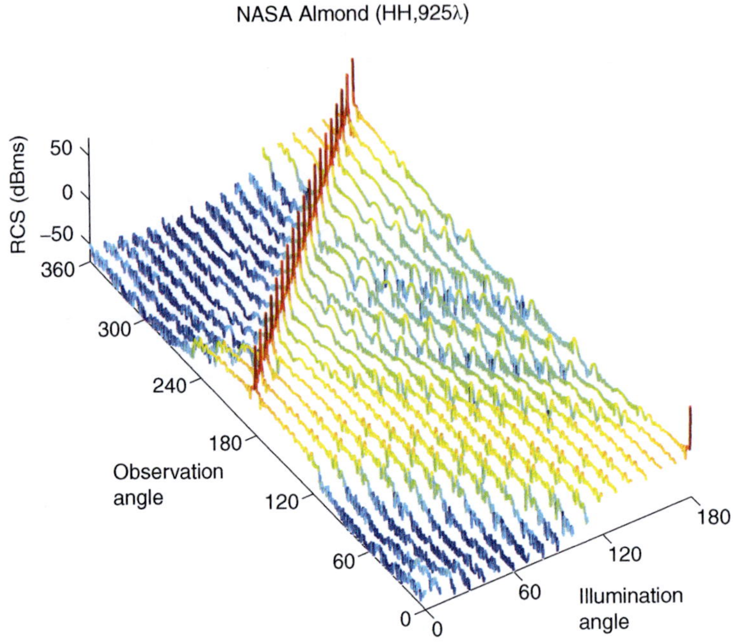

(a)

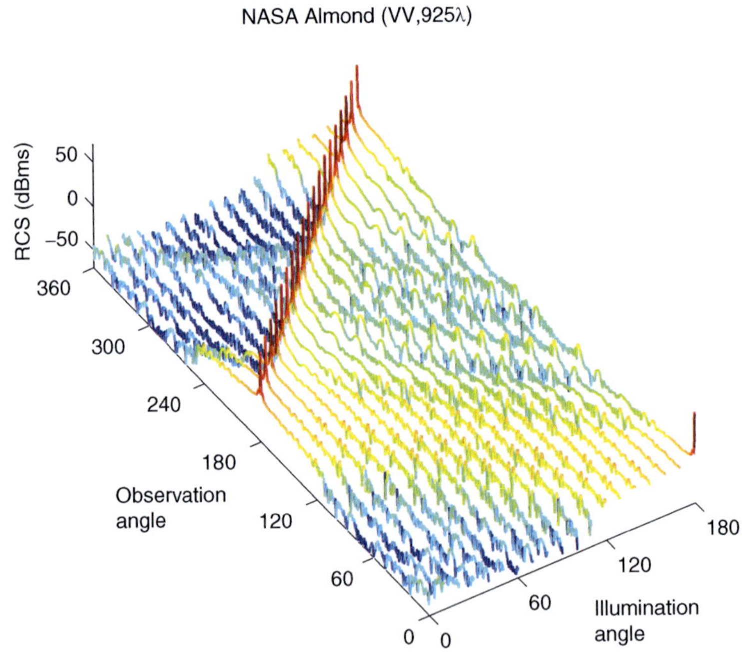

(b)

Figure 4.58 Co-polar bistatic RCS (dBms) of the NASA Almond at 1.1 GHz. RCS is plotted on the x-y plane as a function of the observation and illumination angles when the incident electric field is polarized in (a) ϕ and (b) θ directions.

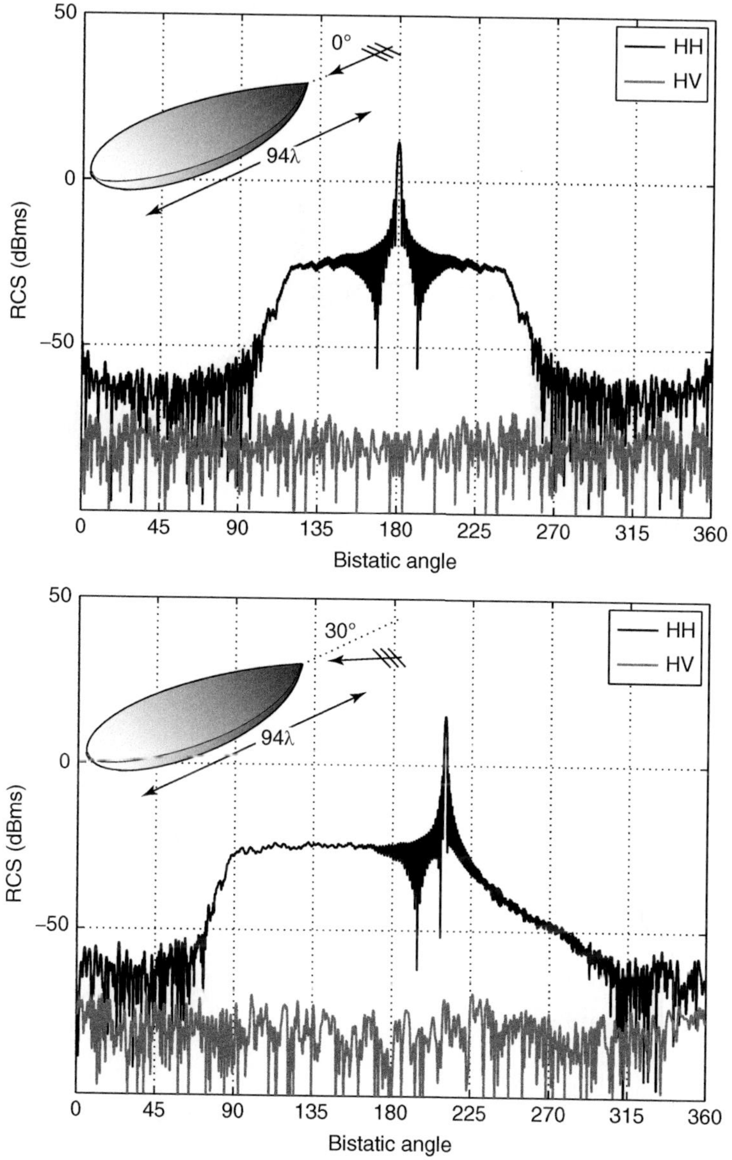

Figure 4.59 Co-polar (HH) and cross-polar (HV) bistatic RCS (dBms) of the NASA Almond at 112.5 GHz. The target is illuminated by plane waves with the electric field polarized horizontally. For numerical solutions, the target is discretized with 2,157,462 unknowns.

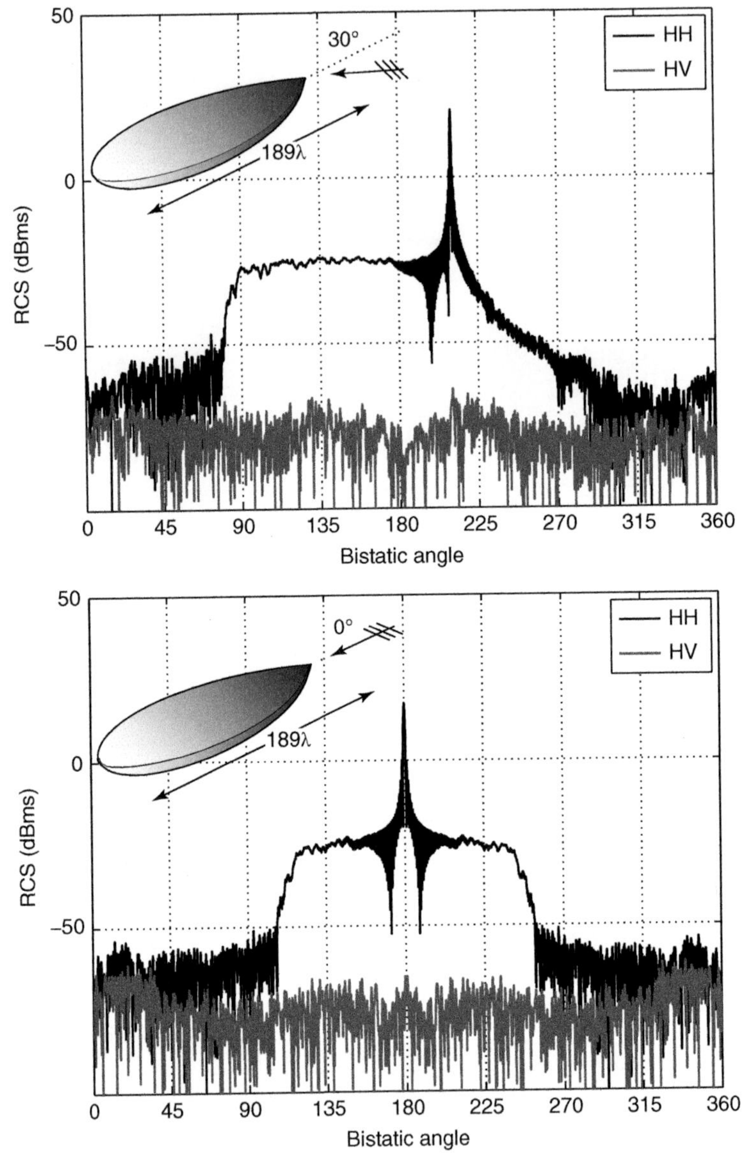

Figure 4.60 Co-polar (HH) and cross-polar (HV) bistatic RCS (dBms) of the NASA Almond at 225 GHz. The target is illuminated by plane waves with the electric field polarized horizontally. For numerical solutions, the target is discretized with 8,629,848 unknowns.

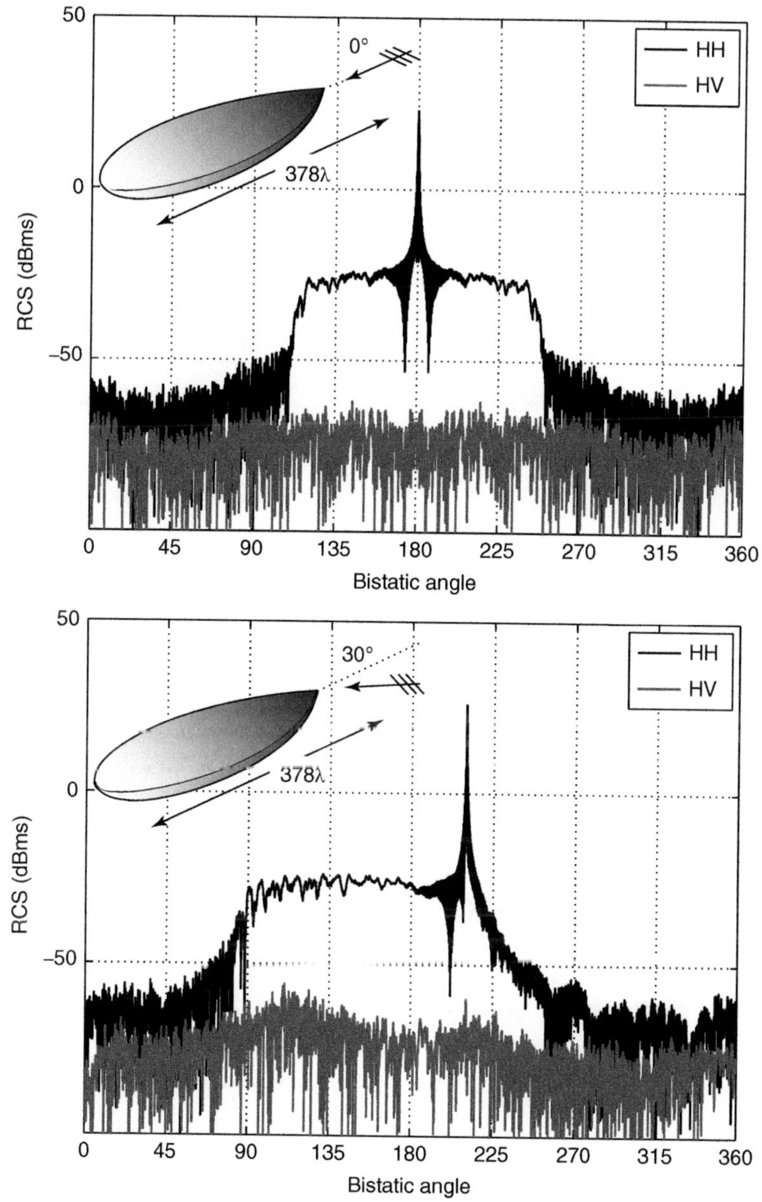

Figure 4.61 Co-polar (HH) and cross-polar (HV) bistatic RCS (dBms) of the NASA Almond at 450 GHz. The target is illuminated by plane waves with the electric field polarized horizontally. For numerical solutions, the target is discretized with 34,519,392 unknowns.

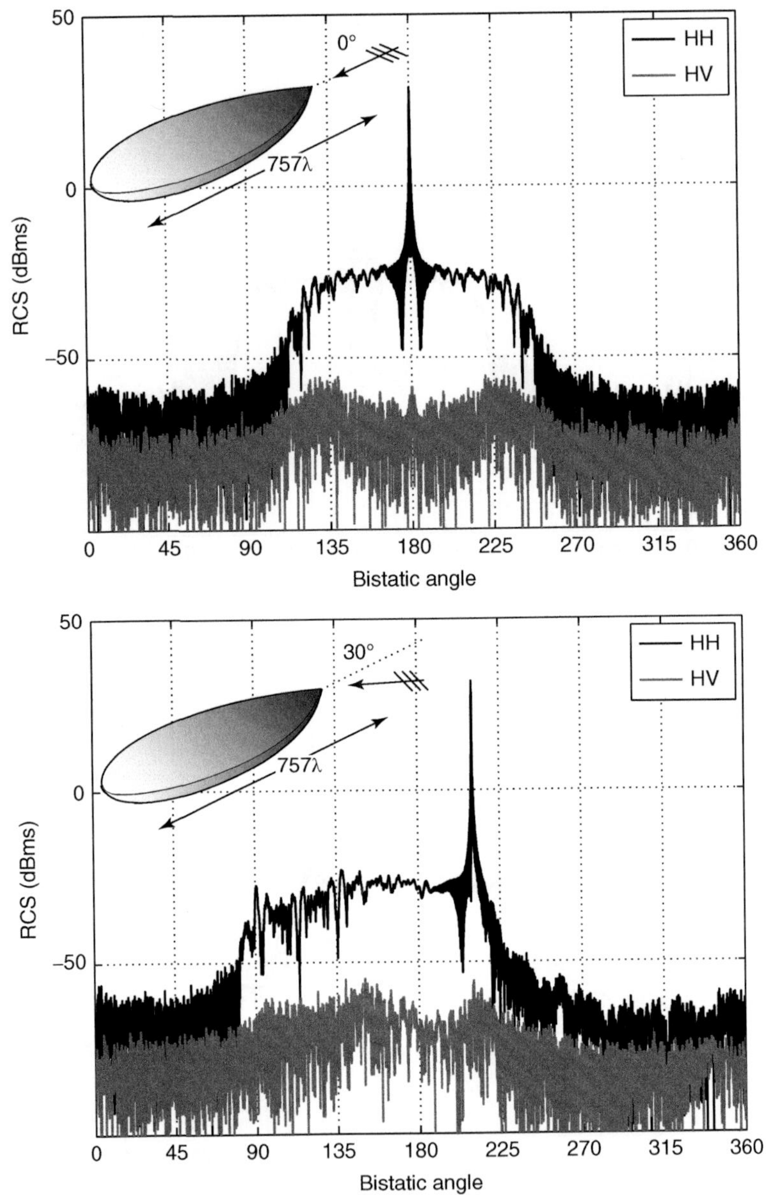

Figure 4.62 Co-polar (HH) and cross-polar (HV) bistatic RCS (dBms) of the NASA Almond at 900 GHz. The target is illuminated by plane waves with the electric field polarized horizontally. For numerical solutions, the target is discretized with 138,077,568 unknowns.

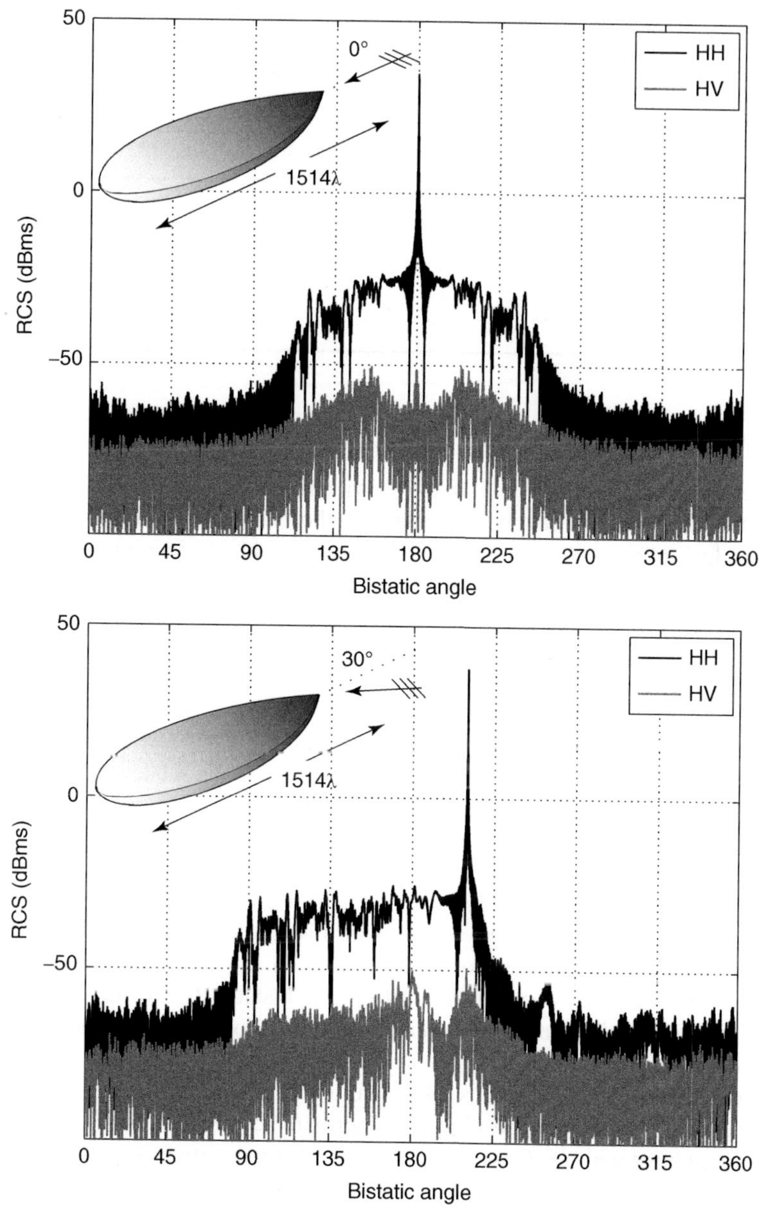

Figure 4.63 Co-polar (HH) and cross-polar (HV) bistatic RCS (dBms) of the NASA Almond at 1.8 THz. The target is illuminated by plane waves with the electric field polarized horizontally. For numerical solutions, the target is discretized with 552,310,272 unknowns. *Source*: Ergül and Gürel 2013 [238]. Reproduced with permission of IEEE.

Table 4.14 Solutions of Increasingly large electromagnetics problems involving the NASA Almond on the S-Nehalem-EX cluster

Frequency (GHz)	Unknowns	Iterations (10^{-3} residual)		Total time (minutes)	
		Exc: 0°	Exc: 30°	Exc: 0°	Exc: 30°
112.5	2,157,462	27	26	8	8
225	8,629,848	31	27	33	31
450	34,519,392	40	32	148	134
900	138,077,568	47	50	634	654
1800	552,310,272	68	70	3269	3396

4.9.4 Flamme

The Flamme depicted in Figure 4.2 is a stealth airborne target [149]. The scaled size of the object is 0.6 m and it involves a cavity at the back (see Chapter 5, Section 5.1). Figures 4.64–4.66 present the solution of scattering problems involving the Flamme at 360 GHz, i.e., when the the size of the object corresponds to approximately 720λ. Discretization of the object with $\lambda/10$ mesh size leads to matrix equations involving 134,741,760 unknowns. The object is located on the *x*-*y* plane such that its nose is directed towards the *x* axis and it is illuminated by various plane waves propagating on the *x*-*y* plane. Both ϕ and θ polarizations of the incident electric field are considered. Figures 4.64 and 4.65 present the number of BiCGStab iterations and processing times (setup, iterative solution, and total) in minutes when the solutions are performed by an 11-level MLFMA on the B-Harpertown cluster. The number of iterations and processing times are plotted with respect to the illumination angle measured from the *x* axis. As opposed to the solutions of the NASA Almond (Figures 4.53 and 4.54), the number of iterations increases with the increasing illumination angle for the Flamme, due to the resonant characteristics of the cavity at the back of the target. Figure 4.66 depicts the co-polar bistatic RCS (dBms) of the Flamme on the *x*-*y* plane as a function of observation and illumination angles. It can be observed that the bistatic RCS of the Flamme is very different from the bistatic RCS of the NASA Almond. Specifically, the Flamme RCS exhibits several peaks in various directions depending on the illumination due to specular reflections.

Figures 4.67–4.69 present solutions of scattering problems involving the Flamme at 440 GHz, i.e., when the size of the object is approximately 880λ. At this frequency, discretization of the object with $\lambda/10$ mesh size leads to matrix equations involving 204,664,320 unknowns. The object is again illuminated by various plane waves propagating on the *x*-*y* plane. Figures 4.67 and 4.68 present the number of BiCGStab iterations and processing times when the solutions are performed by a 12-level MLFMA on the C-Clovertown cluster. The co-polar bistatic RCS (dBms) on the *x*-*y* plane is illustrated in Figure 4.69 for both ϕ and θ polarizations. RCS values are quite similar to those at the lower frequency, depicted in Figure 4.66.

Figures 4.70–4.72 present the solution of scattering problems involving the Flamme at a higher frequency. The object is investigated at 670 GHz, i.e., when its size corresponds

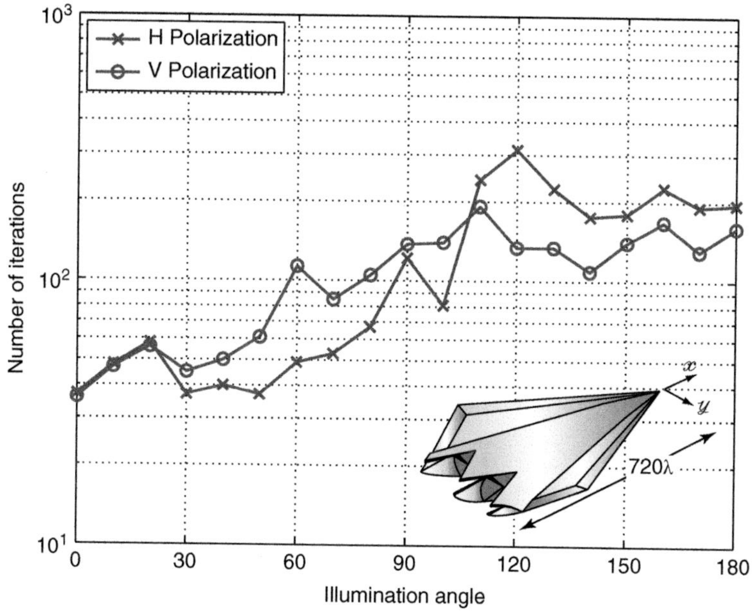

Figure 4.64 Number of BiCGStab iterations (with respect to the illumination angle measured from the x axis) for the solution of scattering problems involving the Flamme at 360 GHz discretized with 134,741,760 unknowns. *Source*: Ergül and Gürel 2011 [230]. Reproduced with permission of IEEE.

to approximately 1340λ. Discretization with $\lambda/10$ triangles leads to matrix equations involving 357,888,000 unknowns. The problems are solved on the C-Clovertown cluster. Figures 4.70–4.72 present the bistatic RCS (dBms) when the target is illuminated by plane waves propagating at 30°, 60°, and 90° angles from the x axis. Both the co-polar and cross-polar RCS is plotted on the x-y plane with respect to the bistatic angle ϕ from 0° to 180°, when the incident electric field is polarized in the θ and ϕ directions. The following observations can be made:

- The Flamme RCS exhibits several peaks in various directions depending on the illumination due to specular reflections from the straight edges and nearly flat surfaces of the target.
- The cross-polar RCS is large and comparable to the co-polar RCS of the Flamme.
- Similar to the NASA Almond, the Flamme exhibits a stealth ability, i.e., the backscattered RCS is significantly lower than the forward-scattered RCS.

Figures 4.73 and 4.74 presents the solution of very large scattering problems involving the Flamme at 820 GHz, i.e., when its size corresponds to approximately 1640λ. The object is discretized with 538,967,040 unknowns and illuminated by plane waves propagating in the $-x$ direction and propagating at 30° angle from the x axis. The electric field is polarized in the ϕ direction. Each problem is solved on the S-Nehalem-EX cluster by using a 13-level MLFMA in approximately 60 hours.

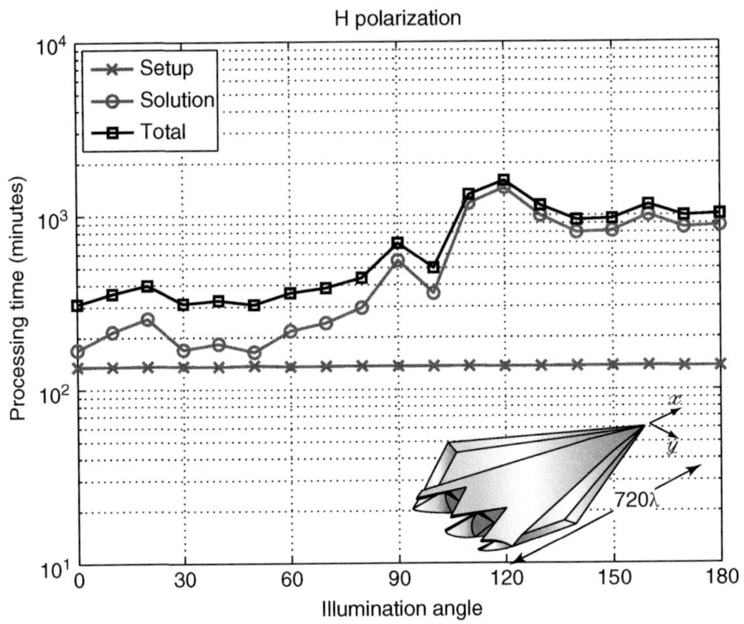

(a)

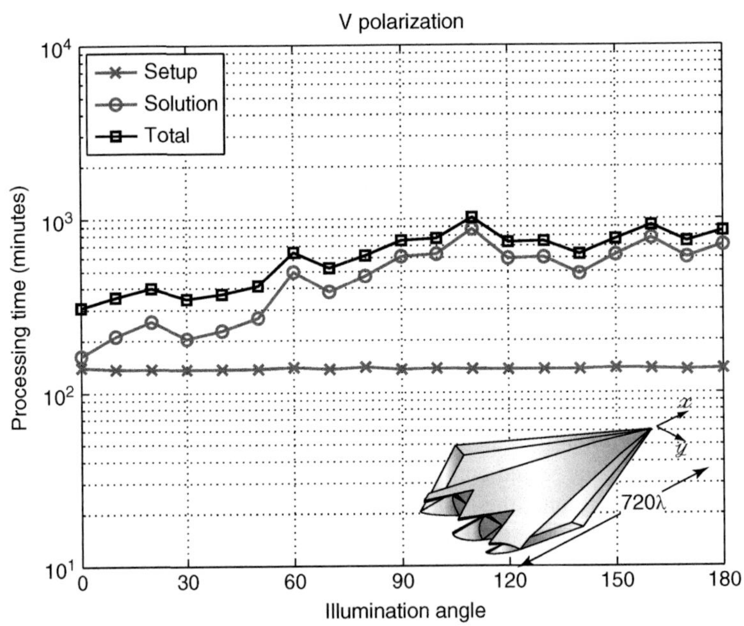

(b)

Figure 4.65 Processing times (with respect to the illumination angle measured from the x axis) for the solution of scattering problems involving the Flamme at 360 GHz discretized with 134,741,760 unknowns (measured on the B-Harpertown cluster). *Source*: Ergül and Gürel 2011 [230]. Reproduced with permission of IEEE.

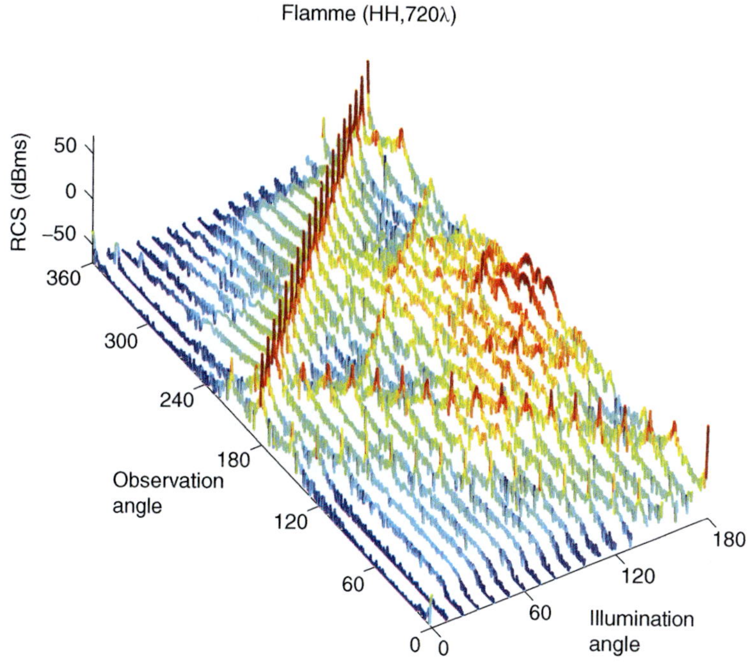

(a)

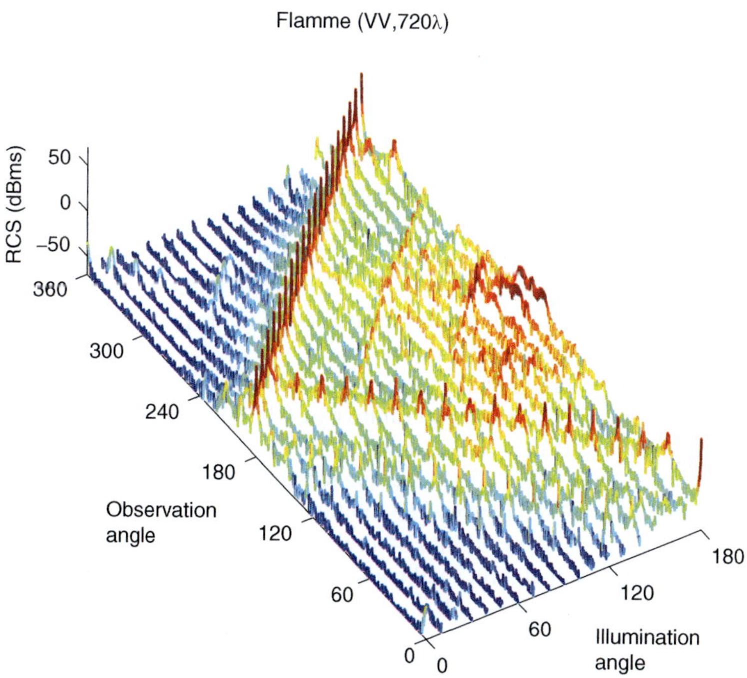

(b)

Figure 4.66 Co-polar bistatic RCS (dBms) of the Flamme at 360 GHz. RCS is plotted on the x-y plane as a function of the observation and illumination angles when the incident electric field is polarized in (a) ϕ and (b) θ directions. *Source*: Ergül and Gürel 2011 [230]. Reproduced with permission of IEEE.

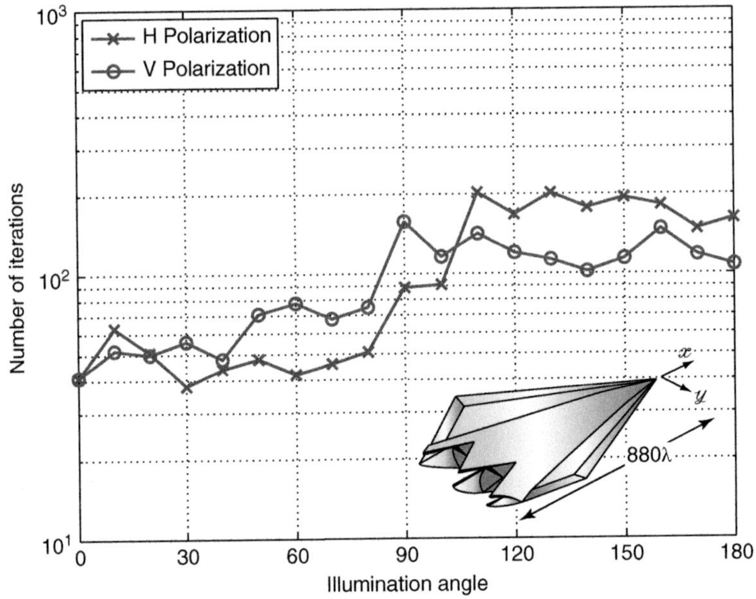

Figure 4.67 Number of BiCGStab iterations (with respect to the illumination angle measured from the x axis) for the solution of scattering problems involving the Flamme at 440 GHz discretized with 204,664,320 unknowns.

4.10 Solutions of Large-Scale Electromagnetics Problems Involving Dielectric Objects

The hierarchical strategy can be extended to large-scale electromagnetics problems involving dielectric objects. Specifically, tree structures constructed for different media can be parallelized efficiently by using the hierarchical strategy. This section presents examples of numerical solutions of very large electromagnetics problems involving dielectric objects. All problems are formulated with the conventional JMCFIE ($\alpha = 0.5$) and solved by using an optimal set of values for the error parameters (as described in the previous section) to achieve maximum 1% error in far-zone electromagnetic fields.

Figure 4.75 presents the solution of a scattering problem involving a dielectric sphere with a radius of 0.3 m in free space. The relative permittivity of the sphere is 2.0 and it is illuminated by a plane wave propagating in the z direction with the electric field polarized in the x direction at 100 GHz. At this frequency, the radius of the sphere is approximately $100\lambda_o$ and the problem is discretized with 67,582,464 unknowns. The solution of the problem, which is performed by using a 10-level MLFMA parallelized into 64 processes on the N-Nehalem cluster, requires 237 iterations (for 0.005 residual error) and a total of 77 hours. The total amount of memory used for the solution is 723 GB. Figure 4.75 depicts bistatic RCS values (in dBms) on the z-x plane as a function of the bistatic angle θ from 0° to 180°. Computational values obtained by using MLFMA are compared with those obtained via analytical Mie-series solutions. RCS

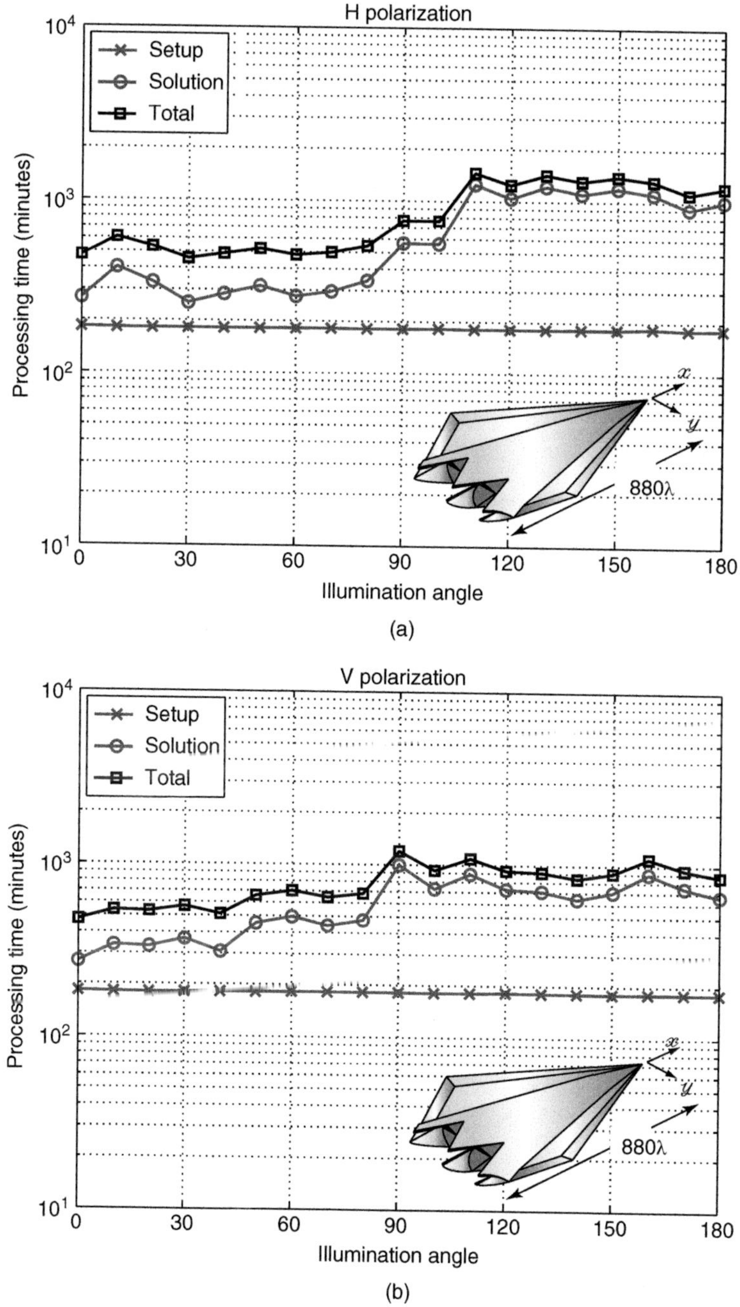

Figure 4.68 Processing times (with respect to the illumination angle measured from the x axis) for the solution of scattering problems involving the Flamme at 440 GHz discretized with 204,664,320 unknowns (measured on the C-Clovertown cluster)

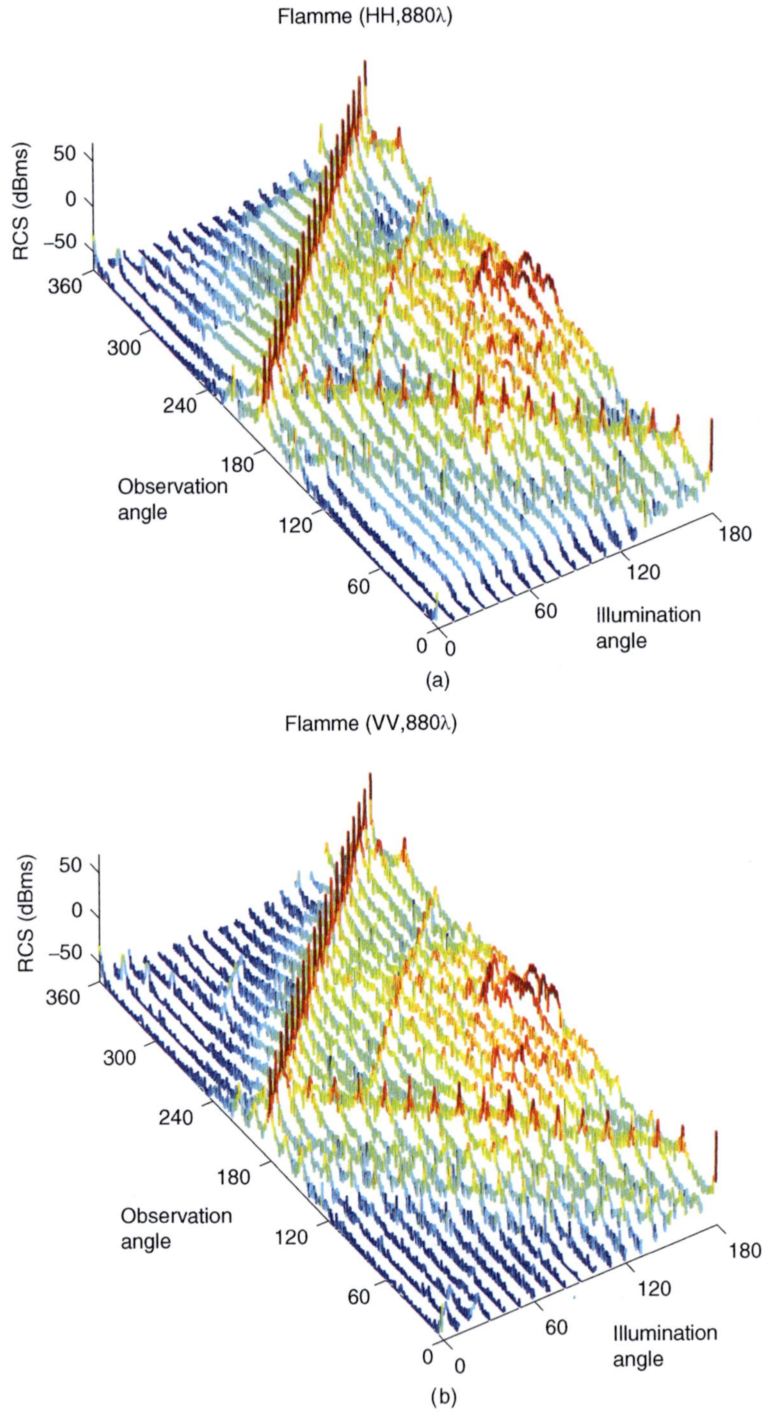

Figure 4.69 Co-polar bistatic RCS (dBms) of the Flamme at 440 GHz. RCS is plotted on the x-y plane as a function of the observation and illumination angles when the incident electric field is polarized in (a) ϕ and (b) θ directions.

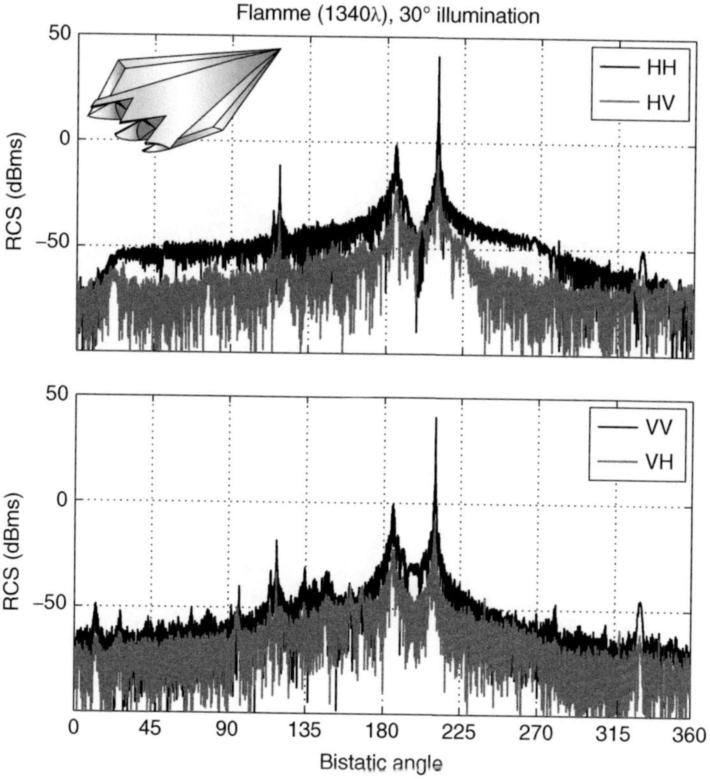

Figure 4.70 Co-polar and cross polar bistatic RCS (dBms) of the Flamme at 670 GHz. The target is illuminated by a plane wave propagating at 30° angle from the x axis. Both θ and ϕ polarizations of the incident electric field are considered. For numerical solutions, the target is discretized with 357,888,000 unknowns.

values around the forward-scattering (0°) and backscattering (180°) directions are also focused in separate plots. It can be observed that the computational values agree very well with the analytical results.

Figure 4.76 presents the solution of an electromagnetics problem involving a hemisphere lens with a radius of 25 mm. The lens has a relative permittivity of 4.8 and it is illuminated by a plane wave propagating in the $-z$ direction (towards its convex surface) at 1.08 THz. At this frequency, the radius of the lens corresponds to approximately $90\lambda_o$. Discretization of the object with $\lambda_o/10$ triangles leads to a matrix equation involving 49,851,936 unknowns. Solution of the problem via 10-level MLFMA requires 105 iterations and a total of 19 hours using 128 processes on the N-Nehalem cluster. The total amount of memory used for the solution is 673 GB. Figure 4.76 depicts the total electric field on the axis of rotation of the lens from $z = -40$ mm to $+40$ mm, where a focussing is observed at around -7 mm.

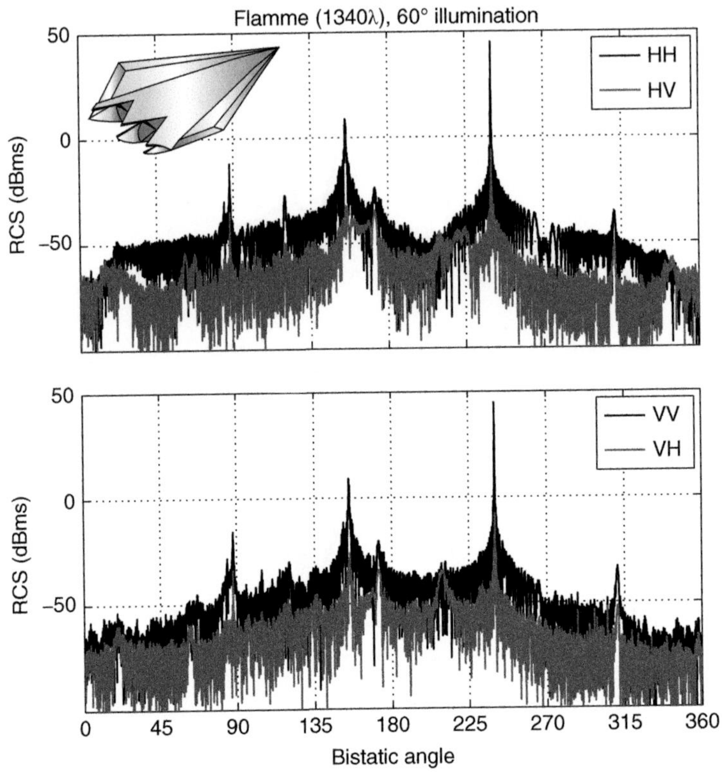

Figure 4.71 Co-polar and cross polar bistatic RCS (dBms) of the Flamme at 670 GHz. The target is illuminated by a plane wave propagating at 60° angle from the x axis. Both θ and ϕ polarizations of the incident electric field are considered. For numerical solutions, the target is discretized with 357,888,000 unknowns.

Figure 4.77 presents the solution of an electromagnetics problem involving a 5-layer periodic structure. Five $2 \times 2 \times 0.41$ cm slabs with 1.6 relative permittivity are placed at 0.5 cm intervals in free space and illuminated by a plane wave at 1.62 GHz. Discretization of the object with $\lambda_o/10$ triangles leads to a matrix equation involving 112,849,200 unknowns. Solution of the problem via 10-level MLFMA requires 102 iterations and a total of 37 hours using 64 processes on the N-Nehalem cluster. The total amount of memory used for the solution is 669 GB. Figure 4.77 depicts the electric field on axis of symmetry from $z = -5$ cm to 5 cm for the inner and outer problems. For the inner problem, the equivalent currents obtained by using MLFMA are allowed to radiate into a homogeneous space with the electrical parameters of the inner medium assumed everywhere. Hence, in this case, equivalent currents should not radiate outside the object and any nonzero field value can be interpreted as the error. Similarly, for the outer problem, equivalent currents radiate into a homogeneous space with the electric

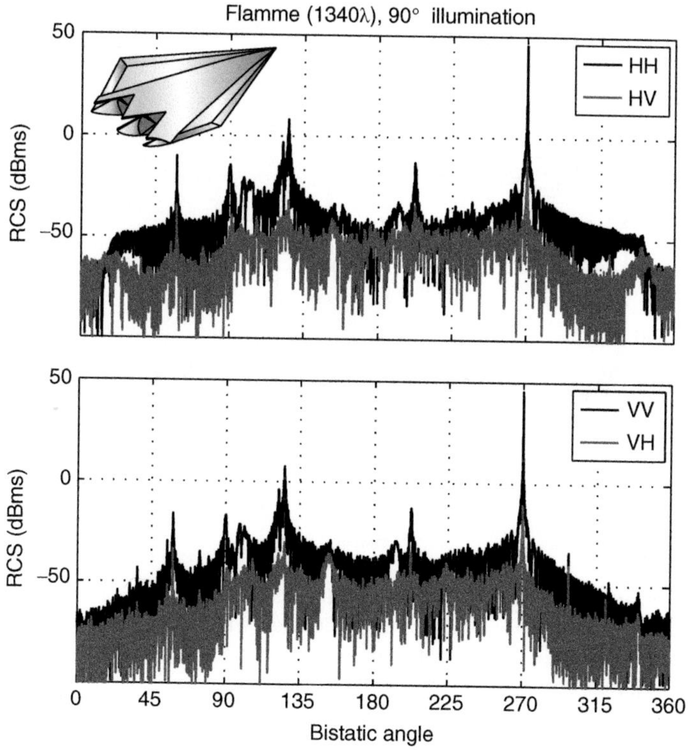

Figure 4.72 Co-polar and cross polar bistatic RCS (dBms) of the Flamme at 670 GHz. The target is illuminated by a plane wave propagating at 90° angle from the x axis. Both θ and ϕ polarizations of the incident electric field are considered. For numerical solutions, the target is discretized with 357,888,000 unknowns.

parameters of the outer medium, and any radiation into the object can be interpreted as error. Note that those solutions of the inner and outer problems can be combined for a complete plot. Figure 4.77 clearly demonstrates the high accuracy of the solution with very low values of unwanted radiations.

Figure 4.78 presents the solution of a scattering problem involving a two-dimensional array of $81 \times 81 = 6561$ lossy dielectric cubes. As depicted in Figure 4.78(a), cubes with edges of 1 μm are periodically arranged on the x-y plane with 2 μm periodicity in both x and y directions. The relative permittivity and conductivity of the cubes are 8.0 and 0.01 S/m, respectively. The array is illuminated by a plane wave propagating in the $-z$ direction with the electric field polarized in the x direction at 300% THz, i.e., when the edges of the cubes are approximately λ_o and the overall size of the structure is approximately $161\lambda_o \times 161\lambda_o \times \lambda_o$. The problem is discretized with 26,926,344 unknowns and solved in a total of 9 hours (including 32 iterations)

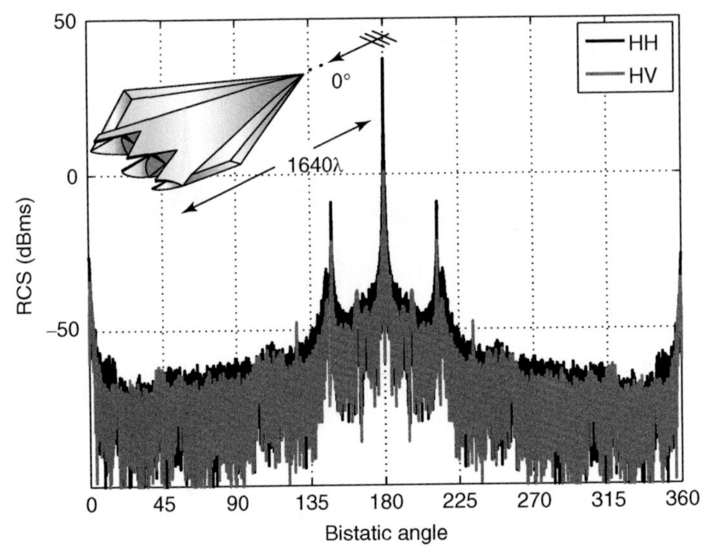

Figure 4.73 Co-polar and cross polar bistatic RCS (dBms) of the Flamme at 820 GHz. The target is illuminated by a plane wave propagating in the $-x$ direction with the electric field polarized in the y direction. For numerical solutions, the target is discretized with 538,967,040 unknowns. *Source*: Gürel and Ergül 2011 [237]. Reproduced with permission of IEEE.

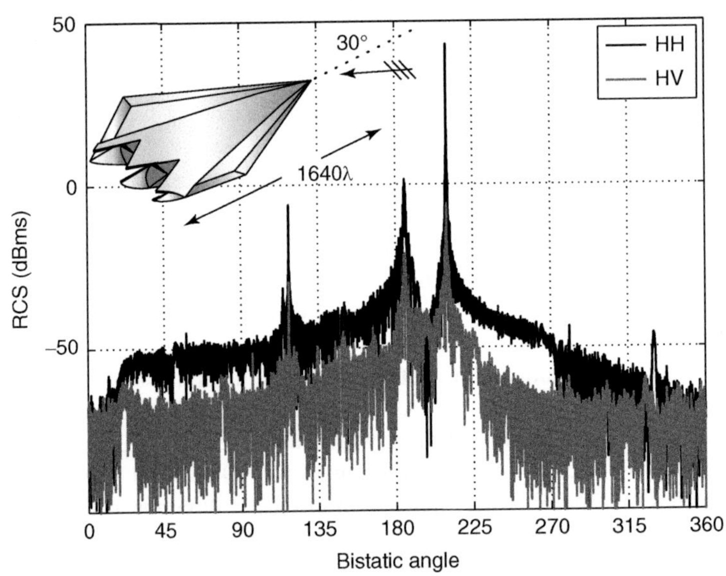

Figure 4.74 Co-polar and cross polar bistatic RCS (dBms) of the Flamme at 820 GHz. The target is illuminated by a plane wave propagating at 30° angle from the x axis with the electric field polarized in the ϕ direction. For numerical solutions, the target is discretized with 538,967,040 unknowns. *Source*: Ergül and Gürel 2013 [238]. Reproduced with permission of IEEE.

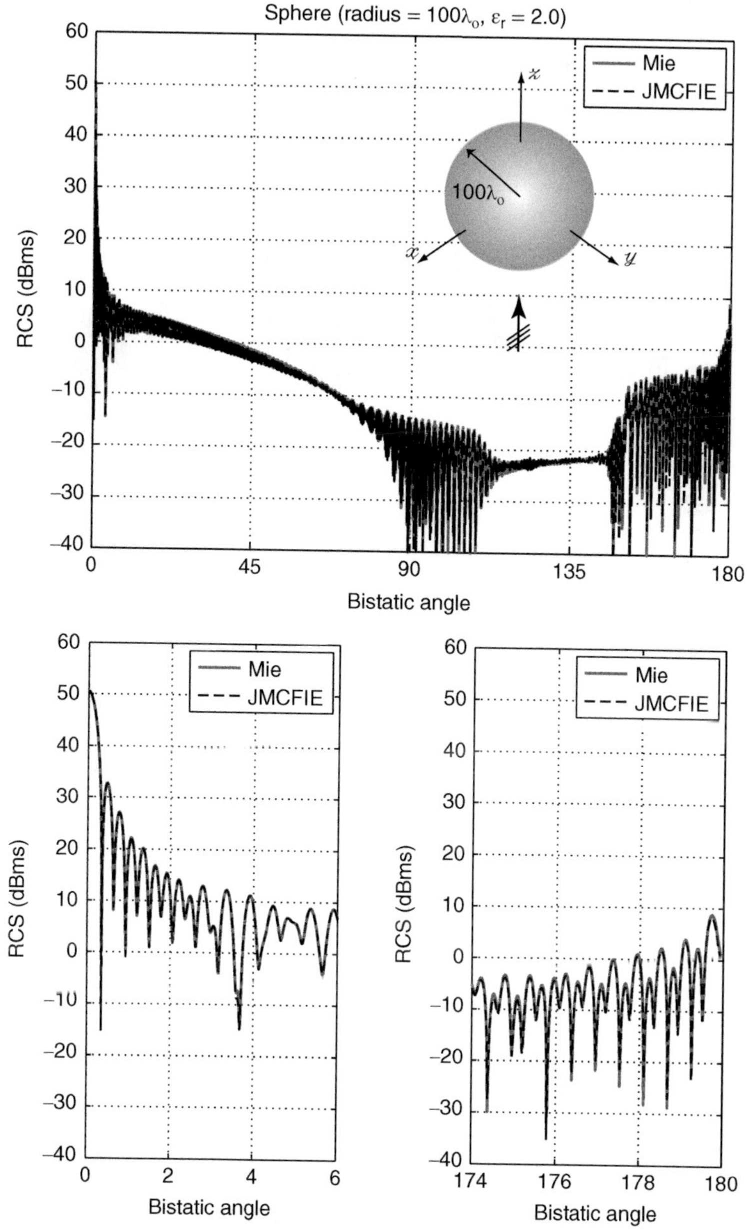

Figure 4.75 Solution of a scattering problem involving a dielectric sphere with a radius of 0.3 m at 100 GHz discretized with 67,582,464 unknowns. RCS (in dBms) is plotted as a function of the bistatic angle from 0° to 180°, where 0° corresponds to the forward-scattering direction. Computational values provided by the parallel MLFMA implementation agree very well with the analytical Mie-series solution. *Source*: Ergül 2011 [232]. Reproduced by permission of the Optical Society of America.

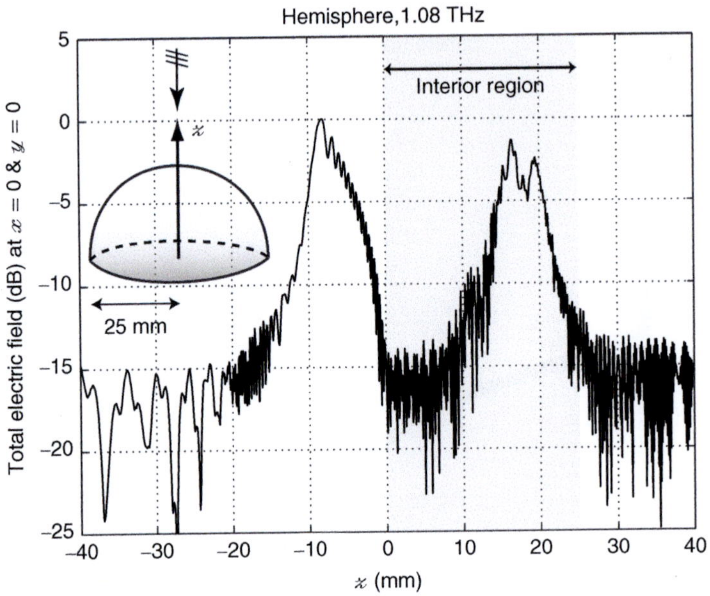

Figure 4.76 Solution of an electromagnetics problem involving a dielectric hemisphere lens with a radius of 25 mm at 1.08 THz. The lens has a relative permittivity of 4.8 and the problem is discretized with 49,851,936 unknowns. The total electric field on the axis of rotation of the lens is plotted from $z = -40$ mm to +40 mm. *Source*: Ergül 2011 [232]. Reproduced by permission of the Optical Society of America.

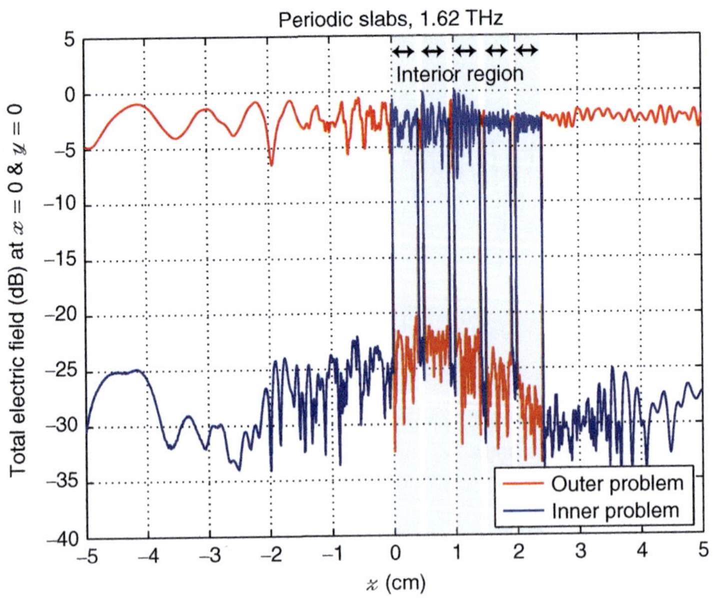

Figure 4.77 Solution of an electromagnetics problem involving five 2 × 2 × 0.41 cm dielectric slabs at 1.62 THz. The structure has a relative permittivity of 1.6 and the problem is discretized with 112,849,200 unknowns. The total electric field on the axis of symmetry is plotted from $z = -5$ cm to 5 cm for the inner and outer problems. *Source*: Ergül 2011 [232]. Reproduced by permission of the Optical Society of America.

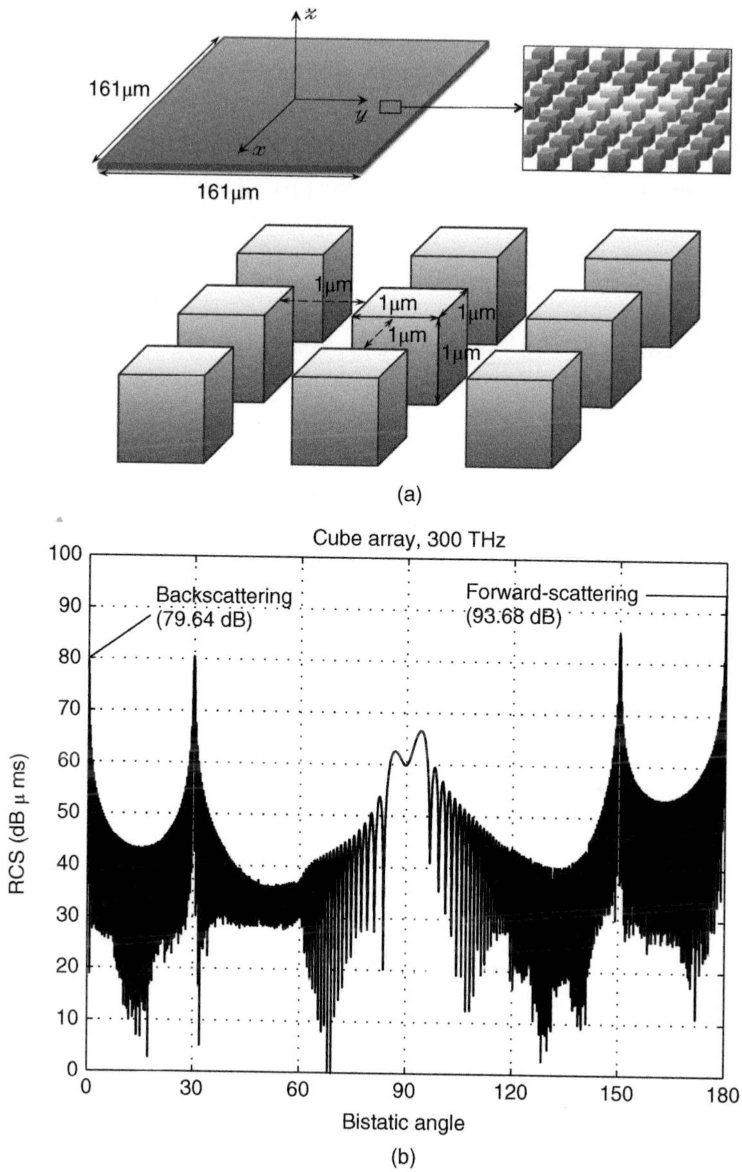

Figure 4.78 Solution of a scattering problem involving an array of $81 \times 81 = 6561$ lossy dielectric cubes. (a) Cubes with edges of 1 μm are periodically arranged on the x-y plane with 2 μm periodicity in both directions. The relative permittivity and conductivity of the cubes are 8.0 and 0.01 S/m, respectively. (b) RCS is plotted on the z-x plane, when the array is illuminated by a plane wave propagating in the $-z$ direction with the electric field polarized in the x direction at 300 THz. *Source*: Ergül 2011 [232]. Reproduced by permission of the Optical Society of America.

using 64 processes and a total of 663 GB memory on the N-Nehalem cluster. Figure 4.78(b) depicts the bistatic RCS (in dBμms) on the z-x plane, demonstrating peaks at around 30° and 150°, in addition to high values in the backscattering and forward-scattering directions.

4.11 Concluding Remarks

On distributed-memory architectures, the hierarchical partitioning strategy enables efficient parallelization of MLFMA. Thanks to the higher parallelization efficiency provided by this strategy, MLFMA can be employed to solve extremely large electromagnetics problems discretized with hundreds of millions of unknowns on moderately large computers. In this chapter, efficiency and accuracy of the developed implementations are demonstrated on various canonical and complicated objects. The next chapter presents more examples to the solution of real-life problems with sequential and parallel implementations of MLFMA.

5

Applications

This chapter consists of various case studies and solutions of real-life electromagnetics problems using MLFMA. The topics are selected based on the authors' experiences on scattering from airborne targets, transmission through metamaterials and photonic crystals, scattering from RBCs, and radiation from log-periodic (LP) antennas and their arrays. Simulation results are presented to demonstrate the capabilities of electromagnetics solvers based on MLFMA.

5.1 Case Study: External Resonances of the Flamme

External resonances can be encountered when a scatterer involves a cavity and the cavity is excited appropriately. Those physical resonances lead to rapid changes in scattering characteristics of the target, which can be visible when electromagnetic fields are investigated with respect to frequency. For example, Figure 5.1 presents the bistatic RCS of the stealth airborne target Flamme [149] from 2 GHz to 11 GHz when the target is illuminated by a plane wave propagating at 140° angle from the nose. The plane wave is vertically polarized and both co-polar (VV) and cross-polar (VH) RCS are considered. It can be observed that the RCS values change rapidly around 7.9 GHz due to the resonance of the cavity located at the back of the Flamme.

To further investigate external resonances of the Flamme, Figure 5.2 presents the electric current density induced on the surface of the target at three different frequencies, i.e., 7.5 GHz, 7.9 GHz, and 8.4 GHz. Comparing Figures 5.2(a) and 5.2(b), one can observe that the value of the current density inside the cavity at the resonance frequency (7.9 GHz) is significantly larger than the value of the current density at the ordinary frequency (7.5 GHz). As depicted in Figure 5.2(c), another external resonance occurs at 8.4 GHz. However, at this frequency, the value of the current density inside the cavity does not increase significantly and the resonance is not visible in the RCS results depicted in Figure 5.1.

In order to verify that the cavity of the Flamme resonates at 7.9 GHz and 8.4 GHz, Figure 5.3 presents the solution of scattering problems involving the cavity alone. The cavity is isolated from the rest of the target and illuminated by the same plane wave at 7.5 GHz, 7.9 GHz, and 8.4 GHz. It can be observed that the value of the current density inside the cavity becomes large at 7.9 GHz and 8.4 GHz due to the resonances.

The Multilevel Fast Multipole Algorithm (MLFMA) for Solving Large-Scale Computational Electromagnetics Problems,
First Edition. Özgür Ergül and Levent Gürel.

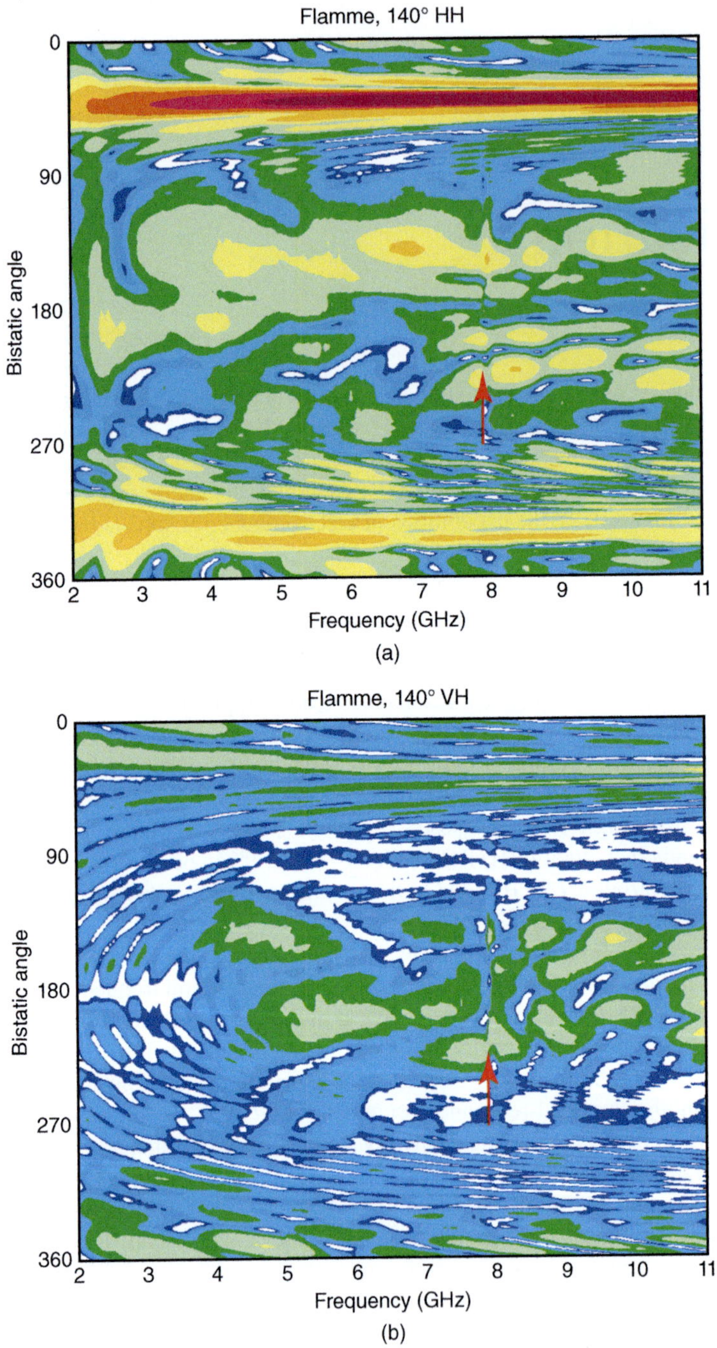

Figure 5.1 (a) Co-polar and (b) cross-polar bistatic RCS of the Flamme from 2 GHz to 11 GHz. The target is illuminated by a plane wave propagating at 140° angle from the nose with the electric field polarized in the vertical direction.

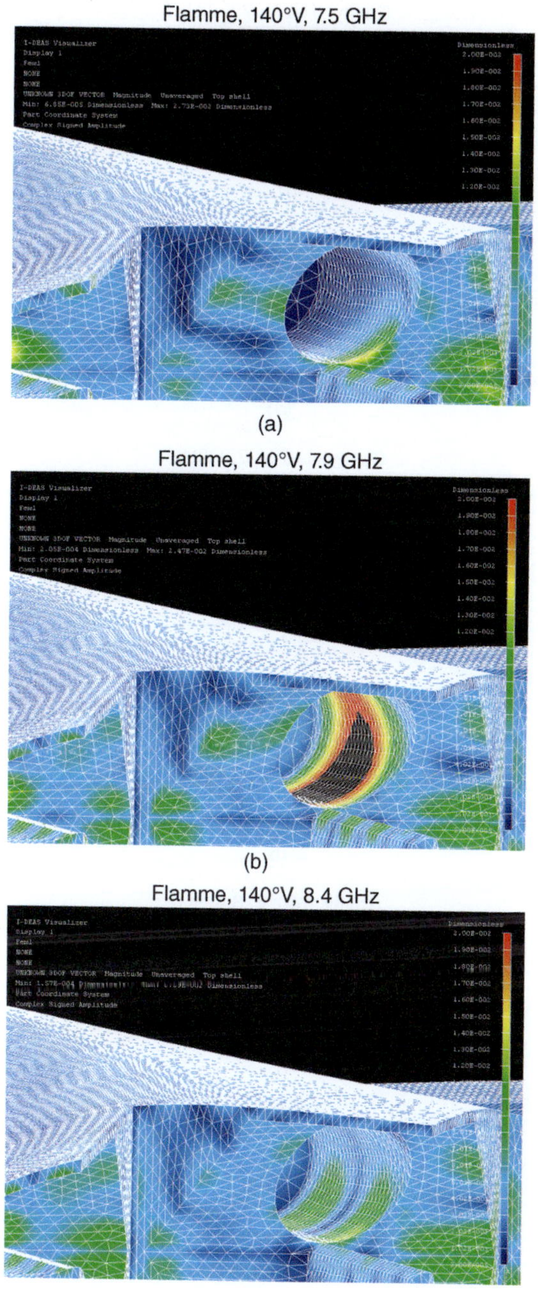

Figure 5.2 Induced electric current on the surface of the Flamme at (a) 7.5 GHz, (b) 7.9 GHz, and (c) 8.4 GHz. The target is illuminated by a plane wave propagating at 140° angle from the nose with the electric field polarized in the vertical direction.

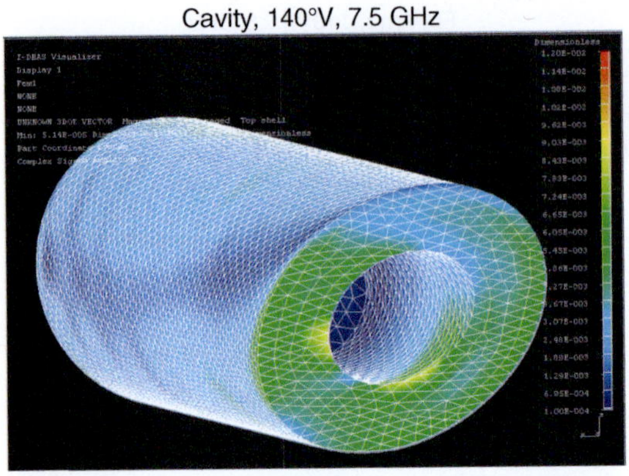

(a)

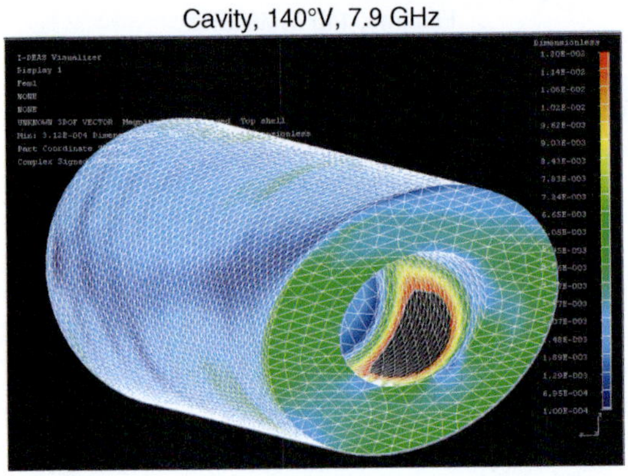

(b)

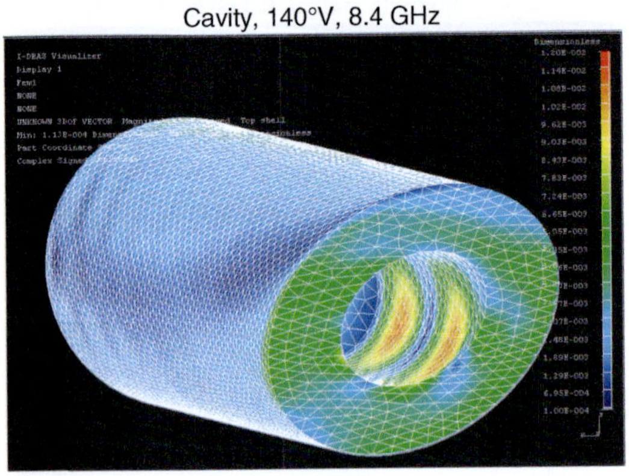

(c)

Figure 5.3 Induced electric current on the surface of the cavity of the Flamme at (a) 7.5 GHz, (b) 7.9 GHz, and (c) 8.4 GHz. The cavity is isolated and illuminated by a plane wave similar to the Flamme.

5.2 Case Study: Realistic Metamaterials Involving Split-Ring Resonators and Thin Wires

Metamaterials are artificial structures that are constructed by periodically arranging unit cells, such as SRRs and thin wires [19]–[25]. Due to resonant natures of those cells, electromagnetic properties of the host medium, i.e., permittivity, permeability, or both, can effectively become negative for some frequencies, which can be useful in diverse applications, such as subwavelength focusing, cloaking, and designing improved antennas. At a resonance frequency, even a single SRR is able to block the transmission of waves, which can be observed as a shadowing effect. This unusual behavior can be explained with the induced negative effective permeability in the medium. Due to the complex effective wavenumber, the propagating power decays rapidly in the transmission region behind the SRR. It is well known that periodical arrangements of thin wires can also be used to block electromagnetic waves, if the spacing between thin wires is small compared to the wavelength and if thin wires are aligned parallel to the electric field of the incident wave. As opposed to SRRs, however, the shadowing effect can be obtained with thin wires over a wide range of frequencies. In the context of metamaterials, the shielding property of thin wires can be explained as the induction of negative effective permittivity in the medium.

Figure 5.4 depicts a 1-layer composite metamaterial (CMM) wall constructed by combining the 18 × 11 SRR array in Figure 3.42 with thin wires. The dimensions of the thin wires are compatible with the dimensions of SRRs, and they exhibit negative effective permittivity in a wide range of frequencies, including the resonance frequency of SRRs, i.e., 100 GHz. By arranging SRRs and thin wires in the same medium, both effective permittivity and effective permeability

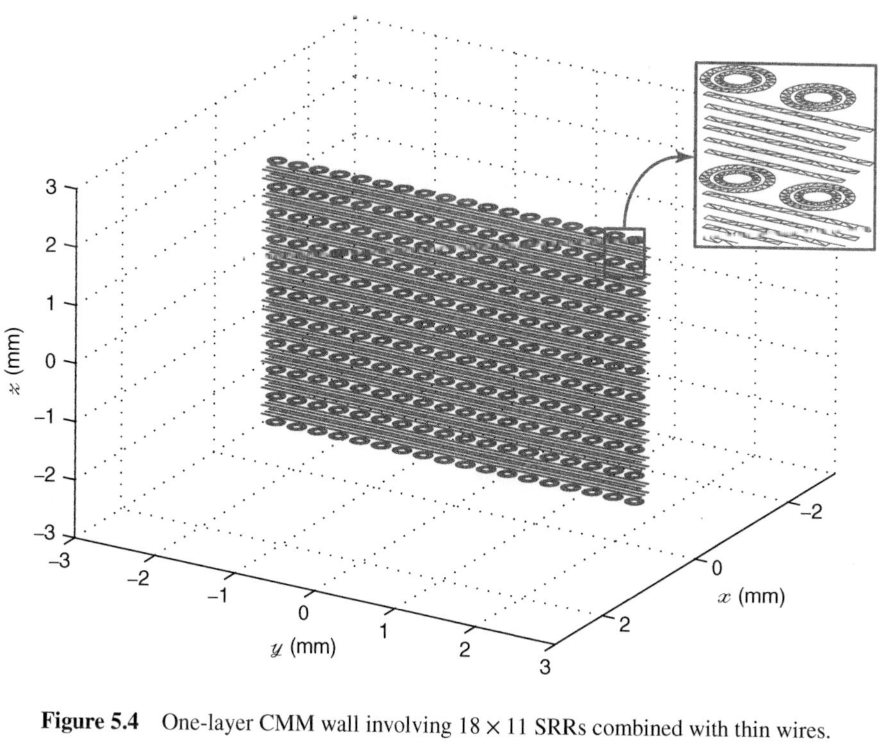

Figure 5.4 One-layer CMM wall involving 18 × 11 SRRs combined with thin wires.

can be simultaneously negative for some frequencies. When SRRs do not resonate, CMM structures are opaque due to the negative effective permittivity induced by thin wires. At resonance frequencies of SRRs, however, CMMs are unexpectedly transparent, which can be explained by the induced double negativity [19].

Figure 5.5 presents the solution of metamaterial problems involving the 1-layer SRR and CMM walls depicted in Figures 3.42 and 5.4, respectively. Both walls are embedded into a

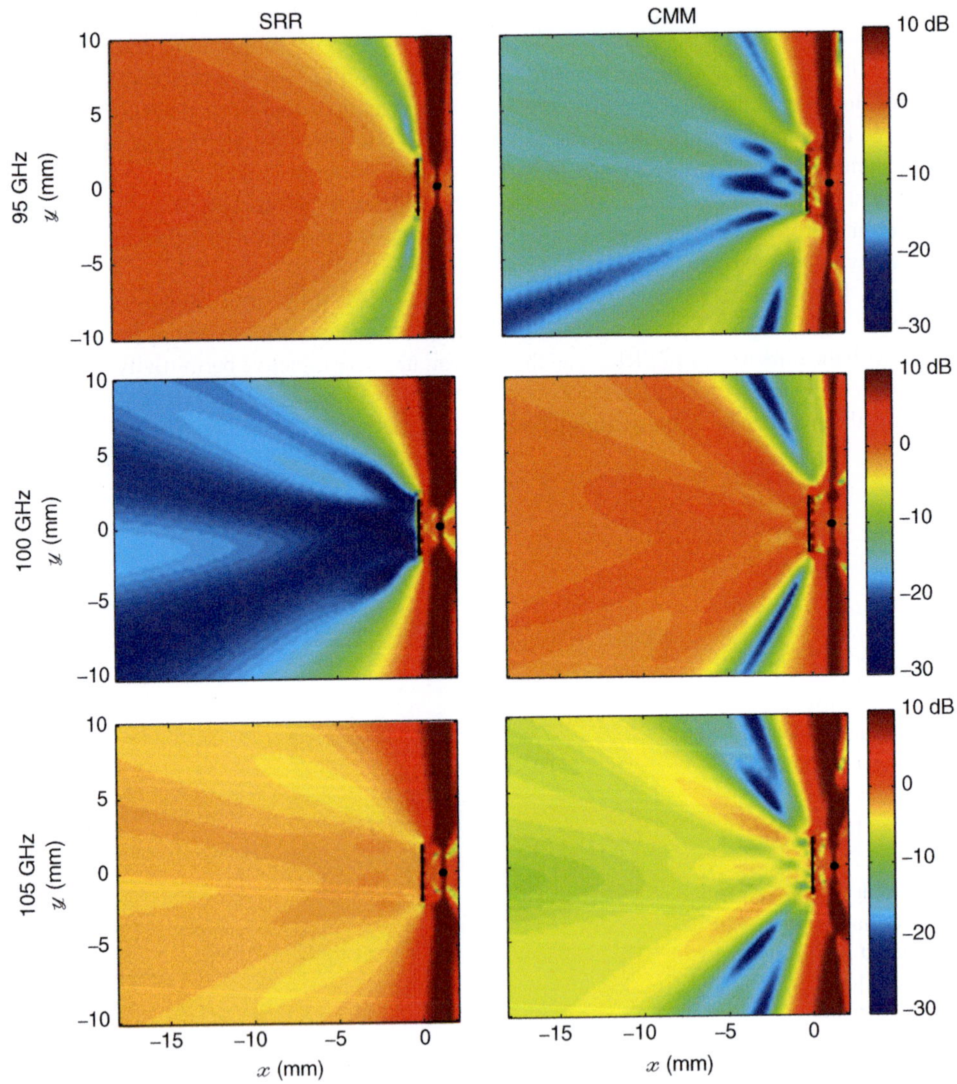

Figure 5.5 Power transmission on the $z = 0$ plane for 1-layer SRR and CMM walls at 95 GHz, 100 GHz, and 105 GHz. A y-directed Hertzian dipole is radiating from $x = 1.2$ mm. *Source*: Gürel *et al.* 2009 [24]. Reproduced with permission of PIER.

homogeneous host medium with a relative permittivity of 4.8 and located at $x = 0$. SRRs are oriented perpendicular to the z direction, while the splits are along the y direction. A y-directed Hertzian dipole is radiating from $x = 1.2$ mm. The problems are formulated with EFIE, discretized with the RWG functions, and solved iteratively using MLFMA. The total electric and magnetic fields (E^{tot} and H^{tot}) are calculated at various points on the $z = 0$ plane. The power transmission at an observation point $r = (x, y, 0)$ is defined as

$$\mathcal{T}(r) = \frac{|\boldsymbol{P}_{av}^{tot}(r)|}{|\boldsymbol{P}_{av}^{inc}(r)|}, \tag{5.1}$$

where P_{av}^{tot} and P_{av}^{inc} represent the total and incident time-average power densities, i.e.,

$$\boldsymbol{P}_{av}^{tot} = \frac{1}{2}\mathrm{Re}\{\boldsymbol{E}^{tot}(r) \times [\boldsymbol{H}^{tot}(r)]^*\} \tag{5.2}$$

$$\boldsymbol{P}_{av}^{inc} = \frac{1}{2}\mathrm{Re}\{\boldsymbol{E}^{inc}(r) \times [\boldsymbol{H}^{inc}(r)]^*\}. \tag{5.3}$$

In Figure 5.5, it can be observed that the power transmission to the left of the SRR wall is almost unity (0 dB) at 95 GHz. Then, at 100 GHz, the transmission drops drastically, and a shadowing effect is observed. At this frequency, the SRR wall is opaque, due to the negative effective permeability stimulated in the medium. The wall becomes transparent again at 105 GHz, although the transmitted power is less than 0 dB. The CMM structure constructed by combining SRRs and TWs is opaque at 95 GHz. At this frequency, the transmission property of the CMM is dominated by the negative effective permittivity dictated by the thin wires. At 100 GHz, however, the structure becomes transparent, and a relatively high power transmission to the left of the array is observed. This unusual behavior of the CMM is a result of double negativity; since SRRs resonate at 100 GHz, both effective permittivity and permeability of the medium become negative at this frequency. The transparency of the CMM wall tapers down at 105 GHz. All computational results in Figure 5.5 agree well with the experimental results in the literature [19]–[23].

Figure 5.6 presents transmission results for a large SRR wall involving 51×29×20 SRRs. The scattering problem is discretized with 2 425 560 unknowns and solved by using MLFMA parallelized into 64 processes. Figures 5.6(a) and 5.6(b) depict the power transmission on the $y = 0$ and $z = 0$ planes, respectively, when the wall is illuminated by a Hertzian dipole at 85 GHz, 100 GHz, and 110 GHz. The area occupied by the wall is indicated with the black frames in the plots. A deep shadowing effect is observed at 100 GHz due to the resonances of the SRRs.

Figures 5.7–5.9 present the solution of scattering problems involving metamaterial walls constructed by using $7 \times 19 \times 19$ SRRs. Figure 5.7(a) depicts a metamaterial wall, where the SRRs have the same orientation. In addition to this ordinary wall, another wall involving $7 \times 19 \times 19$ randomly oriented SRRs, as depicted in Figure 5.7(b), is considered. Discretizations with the RWG functions on $\lambda/100$ triangles lead to matrix equations involving 207 214 unknowns. The walls are illuminated by a Hertzian dipole located at $x = 1.5$ mm. Two orientations (y direction and z direction) of the dipole are considered for investigating the transmission properties for different polarizations. The problems are solved at 100 GHz using a six-level LF-MLFMA. Figure 5.8 present the power transmission as a function of location on the $y = 0$ plane for the metamaterial wall depicted in Figure 5.7(a). One can observe that a shadowing effect occurs only when the Hertzian dipole is oriented in the y direction. For the z orientation,

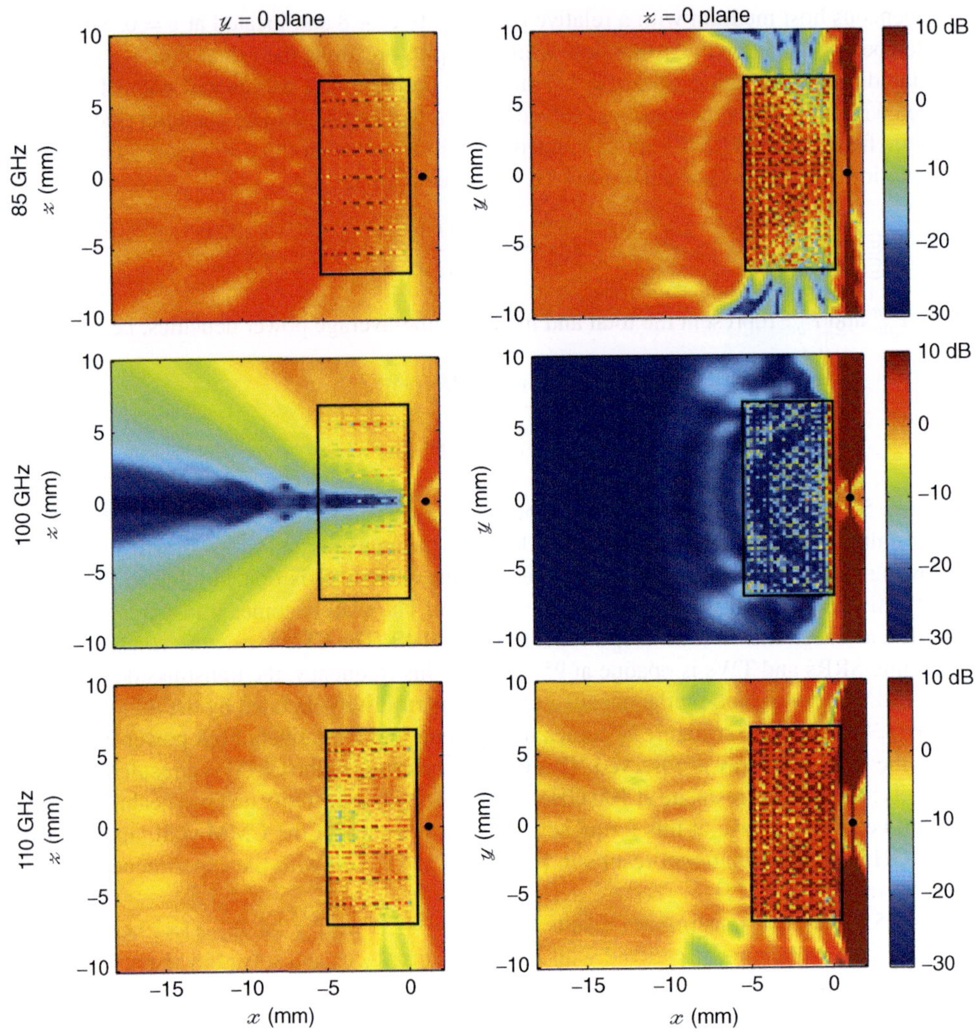

Figure 5.6 Power transmission results for a large SRR wall involving 51×29×20 SRRs on $y = 0$ and $z = 0$ planes. The results are obtained by solving a 2 425 560-unknown problem at 85 GHz, 100 GHz, and 110 GHz. *Source*: Gürel *et al.* 2009 [24]. Reproduced with permission of PIER.

the wall is almost transparent and the power transmission is close to unity in the transmission region on the left of the wall. This is because SRRs do not resonate when the electric field is (almost) perpendicular to their surfaces. However, as depicted in Figure 5.9, this major restriction of the metamaterial structures, i.e., the polarization dependency, can be partially eliminated by using randomly oriented unit cells. Figures 5.9(a) and 5.9(b) shows that the shadowing effect occurs for both polarizations of the dipole, thanks to the randomly oriented SRRs.

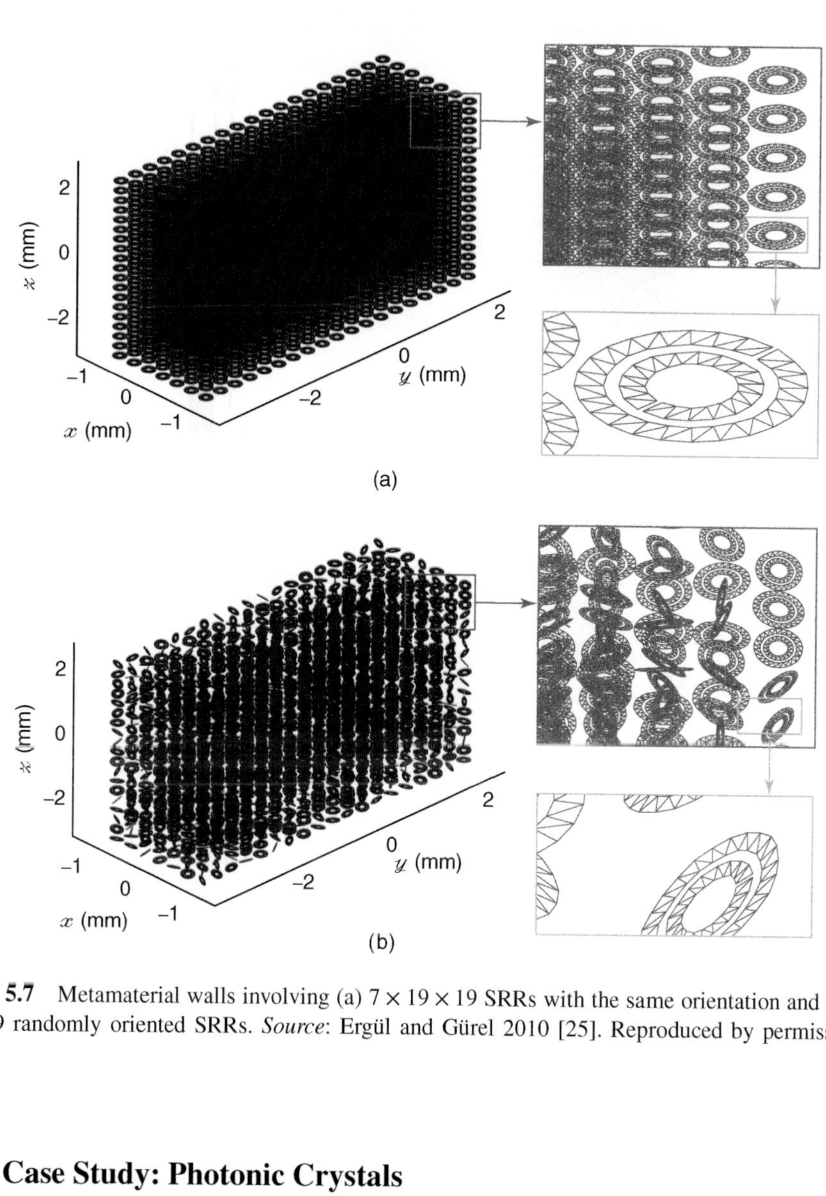

Figure 5.7 Metamaterial walls involving (a) $7 \times 19 \times 19$ SRRs with the same orientation and (b) $7 \times 19 \times 19$ randomly oriented SRRs. *Source*: Ergül and Gürel 2010 [25]. Reproduced by permission of PIER.

5.3 Case Study: Photonic Crystals

Similar to metamaterials, photonic crystals are artificial structures constructed by periodically arranging unit cells [26]–[33]. They exhibit frequency-selective electromagnetic responses, i.e., their electromagnetic transmission properties usually change rapidly as a function of frequency. For example, Figure 5.10(a) depicts a photonic-crystal structure involving periodic dielectric slabs. Depending on the frequency, this relatively simple structure can be transparent and behave like a waveguide or it can be opaque and inhibit the transmission of electromagnetic waves [29]. Due to its frequency-selective property, this structure can be used as a filter in microwave circuits and antenna systems. Another example, namely, a perforated waveguide,

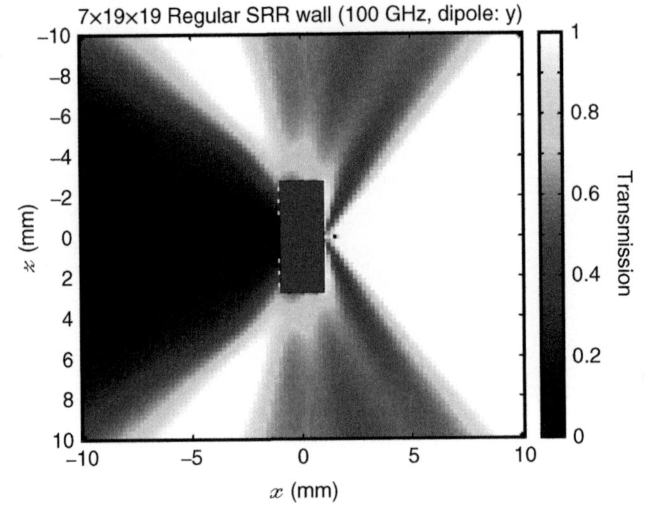

(a)

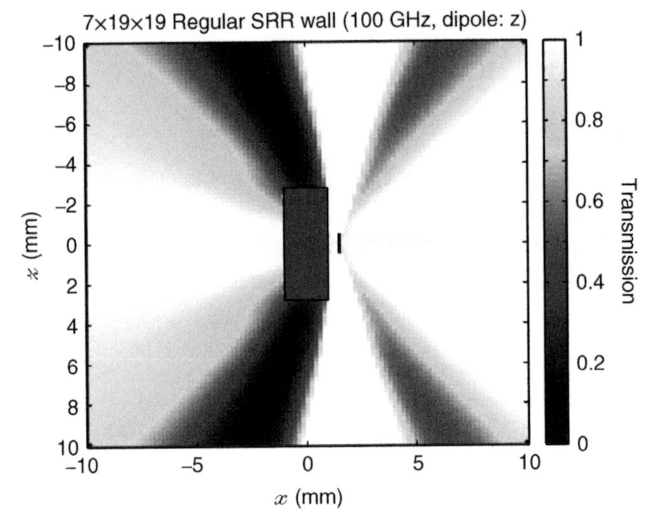

(b)

Figure 5.8 Power transmission as a function of location on the $y = 0$ plane for the metamaterial wall depicted in Figure 5.7(a) when the Hertzian dipole is oriented (a) in the y direction and (b) in the z direction. *Source*: Ergül and Gürel 2010 [25]. Reproduced by permission of PIER.

is depicted in Figure 5.10(b). This structure is also frequency selective, and it can be used to change the direction of electromagnetic waves in a range of frequencies [27]–[32].

Figures 5.11 and 5.12 present the solution of transmission problems involving a periodical arrangement of 0.41 cm × 2 cm × 2 cm dielectric slabs with a relative permittivity of 4.8. The structure involving 10 identical slabs is illuminated by an x-directed Hertzian dipole at 20 GHz, 25 GHz, 30 GHz, and 35 GHz. The problems are formulated with JMCFIE and discretized with

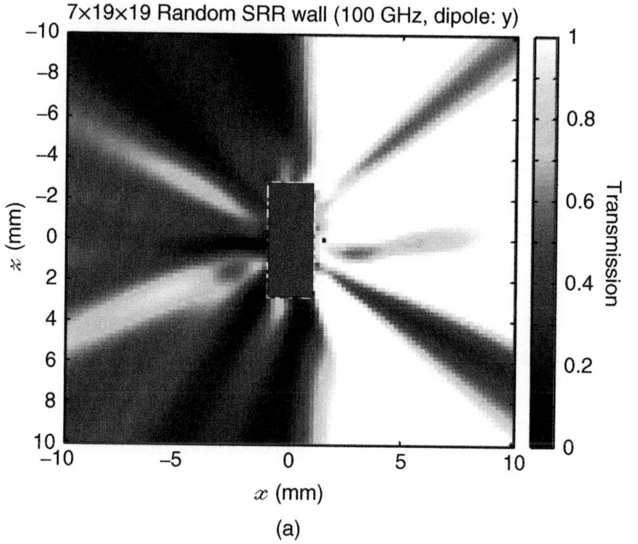

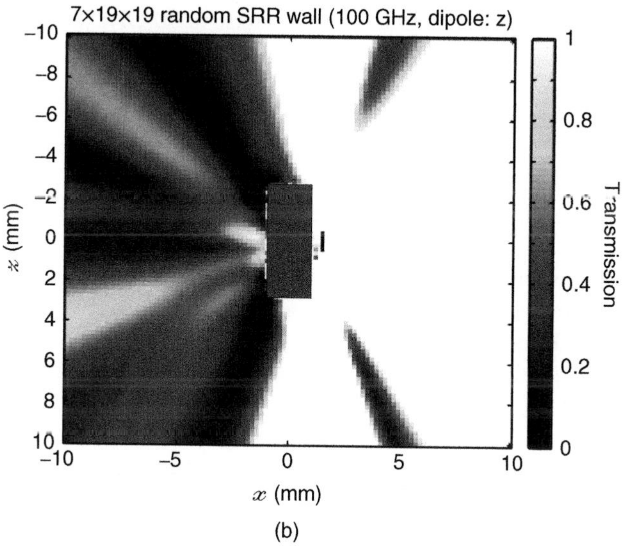

Figure 5.9 Power transmission as a function of location on the $y = 0$ plane for the metamaterial wall depicted in Figure 5.7(b) when the Hertzian dipole is oriented (a) in the y direction and (b) in the z direction. *Source*: Ergül and Gürel 2010 [25]. Reproduced by permission of PIER.

the RWG functions. Matrix equations involving 77 400 unknowns are solved iteratively via MLFMA. Iterative solutions are also accelerated by using the SAI preconditioner. The power transmission defined in (5.1) is computed at various locations in the vicinity of the structure on the z-x plane. Figure 5.11 shows that the dielectric structure behaves like a waveguide at 25 GHz and 30 GHz. At those frequencies, power radiated by the Hertzian dipole is transmitted to the region on the LHS of the structure. At 35 GHz, however, the structure becomes opaque

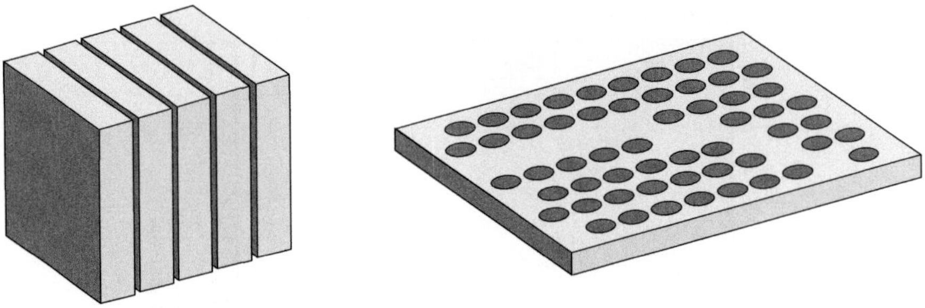

Figure 5.10 Examples of photonic-crystal problems: (a) periodic slabs and (b) perforated waveguide. *Source*: Ergül et al 2011 [33]. Reproduced with permission of IEEE.

and does not transmit the power, as depicted in Figure 5.12(a). Since the structure is constructed by using finite-size slabs, shadowing is not observed for all locations in the transmission region. Specifically, high transmission values are observed even at 35 GHz, especially around the edges of the structure. Nevertheless, the electromagnetic response of the structure at 35 GHz is completely different from its waveguide-like response at 20 GHz and 25 GHz. Finally, as depicted in Figure 5.12(b), the power transmission rises again when the frequency is increased to 35 GHz.

Figure 5.13 presents the solution of transmission problems involving two perforated waveguides with 0.6×26×34 cm and 0.6×29×38 cm dimensions. The structures have a relative permittivity of 12.0 and located in free space. The holes are arranged such that electromagnetic waves can be transmitted from one of the edges to another edge. Specifically, some of the holes in the regular grids are missing to allow the propagation of waves inside the structures. The distance between the centers of the holes is 1 cm, whereas the radius of each hole is 0.29 cm. With these dimensions, the transmission is maximized at 7.6 GHz. The structures are excited by a Hertzian dipole. The problems are formulated with CTF, discretized with 475,782 and 597,462 unknowns, and solved iteratively via MLFMA. Iterations are also accelerated via Schur-complement preconditioners [33]. Figures 5.13(a) and 5.13(b) depict the normalized magnetic field inside and in the vicinity of the structures. It can be observed that the waveguides operate as expected and electromagnetics waves are transmitted efficiently.

5.4 Case Study: Scattering from Red Blood Cells

Scattering from RBCs has attracted the interest of many researchers in the area of biomedical science and engineering [84]–[92]. Scattering cross section (SCS), which is defined similar to RCS, of an RBC can provide essential information on the morphological properties of the cell. Hence, SCS values can be used to diagnose various diseases involving the deformation of RBCs. Using a diagnosis setup, i.e., a flow cytometer, a liquid stream including many RBCs from a blood sample can be passed through a thin tube, and SCS values of each RBC can be measured via an optical detection system. Then, measurement values can be compared with reference values in order to obtain statistical data on the size and shape of RBCs in the blood sample under investigation. This section presents solutions of scattering problems involving ordinary and deformed RBCs for the diagnosis of various diseases.

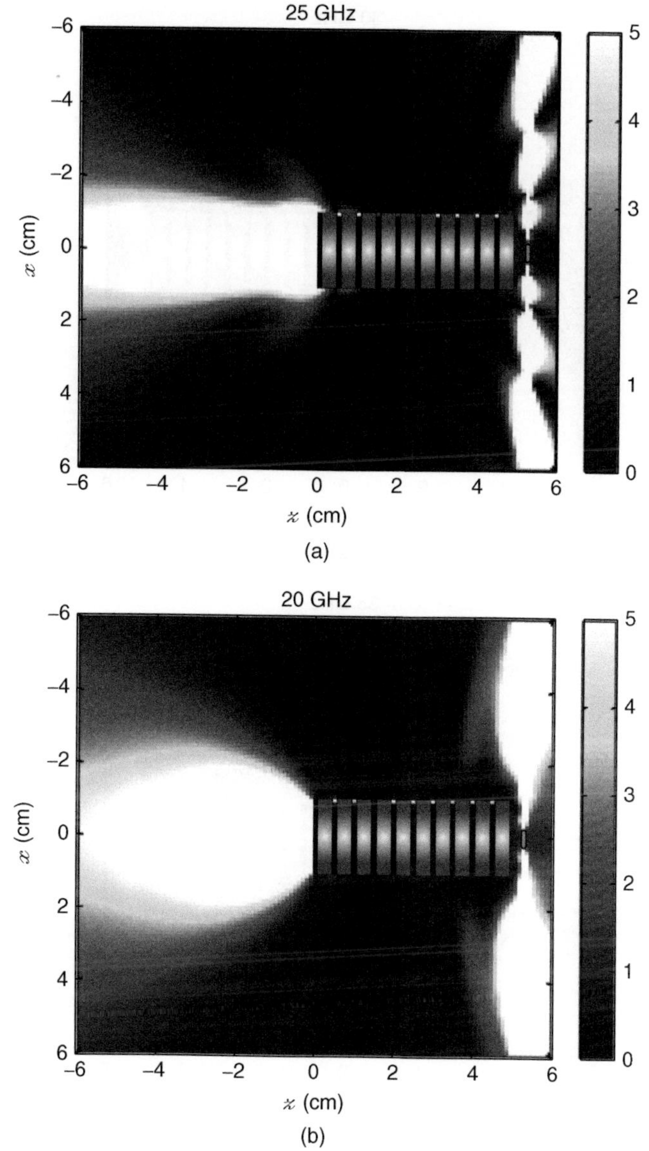

Figure 5.11 Power transmission on the $y = 0$ plane for a 10-layer dielectric structure illuminated by an x-directed Hertzian dipole at (a) 20 GHz and (b) 25 GHz.

Five different RBC shapes are illustrated in Figure 5.14 and listed in Table 5.1. An ordinary RBC depicted in Figure 5.14(a) is modeled by a rotationally symmetric biconcave surface defined as

$$r(\theta, \phi) = a\sin^{q}(\theta) + b, \tag{5.4}$$

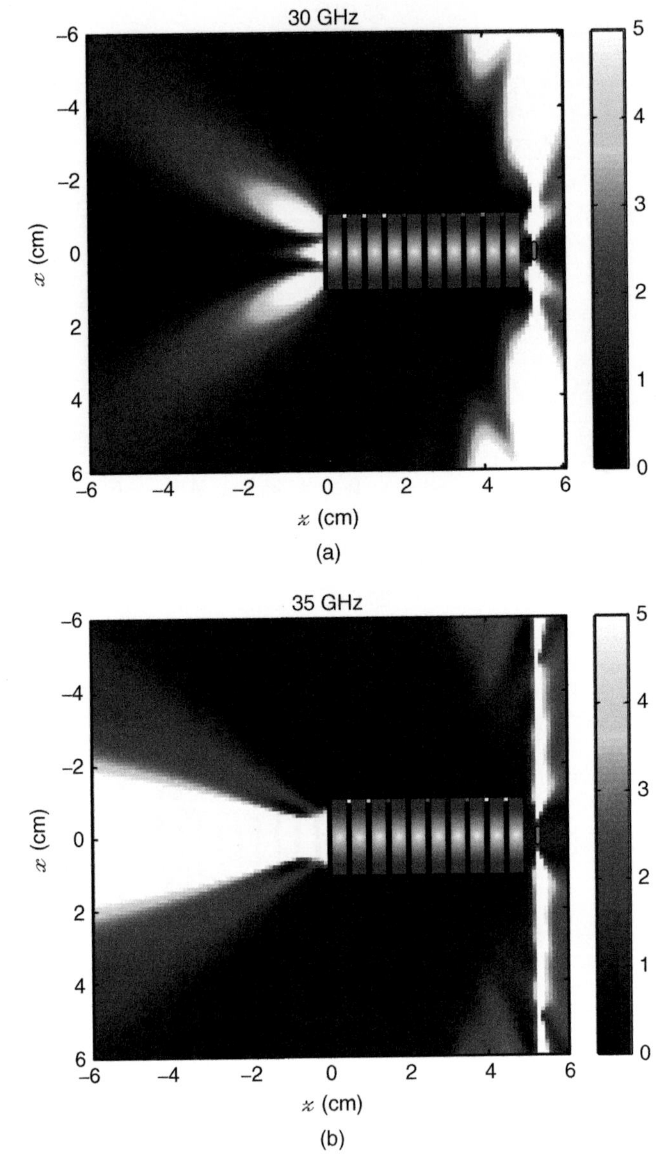

Figure 5.12 Power transmission on the $y = 0$ plane for a 10-layer dielectric structure illuminated by an x-directed Hertzian dipole at (a) 30 GHz and (b) 35 GHz.

where $q = 5$, $a = 3.3$ μm, and $b = 0.55$ μm. With these parameters, the size and volume of the ordinary RBC are 7.70 μm and 80.90 μm^3, respectively. The spherocyte, which is modeled as a dielectric sphere with 5.36 μm diameter, as depicted in Figure 5.14(b), has exactly the same volume as an ordinary RBC. The microcyte and the macrocyte in Figures 5.14(c) and 5.14(d) also have biconcave shapes as defined in (5.4), but their volumes correspond to 5/6 and 4/3 of the volume of an ordinary RBC. Finally, the sickle cell depicted in Figure 5.14(e) also has a

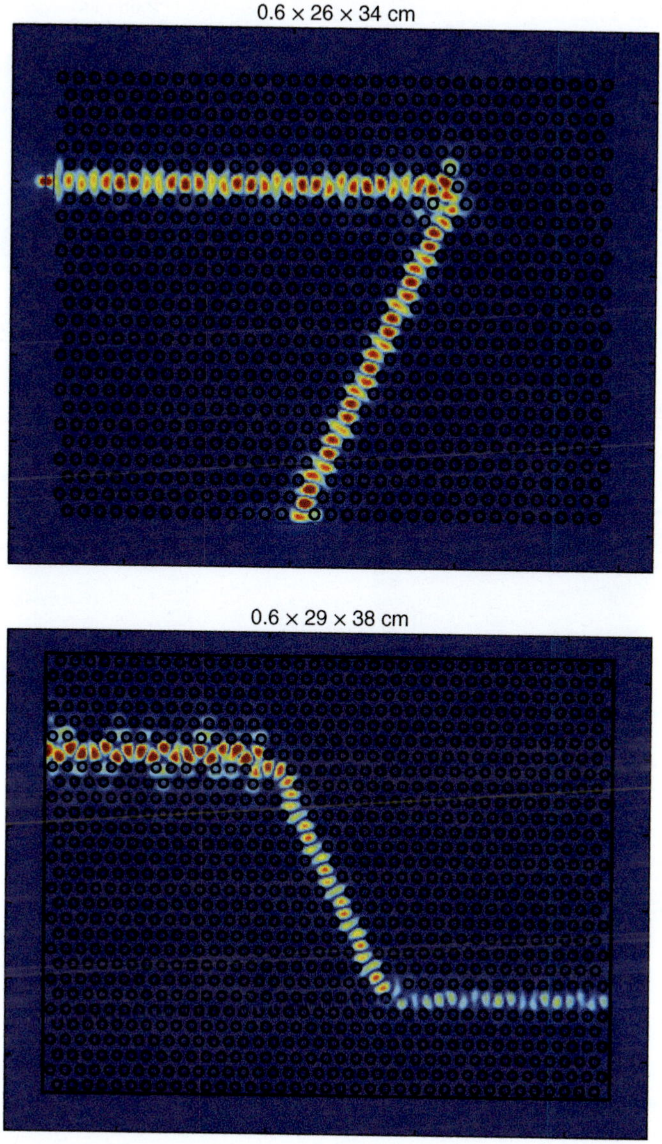

Figure 5.13 Total magnetic field inside and in the vicinity of 0.6×26×34 cm and 0.6×29×38 cm perforated waveguides. *Source*: Ergül et al 2011 [33]. Reproduced with permission of IEEE.

rotationally symmetric shape as the others do, but it is elongated in one dimension. The size of the sickle cell is 20.9 μm, whereas its volume is the same as the volume of an ordinary RBC, i.e., 80.90 μm^3.

RBCs in Figure 5.14 are illuminated by a plane wave propagating in the $-z$ direction with the electric field polarized in the x direction having a unit amplitude. Since the orientation of

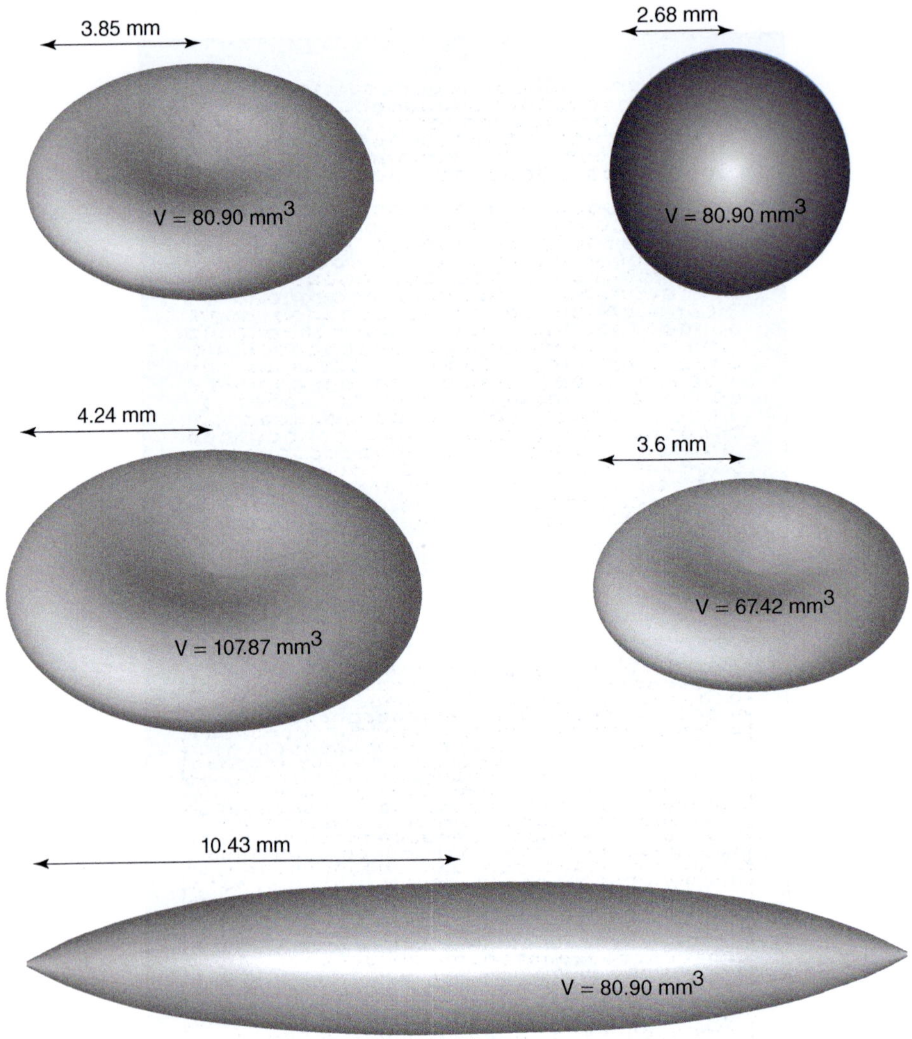

Figure 5.14 Healthy and deformed RBCs. A healthy (ordinary) RBC with a biconcave shape, a spherocyte having exactly the same volume as the ordinary RBC, a microcyte that is smaller than an ordinary RBC with a biconcave shape, a macrocyte that is larger than an ordinary RBC with a biconcave shape, and a sickle cell having exactly the same volume as an ordinary RBC. *Source*: Ergül *et al.* 2010 [92]. Reproduced by permission of SPIE.

an RBC can be arbitrary in the flow cytometry, 13 different cases are investigated for each RBC, except for the spherocyte because it has a fully symmetric geometry. As an example, the default orientation ($\theta_o = 0°, \phi_o = 0°$) of an ordinary RBC is depicted in Figure 5.15. Rotations are performed around x axis (by an amount of θ_o) and z axis (by an amount of ϕ_o), respectively. Specifically, rotations with $\theta_o = \{30°, 60°, 90°\}$ and $\phi_o = \{0°, 30°, 60°, 90°\}$ are considered, in addition to the default case in Figure 5.14. Relative permittivities of RBCs and the host

Table 5.1 Ordinary and deformed RBCs

Red blood cell	Volume (μm^3)	Maximum (μm)	size λ_o	Number of unknowns
Ordinary cell	80.90	7.70	14.1	213 090
Spherocyte	80.90	5.36	9.79	–
Microcyte	67.42	7.24	13.2	188 472
Macrocyte	107.9	8.48	15.5	259 086
Sickle cell	80.90	20.9	38.1	246 828

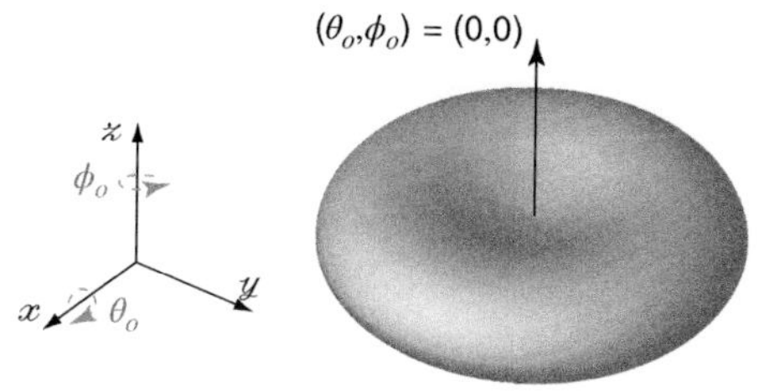

Figure 5.15 Default orientation ($\theta_o = 0°$, $\phi_o = 0°$) of an ordinary RBC.

medium are selected as 1.40 and 1.33, respectively. Numerical simulations are performed at 474 THz, corresponding to the output frequency of a typical Helium-Neon laser. The size of each RBC in terms of the wavelength inside the host medium (λ_o) are listed in Table 5.1. Due to their relatively large sizes, accurate discretizations of RBCs using $\lambda_o/10$ triangles lead to matrix equations involving 180 000–260 000 unknowns, as also listed in Table 5.1. Except for the spherocyte, scattering problems are formulated with JMCFIE and solved iteratively using the BiCGStab algorithm accelerated via MLFMA and 4PBDP. Scattering from the spherocyte is solved exactly by using a Mie-series algorithm for comparisons. Solutions with MLFMA require only 15–20 iterations, and each solution is performed in 3.0–4.5 hours on a 3.6 GHz Intel Xeon processor using 1.5 GB memory.

Figure 5.16 presents the solution of scattering problems involving the RBCs depicted in Figure 5.14. SCS values in the forward-scattering ($\theta = 180°$) and backscattering ($\theta = 0°$) directions are plotted in Figures 5.16(a) and 5.16(b), respectively. Since SCS depends on the orientation of the target, multiple values are obtained for each cell, except for the spherocyte. It can be observe that, depending on the cell type, SCS values change significantly. Considering the results for the ordinary RBC, safe regions are determined as indicated by horizontal lines in Figures 5.16(a) and 5.16(b). For example, an SCS value higher than 44 dB μm^2 or lower than 43 dB μm^2 in the forward-scattering direction indicates a detection of an abnormal RBC, possibly a macrocyte (if greater than 44 dB μm^2) or a microcyte (if smaller than 43 dB μm^2). Note that, although less likely, a sickle cell may also present a low SCS value in the forward-scattering direction. As depicted in Figure 5.16(b), an SCS value smaller than −24 dB μm^2 in the backscattering direction indicates a detection of a macrocyte, a microcyte, or a sickle cell.

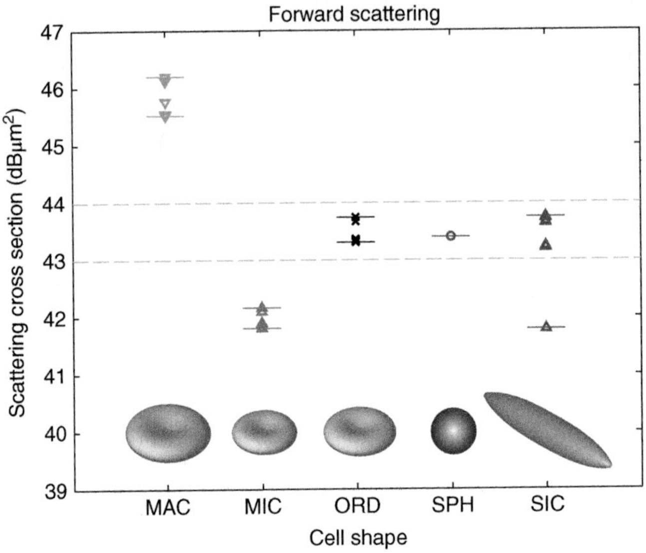

(a)

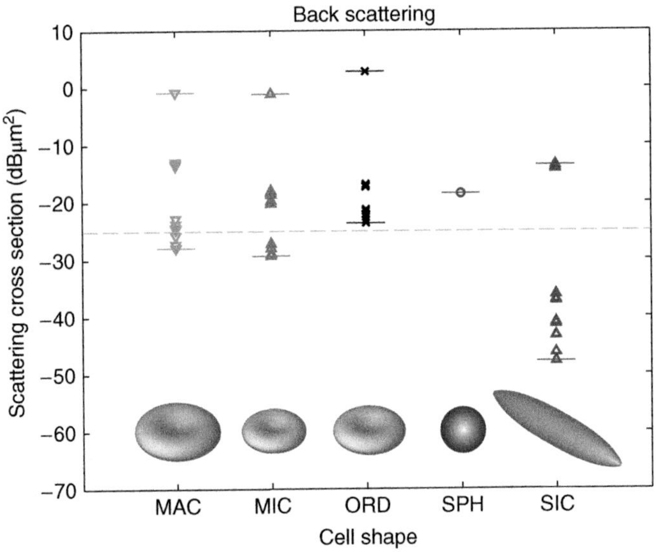

(b)

Figure 5.16 Solutions of scattering problems involving an ordinary RBC (ORD) and four different types of deformed RBCs, i.e., a spherocyte (SPH), a macrocyte (MAC), a microcyte (MIC), and a sickle cell (SIC). (a) SCS in the forward-scattering direction and (b) SCS in the backscattering direction. *Source*: Ergül *et al.* 2010 [92]. Reproduced by permission of SPIE.

SCS values in the forward-scattering and backscattering directions do not provide complete information for diagnosing diseases. For example, using the data in Figures 5.16(a) and 5.16(b), a spherocyte cannot be distinguished from an ordinary (healthy) cell. The required data can be obtained by considering SCS values in a set of side-scattering directions. SCS is sampled on the x-y plane as a function of bistatic scattering angle ϕ from 0° to 360° with 1° intervals. Then, the average SCS value in the side-scattering direction is computed as

$$\mathrm{SCS}_{avg,side} = \sqrt{\frac{1}{360}\sum_{n=1}^{360}\left(|\mathrm{SCS}(\theta = 90°, \phi_n)|^2\right)}, \tag{5.5}$$

where $\phi_n = (n-1)°$ for $n = 1, 2, \ldots, 360$. In the above, $\theta = 90°$ refers to the azimuth plane, i.e., the x-y plane, on which all SCS values in the bistatic ϕ_n directions are averaged. As depicted in Figure 5.17, an average SCS value higher than -15 dB µm^2 or lower than -24 dB µm^2 in the side-scattering direction indicates a detection of an abnormal RBC, such as a macrocyte, a spherocyte, or a sickle cell.

Finally, using all SCS results in Figures 5.16 and 5.17, one can determine strict guidelines to detect deformed RBCs in a diagnosis setup. Figure 5.18 presents a decision chart based on SCS values obtained in the forward-scattering, backscattering, and side-scattering directions. A sufficient number of RBCs are assumed to passed through the setup. The diagnosis of each disease can be described as follows.

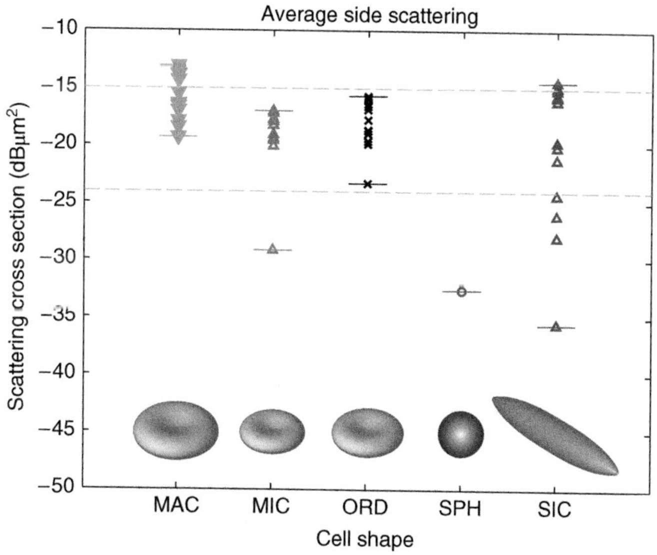

Figure 5.17 Solutions of scattering problems involving an ordinary RBC (ORD) and four different types of deformed RBCs, i.e., a spherocyte (SPH), a macrocyte (MAC), a microcyte (MIC), and a sickle cell (SIC). Average SCS in the side-scattering direction is plotted in dB µm^2 for different orientations of RBCs. *Source*: Ergül *et al.* 2010 [92]. Reproduced by permission of SPIE.

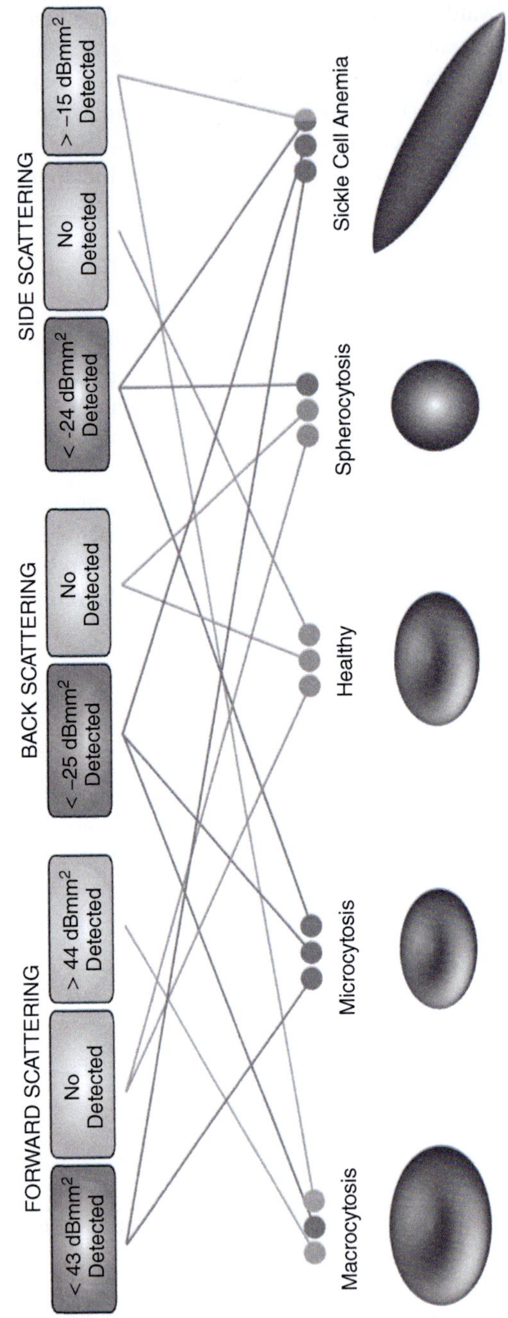

Figure 5.18 A decision chart to diagnose various diseases using SCS data. *Source*: Ergül *et al.* 2010 [92]. Reproduced by permission of SPIE.

- Macrocytosis: Detection of SCS values higher than 44 dB μm^2 in the forward-scattering direction is a major indicator for macrocytosis. SCS values lower than −25 dB μm^2 in the backscattering direction and average SCS values higher than −15 dB μm^2 in the side-scattering direction increase the reliability of the diagnosis.
- Microcytosis: Detection of low SCS values in all directions indicates microcytosis. Specifically, detection of SCS values lower than 43 dB μm^2 in the forward-scattering direction, SCS values lower than −25 dB μm^2 in the backscattering direction, and average SCS values lower than −24 dB μm^2 in the side-scattering direction indicate microcytosis.
- Spherocytosis: Spherocytosis can be diagnosed by the detection of average SCS values lower than −24 dB μm^2 in the side-scattering direction, without any abnormal SCS values in the forward-scattering and backscattering directions.
- Sickle-Cell Anemia: Similar to microcytosis, sickle-cell anemia can be diagnosed by the detection of low SCS values in all directions. On the other hand, sickle-cell anemia also leads to average SCS values higher than −15 dB μm^2 in the side-scattering direction; this can be used to distinguish sickle-cell anemia from microcytosis.

To sum up, by investigating SCS values in the forward-scattering, backscattering, and side-scattering directions, not necessarily limited to conventional cytometer setups, one can able to determine strict guidelines to distinguish deformed RBCs from healthy RBCs and to diagnose related diseases.

5.5 Case Study: Log-Periodic Antennas and Arrays

Log-periodic (LP) antennas are important with their ability to display nearly frequency-independent characteristics over wide bands of frequency [2]–[4]. Unlike LP dipole arrays, which have well-known and time-tested theoretical design recipes, more complicated LP antennas require some degree of correction following their theoretical design according to log-periodic rules. In this section, nonplanar trapezoidal-tooth LP antennas and their circular arrays are considered. Accurate numerical simulations can bridge the gap between approximate designs of LP structures obtained with theoretical design recipes and more precise performance requirements.

Radiation problems involving LP structures are formulated with EFIE, discretized with the RWG functions, and solved iteratively using MLFMA. Although the number of unknowns is relatively low (1000–10 000), acceleration provided by MLFMA is necessary since multiple solutions are required at several frequencies. In many cases, high sampling rate in frequency sweeps is used so that one can properly observe rapid oscillations in radiation patterns and directive gains of LP antennas and their array configurations. This leads to about 100 to 1000 solutions per problem, requiring an efficient algorithm, such as MLFMA, to reach solutions quickly without sacrificing accuracy.

5.5.1 Nonplanar Trapezoidal-Tooth Log-Periodic Antennas

In Figure 5.19, a nonplanar antenna design comprising two planar LP arms with narrow trapezoidal-tooth elements is shown with its apex located at the origin. On each arm, the antenna has 22 teeth, alternating from side to side of the constant-width feed-line, whose lengths, widths, and distances to the source location are determined according to log-periodic

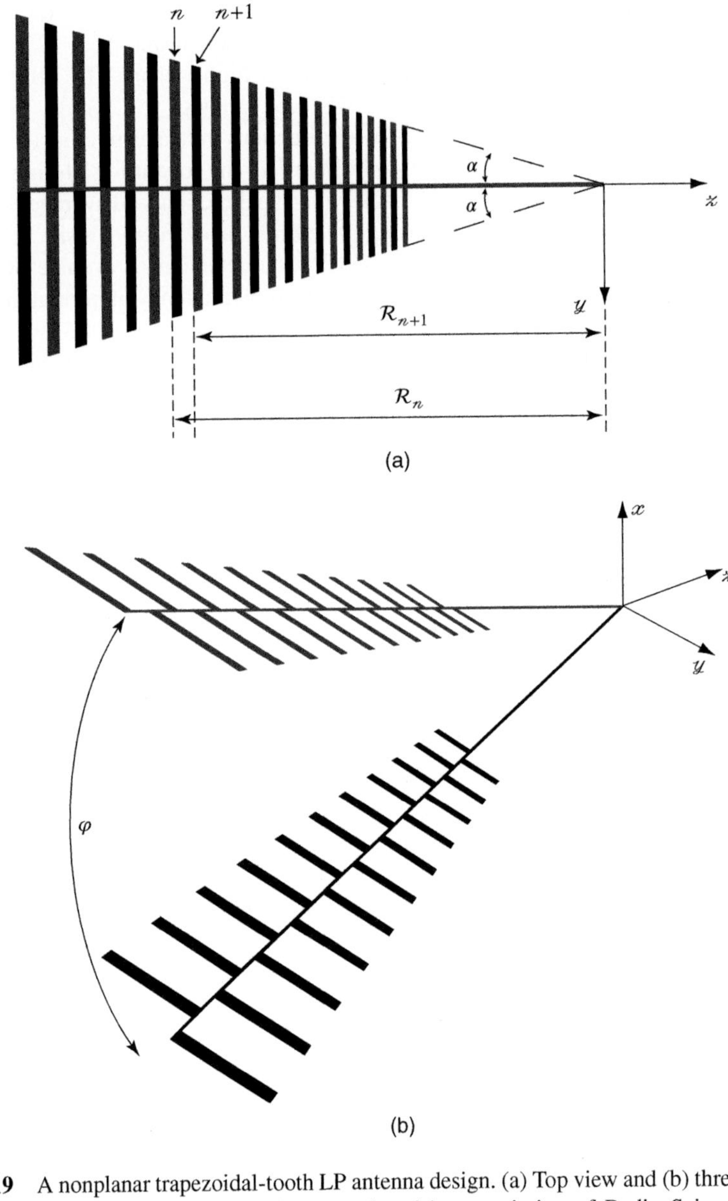

Figure 5.19 A nonplanar trapezoidal-tooth LP antenna design. (a) Top view and (b) three-dimensional view. *Source*: Ergül and Gürel 2005 [6]. Reproduced by permission of Radio Science - American Geophysical Union AGU.

rules. The tooth sequence does not continue up to the apex location, where the source is located, establishing a toothless part of the feed-line to supply the electrical connection. The geometric growth factor

$$\tau = \frac{R_{n+1}}{R_n} \tag{5.6}$$

and the tooth-width ratio

$$\varepsilon = \frac{R_n - w_n}{R_n} \tag{5.7}$$

are selected as 0.95 and 0.98, respectively, where R_n is the distance between the source location and the nth tooth, whereas w_n is the width of the nth tooth. The angle α is 15° so that arms expand from the source location with an angle of 30°, and teeth are attached to center strips with constant widths. If the length of arms is kept constant, increasing or decreasing the angle α shifts the frequency range of operation to lower or higher frequencies, respectively. On the other hand, if it is desirable to keep the frequency range unchanged, then varying the angle α affects the length of the antenna arm.

For the LP antenna in Figure 5.19, the angle between the two arms (φ) is chosen as 45°, which affects the directive gain of the antenna. Specifically, higher values of φ reduce the width of the main beam, but increase the side-lobe level. Lower values of φ suppress side lobes, but increase the beamwidth. Values of φ in the 30°–45° range are observed to provide a trade-off for high directive gain.

The smallest and the largest teeth of the antenna in Figure 5.19 are approximately 9 cm and 27 cm, respectively. A significant majority of the current on an LP antenna exists in the active region, which resides on teeth that are about quarter-wavelength long so that pairs of quasi-symmetric teeth form half-wavelength long radiating elements. Note that the location of the active region depends on the frequency, i.e., for lower and higher frequencies of the operation range, the active region resides on larger and smaller elements, respectively, and partially spills out of the antenna. At intermediate frequencies, for which the active region is completely accommodated on the surface of the antenna, frequency independence is satisfied. In this case, the antenna behaves as if it has an infinite length, since the current out of the active region is negligible. In other words, truncation of the element sequence at both ends of the antenna does not make a difference compared to a theoretical infinite structure. On the other hand, as the active region reaches the end of the antenna and begins to overflow, the antenna becomes dysfunctional and the frequency independence collapses. If the small current outside the active region is ignored, the antenna in Figure 5.19 is expected to operate nearly frequency independent in the 300–800 MHz range. Nevertheless, this simple guess does not consider the overflow of the active region due to its finite but unknown width, which could easily be included in the analysis by using the available information if the design was an LP dipole array [2]. For the design under consideration, this valuable information can only be obtained through numerical simulations.

In the operation frequency range, variation in radiation characteristics with respect to the frequency is directly related to the density of elements on an LP antenna. If τ is close to unity, which means that elements are closely spaced, the active region moves on the antenna smoothly. As a consequence, variation in the radiation characteristics becomes small in the interval $[f_i, f_{i+1}]$, where $f_i = \tau f_{i+1}$. On the other hand, for lower values of τ, frequency independence deteriorates due to the large separation between elements. Note that choosing a small τ value makes it easier to construct the antenna with fewer elements for the same frequency range of operation.

Figure 5.20 shows normalized plots of the three-dimensional far-zone radiation pattern of the LP antenna in Figure 5.19 at various frequencies in the 300–800 MHz range. The main beam points in the z direction, along the apex of the antenna, due to backfire radiation. For

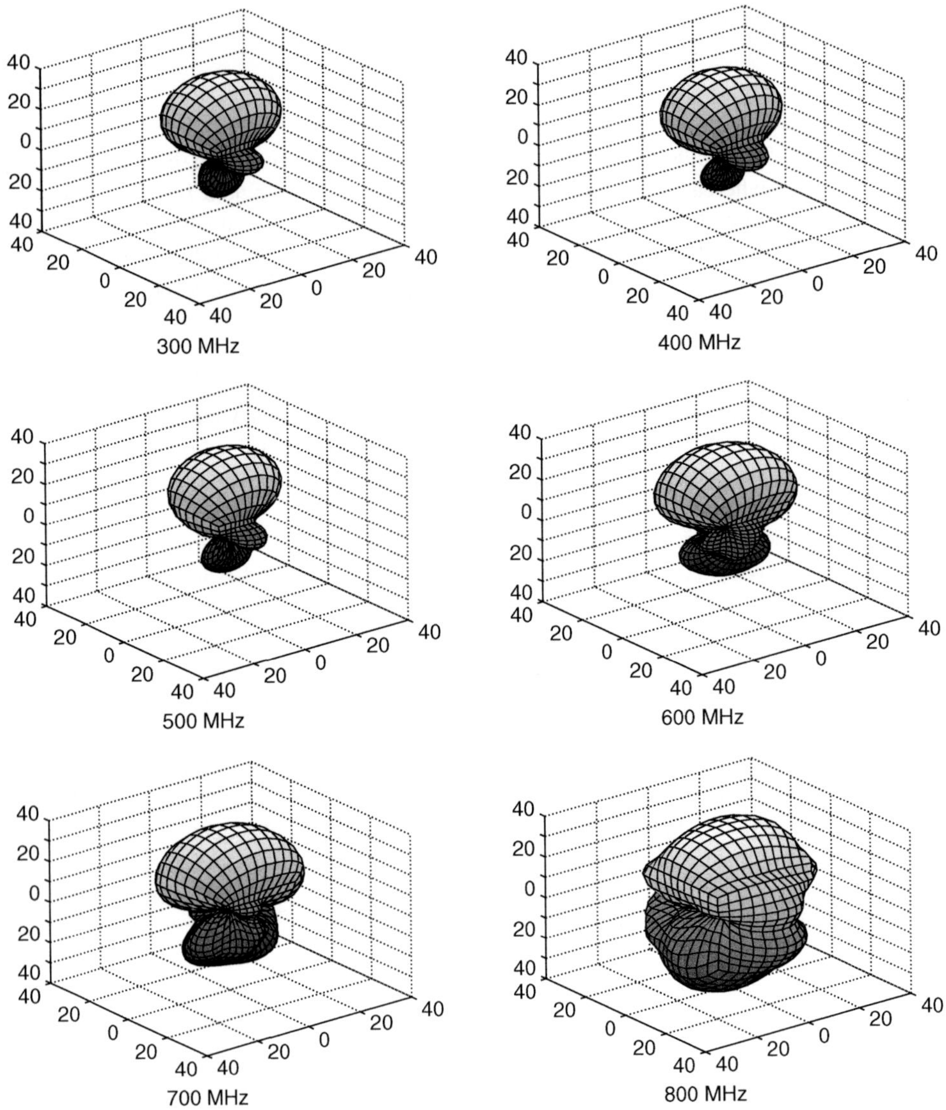

Figure 5.20 Far-zone radiation pattern of the LP antenna design shown in Figure 5.19 for various frequencies in the 300–800 MHz range. The normalized radiated power is presented in the logarithmic scale. *Source*: Ergül and Gürel 2005 [6]. Reproduced by permission of Radio Science - American Geophysical Union AGU.

low frequencies, the radiation pattern seems to be nearly frequency independent. However, the antenna fails to maintain the same radiation pattern for frequencies above 500 MHz. In other words, the antenna does not operate frequency independently in the frequency band it is designed for. Since the radiation pattern of the antenna deteriorates towards the higher end of the frequency band, it is easy to guess that the trouble is associated with the smaller teeth of the antenna. Indeed, plots of the current distribution at various frequencies shown in Figure 5.21 demonstrate that the active region spills out of the antenna for higher frequencies. Both the observation and the diagnosis of this problem, which is not predicted by theoretical design rules, are performed in the simulation environment. The solution is also monitored by simulations: Extra teeth are added one at a time towards the source end of the antenna until the frequency-independence property of the antenna is found satisfactory over the entire 300–800 MHz frequency range.

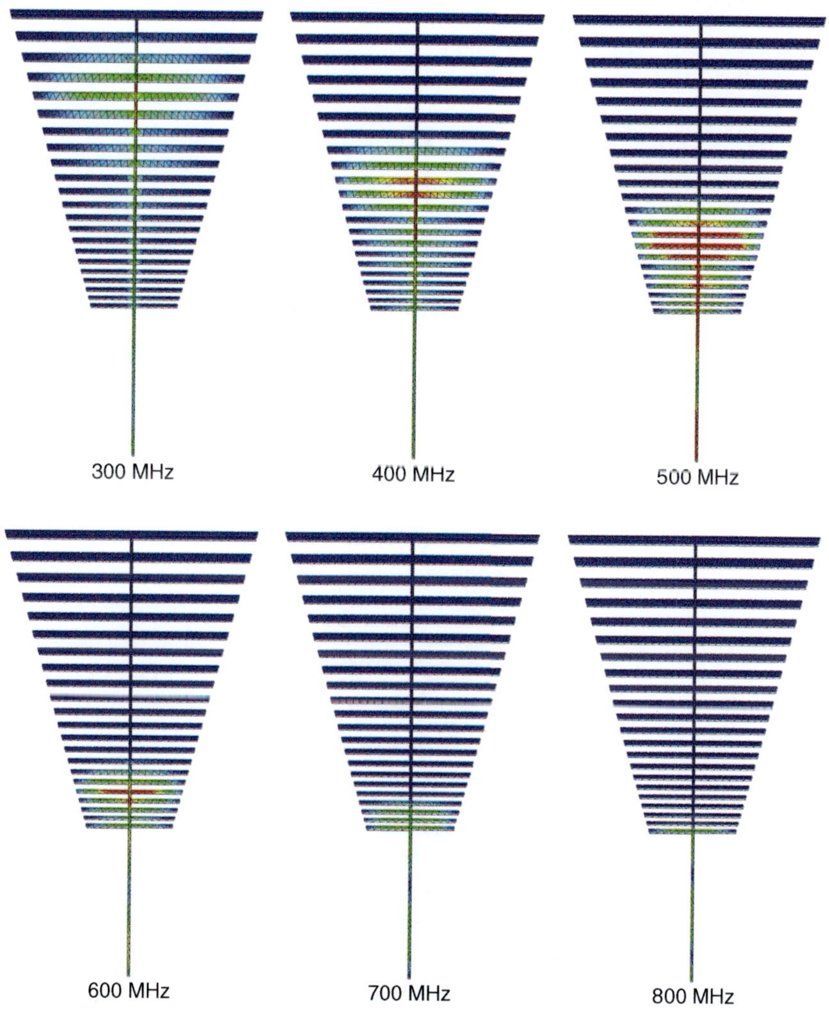

Figure 5.21 Current distribution on the LP antenna design shown in Figure 5.19 at different frequencies. The active region is seen to shift as a function of the frequency and spills out of the antenna at high frequencies. *Source*: Ergül and Gürel 2005 [6]. Reproduced by permission of Radio Science - American Geophysical Union AGU.

The corrected design with 2×38 teeth is shown in Figure 5.22, while radiation-pattern and current-distribution plots for this design are shown in Figures 5.23 and 5.24, respectively. Far-field plots in Figure 5.23 are nearly frequency independent in the desired frequency range. Plots of the current distribution in Figure 5.24 show that the active region is successfully located on the antenna for all frequencies in the desired band, as opposed to the high-frequency spillover demonstrated in Figure 5.21. Finally, for a frequency-independent antenna, the directive gain is expected to remain relatively unchanged over the frequency band of interest. The directive-gain curves in Figure 5.25 clearly demonstrate the improvement in the operation of the modified antenna.

Figure 5.22 Top view of a nonplanar trapezoidal-tooth LP antenna design with 16 teeth added to each arm of the original antenna design shown in Figure 5.19. *Source*: Ergül and Gürel 2005 [6]. Reproduced by permission of Radio Science - American Geophysical Union AGU.

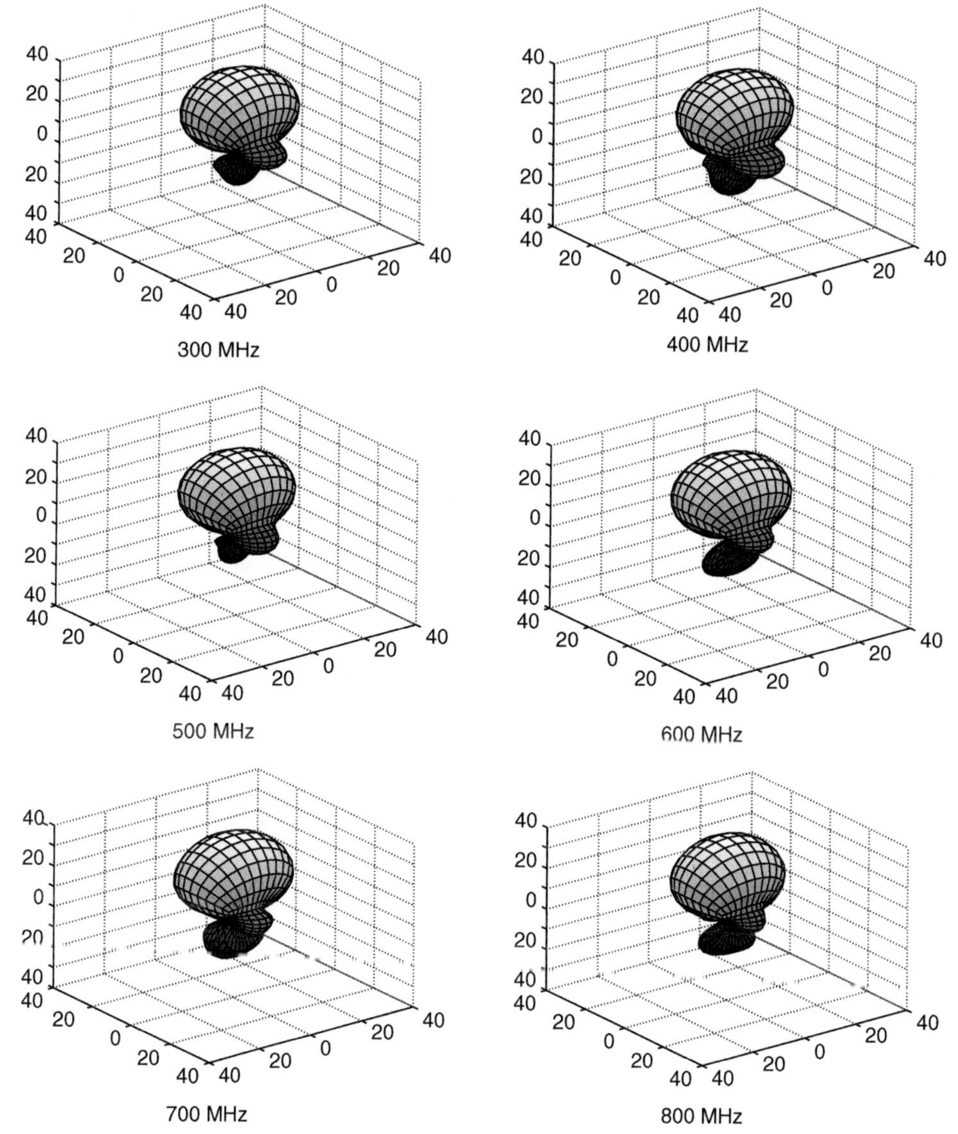

Figure 5.23 Far-zone radiation pattern of the corrected LP antenna design shown in Figure 5.22. *Source*: Ergül and Gürel 2005 [6]. Reproduced by permission of Radio Science - American Geophysical Union AGU.

5.5.2 Circular Arrays of Log-Periodic Antennas

LP antennas can be employed to construct circular arrays with regularly spaced elements, as depicted in Figure 5.26. To preserve frequency independence, the antennas are arranged in a circular form. Theoretically, the radiation intensity of a circular array of identical elements can

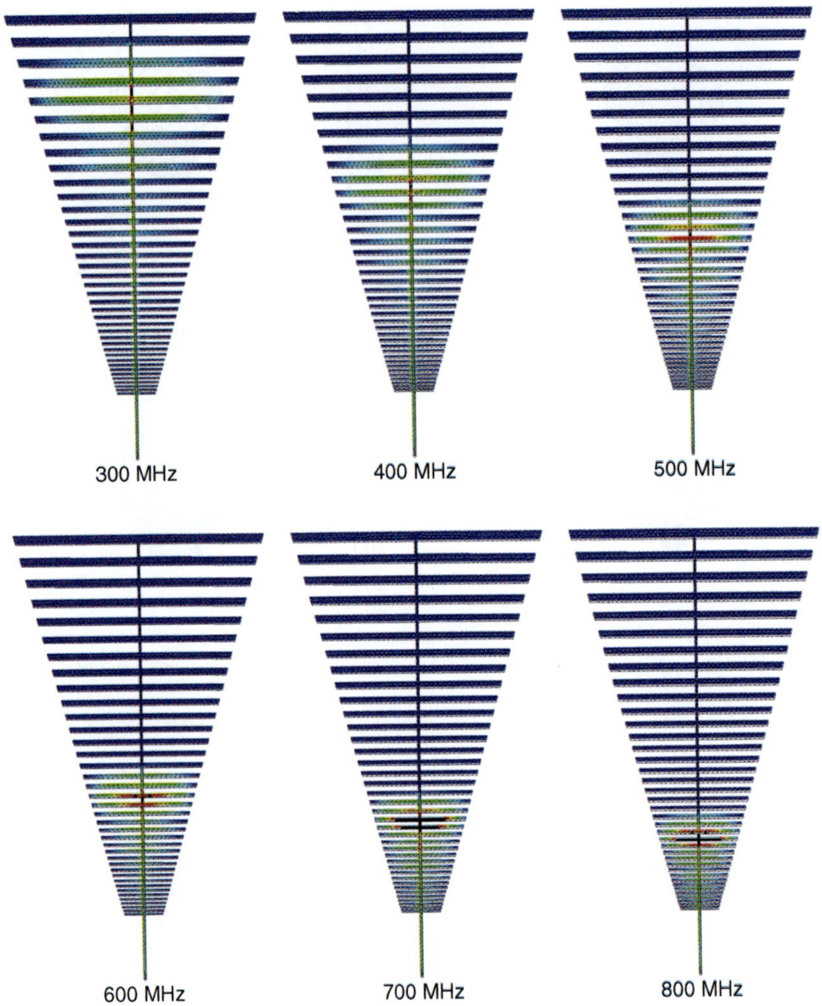

Figure 5.24 Current distribution on the corrected LP antenna design shown in Figure 5.22 at different frequencies. The active region is seen to be successfully contained on the antenna for all frequencies. *Source*: Ergül and Gürel 2005 [6]. Reproduced by permission of Radio Science - American Geophysical Union AGU.

be calculated as [2]

$$P_a^{\infty}(\theta, \phi) = f_{AF}(\theta, \phi) P_e^{\infty}(\theta, \phi), \tag{5.8}$$

where P_e^{∞} and P_a^{∞} represent radiation intensities of a single element and the array, respectively. This theoretical scenario is depicted in Figure 5.27, where the elements are represented by dots

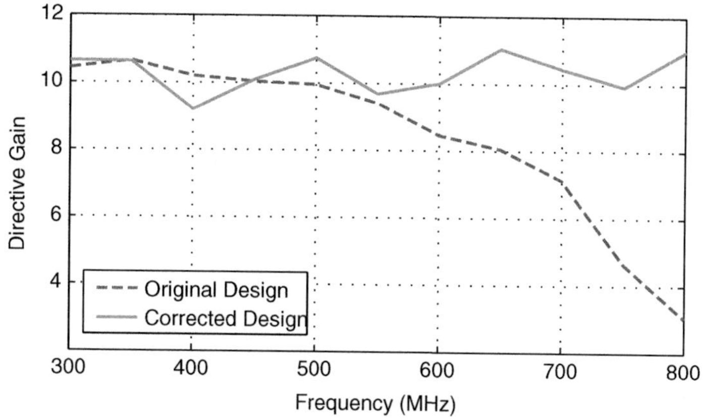

Figure 5.25 Directive gain in the z direction for the original and corrected LP antenna designs shown in Figures 5.19 and 5.22, respectively. *Source*: Ergül and Gürel 2005 [6]. Reproduced by permission of Radio Science - American Geophysical Union AGU.

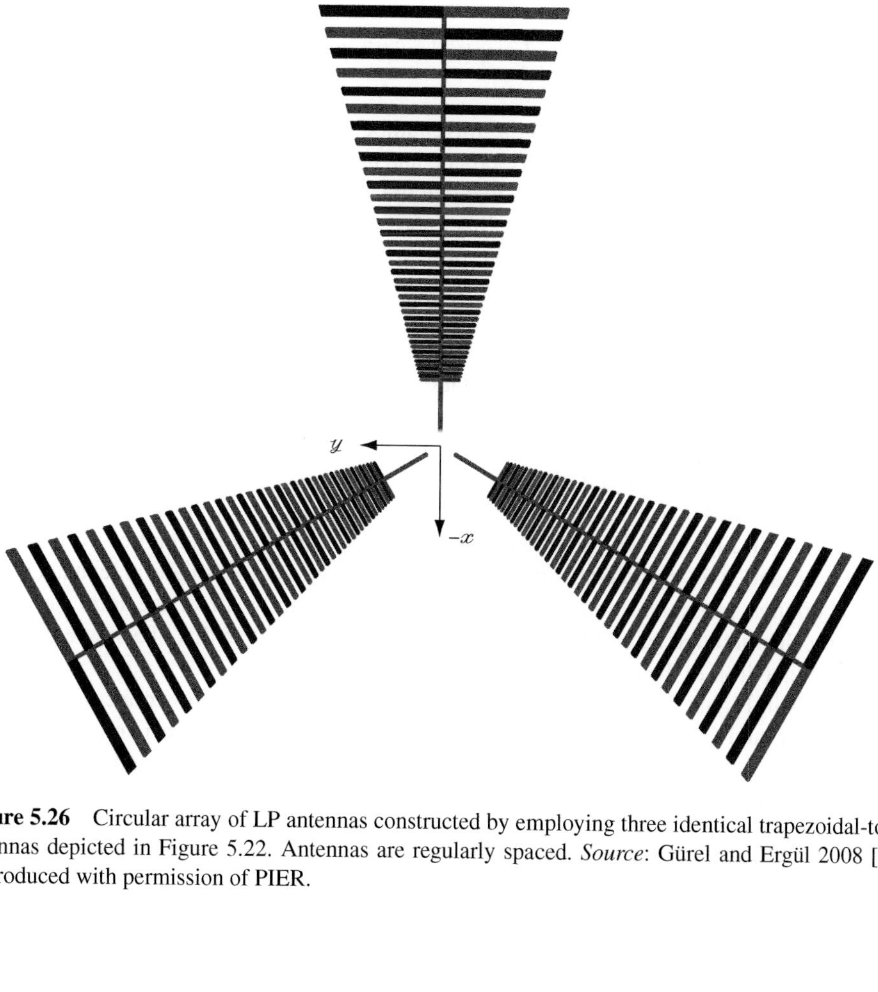

Figure 5.26 Circular array of LP antennas constructed by employing three identical trapezoidal-tooth antennas depicted in Figure 5.22. Antennas are regularly spaced. *Source*: Gürel and Ergül 2008 [10]. Reproduced with permission of PIER.

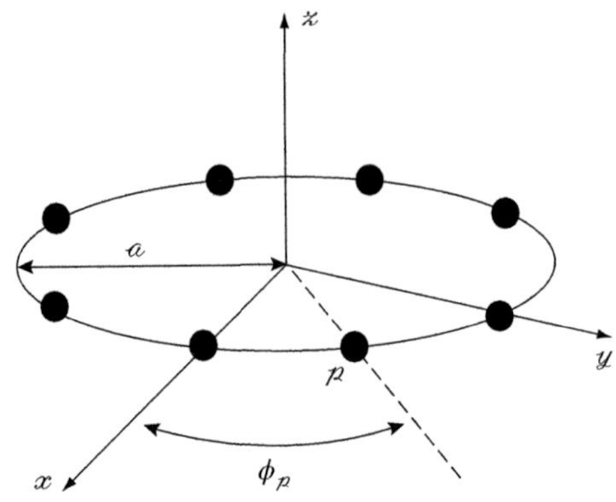

Figure 5.27 Configuration for a circular array, where the array elements are represented by dots on a circle. *Source*: Gürel and Ergül 2008 [10]. Reproduced with permission of PIER.

on a circle. The array factor f_{AF} in (5.8) can be calculated as

$$f_{AF}(\theta, \phi) = \sum_{p=1}^{P} I_p \exp(-ika \sin\theta \cos(\phi - \phi_p)), \tag{5.9}$$

where P is the number of antennas, k is the wavenumber, a is the radius of the circle, I_p is the complex excitation coefficient of the pth element, and ϕ_p is the angle between the pth element and the x axis. For LP arrays, one can assume that the dots in Figure 5.27 represent the center of active regions located on LP antennas. Since the active region appears on those elements that are about a quarter-wavelength long, the circle in Figure 5.27 has a radius that is inversely proportional to k. Consequently, f_{AF} becomes independent of frequency with the elimination of the k factor in the exponential term of (5.9). Finally, if f_e is almost frequency-independent, the radiation intensity of the array f_a should also be independent of frequency to the same degree.

Despite the simple discussion above, simulations show that mutual couplings between LP antennas may not be ignored. To demonstrate this, radiations from the single antenna in Figure 5.22 and the array in Figure 5.26, where only the antenna on the x axis is active, are compared. Considering (5.8) and (5.9) in this case, I_p is zero for the unexcited elements, and the radiation of the array should be the same as the radiation of the single antenna. On the other hand, in Figure 5.28, the directive gain in the $-x$ direction is plotted with respect to frequency, and it is completely different for the array and the single antenna. Due to mutual couplings, unexcited antennas in the array significantly affect the total radiation. Specifically, the directive gain of the array oscillates much more than that of a single element, demonstrating a drop in the quality of frequency independence.

Oscillations in Figure 5.28(b) are directly related to the geometric growth factor (τ) of the LP array. Let f_i and f_{i+1} be two consecutive frequencies, at which the directive gain makes a peak. Then, $f_i/f_{i+1} \approx \tau$, which is equal to 0.95 in this case. Figures 5.29(a) and 5.29(b) show the directive gain when τ becomes 0.98 and 0.85, respectively. It is clearly observed that the oscillation rate depends on τ. In fact, this relation also exists in the case of the single antenna

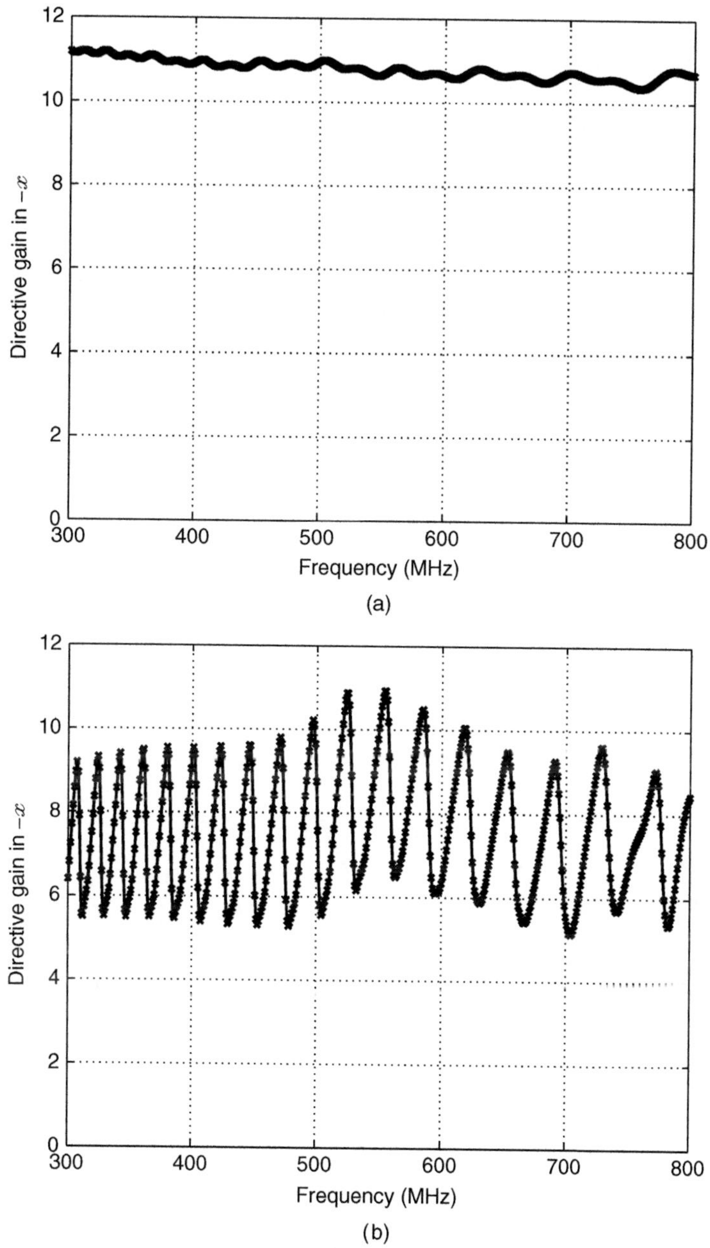

Figure 5.28 Directive gain in the $-x$ direction sampled with 1 MHz resolution for (a) the trapezoidal-tooth LP antenna in Figure 5.22 and (b) the circular array in Figure 5.26 with only the antenna on the x axis active. *Source*: Gürel and Ergül 2008 [10]. Reproduced with permission of PIER.

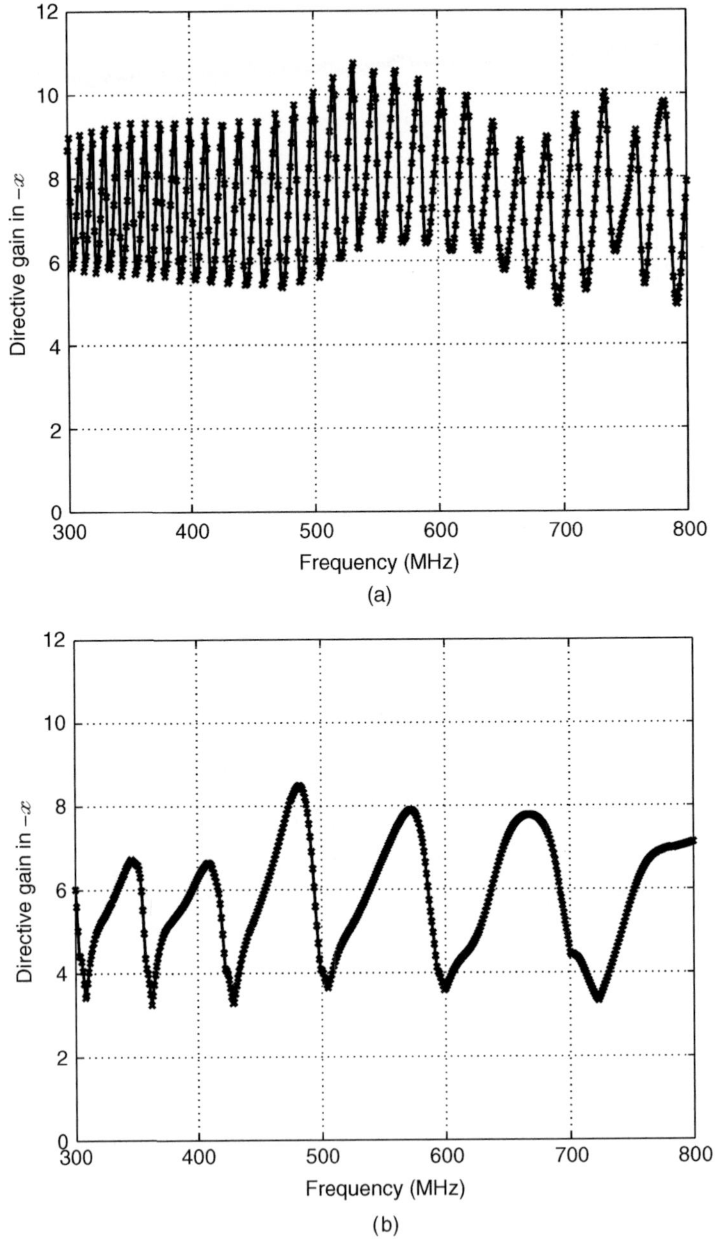

Figure 5.29 Directive gain in the $-x$ direction sampled with 1 MHz resolution for the configuration in Figure 5.26 with only the antenna on the x axis active for (a) $\tau = 0.98$ and (b) $\tau = 0.85$. *Source*: Gürel and Ergül 2008 [10]. Reproduced with permission of PIER.

in Figure 5.28(a). However, the active region on the single antenna moves so smoothly that radiation characteristics change little in intervals $[f_i, f_{i+1}]$ and the oscillation is less visible. Note that, in array configurations, mutual couplings between antennas disturb the localization of the active region, leading to a large effect on radiation characteristics.

As discussed above in Section 5.5.1, frequency independence of an LP structure can be controlled by the geometric growth factor τ. However, as depicted in Figures 5.28 and 5.29, it becomes increasingly difficult to control the variation in radiation characteristics as multiple LP antennas are coalesced in an array configuration. In other words, it might be impossible to obtain the desired level of frequency independence with the highest allowable τ, which must be lower than unity because of nonzero widths of elements. This is mainly due to the complicated mutual couplings among antennas. In order to improve frequency-independence properties of LP arrays, one can optimize the excitations of array elements via genetic algorithms (GAs).

GAs have been successfully employed in many electromagnetics applications [8]. They are especially useful when the optimization space is large and it is difficult to derive an analytical expression for the cost function of the optimization. Therefore, GAs can be employed to improve the frequency independency of LP arrays. The optimization procedure can also be used to provide beam-steering ability to those arrays. Since it is desired to point the main beam in a specific direction, choosing the directive gain as the cost function of the optimization is logical. Specifically, GAs can be used to optimize the excitations of array elements that maximize the directive gain in some given direction (θ, ϕ).

Consider the three-element LP array in Figure 5.26. GAs work on a pool of individuals (citizens), each of which represents a trial combination of optimization variables. An individual suggests values for antenna excitations $I_1 = A_1 \exp \varphi_1$, $I_2 = A_2 \exp \varphi_2$, and $I_3 = A_3 \exp \varphi_3$, where $A_{1,2,3}$ and $\varphi_{1,2,3}$ represent the amplitude and phase, respectively. Without loss of generality, one can set $\varphi_3 = 0$ and there are five variables to be optimized. Parameter values are selected from the optimization space formed by sampling $A_{1,2,3}$ and $\varphi_{1,2}$ in [0,1] and [0°,360°] ranges, respectively. Uniform samplings with intervals of 0.1 in amplitude and 36° in phase are observed to be sufficient, leading to 10 samples for each variable. A one-to-one map is employed to convert values represented by each individual into a single lengthy word of binary numbers, called the chromosome. Each individual also has a degree of success, which is simply the value of the cost function of the optimization, i.e., the directive gain at the optimization angle. A successful individual is defined as the set of excitations I_1, I_2, and I_3, which results in high directive gain. In the beginning of the optimization, individuals can be created randomly. The optimization is then continued as new generations are formed and the pool is modified progressively. There are three important operations to produce a new generation from the old one.

- Crossover: Two successful individuals are selected to exchange some bits of their chromosomes randomly and generate two new individuals called children. There are various crossover schemes to perform these exchanges.
- Mutation: Some of the binary numbers in chromosomes are modified randomly, i.e., 1 changes into 0 and 0 changes into 1.
- Elitism: One or two most successful individuals are preserved in the pool without any modification.

Heuristically, as new generations are produced and the pool evolves, the overall success of the population increases. In the extreme case, after a number of generations, all individuals are the same with the highest possible success. Then, the optimization is completed and any individual gives optimal values via an inverse mapping from the chromosome to excitations. However, it is usually sufficient to interrupt iterations after a number of generations and select the most successful individual in the pool as the optimization result.

Prior to performing an optimization for the LP array, parameters of GAs should be adjusted carefully by examining results obtained at some frequencies. Specifically, different values for the size of the pool, mutation rate, and number of generations should be considered. By checking final results and convergence characteristics of GAs, parameters are selected and fixed, i.e., the same set of parameters is used for all computations at different frequencies. For the optimizations of LP arrays, pools with 20–30 individuals are usually sufficient. The mutation rate for each digit of chromosomes is selected to be 5% and elitism is applied only for the most successful individual. The limit for the number of generations is selected to be 50, i.e., iterations are stopped after the 50th generation. Therefore, the number of trials to complete an optimization at a single frequency is about 1000–1500. A brute-force approach to check all possible combinations of the five optimization variables with the same sampling would require 10^5 trials, which is 100 times larger than that of GAs. With numerous experiments, it has been confirmed that GAs converge to optimized values that are very close to those found by a brute-force approach.

Although GAs reduce the number of trials significantly compared to a brute-force optimization, the calculation of the success for each individual has to be performed efficiently. For any LP array, the number of MLFMA solutions can be kept as low as the number of elements in the array using a superposition technique. Specifically, in each solution, only one antenna is excited with the excitation strength of unity. The radiated field is then calculated and stored in the memory. For the rotationally symmetric array in Figure 5.26, the number of MLFMA solutions per frequency is further reduced to one using the symmetry. Note that the solution by exciting an antenna in the array is different from the solution of an antenna alone, since the former includes mutual couplings between antennas. Whenever it is required to test a set of values for excitations, radiated fields are multiplied by the corresponding coefficients and superposed to obtain the total radiation intensity and directive gain of the whole array.

Figure 5.30 presents the result of the optimization by GAs for the array in Figure 5.26. The directive gain in the $-x$ direction is plotted with respect to frequency for two different cases; when only the antenna on the x axis is excited, and when all antennas are active with optimized excitations. For any frequency, optimization is expected to provide increased directive gain. This is in fact the case; however, optimization is even more effective at those frequencies, where the directive gain is low. Therefore, the optimization by GAs reduces the variation in the directive gain with respect to frequency, and the array becomes more frequency independent.

Since the array in Figure 5.26 is symmetric, i.e., its elements are regularly spaced, it already has some beam-steering ability. Considering the optimization of the directive gain in the $-x$ ($\phi = 180^\circ$) direction, the same directive gain can also be obtained at $\phi = 60^\circ$ and $\phi = 300^\circ$ by simply exchanging excitations among antennas. However, it is also desirable to steer the main beam in any direction within a sector between two antennas. This can again be achieved via optimization with GAs. Figures 5.31 and 5.32 show the results of the optimization. Since it is easier to rotate the antenna than to change the optimization direction, different

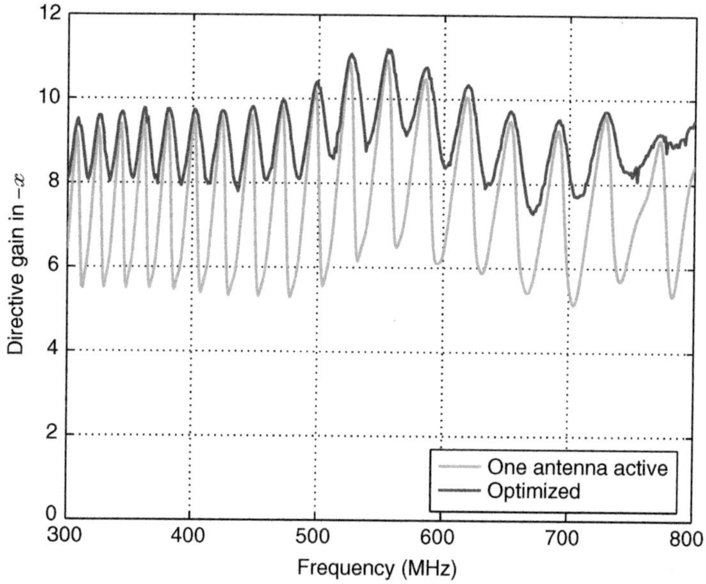

Figure 5.30 Directive gain in the $-x$ direction sampled with 1 MHz resolution for the array in Figure 5.26 when (a) only the antenna on the x axis is active and (b) three antennas are active with optimized excitations. *Source*: Gürel and Ergül 2008 [10]. Reproduced with permission of PIER.

configurations shown at the bottom of Figure 5.31 are used; the optimization angle is fixed at $\phi = 180°$ for all configurations. The array is rotated 10° and 20° in the ϕ direction to test the beam-steering ability in a sector of $2 \times 20° = 40°$. A rotation in the negative direction is not required; due to symmetry, results of an optimization in the $-\phi_0$ direction can be approximated from the optimization in the $+\phi_0$ direction. In Figure 5.32, it can be observed that the optimized directive gain in the $-x$ direction drops as the antenna is rotated. This is because the main beam is not totally controllable, and it is difficult to maintain the directive gain at high levels. This is also evident in Figure 5.31, where the far-zone radiation pattern of the array is plotted for some selected frequencies and for different alignments of the array. The maximum radiation cannot be kept in the $-x$ direction when the antenna is rotated 20°. Nevertheless, the directive gain is larger than five up to 20°. This means that the array in Figure 5.26 can provide a directive gain larger than five in three distinct sectors, each of which has an extent of 40°. Also note that this ability is valid for a wide frequency range from 300 MHz to 800 MHz.

5.5.3 *Circular-Sectoral Arrays of Log-Periodic Antennas*

In sectoral arrays of LP antennas, elements are placed side by side in a circular arrangement, as depicted in Figure 5.33. Such an array with closely spaced elements is observed to provide a wider scanning range and higher directive gain compared to full-circular arrays, e.g., depicted in Figure 5.26. Therefore, sectoral arrays might be preferable depending on the application. On the other hand, sectoral arrays with closely localized elements do not eliminate the

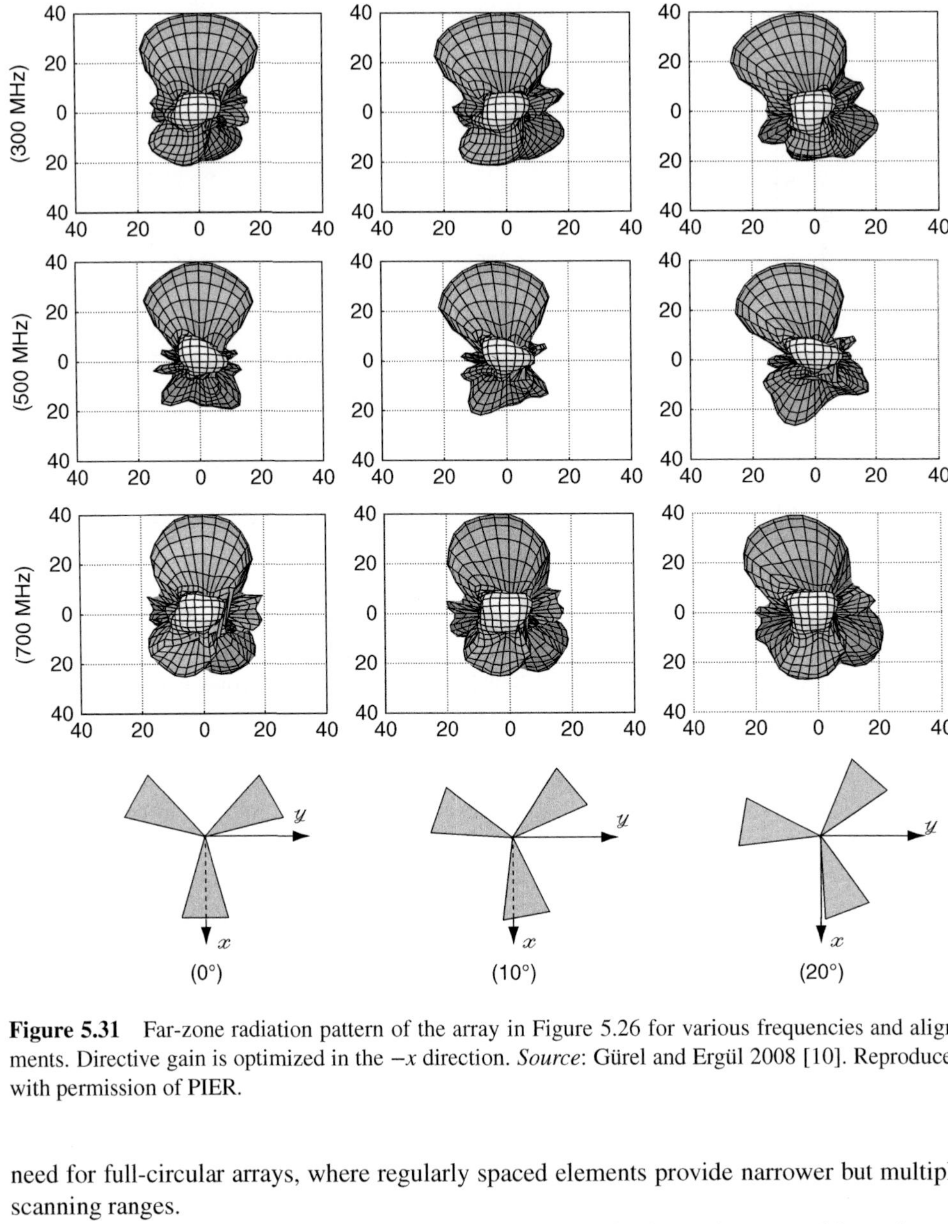

Figure 5.31 Far-zone radiation pattern of the array in Figure 5.26 for various frequencies and alignments. Directive gain is optimized in the $-x$ direction. *Source*: Gürel and Ergül 2008 [10]. Reproduced with permission of PIER.

need for full-circular arrays, where regularly spaced elements provide narrower but multiple scanning ranges.

Figures 5.34 and 5.35 present far-field radiation patterns for three-element and four-element arrays depicted in Figure 5.33, demonstrating beam-steering abilities of arrays for 0°, 10°, 20°, and 30° on the azimuth plane. The directive gain is optimized in the $-x$ direction while rotating arrays from 0° to 30° as depicted in the first column of each figure. Since steering the beam

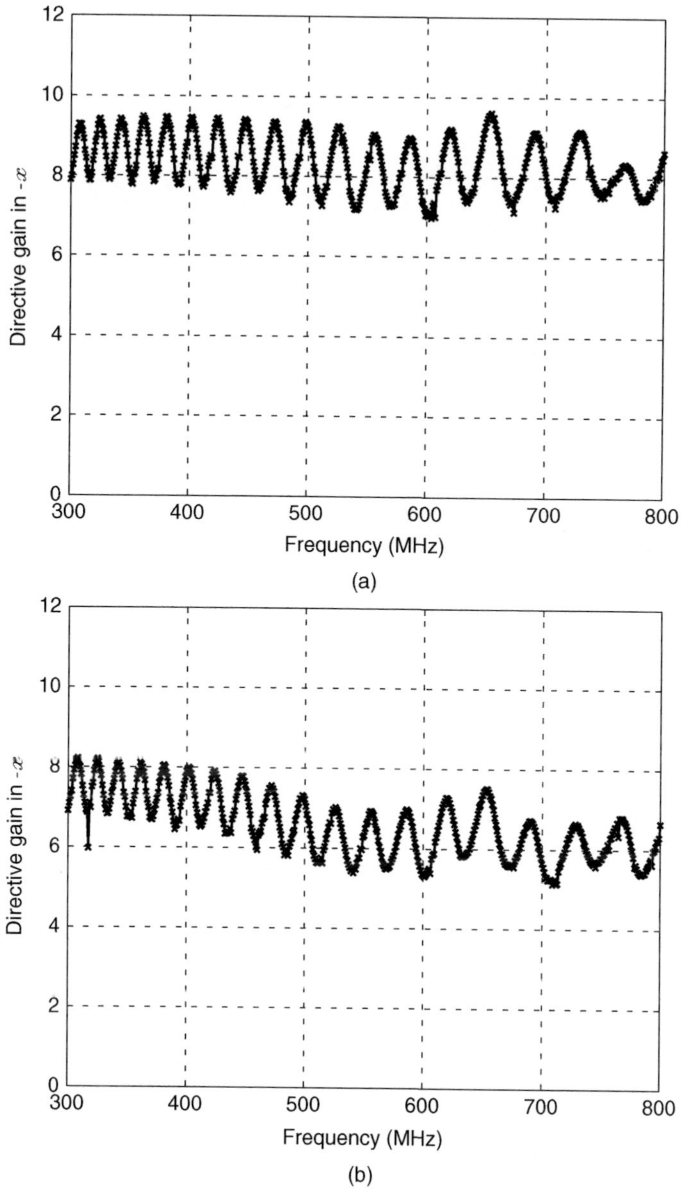

Figure 5.32 Optimized directive gain in the $-x$ direction sampled with 1 MHz resolution when the array in Figure 5.26 is rotated in the ϕ direction for (a) 10° and (b) 20°. *Source*: Gürel and Ergül 2008 [10]. Reproduced with permission of PIER.

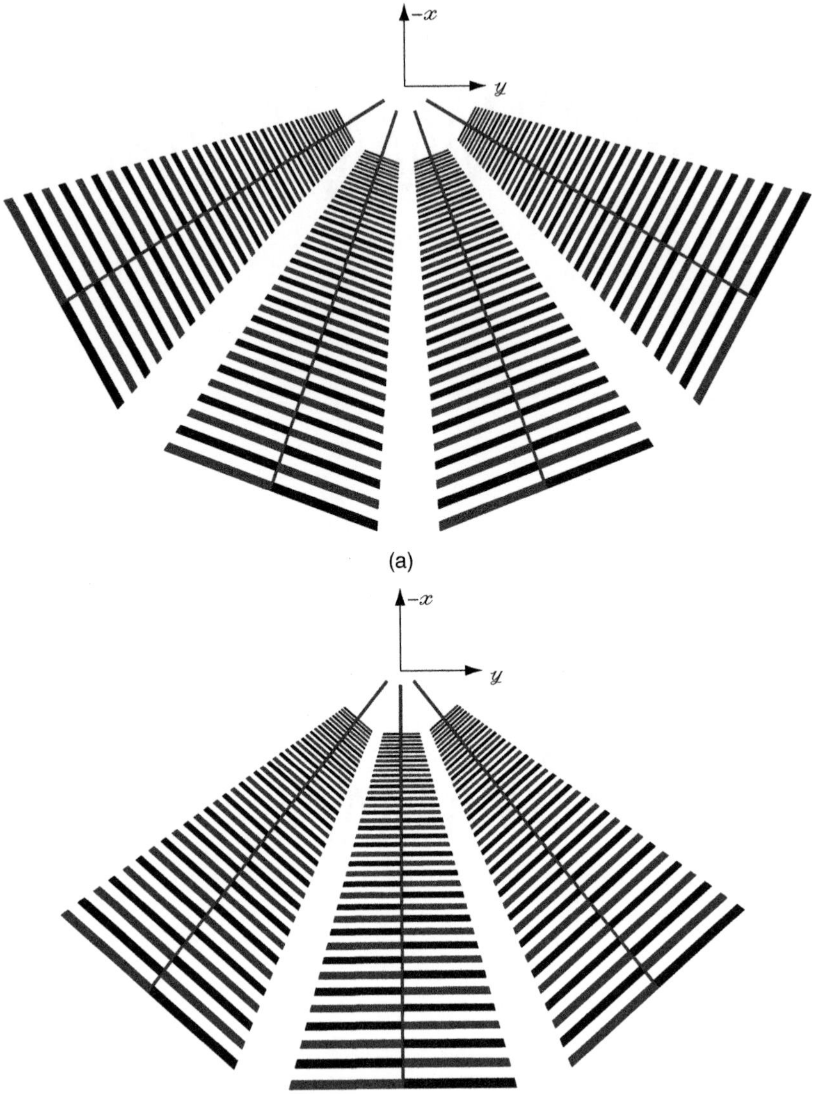

Figure 5.33 Circular-sectoral arrays of closely spaced LP antennas with (a) three elements and (b) four elements. *Source*: Ergül and Gürel 2007 [9]. Reproduced by permission of Radio Science - American Geophysical Union AGU.

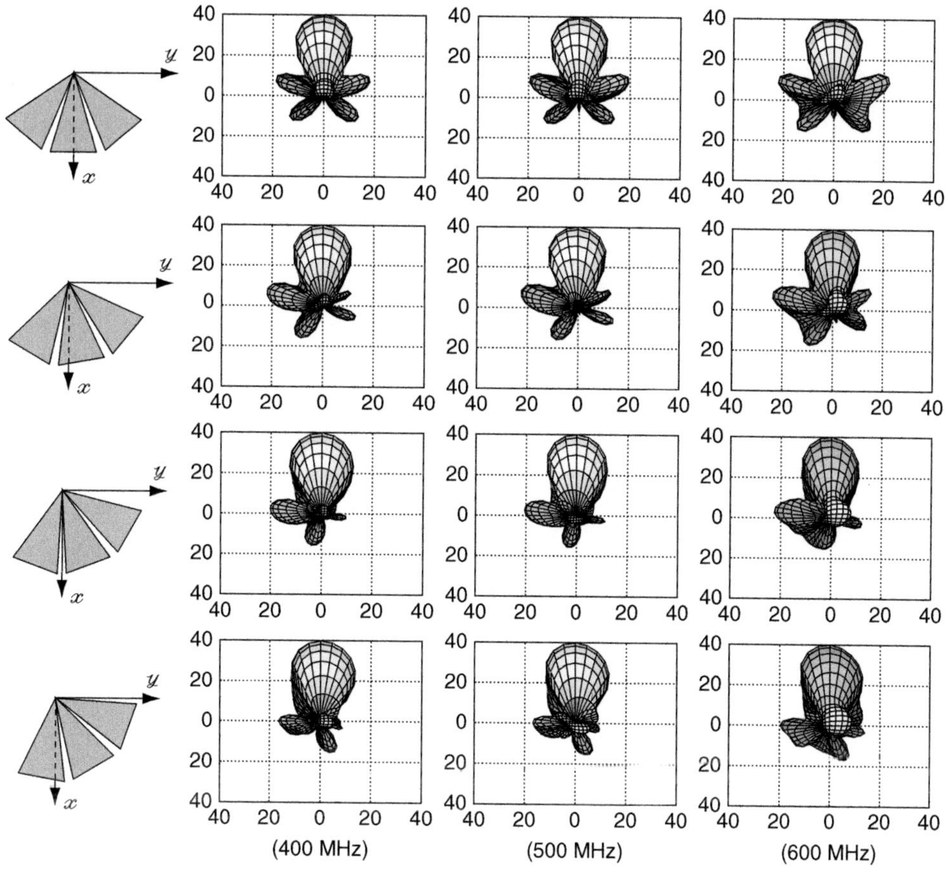

Figure 5.34 Far-zone radiation pattern of the array in Figure 5.33(a) for various frequencies and alignments. The directive gain is optimized in the $-x$ direction. *Source*: Ergül and Gürel 2007 [9]. Reproduced by permission of Radio Science - American Geophysical Union AGU.

at $-10°$, $-20°$, and $-30°$ can also be realized by symmetry, a total scan range of 60° can be achieved. Due to the frequency independence, radiation patterns do not change significantly with respect to the frequency. When the array is rotated from 0° to 30°, the main beam is in the $-x$ direction due to the optimization by GAs. However, as the rotation angle is increased, it becomes difficult to keep the main beam in the $-x$ direction. In other words, although GAs attempt to maximize the directive gain in the $-x$ direction, the main beam may not be pointed in the $-x$ direction for large rotation angles.

For more quantitative information, Figures 5.36(a) and 5.36(b) present the directive gain in the $-x$ direction as a function of the frequency with 25 MHz intervals and for different orientations of the three-element and four-element arrays, respectively. Results of

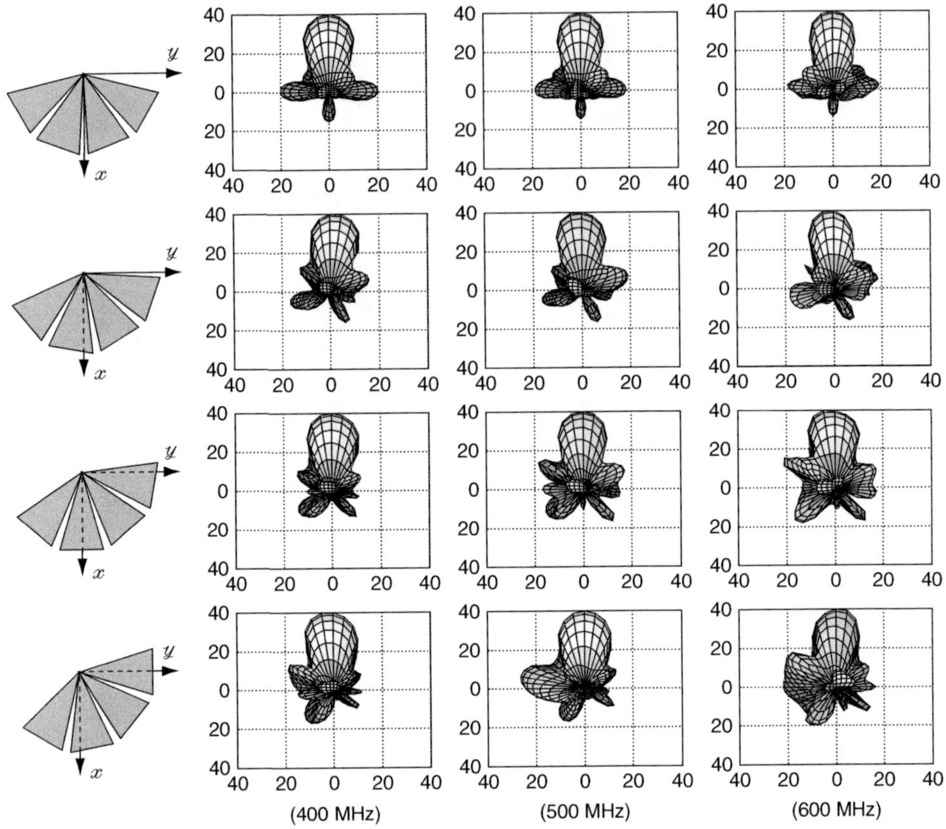

Figure 5.35 Far-zone radiation pattern of the array in Figure 5.33(b) for various frequencies and alignments. The directive gain is optimized in the $-x$ direction. *Source*: Ergül and Gürel 2007 [9]. Reproduced by permission of Radio Science - American Geophysical Union AGU.

optimizations for 40° and 50° orientations are also included. It can be observed that both arrays in Figure 5.33 provide directive gain over 9 in angular sectors of 100°. This is achieved in the frequency range from 300 MHz to 800 MHz, and this range can be extended by adding more teeth to the LP antennas. On the other hand, Figure 5.36 also shows that the directive gain of the four-element array has larger fluctuations compared to the directive gain of the three-element array. This is due to increasing mutual couplings among antennas when the array becomes more populous. In other words, mutual couplings among antennas tend to deteriorate the frequency independence and this becomes more significant as the number of elements in the array is increased.

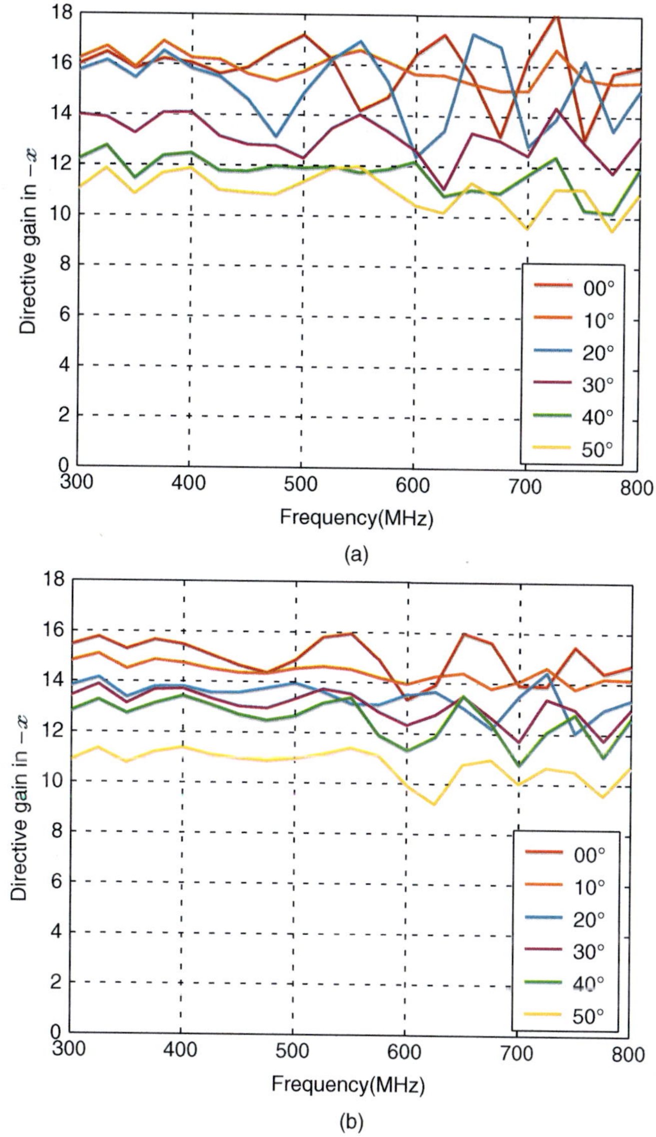

Figure 5.36 The directive gain in the $-x$ direction obtained by the genetic optimization for (a) the three-element array in Figure 5.33(a) and (b) the four-element array in Figure 5.33(b). The arrays are rotated for different angles from 0° to 50° to test the beam-steering ability in a sector of 100°. *Source*: Ergül and Gürel 2007 [9]. Reproduced by permission of Radio Science - American Geophysical Union AGU.

5.6 Concluding Remarks

MLFMA can be used to solve a vast variety of real-life electromagnetics problems. Some examples (case studies) involving airborne targets, metamaterials, photonic crystals, RBCs, and LP antennas are presented in this chapter to demonstrate capabilities of efficient implementations of MLFMA. The results obtained in these and similar simulations can be used to reduce the time and expenses required in building prototypes and carrying out laboratory tests.

Appendix

A.1 Limit Part of the $\mathcal{K}$ Operator

As given in (1.52), the integro-differential operator $\mathcal{K}_u$ is commonly separated into principal-value and limit parts as

$$\mathcal{K}_u\{X\}(r) = \mathcal{K}_{PV,u}\{X\}(r) - \frac{4\pi - \Omega_u}{4\pi}\mathcal{I}^{\times n_u}\{X\}(r), \tag{A.1}$$

where $\Omega_u(r)$ is the solid angle. Specifically, $\Omega_u(r) = 0$ when the observation point r is not located on the source X. If it is located on the source, however, $\Omega_u(r) \in (0, 4\pi)$ is determined by the "shape" of the surface, on which J is defined. If the surface is planar at r, then $\Omega_u(r) = 2\pi$ and

$$\mathcal{K}_u\{X\}(r) = \mathcal{K}_{PV,u}\{X\}(r) - \frac{1}{2}\mathcal{I}^{\times n_u}\{X\}(r). \tag{A.2}$$

Derivation of the limit part of the $\mathcal{K}$ operator is quite straightforward when the source X is locally defined on a planar surface. Consider Figure A.1, where the observation point r approaches a planar surface S_u, which is divided into two parts, i.e., $S_{PV,u}$ and $S_{lim,u}$. Specifically, $S_{lim,u}$ is defined as the infinitesimal circular surface in the vicinity of the projection of the observation point and $S_{PV,u} = S_u - S_{lim,u}$. The $\mathcal{K}$ operator can be written in two parts as

$$\mathcal{K}_u\{X\}(r) = \int dr' X(r') \times \nabla' g_u(r, r') = \mathcal{K}_{PV,u}\{X\}(r) + \mathcal{K}_{lim,u}\{X\}(r), \tag{A.3}$$

where

$$\mathcal{K}_{lim,u}\{X\}(r) = lim_{a\to 0} \int_{S_{\text{lim},u}} dr' X(r') \times \nabla' g_u(r, r') \tag{A.4}$$

$$= \lim_{a\to 0} \int_0^{2\pi} d\phi' \int_0^a d\rho' \rho' X(r') \times \nabla' g_u(r, r'). \tag{A.5}$$

Using

$$\nabla' g_u(r, r') = \nabla' \left(\frac{\exp(ik_u R)}{4\pi R}\right) = \boldsymbol{R}\frac{(1 - ik_u R)}{4\pi R^3}\exp(ik_u R) \tag{A.6}$$

The Multilevel Fast Multipole Algorithm (MLFMA) for Solving Large-Scale Computational Electromagnetics Problems, First Edition. Özgür Ergül and Levent Gürel.

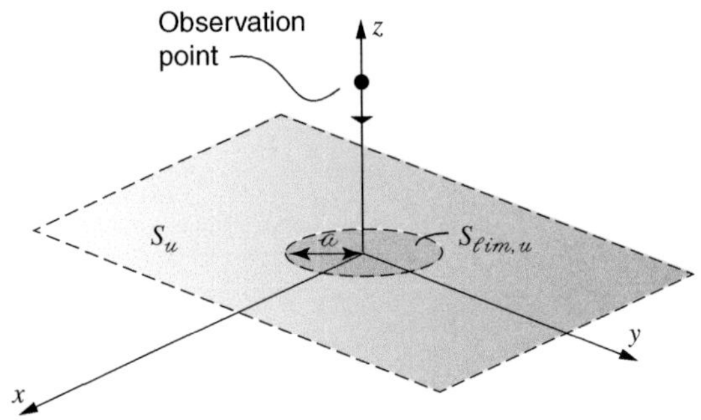

Figure A.1 Observation point approaching a planar surface.

with $R = |\boldsymbol{R}| = |\boldsymbol{r} - \boldsymbol{r}'|$ and considering the limit $a \to 0$,

$$\mathcal{K}_{\text{lim},u}\{\boldsymbol{X}\}(\boldsymbol{r}) \approx \frac{\boldsymbol{X}(\boldsymbol{r})}{4\pi} \times \lim_{a\to 0} \int_0^{2\pi} d\phi' \int_0^a d\rho' \rho' \frac{\boldsymbol{R}}{R^3}$$

$$\approx -\frac{1}{2}\hat{z} \times \boldsymbol{X}(\boldsymbol{r}), \tag{A.7}$$

which can be generalized as

$$\mathcal{K}_{\text{lim},u}\{\boldsymbol{X}\}(\boldsymbol{r}) \approx -\frac{1}{2}\hat{\boldsymbol{n}}_u \times \boldsymbol{X}(\boldsymbol{r}). \tag{A.8}$$

When the observation point approaches a planar surface, the expression in (A.8) indicates that the limit value of the $\mathcal{K}$ operator has a magnitude equal to the half of the source value and has a direction perpendicular to and in the same plane with the source flow. In general, the limit value depends on the solid angle of the surface at the observation point and (A.8) is further generalized as

$$\mathcal{K}_{\text{lim},u}\{\boldsymbol{X}\}(\boldsymbol{r}) \approx -\frac{4\pi - \Omega_u}{4\pi}\hat{\boldsymbol{n}}_u \times \boldsymbol{X}(\boldsymbol{r}). \tag{A.9}$$

Nevertheless, in numerical implementations using planar triangles, the generalized form in (A.9) is never used [125]. For observation points located inside the triangles, $\Omega_u = 2\pi$ and (A.8) is valid. In the limit case, i.e., when the observation point approaches the edge of the source triangle, the solid-angle factor is implicitly included in the near-neighboring interactions [126]. This is discussed in Chapter 1, Section 1.8.1 in the context of the analytical integrals.

A.2 Post Processing

Numerical solutions of electromagnetics problems are usually followed by various post-processing procedures, such as computations of fields at various locations, depending on the application.

A.2.1 Near-Zone Electromagnetic Fields

Finding the coefficients for the electric current on a PEC object, i.e., $\boldsymbol{a}_J[n]$ for $n = 1, 2, \ldots, N$, one can calculate the secondary (scattered or radiated) electric and magnetic fields at any point $\boldsymbol{r}$ as

$$\boldsymbol{E}^{sec}(\boldsymbol{r}) = ik\eta \sum_{n=1}^{N} \boldsymbol{a}_J[n] \int_{S_n} d\boldsymbol{r}' \left[\boldsymbol{b}_n(\boldsymbol{r}') + \frac{1}{k^2}\nabla' \cdot \boldsymbol{b}_n(\boldsymbol{r}')\nabla\right] g(\boldsymbol{r}, \boldsymbol{r}') \tag{A.10}$$

$$\boldsymbol{H}^{sec}(\boldsymbol{r}) = \sum_{n=1}^{N} \boldsymbol{a}_J[n] \int_{S_n} d\boldsymbol{r}' \boldsymbol{b}_n(\boldsymbol{r}') \times \nabla' g(\boldsymbol{r}, \boldsymbol{r}'). \tag{A.11}$$

If the observation point $\boldsymbol{r}$ is close to the object, singularity extractions are required to accurately evaluate the integrals in (A.10) and (A.11). On the other hand, conventional singularity extraction methods must be avoided when the observation point is far from the object and $|\boldsymbol{r}|$ is very large.

For dielectric objects, expansion coefficients of both electric and magnetic currents must be considered to compute the electric and magnetic fields. For example, if the observation point $\boldsymbol{r}$ is outside the object,

$$\begin{aligned}\boldsymbol{E}^{sec}(\boldsymbol{r}) = {} & ik_o\eta_o \sum_{n=1}^{N} \boldsymbol{a}_J[n] \int_{S_n} d\boldsymbol{r}' \left[\boldsymbol{b}_n(\boldsymbol{r}') + \frac{1}{k_o^2}\nabla' \cdot \boldsymbol{b}_n(\boldsymbol{r}')\nabla\right] g_o(\boldsymbol{r}, \boldsymbol{r}') \\ & - \sum_{n=1}^{N} \boldsymbol{a}_M[n] \int_{S_n} d\boldsymbol{r}' \boldsymbol{b}_n(\boldsymbol{r}') \times \nabla' g_o(\boldsymbol{r}, \boldsymbol{r}')\end{aligned} \tag{A.12}$$

$$\begin{aligned}\boldsymbol{H}^{sec}(\boldsymbol{r}) = {} & \sum_{n=1}^{N} \boldsymbol{a}_J[n] \int_{S_n} d\boldsymbol{r}' \boldsymbol{b}_n(\boldsymbol{r}') \times \nabla' g_o(\boldsymbol{r}, \boldsymbol{r}') \\ & + \frac{ik_o}{\eta_o} \sum_{n=1}^{N} \boldsymbol{a}_M[n] \int_{S_n} d\boldsymbol{r}' \left[\boldsymbol{b}_n(\boldsymbol{r}') + \frac{1}{k_o^2}\nabla' \cdot \boldsymbol{b}_n(\boldsymbol{r}')\nabla\right] g_o(\boldsymbol{r}, \boldsymbol{r}'),\end{aligned} \tag{A.13}$$

where the electrical parameters of the host medium are used. On the other hand, if the observation point is inside the object,

$$\begin{aligned}\boldsymbol{E}^{sec}(\boldsymbol{r}) = {} & -ik_i\eta_i \sum_{n=1}^{N} \boldsymbol{a}_J[n] \int_{S_n} d\boldsymbol{r}' \left[\boldsymbol{b}_n(\boldsymbol{r}') + \frac{1}{k_i^2}\nabla' \cdot \boldsymbol{b}_n(\boldsymbol{r}')\nabla\right] g_i(\boldsymbol{r}, \boldsymbol{r}') \\ & + \sum_{n=1}^{N} \boldsymbol{a}_M[n] \int_{S_n} d\boldsymbol{r}' \boldsymbol{b}_n(\boldsymbol{r}') \times \nabla' g_i(\boldsymbol{r}, \boldsymbol{r}')\end{aligned} \tag{A.14}$$

$$\begin{aligned}\boldsymbol{H}^{sec}(\boldsymbol{r}) = {} & -\sum_{n=1}^{N} \boldsymbol{a}_J[n] \int_{S_n} d\boldsymbol{r}' \boldsymbol{b}_n(\boldsymbol{r}') \times \nabla' g_i(\boldsymbol{r}, \boldsymbol{r}') \\ & - \frac{ik_i}{\eta_i} \sum_{n=1}^{N} \boldsymbol{a}_M[n] \int_{S_n} d\boldsymbol{r}' \left[\boldsymbol{b}_n(\boldsymbol{r}') + \frac{1}{k_i^2}\nabla' \cdot \boldsymbol{b}_n(\boldsymbol{r}')\nabla\right] g_i(\boldsymbol{r}, \boldsymbol{r}'),\end{aligned} \tag{A.15}$$

where the electrical parameters of the object is used. In typical computations involving many observation points, one does not need to check whether observation points are located inside or outside the object. Instead, (A.12)–(A.13) and (A.14)–(A.15) can be computed separately and superposed to obtain the secondary fields at all observation points. This is allowed since the expressions in (A.12)–(A.13) and (A.14)–(A.15) are related to the outer and inner problems and they do not contribute to the electromagnetic fields at the observation points inside and outside the object, respectively. A similar procedure can be applied for composite objects involving multiple dielectric and metallic parts. Without checking the positions of observation points, electric and magnetic currents are allowed to radiate using the electrical parameters of each medium.

Figures A.2–A.5 present solutions of various scattering problems involving spherical composite objects located in free space. In Figure A.2, a dielectric sphere of radius $0.5\lambda_o$ is coated with a dielectric shell of radius λ_o. The object is illuminated by a plane wave propagating in the $-z$ direction with the electric field polarized in the x direction. The relative permittivities of the core and shell are 4.0 and 2.0, respectively. The problem is formulated with JMCFIE and solved via MLFMA. Figures A.2(a) and A.2(b) depict the radial (ρ) and angular (ϕ) components of the total electric field on the x-y plane in the vicinity of the object. It can be observed that the angular component of the electric field is continuous across dielectric interfaces. Figures A.3–A.5 present similar results when the material properties of the core and shell are changed. In Figure A.3, the relative permittivities of the core and shell are 2.0 and 4.0, respectively. In Figure A.4, relative permittivity of the shell is 2.0, while the core is PEC. In Figure A.5, the relative permittivity of the shell is 4.0, while the core is again PEC. Note that the magnitude of the electric field becomes vanishingly small inside PEC spheres, as physically expected.

A.2.2 Far-Zone Fields

In the far zone, normalized field quantities are particularly important. These are defined as

$$\begin{aligned}\boldsymbol{E}^{sec,\infty}(\theta,\phi) &= \exp(-ik_o|\boldsymbol{r}|)\lim_{|r|\to\infty}\left\{|\boldsymbol{r}|\boldsymbol{E}^{sec}(\boldsymbol{r})\right\}\\ &= \frac{ik_o\eta_o}{4\pi}\sum_{n=1}^{N}\boldsymbol{a}_J[n]\left(\overline{\boldsymbol{I}}^{3\times3}-\hat{\boldsymbol{k}}\hat{\boldsymbol{k}}\right)\cdot\boldsymbol{V}_n(\theta,\phi)\\ &\quad-\frac{ik_o}{4\pi}\sum_{n=1}^{N}\boldsymbol{a}_M[n]\hat{\boldsymbol{k}}\times\boldsymbol{V}_n(\theta,\phi)\end{aligned} \tag{A.16}$$

and

$$\begin{aligned}\boldsymbol{H}^{sec,\infty}(\theta,\phi) &= \exp(-ik_o|\boldsymbol{r}|)\lim_{|r|\to\infty}\left\{|\boldsymbol{r}|\boldsymbol{H}^{sec}(\boldsymbol{r})\right\}\\ &= \frac{1}{\eta_o}\hat{\boldsymbol{k}}\times\boldsymbol{E}^{sec,\infty}(\theta,\phi),\end{aligned} \tag{A.17}$$

where

$$\boldsymbol{V}_n(\theta,\phi) = \int_{S_n} d\boldsymbol{r}'\boldsymbol{b}_n(\boldsymbol{r}')\exp(-ik_o\hat{\boldsymbol{r}}\cdot\boldsymbol{r}') \tag{A.18}$$

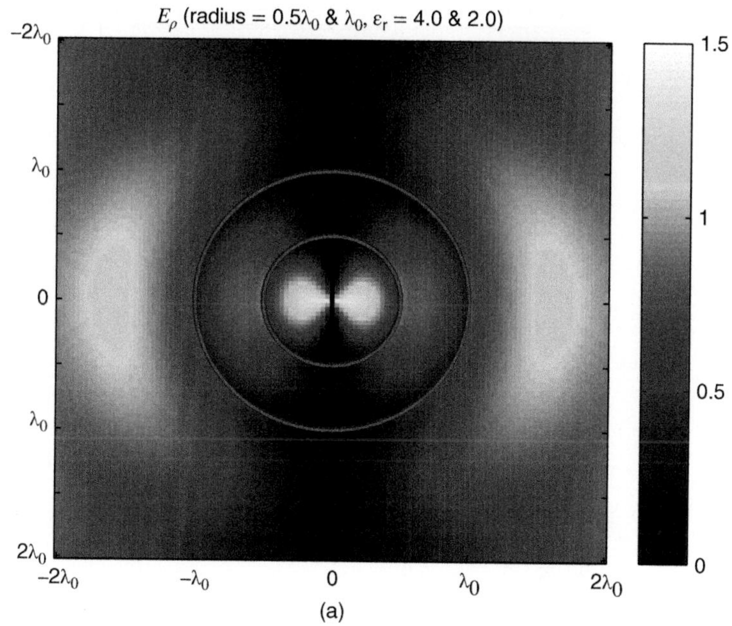

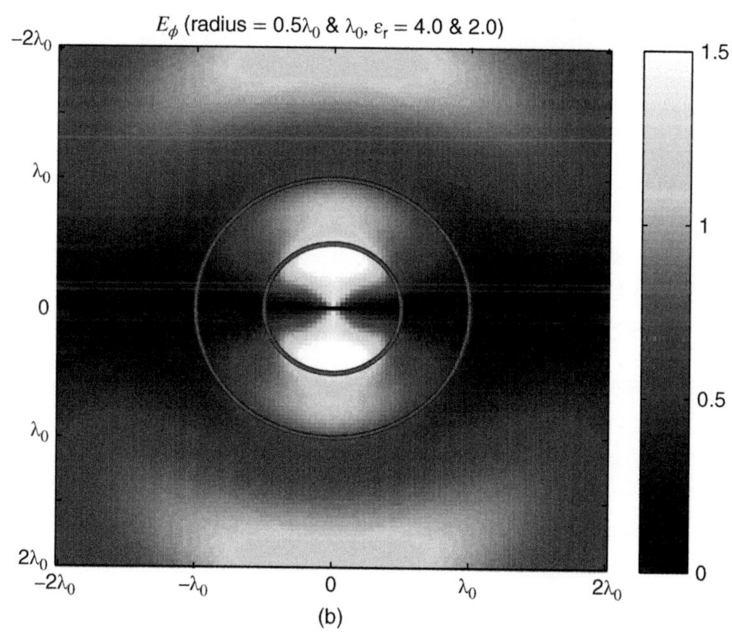

Figure A.2 (a) Radial and (b) angular components of the total electric field in the vicinity of a dielectric sphere of radius $0.5\lambda_o$ coated with a dielectric shell of radius λ_o. The spherical object is illuminated by a plane wave. The relative permittivities of the core and shell are 4.0 and 2.0, respectively.

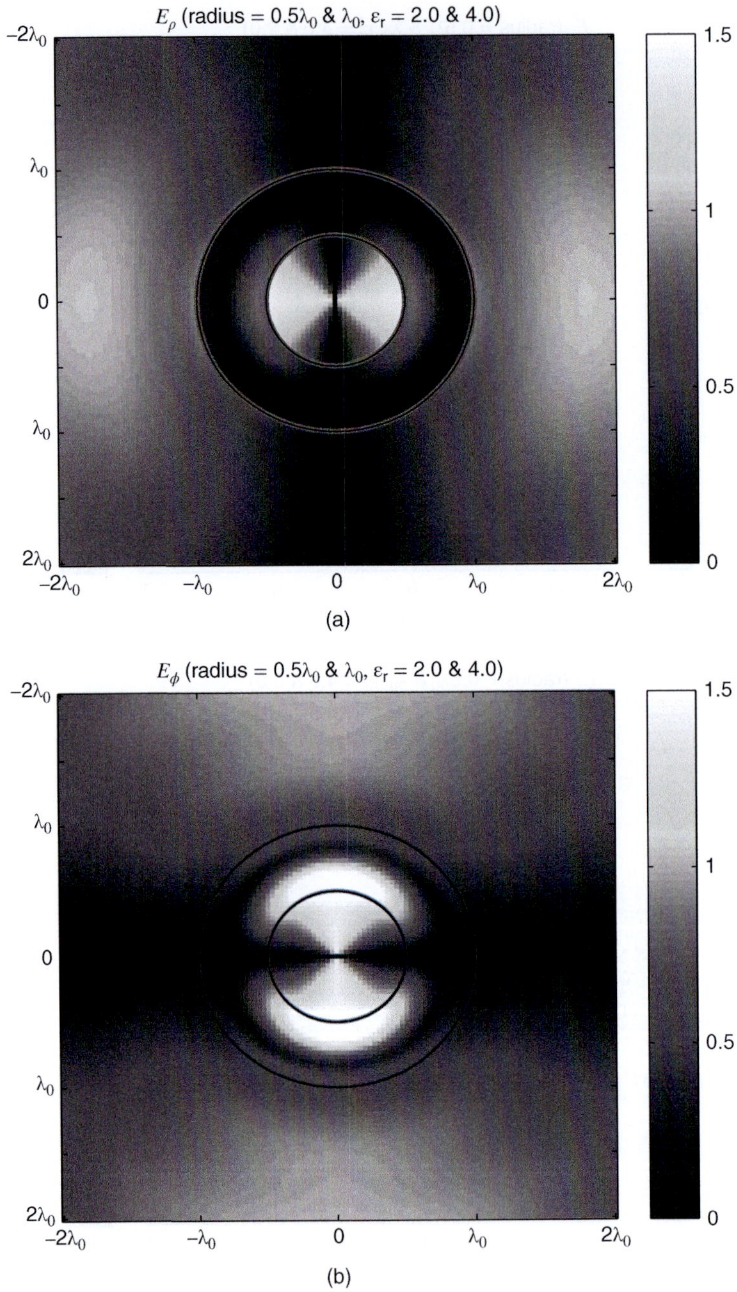

Figure A.3 (a) Radial and (b) angular components of the total electric field in the vicinity of a dielectric sphere of radius $0.5\lambda_o$ coated with a dielectric shell of radius λ_o. The spherical object is illuminated by a plane wave. The relative permittivities of the core and shell are 2.0 and 4.0, respectively.

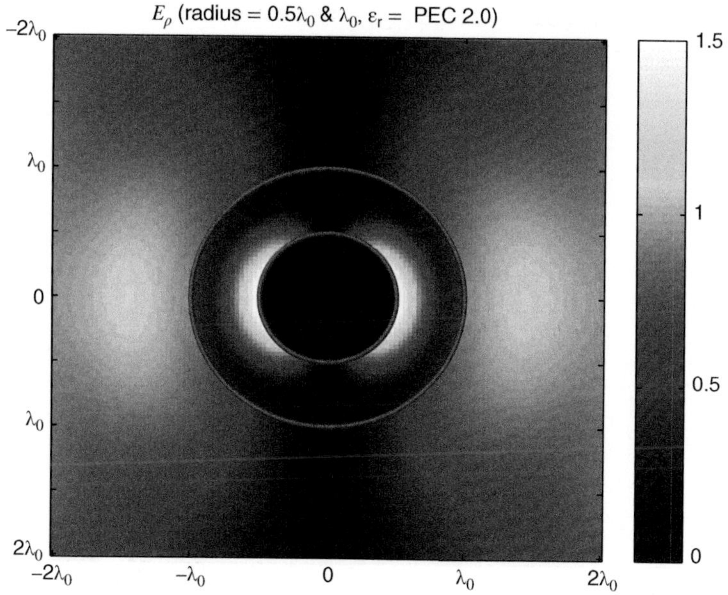

(a)

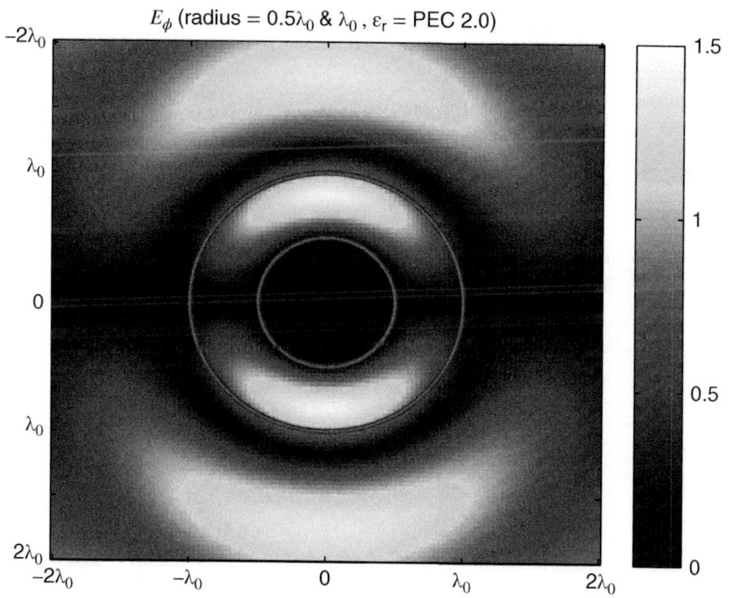

(b)

Figure A.4 (a) Radial and (b) angular components of the total electric field in the vicinity of a PEC sphere of radius $0.5\lambda_o$ coated with a dielectric shell of radius λ_o. The spherical object is illuminated by a plane wave. The relative permittivity of the shell is 2.0.

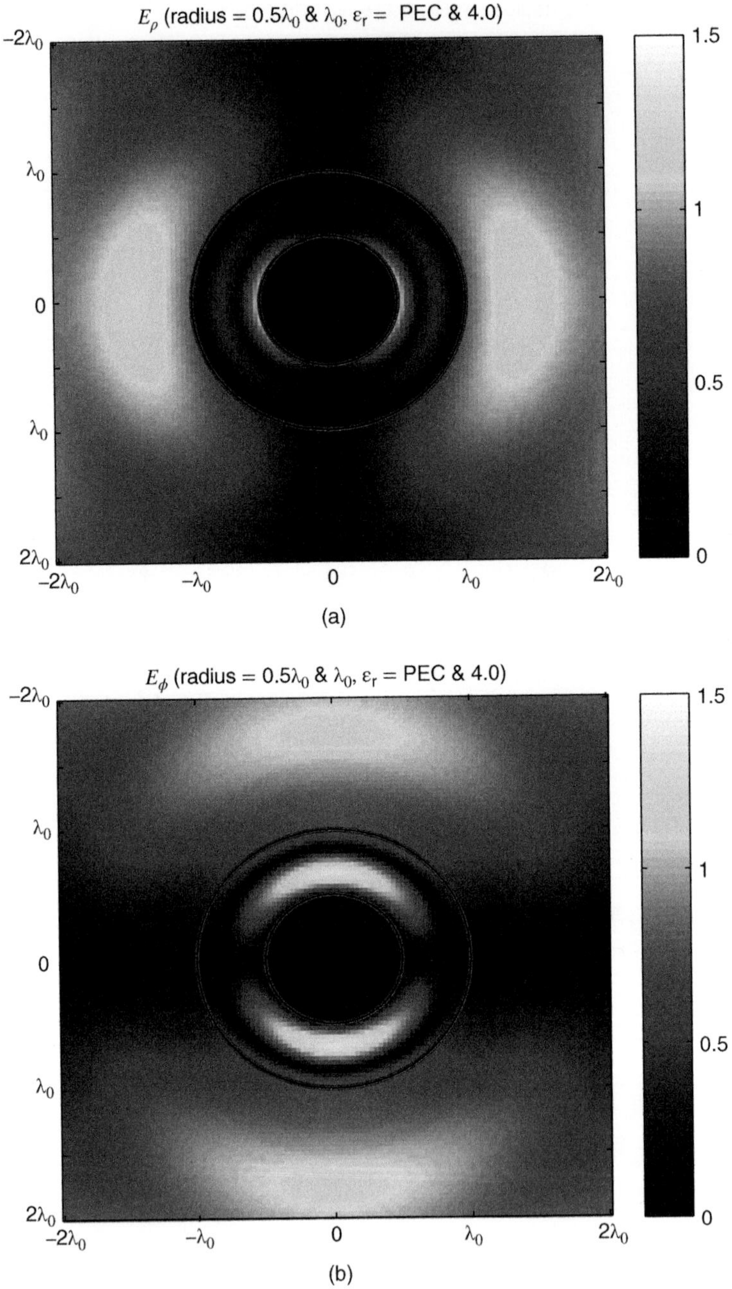

Figure A.5 (a) Radial and (b) angular components of the total electric field in the vicinity of a PEC sphere of radius $0.5\lambda_o$ coated with a dielectric shell of radius λ_o. The spherical object is illuminated by a plane wave and the relative permittivity of the shell is 4.0.

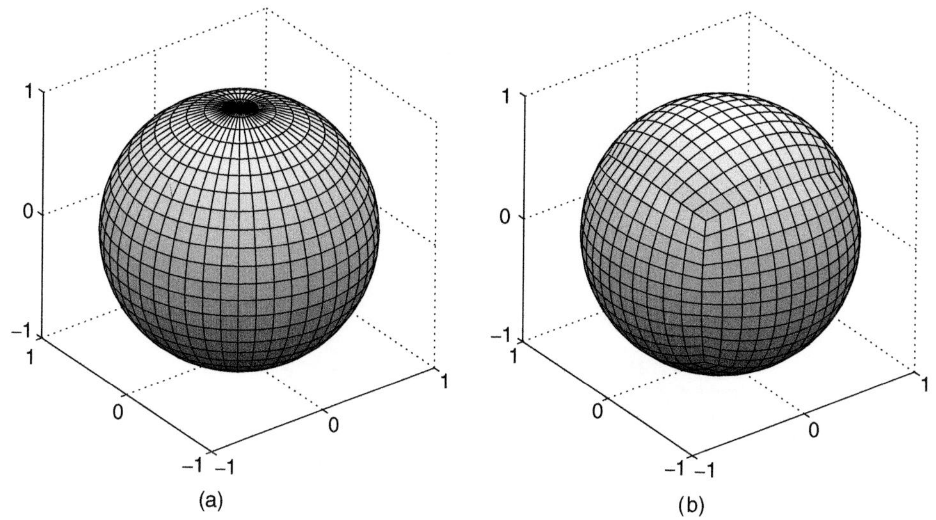

Figure A.6 Discretization of the unit sphere: (a) Sampling with equally spaced points in the θ-ϕ domain and (b) uniform sampling of six biquadratic patches in the u-v domain.

is the vector current moment of the nth basis function. In many cases, far-zone electromagnetic fields need to be calculated at various directions with respect to the angular coordinates θ and ϕ. If the unit sphere is sampled with equally spaced points in the θ-ϕ domain, the resulting discretization is illustrated in Figure A.6(a), where an unnecessary crowding of the sampling points around the north pole and south pole is observed. Since far-zone fields are completely arbitrary, there are no preferred directions requiring denser sampling. Thus, not only is the dense sampling around poles redundant and inefficient, but it also generates misleading (more difficult to read) three dimensional plots. As an alternative, the unit sphere can be modeled with six biquadratic patches, as shown in Figure A.6(b). Each biquadratic patch has a general representation of

$$\boldsymbol{r}(u, v) = \sum_{m=0}^{2} \sum_{n=0}^{2} \boldsymbol{a}_{m,n} u^m v^n \tag{A.19}$$

for $0 \leq u, v \leq 1$, where $\boldsymbol{a}_{m,n}$ represents vector coefficients and $\boldsymbol{r}(u, v)$ is the position vector from the origin pointing to the surface. Then, sampling points can be obtained on each patch by a uniform distribution in the u-v domain. Figure A.6(b) illustrates a discretization with 13×13 points on each patch, producing a total of 864 directions on the unit sphere. In contrast, the discretization in Figure A.6(a), which has comparable fidelity around the equator, contains 33% more sampling directions.

Far-zone scattering properties are usually characterized by RCS defined as

$$\text{RCS}(\theta, \phi) = \frac{4\pi}{\eta_o} \left| E_\theta^{sec,\infty}(\theta, \phi) \right|^2 + \frac{4\pi}{\eta_o} \left| E_\phi^{sec,\infty}(\theta, \phi) \right|^2, \tag{A.20}$$

which depends on the frequency, shape and material properties of the object, excitation (illumination), and observation angle. For a given target and excitation, RCS can be investigated as a function of the frequency and observation angle (bistatic RCS). For example, Figures A.7

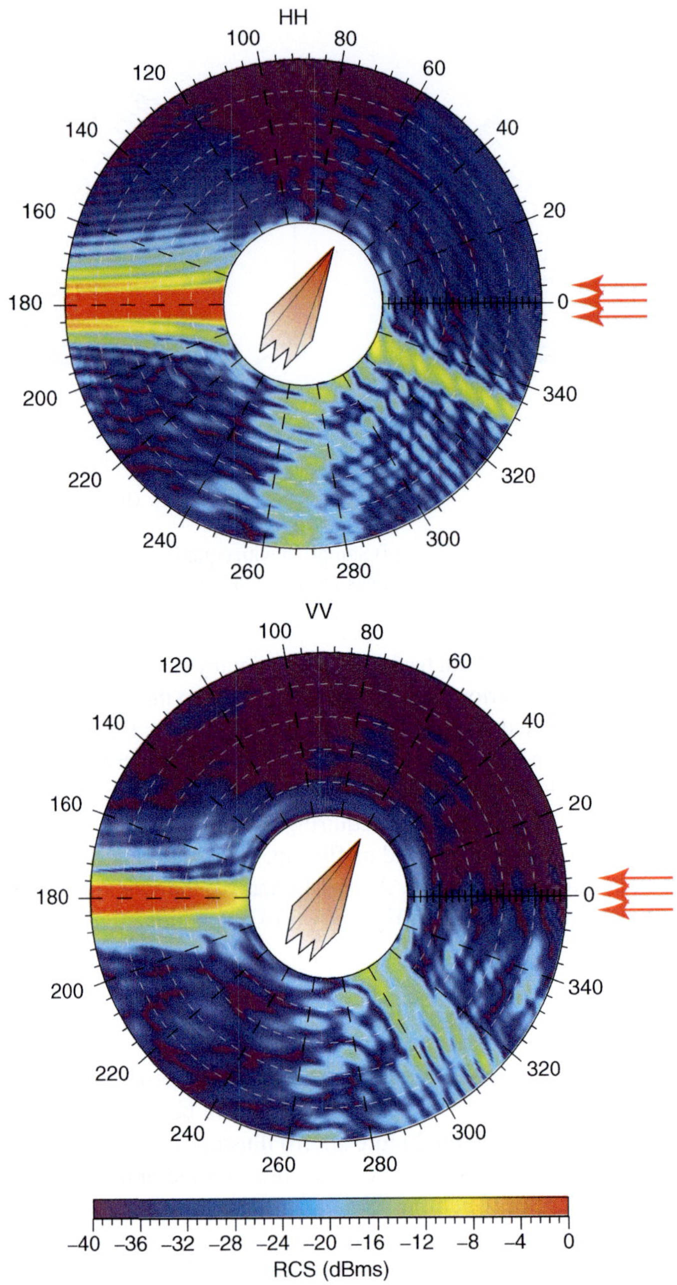

Figure A.7 Co-polar bistatic RCS of the Flamme from 2 GHz to 11 GHz. The target is illuminated by a plane wave propagating at 60° angle from the nose with the electric field polarized horizontally and vertically.

and A.8 illustrate the bistatic co-polar RCS of the stealth airborne target Flamme [149] located on the x-y plane. In Figure A.7, the target is illuminated by a plane wave propagating at 60° angle from the nose. Both horizontal (H) and vertical (V) polarizations are considered. In polar plots, the radial and angular directions represent the frequency and observation angle, respectively. The bistatic RCS is plotted as a function of frequency from 2 GHz to 11 GHz and as a function of observation angle from 0° to 360°. Note that the RCS of the Flamme is very low in the backscattering direction (0°) due to the stealth ability of the target. For the horizontal polarization, a strong reflection from the target at about 330° is also remarkable. Figure A.8 presents similar results when the Flamme is illuminated by a plane wave propagating at 120° angle from the nose. For this excitation, the backscattered RCS of the Flamme is usually at low levels, but it increases at various frequencies, particularly for the horizontal polarization. As shown in Figures A.7 and A.8, two-dimensional RCS plots can provide many useful information on the electromagnetic properties of scatterers.

A major parameter to measure the radiation quality of an antenna is the directive gain in a desired direction (θ_0, ϕ_0), which can be derived as

$$D(\theta_0, \phi_0) = \frac{4\pi P^{\infty}(\theta_0, \phi_0)}{\int_0^{2\pi} d\phi \int_0^{\pi} d\theta \sin\theta P^{\infty}(\theta, \phi)}. \tag{A.21}$$

In (A.21),

$$P^{\infty}(\theta, \phi) = \frac{1}{\eta_o}\left|E_{\theta}^{sec,\infty}(\theta, \phi)\right|^2 + \frac{1}{\eta_o}\left|E_{\phi}^{sec,\infty}(\theta, \phi)\right|^2 \tag{A.22}$$

is the radiation intensity. Using the sampling scheme depicted in A.6(b), the integral in (A.21) can be expressed in the u-v domain for each biquadratic surface as

$$I = \int_0^1 du \int_0^1 dv \left[\left|E_{\theta}^{sec,\infty}(u, v)\right|^2 + \left|E_{\phi}^{sec,\infty}(u, v)\right|^2\right] \varrho(u, v), \tag{A.23}$$

where

$$\varrho(u, v) = \sqrt{j_{uu}(u, v) j_{vv}(u, v) \quad j_{uv}^2(u, v)} \tag{A.24}$$

is the Jacobian and

$$\begin{aligned} j_{uu}(u, v) &= \frac{\partial \mathbf{r}}{\partial u} \cdot \frac{\partial \mathbf{r}}{\partial u} \\ &= \sum_{m_1=1}^{2} \sum_{m_2=1}^{2} \sum_{n_1=0}^{2} \sum_{n_2=0}^{2} \mathbf{a}_{m_1,n_1} \cdot \mathbf{a}_{m_2,n_2} m_1 m_2 u^{m_1+m_2-2} v^{n_1+n_2} \end{aligned} \tag{A.25}$$

$$\begin{aligned} j_{vv}(u, v) &= \frac{\partial \mathbf{r}}{\partial v} \cdot \frac{\partial \mathbf{r}}{\partial v} \\ &= \sum_{m_1=0}^{2} \sum_{m_2=0}^{2} \sum_{n_1=1}^{2} \sum_{n_2=1}^{2} \mathbf{a}_{m_1,n_1} \cdot \mathbf{a}_{m_2,n_2} n_1 n_2 u^{m_1+m_2} v^{n_1+n_2-2} \end{aligned} \tag{A.26}$$

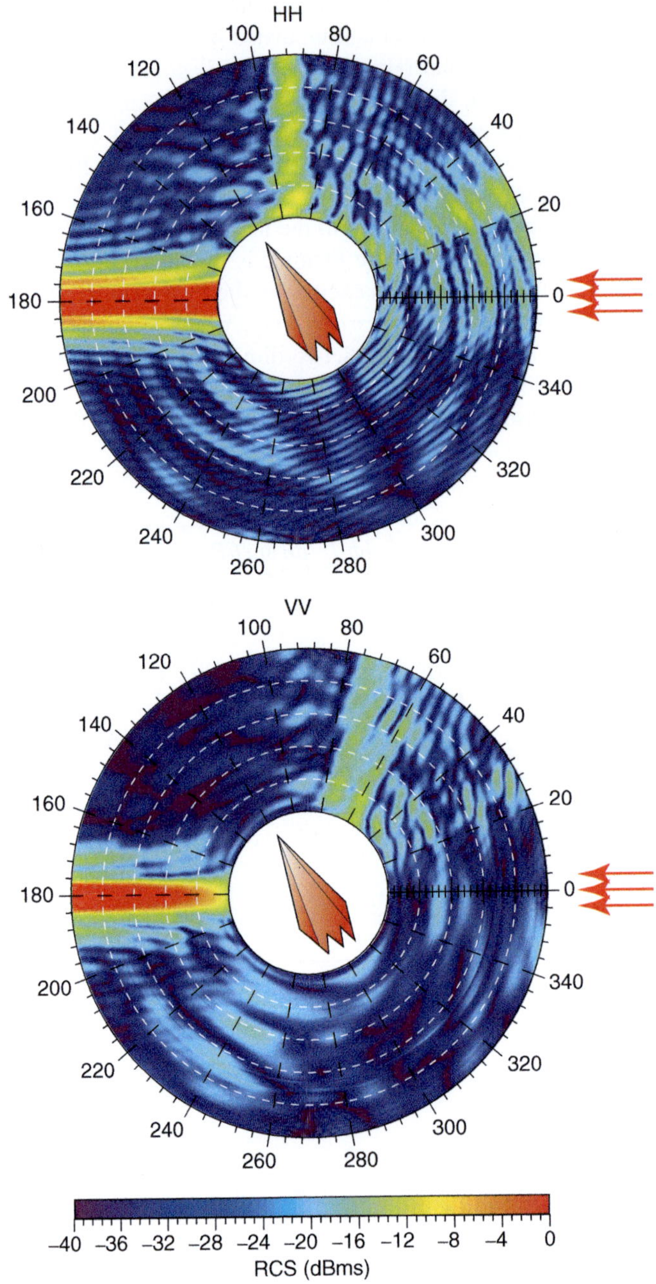

Figure A.8 Co-polar bistatic RCS of the Flamme from 2 GHz to 11 GHz. The target is illuminated by a plane wave propagating at 120° angle from the nose with the electric field polarized horizontally and vertically.

$$
\begin{aligned}
j_{uv}(u,v) &= \frac{\partial \boldsymbol{r}}{\partial u} \cdot \frac{\partial \boldsymbol{r}}{\partial v} \\
&= \sum_{m_1=1}^{2} \sum_{m_2=0}^{2} \sum_{n_1=0}^{2} \sum_{n_2=1}^{2} \boldsymbol{a}_{m_1,n_1} \cdot \boldsymbol{a}_{m_2,n_2} m_1 n_2 u^{m_1+m_2-1} v^{n_1+n_2-1}. \qquad \text{(A.27)}
\end{aligned}
$$

Note that the directive gain is used in Chapter 5, Section 5.5 to compare various designs of LP antennas and their arrays.

A.3 More Details of the Hierarchical Partitioning Strategy

The hierarchical partitioning strategy described in Chapter 4, Section 4.6 provides efficient parallelization of MLFMA. This section presents more details on the implementation of this strategy, particularly considering the organization of communications.

A.3.1 Aggregation/Disaggregation Stages

Figure A.9 depicts a simple example involving a hierarchical partitioning of two consecutive levels among eight processors labeled from 0 to 7. For the sake of brevity, communications involving processor 2 are focused. At level l, boxes (subdomains) and field samples are divided into four and two partitions, respectively. The partitioning is changed at level $l+1$ (assuming that this is more efficient), where boxes and field samples are divided into two and four partitions. To facilitate this rearrangement of partitioning, an intermediate level $l+1/2$ is defined, as also illustrated in Figure A.9. A set of boxes assigned to a processor p at any level l are represented by $\mathcal{N}_l^p$. At actual levels (l and $l+1$), there is no overlapping data between processors. At the intermediate level, however, samples for a single box can be duplicated at some of the neighboring processors.

Table A.1 lists the major operations performed for an aggregation step from level l to level $l+1$ using an intermediate level $l+1/2$. Note that these operations are reversed in a disaggregation step from level $l+1$ to level l. The first loop is constructed over local boxes at the intermediate level. Then, for a given box c, local sub-boxes $Sub\{c\}$ are considered in the second loop. Processors that are in the same column of the intermediate-level partitioning handle

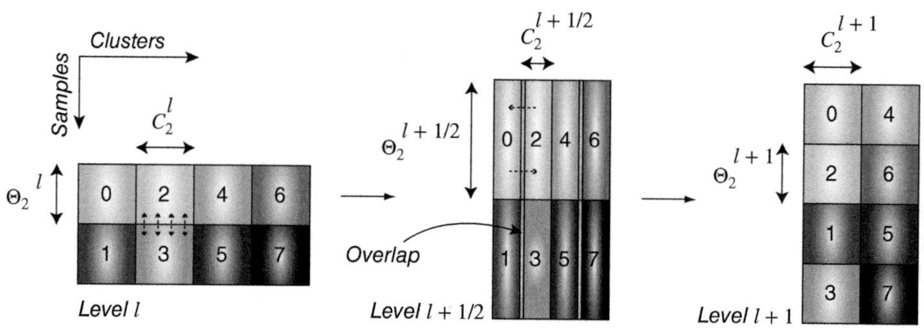

Figure A.9 Partitioning of two consecutive levels among eight processors using the hierarchical partitioning strategy. *Source*: Ergül and Gürel 2011 [239]. Reproduced with permission of IEEE.

Table A.1 Pseudocode of major operations performed by a processor p for an aggregation step from level l to $l+1$

```
do for each box c ∈ N^p_{l+1/2}
    do for each sub-box c_s ∈ Sub{c} & c_s ∈ N^p_l
        get S_p{c_s} from the local aggregation array
        S_p{c_s} → columnwise (neighboring) communications → S_{p+}{c_s}
        S_{p+}{c_s} → local interpolation → S̃_p{c_s}
        divide data into two parts: S̃_p{c_s} = S̃^A_p{c_s} ∪ S̃^B_p{c_s}
        S̃^A_p{c_s} → shift radiation center and add to the local aggregation array
        S̃^B_p{c_s} → shift radiation center and add to a communication array
send and receive data (communication array) to modify the partitioning
```

the same set of boxes and sub-boxes, and this requires a perfect synchronization for efficiency. For a given sub-box c_s, a set of field samples $\mathcal{S}_p\{c_s\}$ is extracted from the aggregation array, and the local data is inflated via one-to-one communications between neighboring processors. Next, the radiated field represented by an inflated set of samples $\mathcal{S}_{p+}\{c_s\}$ is interpolated according to the sample rate of the next level. At this point, the contribution of the sub-box c_s to the radiation of the parent box c can be obtained by shifting the radiation center for the field values at the samples. But, since the partitioning is required to be changed, the inner loop is completed by dividing the samples $\tilde{\mathcal{S}}_p\{c_s\}$ into two parts. Some of the samples, i.e., $\tilde{\mathcal{S}}^A_p\{c_s\}$, are added to the local aggregation array, while the others, i.e., $\tilde{\mathcal{S}}^B_p\{c_s\}$, are added to a communication array to be sent to a corresponding processor. After all boxes at the intermediate level are processed, data collected in the communication arrays are exchanged between pairs of processors to modify the partitioning. In the example depicted in Figure A.9, processor 2 communicates with processor 3 to inflate the local data for each sub-box c_s. When the loops in Table A.1 are completed and the intermediate level is constructed, processor 2 exchanges data with processor 0 to obtain the partitioning at level $l+1$.

A.3.2 Translation Stage

Communications between processors are also required during a translation stage, since boxes are partitioned, and some translations are required among boxes located in different processors. Table A.2 lists the major operations performed for those inter-processor translations. In general, each processor p needs to be paired with all other processors ($p' \neq p$) to exchange data. Given a pair of processors, constructing a single connection between them and exchanging all data at once leads to better performance than pairing the processors repetitively. Hence, as shown in Table A.2, the first loop for inter-processor translations is pairing processors. When a pairing is established, each level is considered one by one in the second loop. At level l, communications are required only if the processors are located in the same row of the partitioning. If this is the case, field samples $\mathcal{S}_{p'}\{c_b\}$ are transferred from processor p' and translations are performed on the receiver side (processor p). For a given box c_b whose field is transferred from processor p', incoming fields are computed for each box c_a that are in the far-field list of c_b and owned by processor p. Then, incoming fields $\mathcal{G}_p\{c_b \rightsquigarrow c_a\}$ are added to the local disaggregation array. Note that processor p also sends field samples $\mathcal{S}_p\{c_b\}$ to p' for inter-processor translations to be performed by p'.

Table A.2 Pseudocode of major operations performed by a processor p for inter-processor translations

```
do for each processor p' ≠ p
    do for each level l (also check if any translation is required or not)
        do for each box c_b that is required to be transferred from p'
            receive S_p'{c_b}
            do for each box c_a owned by p and in the far-field list of c_b
                S_p'{c_b} → translate → G_p{c_b ⇝ c_a}
                add G_p{c_b ⇝ c_a} to the local disaggregation array
        do for each box c_b that is required to be transferred to p'
                send S_p{c_b}
```

When two processors are paired, the amount of communications required for inter-processor translations between them depends on the partitions of the tree structure assigned to the processors. Specifically, processors may need to communicate for some levels while they may not for others. Consequently, for efficient inter-processor translations, one needs to determine the order of pairing among processors carefully. For example, consider the partitioning of two levels among eight processors in Figure A.9. When processor 2 is paired with processor 6, they possibly communicate at levels l and $l + 1$ since they are located in the same row of the partitioning at both levels. Then, at the same time, one can match processor 0 with processor 4, processor 1 with processors 5, and processor 3 with processor 7; because all these pairs involve processors that are located in the same row at both levels. This strategy leads to a very good synchronization among processors and avoids possible delays between the pairing rounds.

A.4 Mie-Series Solutions

Mie-series solutions of scattering problems involving spherical objects are known for more than 100 years [34]. The solution procedure is based on an analytical application of the Maxwell's equations on spherical objects that are illuminated by electromagnetic waves. Although the solution method is named after its developer German physicist Gustav Mie, it is known that the basic theory was presented previously by Ludvig Valentin Lorenz in 1890. Since then, many variations of the theory have been published in journal papers and text books. Mie-series solutions are still used in electromagnetics, especially to test the accuracy of computational solvers.

Mie-series solutions involve the expansion of electromagnetic fields in terms of spherical functions. Among various types of procedures, a rigorous derivation based on Debye potentials is presented in this section. Consider time-harmonic scattering from spherical objects located at the origin in a homogeneous medium extending to infinity. The angular frequency is ω. Total electric and magnetic fields in each homogeneous medium can be separated into known incident and unknown scattered fields. By applying the boundary conditions on spherical surfaces, unknown coefficients for scattered fields can be determined analytically. In theory, electromagnetic fields can only be expanded by infinite numbers of terms. Therefore, an exact solution cannot be generated by using a computer program. However, a finite number of terms would be sufficient to obtain the solution with a desired level of accuracy.

A.4.1 Definitions

Some spherical functions, which are frequently used in Mie-series solutions, are defined as

$$\omega_n(\theta) = \frac{n}{\sin\theta}\left[\cos\theta P_n(\cos\theta) - P_{n-1}(\cos\theta)\right] \tag{A.28}$$

$$\Omega_n(\theta) = \sin\theta\frac{\partial}{\partial\theta}\omega_n(\theta) \tag{A.29}$$

$$= \frac{n}{\sin\theta}\left[(n\cos^2\theta - 1)P_n(\cos\theta)\right.$$
$$\left.-2(n-1)\cos\theta P_{n-1}(\cos\theta) + (n-1)P_{n-2}(\cos\theta)\right] \tag{A.30}$$

$$\gamma_n(kr) = krh_n^{(1)}(kr) \tag{A.31}$$

$$\Gamma_n(kr) = \frac{\partial}{\partial r}\left[rh_n^{(1)}(kr)\right] = \left[krh_{n-1}^{(1)}(kr) - nh_n^{(1)}(kr)\right] \tag{A.32}$$

$$\upsilon_n(kr) = krj_n(kr) \tag{A.33}$$

$$\Upsilon_n(kr) = \frac{\partial}{\partial r}\left[rj_n(kr)\right] = \left[krj_{n-1}(kr) - nj_n(kr)\right], \tag{A.34}$$

where $h_n^{(1)}$ is the spherical Hankel function of the first kind, j_n is the spherical Bessel function of the first kind, and P_n is the Legendre polynomial.

A.4.2 Debye Potentials

Debye potentials at $\boldsymbol{r} = (r, \theta, \phi)$ are defined as

$$\pi_e(\boldsymbol{r}) = \frac{\cos\phi}{\eta}\sum_{n=1}^{\infty} a_n v(kr)\omega_n(\theta) \tag{A.35}$$

$$\pi_m(\boldsymbol{r}) = -\sin\phi\sum_{n=1}^{\infty} b_n v(kr)\omega_n(\theta), \tag{A.36}$$

where v is a spherical Bessel or spherical Hankel function. In (A.35) and (A.36), a_n and b_n are coefficients determined by the boundary conditions. Derivatives of Debye potentials are derived as

$$\frac{\partial}{\partial\theta}\pi_e(\boldsymbol{r}) = \frac{\cos\phi}{\eta\sin\theta}\sum_{n=1}^{\infty} a_n v_n(kr)\Omega_n(\theta) \tag{A.37}$$

$$\frac{\partial}{\partial\phi}\pi_e(\boldsymbol{r}) = -\frac{\sin\phi}{\eta}\sum_{n=1}^{\infty} a_n v_n(kr)\omega_n(\theta) \tag{A.38}$$

$$\frac{\partial}{\partial\theta}\pi_m(\boldsymbol{r}) = -\frac{\sin\phi}{\sin\theta}\sum_{n=1}^{\infty} b_n v_n(kr)\Omega_n(\theta) \tag{A.39}$$

$$\frac{\partial}{\partial\phi}\pi_m(\boldsymbol{r}) = -\cos\phi\sum_{n=1}^{\infty} b_n v_n(kr)\omega_n(\theta). \tag{A.40}$$

A.4.3 Electric and Magnetic Fields

Electric and magnetic fields can be obtained from Debye potentials as

$$E_\theta(\mathbf{r}) = \frac{i}{\omega\epsilon r}\frac{\partial^2}{\partial r\partial\theta} r\pi_e(\mathbf{r}) + \frac{1}{\sin\theta}\frac{\partial}{\partial\phi}\pi_m(\mathbf{r}) \tag{A.41}$$

$$E_\phi(\mathbf{r}) = \frac{i}{\omega\epsilon r\sin\theta}\frac{\partial^2}{\partial r\partial\phi} r\pi_e(\mathbf{r}) - \frac{\partial}{\partial\theta}\pi_m(\mathbf{r}) \tag{A.42}$$

$$H_\theta(\mathbf{r}) = -\frac{i}{\omega\mu r}\frac{\partial^2}{\partial r\partial\theta} r\pi_m(\mathbf{r}) + \frac{1}{\sin\theta}\frac{\partial}{\partial\phi}\pi_e(\mathbf{r}) \tag{A.43}$$

$$H_\phi(\mathbf{r}) = -\frac{i}{\omega\mu r\sin\theta}\frac{\partial^2}{\partial r\partial\phi} r\pi_m(\mathbf{r}) - \frac{\partial}{\partial\theta}\pi_e(\mathbf{r}). \tag{A.44}$$

A.4.4 Incident Fields

Assume that spherical objects are located in a homogeneous space with electrical parameters (ϵ_o, μ_o) and they are illuminated by a plane wave propagating in the z direction with the electric field polarized in the x direction, i.e.,

$$E_\theta^{inc}(\mathbf{r}) = \cos\theta\cos\phi\exp(ik_o r\cos\theta) \tag{A.45}$$

$$E_\phi^{inc}(\mathbf{r}) = -\sin\phi\exp(ik_o r\cos\theta). \tag{A.46}$$

The incident electric and magnetic fields can be expanded in terms of spherical functions at $\mathbf{r} = (r, \theta, \phi)$ as

$$E_\theta^{inc}(\mathbf{r}) = -\frac{\cos\phi}{k_o r\sin\theta}\sum_{n=1}^{\infty} R_n^{(0)}\left[\upsilon_n(k_o r)\omega_n(\theta) - i\Upsilon_n(k_o r)\Omega_n(\theta)\right] \tag{A.47}$$

$$E_\phi^{inc}(\mathbf{r}) = \frac{\sin\phi}{k_o r\sin\theta}\sum_{n=1}^{\infty} R_n^{(0)}\left[\upsilon_n(k_o r)\Omega_n(\theta) - i\Upsilon_n(k_o r)\omega_n(\theta)\right] \tag{A.48}$$

$$H_\theta^{inc}(\mathbf{r}) = \frac{i\sin\phi}{k_o\eta_o r\sin\theta}\sum_{n=1}^{\infty} R_n^{(0)}\left[\Upsilon_n(k_o r)\Omega_n(\theta) + i\upsilon_n(k_o r)\omega_n(\theta)\right] \tag{A.49}$$

$$H_\phi^{inc}(\mathbf{r}) = \frac{i\cos\phi}{k_o\eta_o r\sin\theta}\sum_{n=1}^{\infty} R_n^{(0)}\left[\Upsilon_n(k_o r)\omega_n(\theta) + i\upsilon_n(k_o r)\Omega_n(\theta)\right], \tag{A.50}$$

where

$$R_n^{(0)} = \frac{(-i)^{-n}(2n+1)}{n(n+1)}. \tag{A.51}$$

A.4.5 Perfectly Conducting Sphere

Consider a PEC sphere with a radius of r_o. The scattered electric and magnetic fields outside the sphere at $\boldsymbol{r} = (r, \theta, \phi)$ can be derived as

$$E_\theta^{sca,o}(\boldsymbol{r}) = -\frac{\cos\phi}{k_o r \sin\theta} \sum_{n=1}^{\infty} \left[b_n \gamma_n(k_o r)\omega_n(\theta) - i a_n \Gamma_n(k_o r)\Omega_n(\theta) \right] \tag{A.52}$$

$$E_\phi^{sca,o}(\boldsymbol{r}) = \frac{\sin\phi}{k_o r \sin\theta} \sum_{n=1}^{\infty} \left[b_n \gamma_n(k_o r)\Omega_n(\theta) - i a_n \Gamma_n(k_o r)\omega_n(\theta) \right] \tag{A.53}$$

$$H_\theta^{sca,o}(\boldsymbol{r}) = \frac{i\sin\phi}{k_o \eta_o r \sin\theta} \sum_{n=1}^{\infty} \left[b_n \Gamma_n(k_o r)\Omega_n(\theta) + i a_n \gamma_n(k_o r)\omega_n(\theta) \right] \tag{A.54}$$

$$H_\phi^{sca,o}(\boldsymbol{r}) = \frac{i\cos\phi}{k_o \eta_o r \sin\theta} \sum_{n=1}^{\infty} \left[b_n \Gamma_n(k_o r)\omega_n(\theta) + i a_n \gamma_n(k_o r)\Omega_n(\theta) \right]. \tag{A.55}$$

Applying the boundary conditions on the PEC surface, one can obtain

$$a_n = -R_n^{(0)} \frac{\Upsilon_n(k_o r_o)}{\Gamma_n(k_o r_o)} \tag{A.56}$$

$$b_n = -R_n^{(0)} \frac{\upsilon_n(k_o r_o)}{\gamma_n(k_o r_o)}. \tag{A.57}$$

A.4.6 Dielectric Sphere

Consider a dielectric sphere with a radius of r_o and with electrical parameters (ϵ_i, μ_i). The electric and magnetic fields inside the sphere can be derived as

$$E_\theta^{sca,i}(\boldsymbol{r}) = -\frac{\cos\phi}{k_i r \sin\theta} \sum_{n=1}^{\infty} \left[d_n \upsilon_n(k_i r)\omega_n(\theta) - i c_n \Upsilon_n(k_i r)\Omega_n(\theta) \right] \tag{A.58}$$

$$E_\phi^{sca,i}(\boldsymbol{r}) = \frac{\sin\phi}{k_i r \sin\theta} \sum_{n=1}^{\infty} \left[d_n \upsilon_n(k_i r)\Omega_n(\theta) - i c_n \Upsilon_n(k_i r)\omega_n(\theta) \right] \tag{A.59}$$

$$H_\theta^{sca,i}(\boldsymbol{r}) = \frac{i\sin\phi}{k_i \eta_i r \sin\theta} \sum_{n=1}^{\infty} \left[d_n \Upsilon_n(k_i r)\Omega_n(\theta) + i c_n \upsilon_n(k_i r)\omega_n(\theta) \right] \tag{A.60}$$

$$H_\phi^{sca,i}(\boldsymbol{r}) = \frac{i\cos\phi}{k_i \eta_i r \sin\theta} \sum_{n=1}^{\infty} \left[d_n \Upsilon_n(k_i r)\omega_n(\theta) + i c_n \upsilon_n(k_i r)\Omega_n(\theta) \right]. \tag{A.61}$$

Equations (A.47)–(A.55) can be used to derive the electric and magnetic fields outside the sphere. Applying the boundary conditions and using

$$\upsilon_n(x)\Gamma_n(x) - \gamma_n(x)\Upsilon_n(x) = i, \tag{A.62}$$

one can obtain

$$a_n = R_n^{(0)} \frac{\eta_o^{-1} \upsilon_n(k_o r_o) \Upsilon_n(k_i r_o) - \eta_i^{-1} \upsilon_n(k_i r_o) \Upsilon_n(k_o r_o)}{\eta_i^{-1} \upsilon_n(k_i r_o) \Gamma_n(k_o r_o) - \eta_o^{-1} \gamma_n(k_o r_o) \Upsilon_n(k_i r_o)} \tag{A.63}$$

$$b_n = R_n^{(0)} \frac{\eta_i^{-1} \upsilon_n(k_o r_o) \Upsilon_n(k_i r_o) - \eta_o^{-1} \upsilon_n(k_i r_o) \Upsilon_n(k_o r_o)}{\eta_o^{-1} \upsilon_n(k_i r_o) \Gamma_n(k_o r_o) - \eta_i^{-1} \gamma_n(k_o r_o) \Upsilon_n(k_i r_o)} \tag{A.64}$$

$$c_n = R_n^{(0)} \left(\frac{k_i}{k_o \eta_o} \right) \frac{i}{\eta_i^{-1} \upsilon_n(k_i r_o) \Gamma_n(k_o r_o) - \eta_o^{-1} \gamma_n(k_o r_o) \Upsilon_n(k_i r_o)} \tag{A.65}$$

$$d_n = R_n^{(0)} \left(\frac{k_i}{k_o \eta_o} \right) \frac{i}{\eta_o^{-1} \upsilon_n(k_i r_o) \Gamma_n(k_o r_o) - \eta_i^{-1} \gamma_n(k_o r_o) \Upsilon_n(k_i r_o)}. \tag{A.66}$$

A.4.7 Coated Perfectly Conducting Sphere

Consider a PEC sphere of radius r_s coated with a dielectric shell of radius r_o. The dielectric shell has electrical parameters (ϵ_s, μ_s). The electric and magnetic fields inside the shell can be derived as

$$\begin{aligned} E_\theta^{sca,s}(\boldsymbol{r}) = & -\frac{\cos\phi}{k_s r \sin\theta} \sum_{n=1}^{\infty} \left[f_n^+ \gamma_n(k_s r) \omega_n(\theta) - i e_n^+ \Gamma_n(k_s r) \Omega_n(\theta) \right] \\ & - \frac{\cos\phi}{k_s r \sin\theta} \sum_{n=1}^{\infty} \left[f_n^- \upsilon_n(k_s r) \omega_n(\theta) - i e_n^- \Upsilon_n(k_s r) \Omega_n(\theta) \right] \end{aligned} \tag{A.67}$$

$$\begin{aligned} E_\phi^{sca,s}(\boldsymbol{r}) = & \frac{\sin\phi}{k_s r \sin\theta} \sum_{n=1}^{\infty} \left[f_n^+ \gamma_n(k_s r) \Omega_n(\theta) - i e_n^+ \Gamma_n(k_s r) \omega_n(\theta) \right] \\ & + \frac{\sin\phi}{k_s r \sin\theta} \sum_{n=1}^{\infty} \left[f_n^- \upsilon_n(k_s r) \Omega_n(\theta) - i e_n^- \Upsilon_n(k_s r) \omega_n(\theta) \right] \end{aligned} \tag{A.68}$$

$$\begin{aligned} H_\theta^{sca,s}(\boldsymbol{r}) = & \frac{i \sin\phi}{k_s \eta_s r \sin\theta} \sum_{n=1}^{\infty} \left[f_n^+ \Gamma_n(k_s r) \Omega_n(\theta) + i e_n^+ \gamma_n(k_s r) \omega_n(\theta) \right] \\ & + \frac{i \sin\phi}{k_s \eta_s r \sin\theta} \sum_{n=1}^{\infty} \left[f_n^- \Upsilon_n(k_s r) \Omega_n(\theta) + i e_n^- \upsilon_n(k_s r) \omega_n(\theta) \right] \end{aligned} \tag{A.69}$$

$$\begin{aligned} H_\phi^{sca,s}(\boldsymbol{r}) = & \frac{i \cos\phi}{k_s \eta_s r \sin\theta} \sum_{n=1}^{\infty} \left[f_n^+ \Gamma_n(k_s r) \omega_n(\theta) + i e_n^+ \gamma_n(k_s r) \Omega_n(\theta) \right] \\ & + \frac{i \cos\phi}{k_s \eta_s r \sin\theta} \sum_{n=1}^{\infty} \left[f_n^- \Upsilon_n(k_s r) \omega_n(\theta) + i e_n^- \upsilon_n(k_s r) \Omega_n(\theta) \right]. \end{aligned} \tag{A.70}$$

Equations (A.47)–(A.55) can be used to derive the electric and magnetic fields outside the object. Applying the boundary conditions, one can obtain

$$a_n = R_n^{(0)} \frac{1 + R_n^{(1)} R_n^{(3)}}{R_n^{(5)} + R_n^{(1)} R_n^{(4)}} \tag{A.71}$$

$$b_n = R_n^{(0)} \frac{1 + R_n^{(2)} R_n^{(6)}}{R_n^{(8)} + R_n^{(2)} R_n^{(7)}}, \tag{A.72}$$

where

$$R_n^{(1)} = -\frac{\Upsilon_n(k_s r_s)}{\Gamma_n(k_s r_s)} \tag{A.73}$$

$$R_n^{(2)} = -\frac{\upsilon_n(k_s r_s)}{\gamma_n(k_s r_s)} \tag{A.74}$$

$$R_n^{(3)} = \frac{\eta_s^{-1}\gamma_n(k_s r_o)\Upsilon_n(k_o r_o) - \eta_o^{-1}\upsilon_n(k_o r_o)\Gamma_n(k_s r_o)}{\eta_s^{-1}\upsilon_n(k_s r_o)\Upsilon_n(k_o r_o) - \eta_o^{-1}\upsilon_n(k_o r_o)\Upsilon_n(k_s r_o)} \tag{A.75}$$

$$R_n^{(4)} = \frac{\eta_o^{-1}\gamma_n(k_o r_o)\Gamma_n(k_s r_o) - \eta_s^{-1}\gamma_n(k_s r_o)\Gamma_n(k_o r_o)}{\eta_s^{-1}\upsilon_n(k_s r_o)\Upsilon_n(k_o r_o) - \eta_o^{-1}\upsilon_n(k_o r_o)\Upsilon_n(k_s r_o)} \tag{A.76}$$

$$R_n^{(5)} = \frac{\eta_o^{-1}\gamma_n(k_o r_o)\Upsilon_n(k_s r_o) - \eta_s^{-1}\upsilon_n(k_s r_o)\Gamma_n(k_o r_o)}{\eta_s^{-1}\upsilon_n(k_s r_o)\Upsilon_n(k_o r_o) - \eta_o^{-1}\upsilon_n(k_o r_o)\Upsilon_n(k_s r_o)} \tag{A.77}$$

$$R_n^{(6)} = \frac{\eta_o^{-1}\gamma_n(k_s r_o)\Upsilon_n(k_o r_o) - \eta_s^{-1}\upsilon_n(k_o r_o)\Gamma_n(k_s r_o)}{\eta_o^{-1}\upsilon_n(k_s r_o)\Upsilon_n(k_o r_o) - \eta_s^{-1}\upsilon_n(k_o r_o)\Upsilon_n(k_s r_o)} \tag{A.78}$$

$$R_n^{(7)} = \frac{\eta_s^{-1}\gamma_n(k_o r_o)\Gamma_n(k_s r_o) - \eta_o^{-1}\gamma_n(k_s r_o)\Gamma_n(k_o r_o)}{\eta_o^{-1}\upsilon_n(k_s r_o)\Upsilon_n(k_o r_o) - \eta_s^{-1}\upsilon_n(k_o r_o)\Upsilon_n(k_s r_o)} \tag{A.79}$$

$$R_n^{(8)} = \frac{\eta_s^{-1}\gamma_n(k_o r_o)\Upsilon_n(k_s r_o) - \eta_o^{-1}\upsilon_n(k_s r_o)\Gamma_n(k_o r_o)}{\eta_o^{-1}\upsilon_n(k_s r_o)\Upsilon_n(k_o r_o) - \eta_s^{-1}\upsilon_n(k_o r_o)\Upsilon_n(k_s r_o)}. \tag{A.80}$$

Let $r_s \neq r_o$, $\epsilon_s = \epsilon_o$, and $\mu_s = \mu_o$. Then,

$$a_n \to R_n^{(0)} \frac{R_n^{(1)} R_n^{(3)}}{R_n^{(5)}} = -R_n^{(0)} \frac{\Upsilon_n(k_o r_o)}{\Gamma_n(k_o r_o)} \tag{A.81}$$

$$b_n \to R_n^{(0)} \frac{R_n^{(2)} R_n^{(6)}}{R_n^{(8)}} = -R_n^{(0)} \frac{\upsilon_n(k_o r_o)}{\gamma_n(k_o r_o)}, \tag{A.82}$$

which are in agreement with the expressions in (A.56) and (A.57) for a PEC sphere in a homogeneous space.

A.4.8 Coated Dielectric Sphere

Consider a dielectric sphere of radius r_s coated with a dielectric shell of radius r_o. The core and shell have electrical parameters (ϵ_i, μ_i) and (ϵ_s, μ_s), respectively. Once again, (A.47)–(A.55) can be used to derive the electric and magnetic fields outside the sphere. Similarly, Equations (A.58)–(A.61) and (A.67)–(A.70) can be used to derive the electromagnetic fields inside the core and shell, respectively. Using the boundary conditions, one can obtain

$$a_n = R_n^{(0)} \frac{1 + R_n^{(1)} R_n^{(3)}}{R_n^{(5)} + R_n^{(1)} R_n^{(4)}} \tag{A.83}$$

$$b_n = R_n^{(0)} \frac{1 + R_n^{(2)} R_n^{(6)}}{R_n^{(8)} + R_n^{(2)} R_n^{(7)}}, \tag{A.84}$$

where

$$R_n^{(1)} = \frac{\eta_s^{-1} \gamma_n(k_s r_s) \Upsilon_n(k_i r_s) - \eta_i^{-1} \upsilon_n(k_i r_s) \Gamma_n(k_s r_s)}{\eta_i^{-1} \upsilon_n(k_i r_s) \Upsilon_n(k_s r_s) - \eta_s^{-1} \upsilon_n(k_s r_s) \Upsilon_n(k_i r_s)} \tag{A.85}$$

$$R_n^{(2)} = \frac{\eta_i^{-1} \gamma_n(k_s r_s) \Upsilon_n(k_i r_s) - \eta_s^{-1} \upsilon_n(k_i r_s) \Gamma_n(k_s r_s)}{\eta_s^{-1} \upsilon_n(k_i r_s) \Upsilon_n(k_s r_s) - \eta_i^{-1} \upsilon_n(k_s r_s) \Upsilon_n(k_i r_s)} \tag{A.86}$$

$$R_n^{(3)} = \frac{\eta_s^{-1} \upsilon_n(k_s r_o) \Upsilon_n(k_o r_o) - \eta_o^{-1} \upsilon_n(k_o r_o) \Upsilon_n(k_s r_o)}{\eta_s^{-1} \gamma_n(k_s r_o) \Upsilon_n(k_o r_o) - \eta_o^{-1} \upsilon_n(k_o r_o) \Gamma_n(k_s r_o)} \tag{A.87}$$

$$R_n^{(4)} = \frac{\eta_o^{-1} \gamma_n(k_o r_o) \Upsilon_n(k_s r_o) - \eta_s^{-1} \upsilon_n(k_s r_o) \Gamma_n(k_o r_o)}{\eta_s^{-1} \gamma_n(k_s r_o) \Upsilon_n(k_o r_o) - \eta_o^{-1} \upsilon_n(k_o r_o) \Gamma_n(k_s r_o)} \tag{A.88}$$

$$R_n^{(5)} = \frac{\eta_o^{-1} \gamma_n(k_o r_o) \Gamma_n(k_s r_o) - \eta_s^{-1} \gamma_n(k_s r_o) \Gamma_n(k_o r_o)}{\eta_s^{-1} \gamma_n(k_s r_o) \Upsilon_n(k_o r_o) - \eta_o^{-1} \upsilon_n(k_o r_o) \Gamma_n(k_s r_o)} \tag{A.89}$$

$$R_n^{(6)} = \frac{\eta_o^{-1} \upsilon_n(k_s r_o) \Upsilon_n(k_o r_o) - \eta_s^{-1} \upsilon_n(k_o r_o) \Upsilon_n(k_s r_o)}{\eta_o^{-1} \gamma_n(k_s r_o) \Upsilon_n(k_o r_o) - \eta_s^{-1} \upsilon_n(k_o r_o) \Gamma_n(k_s r_o)} \tag{A.90}$$

$$R_n^{(7)} = \frac{\eta_s^{-1} \gamma_n(k_o r_o) \Upsilon_n(k_s r_o) - \eta_o^{-1} \upsilon_n(k_s r_o) \Gamma_n(k_o r_o)}{\eta_o^{-1} \gamma_n(k_s r_o) \Upsilon_n(k_o r_o) - \eta_s^{-1} \upsilon_n(k_o r_o) \Gamma_n(k_s r_o)} \tag{A.91}$$

$$R_n^{(8)} = \frac{\eta_s^{-1} \gamma_n(k_o r_o) \Gamma_n(k_s r_o) - \eta_o^{-1} \gamma_n(k_s r_o) \Gamma_n(k_o r_o)}{\eta_o^{-1} \gamma_n(k_s r_o) \Upsilon_n(k_o r_o) - \eta_s^{-1} \upsilon_n(k_o r_o) \Gamma_n(k_s r_o)}. \tag{A.92}$$

Let $r_s \neq r_o$, $\epsilon_s = \epsilon_o \neq \epsilon_i$, and $\mu_s = \mu_o \neq \mu_i$. Then,

$$a_n \to \frac{R_n^{(0)}}{R_n^{(1)} R_n^{(4)}} = R_n^{(0)} \frac{\eta_i^{-1} \upsilon_n(k_i r_s) \Upsilon_n(k_o r_s) - \eta_o^{-1} \upsilon_n(k_o r_s) \Upsilon_n(k_i r_s)}{\eta_o^{-1} \gamma_n(k_o r_s) \Upsilon_n(k_i r_s) - \eta_i^{-1} \upsilon_n(k_i r_s) \Gamma_n(k_o r_s)} \tag{A.93}$$

$$b_n \to \frac{R_n^{(0)}}{R_n^{(2)} R_n^{(7)}} = R_n^{(0)} \frac{\eta_o^{-1} \upsilon_n(k_i r_s) \Upsilon_n(k_o r_s) - \eta_i^{-1} \upsilon_n(k_o r_s) \Upsilon_n(k_i r_s)}{\eta_i^{-1} \gamma_n(k_o r_s) \Upsilon_n(k_i r_s) - \eta_o^{-1} \upsilon_n(k_i r_s) \Gamma_n(k_o r_s)}, \tag{A.94}$$

which are in agreement with the expressions in (A.63) and (A.64) for a dielectric sphere in a homogeneous domain.

A.4.9 Far-Field Expressions

Using the large-argument form of the spherical Hankel function, i.e.,

$$\lim_{x\to\infty} h_n^{(1)}(x) \approx (-i)^{n+1}\frac{\exp(ix)}{x} \tag{A.95}$$

$$\lim_{x\to\infty} \Gamma_n(x) \approx \left[(-i)^n \exp(ix) - n(-i)^{n+1}\frac{\exp(ix)}{x}\right] \approx (-i)^n \exp(ix), \tag{A.96}$$

one can derive

$$\begin{aligned} E_\theta^\infty(\theta,\phi) &= \lim_{r\to\infty}\left\{\left[r\exp(-ik_o r)\right] E_\theta^{sca,o}\right\} \\ &= \frac{i\cos\phi}{k_o\sin\theta}\sum_{n=1}^{\infty}(-i)^n\left[a_n\Omega_n(\theta)+b_n\omega_n(\theta)\right] \end{aligned} \tag{A.97}$$

$$\begin{aligned} E_\phi^\infty(\theta,\phi) &= \lim_{r\to\infty}\left\{\left[r\exp(-ik_o r)\right] E_\phi^{sca,o}\right\} \\ &= -\frac{i\sin\phi}{k_o\sin\theta}\sum_{n=1}^{\infty}(-i)^n\left[a_n\omega_n(\theta)+b_n\Omega_n(\theta)\right] \end{aligned} \tag{A.98}$$

$$\mathrm{RCS}(\theta,\phi) = 4\pi\left[\left|E_\theta^\infty(\theta,\phi)\right|^2 + \left|E_\phi^\infty(\theta,\phi)\right|^2\right]. \tag{A.99}$$

When $\theta = 0°$ (backscattering direction),

$$\frac{\omega_n(\theta)}{\sin\theta} \to -\frac{n(n+1)}{2} \tag{A.100}$$

$$\frac{\Omega_n(\theta)}{\sin\theta} \to -\frac{n(n+1)}{2} \tag{A.101}$$

$$E_\theta^\infty(\theta,\phi) \to -\frac{i\cos\phi}{k_o}\sum_{n=1}^{\infty}\frac{n(n+1)}{2}(-i)^n\left(a_n+b_n\right) \tag{A.102}$$

$$E_\phi^\infty(\theta,\phi) \to \frac{i\sin\phi}{k_o}\sum_{n=1}^{\infty}\frac{n(n+1)}{2}(-i)^n\left(a_n+b_n\right), \tag{A.103}$$

and

$$\mathrm{RCS}(\theta=0°) = \frac{\pi}{k_o^2}\left|\sum_{n=1}^{\infty} n(n+1)(-i)^n\left(a_n+b_n\right)\right|^2. \tag{A.104}$$

Similarly, when $\theta = 180°$ (forward-scattering direction),

$$\frac{\omega_n(\theta)}{\sin\theta} \to \frac{n(n+1)}{2}(-1)^n \tag{A.105}$$

$$\frac{\Omega_n(\theta)}{\sin\theta} \to -\frac{n(n+1)}{2}(-1)^n \tag{A.106}$$

$$E_\theta^\infty(\theta,\phi) \to -\frac{i\cos\phi}{k_o}\sum_{n=1}^{\infty}\frac{n(n+1)}{2}(i)^n\left(a_n-b_n\right) \tag{A.107}$$

$$E_\phi^\infty(\theta,\phi) \to -\frac{i\sin\phi}{k_o}\sum_{n=1}^{\infty}\frac{n(n+1)}{2}(i)^n\left(a_n-b_n\right), \tag{A.108}$$

and

$$\mathrm{RCS}(\theta=180^\circ)=\frac{\pi}{k_o^2}\left|\sum_{n=1}^{\infty}n(n+1)(i)^n\left(a_n-b_n\right)\right|^2. \tag{A.109}$$

A.5 Electric-Field Volume Integral Equation

Volume integral equations can be used to formulate electromagnetics problems involving dielectric objects. Major advantages and disadvantages of volume formulations compared to surface formulations can be listed as follows:

- For thin dielectric objects, volume integral equations produce better-conditioned matrix equations than surface integral equations.
- Volume integral equations do not breakdown for low-contrast objects, as opposed to conventional surface integral equations.
- Heterogeneous dielectric objects can easily be modeled with volume integral equations, which allow different electrical parameters for different discretization elements.
- Volume integral equations usually require more numbers of unknowns than surface integral equations for a given problem.

This section presents a volume integral equation, namely, EFVIE, for nonmagnetic dielectric objects.

Consider a dielectric object with a relative permittivity of ϵ_r located in free space. An equivalent electric current can be defined as

$$\boldsymbol{J}(\boldsymbol{r}) = i\omega\kappa(\boldsymbol{r})\boldsymbol{D}(\boldsymbol{r}) = i\omega\kappa(\boldsymbol{r})\epsilon_0\epsilon_r(\boldsymbol{r})\boldsymbol{E}(\boldsymbol{r}), \tag{A.110}$$

where ϵ_0 is the permittivity of free space and

$$\kappa(\boldsymbol{r}) = \frac{1-\epsilon_r(\boldsymbol{r})}{\epsilon_r(\boldsymbol{r})} \tag{A.111}$$

is the contrast at the observation point $\boldsymbol{r}$. When $\boldsymbol{r}$ is located outside the object, $\kappa(\boldsymbol{r})=0$, hence $\boldsymbol{J}(\boldsymbol{r})=0$. The total electric field in (A.110) can be written as

$$\boldsymbol{E}(\boldsymbol{r}) = \boldsymbol{E}^{inc}(\boldsymbol{r}) + ik_0\eta_0\int_D d\boldsymbol{r}'\boldsymbol{J}(\boldsymbol{r}')g_0(\boldsymbol{r},\boldsymbol{r}') - \frac{i\eta_0}{k_0}\nabla\int_D d\boldsymbol{r}'\boldsymbol{J}(\boldsymbol{r}')\cdot\nabla' g_0(\boldsymbol{r},\boldsymbol{r}'), \tag{A.112}$$

where $\boldsymbol{E}^{inc}$ is the incident electric field created by external sources, $k_0 = 2\pi/\lambda_0 = \omega\sqrt{\mu_0\epsilon_0}$ is the wave number, and $\eta_0 = \sqrt{\frac{\mu_0}{\epsilon_0}}$ is the wave impedance in free space. Using (A.112) in (A.110), one can obtain [36]

$$\frac{\boldsymbol{D}(\boldsymbol{r})}{\epsilon_0\epsilon_r(\boldsymbol{r})} - ik_0\eta_0\int_D d\boldsymbol{r}' i\omega\kappa(\boldsymbol{r}')\boldsymbol{D}(\boldsymbol{r}')g_0(\boldsymbol{r},\boldsymbol{r}')$$
$$+\frac{i\eta_0}{k_0}\nabla\int_D d\boldsymbol{r}' i\omega\kappa(\boldsymbol{r}')\boldsymbol{D}(\boldsymbol{r}')\cdot\nabla' g_0(\boldsymbol{r},\boldsymbol{r}') = \boldsymbol{E}^{inc}(\boldsymbol{r}), \quad \text{(A.113)}$$

which is called EFVIE. Note that, when the contrast of the object is low, i.e., $\kappa(\boldsymbol{r}) \approx 0$, the equivalent electric current can be approximated as

$$\boldsymbol{J}(\boldsymbol{r}) = i\omega\kappa(\boldsymbol{r})\epsilon_0\epsilon_r(\boldsymbol{r})\boldsymbol{E}^{inc}(\boldsymbol{r}) = i\omega[1-\epsilon_r(\boldsymbol{r})]\epsilon_0\boldsymbol{E}^{inc}(\boldsymbol{r}), \quad \text{(A.114)}$$

which is known as the Born approximation.

For numerical solutions of EFVIE, the quantity $i\omega\boldsymbol{D}$ can be expanded in a series of basis functions, i.e.,

$$i\omega\boldsymbol{D}(\boldsymbol{r}) = \sum_{n=1}^{N} a_n\boldsymbol{b}_n(\boldsymbol{r}), \quad \text{(A.115)}$$

and the equation can be tested by using a set of testing functions $\boldsymbol{t}_m$ for $m = 1, 2, \ldots, N$. This leads to $N \times N$ dense matrix equations in the form of

$$\overline{\boldsymbol{Z}}^{\text{EFVIE}} \cdot \boldsymbol{a} = \boldsymbol{w}^{\text{EFVIE}}, \quad \text{(A.116)}$$

where the matrix elements and the elements of the RHS vector are derived as

$$\overline{\boldsymbol{Z}}^{\text{EFVIE}}[m,n] = \frac{1}{i\omega\epsilon_0}\int_{D_m} d\boldsymbol{r}\frac{1}{\epsilon_r(\boldsymbol{r})}\boldsymbol{t}_m(\boldsymbol{r})\cdot\boldsymbol{b}_n(\boldsymbol{r})$$
$$- i\omega\mu_0\int_{D_m} d\boldsymbol{r}\boldsymbol{t}_m(\boldsymbol{r})\cdot\int_{D_n} d\boldsymbol{r}'\kappa(\boldsymbol{r}')\boldsymbol{b}_n(\boldsymbol{r}')g_0(\boldsymbol{r},\boldsymbol{r}')$$
$$- \frac{1}{i\omega\epsilon_0}\int_{D_m} d\boldsymbol{r}\boldsymbol{t}_m(\boldsymbol{r})\cdot\nabla\int_{D_n} d\boldsymbol{r}'\kappa(\boldsymbol{r}')\boldsymbol{b}_n(\boldsymbol{r}')\cdot\nabla' g_0(\boldsymbol{r},\boldsymbol{r}') \quad \text{(A.117)}$$

and

$$\boldsymbol{w}^{\text{EFVIE}}[m] = \int_{D_m} d\boldsymbol{r}\boldsymbol{t}_m(\boldsymbol{r})\cdot\boldsymbol{E}^{inc}(\boldsymbol{r}), \quad \text{(A.118)}$$

respectively.

EFVIE can be discretized with the Schaubert-Wilton-Glisson (SWG) functions [36]. Each SWG function is defined in a pair of tetrahedra sharing a common surface. The nth SWG function with a spatial support of $D_n = D_{n1} + D_{n2}$ can be written as

$$\boldsymbol{f}_n^{\text{SWG}}(\boldsymbol{r}) = \begin{cases} \dfrac{A_n}{3V_{n1}}(\boldsymbol{r}-\boldsymbol{r}_{n1}), & \boldsymbol{r}\in D_{n1} \\ \dfrac{A_n}{3V_{n2}}(\boldsymbol{r}_{n2}-\boldsymbol{r}), & \boldsymbol{r}\in D_{n2} \\ 0, & \boldsymbol{r}\notin D_n, \end{cases} \quad \text{(A.119)}$$

where A_n represents the area of the common surface, whereas V_{n1} and V_{n2} are volumes of the first (D_{n1}) and second (D_{n2}) tetrahedron, respectively. In (A.119), $\boldsymbol{r}_{n1}$ and $\boldsymbol{r}_{n2}$ are the coordinates of the two vertices opposite of the common surface. The charge density represented by the nth SWG function can be derived as

$$\nabla \cdot \boldsymbol{f}_n^{\mathrm{SWG}}(\boldsymbol{r}) = \begin{cases} \dfrac{A_n}{V_{n1}}, & \boldsymbol{r} \in D_{n1} \\ -\dfrac{A_n}{V_{n2}}, & \boldsymbol{r} \in D_{n2} \\ 0, & \boldsymbol{r} \notin D_n. \end{cases} \tag{A.120}$$

Consider the interaction of two half SWG functions $\boldsymbol{t}_{ma}^{\mathrm{SWG}}$ and $\boldsymbol{b}_{nb}^{\mathrm{SWG}}$ associated with the ath tetrahedron of the mth surface and bth tetrahedron of the nth surface, respectively. One can obtain

$$\begin{aligned} \overline{\boldsymbol{Z}}^{\mathrm{EFVIE}}[m,n,a,b] &= \frac{C_{ma,nb}}{i\omega\epsilon_0}\int_{D_{ma}} d\boldsymbol{r}\frac{1}{\epsilon_r(\boldsymbol{r})}(\boldsymbol{r}-\boldsymbol{r}_{ma})\cdot(\boldsymbol{r}-\boldsymbol{r}_{nb}) \\ &- i\omega\mu_0 C_{ma,nb}\int_{D_{ma}} d\boldsymbol{r}(\boldsymbol{r}-\boldsymbol{r}_{ma})\cdot\int_{D_{nb}} d\boldsymbol{r}'\kappa(\boldsymbol{r}')(\boldsymbol{r}'-\boldsymbol{r}_{nb})g_0(\boldsymbol{r},\boldsymbol{r}') \\ &- \frac{C_{ma,nb}}{i\omega\epsilon_0}\int_{D_{ma}} d\boldsymbol{r}(\boldsymbol{r}-\boldsymbol{r}_{ma})\cdot\nabla\int_{D_{nb}} d\boldsymbol{r}'\kappa(\boldsymbol{r}')(\boldsymbol{r}'-\boldsymbol{r}_{nb})\cdot\nabla' g_0(\boldsymbol{r},\boldsymbol{r}'), \end{aligned} \tag{A.121}$$

where

$$C_{ma,nb} = \frac{A_m A_n}{9V_{ma}V_{nb}}\gamma_{ma}\gamma_{nb} \tag{A.122}$$

and $\gamma_{nb}, \gamma_{ma} = \pm 1$, depending on the direction of the basis and testing functions in tetrahedra.

At this stage, assume that the permittivity is constant inside each tetrahedron. Then,

$$\begin{aligned} \overline{\boldsymbol{Z}}^{\mathrm{EFVIE}}[m,n,a,b] &= \frac{C_{ma,nb}}{i\omega\epsilon_0\epsilon_{r,nb}}\int_{D_{ma}} d\boldsymbol{r}(\boldsymbol{r}-\boldsymbol{r}_{ma})\cdot(\boldsymbol{r}-\boldsymbol{r}_{nb}) \\ &- i\omega\mu_0\kappa_{nb}C_{ma,nb}\int_{D_{ma}} d\boldsymbol{r}(\boldsymbol{r}-\boldsymbol{r}_{ma})\cdot\int_{D_{nb}} d\boldsymbol{r}'(\boldsymbol{r}'-\boldsymbol{r}_{nb})g_0(\boldsymbol{r},\boldsymbol{r}') \\ &- \frac{C_{ma,nb}\kappa_{nb}}{i\omega\epsilon_0}\int_{D_{ma}} d\boldsymbol{r}(\boldsymbol{r}-\boldsymbol{r}_{ma})\cdot\nabla\int_{D_{nb}} d\boldsymbol{r}'(\boldsymbol{r}'-\boldsymbol{r}_{nb})\cdot\nabla' g_0(\boldsymbol{r},\boldsymbol{r}'). \end{aligned} \tag{A.123}$$

For the third term in (A.123), the divergence theorem and the properties of the SWG functions can be used to arrive at

$$\int_{D_{nb}} d\boldsymbol{r}'(\boldsymbol{r}'-\boldsymbol{r}_{nb})\cdot\nabla' g_0(\boldsymbol{r},\boldsymbol{r}') = \frac{3V_{nb}}{A_n}\int_{S_n} d\boldsymbol{r}' g_0(\boldsymbol{r},\boldsymbol{r}') - 3\int_{D_{nb}} d\boldsymbol{r}' g_0(\boldsymbol{r},\boldsymbol{r}'). \tag{A.124}$$

Inserting (A.124) in (A.123), and using the divergence theorem again,

$$\overline{\boldsymbol{Z}}^{\mathrm{EFVIE}}[m,n,a,b] = \frac{C_{ma,nb}}{i\omega\epsilon_0\epsilon_{r,nb}}\int_{D_{ma}} d\boldsymbol{r}(\boldsymbol{r}-\boldsymbol{r}_{ma})\cdot(\boldsymbol{r}-\boldsymbol{r}_{nb})$$

(continued)

$$- i\omega\mu_0\kappa_{nb}C_{ma,nb}\int_{D_{ma}} d\boldsymbol{r}(\boldsymbol{r}-\boldsymbol{r}_{ma})\cdot\int_{D_{nb}} d\boldsymbol{r}'(\boldsymbol{r}'-\boldsymbol{r}_{nb})g_0(\boldsymbol{r},\boldsymbol{r}')$$

$$-\frac{9C_{ma,nb}\kappa_{nb}}{i\omega\epsilon_0}\int_{D_{ma}} d\boldsymbol{r}\int_{D_{nb}} d\boldsymbol{r}'g_0(\boldsymbol{r},\boldsymbol{r}')$$

$$+\frac{A_n\kappa_{nb}\gamma_{nb}\gamma_{ma}}{i\omega\epsilon_0 V_{nb}}\int_{S_m} d\boldsymbol{r}\int_{D_{nb}} d\boldsymbol{r}'g_0(\boldsymbol{r},\boldsymbol{r}')$$

$$+\frac{A_m\kappa_{nb}\gamma_{nb}\gamma_{ma}}{i\omega\epsilon_0 V_{ma}}\int_{D_{ma}} d\boldsymbol{r}\int_{S_n} d\boldsymbol{r}'g_0(\boldsymbol{r},\boldsymbol{r}')$$

$$-\frac{\kappa_{nb}\gamma_{nb}\gamma_{ma}}{i\omega\epsilon_0}\int_{S_m} d\boldsymbol{r}\int_{S_n} d\boldsymbol{r}'g_0(\boldsymbol{r},\boldsymbol{r}'). \tag{A.125}$$

Finally, (A.125) can be written in terms of basic integrals as

$$\overline{\boldsymbol{Z}}^{\text{EFVIE}}[m,n,a,b] = \frac{C_{ma,nb}}{i\omega\epsilon_0\epsilon_{r,nb}}\left\{(x_{ma}x_{nb}+y_{ma}y_{nb}+z_{ma}z_{nb})V_{ma}+I_{17}+I_{18}+I_{19}\right.$$
$$\left.-(x_{ma}+x_{nb})I_{14}-(y_{ma}+y_{nb})I_{15}-(z_{ma}+z_{nb})I_{16}\right\}$$
$$- i\omega\mu_0\kappa_{nb}C_{ma,nb}\left\{(x_{ma}x_{nb}+y_{ma}y_{nb}+z_{ma}z_{nb})I_1+I_8+I_9+I_{10}\right.$$
$$\left.-x_{nb}I_2-y_{nb}I_3-z_{nb}I_4-x_{ma}I_5-y_{ma}I_6-z_{ma}I_7\right\}$$
$$-\frac{9C_{ma,nb}\kappa_{nb}}{i\omega\epsilon_0}\left\{I_1-\frac{V_{ma}}{A_m}I_{11}-\frac{V_{nb}}{A_n}I_{12}+\frac{V_{ma}V_{nb}}{A_mA_n}I_{13}\right\}, \tag{A.126}$$

where

$$I_{\{1,2,3,4\}} = \int_{V_{ma}} d\boldsymbol{r}\,\{1,x,y,z\}\int_{V_{nb}} d\boldsymbol{r}'g_0(\boldsymbol{r},\boldsymbol{r}') \tag{A.127}$$

$$I_{\{5,6,7\}} = \int_{V_{ma}} d\boldsymbol{r}\int_{V_{nb}} d\boldsymbol{r}'\,\{x',y',z'\}\,g_0(\boldsymbol{r},\boldsymbol{r}') \tag{A.128}$$

$$I_8 = \int_{V_{ma}} d\boldsymbol{r}x\int_{V_{nb}} d\boldsymbol{r}'x'g_0(\boldsymbol{r},\boldsymbol{r}') \tag{A.129}$$

$$I_9 = \int_{V_{ma}} d\boldsymbol{r}y\int_{V_{nb}} d\boldsymbol{r}'y'g_0(\boldsymbol{r},\boldsymbol{r}') \tag{A.130}$$

$$I_{10} = \int_{V_{ma}} d\boldsymbol{r}z\int_{V_{nb}} d\boldsymbol{r}'z'g_0(\boldsymbol{r},\boldsymbol{r}') \tag{A.131}$$

$$I_{11} = \int_{S_m} d\boldsymbol{r}\int_{D_{nb}} d\boldsymbol{r}'g_0(\boldsymbol{r},\boldsymbol{r}') \tag{A.132}$$

$$I_{12} = \int_{D_{ma}} d\boldsymbol{r}\int_{S_n} d\boldsymbol{r}'g_0(\boldsymbol{r},\boldsymbol{r}') \tag{A.133}$$

$$I_{13} = \int_{S_m} d\boldsymbol{r}\int_{S_n} d\boldsymbol{r}'g_0(\boldsymbol{r},\boldsymbol{r}') \tag{A.134}$$

$$I_{\{14,15,16\}} = \int_{V_{ma}} d\boldsymbol{r} \{x, y, z\} \tag{A.135}$$

$$I_{\{17,18,19\}} = \int_{V_{ma}} d\boldsymbol{r} \left\{x^2, y^2, z^2\right\}. \tag{A.136}$$

The inner integrals in (A.127)–(A.136) can be calculated accurately by using singularity extraction methods [143]. Note that the surface integrals in I_{11}, I_{12}, and I_{13} can be omitted when electrical parameters (in this case the permittivity) do not change across S_m or S_n. This is because contributions from two tetrahedra on the two sides of the surface cancel each other.

A.6 Calculation of Some Special Functions

In this section, fast and efficient computations of some special functions that may be required for various implementations of MLFMA are presented.

A.6.1 *Spherical Bessel Functions*

Spherical Bessel and Hankel functions of order t are calculated as

$$j_t(x) = \sqrt{\frac{\pi}{2x}} J_{t+1/2}(x) \tag{A.137}$$

$$h_t^{(1,2)}(x) = \sqrt{\frac{\pi}{2x}} H_{t+1/2}^{(1,2)}(x), \tag{A.138}$$

where $J_{t+1/2}$ and $H_{t+1/2}^{(1,2)}$ are the ordinary Bessel and Hankel functions that are available in many numerical libraries. Note that a recursive calculation of a Bessel or Hankel function of order t requires the calculation of the functions with lower orders $t' < t$. Hence, for a given value of t, numerical libraries usually provide all orders up to t simultaneously.

A.6.2 *Legendre Functions*

Normalized Legendre functions are defined as

$$R_t^u(\cos\theta) = \sqrt{\frac{(t-u)!}{(t+u)!}} P_t^u(\cos\theta) \tag{A.139}$$

for $|u| \leq t$. Starting with

$$R_0^0(\cos\theta) = 1 \tag{A.140}$$

$$R_1^0(\cos\theta) = \cos\theta \tag{A.141}$$

$$R_1^1(\cos\theta) = -\frac{1}{\sqrt{2}}\left(1-\cos^2\theta\right)^{1/2} = -\frac{1}{\sqrt{2}}\sin\theta \tag{A.142}$$

$$R_1^{-1}(\cos\theta) = \sqrt{2}P_1^{-1}(\cos\theta) = -\frac{1}{\sqrt{2}}P_1^1(\cos\theta) = -R_1^1(\cos\theta), \tag{A.143}$$

a recursion can be constructed as follows:

- $|u| \neq t$

$$\begin{aligned}
R_t^u(\cos\theta) &= \sqrt{\frac{(t-u)!}{(t+u)!}}P_t^u(\cos\theta) \\
&= \cos\theta\frac{(2t-1)}{(t-u)}\sqrt{\frac{(t-u)}{(t+u)}}\sqrt{\frac{(t-1-u)!}{(t-1+u)!}}P_{t-1}^u(\cos\theta) \\
&\quad - \frac{(t+u-1)}{(t-u)}\sqrt{\frac{(t-u)}{(t+u)}}\sqrt{\frac{(t-1-u)}{(t-1+u)}}\sqrt{\frac{(t-2-u)!}{(t-2+u)!}}P_{t-2}^u(\cos\theta) \\
&= \cos\theta\frac{(2t-1)}{\sqrt{(t-u)(t+u)}}R_{t-1}^u(\cos\theta) \\
&\quad - \frac{\sqrt{(t-1+u)(t-1-u)}}{\sqrt{(t-u)(t+u)}}R_{t-2}^u(\cos\theta).
\end{aligned} \tag{A.144}$$

- $u = t$

$$\begin{aligned}
R_t^u(\cos\theta) = R_t^t(\cos\theta) &= \frac{1}{\sqrt{(2t)!}}P_t^t(\cos\theta) \\
&= \frac{1}{\sqrt{(2t)!}}(-1)^t(2t-1)!!(1-\cos^2\theta)^{t/2} \\
&= \frac{1}{\sqrt{(2t)!}}(-1)^t\frac{(2t)!}{2^t t!}\left(1-\cos^2\theta\right)^{t/2} \\
&= -\sqrt{\frac{(2t-1)}{2t}}\left(1-\cos^2\theta\right)^{1/2}R_{t-1}^{t-1}(\cos\theta).
\end{aligned} \tag{A.145}$$

- $u = -t$

$$\begin{aligned}
R_t^u(\cos\theta) = R_t^{-t}(\cos\theta) &= \sqrt{(2t)!}P_t^{-t}(\cos\theta) = \frac{\sqrt{(2t)!}}{(2t)!}(-1)^t P_t^t(\cos\theta) \\
&= \frac{\sqrt{(2t)!}}{(2t)!}\frac{(2t)!}{2^t t!}\left(1-\cos^2\theta)\right)^{t/2} = \frac{\sqrt{(2t)!}}{2^t t!}\left(1-\cos^2\theta\right)^{t/2} \\
&= \sqrt{\frac{(2t-1)}{2t}}\left(1-\cos^2\theta)\right)^{1/2}R_{t-1}^{-t+1}(\cos\theta).
\end{aligned} \tag{A.146}$$

In the recursive formulas (A.144)–(A.146), $R_t^u(\cos\theta) = 0$, when $u > t$ or $u < -t$.

A.6.3 Gradient of Multipole-to-Monopole Shift Functions

LF-MLFMA implementations using MFIE require the gradient of multipole-to-monopole shift functions, which can be calculated as

$$\begin{aligned}
\nabla\beta_{0,0,t,u}(k,\boldsymbol{r}) &= \sqrt{4\pi}\nabla\left(j_t(kr)Y_{t,u}(\theta,\phi)\right)\\
&= \hat{\boldsymbol{r}}\sqrt{4\pi}\frac{\partial j_t(kr)}{\partial r}Y_{t,u}(\theta,\phi) + \hat{\boldsymbol{\theta}}\sqrt{4\pi}\frac{j_t(kr)}{r}\frac{\partial Y_{t,u}(\theta,\phi)}{\partial\theta}\\
&+ \hat{\boldsymbol{\phi}}\sqrt{4\pi}\frac{j_t(kr)}{r\sin\theta}\frac{\partial Y_{t,u}(\theta,\phi)}{\partial\phi}\\
&= \hat{\boldsymbol{r}}\sqrt{4\pi}(-1)^t\left\{\frac{t}{r}j_t(r) - kj_{t+1}(kr)\right\}Y_{t,u}(\theta,\phi)\\
&+ \hat{\boldsymbol{\theta}}\sqrt{4\pi}(-1)^t\frac{j_t(kr)}{r}\sqrt{\frac{(2t+1)}{4\pi}\frac{(t-u)!}{(t+u)!}}\exp(iu\phi)\frac{\partial P_t^u(\cos\theta)}{\partial\theta}\\
&+ \hat{\boldsymbol{\phi}}\sqrt{4\pi}(-1)^t\frac{j_t(kr)}{r\sin\theta}iuY_{t,u}(\theta,\phi)\\
&= \hat{\boldsymbol{r}}\sqrt{4\pi}(-1)^t\left\{\frac{t}{r}j_t(kr) - kj_{t+1}(kr)\right\}Y_{t,u}(\theta,\phi)\\
&+ \hat{\boldsymbol{\theta}}\sqrt{4\pi}(-1)^t\frac{j_t(kr)}{r\sin\theta}\sqrt{\frac{(2t+1)(t+1-u)(t+1+u)}{(2t+3)}}Y_{t+1,u}(\cos\theta)\\
&- \hat{\boldsymbol{\theta}}\sqrt{4\pi}(-1)^t\frac{j_t(kr)\cos\theta}{r\sin\theta}(t+1)Y_t^u(\cos\theta)\\
&+ \hat{\boldsymbol{\phi}}\sqrt{4\pi}(-1)^t\frac{j_t(kr)}{r\sin\theta}iuY_{t,u}(\theta,\phi).
\end{aligned} \tag{A.147}$$

A.6.4 Gaunt Coefficients

Gaunt coefficients $G_{t,u,t',u',t'',u''}$, which are required in LF-MLFMA implementations, are calculated for $|u| \leq t$, $|u'| \leq t'$, $|u''| \leq t''$, $|t-t'| \leq t'' \leq |t+t'|$, $u'' = u - u'$, and when $t + t' + t''$ is even. They can be derived as

$$\begin{aligned}
G_{t,u,t',u',t'',u''} &= (-1)^u\sqrt{\frac{(2t+1)(2t'+1)(2t''+1)}{4\pi}}\\
&\begin{pmatrix} t & t' & t'' \\ 0 & 0 & 0 \end{pmatrix}\begin{pmatrix} t & t' & t'' \\ -u & u' & u'' \end{pmatrix}
\end{aligned} \tag{A.148}$$

in terms of Wigner-3j symbols. In order to calculate Wigner-3j symbols, one can start with

$$\begin{pmatrix} t & t' & t+t' \\ u & u' & u'' \end{pmatrix} = \frac{(-1)^{t-t'-u''}\left[(2t)!(2t')!(t+t'+u'')!(t+t'-u'')!\right]^{1/2}}{\left[(2t+2t'+1)!(t+u)!(t-u)!(t'+u')!(t'-u')!\right]^{1/2}} \tag{A.149}$$

and construct a recursion for $|t - t'| \le t'' < |t + t'|$ as

$$\begin{pmatrix} t & t' & t'' \\ u & u' & u'' \end{pmatrix} = -\frac{a}{b} \begin{pmatrix} t & t' & t''+2 \\ u & u' & u'' \end{pmatrix} - \frac{c}{b} \begin{pmatrix} t & t' & t''+1 \\ u & u' & u'' \end{pmatrix}, \tag{A.150}$$

where

$$\begin{aligned} a = {} & (t''+1)\left\{\left[(t''+2)^2 - (t-t')^2\right]\right. \\ & \left.\left[(t+t'+1)^2 - (t''+2)^2\right]\left[(t''+2)^2 - (u'')^2\right]\right\}^{1/2} \end{aligned} \tag{A.151}$$

$$\begin{aligned} b = {} & (t''+2)\left\{\left[(t''+1)^2 - (t-t')^2\right]\right. \\ & \left.\left[(t+t'+1)^2 - (t''+1)^2\right]\left[(t''+1)^2 - (u'')^2\right]\right\}^{1/2} \end{aligned} \tag{A.152}$$

$$c = -(2t''+3)\left[t(t+1)u'' - t'(t'+1)u'' - (t''+1)(t''+2)(u'-u)\right]. \tag{A.153}$$

References

[1] W. C. Chew, J.-M. Jin, E. Michielssen, and J. Song, *Fast and Efficient Algorithms in Computational Electromagnetics.* Boston: Artech House, 2001.

[2] C. A. Balanis, *Antenna Theory: Analysis and Design.* New York: John Wiley & Sons, Inc., 1997.

[3] W. L. Stutzman and G. A. Thiele, *Antenna Theory and Design.* New York: John Wiley & Sons, Inc., 1998.

[4] J. D. Kraus and R. J. Marhefka, *Antennas for All Applications.* New York: McGraw-Hill, 2002.

[5] W.-J. Zhao, L.-W. Li, and Y.-B. Gan, "Efficient analysis of antenna radiation in the presence of airborne dielectric radomes of arbitrary shape," *IEEE Trans. Antennas Propag.*, vol. 53, no. 1, pp. 442–9, Jan. 2005.

[6] Ö. Ergül and L. Gürel, "Nonplanar trapezoidal-tooth log-periodic antennas: design and electromagnetic modelling," *Radio Sci.*, vol. 40, RS5010, Oct. 2005.

[7] S. R. Saunders and A. A. Zavala, *Antennas and Propagation for Wireless Communication Systems.* Chichester: John Wiley & Sons, Ltd, 2007.

[8] R. L. Haupt and D. H. Werner, *Genetic Algorithms in Electromagnetics.* New Jersey: John Wiley & Sons, Inc., 2007.

[9] Ö. Ergül and L. Gürel, "Modeling and synthesis of circular-sectoral arrays of log-periodic antennas using multilevel fast multipole algorithm and genetic algorithms," *Radio Sci.*, vol. 42, RS3018, Jun. 2007.

[10] L. Gürel and Ö. Ergül, "Design and simulation of circular arrays of trapezoidal-tooth log-periodic antennas via genetic optimization," *Prog. Electromagn. Res.*, vol. 85, pp. 243–60, 2008.

[11] J.-M. Jin and D. J. Riley, *Finite Element Analysis of Antennas and Arrays.* New Jersey: John Wiley & Sons, Inc., 2009.

[12] R. C. Hansen, *Phased Array Antennas.* New Jersey: John Wiley & Sons, Inc., 2009.

[13] J. M. Taboada, J. Rivero, F. Obelleiro, M. G. Araujo, and L. Landesa, "Method-of-moments formulation for the analysis of plasmonic nano-optical antennas," *J. Opt. Soc. Am. A*, vol. 28, no. 7, pp. 1341–8, Jun. 2011.

[14] E. Knott, J. F. Shaeffer, and M. T. Tuley, *Radar Cross Section.* Boston: Artech House, 1993.

[15] M. Born and E. Wolf, *Principles of Optics: Electromagnetic Theory of Propagation Interference and Diffraction of Light.* Cambridge: Cambridge University Press, 1999.

[16] Z.-H. Cho, J. P. Jones, and M. Singh, *Foundations of Medical Imaging.* New York: John Wiley & Sons, Inc., 1993.

[17] T. Rappaport, *Wireless Communications: Principles and Practice.* New Jersey: Prentice Hall, 2001.

The Multilevel Fast Multipole Algorithm (MLFMA) for Solving Large-Scale Computational Electromagnetics Problems, First Edition. Özgür Ergül and Levent Gürel.

[18] J. J. Greffet, "Nanoantennas for light emission," *Science*, vol. 308, pp. 1561–3, Jun. 2005.
[19] R. A. Shelby, D. R. Smith, and S. Schultz, "Experimental verification of a negative index of refraction," *Science*, vol. 292, pp. 77–9, Apr. 2001.
[20] G. V. Eleftheriades and K. G. Balmain, *Negative-Refraction Metamaterials: Fundamental Principles and Applications.* New Jersey: John Wiley & Sons, Inc., 2005.
[21] N. Engheta and R. W. Ziolkowski, "A positive future for double-negative metamaterials," *IEEE Trans. Microw. Theory Tech.*, vol. 53, no. 4, pp. 1535–6, Apr. 2005.
[22] M. Gokkavas, K. Güven, I. Bulu, K. AydÇn, R. S. Penciu, M. Kafesaki, C. M. Soukoulis, and E. Özbay, "Experimental demonstration of a left-handed metamaterial operating at 100 GHz," *Phys. Rev. B.*, vol. 73, no. 19, no. 193103, May 2006.
[23] D. Schurig, J. J. Mock, B. J. Justice, S. A. Cummer, J. B. Pendry, A. F. Starr, and D. R. Smith, "Metamaterial electromagnetic cloak at microwave frequencies," *Science*, vol. 314, pp. 977–80, Nov. 2006.
[24] L. Gürel, Ö. Ergül, A. Ünal, and T. Malas, "Fast and accurate analysis of large metamaterial structures using the multilevel fast multipole algorithm," *Prog. Electromagn. Res.*, vol. 95, pp. 179–98, 2009.
[25] Ö. Ergül and L. Gürel, "Efficient solutions of metamaterial problems using a low-frequency multilevel fast multipole algorithm," *Prog. Electromagn. Res.*, vol. 108, pp. 81–99, 2010.
[26] J. D. Joannopoulos, S. G. Johnson, J. N. Winn, and R. D. Meade, *Photonic Crystals, Molding the Flow of Light.* New Jersey: Princeton Univ. Press, 2008.
[27] A. Mekis, J. C. Chen, I. Kurland, S. Fan, P. R. Villeneuve, and J. D. Joannopoulos, "High transmission through sharp bends in photonic crystal waveguides," *Phys. Rev. Lett.*, vol. 77, no. 18, pp. 3787–90, Oct. 1996.
[28] S. G. Johnson, S. Fan, P. R. Villeneuve, and J. D. Joannopoulos, "Guided modes in photonic crystal slabs," *Physical Review B*, vol. 60, no. 8, pp. 5751–8, 1999.
[29] P. Loschialpo, D. W. Forester, and J. Schelleng, "Anomalous transmission through near unit index contrast dielectric photonic crystals," *J. Appl. Phys.*, vol. 86, no. 10, pp. 5342–7, Nov. 1999.
[30] A. Talneau, P. Lalanne, M. Agio, and C. M. Soukoulis, "Low-reflection photonic-crystal taper for efficient coupling between guide sections of arbitrary widths," *Opt. Lett.*, vol. 27, no. 17, pp. 1522–4, Sep. 2002.
[31] S. Boscolo and M. Midrio, "Three-dimensional multiple-scattering technique for the analysis of photonic-crystal slabs," *J. Lightwave Technol.*, vol. 22, no. 12, pp. 2778–86, Dec. 2004.
[32] D. Pissoort, E. Michielssen, D. V. Ginste, and F. Olyslager, "Fast-multipole analysis of electromagnetic scattering by photonic crystal slabs," *J. Lightwave Technol.*, vol. 25, no. 9, pp. 2847–63, Sep. 2007.
[33] Ö. Ergül, T. Malas, and L. Gürel, "Analysis of dielectric photonic-crystal problems with MLFMA and Schur-complement preconditioners," *J. Lightwave Technol.*, vol. 29, no. 6, pp. 888–97, Mar. 2011.
[34] G. Mie, "Contributions to the optics of turbid media, particularly of colloidal metal solutions," in German, *Ann. Phys.*, vol. 25, no. 3, pp. 377–445, 1908.
[35] J. Song, C.-C. Lu, and W. C. Chew, "Multilevel fast multipole algorithm for electromagnetic scattering by large complex objects," *IEEE Trans. Antennas Propag.*, vol. 45, no. 10, pp. 1488–93, Oct. 1997.
[36] D. H. Schaubert, D. R. Wilton, and A. W. Glisson, "A tedrahedral modeling method for electromagnetic scattering by arbitrarily shaped inhomogeneous dielectric bodies," *IEEE Trans. Antennas Propag.*, vol. 32, no. 1, pp. 77–85, Jan. 1984.
[37] J. P. Kottmann and O. J. F. Martin, "Accurate solution of the volume integral equation for high-permittivity scatterers," *IEEE Trans. Antennas Propag.*, vol. 48, no. 11, pp. 1719–26, Nov. 2000.
[38] K. Sertel and J. L. Volakis, "Multilevel fast multipole method solution of volume integral equations using parametric geometry modeling," *IEEE Trans. Antennas Propag.* vol. 52, no. 7, pp. 1686–92, Jul. 2004.

[39] X.-C. Nie, N. Yuan, L.-W. Li, Y.-B. Gan, and T. S. Yeo, "A fast combined field volume integral equation solution to EM scattering by 3-D dielectric objects of arbitrary permittivity and permeability," *IEEE Trans. Antennas Propag.* vol. 54, no. 3, pp. 961–9, Mar. 2006.

[40] J. De Zaeytijd, I. Bogaert, and A. Franchois, "An efficient hybrid MLFMA-FFT solver for the volume integral equation in case of sparse 3D inhomogeneous dielectric scatterers," *J. Comput. Phys.*, vol. 227, no. 14, pp. 7052–68, Jul. 2008.

[41] T. Sarkar, E. Arvas, and S. Rao, "Application of FFT and the conjugate gradient method for the solution of electromagnetic radiation from electrically large and small conducting bodies," *IEEE Trans. Antennas Propag.* vol. 34, no. 5, pp. 635–40, May 1986.

[42] J. M. Jin and J. L. Volakis, "A biconjugate gradient FFT solution for scattering by planar plates," *Electromagn.*, vol. 12, no. 1, pp. 105–19, 1992.

[43] H. Gan and W. C. Chew, "A discrete BCG-FFT algorithm for solving 3-D inhomogeneous scatterer problems," *J. Electromagn. Waves Appl.*, vol. 9, no. 10, pp. 1339–57, 1995.

[44] Q. H. Liu, Z. Q. Zhang, and X. M. Xu, "The hybrid extended Born approximation and CG-FFT method for electromagnetic induction problems," *IEEE Trans. Geosci. Remote Sensing*, vol. 39, no. 2, pp. 347–55, Feb. 2001.

[45] A. E. Yilmaz, D. S. Weile, H. -M. Jin, and E. Michielssen, "A hierarchical FFT algorithm (HIL-FFT) for the fast analysis of transient electromagnetic scattering phenomena," *IEEE Trans. Antennas Propag.* vol. 50, no. 7, pp. 971–82, Jul. 2002.

[46] N. Yuan, T. S. Yeo, X. C. Nie, L. W. Li, and Y. B. Gan, "Analysis of scattering from composite conducting and dielectric targets using the precorrected-FFT algorithm," *J. Electromagn. Waves Appl.*, vol. 17, no. 3, pp. 499–515, 2003.

[47] E. Bleszynski, M. Bleszynski, and T. Jaroszewicz, "AIM: Adaptive integral method for solving large-scale electromagnetic scattering and radiation problems," *Radio Sci.*, vol. 31, no. 5, pp. 1225–51, 1996.

[48] L. Hu, L.-W. Li, and R. Mittra, "Electromagnetic scattering by finite periodic arrays using the characteristic basis function and adaptive integral methods," *IEEE Trans. Antennas Propag.* vol. 58, no. 9, pp. 3086–90, Sep. 2010.

[49] W. C. Chew, T. J. Cui, and J. M. Song, "A FAFFA-MLFMA algorithm for electromagnetic scattering," *IEEE Trans. Antennas Propag.*, vol. 50, no. 11, pp. 1641–9, Nov. 2002.

[50] T. J. Cui, W. C. Chew, G. Chen, and J. M. Song, "Efficient MLFMA, RPFMA, and FAFFA algorithms for EM scattering by very large structures," *IEEE Trans. Antennas Propag.*, vol. 52, no. 3, pp. 759–70, Mar. 2004.

[51] C. Müller, *Foundations of the Mathematical Theory of Electromagnetic Waves.* New York: Springer, 1969.

[52] A. J. Poggio and E. K. Miller, "Integral equation solutions of three-dimensional scattering problems," in *Computer Techniques for Electromagnetics*, R. Mittra, Ed. Oxford: Pergamon Press, 1973, Chap. 4.

[53] Y. Chang and R. F. Harrington, "A surface formulation for characteristic modes of material bodies," *IEEE Trans. Antennas Propag.*, vol. 25, pp. 789–95, Nov. 1977.

[54] T. K. Wu and L. L. Tsai, "Scattering from arbitrarily-shaped lossy dielectric bodies of revolution," *Radio Sci.*, vol. 12, pp. 709–18, Sep.-Oct. 1977.

[55] J. R. Mautz and R. F. Harrington, "H-field, E-field, and combined field solutions for conducting bodies of revolution," *AEÜ*, vol. 32, no. 4, pp. 157–64, Apr. 1978.

[56] J. R. Mautz and R. F. Harrington, "Electromagnetic scattering from a homogeneous material body of revolution," *AEÜ*, vol. 33, pp. 71–80, Feb. 1979.

[57] S. M. Rao, D. R. Wilton, and A. W. Glisson, "Electromagnetic scattering by surfaces of arbitrary shape," *IEEE Trans. Antennas Propag.*, vol. 30, no. 3, pp. 409–18, May 1982.

[58] K. Umashankar, A. Taflove, and S. M. Rao, "Electromagnetic scattering by arbitrary shaped three-dimensional homogeneous lossy dielectric objects," *IEEE Trans. Antennas Propag.*, vol. 34, no. 6, pp. 758–66, Jun. 1986.

[59] W. V. T. Rusch and R. J. Pogorzelski, "A mixed-field solution for scattering from composite bodies," *IEEE Trans. Antennas Propag.*, vol. 34, no. 7, pp. 955–8, Jul. 1986.

[60] R. F. Harrington, "Boundary integral formulations for homogeneous material bodies," *J. Electromagn. Waves Applicat.*, vol. 3, no. 1, pp. 1–15, 1989.

[61] N. Morita, N. Kumagai, and J. R. Mautz, *Integral Equation Methods for Electromagnetics.* Boston: Artech House, 1990.

[62] S. M. Rao and D. R. Wilton, "E-field, H-field, and combined field solution for arbitrarily shaped three-dimensional dielectric bodies," *Electromagn.*, vol. 10, no. 4, pp. 407–21, Oct.-Dec. 1990.

[63] S. M. Rao, C. C. Cha, R. L. Cravey, and D. Wilkes, "Electromagnetic scattering from arbitrary shaped conducting bodies coated with lossy materials of arbitrary thickness," *IEEE Trans. Antennas Propag.*, vol. 39, pp. 627–31, no. 5, May 1991.

[64] A. Helaly and H. M. Fahmy, "Combined-field integral equation," *Electron. Lett.*, vol. 29, no. 19, pp. 1678–9, Sep. 1993.

[65] R. E. Hodges and T. Rahmat-Samii, "An iterative current-based hybrid method for complex structures," *IEEE Trans. Antennas Propag.*, vol. 45, no. 2, pp. 265–76, Feb. 1997.

[66] G. C. Hsiao and R. E. Kleinman, "Mathematical foundations for error estimation in numerical solutions of integral equations in electromagnetics," *IEEE Trans. Antennas Propag.*, vol. 45, no. 3, pp. 316–28, Mar. 1997.

[67] A. F. Peterson, S. L. Ray, and R. Mittra, *Computational Methods for Electromagnetics.* New York: IEEE Press, 1998.

[68] X.-Q. Sheng, J.-M. Jin, J. Song, W. C. Chew, and C.-C. Lu, "Solution of combined-field integral equation using multilevel fast multipole algorithm for scattering by homogeneous bodies," *IEEE Trans. Antennas Propag.*, vol. 46, no. 11, pp. 1718–26, Nov. 1998.

[69] P. Ylä-Oijala, "Application of a novel CFIE for electromagnetic scattering by dielectric objects," *Microw. Opt. Technol. Lett.*, vol. 35, no. 10, pp. 3–5, Oct. 2002.

[70] S. Chakraborty and V. Jandhyala, "Accurate computation of vector potentials in lossy media", *Microw. Opt Technol. Lett.*, vol. 36, pp. 359–63, Mar. 2003.

[71] P. Ylä-Oijala and M. Taskinen, "Application of combined field integral equation for electromagnetic scattering by dielectric and composite objects," *IEEE Trans. Antennas Propag.*, vol. 53, no. 3, pp. 1168–73, Mar. 2005.

[72] Y.-H. Chu and W. C. Chew, "A robust surface-integral-equation formulation for conductive media," *Microw. Opt. Technol. Lett.*, vol. 46, no. 2, pp. 109–14, Jul. 2005.

[73] P. Ylä-Oijala and M. Taskinen, "Well-conditioned Müller formulation for electromagnetic scattering by dielectric objects," *IEEE Trans. Antennas Propag.*, vol. 53, no. 10, pp. 3316–23, Oct. 2005.

[74] P. Ylä-Oijala, M. Taskinen, and S. Järvenpää, "Surface integral equation formulations for solving electromagnetic scattering problems with iterative methods," *Radio Sci.*, vol. 40, RS6002, Nov. 2005.

[75] P. Ylä-Oijala, M. Taskinen, and J. Sarvas, "Surface integral equation method for general integral equation method for general composite metallic and dielectric structures with junctions," *Prog. Electromagn. Res.*, vol. 52, pp. 81–108, 2005.

[76] L. Gürel and Ö. Ergül, "Extending the applicability of the combined-field integral equation to geometries containing open surfaces," *IEEE Antennas Wireless Propag. Lett.*, vol. 5, pp. 515–16, 2006.

[77] P. Ylä-Oijala and M. Taskinen, "Improving conditioning of electromagnetic surface integral equations using normalized field quantities," *IEEE Trans. Antennas Propag.*, vol. 55, no. 1, pp. 178–85, Jan. 2007.

[78] P. Ylä-Oijala, "Numerical analysis of combined field integral equation formulations for electromagnetic scattering by dielectric and composite objects," *Prog. Electromagn. Res. C,* vol. 3, pp. 19–43, 2008.

[79] W.-D. Li, H.-X. Zhou, and T. Weiland, "Overlapped domain decomposition method for modified N-Mller formulation," *IEEE Antennas Wireless Propag. Lett.*, vol. 7, pp. 710–13, 2008.
[80] Ö. Ergül and L. Gürel, "Comparison of integral-equation formulations for the fast and accurate solution of scattering problems involving dielectric objects with the multilevel fast multipole algorithm," *IEEE Trans. Antennas Propag.*, vol. 57, no. 1, pp. 176–87, Jan. 2009.
[81] Ö. Ergül and L. Gürel, "Fast and accurate solutions of scattering problems involving dielectric objects with moderate and low contrasts," in *Proc. Computational Electromagnetics Workshop*, 2007, pp. 59–64.
[82] Ö. Ergül and L. Gürel, "Iterative solutions of hybrid integral equations for coexisting open and closed surfaces," *IEEE Trans. Antennas Propag.*, vol. 57, no. 6, pp. 1751–8, Jun. 2009.
[83] Ö. Ergül and L. Gürel, "Efficient solution of the electric and magnetic current combined-field integral equation with the multilevel fast multipole algorithm and block-diagonal preconditioning," *Radio Sci.*, vol. 44, RS6001, Nov. 2009.
[84] J. M. Steinke and A. P. Shepherd, "Comparison of Mie theory and the light scattering of red blood cells," *Appl. Opt.*, vol. 27, no. 19, pp. 4027–33, Oct. 1988.
[85] A. M. K. Nilsson, P. Alsholm, A. Karlsson, and S. Andersson-Engels, "T-matrix computations of light scattering by red blood cells," *Appl. Opt.*, vol. 37, no. 13, pp. 2735–48, May 1998.
[86] M. Hammer, D. Schweitzer, B. Michel, E. Thamm, and A. Kolb, "Single scattering by red blood cells," *Appl. Opt.*, vol. 37, no. 31, pp. 7410–18, Nov. 1998.
[87] S. V. Tsinopoulos, E. J. Sellountos, and D. Polyzos, "Light scattering by aggregated red blood cells," *Appl. Opt.*, vol. 41, no. 7, pp. 1408–17, Mar. 2002.
[88] J. He, A. Karlsson, J. Swartling, and S. Andersson-Engels, "Light scattering by multiple red blood cells," *J. Opt. Soc. Am. A*, vol. 21, no. 10, pp. 1953–61, Oct. 2004.
[89] A. Karlsson, J. He, J. Swartling, and S. Andersson-Engels, "Numerical simulations of light scattering by red blood cells," *IEEE Trans. Biomed. Eng.*, vol. 52, no. 1, pp. 13–18, Jan. 2005.
[90] T. W. Lloyd, J. M. Song, and M. Yang, "Numerical study of surface integral formulations for low-contrast objects," *IEEE Antennas Wireless Propag. Lett.*, vol. 4, pp. 482–5, 2005.
[91] J. Q. Lu, P. Yang, and X.-H. Hu, "Simulations of light scattering from a biconcave red blood cell using the finite-difference time-domain method," *J. Biomed. Opt.*, vol. 10, no. 024022, Mar.-Apr. 2005.
[92] Ö. Ergül, A. Arslan-Ergül, and L. Gürel, "Computational study of scattering from healthy and diseased red blood cells using surface integral equations and the multilevel fast multipole algorithm," *J. Biomed. Opt.*, vol. 15, no. 4, 045004-1–045004-8, Jul./Aug. 2010.
[93] D. A. Hill, "Electromagnetic scattering by buried objects of low contrast," *IEEE Trans. Geosci. Remote Sen.*, vol. 26, no. 2, pp. 195–203, Mar. 1988.
[94] E. S. Thiele, "Scattering of electromagnetic radiation by complex microstructures in the resonant regime," Ph.D. thesis, University of Pennsylvania, 1998.
[95] P. M. Goggans and A. Glisson, "A surface integral equation formulation for low contrast scatterers based on radiation currents," *ACES J.*, vol. 10, pp. 15–18, Mar. 1995.
[96] Ö. Ergül and L. Gürel, "Stabilization of integral-equation formulations for the accurate solution of scattering problems involving low-contrast dielectric objects," *IEEE Trans. Antennas Propag.*, vol. 56, no. 3, pp. 799–805, Mar. 2008.
[97] Ö. Ergül and L. Gürel, "Novel electromagnetic surface integral equations for highly accurate computations of dielectric bodies with arbitrarily low contrasts," *J. Comput. Phys.*, vol. 227, no. 23, pp. 9898–9912, Dec. 2008.
[98] J. R. Mautz and R. F. Harrington, "An E-field solution for a conducting surface small or comparable to the wavelength," *IEEE Trans. Antennas Propag.*, vol. 32, no. 4, pp. 330–8, Apr. 1984.
[99] J.-S. Zhao and W. C. Chew, "Integral equation solution of Maxwell's equations from zero frequency to microwave frequencies," *IEEE Trans. Antennas Propag.*, vol. 48, no. 10, pp. 330–9, Oct. 2000.

[100] H. Contopanagos, B. Dembart, M. Epton, J. J. Ottusch, V. Rokhlin, J. L. Visher, and S. M. Wandzura "Well-conditioned boundary integral equations for three-dimensional electromagnetic scattering," *IEEE Trans. Antennas Propag.*, vol. 50, no. 12, pp. 1824–30, Dec. 2002.
[101] F. P. Andriulli and E. Michielssen, "A regularized combined field integral equation for scattering from 2-D perfect electrically conducting objects," *IEEE Trans. Antennas Propag.*, vol. 55, no. 9, pp. 2522–9, Sep. 2007.
[102] Y. Zhang, T. J. Cui, W. C. Chew, and J.-S. Zhao, "Magnetic field integral equation at very low frequencies," *IEEE Trans. Antennas Propag.*, vol. 51, no. 8, pp. 1864–71, Aug. 2003.
[103] M. Burton and S. Kashyap, "A study of a recent moment-method algorithm that is accurate to very low frequencies," *ACES J.*, vol. 10, no. 3, pp. 58–68, Nov. 1995.
[104] L. Gürel and Ö. Ergül, "Comparisons of FMM implementations employing different formulations and iterative solvers," in *Proc. IEEE Antennas and Propagation Soc. Int. Symp.*, vol. 1, 2003, pp. 19–22.
[105] W.-L. Wu, A. W. Glisson, and D. Kajfez, "A study of two numerical solution procedures for the electric field integral equation at low frequency," *ACES J.*, vol. 10, no. 3, pp. 69–80, Nov. 1995.
[106] G. Vecchi, "Loop-star decomposition of basis functions in the discretization of the EFIE," *IEEE Trans. Antennas Propag.*, vol. 47, no. 2, pp. 339–46, Feb. 1999.
[107] J.-F. Lee, R. Lee, and R. J. Burkholder, "Loop star basis functions and a robust preconditioner for EFIE scattering problems," *IEEE Trans. Antennas Propag.*, vol. 51, no. 8, pp. 1855–63, Aug. 2003.
[108] M. Taskinen and P. Ylä-Oijala, "Current and charge integral equation formulation," *IEEE Trans. Antennas Propag.*, vol. 54, no. 1, pp. 58–67, Jan. 2006.
[109] Z. G. Qian and W. C. Chew, "An augmented EFIE for high speed interconnect analysis," *Microw. Opt. Tech. Lett.*, vol. 50, no. 10, pp. 2658–62, Oct. 2008.
[110] R. J. Adams, "Physical and analytical properties of a stabilized electric field integral equation," *IEEE Trans. Antennas Propag.*, vol. 52, no. 2, pp. 362–72, Feb. 2004.
[111] R. J. Adams and N. J. Champagne, "A numerical implementation of a modified form of the electric field integral equation," *IEEE Trans. Antennas Propag.*, vol. 52, no. 9, pp. 2262–6, Sep. 2004.
[112] F. P. Andriulli, K. Cools, H. Bagcζ, F. Olyslager, A. Buffa, S. Christiansen, and E. Michielssen, "A multiplicative Calderon preconditioner for the electric field integral equation," *IEEE Trans. Antennas Propag.*, vol. 56, no. 8, pp. 2398–2412, Aug. 2008.
[113] M. B. Stephanson and J. F. Lee, "Preconditioned electric field integral equation using Calderon identities and dual loop/star basis functions," *IEEE Trans. Antennas Propag.*, vol. 57, no. 4, pp. 1274–9, Apr. 2009.
[114] S. Yan, J. M. Jin, and Z. P. Nie, "EFIE Analysis of low-frequency problems with loop-star decomposition and Calderon multiplicative preconditioner," *IEEE Trans. Antennas Propag.*, vol. 58, no. 3, pp. 857–67, Mar. 2010.
[115] J. Peeters, K. Cools, I. Bogaert, F. Olyslager, and D. De Zutter, "Embedding Calderon multiplicative preconditioners in multilevel fast multipole algorithms," *IEEE Trans. Antennas Propag.*, vol. 58, no. 4, pp. 1236–50, Apr. 2010.
[116] S. Yan, J. M. Jin, and Z. P. Nie, "Analysis of electrically large problems using the augmented EFIE with a Calderon preconditioner," *IEEE Trans. Antennas Propag.*, vol. 59, no. 6, pp. 2303–14, Jun. 2011.
[117] R. F. Harrington, *Field Computation by Moment Methods.* New York: Macmillan, 1968.
[118] W. C. Gibson, *The Method of Moments in Electromagnetics.* Boca Raton, FL: Chapman and Hall/CRC, 2007.
[119] D. R. Wilton and J. E. Wheeler III, "Comparison of convergence rates of the conjugate gradient method applied to various integral equation formulations," *Prog. Electromagn. Res.*, pp. 131–58, 1991.
[120] P. Ylä-Oijala, M. Taskinen, and S. Järvenpää, "Analysis of surface integral equations in electromagnetic scattering and radiation problems," *Eng. Anal. Boundary Elem.*, vol. 32, no. 3, pp. 196–209, Mar. 2008.

[121] Ö. Ergül and L. Gürel, "Discretization error due to the identity operator in surface integral equations," *Comput. Phys. Comm.*, vol. 180, no. 10, pp. 1746–52, Oct. 2009.

[122] L. Gürel and Ö. Ergül, "Contamination of the accuracy of the combined-field integral equation with the discretization error of the magnetic-field integral equation," *IEEE Trans. Antennas Propag.*, vol. 57, no. 9, pp. 2650–7, Sep. 2009.

[123] J. M. Rius, E. Ubeda, and J. Parron, "On the testing of the magnetic field integral equation with RWG basis functions in method of moments," *IEEE Trans. Antennas Propag.*, vol. 49, pp. 1550–3, Nov. 2001.

[124] L. Gürel and Ö. Ergül, "Singularity of the magnetic-field integral equation and its extraction," *IEEE Antennas Wireless Propag. Lett.*, vol. 4, pp. 229–32, 2005.

[125] Ö. Ergül and L. Gürel, "Solid-angle factor in the magnetic-field integral equation," *Microw. Opt. Technol. Lett.*, vol. 45, no. 5, pp. 442–56, Jun. 2005.

[126] Ö. Ergül and L. Gürel, "Improved testing of the magnetic-field integral equation," *IEEE Microw. Wireless Comp. Lett.*, vol. 15, no. 10, pp. 615–17, Oct. 2005.

[127] Ö. Ergül and L. Gürel, "The use of curl-conforming basis functions for the magnetic-field integral equation," *IEEE Trans. Antennas Propag.*, vol. 54, no. 7, pp. 1917–26, Jul. 2006.

[128] Ö. Ergül and L. Gürel, "Improving the accuracy of the magnetic field integral equation with the linear-linear basis functions," *Radio Sci.*, vol. 41, RS4004, Jul. 2006.

[129] Ö. Ergül and L. Gürel, "Linear-linear basis functions for MLFMA solutions of magnetic-field and combined-field integral equations," *IEEE Trans. Antennas Propag.*, vol. 55, no. 4, pp. 1103–10, Apr. 2007.

[130] R. D. Graglia, D. R. Wilton, and A. F. Peterson, "Higher order interpolatory vector bases for computational electromagnetics," *IEEE Trans. Antennas Propag.*, vol. 45, no. 3, pp. 329–42, Mar. 1997.

[131] J. Wang and J. P. Webb, "Hierarchical vector boundary elements and p-adaptation for 3-D electromagnetic scattering," *IEEE Trans. Antennas Propag.*, vol. 45, no. 12, pp. 1869–79, Dec. 1997.

[132] W. Cai, T. Yu, H. Wang, and Y. Yu, "High-order mixed RWG basis functions for electromagnetic applications," *IEEE Trans. Microw. Theory Tech.*, vol. 49, no. 7, pp. 1295–1303, Jul. 2001.

[133] L. C. Trintinalia and H. Ling, "First order triangular patch basis functions for electromagnetic scattering analysis," *J. of Electromagn. Waves Applicat.*, vol. 15, no. 11, pp. 1521–37, 2001.

[134] E. Ubeda and J. M. Rius, "MFIE MOM-formulation with curl-conforming basis functions and accurate kernel integration in the analysis of perfectly conducting sharp-edged objects," *Microw. Opt. Technol. Lett.*, vol. 44, no. 4, pp. 354–8, Feb. 2005.

[135] E. Ubeda and J. M. Rius, "Monopolar divergence-conforming and curl-conforming low-order basis functions for the electromagnetic scattering analysis," *Microw. Opt. Technol. Lett.*, vol. 46, no. 3, pp. 237–41, Aug. 2005.

[136] E. Ubeda and J. M. Rius, "Novel monopolar MFIE MoM-discretization for the scattering analysis of small objects," *IEEE Trans. Antennas Propag.*, vol. 54, no. 1, pp. 50–7, Jan. 2006.

[137] L. Zhang, A. Deng, M. Wang, and S. Yang, "Numerical study of a novel MFIE formulation using monopolar RWG basis functions," *Comput. Phys. Comm.*, vol. 181, no. 1, pp. 52–60, Jan. 2010.

[138] A. Deng, L. Zhang, and M. Wang, "Application of the combined RWG basis functions to the MFIE for the scattering analysis of small sharp-edged objects," *Eng. Anal. Boundary Elem.*, vol. 34, no. 1, pp. 17–22, Jan. 2010.

[139] L. Zhang, A. Deng, M. Wang, and S. Yang, "Investigation of two formulations using curl-conforming basis functions for the MFIE," *Eng. Anal. Boundary Elem.*, vol. 35, no. 3, pp. 383–8, Mar. 2011.

[140] C. P. Davis and K. F. Warnick, "High-order convergence with a low-order discretization of the 2-D MFIE," *IEEE Antennas Wireless Propag. Lett.*, vol. 3, pp. 355–8, 2004.

[141] L. Gürel, K. Sertel, and İ. K. Şendur, "On the choice of basis functions to model surface electric current densities in computational electromagnetics," *Radio Sci.*, vol. 34, no. 6, pp. 1373–87, Nov.-Dec. 1999.

[142] D. A. Dunavant, "High degree efficient symmetrical Gaussian quadrature rules for the triangle," *Int. J. Numer. Meth. Eng.*, vol. 21, pp. 1129–48, 1985.
[143] D. R. Wilton, S. M. Rao, A. W. Glisson, D. H. Schaubert, O. M. Al-Bundak, and C. M. Butler, "Potential integrals for uniform and linear source distributions on polygonal and polyhedral domains," *IEEE Trans. Antennas Propag.*, vol. 32, pp. 276–81, Mar. 1984.
[144] R. D. Graglia, "On the numerical integration of the linear shape functions times the 3-D Green's function or its gradient on a plane triangle," *IEEE Trans. Antennas Propag.*, vol. 41, no. 10, pp. 1448–55, Oct. 1993.
[145] K. Tap, "Complex source point beam expansion for some electromagnetic radiation and scattering problems," Ph.D. thesis, The Ohio State University, 2007.
[146] V. Rokhlin, "Rapid solution of integral equations of scattering theory in two dimensions," *J. Comput. Phys.*, vol. 86, no. 2, pp. 414–39, Feb. 1990.
[147] R. Coifman, V. Rokhlin, and S. Wandzura, "The fast multipole method for the wave equation: a pedestrian prescription," *IEEE Antennas Propag. Mag.*, vol. 35, no. 3, pp. 7–12, Jun. 1993.
[148] N. Geng, A. Sullivan, and L. Carin, "Fast multipole method for scattering from an arbitrary PEC target above or buried in a lossy half space," *IEEE Trans. Antennas Propag.*, vol. 49, no. 5, pp. 740–8, May 2001.
[149] L. Gürel, H. Bağcζ, J. C. Castelli, A. Cheraly, and F. Tardivel, "Validation through comparison: measurement and calculation of the bistatic radar cross section (BRCS) of a stealth target," *Radio Sci.*, vol. 38, no. 3, Jun. 2003.
[150] M. F. Gyure and M. A. Stalzer, "A prescription for the multilevel Helmholtz FMM," *IEEE Comput. Sci. Eng.*, vol. 5, no. 3, pp. 39–47, Jul.–Sep. 1998.
[151] H.-Y. Chao, J.-S. Zhao, and W. C. Chew, "Application of curvilinear basis functions and MLFMA for radiation and scattering problems involving curved PEC structures," *IEEE Trans. Antennas Propag.*, vol. 51, no. 2, pp. 331–6, Feb. 2003.
[152] L. Li, J. He, Z. Liu, X. Dong, and L. Carin, "MLFMA analysis of scattering from multiple targets in the presence of a half-space," *IEEE Trans. Antennas Propag.*, vol. 51, no. 4, pp. 810–19, Apr. 2003.
[153] Z. Liu, R. J. Adams, and L. Carin, "Well-conditioned MLFMA formulation for closed PEC targets in the vicinity of a half space," *IEEE Trans. Antennas Propag.*, vol. 51, no. 10, pp. 2822–9, Oct. 2003.
[154] K. C. Donepudi, J.-M. Jin, and W. C. Chew, "A higher order multilevel fast multipole algorithm for scattering from mixed conducting/dielectric bodies," *IEEE Trans. Antennas Propag.*, vol. 51, no. 10, pp. 2814–21, Oct. 2003.
[155] M. S. Tong, W. C. Chew, and M. J. White, "Multilevel fast multipole algorithm for acoustic wave scattering by truncated ground with trenches," *J. Acoust. Soc. Am.*, vol. 123, no. 5, pp. 2513–21, 2008.
[156] P. Wang, Y. J. Xie, and R. Yang, "Analysis of electrical large radiation problem with precorrected multilevel fast multipole algorithm," *IEEE Antennas Wireless Propag. Lett.*, vol. 6, pp. 340–3, 2007.
[157] M. S. Tong and W. C. Chew, "Multilevel fast multipole algorithm for elastic wave scattering by large three-dimensional objects," *J. Comput. Phys.*, vol. 229, no. 3, pp. 921–32, Feb. 2009.
[158] S. Velamparambil, W. C. Chew, and J. Song, "10 million unknowns: Is it that big?," *IEEE Antennas Propag. Mag.*, vol. 45, no. 2, pp. 43–58, Apr. 2003.
[159] S. Koc, J. M. Song, and W. C. Chew, "Error analysis for the numerical evaluation of the diagonal forms of the scalar spherical addition theorem," *SIAM J. Numer. Anal.*, vol. 36, no. 3, pp. 906–21, 1999.
[160] A. Brandt, "Multilevel computations of integral transforms and particle interactions with oscillatory kernels," *Comput. Phys. Comm.*, vol. 65, pp. 24–38, Apr. 1991.
[161] R. Jacob-Chien and B. K. Alpert, "A fast spherical filter with uniform resolution," *J. Comput. Phys.*, vol. 136, no. 2, pp. 580–4, Sep. 1997.

[162] J. Sarvas, "Performing interpolation and anterpolation entirely by fast Fourier transform in the 3-D multilevel fast multipole algorithm", *SIAM J. Numer. Anal.*, vol. 41, no. 6, pp. 2180–96, Nov. 2003.

[163] O. M. Bucci, C. Gennarelli, and C. Savarese, "Optimal interpolation of radiated fields over a sphere," *IEEE Trans. Antennas Propag.*, vol. 39, no. 11, pp. 1633–43, Nov. 1991.

[164] Ö. Ergül and L. Gürel, "Enhancing the accuracy of the interpolations and anterpolations in MLFMA," *IEEE Antennas Wireless Propag. Lett.*, vol. 5, pp. 467–70, 2006.

[165] Ö. Ergül, I. van den Bosch, and L. Gürel, "Two-step Lagrange interpolation method for the multilevel fast multipole algorithm," *IEEE Antennas Wireless Propag. Lett.*, vol. 8, pp. 69–71, 2009.

[166] J. Song and W. C. Chew, "Interpolation of translation matrix in MLFMA," *Microw. Opt. Technol. Lett.*, vol. 30, no. 2, pp. 109–14, Jul. 2001.

[167] Ö. Ergül and L. Gürel, "Optimal interpolation of translation operator in multilevel fast multipole algorithm," *IEEE Trans. Antennas Propag.*, vol. 54, no. 12, pp. 3822–6, Dec. 2006.

[168] R. E. Hodges and Y. Rahmat-Samii, "The evaluation of MFIE integrals with the use of vector triangle basis functions," *Microw. Opt. Technol. Lett.*, vol. 14, no. 1, pp. 9–14, Jan. 1997.

[169] P. Ylä-Oijala and M. Taskinen, "Calculation of CFIE impedance matrix elements with RWG and $\hat{n}\times$RWG functions," *IEEE Trans. Antennas Propag.*, vol. 51, no. 8, pp. 1837–46, Aug. 2003.

[170] M. L. Hastriter, S. Ohnuki, and W. C. Chew, "Error control of the translation operator in 3D MLFMA," *Microw. Opt. Technol. Lett.*, vol. 37, no. 3, pp. 184–8, May 2003.

[171] S. Ohnuki and W. C. Chew, "Truncation error analysis of multipole expansions," *SIAM J. Sci. Comput.*, vol. 25, no. 4, pp. 1293–1306, 2003.

[172] S. Ohnuki and W. C. Chew, "Numerical accuracy of multipole expansion for 2D MLFMA," *IEEE Trans. Antennas Propag.*, vol. 51, no. 8, pp. 1883–89, Aug. 2003.

[173] D. V. Ginste, L. Knockaert, and D. De Zutter, "Error control in the perfectly matched layer based multilevel fast multipole algorithm," *J. Comput. Phys.*, vol. 228, no. 13, pp. 4811–22, Jul. 2009.

[174] L. Greengard, J. Huang, V. Rokhlin, and S. Wadzura, "Accelerating fast multipole methods for the Helmholtz equation at low frequencies," *IEEE Comput. Sci. Eng.*, vol. 5, pp. 32–8, Jul.–Sep. 1998.

[175] L. J. Jiang and W. C. Chew, "Low-frequency fast inhomogeneous plane-wave algorithm (LF-FIPWA)," *Microw. Opt. Technol. Lett.*, vol. 40, no. 2, pp. 117–22, Jan. 2004.

[176] E. Darve and P. Have, "A fast multipole method for Maxwell equations stable at all frequencies," *Phil. Trans. R. Soc. Lond. A*, vol. 362, pp. 603–28, Mar. 2004.

[177] H. Wallen and J. Sarvas, "Translation procedures for broadband MLFMA," *Prog. Electromagn. Res.*, vol. 55, pp. 47–78, 2005.

[178] W. C. Chew, S. Koc, J. M. Song, C. C. Lu, and E. Michielssen, "A succinct way to diagonalize the translation matrix in three dimensions," *Microw. Opt. Technol. Lett.*, vol. 15, no. 3, pp. 144–7, Jun. 1997.

[179] J.-S. Zhao and W. C. Chew, "Three dimensional multilevel fast multipole algorithm from static to electrodynamic," *Microw. Opt. Technol. Lett.*, vol. 26, no. 1, pp. 43–8, Jul. 2000.

[180] J.-S. Zhao and W. C. Chew, "Applying matrix rotation to the three-dimensional low-frequency multilevel fast multipole algorithm," *Microw. Opt. Technol. Lett.*, vol. 26, no. 2, pp. 105–10, Jul. 2000.

[181] J.-S. Zhao and W. C. Chew, "Applying LF-MLFMA to solve complex PEC structures," *Microw. Opt. Technol. Lett.*, vol. 28, no. 3, pp. 155–60, Feb. 2001.

[182] Y.-H. Chu and W. C. Chew, "A multilevel fast multipole algorithm for electrically small composite structures," *Microw. Opt. Technol. Lett.*, vol. 43, no. 3, pp. 202–7, Nov. 2004.

[183] L. J. Jiang and W. C. Chew, "A mixed-form fast multipole algorithm," *IEEE Trans. Antennas Propag.*, vol. 53, no. 12, pp. 4145–56, Dec. 2005.

[184] I. Bogaert, J. Peeters, and F. Olyslager, "A nondirective plane wave MLFMA stable at low frequencies," *IEEE Trans. Antennas Propag.*, vol. 56, no. 12, pp. 3752–67, Dec. 2008.

[185] I. Bogaert and F. Olyslager, "A low frequency stable plane wave addition theorem," *J. Comput. Phys.*, vol. 228, no. 4, pp. 1000–16, Mar. 2009.

[186] R. Barrett, M. Berry, T. Chan, J. Demmel, J. Donato, J. Dongarra, V. Eijkhout, R. Pozo, C. Romine, and H. van der Vorst, *Templates for the Solution of Linear Systems: Building Blocks for Iterative Methods.* Philadelphia: SIAM, 1994.

[187] A. Greenbaum, *Iterative Methods for Solving Linear Systems.* Philadelphia: SIAM, 1997.

[188] H. van der Vorst, "Bi-CGSTAB: A fast and smoothly converging variant of Bi-CG for the solution of nonsymmetric linear systems," *SIAM J. Sci. Stat. Comput.*, vol. 13, no. 2, pp. 631–44, Mar. 1992.

[189] P. Sonneveld, "CGS: A fast Lanczos-type solver for nonsymmetric linear systems," *SIAM J. Sci. Stat. Comput.*, vol. 10, no. 1, pp. 36–52, Jan. 1989.

[190] Y. Saad and M. H. Schultz, "GMRES: A generalized minimal residual algorithm for solving nonsymmetric linear systems," *SIAM J. Sci. Stat. Comput.*, vol. 7, no. 3, pp. 856–69, 1986.

[191] C. C. Paige and M. A. Saunders, "LSQR: An algorithm for sparse linear equations and sparse least squares," *ACM Trans. Math. Software*, vol. 8, no. 1, pp. 43–71, Mar. 1982.

[192] C. C. Paige and M. A. Saunders, "LSQR: Sparse linear equations and least squares problems," *ACM Trans. Math. Software*, vol. 8, no. 2, pp. 195–209, Jun. 1982.

[193] A. Bjorck, T. Elfving, and Z. Strakos, "Stability of conjugate gradient and Lanczos methods for linear least squares problems," *SIAM J. Matrix Anal. Appl.*, vol. 19, no. 3, pp. 720–36, Jul. 1998.

[194] R. W. Freund and N. M. Nachtigal, "QMR: A quasi-minimal residual method for non-Hermitian linear systems," *Numerische Mathematik*, vol. 60, pp. 315–38, 1991.

[195] R. W. Freund, "A transpose-free quasi-minimal residual algorithm for non-hermitian linear systems," *SIAM J. Sci. Comput.*, vol. 14, no. 2, pp. 470–82, Mar. 1993.

[196] Y.-F. Jing, B. Carpentieri, and T.-Z. Huang, "Experiments with lanczos biconjugate a-orthonormalization methods for MoM discretizations of Maxwell's equations," *Prog. Electromagn. Res.*, vol. 99, 427–51, 2009.

[197] Ö. Ergül and L. Gürel, "Efficient solution of the electric-field integral equation using the iterative LSQR algorithm," *IEEE Antennas Wireless Propag. Lett.*, vol. 7, pp. 36–9, 2008.

[198] Ö. Ergül and L. Gürel, "Improving iterative solutions of the electric-field integral equation via transformations into normal equations," *J Electromagn. Waves Appl.*, vol. 24, no. 16, pp. 2129–38, 2010.

[199] S. Balay, K. Buschelman, V. Eijkhout, W. D. Gropp, D. Kaushik, M. G. Knepley, L. C. McInnes, B. F. Smith, and H. Zhang, *PETSc Users Manual*, Argonne National Laboratory, 2004.

[200] M. Benzi, "Preconditioning techniques for large linear systems: a survey," *J. Comput. Phys.*, vol. 182, pp. 418–77, 2002.

[201] E. Cuthill and J. McKee, "Reducing the bandwidth of sparse symmetric matrices," in *Proc. 24th Nat. Conf. ACM*, pp. 157–72, 1969.

[202] E. Ubeda, J. M. Rius, and J. Romeu, "Preconditioning techniques in the analysis of finite metamaterial slabs," *IEEE Trans. Antennas Propag.*, vol. 54, no. 1, pp. 265–8, Jan. 2006.

[203] K. Sertel and J. L. Volakis, "Incomplete LU preconditioner for FMM implementations," *Microw. Opt. Technol. Lett.*, vol. 26, no. 4, pp. 265–7, Aug. 2000.

[204] J. Lee, J. Zhang, and C.-C. Lu, "Incomplete LU preconditioning for large scale dense complex linear systems from electromagnetic wave scattering problems," *J. Comput. Phys.*, vol. 185, no. 1, pp. 158–75, Feb. 2003.

[205] Y. Saad, *Iterative Methods for Sparse Linear Systems.* Philadelphia: SIAM, 2003.

[206] T. Malas and L. Gürel, "Incomplete LU preconditioning with multilevel fast multipole algorithm for electromagnetic scattering," *SIAM J. Sci. Comput.*, vol. 29, no. 4, pp. 1476–94, Jun. 2007.

[207] J. Lee, J. Zhang, and C.-C. Lu, "Sparse inverse preconditioning of multilevel fast multipole algorithm for hybrid integral equations in electromagnetics," *IEEE Trans. Antennas and Propag.*, vol. 52, no. 9, pp. 2277–87, Sep. 2004.

[208] B. Carpentieri, I. S. Duff, L. Giraud, and G. Sylvand, "Combining fast multipole techniques and an approximate inverse preconditioner for large electromagnetism calculations," *SIAM J. Sci. Comput.*, vol. 27, no. 3, pp. 774–92, 2005.

[209] A. J. Hesford and W. C. Chew, "On preconditioning and the eigensystems of electromagnetic radiation problems," *IEEE Trans. Antennas and Propag.*, vol. 56, no. 8, pp. 2413–20, Aug. 2008.

[210] T. Malas and L. Gürel, "Accelerating the multilevel fast multipole algorithm with the sparse-approximate-inverse (SAI) preconditioning," *SIAM J. Sci. Comput.*, vol. 31, no. 3, pp. 1968–84, Mar. 2009.

[211] Ö. Ergül, T. Malas, and L. Gürel, "Solutions of large-scale electromagnetics problems using an iterative inner-outer scheme with ordinary and approximate multilevel fast multipole algorithms," *Prog. Electromagn. Res.*, vol. 106, pp. 203–23, 2010.

[212] J. van den Eshof, G. L. G. Sleijpen, and M. B. van Gijzen, "Relaxation strategies for nested Krylov methods," *J. Comput. Appl. Math.*, vol. 177, no. 2, pp. 347–65, May 2005.

[213] H. van der Vorst and C. Vuik, "GMRESR: a family of nested GMRES methods," *Numer. Linear Algebra Appl.*, vol. 1, no. 4, pp. 369–86, Jul. 1994.

[214] K. C. Donepudi, J.-M. Jin, S. Velamparambil, J. Song, and W. C. Chew, "A higher order parallelized multilevel fast multipole algorithm for 3-D scattering," *IEEE Trans. Antennas Propag.*, vol. 49, no. 7, pp. 1069–78, Jul. 2001.

[215] M. L. Hastriter, "A study of MLFMA for large-scale scattering problems," Ph.D. thesis, University of Illinois at Urbana-Champaign, 2003.

[216] G. Sylvand, "Performance of a parallel implementation of the FMM for electromagnetics applications," *Int. J. Numer. Meth. Fluids*, vol. 43, pp. 865–79, 2003.

[217] S. Velamparambil and W. C. Chew, "Analysis and performance of a distributed memory multilevel fast multipole algorithm," *IEEE Trans. Antennas Propag.*, vol. 53, no. 8, pp. 2719–27, Aug. 2005.

[218] L. Gürel and Ö. Ergül, "Fast and accurate solutions of extremely large integral-equation formulations discretised with tens of millions of unknowns," *Electron. Lett.*, vol. 43, no. 9, pp. 499–500, Apr. 2007.

[219] Ö. Ergül and L. Gürel, "Hierarchical parallelisation strategy for multilevel fast multipole algorithm in computational electromagnetics," *Electron. Lett.*, vol. 44, no. 1, pp. 3–5, Jan. 2008.

[220] X.-M. Pan and X.-Q. Sheng, "A sophisticated parallel MLFMA for scattering by extremely large targets," *IEEE Antennas Propag. Mag.*, vol. 50, no. 3, pp. 129–38, Jun. 2008.

[221] Ö. Ergül and L. Gürel, "Efficient parallelization of the multilevel fast multipole algorithm for the solution of large-scale scattering problems," *IEEE Trans. Antennas Propag.*, vol. 56, no. 8, pp. 2335–45, Aug. 2008.

[222] J. Fostier and F. Olyslager, "An asynchronous parallel MLFMA for scattering at multiple dielectric objects," *IEEE Trans. Antennas Propag.*, vol. 56, no. 8, pp. 2346–55, Aug. 2008.

[223] J. Fostier and F. Olyslager, "Provably scalable parallel multilevel fast multipole algorithm," *Electron. Lett.*, vol. 44, no. 19, pp. 1111–13, Sep. 2008.

[224] J. Fostier and F. Olyslager, "Full-wave electromagnetic scattering at extremely large 2-D objects," *Electron. Lett.*, vol. 45, no. 5, pp. 245–6, Feb. 2009.

[225] Ö. Ergül and L. Gürel, "A hierarchical partitioning strategy for an efficient parallelization of the multilevel fast multipole algorithm," *IEEE Trans. Antennas Propag.*, vol. 57, no. 6, pp. 1740–50, Jun. 2009.

[226] J. M. Taboada, L. Landesa, F. Obelleiro, J. L. Rodriguez, J. M. Bertolo, M. G. Araujo, J. C. Mourino, and A. Gomez, "High scalability FMM-FFT electromagnetic solver for supercomputer systems," *IEEE Antennas Propag. Mag.*, vol. 51, no. 6, pp. 21–8, Dec. 2009.

[227] J. M. Taboada, M. G. Araujo, J. M. Bertolo, L. Landesa, F. Obelleiro, and J. L. Rodriguez, "MLFMA-FFT parallel algorithm for the solution of large-scale problems in electromagnetics," *Prog. Electromagn. Res.*, vol. 105, pp. 15–30, 2010.

[228] M. G. Araujo, J. M. Taboada, F. Obelleiro, J. M. Bertolo, L. Landesa, J. Rivero, and J. L. Rodriguez, "Supercomputer aware approach for the solution of challenging electromagnetic problems," vol. 101, pp. 241–56, 2010.

[229] J. Fostier and F. Olyslager, "An open-source implementation for full-wave scattering by million-wavelength-size objects," *IEEE Antennas Propag. Mag.*, vol. 52, no. 5, pp. 23–34, Oct. 2010.

[230] Ö. Ergül and L. Gürel, "Rigorous solutions of electromagnetics problems involving hundreds of millions of unknowns," *IEEE Antennas Propag. Mag.*, vol. 53, no. 1, pp. 18–26, Feb. 2011.

[231] Ö. Ergül, "Parallel implementation of MLFMA for homogeneous objects with various material properties," *Prog. Electromagn. Res.*, vol. 121, pp. 505–20, 2011.

[232] Ö. Ergül, "Solutions of large-scale electromagnetics problems involving dielectric objects with the parallel multilevel fast multipole algorithm," *J. Opt. Soc. Am. A.*, vol. 28, no. 11, pp. 2261–8, Nov. 2011.

[233] M. Snir, S. Otto, S. Huss-Lederman, D. Walker, and J. Dongarra, *MPI: The Complete Reference (Volume 1, The MPI Core).* Cambridge: MIT Press, 1998.

[234] A. K. Dominek, M. C. Gilreath, and R. M. Wood, "Almond test body," *United States Patent*, no. 4809003, 1989.

[235] L. Gürel and Ö. Ergül, "Hierarchical parallelization of the multilevel fast multipole algorithm (MLFMA)," in *IEEE Proceedings*, vol. 101, no. 2, pp. 332–341, Feb. 2013.

[236] Ö. Ergül and L. Gürel, "Accuracy: the frequently overlooked parameter in the solution of extremely large problems," in *Proc. European Conf. on Antennas and Propagation (EuCAP)*, vol. 101, no. 2, pp. 3928–3931.

[237] L. Gürel and Ö. Ergül, "Accuracy and efficiency considerations in the solution of extremely large electromagnetics problems," in *Proc. Computational Electromagnetics International Workshop*, 2011, 136–139.

[238] Ö. Ergül and L. Gürel, "Accurate solutions of extremely large integral-equation problems in computational electromagnetics," *IEEE Proceedings*, vol. 101, no. 2, pp. 342–349, Feb. 2013.

[239] Ö. Ergül and L. Gürel, "Advanced partitioning and communication strategies for the efficient parallelization of the multilevel fast multipole algorithm," *Proc. IEEE Antennas and Propagation Soc. Int. Symp.*, 2010.

Index

The Multilevel Fast Multipole Algorithm (MLFMA) for Solving Large-Scale Computational Electromagnetics Problems, First Edition. Özgür Ergül and Levent Gürel.

Printed and bound by CPI Group (UK) Ltd, Croydon, CR0 4YY
17/09/2026
14964269-0005